# Lecture Notes in Computer Science 4495

*Commenced Publication in 1973*
Founding and Former Series Editors:
Gerhard Goos, Juris Hartmanis, and Jan van Leeuwen

John Krogstie Andreas Opdahl
Guttorm Sindre (Eds.)

# Advanced Information Systems Engineering

19th International Conference, CAiSE 2007
Trondheim, Norway, June 11-15, 2007
Proceedings

Volume Editors

John Krogstie
Guttorm Sindre

Norwegian University of Science and Technology
Dept. of Computer and Information Science
Sem Sælands vei 7-9, 7491 Trondheim, Norway
E-mail: {krogstie, guttors}@idi.ntnu.no

Andreas Opdahl
University of Bergen
Department of Information Science and Media Studies
Fosswinckelsgate 6, 5007 Bergen, Norway
E-mail: Andreas.Opdahl@uib.no

Library of Congress Control Number: 2007928349

CR Subject Classification (1998): H.2, H.3-5, J.1, K.4.3-4, K.6, D.2, I.2.11

LNCS Sublibrary: SL 3 – Information Systems and Application, incl. Internet/Web and HCI

ISSN 0302-9743
ISBN-10 3-540-72987-9 Springer Berlin Heidelberg New York
ISBN-13 978-3-540-72987-7 Springer Berlin Heidelberg New York



Springer is a part of Springer Science+Business Media

springer.com


Printed in Germany

Typesetting: Camera-ready by author, data conversion by Scientific Publishing Services, Chennai, India
Printed on acid-free paper SPIN: 12074839 06/3180 5 4 3 2 1 0

# Preface

CAiSE 2007 was the 19th in the series of International Conferences on Advanced Information Systems Engineering. This year's conference was located in Trondheim and hosted by the Norwegian University of Science and Technology, with the aim of bringing together practitioners and researchers in the field of information systems engineering. The CAiSE series thereby returned to the city where the third CAiSE conference was held in 1991.

Since the first CAiSE was organized in Stockholm in 1989, CAiSE has grown to become one of the most prestigious international conferences at the intersection between information systems, software engineering, database technology and other related fields. The CAiSE conferences present basic and applied research results from academia alongside keynotes and research presentations from industry.

The special theme of CAiSE 2007 was "Ubiquitous Information Systems Engineering," reflecting that modern information systems often span activities performed in several organizations and at different geographical locations. They often support the untethered mobility of their users. Already today, these systems have a large impact on the everyday life of individuals and organizations. As we move towards ambient, pervasive and ubiquitous computing, this impact will increase significantly.

While CAiSE 2007 invited general submissions on the development, maintenance, procurement and use of information systems, submissions dealing with aspects related to information systems engineering in ubiquitous environments were especially welcome. The response was overwhelming. In all, 301 papers were submitted, which is a new record for CAiSE conferences. After all submissions had been carefully assessed by three independent reviewers, the Program Committee meeting selected 40 top-quality papers, resulting in an acceptance rate of around 13%. Several other high-quality papers were selected for the CAiSE Forum, a tradition initiated at CAiSE 2003 in Velden to stimulate open discussions of high-quality on-going research.

The success of CAiSE 2007 is also evidenced by the many top-quality workshops that were arranged as CAiSE pre-conference events. The longest running, the REFSQ series on Requirements Engineering: Foundation for Software Quality, was organized for the 13th time in Trondheim. Over the years it has evolved into a Working Conference, this year with its own LNCS proceedings published by Springer. Other workshops associated with CAiSE have almost equally long histories. The EMMSAD 2007 Workshop on Exploring Modelling Methods for Information Systems Analysis and Design was organized for the 12th time. The AOIS 2007 Workshop on Agent-Oriented Information Systems was organized for the 17th time and was associated with CAiSE for the eighth time. Other high-quality international workshops this year were BPMDS 2007 on Business

Process Modelling, Development, and Support, BUSITAL 2007 on Business IT alignment and WISM 2007 on Web Information Systems Modelling.

The special theme of CAiSE 2007 was high-lighted by an additional workshop on Ubiquitous Mobile Information and collaboration systems, UMICS 2007, and by three industrial keynote speeches: Ora Lassila of Nokia Research, UK, on "Setting Your Data Free: Thoughts on Information Interoperability," Pekka Abrahamsson of VTT, Finland, on "Agile Software Development of Mobile Information Systems" and Christen Krogh of Opera Software, Norway, with a talk titled "40 Million Users, 300 Engineers, 40 Enterprise Customers, 7 Development Locations, and 1 CVS - Lessons Learned Through Design, Development and Deployment of the Opera Browser."

Contact with industry was emphasized through two one-day industrial seminars on Agile Methods in Practice (organized by Torgeir Dingsøyr) and on Interoperability in the Public Sector (organized by Arne-Jørgen Berre). As usual, a doctoral consortium was also organized in conjunction with CAiSE, giving research students an opportunity to present and discuss their PhD topics and plans face to face with internationally leading researchers in their fields.

Last, but not least, CAiSE 2007 was also an occasion to honor one of the founding fathers of the CAiSE series and Organizing Chair of the 1991 conference, Professor Arne Sølvberg, who celebrated his 67th birthday in 2007. This is the usual retirement age in Norway, although Arne has promised to be working until 70 (at least)! A symposium to honor Professor Sølvberg was arranged before CAiSE 2007 as an additional pre-conference event.

As the organizers of the CAiSE 2007 main conference, we deeply thank the many members of the CAiSE 2007 Program Committee and the additional reviewers for making the reviewing process so thorough and smooth. We equally deeply thank the Chairs and other committee members involved in the many additional events associated with CAiSE 2007. We also want to thank Richard van de Stadt for his excellent technical support during the various stages of evaluating papers and preparing the proceedings. We also wish to thank all the local organizers at the Norwegian University of Science and Technology (NTNU) for their hard work and devotion. Finally, we would like to thank the conference gold sponsors Google, SINTEF and Telenor, institutional sponsor ERCIM, local sponsors The City of Trondheim and the NTNU, as well as our collaborators, the University of Bergen and The Norwegian Computer Society.

March 2007

John Krogstie
Andreas L. Opdahl
Guttorm Sindre

# Organization

| | |
|---|---|
| Advisory Committee | Janis Bubenko Jr<br>Royal Institute of Technology, Sweden<br>Colette Rolland<br>Université Paris 1 - Sorbonne, France<br>Arne Sølvberg<br>Norwegian University of Science and Technology, Norway |
| General Chair | John Krogstie<br>Norwegian University of Science and Technology, Norway |
| Program Chairs | Andreas Opdahl<br>University of Bergen, Norway<br>Guttorm Sindre<br>Norwegian University of Science and Technology, Norway |
| Organization Chair | Hallvard Trætteberg<br>Norwegian University of Science and Technology, Norway |
| Forum Chairs | Johann Eder<br>University of Vienna, Austria<br>Stein Løkke Tomassen<br>Norwegian University of Science and Technology, Norway |
| Workshop Chairs | Barbara Pernici<br>Politecnico di Milano, Italy<br>Jon Atle Gulla<br>Norwegian University of Science and Technology, Norway |
| Tutorial Chair | Terje Brasethvik<br>Norwegian University of Science and Technology, Norway |
| Doctoral Consortium Chairs | Moira Norrie<br>ETH Zürich, Switzerland<br>Jon Espen Ingvaldsen<br>Norwegian University of Science and Technology, Norway<br>Renate Kristiansen<br>Norwegian University of Science and Technology, Norway |

| | |
|---|---|
| Webmasters | Rune Molden<br>Norwegian University of Science and Technology, Norway<br>Lillian Hella<br>Norwegian University of Science and Technology, Norway |
| Industrial Chairs | Arne Jørgen Berre<br>SINTEF<br>Torgeir Dingsøyr<br>SINTEF |
| Sponsor Chair | Babak Amin Farschhian<br>Telenor, Norway |

## Program Committee

| | |
|---|---|
| Jan Øyvind Aagedal | Norway |
| Wil van der Aalst | The Netherlands |
| Pär Ågerfalk | Ireland |
| Luciano Baresi | Italy |
| Zohra Bellahsene | France |
| Giuseppe Berio | Italy |
| Claudio Bettini | Italy |
| Nacer Boudjlida | France |
| Mokrane Bouzeghoub | France |
| Svein E. Bratsberg | Norway |
| Sjaak Brinkkemper | The Netherlands |
| Silvana Castano | Italy |
| Jaelson Castro | Brazil |
| João Falcão e Cunha | Portugal |
| Monica Divitini | Norway |
| Dov Dori | Israel |
| Eric Dubois | Luxembourg |
| Johann Eder | Austria |
| David Embley | USA |
| Joerg Evermann | New Zealand |
| Xavier Franch | Spain |
| Paolo Giorgini | Italy |
| Claude Godart | France |
| Jaap Gordijn | The Netherlands |
| Peter Green | Australia |
| Terry Halpin | USA |
| Manfred Hauswirth | Ireland |
| Brian Henderson-Sellers | Australia |
| Patrick Heymans | Belgium |
| Matthias Jarke | Germany |

| | |
|---|---|
| Manfred Jeusfeld | The Netherlands |
| Paul Johannesson | Sweden |
| Henk Jonkers | The Netherlands |
| Håvard Jørgensen | Norway |
| Roland Kaschek | New Zealand |
| Marc Lankhorst | The Netherlands |
| Julio Leite | Brazil |
| Michel Lemoine | France |
| Michel Leonard | Switzerland |
| Pericles Locoupolous | UK |
| Kalle Lyytinen | USA |
| Neil Maiden | UK |
| Michele Missikoff | Italy |
| Haris Mouratidis | UK |
| John Mylopoulos | Canada |
| Moira Norrie | Switzerland |
| Andreas Oberweis | Germany |
| Antoni Olivé | Spain |
| Hervé Panetto | France |
| Jeffrey Parsons | Canada |
| Oscar Pastor Lopez | Spain |
| Barbara Pernici | Italy |
| Anne Persson | Sweden |
| Michaël Petit | Belgium |
| Yves Pigneur | Switzerland |
| Geert Poels | Belgium |
| Klaus Pohl | Germany |
| Erik Proper | The Netherlands |
| Jolita Ralyte | Switzerland |
| Björn Regnell | Sweden |
| Colette Rolland | France |
| Michael Rosemann | Australia |
| Gustavo Rossi | Argentina |
| Matti Rossi | Finland |
| Kevin Ryan | Ireland |
| Motoshi Saeki | Japan |
| Camille Salinesi | France |
| Tony C. Shan | USA |
| Monique Snoeck | Belgium |
| Arnor Solberg | Norway |
| Erlend Stav | Norway |
| Janis Stirna | Sweden |
| Alistair Sutcliffe | UK |
| David Taniar | Australia |
| Bernhard Thalheim | Germany |

| | |
|---|---|
| Aphrodite Tsalgatidou | Greece |
| Olegas Vasilecas | Lithuania |
| Yair Wand | Canada |
| Roel Wieringa | The Netherlands |
| Petia Wohed | Sweden |
| Carson Woo | Canada |
| Eric Yu | Canada |
| Didar Zowghi | Australia |

## Additional Referees

Franz Acherman
Ole Agesen
Xavier Alvarez
Davide Ancona
Joaquim Aparício
João Araújo
Ulf Asklund
Dharini Balasubramaniam
Carlos Baquero
Luís Barbosa
Lodewijk Bergmans
Joshua Bloch
Noury Bouraqadi
Johan Brichau
Fernando Brito e Abreu
Pim van den Broek
Kim Bruce
Luis Caires
Giuseppe Castagna
Barbara Catania
Walter Cazzola
Shigeru Chiba
Tal Cohen
Aino Cornils
Erik Corry
Juan-Carlos Cruz
Gianpaolo Cugola
Padraig Cunningham
Christian D. Jensen
Silvano Dal-Zilio
Wolfgang De Meuter
Kris De Volder
Giorgio Delzanno
David Detlefs
Anne Doucet
Rémi Douence
Jim Dowling
Karel Driesen
Sophia Drossopoulou
Stéphane Ducasse
Natalie Eckel
Marc Evers
Johan Fabry
Leonidas Fegaras
Luca Ferrarini
Rony Flatscher
Jacques Garrigue
Marie-Pierre Gervais
Miguel Goulão
Thomas Gschwind
Pedro Guerreiro
I. Hakki Toroslu
Görel Hedin
Christian Heide Damm
Roger Henriksson
Martin Hitz
David Holmes
James Hoover
Antony Hosking
Cengiz Icdem
Yuuji Ichisugi
Anders Ive
Hannu-Matti Järvinen
Andrew Kennedy
Graham Kirby
Svetlana Kouznetsova
Kresten Krab Thorup
Reino Kurki-Suonio
Thomas Ledoux
Yuri Leontiev
Cristina Videira Lopes
David Lorenz
Steve MacDonald
Ole Lehrmann Madsen
Eva Magnusson
Margarida Mamede
Klaus Marius Hansen
Kim Mens
Tom Mens
Isabella Merlo
Marco Mesiti
Thomas Meurisse
Mattia Monga
Sandro Morasca
M. Murat Ezbiderli
Oner N. Hamali
Hidemoto Nakada
Jacques Noye
Deniz Oguz
José Orlando Pereira
Alessandro Orso
Johan Ovlinger
Marc Pantel
Jean-François Perrot
Patrik Persson
Frédéric Peschanski
Gian Pietro Picco
Birgit Pröll
Christian Queinnec
Osmar R. Zaiane
Barry Redmond
Sigi Reich

Arend Rensink
Werner Retschitzegger
Nicolas Revault
Matthias Rieger
Mario Südholt
Paulo Sérgio Almeida
Ichiro Satoh
Tilman Schaefer
Jean-Guy Schneider
Pierre Sens
Veikko Seppänen
Magnus Steinby
Don Syme
Tarja Systä
Duane Szafron
Yusuf Tambag
Kenjiro Taura
Michael Thomsen
Sander Tichelaar
Mads Torgersen
Tom Tourwé
Arif Tumer
Ozgur Ulusoy
Werner Van Belle
Vasco Vasconcelos
Karsten Verelst
Juha Vihavainen
John Whaley
Mario Wolzko
Mikal Ziane
Gabi Zodik
Elena Zucca

## Gold Sponsors

## Institutional Sponsor

## Local Sponsors

## Additional Collaborators

# Table of Contents

## Strategic Alignment



## Service-oriented Architecture II



## Requirements I



## Process Modelling I

## Participative Modelling



## Process-Aware Information Systems

# Agile Software Development of Mobile Information Systems

Pekka Abrahamsson

VTT Technical Research Centre of Finland,
P.O.Box 1100, FIN-90571 Oulu, Finland
pekka.abrahamsson@vtt.fi

**Abstract.** Agile software development methods are quickly being adopted by the software industry. Concerns have been raised whether agile methods are suitable for any given information systems development domain. Indeed, quite little is known empirically about the validity of agile methods in most of the industrial domains. Mobile information systems present no exception in this sense. Yet, they are subject to frequent requirements changes in terms of changing business needs and technology, and their market is highly competitive. Moreover, most of these systems are far away from so called agile home ground. This talk presents the need for agile methods in the focal domain, identifies their shortcomings on the basis of three large-scale case studies from industry. All of the cases deal with the development of mobile information system and come from Nokia, F-Secure and Philips. The talk also discusses the possible strategies for deploying agile solutions in practice.

**Keywords:** Agile software development, case study, mobile information systems.

## 1 Introduction

Agile software development methods have emerged rapidly since the mid 1990's. The roots of agile methods are placed far beyond the last decade, however [1]. Industry has been keen on adopting agile solutions in recent years. Large software corporations such as Microsoft and SAP have announced publicly of their plans to adopt agile methods. Information systems build in embedded devices such as cars, telecom systems or consumer electronics systems present no exception to this. Large companies such as Philips, Nokia, British Telecom, to name a few, are either already adopting or plan to adopt agile software solutions as their means to tackle software related challenges. A recent Forrester survey[1] suggested that one out of seven software companies already use agile methods.

Agile methods have been criticized for the lack of solid, scientifically valid empirical data to back up their claims [2, 3]. Yet, the situation is far from unique in

[1] Corporate IT Leads The Second Wave Of Agile Adoption, 30 November 2005, Forrester Research, http://www.forrester.com/Research/Document/Excerpt/0,7211,38334,00.html

J. Krogstie, A.L. Opdahl, and G. Sindre (Eds.): CAiSE 2007, LNCS 4495, pp. 1–4, 2007.

the fields of software and information systems development. Fenton [4] argued that while software professional often seek for rational basis for making a decision about which development method they should adopt, the basis for such rationalization is completely missing. Fenton went as far as claiming that the "methods introduced continue to be based on more on faith than on an empirical data". Based on recent industrial attention, we can suggest that the lack of data has not slowed down the adoption of agile methods.

There is no agreed definition of agile methods in information systems or software engineering fields. In industrial engineering science the situation is quite different. They have proposed 17 competing definitions of the concept of agile manufacturing [5]. In software engineering, the concept of agile software is most often related to the elements presented in the Agile Manifesto (http://www.agilemanifesto.org). To some extent, we can assert that it is the most accepted conceptualization of agile software development as it has received more than 5000 independent signatories over a few years time. This holds, however, no scientific meaning. Rather, it shows something about the popularity of these methods. In information systems field, the concept of agility emerged in the late 1990's with studies about internet-speed development [e.g., 6]. Yet, the very concept of agile development still denotes to "more formal than hacking and less formal than traditional methods" [7, p.29].

Mobile information systems, whether they are operated as a stand alone application in mobile terminals or as an access provider to a back end system, are restricted by terminal constraints. Some of these constraints are screen size, keyboard, memory constraints and the battery power. The devices, however, are developing technologically very rapidly and most of the modern handheld devices called smart phones have the processing power of regular computers. Thus, the restrictions that were in place a few years ago do not hold within the next coming years. Moreover, the mobile telecommunications industry has shown to be comprised of a highly competitive, uncertain and dynamic environment [8]. While, so far, mobile commerce applications have not been very successful, telecommunications companies believe a change in short term due to the adoption of 3G technologies. This should lead to a widespread adoption of mobile services in combination with mobile commerce applications [9]. Rather than technology driven, mobile information systems today are described as location-aware service providers, which bring connectivity and mobility to a new level. This is still relatively poorly captured in contemporary mobile service offering, which to a certain extent explains the slower-than-expected adoption of mobile services.

Mobile information systems are growing in terms of their size and complexity. This calls for attention in developing information systems development methods that meet the needs of volatile business environment. Also, it can be argued, that certain type agility is required in order to survive in global competitive marketplace. The characteristic of agility should take place in all operational levels of a corporation or a network of multiple corporations. This is not easily achieved and calls for broader approaches than those that are currently available. This is not in accordance with current development of agile software solutions, however. Agile methods purposefully strive for a minimalist set of activities and artifacts.

It is proposed that a success factor explaining the acceptance of agile software development methods by industry is the explicit attention to a set of concrete practices. At a second layer, the identification of principles, called the agile principles, place meaningful value for these practices. They thus provide a rationale for executing a practice rather than just imposing it as such. Also, by placing an explicit focus on practices, the development teams are faced with an immediate change of behavior, which is bound to lead to concrete and visible results. As an example, this has been a key challenge in the software process improvement literature. Software process changes do not necessarily lead to changes in the behavior of the people executing the processes since most often the processes are not followed as suggested [10]. A visible testimonial of this is the lack of studies addressing return-of-investment (ROI) of software process improvements [11]. Indeed, van Soligen [11] found only eight studies in the literature with explicit consideration of ROI in the studied software process improvement initiatives. A literature search of IEEE database reveals more than 1000 studies in the area. Thus, it can be argued that 0,21% or less of published software process improvement studies have some ROI values to present.

A series of case studies of the use of agile software development methods in the area of mobile information systems development are presented. The case studies come from Nokia, F-Secure and Philips. These cases provide evidence on the applicability of agile solutions in a specific type of industrial domain, ROI impact on the case organizations in terms of developer satisfaction, customer satisfaction, product quality, time-to-market and development costs. The case studies also improve our understanding of the use of different strategies that were exploited in the case organizations to deploy agile solutions in practice.

The cases are part of ITEA-AGILE (http://www.agile-itea.org) research project, which studied the use of agile methods in the area of embedded systems development. The case material is presented in the House of Agile (http://www.houseofagile.org), which is an interactive web portal for embedded agile development.

## References


1. Larman, C., Basili, V.R.: Iterative and incremental development: A brief history. IEEE Software 20, 47–56 (2003)
2. Melnik, G., Williams, L., Geras, A.: Empirical Evaluation of Agile Processes. In: Wells, D., Williams, L. (eds.) Lecture Notes in Computer Science, vol. 2418, Springer, Heidelberg (2002)
3. Lindvall, M., Basili, V.R., Boehm, B.W., Costa, P., Dangle, K., Shull, F., Tesoriero, R., Williams, L.A., Zelkowitz, M.V.: Empirical findings in agile methods. In: Wells, D., Williams, L. (eds.) Lecture Notes in Computer Science, vol. 2418, Springer, Heidelberg (2002)
4. Fenton, N.: Viewpoint Article: Conducting and presenting empirical software engineering. Empirical Software Engineering 6, 195–200 (2001)
5. Iskanius, P.: An agile supply chain for a project-oriented steel product network, in Department of Industrial Engineering, University of Oulu, Oulu (2006)
6. Baskerville, R., Levine, L., Pries-Heje, J., Ramesh, B., Slaughter, S.: How Internet companies negotiate quality. IEEE Computer 5, 51–57 (2001)
7. Baskerville, R., Balasubramaniam, R., Levine, L., Pries-Heje, J.: High-speed software development practices: What works, what doesn't. IT Professional 4, 29–36 (2006)

8. Lal, D., Pitt, D.C., Beloucif, A.: Restructuring in European telecommunications: Modeling the evolving market. European Business Review 3, 152–156 (2001)
9. Blazevic, V., Lievens, A., Klein, E.: Antecedents of project learning and time-to-market during new mobile service development. International Journal of Service Industry Management 1, 120–147 (2003)
10. Truex, D.P., Baskerville, R., Travis, J.: Amethodological systems development: The deferred meaning of systems development methods. Accounting, Management and Information Technology 10, 53–79 (2001)
11. van Solingen, R.: Measuring the ROI of Software Process Improvement. IEEE Software 3, 32–38 (2004)

# Modal Aspects of Object Types and Part-Whole Relations and the *de re/de dicto* Distinction

Giancarlo Guizzardi

Federal University of Espírito Santo (UFES), Vitória, Brazil
Laboratory for Applied Ontology, ISTC-CNR, Trento, Italy
guizzardi@loa-cnr.it

**Abstract.** In a series of publications, we have proposed a foundational system of ontological categories which has been used to evaluate and improve the quality of conceptual modeling languages and models. In this article, we continue this work by employing theories from Formal Ontology, Cognitive Psychology and Philosophical Logic to systematically investigate some important modal aspects of the ontological categories represented in structural conceptual models. In particular, we focus on *Object Types* and *Part-Whole Relations*, formally characterizing some modal properties that motivate the proposal of a number of distinctions within these categories. In addition, we show how two types of modality known in philosophical logic (*de re/de dicto* modality) can be used to address some subtle issues that appear in conceptual diagrams when different sorts of object types and part-whole relations are combined.

## 1 Introduction

In recent years, there has been a growing interest in the application of *Foundational Ontologies*, i.e., formal ontological theories in the philosophical sense, for providing real-world semantics for conceptual modeling languages, and theoretically sound foundations and methodological guidelines for evaluating and improving the individual models produced using these languages. This increasing interest can be noticed by the growth of the number of publications dedicated to the subject, including books [13], journal issues [17] and articles published at this forum [1,6]. However, by looking at these publications, one may notice that there is an issue of substantial importance in Formal Ontology but which has been given relative little attention in that community, namely, the examination of the modal properties of the ontological categories represented in the constructs of these languages.

In this article we continue our work on developing ontological foundations for conceptual modeling [4-6]. The objective here is to employ theories from Formal Ontology, Cognitive Psychology and Philosophical Logic to systematically investigate some important modal properties of structural conceptual models. In section 2, we give a brief presentation of a system of Quantified Modal Logics used in the remaining sections. In section 3, we revisit our theory of Object Types (e.g., Kinds, Roles, States, Mixins) presented in [6] focusing on some modal aspects of these categories, and formally characterizing these aspects with the system presented in section 2.

J. Krogstie, A.L. Opdahl, and G. Sindre (Eds.): CAiSE 2007, LNCS 4495, pp. 5–20, 2007.

In section 4, we revisit a theory presented in [4] elaborating on distinctions between mereological (parthood) relations motivated by different modal properties governing the relations between parts and wholes[1]. In section 5, we present the main contribution of this paper, namely, to formally elaborate on some subtle issues regarding the distinction between *de re* and *de dicto* modality, which are manifest in conceptual diagrams when the categories presented in section 4 and 5 are combined. Section 6 briefly discusses related work. Finally, section 7 presents some final considerations.

## 2 A Brief Presentation of a System of Quantified Modal Logics

In order to present a formal characterizations of the notions discussed in this article we make use of a language L of quantified modal logics with identity. The alphabet of L contains the traditional operators $\wedge$ (conjunction), $\vee$ (disjunction), $\neg$ (negation), $\rightarrow$ (conditional), $\leftrightarrow$ (biconditional), $\forall$(universal quantification), $\exists$ (existential quantification), with the addition of the equality operator =, the uniqueness existential quantification operator $\exists!$, and the modal operators $\Box$ (necessity) and $\Diamond$ (possibility). The following holds for these three latter operators: (1) $\Diamond A =_{def} \neg\Box\neg A$; (2) $\Box A =_{def} \neg\Diamond\neg A$ and (3) $\exists! x\ A =_{def} \exists y \forall x\ (A \leftrightarrow (x = y))$.

A Model-Theoretic semantics for this language can be given by defining an interpretation function δ that assigns values to the non-logical constants of the language and a *model structure* M. In this language M has a structure <W,D> where W is a non-empty set of worlds and D is a non-empty domain of objects. The domain D of quantification is that of *possibilia*, which includes all possible entities independent of their actual existence. Therefore we shall quantify over a constant domain in all possible worlds. Informally, we can state that the truth of formulas involving the modal operators can be defined such that the semantic value of formula $\Box A$ is true in world w iff A is true in every world w' accessible from w. Likewise, the semantic value of formula $\Diamond A$ is true in world w iff A is true in at least one world w' accessible from w.

There are alternative interpretations regarding the ontological status of possible worlds and a full discussion of the topic is outside the scope of this article. Here, unless explicitly mentioned, we take worlds to represent maximal states of affairs (states of the world) which can be factual (i.e., obtaining in reality) or counterfactual. An alternative interpretation which also appears in the article is that of worlds as histories, i.e., as causally connected sequences of world snapshots (state of affairs), which again, can be either factual or counterfactual. Moreover, we take all worlds to be equally accessible and therefore we omit the accessibility relation from the model structure. As a result we have the simplest language of quantified modal logic (QS5). For a full presentation of such a system one should refer to [2].

Finally, in order to simplify the presentation of the formulas throughout the article we make use a restricted quantification scheme following the notation proposed in [15]: **(i) (∀S,x)** A and **(ii) (∃S,x) A**, which can be read as *for every instance of S, A holds* and *there is an instance of S such that A holds*, respectively. In other words, (i)

[1] The theory proposed in [4] and elaborated in [5] discusses a number of other properties of part-whole relations. Here, due to the scope and objectives of this article we focus solely on modally related properties.

and (ii) are meta-linguistic abbreviations to the formulas ($\forall$x S(x) $\rightarrow$ A) and ($\exists$x S(x) $\wedge$ A), respectively, i.e., they conform to the so-called Fregean analysis of restricted quantification.

# 3 Modal Distinctions in a Theory of Object Types

In the practice of conceptual modeling, a set of primitives is often used to represent distinctions in different sorts of *Object Types* (Kind, Role, State, Mixin, among others). However, most conceptual modeling languages do not offer methodological support for helping their uses to decide how to represent elements that denote general terms in a given domain (viz. Person, Student, Red Thing, Physical Thing, Deceased Person, Customer) and, hence, modeling choices are often made in an ad hoc manner. Additionally, an inspection of the literature shows that there has been traditionally much disagreement on the meaning of these categories (for extended discussion on this see [6]).

In [6], we propose a philosophically and psychologically well-founded theory of types for conceptual modeling and a UML modeling profile based on this theory[2]. In the remaining of this section we briefly revisit this theory. However, the focus here is on the modal properties that motivate the distinctions populating this *"Typology of Object Types"*, as well as on the formal characterization of these distinctions using the system of modal logics presented in section 2. In addition, we focus here on a subset of these distinctions, namely, on ***Kinds***, ***Roles, Phases,*** and ***RoleMixins***, which are the most relevant ones for the purposes of this article.

The categories forming this typology that we discuss here are depicted in Figure 1.a. As it can be observed, a fundamental distinction between Object Types is made between *Sortal* and *Mixin Types.* Sortals are sorts of types that carry *principles of identity*, *individuation* and *counting* for their instances. A principle of identity is a principle for which we can judge whether two individuals are the same. A principle of counting, in contrast, is one that supports individuation and counting of individuals. To illustrate this point, let us make use of the following thought experiment. Suppose someone is presented with a *red entity* (e.g., a red shirt) at time $t_1$ and asked the following question: "Exactly how many *red entities* do you see in front of you?". Now, suppose that a part (e.g., one sleeve) of this red entity is extracted and destroyed at a time $t_2$, and an additional question is asked: "Is the red entity you are seeing now ($t_2$) the same you saw before (in $t_1$)?" Notice that none of the questions can receive a determinate answer (an answer with a determinate truth-value): (i) Should a red shirt be counted as one or should the shirt, the two sleeves, and two pockets be counted separately so that we have five reds? The problem in this case is not that one would not know how to finish the counting but that one would not know how to start, since arbitrarily many *subparts of a red thing are still red*; (ii) How can one know if extracting a piece of the entity alters the identity of that entity? How can one know, for example, if having that piece is an essential property of that entity? The problem in both cases is the type *Red* does not supply principles based on which these questions can be given determinate answers. Now, notice that if *(red) entity* is replaced in these

[2] This theory as presented in [6] can be seen as the conceptual modeling extension of the Onto-Clean methodology [3].

questions by *(red) shirt*, determinate answers can be given to all these questions. Types such as Shirt (but also Person, Car, Dog, Student) are examples of *Sortal Types*. In contrast, types such as Red (but also Thing, Tall, Heavy and Insured Item) are named *Characterizing Types*, *Attributions* or *Mixins*, since they only attribute properties to (characterize) individuals which have already being individuated by sortal-supplied principles.

The statement that the identity of an individual can only be traced in connection with a sortal type, which carries a principle of individuation and identity to the particulars it collects amounts to one of the best-supported theories in the philosophy of language [10,15], and one that finds strong empirical support in cognitive psychology [5]. Moreover, the distinction between sortals and mixins is reflected in natural language in the distinction between common nouns and other general terms (e.g., adjectives, verbs), respectively. Finally, as discussed in [3,5,6], the role of (sortal-supplied) identity principles is explicitly defended in conceptual modeling as a method for deriving stable and ontologically sound taxonomic structures.

A principle of identity must apply to an individual in all possible situations. For this reason, principles of identity must be supplied by types that are also instantiated by their instances in all possible situations, i.e., type whose instances cannot cease to instantiate without ceasing to exist. This meta-property of types is named *Modal Constancy* or *rigidity* and can be formally characterized as in the following *formula schema*:

**Definition 1 (Rigidity).** A type T is rigid if for every instance $x$ of T, $x$ is necessarily (in the modal sense) an instance of T. In other words, if $x$ instantiates T in a given world $w$, then $x$ must instantiate T in every possible world $w'$: **(1).** $\mathbf{R(T) =_{def} \Box(\forall x\ T(x) \rightarrow \Box(T(x)))}$. ■

We have that only *rigid sortals* can supply principles of identities for their instances. A rigid sortal type that supplies a principle of identity for its instances is named here a *Substance Sortal* or a ***Kind***. This notion of Kind as presented here (also sometimes termed *Natural Kind*) is associated with the notion of *Essence* in the philosophical literature. More specifically, a Kind is a type defining all the essential properties for the individuals it classifies. Examples of types typically modeled as Kinds include *Person*, *Planet*, *Gold*, *Water*, *Lepidopteron* and *City*.

Within the category of sortals, we also have types that apply to their instances only *contingently* (i.e., possibly only in certain situations). Examples include types such as *Adolescent*, *Student*, *Employee*, *Philosopher*, *Deceased*, *Customer* and *Caterpillar*. Sortals that possibly apply to an individual only during a certain phase of its existence are named *Phased-Sortals*. Contrary to kinds, phased-sortals are *anti-rigid* types:

**Definition 2 (Anti-rigidity).** A types T is anti-rigid if for every instance $x$ of T, $x$ is *possibly* (in the modal sense) not an instance of T. In other words, if $x$ instantiates T in a given world $w$, then there is a possible world $w'$ in which $x$ does not instantiate T: **(2).** $\mathbf{AR(T) =_{def} \Box(\forall x\ T(x) \rightarrow \Diamond(\neg T(x)))}$. ■

Being anti-rigid, phased-sortals cannot *supply* a principle of identity for their instances. However, since they are sortals, they must *carry* a principle of identity, which they inherit from a Kind. Therefore, we have that every phase-sortal PS must be a *subtype* of Kind such that PS inherits the principle of identity supplied by K. In other words, every instance of PS is necessarily a K and, thus, obeys the principle of identity supplied by K. For example, for an individual John instance of Student, we can easily imagine John moving in and out of the Student type, while being the same individual, i.e. without losing his identity. This is because the principle of identity that applies to instances of Student and, in particular, that can be applied to John, is the one which is supplied by the kind Person of which the phase-sortal Student is a subtype.

If PS is a phased-sortal and K is the Kind specialized by PS, there is a *specialization condition* φ such that *x* is an instance of PS iff *x* is an instance of K that satisfies φ [15]. A further clarification on the different types of specialization conditions allows us to distinguish between two different types of phased-sortals which are of great importance to the practice of conceptual modeling, namely, *Phases* and *Roles*. **Phases** constitute possible stages in the history of a Kind. Examples include: (a) Alive and Deceased: as possible stages of a Person; (b) Catterpillar and Butterfly of a Lepidopteran; (c) Town and Metropolis of a City; (d) Boy, Male Teenager and Adult Male of a Male Person.

***Roles*** differ from phases with respect to the specialization condition φ. For a phase Ph, φ represents a condition that depends solely on intrinsic properties of Ph. For instance, one might say that if John is a Living Person then he is a Person who has the property of being alive or, if Spot is a Puppy then it is a Dog who has the property of being less than one year old. For a role Rl, conversely, φ depends on extrinsic (relational) properties of Rl. For example, one might say that if John is a Student then John is a Person who is enrolled in some educational institution, if Peter is a Customer then Peter is a Person who buys a Product *x* from a Supplier *y*, or if Mary is a Patient than she is a Person who is treated in a certain medical unit. In other words, an entity plays a role in a certain context, demarcated by its relation with other entities. This meta-property of Roles is named *Relational Dependence* and can be formally characterized as follows:

**Definition 3 (Relational Dependence).** A type T is relationally dependent on another type P via relation R iff for every instance *x* of T there is an instance *y* of P such that *x* and *y* are related via R: **(3).** $\mathbf{R(T,P,R) =_{def} \square(\forall x\ T(x) \rightarrow \exists y\ P(y) \land R(x,y))}$. ■

Mixins (i.e., non-sortals) are types that classify entities that belong to different Kinds, i.e., that obey different principles of identity. As with the category of sortals, mixins can also be rigid or anti-rigid. One type of mixin of great interest in conceptual modeling is the so-called ***RoleMixin***. For example, take the type *Insured Item*. This type can have as instances entities such as Boats, Cars, Persons, Houses, Work of Art, among others, clearly belonging to different kinds. In addition, instances of this type are only so contingently (an entity can be insured in one situation and not in another one). Finally, an Insured Item is defined in a certain context that includes types such as Insurance Policy and Insurance Agency. Thus, the type Insured Item is an example of a role mixin, i.e., an *anti-rigid* and *relationally dependent* mixin.

The discussion of this section is summarized in figures 1.a below. In this figure, we use the notational shortcuts **R+** and **R-** to represent the meta-properties or rigidity and anti-rigidity, respectively and **D** **(-/+)** to represent the meta-property of relational (in)dependence. In summary, ***Kinds*** are *rigid*, *independent sortals* that supply a principle of identity for their instances; ***Phases*** are *independent anti-rigid sortals*; ***Roles*** are *anti-rigid* and *relationally dependent sortals*, and ***RoleMixins*** are *anti-rigid* and *relationally dependent non-sortals*. In this article, we use the stereotypes «Kind», «Role», «Phase», and «RoleMixin» to decorate classes in a UML conceptual model (see figure 1.b) representing these distinctions among object types. It is important to emphasize that UML is used here only for the sake of exemplification, and that the issues addressed here are present in all major conceptual modeling languages.

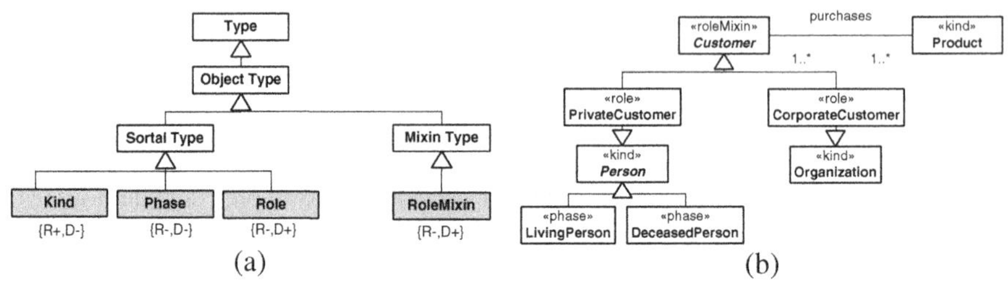


**Fig. 1.a (left)** Ontological Distinctions among Object Types motivated by Modal Meta-Properties;.

**Fig. 1.b** Example of use of a modeling profile based on these distinctions.

# 4 Modal Distinctions in Part-Whole Relations

Parthood is a relation of significant importance in conceptual modeling, being present in practically all conceptual modeling languages (e.g., OML, UML, EER). Nonetheless, in many of these languages, the concepts of part and whole are understood only intuitively, or are based on the very minimal axiomatization that these notions require, namely, that of a strict partial order (the so-called *Ground Mereology*). However, an important aspect to be addressed by any conceptual theory of parthood is to stipulate the different status that parts can have w.r.t. the whole they compose. As discussed by [14], many of the issues regarding this point cannot be clarified without considering *modality*. One of these issues refers to the notion of *separability*.

In order to formally define separability, we first define some notions related to the topic of *ontological dependence*. In particular, the relations of existential and generic dependence discussed in the sequel are strongly based on those defined in [8].

**Definition 4 (existential dependence).** Let the predicate ε denote existence. We have that an individual *x* is *existentially dependent* on another individual *y* (symbolized as *ed(x,y)*) iff, as a matter of necessity, y must exist whenever x exists, or formally **(4).**
$\mathbf{ed(x,y) =_{def} \Box(\varepsilon(x) \rightarrow \varepsilon(y))}$. ■

With definition 4 we can propose the concept of an essential part as follows[3]

**Definition 5 (essential part).** An individual *x* is an essential part of another individual *y* iff, *y* is existentially dependent on *x* and *x* is, necessarily, a part of *y*: $EP(x,y) =_{def} ed(y,x) \wedge \Box(x \leq y)$. This is equivalent to stating that $EP(x,y) =_{def} \Box(\varepsilon(y) \rightarrow \varepsilon(x)) \wedge \Box$

[3] Following [14] we use the symbols ≤ and < to represent parthood and proper parthood, respectively, and we have that $(x \leq y) =_{def} (x < y) \vee (x = y)$.

$(x \leq y)$, which is, in turn, equivalent to $EP(x,y) =_{def} \Box(\varepsilon(y) \rightarrow \varepsilon(x) \wedge (x \leq y))$. We adopt here the *mereological continuism* defended by [14], which states that the part-whole relation should only be considered to hold among existents, i.e., $\forall x,y\ (x \leq y) \rightarrow \varepsilon(x) \wedge \varepsilon(y)$. As a consequence, we can have this definition in its final simplification **(5). $EP(x,y) =_{def} \Box(\varepsilon(y) \rightarrow (x \leq y))$.** ■

Figures 2.a and 2.b below depict examples of essential parts. In figure 2.a, every person has a brain as part, and in every world that the person exists, the very same brain exists and is a part of that person. In figure 2.b, we have an analogous example: a car has a chassis as an essential part, thus, the part-whole relation between car and chassis holds in every world that the car exists. To put in a different way, if the chassis is removed, the car ceases to exist as such, i.e., it looses its identity.

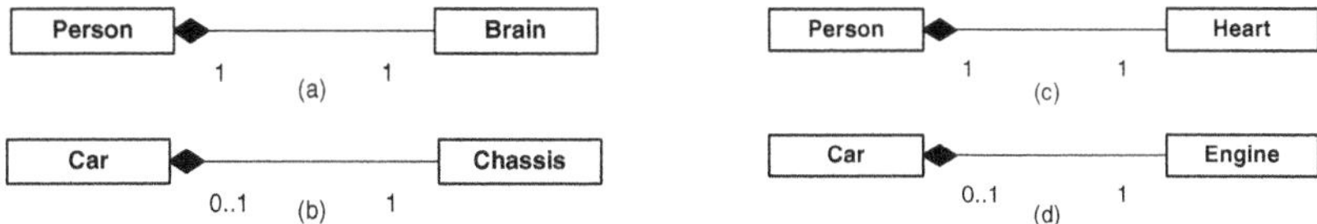


**Fig. 2. (a-b)** Wholes and their Essential parts; (c-d) Wholes and their Mandatory parts

The UML notation used in figure 2 highlights a problem that exists in practically all conceptual modeling languages. In order to discuss this problem, let us examine the models represented in figures 2.c and 2.d. According to the UML semantics, the models of figure 2.a and 2.c convey exactly the same kind of information. However, this is not the case, in general, in this domain in reality. Typically, the relation between a person and his brain is not of the same nature as the relation between a person and his heart. Differently from the former, a particular heart is not an essential part of a person, i.e., it is not the case that for every person x there is a heart y, such that in every possible circumstance y is part of x. For instance, the fact that an individual John had the same heart during his entire lifetime is only accidental. With the advent of heart transplants, one can easily imagine a counterfactual in which John had been transplanted a different heart. An analogous argument can be made in the case of figure 2.d. Although every car needs an engine, it certainly does not have to be the same engine in every possible world.

The difference in the underlying real-world semantics in the cases of figure 2.a and 2.c are made explicit if we consider their corresponding formal characterization. In the case of fig.2.a, since it is a case of essential parthood, we have that: **(figure 2.a) $\Box((\forall Person,x)(\exists!Brain,y)\ \Box(\varepsilon(x) \rightarrow (y < x)))$,** whereas in the case of figure 2.c, the corresponding axiomatization is **(figure 2.c) $\Box((\forall Person,x)\ \Box(\varepsilon(x) \rightarrow (\exists!Heart,y)(y < x)))$.** A similar distinction can be made for the case of figures 2.b and 2.d: **(figure 2.b) $\Box((\forall Car,x)(\exists!Chassis,y)\ \Box(\varepsilon(x) \rightarrow (y < x)))$** and **(figure 2.d) $\Box((\forall Car,x)\ \Box(\varepsilon(x) \rightarrow (\exists!Engine,y)(y < x)))$.**

In cases such as those depicted in the specifications of figures 2.c and 2.d, an individual is not specifically dependent of another individual, but *generically dependent* of any individual that instantiates a given type. The concept of generic dependence is defined as follows:

**Definition 6 (generic dependence).** An individual *y* is *generic dependent* of a type T iff, whenever y exists it is necessary that an instance of T exists. This can be formally characterized by the following *formula schema*: **(6). GD(y,T)** $=_{def}$ $\Box(\varepsilon(y) \rightarrow \exists T,x\ \varepsilon(x))$. ■

We name individuals such as the instances of Heart and Engine in figures 2.c and 2.d, respectively, *mandatory parts*:

**Definition 7 (mandatory part).** An individual *x* is a mandatory part of another individual *y* iff, *y* is generically dependent of an type T that *x* instantiates, and y has, necessarily, as a part an instance of T*:* **(7). MP(T,y)** $=_{def}$ $\Box(\varepsilon(y) \rightarrow (\exists T,x)(x < y))$. ■

In order to represent the ontological distinction between essential and mandatory parts, we propose an extension to the UML notation used in the examples for the remaining of this paper. We assume that the minimum cardinality of **1** in the association end corresponding to the part represents a *mandatory part-whole* relation. To represent the case of an *essential part-whole* relation, we propose to extend the current UML aggregation notation by defining the Boolean meta-attribute **essential**.

When the meta-attribute *essential* equals *true* then the minimum cardinality in the association end corresponding to the part must also be **1**. This is expected to be the case, since essential parthood can be seen as a limit case of mandatory parthood. When *essential* equals *false*, the tagged value textual representation can be omitted. This extended notation is exemplified in figure 3 below.

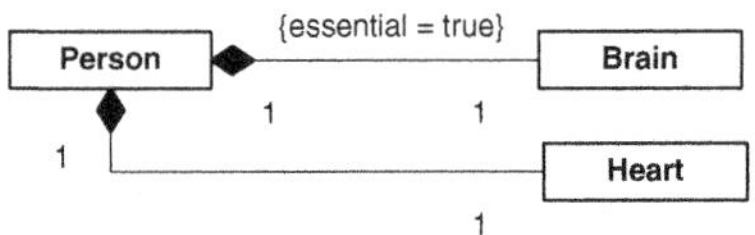


**Fig. 3.** Extensions to the UML notation to distinguish between *essential* and *mandatory* parts

We emphasize that the particular examples chosen to illustrate the distinction between *essential* and *mandatory* parts are used here for illustration purposes only. For example, when modeling *brain* as an essential part of persons and *heart* as a mandatory one, we are not advocating that this is a general ontological choice that should be countenanced in all conceptualizations. Conversely, the intention is to make explicit the consequences of this modeling choice, and to advocate for the need of explicitly differentiating between these two modes of parthood. The choice itself, however, is always left to the model designer and is conceptualization-dependent.

Up to this moment, we have interpreted possible worlds as maximal state of affairs, which can be factual or counterfactual. In other words, we have assumed a branching structure of time, and each world is taken at a time interval in a (factual or counterfactual) time branch. An alternative is to interpret possible worlds as histories, i.e., as the sum of all state of affairs in a given time branch. In this alternative conception of worlds, we can examine the possible relations between the lifespan of wholes and parts in different types of parthood relations. For instance, figure 4.a illustrates the possible relations between the lifespan of a whole and one of its essential parts.

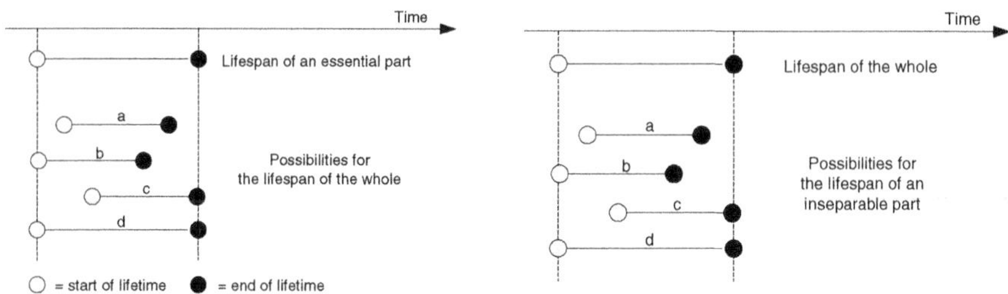


**Fig. 4.** Possible relations between the life spans of an individual whole and: **(a-left)** one of its *essential parts;* **(b)** one of its *inseparable parts.*

This figure illustrates the true possibilities for, for instance, the relation between a chassis and a car as depicted in figure 2.b. In this case, the lifetime of the chassis is completely independent from the lifetime of any of the cars it happens to be a part of. Actually, as represented in figure 2.b, a chassis does not even have to be connected to a car (whole). This is a case of, what we term, *essential part with optional whole*.

Conversely, if we analyze the relation between a brain and a person, we come to the conclusion that the lifespan (d) in figure 4.a is the only real possibility in this case. That is to say that the lifespan of a person and her brain should necessarily coincide. This is because, in this case, a brain is also existentially dependent on its host. Whenever we have the situation that a part is existentially dependent on the whole it composes, we name it an *inseparable part:*

**Definition 8 (inseparable part).** An individual *x* is an inseparable part of another individual *y* iff, *x* is existentially dependent on *y*, and *x* is, necessarily, a part of *y:* **(6). $\mathbf{IP(x,y) =_{def} \Box(\varepsilon(x) \rightarrow (x \leq y))}$.** ■

The possible relations between the life spans of an inseparable part and its (essential) whole are depicted in figure 4.b. The case of an essential and inseparable part is shown in figure 5 below.

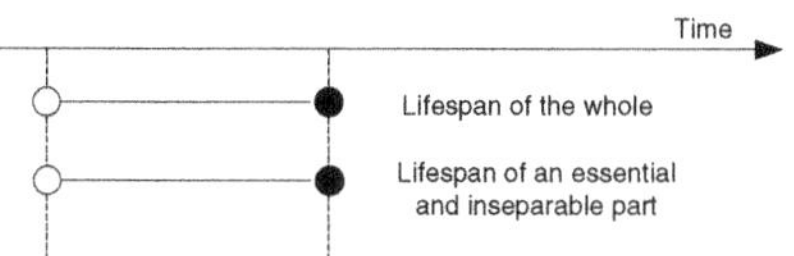


**Fig. 5.** Possible relations between the life spans of an individual whole and one of its *essential and inseparable parts*

Figure 4.b does not represent all the possibilities for, for instance, the relation between a heart and its bearer (figure 2.c), since the heart of person is not an inseparable part of a person and, hence, their life spans can be completely independent. A heart can pre-exist its bearer as well as survive its death. Nonetheless, a heart must be part of *a* person, only not necessarily the same person in all possible circumstances. For these cases, of generic dependence from the part to a whole, we use the term *parts with mandatory wholes*:

**Definition 9 (mandatory whole).** An individual *y* is a mandatory whole for another individual *x* iff, *x* is generically dependent on a type T that *y* instantiates, and *x* is, necessarily, part of an individual instantiating T: **(7). MW(T,x)** $=_{def}$ **□(ε(x) → (∃T,y)(x < y))).** ■

Once more, the distinction between inseparable parts and parts with mandatory wholes is neglected in practically all conceptual modeling languages. For this reason, we propose to extend the current UML aggregation notation with the Boolean meta-attribute **inseparable** to represent inseparable parts. When *inseparable* is equal to *true*, the minimum cardinality constraint in the association end corresponding to the whole type must be at least **1**. If *inseparable* is equal to *false*, the tagged value textual representation can be omitted. A UML class representing a whole type involved in an aggregation relation with minimum cardinality constraint of at least **1** in its association end represents a type whose instances are mandatory wholes.

## 5 The *de re/de dicto* Modal Distinction

In the previous section, we have presented a distinction between parthood relations w.r.t. ontological dependence containing two possible subtypes: (i) *essential parts*: characterized by existential dependence from the whole to a part; (ii) *mandatory parts*: characterized by generic constant dependence from the whole to the type a part instantiates.

As mentioned in the previous section, the relations between a person and her brain, on one hand, and a person and her heart, on the other, can exemplify part-whole relations of sort (i) and (ii), respectively. These two situations taking the human body as an example are depicted in figure 6 together with their corresponding modal logics formalizations. For the sake of simplicity, we formalize in this case only the axioms w.r.t. the relation from the whole to the part. All other axioms are omitted.

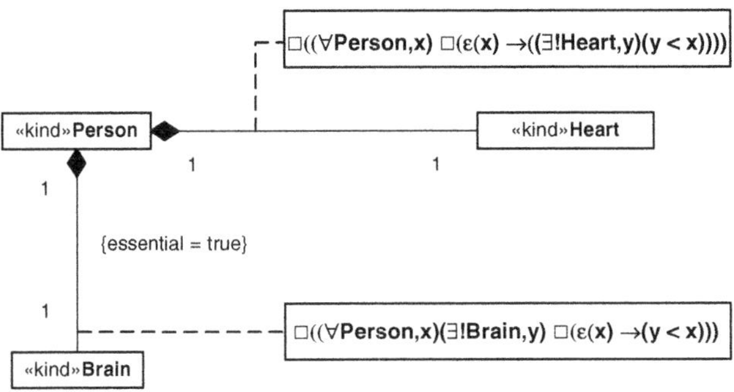


**Fig. 6.** Representation of essential and mandatory parthood in a model of the human body

In all examples used in section 4, the object types representing wholes are *Kinds*. Let us now investigate how these different sorts of necessary parthood relations can be used to characterize non-rigid types, such as *Roles, Phases* or *Role Mixins*. Suppose, for instance, the situation depicted in figure 7. The figure illustrates the relation between a Boxer and one of his hands. What the picture attempts at representing is the statement that "every boxer must have a hand". This relation is certainly not one of mandatory parthood, since it is not the case that a Boxer depends generically on the type hand but specifically on one particular hand[4]. It thus appears to be the case that this relation is one of essential parthood. However, this is not true either. If a hand were to be considered an essential part of a particular boxer then the corresponding formula represented in figure 7 should be valid. To show that this is not the case, suppose the following: let John be a boxer in world w and let x be John's hand in w. What the formula in figure 7 states is that in every world w' in which John exists, x must be part of John in w'. This formula is clearly falsifiable. One just have to imagine a world w'', in which John exists without being a boxer and without having x as his hand (supposed that x has been tragically amputated in w''). This problem arises from the ambiguity of the word "must" in "every boxer must have a hand". Intuitively, the situation that this model intended to express is the valid statement that "For every Person *x*, there is a hand *y*, such that *in every world that x is a Boxer*, *y* is a hand of *x*".

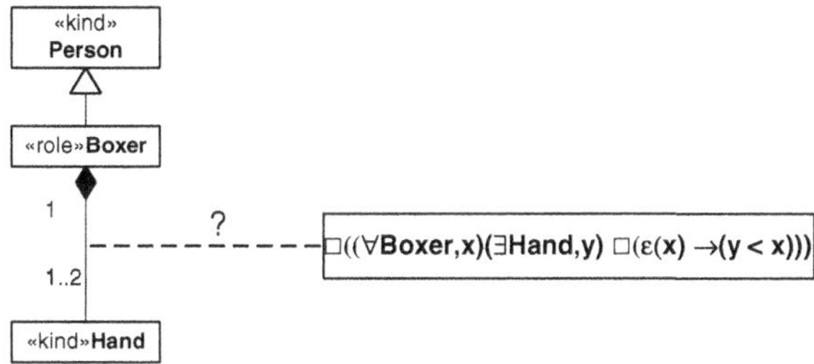


**Fig. 7.** Problems in the representation of specifically dependent parts for anti-rigid types

In the example of figure 7, Boxer cannot have essential properties and, in particular, cannot have essential parts, since it is an anti-rigid type. In other words, if "to be a boxer" is consider as a property, it is not an essential property itself of any individual. However, this situation can be understood in terms of the philosophical distinction between *de re* and *de dicto* modality. Take the following two sentences: (i) *The queen of the Netherlands is necessarily queen*; (ii) *The number of planets in the solar system is necessarily odd*. In the *de re* reading, the first sentence expresses that a certain individual (Beatrix) is necessarily queen. This is clearly false, since we can conceive a different world in which Beatrix decides to abdicate the throne. However, in the *de dicto* reading the sentence simply expresses that it is necessarily true that in any circumstance whoever is the Dutch queen is a queen. The second sentence works in the converse manner. In the *de re* reading the sentence (ii) expresses that a certain number (9) is necessarily odd. This is indeed necessarily true. The *de dicto* reading of the sentence however is false. It is not necessarily the case that the number of planets in the solar system is odd. We can imagine a counterfactual situation in which the solar system has, for instance, 8 or 10 planets. The Latin expressions *de re* represents a

[4] We are here not considering the possibility of hand transplants. Once more, the point of the argumentation is not the specific example.

modality which refers to a property of the thing itself (*res*), whereas *de dicto* represents a modality that refers to an expression (*dictum*). This is made explicit in the logical rendering of the possible readings of these two expressions: (iii-a) *de re* (false): $\forall$x QueenOfTheNetherlands(x) $\rightarrow \Box$(Queen(x)); (iii-b) *de dicto* (true): $\Box$($\forall$x QueenOfTheNetherlands(x) $\rightarrow$ Queen(x)); (iv-a) *de re* (true): $\forall$x NumberOfPlanets(x) $\rightarrow \Box$(Odd(x)); (iv-b) *de dicto* (false): $\Box$($\forall$x NumberOfPlanets(x) $\rightarrow$ Odd(x)).

Take now the expression "every boxer has necessarily a hand". Once more, this expression is true only in one of the readings, namely, the *de dicto* reading. Whilst it is the case that the expression "In any circumstance, whoever is boxer has at least one hand" is necessarily true, it is false that "If someone is a boxer than he has at least a hand in every possible circumstance". Figure 8, expresses a correct representation of this situation in the *de dicto* modality.

We now have expressed three different types of dependency relations between wholes and parts: (i) specific dependence with *de re* modality; (ii) generic dependence with *de re* modality; (iii) specific dependence with *de dicto* modality. The remaining option is, of course, conceivable, i.e., generic dependence with *de dicto* modality. This situation can be captured by the following formula **(v)** $\Box(\forall \mathbf{A,x}\ \Box(\varepsilon(\mathbf{x}) \wedge \mathbf{A(x)} \rightarrow \exists \mathbf{y}\ \mathbf{B(y)} \wedge (\mathbf{y} < \mathbf{x})))$, in which A represents the (anti-rigid) whole and B represents the part. In this formula, the predicate B is used as what we term here a *guard predicate*. Intuitively, this predicate "selects" those worlds, in which the parthood relation must hold. The same holds for the predicate Boxer in figure 8.

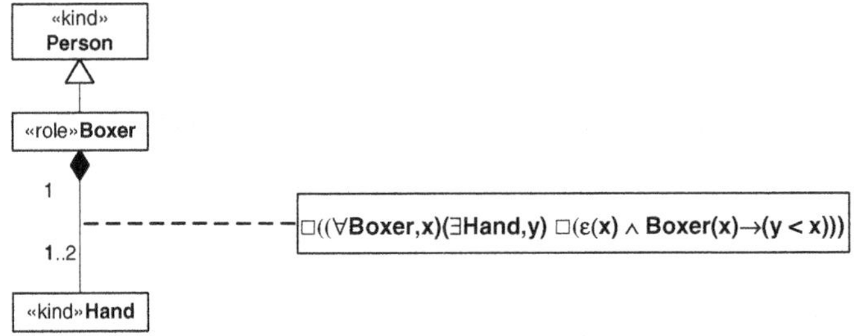


**Fig. 8.** Correct representation of specifically dependent parts of anti-rigid types

We have seen that essential properties, i.e., specific dependence expressed in terms of the *de re* modality, can only be expressed for rigid types. For anti-rigid types (roles, phases, role mixins), only the corresponding *de dicto* modality can be applied. Nonetheless, it is also true that for every *de re* statement regarding an individual *x*, we can express a corresponding *de dicto* one, by using as guard predicate the substance sortal that x instantiates. For instance, if it is true that "The number of planets in the solar system (9) is essentially odd" then it is also true that "In any circumstance, if 9 is a number then 9 is odd". We therefore could rephrase the formulas in figure 6 as follows: **(vi)** $\Box((\forall \mathbf{Person,x})(\exists!\mathbf{Heart,y})\ \Box(\varepsilon(\mathbf{x}) \wedge \mathbf{person(x)} \rightarrow (\mathbf{y} < \mathbf{x})))$ and **(vii)** $\Box((\forall \mathbf{Person,x})\ \Box(\varepsilon(\mathbf{x}) \wedge \mathbf{person(x)} \rightarrow (\exists!\mathbf{Heart,y})(\mathbf{y} < \mathbf{x})))$. Since Person is a kind (rigid type), everything that is person is necessarily a person. In other words, the predicate *person* is modally constant, and for every object selected by the universal quantifier, *person* must be true for this object in every possible world. Consequently, (vi) and (vii) are logically equivalent to their counterparts in figure 6.

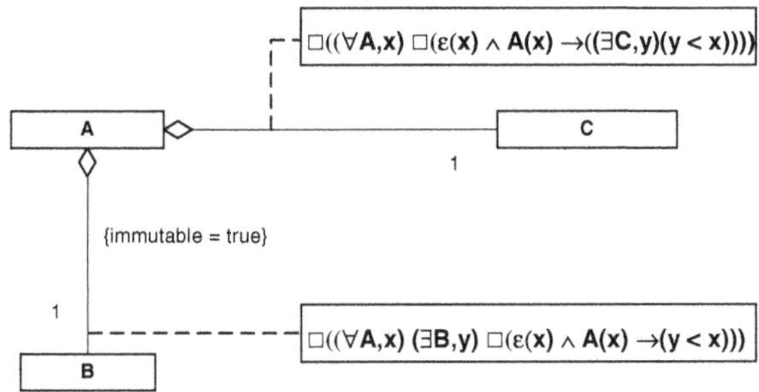


**Fig. 9.** General representation for Immutable and Mandatory parts

In order to achieve a uniform axiomatization, we therefore propose the following formula schemas depicted in figure 9, which must hold irrespective of the type representing the whole being rigid or anti-rigid sortals. If the type A is rigid then A(x) is necessarily true (if true) and the antecedent ($\varepsilon(x) \wedge A(x)$) can be expressed only by ($\varepsilon(x)$). In this case, the B's are truly essential parts of A's. We refrain from using the term *essential part* for the cases in which a mere de dicto modality is expressed. Therefore, for the case of specific dependence from instances of anti-rigid types to theirs part we adopt the term *immutable part* instead. Of course, every essential part is also immutable. Generalization axioms analogous to those in figure 9 can be produced for the case of inseparable and mandatory wholes. Figure 10 depicts a representation of inseparable parts and mandatory wholes, in which guard predicates are included to produce generalizations of the axioms in definitions 6 and 7 that are suitable for the cases of both rigid and anti-rigid types.

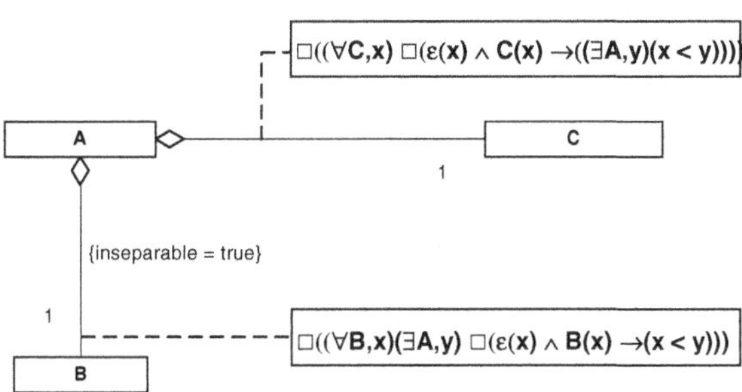


**Fig. 10.** A general representation scheme for Inseparable Parts and Mandatory Wholes

## 6 Related Work

Two of the works the are closest to ours in objectives w.r.t. establishing a foundation for part-whole relations in conceptual modeling are the pioneering works of James Odell reported in *"Six Different Kinds of Composition"* [11], and of Brian Henderson-Sellers and Colleagues reported in a series of articles that includes [7, 12]. There are a number of important issues in which our approach differs from these two proposals regarding a number of ontological aspects of part-whole relations (e.g., constitution versus parthood, emergent properties, transitivity of parthood, among others). A fuller comparison between our proposal and these two approaches (among others) can be found in [5]. Here, we concentrate exclusively on the notions directly related to modal aspects of part-whole modeling and, in particular, on the treatment of the notion of *separability* between parts and wholes.

In his article, Odell has proposed an adaptation of the taxonomy of part-whole relations proposed by Winston, Chaffin and Herrman [16] (henceforth WCH) for the purpose of modeling object-oriented systems. Following WCH, Odell employs a notion of separability as one of the criterion for differentiating between six kinds of part-whole relations. However, this notion employed by Odell (and inherited from WCH) is not a modal notion, but one of physical entanglement. For instance, on page 4 of his article, Odell proposes that the difference between *place-area* (e.g., Everglades-Florida) and *portion-object* (e.g., slice-pie) compositions is that only the former is constituted solely by inseparable parts. Now, if separability is taken in an ontologically meaningful modal sense, there is nothing in the place-area composition relation that requires the parts to be inseparable. For instance, the province of Trentino-Alto Adige is a (place-area) part of Italy, but not an inseparable part, since there are possible worlds (namely before 1921), in which it belonged to the Austrian-Hungarian Empire.

In a different perspective, *contra* Henderson-Sellers and colleagues [7], we take *Lifetime dependency* to be a defining feature of those part-whole relations with essential and/or inseparable parts. In this sense, we disagree with examples such as the one used by the authors to justify the existence of parts that are separable, but share the same destruction as the whole: *"a car wheel is independent of the car but if the wheel is in the car during the car's destruction then it is also destroyed"*. In this case, the wheel is clearly separable from the car, it just happened to be the same event that caused the destruction of both objects (had the wheel been separated from the car, the car's destruction would not propagate to the wheel; the wheel can clearly exist in possible worlds in which the car does not exist). In other words, the lifetime coincidence of two separable objects is merely a contingent fact.

This confusion seems to be motivated by an object-oriented programming bias towards conceptual modeling. Traditionally, in OO programming languages, an object can be made responsible for the destruction of other objects as a procedure for memory de-allocation named *garbage collecting*. Thus, it can be warranted that an object X should trigger the destruction of other objects coupled with X in the moment of its destruction, even if the coupling is merely a contingent one.

Finally, it is important to highlight that none of these approaches investigate the modal properties of Object Types. As a consequence, they also do not establish a systematic relation between the different modal properties of part-whole relations and of the object types they are attached to. Here, in contrast, by exposing some subtle ntions that arise when these categories are combined, we can derive practical modeling constraints for the construction of ontologically well-founded conceptual models.

## 7 Final Considerations

The main objective of this article is to demonstrate the importance of some modal notions in capturing the real-world semantics of some of the conceptual modeling most important constructs, namely, the ones representing the notions of *object types* and *part-whole relations*. The article offers a new formal characterization of the modal aspects of the theory originally proposed in [4] by using a systems of quantified modal logics. Moreover, it shows how the formal characterization of the aforementioned ontological categories makes explicit some subtle issues regarding the *de re* and *de dicto* modalities in conceptual modeling diagrams which are reflected in two different modes of necessary parthood.

The different types of object types (Kind, Role, Phase and Role Mixin) and part-whole relations (essential, immutable, inseparable and mandatory parts, and mandatory wholes) which result from this analysis, as well as the constraints on how they can be combined (e.g., if a part-whole relation is of type Essential then the whole type must be of type Kind) can be used to analyze and re-design the metamodel of current conceptual modeling languages. An example of an ontologically well-founded language redesigned in this manner is the version of UML proposed in [5]. An example of similar approach towards an extension of ORM using the same ontology proposed in [5] can be found in [9].

It is important to emphasize that the focus of this article is not on aspects of formalization *per se* but on how some philosophical issues regarding modality can be used to: (i) illuminate the real-world semantics of conceptual modeling constructs; (ii) justify the proposal of more elaborated extensions of these constructs capturing ontological distinctions within the represented categories; (iii) provide some methodological guidelines for helping the user of the language in choosing the most suitable constructs for representing the elements in the universe of discourse according to his own conceptualization. As an example of (iii), if in a given conceptualization the concept *Person* is taken to be anti-rigid (for instance, in a Legal Ontology, only a conscious entity fully responsible for her acts may be considered to be a person), the model designer knows that this concept should be modeled as a phase, not as a kind. Moreover, since phases are always defined in a phase partition and as a subtype of kind, the designer knows that there are other phases (e.g, *UncounciousHumanBeing*) that are subsumed by the same kind (e.g., *HumanBeing*) that are missing in the model. Still on this example, if we have that in two different models a concept represented by the same lexical label (e.g., Person) but with incompatible modal meta-properties (e.g., Person-as-Phase and Person-as-Kind), we have a formal ground for justifying that they are actually different concepts, and for studying what exactly is the relation between them (e.g., Person-as-Kind is equivalent to Human Being). This feature makes an approach such as this one also relevant for the tasks of model integration and semantic interoperability.

## References


1. Evermann, J.: The Association Construct in Conceptual Modelling - An Analysis Using the Bunge Ontological Model. In: Proceedings of 17th CAiSE, Portugal, pp. 33–47 (2005)
2. Fitting, M., Mendelsohn, R.L.: First-Order Modal Logic. Kluwer Publishers, Boston (1999)

3. Guarino, N., Welty, C.: An Overview of OntoClean. In: Staab, S., Studer, R. (eds.) Handbook on Ontologies, pp. 151–159. Springer, Heidelberg (2004)
4. Guizzardi, G., Herre, H., Wagner, G.: Towards Ontological Foundations for UML Conceptual Models. In: Proceedings of the 1st ODBASE, USA, 2002, p. 1100-1117 ( 2002)
5. Guizzardi, G.: Ontological Foundations for Structural Conceptual Models, PhD thesis, University of Twente, The Netherlands (2005)
6. Guizzardi, G., Wagner, G., Guarino, N., van Sinderen, M.: An Ontologically Well-Founded Profile for UML Conceptual Models. In: Proc. of the 16th CAiSE, Latvia, pp.112–126 ( 2004)
7. Henderson-Sellers, B., Barbier, F.: What Is This Thing Called Aggregation? In: proceedings of Technology of Object-Oriented Languages and Systems Europe'99, Nancy, France, pp. 236–250. IEEE Computer Society Press, Washington (ISBN: 0-7695-0275-X) (June 7-10, 1999)
8. Husserl, E.: Logical Investigations, Routledge, London, 1970 (original 1900/1901)
9. Keet, C.M.: Part-Whole Relations in Object-Role Models, International Workshop on Object-Role Modeling (ORM'2006), Montpellier, France, pp.1116-1127 (2006)
10. McNamara, J.: A Border Dispute, the Place of Logic in Psychology. M.I.T. Press, Cambridge (1986)
11. Odell, J.J.: Six Different Kinds of Composition. In: Advanced Object-Oriented Analysis and Design Using UML, Cambridge University Press, New York (1998)
12. Opdahl, A., Henderson-Sellers, B., Barbier, F.: Ontological Analysis of whole-part relationships in OO-models. Information and Software Technology 43, 387–399 (2001)
13. Rosemann, M., Green, P.: Ontologies and Business Systems Analysis, IDEA, USA (2005)
14. Simons, P.M.: Parts. An Essay in Ontology. Clarendon Press, Oxford (1987)
15. van Leeuwen, J.: Individuals and sortal concepts : an essay in logical descriptive metaphysics, PhD Thesis, University of Amsterdam (1991)
16. Winston, M.E., Chaffin, R., Herrman, D.: A taxonomy of part-whole relations. Cognitive Science 11, 417–444 (1987)
17. Wyssusek, B.: On Ontological Foundations of Conceptual Modeling, Scandinavian Journal of Information Systems, Vol.18, No. 1, ISSN 0905-0167 (2006)

# Change Detection in Ontologies Using DAG Comparison

Johann Eder[1] and Karl Wiggisser[2]

[1] University of Vienna, Dep. of Knowledge and Business Engineering
johann.eder@univie.ac.at
[2] Klagenfurt University, Dep. of Informatics-Systems
wiggisser@isys.uni-klu.ac.at

**Abstract.** Ontologies are shared conceptualizations of a domain. As this domain may change over the time, the ontology has to evolve as well. Additionally, for many applications, it is important to know which version of an ontology was valid at a certain point in time. Several ontology version management systems address this problem. If a user is confronted with different versions of an ontology it is often necessary to identify the changes. We present an efficient graph based approach for change detection between two versions of an ontology based on structural comparisons. The result is a change script transforming the old to the new version. Furthermore, we present an extensive evaluation of the prototype implementation of the change detection system.

## 1 Introduction

An ontology is *an explicit specification of a conceptualization* [1]. Ontologies are seen as important technique for semantic data processing, and in particular for interoperability. They represent knowledge about a certain real world domain. But as the real world tends to change, the ontologies have to change as well. Knowledge about these changes is mandatory to correctly interpret data or documents which were based on the semantics defined in the changed ontology. Furthermore, the correct comparison of data and documents from different points in time, based on different versions of an ontology is only possible if the differences between these versions are known. E. g. when analyzing the development of unemployment rate in the European Union over the last 30 years one has to be aware that both the European Union and the formula for computing the rates changed considerably over this period of time.

Changes between versions of an ontology might not be explicitly available. Frequently, only the different versions are available, but a change history is missing. To help in this situation is the ambition of the work presented here. In particular, we focus on the following problem: Given two versions of an ontology we want to derive an edit script, i. e. a series of change operations, which is able to transform one version into the other. This edit script is then an explicit representation of changes which occurred between the versions of the ontology.

J. Krogstie, A.L. Opdahl, and G. Sindre (Eds.): CAiSE 2007, LNCS 4495, pp. 21–35, 2007.

Our change detection system is based on the structure of the ontology only. One might argue that the important changes in ontologies are changes in the semantics. We assume that every semantic change has to be represented by a structural change, as otherwise two identical representations will have different semantics. There might be structural changes which are not semantic changes. Examples for such changes are representational variations for performance increase. Thus, a fast and reliable algorithm for identifying and describing structural changes is a good start for analyzing the changes in the semantics.

In this paper we present our graph based algorithm for a semiautomatic change detection between two versions of an ontology in detail. Based on an extensive evaluation of the algorithm (Sect. 5) we claim that it is very efficient in terms of both speed and precision.

## 2 Related Work

In [2,3] we presented a graph based approach for ontology versioning. Incorporating changes in such a temporal ontology is easy if one knows all changes, but can be a very complex task, if the differences are not previously known. This is particularly important for users of ontologies who have access to the latest version, but do not have a representation of the changes since their last download. There are approaches for ontology comparison published, e. g. [4,5,6]. Among them, PromptDiff, as a part of the Protege framework [7], and OntoView, a web based system, are the best known. However, to the best of our knowledge, there are no evaluation figures for these systems published. Ontology matching/alignment/merging systems like GLUE [8], Cato [9] or Chimaera [10], although somehow related to our change-detection problem, in fact address a different issue. They are more intended to find the semantic overlapping of two or more *different independently developed ontologies*, whereas our approach is designed to find changes in *two versions* of the *same ontology*.

In [11], we presented a brief first sketch of our concepts without detailed description of the algorithm. It is an extension of an algorithm, successfully applied for identifying changes in dimension structures of data warehouses [12,13]. The major challenges were the far more complex (data) structure of ontologies and the usage of ontology specific information (in particular various forms of relationships) for further improving the accuracy of the applied heuristics.

Graph matching and graph comparison is a long known problem. Because the graph isomorphism problem is in $\mathcal{NP}$ [14], there are several approaches comparing two graphs and/or determine their edit distance using some heuristics or restricting the data structure. For instance, the approaches presented in [15,16,17,18,19,20,21] are only some of them. We evaluated these algorithms but they all have some shortcomings which make them either completely unusable for our purpose or at least very hard to adapt to our problem. Some of the approaches are defined on undirected graphs and others are missing operations essential for our purpose. If an adaption was possible, main advantages of the algorithms would have vanished.

To the best of our knowledge there is no algorithm which can easily be adapted to calculate the edit operations between two DAGs (directed acyclic graphs) as we need them for our ontology versioning system. So we developed a new algorithm inspired by the tree comparison algorithm of Chawathe et al. [22]. It is built upon the same principles, but with major enhancements to support the comparison of directed acyclic graphs. The renaming detection component is adapted from our previous work in this area [12].

## 3 Ontology Graphs and Graph Operations

An ontology can be seen as a graph where the concepts are represented by nodes and semantic relations between concepts by edges. A node consists of an unique *id*, a *label* which represents the concept's name, an object holding implementation-dependent *attributes* (e. g. some comment or description of the concept) and a set of *slots*, a concept we will describe later. The ontology's relations are represented by edges. An edge consists of two nodes (*parent* and *child*) and an *edgetype*, which represents the type of the relation. Common ontological relations like generalization (`IS-A`) or aggregation (`PART-OF`) typically build up a *directed acyclic graph* (DAG). Other relations, e. g. `IS-FRIEND-OF`, may also create cycles in the graph. The user may explicitly define edge types as *acyclic*, i. e. not creating cycles in the graph. All other edges are per default treated as *possibly cyclic*, thus may build up cyclic graphs. For our approach, we assume the ontology graph to be a *rooted directed acyclic graph* (RDAG), i. e. a DAG with exactly one node not having any parents.

To transform an arbitrary digraph, representing an ontology version, to such a RDAG we perform the following steps: First, we assume that there is one single root in the graph, i. e. there is only one node in the graph, not having any parents. If such a root is not present, we create a new node $root_v$, which becomes the *virtual root* of the graph by creating a *vroot* edge from $root_v$ to each node $x$ in the graph not having any parents yet. Next we eliminate cycles. For that purpose, we assume that every node is connected to the root via a path consisting only of edges defined to be acyclic. This will hold for many ontologies, because they often comprise a generalization hierarchy. For all nodes $x$ not satisfying this requirement, we create a *vroot* relation from the root to $x$. As a last step, we create the so called *slots*, representing cyclic edges. Each slot has a *name* and a *type*. For each relation from a node *parent* to node *child*, with relation type $edgeType$, where $edgeType$ is not defined as acyclic, we add the slot with name $edgeType$ of type *child* to the node *parent* and remove the edge between *parent* and *child*. With this steps we can transform any graph into a RDAG. When we assume all relations in an ontology to be directed, i. e. we can determine the start and the end of a relation, this transformation is lossless and unique and can be reversed by replacing the slots of each node with the respective edge and removing the *vroot* node and edges.

Figure 1 shows an example for such a transformation. On the left the original cyclic graph is shown. The relations `IS A` and `PART OF` are defined to be acyclic.

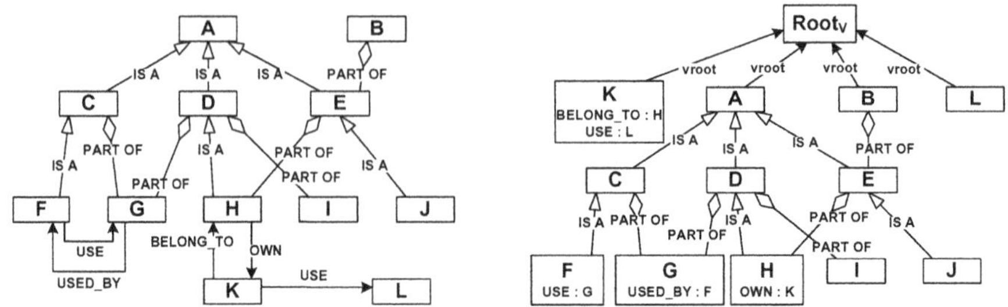


**Fig. 1.** Transformation of arbitrary graph to RDAG

All other relations may build cycles. There is no single root. In the right, the resulting RDAG is shown. A new node $Root_v$ is created and the nodes $A$ and $B$ are attached to it with a *vroot* relation. Now all nodes except $K$ and $L$ are connected to the root via a path, consisting only of acyclic edges. So these two nodes are also connected with a *vroot* relation. For edges like *use* or *belong_to* slots are inserted to the respective nodes, e. g. $K$.

With the operations defined below we can transform any two RDAG into each other. We represent the old version of an ontology with the graph $v_{old}$ and a new version of the same ontology with the graph $v_{new}$. Our goal is to find the differences between two ontology versions in terms of graph operations. We present an algorithm which calculates a so called edit script, which is a sequence of graph operations that transform $v_{old}$ into $v_{new}$. This edit script acts as representation for the changes between the ontologies and thus enables us to incorporate changes of the ontology into virtually any ontology versioning system, for instance like proposed in [2]. The operations defined on the ontology graph are:

- $InsertNode(name, attributes, slots, parents)$ inserts a new node with the label *name*, the attributes *attributes*, and set of *slots* to all *parents*. The set *parents* holds pairs of nodes and edge types (*parent*, *type*), meaning the edge from *parent* to the new node to be of type *type*.
- $DeleteNode(node)$ deletes *node* from the graph. A node can only be deleted, if it does not have any children. With the node, all its incident edges are deleted as well.
- $InsertEdge(parent, child, type)$ creates an edge of type *type* from *parent* to *child*. The new edge must not close a cycle in the graph.
- $DeleteEdge(parent, child)$ deletes the edge from *parent* to *child*.
- $InsertSlot(node, slot)$ adds a new *slot* consisting of name and type to *node*.
- $DeleteSlot(node, slot)$ removes the *slot* from the *node*.
- $UpdateNode(node, attributes)$ changes the attributes of *node* to *attributes*.
- $RenameNode(node, name)$ changes the name of *node* to *name*.
- $ChangeEdgeType(parent, child, type)$ changes the type of the edge between *parent* to *child* to *type*.

# 4 The Comparison Algorithm

Our graph comparison algorithm is inspired by the tree comparison algorithm of Chawathe et al. [22]. Although the algorithm works quite well on ordered trees, it has some shortcomings for our purpose: (i) It is defined on trees and not on RDAG structures. (ii) The renaming of nodes is not supported. (iii) It depends on the ordering of the nodes' children. (iv) It does not support typed edges.

Our approach is built upon the same principles, especially when calculating the node matching between two graphs, but includes some major changes in order to support the comparison of directed acyclic graphs. The renaming detection component is adapted from our previous work in this area [12]. The algorithm is based on the assumption that ontologies do not change very much from one version to the next. This is also supported by [4].

## 4.1 The Longest Common Subsequence

A *Subsequence* of a string is any string obtained by deleting zero or more symbols from the given string. A *Common Subsequence* of two strings $A$ and $B$ is a subsequence of both [24]. The *Longest Common Subsequence*(LCS) of two strings $A$ and $B$ is a common subsequence of $A$ and $B$ such that there is no common subsequence of $A$ and $B$ which contains more symbols. Note that the LCS is not necessarily unique, but there can be different common subsequences of maximal length. Efficient algorithms calculating the LCS are for instance given in [24,25].

The concept of subsequences can easily be extended from strings to sequences of objects of any type. For that purpose we also need to specify a comparison function, which determines whether two elements stemming from either of the sequences are equal. We define the function $LCS$ as follows: $LCS(A, B, equal)$, where $A$ and $B$ are sequences of objects of the same type and $equal(a, b)$ is a function which decides the equality of the objects $a$ and $b$ and returns either `true` or `false`. The function returns a sequence of object pairs $\langle(a_1, b_1), \dots, (a_n, b_n)\rangle$ with the following properties: (i) $a_i \in A$ and $b_i \in B$ (ii) $equal(a_i, b_i) =$ `true` (iii) $\langle a_1, \dots, a_n\rangle$ is a subsequence of $A$ and $\langle b_1, \dots, b_n\rangle$ is a subsequence of $B$ (iv) There is no longer sequence of object pairs which fulfils (i)–(iii).

We use this $LCS$-function to efficiently compare sequences of graph nodes during node matching (see Sec. 4.3).

## 4.2 Node Matching

The first step, when comparing two graphs is to find a good matching between them, i. e. finding nodes which represent the same concept in both graphs.

As for ontologies the concept's name often acts as key, we defined the node's name to be the primary key for matching nodes. That means two nodes cannot match if their names differ. Furthermore, we do not expect nodes to change their hierarchical position within the graph, i. e. leaf nodes will rarely become inner nodes and vice versa. The third assumption we act on is that the attributes and descendants, i. e. edges and slots, will not change very much. Thus, we define a function $similar(x, y)$ returning `true` or `false`, which compares two concepts $x$ and $y$ to take into account the following properties (compare [22,26]):

1. $x.name = y.name$: Two concepts can only match if their name is equal.
2. The function $compare(x, y)$ compares all attributes of $x$ and $y$ for similarity and returns a number between 0 (no similarity) and 1 (identical).
3. $commonSlotsRatio(x, y) = \frac{commonSlots(x,y)}{maxSlots(x,y)}$, where $commonSlots(x, y)$ is the number of slots appearing in both, $x$ and $y$. The function $maxSlots(x, y)$ is the maximum number of slots of $x$ and $y$. Thus, $commonSlotsRatio(x, y)$ returns a value between 0 and 1. If none of the nodes contains slots, the function is defined to return 1.
4. $commonLeavesRatio(x, y) = \frac{commonLeaves(x,y)}{maxLeaves(x,y)}$, with $commonLeaves(x, y)$ calculating the number of common, i.e. matched, leaf descendants of $x$ and $y$ and $maxLeaves(x, y)$ gives the maximum number of leaf descendants of $x$ and $y$. So $commonLeavesRatio(x, y)$ returns a number between 0 and 1. If one of the nodes is a leaf, the function is defined to return 1.

The user may configure the influence for each of these similarity measures, depending on the expected changes. For instance, if the ontology's structure has remained stable but the comments for many concepts changed, the administrator can pay more attention to common slots and leaves than to attributes. Only the criterion of equal names is mandatory. The function $similar(x, y)$ returns `true` iff the two concepts have the same name and each of the above comparison functions returns a value greater than the user defined threshold.

### 4.3 Matching Algorithm

Chawathe et al.'s matching algorithm relies on the ordering of children. Ontologies do not have such an ordering, but a defined order dramatically reduces the complexity during the matching. Therefore, we first sort the nodes' children alphabetically by their name. Then, for each of the graphs we build a list of leaves, traversing the graph from left to right. From these two node lists we build the Longest Common Subsequence, with the function $similar(x, y)$ as equality check. This gives a set of matchings $\mathbb{M}$. A matching is a pair of nodes $(n_i, n_j)$, with $n_i \in v_{old}$ and $n_j \in v_{new}$ which represent the same concept in both versions. We do the same for all inner nodes and add the resulting pairs to $\mathbb{M}$. As a last step during matching calculation, we build a list of still unmatched nodes for each of the graphs and run the LCS algorithm again. The alphabetic sorting of nodes will significantly reduce the effort for the LCS. Figure 2 shows the pseudocode for the matching calculation.

The first run of LCS will match all similar leaves. The second run of LCS does the matching of all similar inner nodes, which depends on the matchings of leaves. The third run of LCS will match nodes, which can either be inner nodes or leaves. The execution order takes into account our assumption that inner nodes seldom will become leaf nodes and vice versa. This approach is a heuristic one and errors may occur. Thus, in production environments, the administrator

**Function** $calculateMatching(v_{old}, v_{new})$

1. Let matching set $\mathbb{M} = \emptyset$;
2. Let $L_o(L_n)$ be the list of leaves, when traversing $v_{old}(v_{new})$ from left to right;
3. Let $\mathbb{M} = \mathbb{M} \cup LCS(L_o, L_n, similar)$
4. For each unmatched node $x$ in $L_o$ if there is an unmatched node $y$ in $L_n$ with $similar(x, y)$: $\mathbb{M} = \mathbb{M} \cup \{(x, y)\}$;
5. Repeat steps 2 – 5 for inner nodes;
6. Repeat steps 2 – 5 for all still unmatched nodes;
7. Let the user acknowledge and correct $\mathbb{M}$
8. Return $\mathbb{M}$;

**Fig. 2.** Pseudocode for the calculation of the matching set

must have the possibility to modify the results of the algorithm, i. e. break up matchings or establish matchings not detected by the system.

Of course, this matching order does not guarantee the best matching, i. e. the matching with the minimum differences. Consider two concepts $a$ and $a'$ within the same ontology version which have the same label. They are quite similar to each other such that $similar(a, a')$ returns `true`, but they are not equal. Now in version 1 of the ontology, when traversing the graph, they appear in order $a$ and later $a'$ but in the traversal of version 2 they appear in order $a'$ and then $a$. As we build the LCS of these traversal sequences and $similar(a, a')$ gives `true`, $a$ from version 1 will be matched to $a'$ from version 2 and vice versa. Thus, this is not the best matching. But as we assume that such situations seldom occur and the calculation a perfect matching is considered to be in $\mathcal{NP}$ [14], we think the result is reasonably good with respect to the gained performance.

### 4.4 Renaming Detection

As the name of a concept is the primary matching criterion, all nodes that cannot be matched could possibly have been renamed. So in the next step, we try to find pairs of nodes, which differ in their names but are so similar with respect to their attributes and structure that they may represent the same concept nonetheless.

Theoretically, each unmatched node from the old graph could have been renamed to any unmatched node in the new graph. To reduce complexity, we only consider node pairs under matched parents as possible renamings. But as this rule may foreclose many renamings to be detected, when, for instance, a new prefix is added to every node, in this phase we also consider possible renamed parents as matched parents. So the renaming of node $w \in v_{old}$ to $x \in v_{new}$ can be detected iff there is at least one parent of $w$ that is matched to a parent $x$ or possible renamed to a parent of $x$. Pairs of nodes, which are possibly renamed are stored in the set *possibleRenamings*. We sort this set according to the hierarchical position of the node such that pairs containing only leaves come first, then the pairs containing only inner nodes, and last the mixed pairs.

Next we calculate the similarity for all possibly renamed pairs, again using the function $similar(w, x)$, but now of course neglecting the different node name. If the similarity of a pair is greater than a user defined threshold, it is considered as

**Function** *calculateRenamings*$(v_{old}, v_{new}, \mathbb{M})$

1. Let *edit script* $\mathbb{E} = \emptyset$; *likelyRenamings* $\mathbb{L} = \emptyset$; *unlikelyRenamings* $\mathbb{U} = \emptyset$
2. Let possible renamings $\mathbb{P} = \{(w, x) | \nexists (w, \_) \in \mathbb{M} \wedge \nexists (\_, x) \in \mathbb{M} \wedge \exists u \in w.parents, v \in x.parents : (u, v) \in \mathbb{M} \vee (u, v) \in \mathbb{P}\}$;
3. Reorder $\mathbb{P}$: Leaf pairs $\longrightarrow$ Inner node pairs $\longrightarrow$ Mixed Pairs;
4. For all pairs $(w, x) \in \mathbb{P}$
   (a) If $similar(w, x) : \mathbb{L} = \mathbb{L} \cup \{(w, x)\}$;
   (b) Else $\mathbb{U} = \mathbb{U} \cup \{(w, x)\}$;
5. Let the user acknowledge and correct the renamings;
6. For each renaming pair $(x, y)$ the user acknowledged
   (a) $\mathbb{M} = \mathbb{M} \cup \{(w, x)\}$;
   (b) $\mathbb{E} = \mathbb{E} \cup \{RenameNode(w, x.name)\}$;
7. Return $\mathbb{E}$;

**Fig. 3.** Pseudocode for the calculation of the node renamings

a *likely renaming*, otherwise we call it an *unlikely renaming*. Of course, each node can only appear in one likely renaming. In case that a node is contained in more than one pair with adequate similarity, the pair with the highest similarity is chosen to be the likely renaming, all others become unlikely renamings. As each renaming results in a matching, when comparing inner nodes, likely renamings are treated as common leaves and contribute to the similarity of inner nodes.

This approach is a heuristic one. For instance, for a node that was renamed and attached to totally different parents, no renaming will be detected, but the node will remain unmatched, even after renaming detection. Thus, in production environments, the user will have to acknowledge or correct the detected renamings. For each acknowledged renaming $(w, x)$, a $RenameNode(w, x.name)$ operation is appended to the edit script and immediately applied on $v_{old}$. The pseudocode for the renaming detection is shown in Fig. 3.

### 4.5 Comparing Two DAGs

After matching and renaming detection, we have all preliminaries for the change detection algorithm. This is split into five phases, each responsible for finding a particular set of operations. For the description of these phases, we need to introduce the *partner* of a node $x$, which is the node $y$ to which $x$ is matched. Thus $x$ and $y$ have to stem from different graphs. Each of the operations detected during the comparison is immediately applied to $v_{old}$. So, during the comparison $v_{old}$ is transformed into $v_{new}$, and when finished, both graphs are identical.

**Insert Phase.** Let $x$ be the current node when traversing $v_{new}$ in topological order. If $x$ is not matched yet, it must have been inserted. Let $\mathbb{Y}$ be the set of parents of $x$ combined with the type of the edge to $x$. Then $\mathbb{Z}$ is the set of partners of the nodes in $\mathbb{Y}$, combined with the respective edge type. As we traverse the graph in topological order, we can be sure that every parent of $x$ has already been visited and thus must have a partner in $v_{old}$. We now can easily create the appropriate $InsertNode(x.name, x.attributes, x.slots, \mathbb{Z})$ operation.

**Update Phase.** Let $x$ be the current node when traversing $v_{new}$ in topological order and $w$ its partner. If the attributes of $x$ and $w$ differ, we append an $UpdateNode(w, x.attributes)$ operation to the edit script.

**Slot Changing Phase.** Let $x$ be the current node when traversing $v_{new}$ in topological order and $w$ its partner. For every slot $s_n$ contained in $x$ but not in $w$, we append an $InsertSlot(w, s_n)$ to the edit script. For every slot $s_o$ contained in $w$ but not in $x$, we append an $DeleteSlot(w, s_o)$ to the edit script.

**Edge Changing Phase.** Let $x$ be the current node when traversing $v_{new}$ in topological order and $w$ its partner. Let $\mathbb{Y}$ be the set of parents of $x$, $\mathbb{V}$ be the set of parents of $w$, each of them combined with the respective edge type from parent to child. We now have to check, whether every node in $\mathbb{Y}$ has a partner in $\mathbb{V}$ and vice versa, and whether all edges are of the correct edge type. For every edge $e$ from $y \in \mathbb{Y}$ to $x$ where the partner of $y$ is not in $\mathbb{V}$, we append an $InsertEdge(y.partner, w, e.type)$ to the edit script. For every edge from $v \in \mathbb{V}$ to $w$ where the partner of $v$ is not in $\mathbb{Y}$, we append a $DeleteEdge(v, w)$ to the edit script. For every edge $e_o$ from $v \in \mathbb{V}$ to $w$ where exists an edge $e_n$ from $v.partner \in \mathbb{Y}$ to $x$ and $e_o.type \neq e_n.type$ we append an $ChangeEdgeType(v, w, e_n.type)$ to the edit script.

**Delete Phase.** Let $w$ be the current node when traversing $v_{old}$ in post-order. If $w$ is not matched, it has been deleted. Thus, we append a $DeleteNode(w)$ operation to the edit script.

**Complete Algorithm.** Figure 4 shows the pseudocode for the complete change detection algorithm (compare [22]). The function $compareOntologies(v_{old}, v_{new})$ takes two versions of an ontology as input and returns an edit script $\mathbb{E}$, which represents the differences between them. First, the Matchings and Renamings are detected. Then, the Inserts, Updates, Edge and Slot changes can be found during one topological graph traversal. The delete phase needs one additional post-order traversal. Every operation that is appended to the edit script, is immediately applied to $v_{old}$, thus the graph is transformed to be equal to $v_{new}$.

### 4.6 Complexity Analysis

After having presented the complete algorithm, we now give a short complexity analysis. Let $n$ be the number of nodes, $n_l$ be the number leaves, $n_i$ the number inner nodes in the graph and $d$ the number of differences between the two version graphs. Let $p$ be the average number of parents of a node and $c$ be the average number of children of a node. Typically $d \ll n$, $p \ll n_i$ and $c \ll n$.

In the matching phase for each of the matching types (leaves, inner nodes, mixed), we have to do an LCS run, which is in $O(n \cdot d)$ and then compare the unmatched nodes with each other, which is in $O(d^2)$. Thus, number of node comparisons is in $O(n \cdot d + d^2)$. But as comparing two inner nodes requires building the intersection of the contained leaves sets, of course $similar(x, y)$ is

**Function** *compareOntologies*$(v_{old}, v_{new})$

1. Edit script $\mathbb{E} = \emptyset$;
2. Matching set $\mathbb{M} = calculateMatching(v_{old}, v_{new})$;
3. Renamings $\mathbb{R} = calculateRenamings(v_{old}, v_{new}, \mathbb{M})$;
4. $\mathbb{E} = \mathbb{E} \cup \mathbb{R}$;
5. Let $x$ be the current node in topological traversal of $v_{new}$;
   (a) If $x$ has no partner
      i. $\mathbb{Z} = \{(p.partner, type) | (p, x, type)$ is an edge in $v_{new}\}$;
      ii. $\mathbb{E} = \mathbb{E} \cup \{InsertNode(x.name, x.attributes, x.slots, \mathbb{Z})\}$;
      iii. $w = InsertNode(x.name, x.attributes, x.slots, \mathbb{Z})$ applied to $v_{old}$;
      iv. $\mathbb{M} = \mathbb{M} \cup \{(w, x)\}$;
   (b) Else
      i. Let $w$ be the partner of $x$;
      ii. If $w.attributes \neq x.attributes$ $\mathbb{E} = \mathbb{E} \cup \{UpdateNode(w, x.attributes)\}$;
      iii. If $commonSlotsRatio(w, x) \neq 1$
         A. For each slot $s_o$ in $w$ which is not in $x$: $\mathbb{E} = \mathbb{E} \cup \{DeleteSlot(w, s_o)\}$;
         B. For each slot $s_n$ in $x$ which is not in $w$: $\mathbb{E} = \mathbb{E} \cup \{InsertSlot(w, s_n)\}$;
      iv. Let $\mathbb{Y} = \{(y, type_1) | (u, x, type_1)$ is an edge in $v_{new}\}$;
      v. Let $\mathbb{U} = \{(u, type_2) | (u, w, type_2)$ is an edge in $v_{old}\}$;
      vi. For each pair $(y, type_1) \in \mathbb{Y}$ with $u = y.partner$, $(u, type_2) \in \mathbb{U}$ and $type_1 \neq type_2$: $\mathbb{E} = \mathbb{E} \cup \{ChangeEdgeType(u, w, type_1)\}$;
      vii. For each pair $(y, type_1) \in \mathbb{Y}$ with $u = y.partner$, $(u, \_) \notin \mathbb{U}$: $\mathbb{E} = \mathbb{E} \cup \{InsertEdge(u, w, type_1)\}$;
      viii. For each pair $(u.type_2) \in \mathbb{U}$ with $y = u.partner$, $(y, \_) \notin \mathbb{Y}$: $\mathbb{E} = \mathbb{E} \cup \{DeleteEdge(u, w)\}$;
6. Let $w$ be the current node in a bottom-up traversal of $v_{old}$;
   (a) If $w$ has no partner
      i. $\mathbb{E} = \mathbb{E} \cup \{DeleteNode(w)\}$;
      ii. Delete $w$ from $v_{old}$;
7. Return $\mathbb{E}$;

**Fig. 4.** Pseudocode for comparing two ontology graphs

in $O(n_l)$ for the inner nodes $x$ and $y$. Thus, the complete matching phase is in $O(n_i \cdot n_l \cdot d + n_l \cdot d^2)$.

All nodes that could not be matched (bound by $d$) could have been renamed. This gives at most $d^2$ node pairs to be compared. For similarity determination of inner nodes we again have to compare the common leaves, which is in $O(n_l)$, thus renaming detection is in $O(n_l \cdot d^2)$.

The detection of the remaining graph differences (i.e. Inserts, Deletes, Updates, Edge and Slot changes) requires at least two graph traversals ($\in O(n)$). The number of edit operations to be found here is in $O(d)$. As for the Inserts and Edge Changes for each node its parents have to be checked, the overall complexity here is in $O(n \cdot p + d \cdot p)$.

The overall complexity for our algorithm is the combination of matching, renaming and difference detection and thus $O(n_i \cdot n_l \cdot d + n_l \cdot d^2) \cup O(n_l \cdot d^2) \cup O(n \cdot p + d \cdot p)$. As $O(n_l \cdot d^2) \subset O(n^2 \cdot d + n \cdot d^2)$, $O(n \cdot p + d \cdot p) \subset O(n^2 \cdot d + n \cdot d^2)$ and $O(n_i \cdot n_l \cdot d + n_l \cdot d^2) \subset O(n^2 \cdot d + n \cdot d^2)$, the overall complexity is $O(n^2 \cdot d + n \cdot d^2)$.

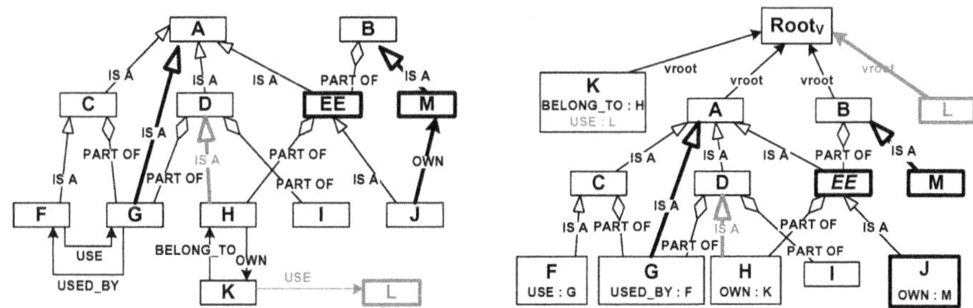


**Fig. 5.** Graph and RDAG representing the new ontology version

1. *RenameNode(E, "EE")* 2. *InsertNode("M", attributes, ∅, (B, IS_A))*
3. *InsertSlot(J, (OWN, M))* 4. *DeleteSlot(K, (USE, L)))* 5. *InsertEdge(A, G, IS_A)*
6. *RemoveEdge(D, H)* 7. *DeleteNode(L)*

**Fig. 6.** Edit script between the two versions

### 4.7 Example Edit Script

Remember the graph and the corresponding RDAG given in Fig. 1. In Fig. 5 we give a subsequent version of this ontology, represented again by the ontology graph and a corresponding RDAG. Elements which have been deleted between the two versions are depicted with strong grey lines. Elements which have changed or have been inserted are depicted with strong black lines. So, when looking at the left graph, we can see that the concept $L$ and the generalization between $D$ and $H$ were deleted. We introduced a new concept $M$, which is a subconcept of $B$. Moreover, the concept $G$ became a subconcept of $A$ and a relation of type *own* was introduced between $J$ and $M$. Finally the concept $E$ was renamed to $EE$. The same changes can be seen in the RDAG in the right side of Fig. 5. When applying our change detection algorithm, the edit script shown in Fig. 6 is generated, including all differences between the two graphs.

### 4.8 Structure Versus Semantics

Our approach compares ontology versions solely on the syntactic level, i.e. it performs a structural comparison. With *structural comparison* in this context we address both: (1) Comparing the structure of the graph (i. e. the relations between concepts) and (2) comparing the structure of the concepts (i. e. their name and attributes). So we also have to discuss, whether this mere structural comparison can provide appropriate means to detect ontology changes.

If the semantics of an ontology changes between two versions, then the structure will change as well. Otherwise, there is no representation of the semantic changes and the changes can neither be detected by structural nor by semantic comparison algorithms. If the structure of an ontology (i.e. its representation) changes, then these structural changes might constitute semantic changes. There are cases where the structure is changed without an underlying semantic change,

e. g. for improving the representation for performance reasons, for better understandability, etc. So probably not all structural changes identified by our approach will also constitute semantic changes. Note that we do not compare two independently developed ontologies but rather two different versions of the same ontology. So the general difficulties of schema integration [27], where frequently the same semantical concepts differ considerable in their representations do not apply. This problem is covered by ontology matching techniques as e. g. [8,9,10].

We can conclude that every semantic change is represented by structural changes. Since semantic comparisons are usually more complex than structural analysis, we expect a facilitation and acceleration of identifying semantic changes if it can be restricted to the structural changes resulting from our change detection procedure. So the edit script which is the output of our algorithm can be input to a procedure for identifying and characterizing changes in the semantics.

## 5 Implementation and Evaluation

### 5.1 Evaluation Environment

For evaluating our approach we need two versions of an ontology to compare them. As we do not have enough ontologies with defined differences available, we simulated them by random DAG. The differences between two versions are also generated randomly.

The generated graphs have the following structure: As an ontology often describes – among other relations – a sort of a taxonomy we divide the graph into levels. A graph may have up to seven levels, depending on its size. After creating the nodes for each level, we define hierarchical relations between them. These hierarchical relations create a directed acyclic graphs. Each node can have one ($p = 0.7$), two ($p = 0.15$) or three ($p = 0.15$) parents from the parent or the grandparent level. Due to the layered structure of the graph, we easily can prevent creating cycles in the graph, by only creating edges from higher levels to lower levels. Cyclic graphs are represented by slots. About a quarter of the created nodes get up to five slots.

The number of generated differences is given as percentage of the number of nodes. Thus, for a graph with size 1000 and change rate of 10%, we generate an edit script consisting of about 100 operations. Within the generated edit script the operations are distributed as follows: Insert Node 10% Delete Node 5% Rename Node 5%, Update Node 10%, Insert Edge 15%, Remove Edge 15%, Insert Slot 15%, Remove Slot 15%, and Change Edge Type 10%. As the graphs are representing ontologies, and ontologies tend not to change very much [4], we tested our algorithm with difference rates of not more than 10%.

### 5.2 Evaluation Results

In the evaluation of the algorithm there are two major points to look at: *calculation time* and *correctness*. Figure 7 shows the average overall calculation

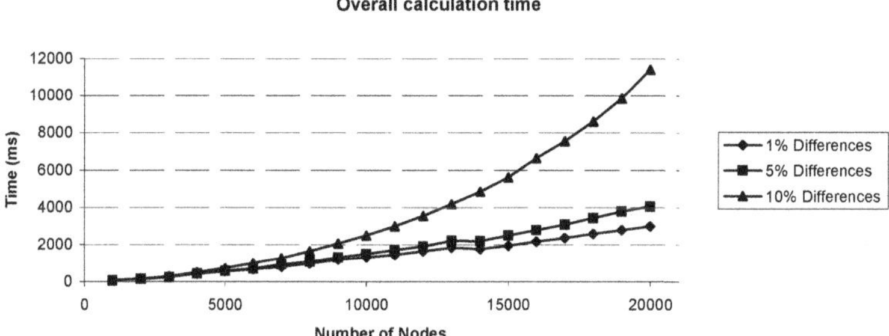


**Fig. 7.** Overall Calculation Time

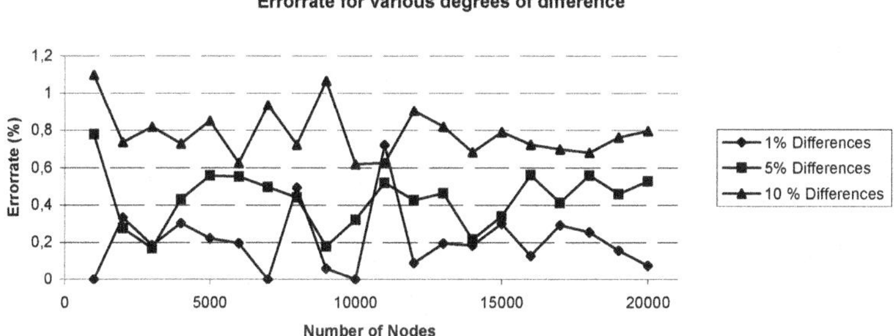


**Fig. 8.** Percentage of Errors in the Result

time for the algorithm for each difference rate. As expected from the complexity analysis, the chart shows quadratic complexity. Figure 8 shows the percentage of errors in the detected edit scripts with respect to the generated differences. This error rate does not only cover the absolute number of found edit operations but each operation is checked for its correctness against the differences created before the comparison. It can be seen that the error rate grows with the percentage of difference but does not strongly depend on the number of nodes in the graph. We also calculated the overall average error rates which are about 0.21% for 1% differences, 0.43% for 5% differences and 0.78% for 10% differences.

When taking a closer look on the generated errors, we see that many of the wrong operations are a direct result of errors in the matching and renaming detection. Because, if the algorithm does not match two nodes which represent the same concept or matches two nodes which do not represent the same concept, the algorithm will produce a series of operations (insert, delete, update, change edges and slots) for reestablishing what seems to be the correct structure. Thus, when using this approach in a production environment, we have to ask the user to acknowledge all detected matchings and renamings for correctness and therefore can foreclose many of the resulting errors.

## 6 Conclusion

We presented an approach for computing an edit script which explicitly represents the structural changes between two versions of an ontology. The approach is based on an efficient graph comparison algorithm.

Our approach can be used if only snapshots of an ontology are available but not a change history. It supports ontology administrators and ontology integrators to identify which changes have taken place between the two given versions. The result can, for an example, be applied to mark data where the underlying semantics changed. It can also provide the input for feeding a temporal ontology where several versions of an ontology together with the mappings between them are represented. Also semantic change analysis can be accelerated when it has to consider only the structural changes identified by structural comparison.

We tested the algorithm exhaustively, and honestly, we were surprised by its good performance figures, both in terms of time and precision.

## Acknowledgements

We would like to thank Mark Musen (SMI, Stanford) who provided us a version history of the BioSTORM ontology [23] for our experiments.

The work on this project was partially supported by the project GATIB - Genome Austrian Tissue bank within the Austrian Genome Program GEN-AU, and by the EU Network of Excellence INTEROP.

## References

1. Gruber, T.: A Translation Approach to Portable Onotology Specifications. Knowledge Acquisition 5(2) (1993)
2. Eder, J., Koncilia, C.: Modelling Changes in Ontologies. In: Proc. of On The Move - Federated Conferences. LNCS, vol. 3292, Springer, Heidelberg (2004)
3. Eder, J., Koncilia, C.: Interoperability in Temporal Ontologies. In: Proc. of the Open Interop Workshop on Enterprise Modelling and Ontologies for Interoperability (2005)
4. Noy, N., Musen, M.: PromptDiff: A fixed-point algorithm for comparing ontology versions. In: Proc. of the Nat'l Conf. on Artificial Intelligence (2002)
5. Klein, M., Fensel, D., Kiryakov, A., Ognyavov, D.: Ontology versioning and change detection on the Web. In: Knowledge Engineering and Knowledge Management. Ontologies and the Semantic Web, 13th Int'l Conf (2002)
6. Mostowfi, F., Fotouhi, F.: Change in Ontology and Ontology of Change. In: Proc. of Workshop on Ontology Management: Searching, Selection, Ranking, and Segmentation (2005)
7. Noy, N., Kunnatur, S., Klein, M., Musen, M.: Tracking Changes During Ontology Evolution. In: Proc. of the 3rd Int'l Conf. on the Semantic Web (2004)
8. Doan, A., Madhavan, J., Domingos, P., Halevy, A.Y.: Ontology matching: A machine learning approach. In: Staab, S., Studer, R. (eds.) Handbook on Ontologies, pp. 385–404. Springer, Heidelberg (2004)

9. Felicíssimo, C.H., Breitman, K.K.: Taxonomic ontology alignment – an implementation. In: Workshop em Engenharia de Requisitos. pp. 152–163 (2004)
10. McGuinness, D., Fikes, R., Rice, J., Wilder, S.: An environment for merging and testing large ontologies. In: Proc. of the 7th Int'l Conf. on Principles of Knowledge Representation and Reasoning. pp.483–493 (2000)
11. Eder, J., Wiggisser, K.: Detecting Changes in Ontologies via DAG Comparison. In: Proc. of the Open Interop Workshop on Enterprise Modelling and Ontologies for Interoperability (2006)
12. Eder, J., Koncilia, C., Wiggisser, K.: A Tree Comparison Approach to Detect Changes in Data Warehouse Structures. In: Proc. of the 7th Int'l Conf. on Data Warehousing and Knowledge Discovery. pp.1–10 (2005)
13. Eder, J., Wiggisser, K.: A DAG Comparison Algorithm and Its Application to Temporal Data Warehousing. In: Advances in Conceptual Modeling – Theory and Practice, ER Workshops 2006. pp.217–226 (2006)
14. Garey, M., Johnson, D.: Computers and Intractability – A Guide to the Theory of NP-Completeness. W.H. Freeman and Company, New York (1979)
15. Wang, J.T.L., Zhang, K., Chirn, G.W.: Algorithms for approximate graph matching. Information Sciences 82(1-2), 45–74 (1995)
16. Zhang, K., Wang, J., Sasha, D.: On the editing distance between undirected acyclic graphs. Int'l Journal of Foundations of Computer Science 7(1), 43–58 (1996)
17. Messmer, B., Bunke, H.: A new algorithm for error-tolerant subgraph isomorphism detection. IEEE Trans. on Pattern Analysis and Machine Intelligence 20, 493–505 (1998)
18. Shoubridge, P., Kraetzl, M., Ray, D.: Detection of abnormal change in dynamic networks. In: Proc. of Information Decision and Control, IEEE Inc. pp.557–562 (1999)
19. Cordella, L., Foggia, P., Sansone, C., Vento, M.: Perfomance evaluation of the vf graph matching algorithm. In: Proc. of the 10th Int'l Conf. on Image Analysis and Processing. pp.1172–1177 (1999)
20. Hlaoui, A., Wang, S.: A new algorithm for inexact graph matching. In: Proc. of the 16th Int'l Conf. on Pattern Recognition (ICPR'02) - Vol. 4 (2002)
21. Gori, M., Maggini, M., Sarti, L.: Exact and approximate graph matching using random walks. IEEE Trans. on Pattern Analysis and Machine Intelligence 27(7), 1100–1111 (2005)
22. Chawathe, S., Rajaraman, A., Garcia-Molina, H., Widom, J.: Change detection in hierarchically structured information. In: Proc. of the ACM SIGMOD Int'l Conf. on Management of Data. pp.493–504 (1996)
23. Crubzy, M., O'Connor, M., Buckeridge, D., Pincus, Z., Musen, M.: Ontology-centered syndromic surveillance for bioterrorism. IEEE Intelligent Systems 20(5), 26–35 (2005)
24. Myers, E.: An O(N D) Difference Algorithm and Its Variations. Algorithmica 1(2), 251–266 (1986)
25. Bergroth, L., Hakonen, H., Väisänen, H.: New Refinement Techniques for Longest Common Subsequence Algorithms. In: String Processing and Information Retrieval, Proceedings. pp.287–303 (2003)
26. Zhang, L.: On matching nodes between trees. Tech. Rep. 2003–2067, HP Labs (2003)
27. Halevy, A.Y.: Structures, semantics and statistics. In: Proc. of the 13 th Int'l Conf. on Very Large Data Bases.pp. 4–6 (2004)

# Automatic Generation of Model Translations

Paolo Papotti and Riccardo Torlone

Università Roma Tre
Roma, Italy
{papotti,torlone}@dia.uniroma3.it

**Abstract.** The translation of information between heterogeneous representations is a long standing issue. With the large spreading of cooperative applications fostered by the advent of the Internet the problem has gained more and more attention but there are still few and partial solutions. In general, given an information source, different translations can be defined for the same target model. In this work, we first identify general properties that "good" translations should fulfill. We then propose novel techniques for the automatic generation of model translations. A translation is obtained by combining a set of basic transformations and the above properties are verified locally (at the transformation level) and globally (at the translation level) without resorting to an exhaustive search. These techniques have been implemented in a tool for the management of heterogeneous data models and some experimental results support the effectiveness and the efficiency of the approach.

## 1 Introduction

### 1.1 Goal and Motivations

In today's world of communication, information needs to be shared and exchanged continuously but organizations collect, store, and process data differently, making this fundamental process difficult and time-consuming. There is therefore a compelling need for effective methodologies and flexible tools supporting the management of heterogeneous data and the automatic translations from one system to another.

In this scenario, we are involved into a large research project at Roma Tre University whose goal is the development of a tool supporting the complex tasks related to the translation of data described according to a large variety of formats and data models [1,2,3]. These include the majority of the formats used to represent data in current applications: semi-structured models, schema languages for XML, specific formats for, e.g., scientific data, as well as database and conceptual data models. In this paper, we focus our attention on the problem of the automatic generation of "good" data model translations.

We start observing that, in general, given a data source, different translations can be defined for the same target model. To clarify this aspect, let us consider the example in Figure 1 where the relational schema $a$ is translated into an XML

J. Krogstie, A.L. Opdahl, and G. Sindre (Eds.): CAiSE 2007, LNCS 4495, pp. 36–50, 2007.

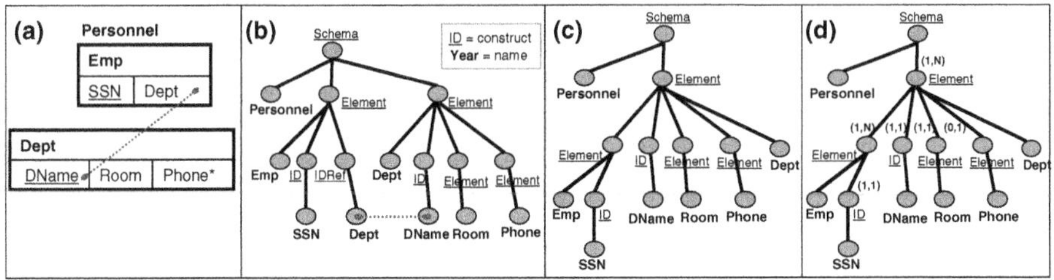


**Fig. 1.** The translation of a relational schema into an XML based model

based structure. Actually, several solutions are possible since it is well known that different strategies can be followed [4]. We report just three of them.

In our example, we can choose between a nested-based (schemas $c$ and $d$) and a flat-based (schema $b$) structure. The latter can be easily generated, but the schema we obtain is probably not desirable in a model with nesting capabilities. Moreover, the question arises whether we want to force "model" constraints like the presence of an order or the absence of duplicates. The second point is that differences between translations are not only structural. For instance, schemas $c$ and $d$ are similar but they have significant differences in the schema semantics, since $d$ also includes cardinality constraints on the elements. Finally, the efficiency of the translation is clearly an issue [5,6].

In order to tackle this problem, in this paper we first identify general properties that the translations should fulfill and investigate the conditions under which a translation can be considered better than another. We then propose efficient methods for the automatic generation of schema and data translations from one model to another. We also show experimental results obtained with a tool for the management of heterogeneous data models in which the proposed methods have been implemented.

### 1.2 Related Works and Organization

The problem of model translation is one of the issues that arises when there is the need to combine heterogeneous sources of information. Many studies can be found on this problem. For instance, translations between specific pairs of models have been deeply investigated [4,7] and are widely supported in commercial products. Our goal is more general: the development of a flexible framework able to automatically translate between data models that, in principle, are not fixed a priori. In recent years, an aspect of the translation problem that has been deeply studied is data exchange [8], where the focus is on the translation of data between two fixed schemas, given a set of correspondences between the elements. Recently, the problem has been set in the general framework of *model management* [9], where a set of generic operators is introduced to cope, in a uniform way, various metadata related problems. One of them is the ModelGen operator that corresponds to the problem tackled in this paper. An early approach to

ModelGen for conceptual data models was proposed by Atzeni and Torlone [2] with a tool based on an internal metamodel and a library of transformations. Following works [3] and similar approaches [10,19] has been presented in the last years. Currently, there are two active projects working on this subject. Atzeni et. al [1] recently provided a comprehensive solution based on a relational dictionary of schemas, models and translation rules. Their approach however does not consider the automatic generation of translations. The approach of Bernstein et al. [11] is also rule-based and it introduces incremental regeneration of instance mappings when source schema changes. A detailed description of their approach has not yet appeared. Our contribution is orthogonal to both projects. In previous works of ours [3] we have focused our attention to the management of Web information and we have proposed a general methodology for the translation of schema and data between data models. In this paper, the focus is on the *automatic generation* of translations based on the *ranking* of the possible solutions. MOF [12] is an industry-standard framework where models can be exchanged and transformed between different formats and provides a uniform syntax for model transformation. Our approach is complementary: we provide methods to automatically perform translations between models, possibly expressed in a MOF-compliant way.

The rest of the paper is organized as follows. In Section 2 we provide the needed background and, in Section 3, we investigate the general properties of model translations. In Section 4 we present the algorithms for the automatic generation of translation and, in Section 5, we provide some experimental results. Finally, in Section 6, some conclusions are drawn and future work is sketched.

## 2 Background

### 2.1 Translations, Metamodel and Patterns

We identify four levels of abstraction: (1) *data* (or *instances*) organized according to a variety of (semi) structured formats (relational tables, XML documents, HTML files, scientific data, and so on); (2) *schemas*, which describe the structure of the instances (a relational schema, a DTD, an XML Schema or one of its dialects, etc.); (3) *(data) models*, that is, formalisms for the definition of schemas (e.g., the relational model, the XML Schema model or a conceptual model like the ER model), and (4) a *metamodel*, that is, a general formalism for the definition of models.[1]

In this framework, a *schema translation* from a source model $M_s$ to a target model $M_t$ is a function $\sigma : \mathcal{S}(M_s) \rightarrow \mathcal{S}(M_t)$, where $\mathcal{S}(M)$ denotes the set of schemas of $M$, and $\mathcal{I}(S)$ the set of instances of $S$. If $S \in \mathcal{S}(M_s)$ then $\sigma(S)$ is called the $\sigma$*−translation* of $S$ (this corresponds to the ModelGen operator [9]). Similarly, a *data translation* from a source schema $S_s$ to a target schema $S_t$ is a function $\delta : \mathcal{I}(S_s) \rightarrow \mathcal{I}(S_t)$.

[1] We refer to a "database" terminology; in other works (e.g., [9,12]), a schema is called *model*, a model is called *metamodel*, and a metamodel is called *metametamodel*.

As others [1,2,3,10,11], our approach is based on a unifying *metamodel* made of a set of *metaprimitives* each of which captures similar constructs of different data models. More precisely, a metaprimitive represents a set of constructs that implement, in different data models, the same basic abstraction principle [14]. For instance, a set of objects is represented by a class in ODL and by an entity in the Entity-Relationship model. Clearly, metaprimitives can be combined. We will call a specific combination of metaprimitives a *pattern*. In this framework, a model is defined by means of: (i) set of *primitives*, each of which is classified according to a metaprimitive of the metamodel, and (ii) a set of patterns over the given primitives.

As an example, the table in Figure 2 describes a set of models. The first column contains a set of possible patterns over the metaprimitives of the metamodel. Each pattern has a metaprimitive $m$ as root and a collection of metaprimitives that are used as components of $m$ ("*" means 0 or more times). In the other columns of the table different models are defined by listing the patterns used and the names given to them in the model. For instance, the relational model is defined by means of a set having the table pattern (which correspond to the metaprimitive relation) which is composed by a number of attribute constructs, one key and, possibly, a foreign key.

| | XmlSchema | DTD | ODL | Relational | ER |
|---|---|---|---|---|---|
| **Element** | *element* | *element/entity* | - | - | - |
| Domain | √ | √ | - | - | - |
| AttributeOfElement* | √ | √ | - | - | - |
| Key | √ | √ | - | - | - |
| Cardinality | √ | √ | - | - | - |
| **Object** | - | - | *class* | - | *entity* |
| Key | - | - | √ | - | √ |
| Attribute* | - | - | √ | - | √ |
| Relationship | - | - | √ | - | √ |
| **Relation** | - | - | - | *table* | - |
| Attribute* | - | - | - | √ | - |
| Key | - | - | - | √ | - |
| Foreign key | - | - | - | √ | - |
| **Domain** | *type* | *type* | *type* | *type* | *type* |
| Struct* | - | - | √ | - | - |
| Restriction | √ | - | √ | √ | √ |
| List | √ | √ | - | - | - |
| Extension | √ | - | - | - | - |
| **Key** | *key* | *key* | *key* | *key* | *key* |
| Domain | √ | √ | √ | √ | √ |
| **Attribute** | - | - | *attribute* | *attribute* | *attribute* |
| Domain | - | - | √ | √ | √ |
| ForeignKey | - | - | - | √ | - |
| Cardinality | - | - | √ | - | √ |
| **Relationship** | - | - | *relationship* | - | *relationship* |
| Cardinality* | - | - | - | - | √ |
| AttributeOfObject | - | - | - | - | √ |
| RelationshipType | - | - | √ | - | - |
| ... | | | | | |

**Fig. 2.** A set of models described by patterns

A pattern corresponds to a context free grammar that makes use of an alphabet denoting the primitives of the metamodel. We call a string of this grammar a *structure*. A *schema* can be obtained by associating names to the symbols of a structure. For instance, the schema (b) in Figure 1 is obtained by adding the nodes in bold to the rest of the tree, which corresponds to the underlying structure.

### 2.2 A Transformational Approach

In [3], we have introduced a general methodology for model translation based on three main steps: (1) the source schema $S$ is first represented in terms of the metamodel so that it can be compared with the target data model definition; (2) source schema and target model may share some constructs (metaprimitives), but others must be translated or eliminated to obtain a schema consistent with the target data model. This operation is performed on $S$: the system tries to translate the metaprimitives of $S$ into metaprimitives of the target model or, if the translation fail, it removes them; (3) a rewriting of the generated schema in terms of the target model syntax is executed.

The translation step, which is the fundamental phase of the process, takes as input a schema expressed in terms of (patterns of) metaprimitives. As the number of metaprimitives is limited, it is possible to define a library of basic and "generic" transformations that can be composed to build more complex translations. These basic transformations implement rather standard translations between metaprimitives (e.g., from a relation to an element or from a n-ary aggregation to a binary one). Representatives of such transformation have been illustrated in [3].

Actually, each basic transformation $b$ has two components: a schema translation $\sigma$ and a data translation $\delta$. Its behavior can be conveniently represented by a *signature* $b[\mathbf{P}_{in} : \mathbf{P}_{out}]$, that is, an abstract description of the set of patterns $\mathbf{P}_{in}$ on which $p$ operates and of the set of patterns $\mathbf{P}_{out}$ introduced by $p$. Note that this description makes the approach independent of the actual implementation of the various transformations. As an example, the signature of an unnesting transformation that transforms each nested element into a set of flat elements related by foreign keys is the following:

$$\begin{aligned} &b[\{ComplexElement(ComplexElement+, AtomicElement*, Domain)\} : \\ &\quad \{ComplexElement(Key+, ForeignKey*, AtomicElement*, Domain)\}] \end{aligned}$$

It turns out that the effect of a transformation with signature $b[\mathbf{P}_{in} : \mathbf{P}_{out}]$ over a structure that makes use of a set of primitives $\mathbf{P}$ is a structure using the primitives $(\mathbf{P} - \mathbf{P}_{in}) \bigcup \mathbf{P}_{out}$.

## 3 Transformations and Translation

In this section, we first investigate general properties of basic transformations and then introduce properties for complex translations.

### 3.1 Properties of the Basic Transformations

Several properties characterizing the correctness, the effectiveness and the efficiency of a basic transformation can be defined, and this issue has been largely debated in the literature (see for instance [2]). Among them, we have focused our attention into the properties that follow.

The first property states a consistency relationship between the schema translation and the data translation which compose a basic transformation.

**Definition 1.** *A basic transformation* $b = (\sigma, \delta)$ *is* consistent *if for each schema* $S \in \mathcal{S}(M_s)$ *and for each instance* $I$ *of* $S$*,* $\delta(I)$ *is an instance of* $\sigma(S)$*.*

A key aspect for a schema transformation is its "correctness", that is, the fact that the output schema is somehow equivalent to the input one. The equivalence of two schemas is a widely debated topic in literature, and all the approaches rely on the ability of the target schema to represent the same information of the source one [15,16,17]. In other words, all data associated with the input schema can be recovered from the data associated with the output schema. This notion has been named *equivalence preserving* or *information preserving* and have been formalized by means of the following properties:

- a data translation $\delta$ from $S_s$ to $S_t$ is *query preserving* w.r.t. a query language $\mathcal{L}$ if there exists a computable function $F : \mathcal{L} \rightarrow \mathcal{L}$ such that for any query $Q \in \mathcal{L}$ over $S_s$ and any $I \in \mathcal{I}(S_s)$, $Q(I) = F(Q)(\delta(I))$;
- a data translation $\delta$ from $S_s$ to $S_t$ is *invertible* if there exists an inverse $\delta^{-1}$ of $\delta$ such that, for each instance $I \in \mathcal{I}(S_s)$, $\delta^{-1}(\delta(I)) = I$.

In our context, the property that actually guarantees the equivalence depends on the internal model used to represent the schemas. In [1] the internal model is based on a relational dictionary, and it has been shown that calculus dominance and query dominance are equivalent for relational settings [16]. In contrast, invertibility and query preservation do not necessarily coincide for XML mappings and query languages [18,15]. In the following, we will refer to a notion of "equivalence preserving" that relies on query preservation.

As we have shown in the introduction, even if we assume that all the transformations preserve the above properties, there are transformations that are preferable than others. Different issues can be considered in this respect: redundancy, ease of update maintenance, ease of query processing with respect to certain workload, and so on. We therefore assume that a preference relationship can be defined over the basic transformations according to one or more of these aspects. One important point is that this preference relationship depends, in many cases, on the target model. For instance, a translation to an object model (with both relationships and generalizations) that is able to identify generalization hierarchies between classes is preferable to a translation that only identifies generic relationships between them. This is not true if the target is the relational model.

First of all, we say that two basic transformations $b[\mathbf{P}_{in} : \mathbf{P}_{out}]$ and $b'[\mathbf{P}'_{in} : \mathbf{P}'_{out}]$ are *comparable* if either $\mathbf{P}_{in} \subseteq \mathbf{P}'_{in}$ or $\mathbf{P}'_{in} \subseteq \mathbf{P}_{in}$.

**Definition 2.** *Given a set $L$ of basic transformations and a target data model $M_t$ a* preference relationship $>_{M_t}$ towards $M_t$ *is a poset over comparable transformations in $L$. Given two comparable basic transformations $b_1$ and $b_2$ in $L$, we say that $b_1$ is* preferable *to $b_2$ w.r.t. $M_t$ if $b_1 >_{M_t} b_2$.*

*Example 1.* Let $P_1$ be a pattern denoting an Entity, $P_2$ be a pattern denoting a Relationship, and $P_3$ a pattern denoting a Generalization. Given the basic transformations $b_1$ with signature $[\{P_3\}{:}\{P_1, P_2\}]$, which translates generalizations into entities and relationships, and $b_2[\{P_3\}{:}\{P_1\}]$, which simply translates generalizations into entities. Then, we can state that $b_1 >_{M_t} b_2$ if $M_t$ is a model with entities and relationships. The rationale under this assertion is that $b_1$ takes more advantage than $b_2$ of the *expressiveness* of the target data model.

In Section5, we will concretely specify a specific preference relationship that is suitable for our purposes.

We finally define a property for the evaluation of the performance of a basic transformation. The best way to measure the effective cost of a transformation would be the evaluation of its execution at runtime. Obviously we would prefer to not actually execute basic transformations on instances to compare their performance. Since execution time optimization is not our primary goal, an estimation of each basic transformation complexity is a reasonable solution. In particular, we assume that the designer provides a specification of the complexity of the algorithm with respect to the size of the database. For instance, the complexity of the unnesting transformation described in the previous is linear with respect to the database. We denote the complexity for a basic transformation $b_x$ with $c(b_x)$.

**Definition 3.** *Given a set $L$ of basic transformations an* efficiency relationship *$\succ$ is a total order over $L$ such that $b_i \succ b_j$ if $c(b_i) > c(b_j)$. Given two basic transformations $b_i$ and $b_j$, we say that $b_i$ is more* efficient *than $b_j$ if $b_j \succ b_i$.*

### 3.2 Properties of Translations

We have just defined some *local* properties of transformations, but we would like to study also *global* properties of entire translations.

It has been observed that, in the transformational approach, if every transformation $b_i$ in the library is equivalence preserving, the information preservation for the sequential application of two or more transformations is guaranteed by construction [2,17]. The same guarantee applies for the schema validation: the generated output schema cannot contain primitives that are not allowed in the target data model. It turns out that a translation $t = b_1, \ldots, b_n$ from $M_s$ to $M_t$ is *consistent* if each $b_i$ in $\sigma$ is consistent.

We now extend the preference property to translations.

**Definition 4.** *Given two translations $t = b_1, \ldots, b_m$ and $t' = b'_1, \ldots, b'_n$ and a target model $M_t$, $t >_{M_t} t'$ if: (i) there exists a transformation $b_i$ in $t$ such that $b_i >_{M_t} b'_j$, for some transformation $b'_j$ in $t'$, and (ii) there is no transformation $b'_k$ in $t'$ such that $b'_k >_{M_t} b_l$ for some transformation $b_l$ in $t$.*

It is easy to show that the above relationship is a poset over the set of all possible translations.

*Example 2.* Assume that $L$ contains three basic transformations: $b_1$ with signature $[\{P_3\}{:}\{P_1, P_2\}]$, which translates generalizations ($P_3$) in entities and relationships ($P_1$ and $P_2$ respectively), $b_2$ with $[\{P_1, P_2\}{:}\{P4\}]$, which translates entities and relationships into elements ($P_4$), and $b_3[\{P_3\}{:}\{P_1\}]$, which translates the generalizations into entities. Consider a target data model $M_t$ with just the element pattern (a subset of DTD). If we assume the preference discussed in Example 1 ($b_1 >_{M_t} b_3$), and consider the following translations: $t_1 = b_3, b_2$, $t_2 = b_2, b_3$, and $t_3 = b_1, b_2$, then we have that $t_3 >_{M_t} t_2$ and $t_3 >_{M_t} t_1$.

The efficiency of a translation can be defined with different levels of granularity. In [3] we proposed a preliminary evaluation based on the *length* of the translation, that is, the number of basic transformations that composed the actual solution. We now extended the definition: the efficient solution is the translation $t$ that globally minimize the cost of the basic transformations $b_1, \ldots, b_n$ that compose $t$.

**Definition 5.** *Given two schema translations $t = b_1, \ldots, b_m$ and $t' = b'_1, \ldots, b'_n$, $t$ is more* efficient *than $t'$ if* $\max_{j=1}^{m} c(b_j) < \max_{i=1}^{n} c(b'_i)$.

If we consider the data model translation problem (a translation of source schema $S$ into a target model $M_t$ given a library $L$ of basic transformations $\{b_1, \ldots, b_n\}$), it turns out that the above properties lead to two different classifications of the possible solutions for the problem based on the orthogonal notions of efficiency and preferability.

**Definition 6.** *A translation $t$ is* optimal *if there is no other translation that is more preferable and more efficient than $t$.*

Note that, in general, several optimal translations can exist. We will see in the next section how the classification of translations can be the basis for an automatic ranking of the solutions and for an efficient algorithm that retrieves solutions without generating all the alternatives.

## 4 Automatic Generation of Translation

In this section we present two approaches to the problem of the automatic generation of translations. The former is based on an exhaustive search, the latter relies on a best-first technique and is much more efficient.

### 4.1 Computing and Ranking all Translations

In [3], we have proposed a basic strategy that follows a greedy approach: given a source schema $S_s$, a target model $M_t$ and a library of basic transformations $L$, this method applies exhaustively the following rule over a working schema $S$, which initially coincides with $S_s$, and an initially empty sequence of transformations $t$.

if (a) the $S$ makes use of a pattern $P$ that is not allowed in the target model, and (b) there exists a transformation $b(\sigma, \delta) \in L$ whose effect is the removal of $P$ and (possibly) the introduction of patterns allowed in the target model, then append $b$ to $t$ and set $S$ to $\sigma(S)$.

When condition (a) fails, the process terminates successfully and the sequence $t$ is a solution for the data translation problem.

This simple method can be extended to an algorithm ComputeAllSolutions that generates all the possible solutions for $S_s$. It is possible to show that this algorithm is complete in the sense that every valid translation from the $S_s$ to the target data model is in the solution set $T$. This algorithm is shown in Figure 3.

**Input:** A schema $S_s$, a target model $M_t$ and a library of basic transformations $L = \{b_1, \ldots, b_n\}$.
**Output:** A set of all translations $T = \{t_1, \ldots, t_n\}$ of $S_s$ into $M_t$ (each $t_i$ is a sequence of basic transformations in $L = \{b_1, \ldots, b_n\}$).
**begin**
(1) Set $t$ to the empty translation and $st$ to the structure of $S_s$;
(2) Add $(t, st)$ to the set of possible solution $Sol$;
(3) **while**, for each $s \in Sol$, there is a pattern $P$ in $st$ s.t. $P$ is not allowed in $M_t$ **do**
(4) let $B$ denote the set of all $(b_p, t', st'_i)$ branches under $st$ such that:
(a) $b_p$ is a basic transformation whose input signature matches $P$,
(b) $t'$ is a copy of the actual $t$,
(c) $st'_i$ is the resulting structure after $b_p$'s application;
(5) for each branch $(b_p, t', st'_i)$ in $B$:
(6) **if** $b_p \in t'$:
(7) **then** discard $(b_p, t', st'_i)$;
(8) **else** append $b_p$ to $t'$ and add $(t', st'_i)$ to $Sol$.
**end while**
(9) Add the valid translation $t$ to the solution's set $T$.
(10) Return $T$.
**end**

**Fig. 3.** ComputeAllSolutions algorithm

Observe that in step (3), we allow the search of a basic transformation to consider any pattern of the input structure $st$. Since there could be basic transformations that are commutative (the result of the translation could not depend on the order of the basic transformations) we can have solutions that are different but composed by the same basic transformations set. For example, translating to the *DTD* data model, the basic transformation $b_x$ that removes a metaprimitive that cannot be transformed, as the Namespace construct, can be added to the translation at any place in the sequence: $b_i, b_j, b_x \equiv b_x, b_i, b_j \equiv b_i, b_x, b_j$.

We can have translations in $T$ that are equivalent, but notice that we only have translations in which the same transformation does not appear more than once. Consider steps 5-8: if a basic transformation has been already added to $t$, we discard the branch. This choice prevents also loops in case of pairs of basic transformations that add and remove the same metaprimitive (of a pattern of them).

This algorithm captures all the possible valid translations, including optimal ones. Consider now the efficiency and preferability issues. We can use them to rank the solutions in $T$ and expose to the user only the translations that are in the optimal set. Given the set of solutions $T = \{t_1, \ldots, t_n\}$ we can order $T$ according to $>_{M_t}$ and according to $\succ$: we get two ordered lists of transformations. The optimal set is the union of the top elements in these lists. We show experimentally next that in the optimal set there are often several solutions. This is rather intuitive: some solutions are better in terms of efficiency, others are better in terms of preferability.

Note that, even if the solution's generation and ranking are efficient, an exhaustive exploration of the search space is required. In particular, if the set of basic transformations is large, an exhaustive search can be expensive, since the complexity of the algorithm depends exponentially on the size of the library. Another approach it is introduced next to overcome this limit.

### 4.2 Best-First Search Algorithm

The approach we have followed to limit the search is based on the $A^*$ strategy. This algorithm avoids the expansion of paths of the tree that are already expensive and returns as first result the best solution with respect to an appropriate function $f(n)$ that *estimates* the total cost of a solution. Specifically, the value of $f(n)$ for a node $n$ is obtained as the sum of $g(n)$, the cost of the path so far, and the estimated cost $h(n)$ from $n$ to a valid solution. This function is rather difficult to define in our context. Indeed, the $A*$ search on a tree grants the best solution only if $h(n)$ is admissible, that is, only if for every node $n$, $h(n) \leq h^*(n)$, where $h^*(n)$ is the real cost to reach the solution from $n$. We have followed here a practical solution: $h(n)$ is defined as a piecewise function that returns zero if the current structure associated with $n$ is empty, and it is proportional to "distance" to the target, that is, the number of patterns of the current structure not occurring in the target model when the set is not empty.

The crucial point is the identification of some heuristics able to limit the search space and generate efficiently good solutions. For each node we consider two properties:

- the size, $l(n)$, that is the length of the current translation in terms of the number of basic transformations that compose it;
- the recall, $r(n)$, which corresponds to the number of patterns of the target model not occurring in the current structure associated with $n$.

To this end, we have defined a cost function for each node $n$ in the search tree as the linear combination of the two functions: $g(n) = w_1 \times l(n) + w_2 \times r(n)$.

The weights $w_1$ and $w_2$ have been chosen such that the recall is privileged to the length of a translation. Intuitively, modifying the $w_1$ and $w_2$ we can choose a solution with a better efficiency or a better preferability. Notice that we cannot use the complexity values for the cost function $g(n)$, even if it more precise than the translation size: we must use a value that is comparable to the heuristic $h(n)$, that it is based on the number of patterns, since we cannot know *a priori* the complexity of translation.

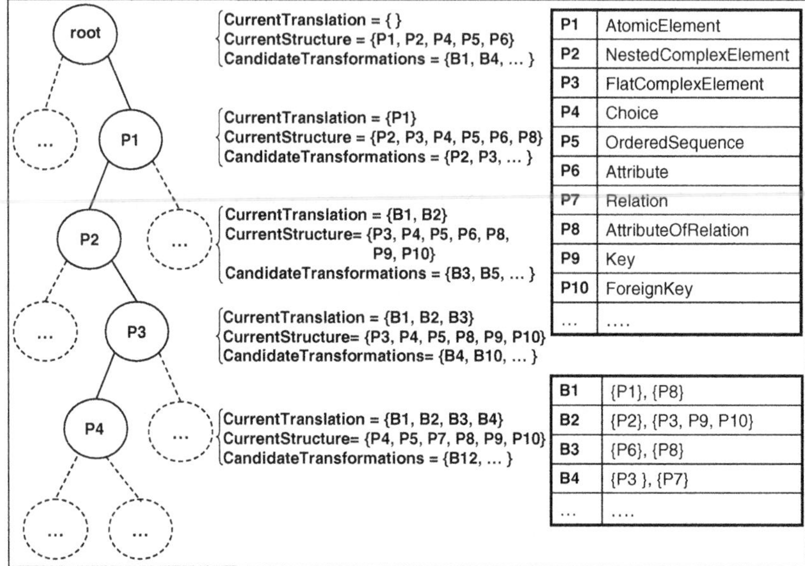


**Fig. 4.** The search space for the translation

Let us introduce in Figure 4 a practical example to show, in more detail, how the algorithm proceeds. In the example we refer to metaprimitives and patterns with same notation *PX*, see for instance *P2*, that is a ComplexElement that nests another ComplexElement. The example refers to the translation of an XML Schema and its data set into the relational model. Each node of the tree corresponds to a possible state reached by the execution of the algorithm. The first observation is that for each node there could be many candidate transformations and so in principle, to find the optimal solution according to the properties discussed in Section 3, all the possible alternatives should be evaluated. Each node has associated three sets: (i) the set of *PX* (metaprimitives or patterns) involved in the current structure, the *current structure*, (ii) the set of *candidate transformations*, and (iii) the set of transformations collected in the preceding states, that is the *current translation*. As we said, the current structure is initialized to the metaprimitive used in the source schema. For each candidate basic transformation $b$ of a node $n$, there is a child whose structure is the effect of $b$

on the current structure of $n$. A leaf of the tree corresponds to a state in which no candidate transformation exists: if the current structure is not valid for the target model it is a *failing* state. The algorithm stops when the current structure is valid for the target model definition, this would be a *successful* state or a *solution*. In the example at hand, a successful solution is composed by the set of following transformations.

- Atomic elements have been changed into relational attributes.
- Nested elements have been unnested.
- XML attributes have been changed into relational attributes.
- Complex flat elements have been turned into relations with keys.
- Choices have been implemented by means of separate relations.
- Order of sequences have been coded into relational attributes.

The time complexity of the algorithm strictly depends on the heuristic. It is exponential in the worst case, but is polynomial when the heuristic function $h(n)$ meets the condition $|h(n) - h^*(n)| \leq O(\log h^*(n))$, where $h^*(n)$ is the exact cost to get from $n$ to the goal. This condition cannot be guaranteed in our context, since a basic transformation could remove or transform more than just one pattern from the input structure, but in the average case it is polynomial as we show experimentally shortly. More problematic than time complexity can be the memory usage for this algorithm. We have also tried variants of $A^*$ able to cope with this issue, such as memory-bounded $A^*(MA^*)$, but we never hit memory limit even in the original implementation, since the heuristic we used removes effectively not promising paths so that the algorithm does not expand them.

## 5 Testing the Approach

To evaluate the effectiveness of our approach, we have implemented the translation process within a prototype and we have conducted a series of experiments on several schemas of different data models that vary in terms of dimension and complexity.

In our tests, we have used a specific preference relationship. Consider again the example in Figure 1: the output schema $d$ takes more advantage of the expressiveness of the target model than schemas $b$ or $c$. It follows that the translation to schema $d$ it is preferable to the other solutions because it also includes the information on cardinalities, which is a construct available in the target data model. For a translation $t_x$ on an input schema $S$, we denote the number of distinct valid patterns in the target schema by $|t_x(S)|$ and we assume that $t_1 >_{M_t} t_2$ if $|t_1(S)| > |t_2(S)|$.

The results of these experiments are summarized in Figure 5. Each row presents quantitative results for a schema translated from a source model (first column) to a target model (fourth column). The actual number of metaprimitives occurring in the schema is in the second column $|\mathbf{S}|$. The size complexity is estimated by the number of distinct metaprimitives involved in the source schema and it is reported in the third column $|\mathbf{C}|$.

| Source Schema | $\lvert S\rvert$ | $\lvert C\rvert$ | Target Model | Num. of sol. | Alg. time | Sol. with min. size | Min. size | Sol. with max. preferabi. | Max. pref. | Number of optimal sol. | A* First Solution | |
|---|---|---|---|---|---|---|---|---|---|---|---|---|
| | | | | | | | | | | | Size | Pref |
| XMLSCHEMA | 16 | 4 | ODL | 384 | 2.5 | 2 | 3 | 16 | 4 | 18 | 4 | 4 |
| XMLSCHEMA | 16 | 4 | ER | 367 | 3.4 | 1 | 1 | 4 | 4 | 5 | 2 | 4 |
| XMLSCHEMA | 35 | 5 | DTD | 3 | 2.4 | 1 | 1 | 3 | 4 | 1 | 1 | 4 |
| XMLSCHEMA | 35 | 5 | RELATIONAL | 352 | 2.5 | 2 | 3 | 176 | 5 | 2 | 3 | 5 |
| RELATIONAL | 20 | 5 | XMLSCHEMA | 8 | 1.4 | 2 | 3 | 1 | 5 | 1 | 4 | 5 |
| RELATIONAL | 20 | 5 | ER | 2 | 0.4 | 1 | 1 | 6 | 5 | 1 | 1 | 5 |
| RELATIONAL | 20 | 5 | DTD | 6 | 1.3 | 2 | 3 | 16 | 5 | 18 | 4 | 5 |
| RELATIONAL | 42 | 6 | XMLSCHEMA | 24 | 4.2 | 1 | 4 | 24 | 6 | 1 | 6 | 6 |
| RELATIONAL | 42 | 6 | ODL | 8 | 4.2 | 1 | 3 | 4 | 6 | 5 | 3 | 5 |
| DTD | 27 | 6 | XMLSCHEMA | 121 | 8.5 | 1 | 2 | 89 | 5 | 1 | 2 | 5 |
| RELATIONAL | 54 | 8 | ODL | 13 | 5.1 | 1 | 2 | 4 | 7 | 5 | 2 | 6 |
| RELATIONAL | 54 | 8 | DTD | 6 | 5.4 | 1 | 6 | 6 | 6 | 7 | 7 | 5 |

**Fig. 5.** Experimental results

The first result we report for each translation is the total number of valid possible solutions. It turns out that even for a simple source schema (e.g. a DTD with a few distinct metaprimitives) there are a lot of possible solutions. For instance, valid translations from an XML schema into ODL, as well as into the ER or the relational model, are hundreds. As we said, many of these solutions differ only by the order in which basic transformations occur in the translation, but we still want a criterium to limit the search as discussed in the previous section. Following columns report: (i) the number of solutions with minimal size and the corresponding size value, and (ii) the number of preferable solutions and the corresponding value. We reported the size results instead of the complexity ones to compare them with the $A^*$ results. The optimal solutions are those both efficient and preferable. It turns out that in several cases optimal solutions do not exists and this confirms the intuition that longer solutions may be more "accurate". We also report the overall time (in minutes) the exhaustive search algorithm took to find all the possible solutions for a translation. Notice that this value increases with the number of distinct metaprimitives of the source schema, but in a rather irregular way, since it depends also on the library of transformations and the target model complexity. Indeed, the problem depends on the complexity of the models involved. The required time to find all the solutions becomes critic with models that present more than five distinct metaprimitives. Conversely, the translations involving a restricted number of metaprimitives require a rather limited effort and increase linearly with the complexity of the source schema. Finally, notice in the last two columns the results for the $A^*$ implementation. The first column reports the size of the first solution returned and the second one its preferability. The solution coincides with the optimal for translations between XML based models, whereas in the other cases preferability is close or equal to the best value. We expected this result, since we have tuned the heuristics to maximize the preferability of the solution.

We further evaluated the performance of our best-first search algorithm aggregating results under four main scenarios: (i) translations between database models (Relational, ER, ODL), (ii) between XML based models (XSD, DTD), (iii) from XML based models to database ones and (iv) vice versa. Figure 6 shows the performance of the $A^*$ algorithm on the four scenarios with increasing schema complexity on the x-axis. We ran several examples for each scenario, using schemas with different structures, and reporting average results on the charts. The chart with the size analysis is on the left hand side of the figure. As expected, the number of transformations increases with the number of distinct metaprimitives in the source schema. The linearity in (iii) and (iv) represents the heterogeneity between the two classes of models: for almost each metaprimitive the system uses a transformation from the library. Lines (i) and (ii), instead, show that translations between models that share many metaprimitives require only few basic transformations: this fact strongly depends on the quality of the metamodel implemented in the system. The chart on the right hand side of the figure shows the results of the preferability analysis. The algorithm scales well for the four scenarios: the number of metaprimitives in the target schemas increases with the increasing complexity of the input schemas. In particular, in scenarios (i) and (iv) the system translates with almost linear results: these performances depend on the low complexity of the considered source models, that is, it is easier to find a preferable solution translating a schema from a database model, which presents less constructs and less patterns than an XML data model.

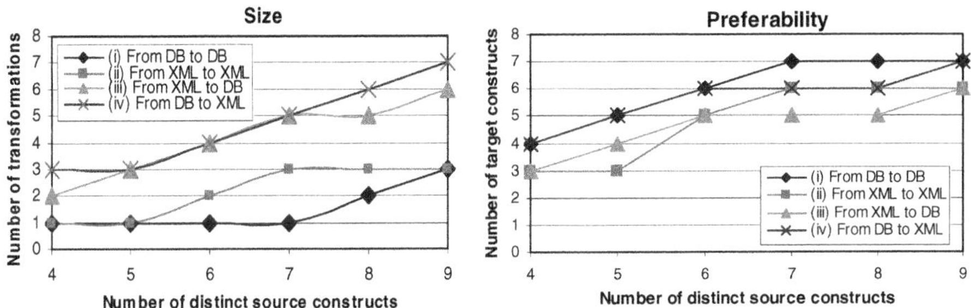


**Fig. 6.** $A^*$ search strategy analysis

## 6 Conclusions

In this paper, we have presented new techniques, and supporting results, for the automatic generation of data model translations. We have defined a number of properties for evaluating the quality of the translations and we have experimented them evaluating the translations generated by a prototype. It turned out that the system can generate all the optimal solutions with respect to these properties, but this usually requires significant effort. However, by adopting a best-first search algorithm and appropriate heuristics, a solution can be retrieved efficiently and it is usually optimal in terms of preferability.

In the future, we plan to investigate and develop supporting tools for the design of information preserving transformations. Our work is now focusing on a formal language to express transformation between data models definitions and a graph-based, high-level notation to easily define them.

## References

1. Atzeni, P., Cappellari, P., Bernstein, P.A.: Model-independent schema and data translation. In: EDBT. pp.368–385 (2006)
2. Atzeni, P., Torlone, R.: Management of multiple models in an extensible database design tool. In: EDBT, pp. 79–95 ( 1996)
3. Papotti, P., Torlone, R.: Heterogeneous data translation through xml conversion. J. Web. Eng. 4(3), 189–204 (2005)
4. Fernandez, M.F., Kadiyska, Y., Suciu, D., Morishima, A., Tan, W.C.: Silkroute: A framework for publishing relational data in xml. ACM Trans. Database Syst. 27(4), 438–493 (2002)
5. Bohannon, P., Freire, J., Roy, P., Siméon, J.: From xml schema to relations: A cost-based approach to xml storage. In: ICDE (2002)
6. Ramanath, M., Freire, J., Haritsa, J.R., Roy, P.: Searching for efficient xml-to-relational mappings. In: Xsym, pp. 19–36 ( 2003)
7. Shu, N.C., Housel, B.C., Taylor, R.W., Ghosh, S.P., Lum, V.Y.: EXPRESS: A Data EXtraction, Processing, amd REStructuring System. ACM TODS 2(2), 134–174 (1977)
8. Fagin, R., Kolaitis, P.G., Miller, R.J., Popa, L.: Data Exchange: Semantics and Query Answering. TCS 336(1), 89–124 (2005)
9. Bernstein, P.A.: Applying model management to classical meta data problems. In: CIDR. pp. 209–220 (2003)
10. Bowers, S., Delcambre, L.M.L.: The uni-level description: A uniform framework for representing information in multiple data models. In: ER, pp. 45–58 ( 2003)
11. Bernstein, P.A., Melnik, S., Mork, P.: Interactive schema translation with instance-level mappings. In: VLDB. pp.1283–1286 (2005)
12. MOF: OMG's MetaObject Facility. `http://www.omg.org/mof/` (2006)
13. Lenzerini, M.: Data Integration: A Theoretical Perspective. In: PODS. pp. 233–246 (2002)
14. Hull, R., King, R.: Semantic database modeling: Survey, applications, and research issues. ACM Comput. Surv. 19(3), 201–260 (1987)
15. Bohannon, P., Fan, W., Flaster, M., Narayan, P.P.S.: Information preserving xml schema embedding. In: VLDB (2005)
16. Hull, R.: Relative information capacity of simple relational database schemata. SIAM J. Comput. 15(3), 856–886 (1986)
17. Miller, R.J., Ioannidis, Y.E., Ramakrishnan, R.: Schema equivalence in heterogeneous systems: bridging theory and practice. Inf. Syst. 19(1), 3–31 (1994)
18. Arenas, M., Libkin, L.: A normal form for xml documents. ACM Trans. Database Syst. 29, 195–232 (2004)
19. Kensche, D., Quix, C., Chatti, M.A., Jarke, M.: GeRoMe: A Generic Role Based Metamodel for Model Management. In: OTM Conferences. pp.1206–1224 (2005)

# Handling Instance Correspondence in Inter-organisational Workflows

Xiaohui Zhao[1], Chengfei Liu[1], Yun Yang[1], and Wasim Sadiq[2]

[1] Faculty of Information and Communication Technologies
Swinburne University of Technology
Melbourne, Victoria, Australia
{xzhao, cliu, yyang}@ict.swin.edu.au
[2] SAP Research
Brisbane, Australia
wasim.sadiq@sap.com

**Abstract.** As business collaboration involves multiple business processes from different participating organisations, it becomes a challenging issue to manage the complex correspondence between instances of these business processes. Yet very limited support has been provided by inter-organisational workflow research. In this paper, we develop a formal method to specify instance correspondence based on a novel correspondence Petri net model. In this method, cardinality parameters are defined to represent cardinality relationships between collaborating business processes at build time, while correlation structures are designed to characterise correspondence between collaborating business process instances at run time. Corresponding algorithms are also developed to generate the correspondence Petri nets for collaborative business processes, and to trace instance correlation on the fly using the generated Petri nets.

**Keywords:** Inter-organisational workflow management, workflow instance correspondence, correspondence Petri net.

## 1 Introduction

In recent years, organisations have been undergoing a thorough transformation towards highly flexible and agile collaborations [1]. Organisations are required to dynamically create and manage collaborative business processes to grasp market opportunities [2]. A collaborative business process involves multiple parties and their business processes, thus it inevitably brings new challenges to workflow choreography and orchestration. One of the most pressing issues in this context is the instance correspondence.

Complex instance correspondences may exist at both build time and run time. Here, we characterise instance correspondences in terms of cardinality and correlations. Thereby, we can define and represent statical and dynamic correspondence when modelling and executing a collaborative business process.

J. Krogstie, A.L. Opdahl, and G. Sindre (Eds.): CAiSE 2007, LNCS 4495, pp. 51–65, 2007.

Some research efforts were put in this field. Multiple workflow instantiation was discussed by Dumas and ter Hofstede [3], using UML activity diagrams. Later they extended their work to service interactions [4]. van der Aalst et al. [5] deployed coloured Petri nets to represent multiple workflow cases in workflow patterns, and implemented it in the YAWL system [6]. Guabtni and Charoy [7] extended the multiple instantiation patterns and classified multiple workflow instantiation into parallel and iterative instances. However, most of the above research focus on interaction patterns, and sidestep the instance correspondence issue in collaborative business processes. WS-BPEL (previously BPEL4WS) [8] uses its own correlation set to combine workflow instances, which have same values on specified message fields. However, WS-BPEL defines a business process in terms of a pivot organisation. This results in that a WS-BPEL business process only represents the interaction behaviours of the pivot organisation with its neighbouring organisations. This feature limits its application for complex business collaborations, which are likely to include interactions beyond neighbouring organisations.

Aiming to address this issue, this paper proposes a method to support instance correspondences from an organisation-oriented view. In our method, cardinality parameters are developed to characterise cardinality relationships between collaborating business processes at build time. Besides, a correlation structure is combined with each instance to trace dynamic workflow correlations at run time. In addition, we formalise this method with a novel correspondence Petri net to describe instance correspondence precisely.

The rest of this paper is organised as follows: Section 2 analyses the instance correspondence within collaborative business processes with a motivating example. In Section 3, we discuss workflow cardinality and correlation issues in business collaboration context. In Section 4, we establish a novel correspondence Petri net model with extensions on workflow cardinality and correlation as our formal method. In Section 5, we develop algorithms to illustrate how to model collaborative business processes and manage run time executions of business collaborations with these special Petri nets. Section 6 concludes this paper and indicates our future work.

## 2 Motivating Example

Figure 1 illustrates a business collaboration scenario, where a retailer may initiate a product-ordering process instance that orders products from a manufacturer. The manufacturer may use a production process instance to receive orders from retailers. Once it obtains enough orders, the production instance may start making goods in bulk. At the same time, the manufacturer may assign several shippers to handle goods delivery. These shippers arrange their goods transfer according to their transfer capability and route optimisation etc. Finally, these shipping instances send goods to the proper retailers according to these correlations.

From this collaboration scenario, we see that an instance of one business process is likely to interact with multiple instances of another business process. For example, one production instance may correspond to multiple product-ordering instances, and multiple shipping instances may correspond to multiple product-ordering instances.

In contrast to such complex quantitative relationship, most current workflow modelling approaches simply assume and support a one-to-one relationship between business process instances. Although such requirements are quite common in B2B collaborations, they are primarily supported by enterprise applications internally and not adequately by workflow management systems.

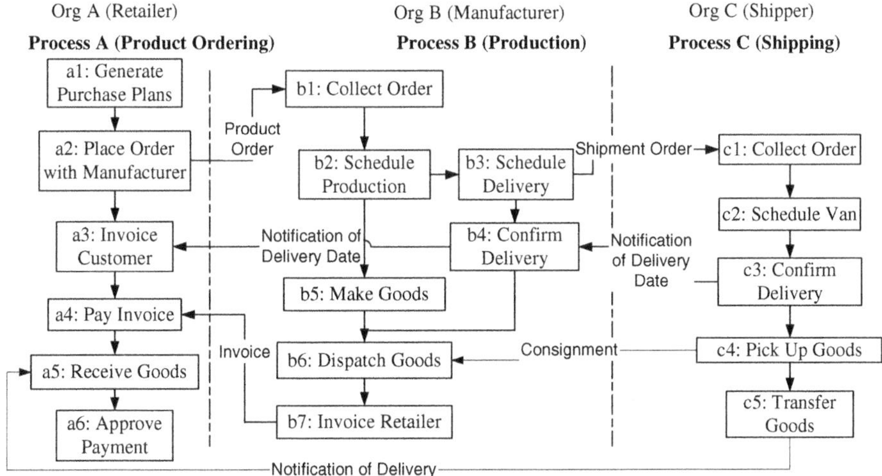


**Fig. 1.** Motivating example

At run time, instance correspondences are subject to the correlations between instances of different business processes. These correlations result from the underlying business semantics of interactions. In real cases, such correlations may be realised by real interactions (direct) or passing unique identifiers (indirect), such as order number. Sometimes, real interactions between instances may be triggered by time duration, external events etc. In this example, the manufacturer's production instance is correlated with retailers' product-ordering instances during the real interaction of receiving orders from retailers. Afterwards, the manufacturer contacts shippers for booking deliveries. At the same time, the manufacturer also passes the order numbers to proper shippers. With these order numbers, shippers' shipping instances are indirectly correlated with retailers' product-ordering instances. Following these correlations, shippers can pick up produced goods from the manufacturer, and then transfer them to proper retailers.

From the above discussion, we see that workflow correlations combine business interactions into a meaningful collaboration. Some existing approaches provide primitive support for correlation handling, such as message correlations in WS-BPEL. As discussed in Section 1, a WS-BPEL business process generated for a retailer cannot cover the interactions between the manufacturer and shippers, not to mention the correlations between their production and shipping instances.

## 3 Cardinality and Correlation Issues in Business Collaboration

In a collaborative business process, each participating organisation may play a specific role and only care about its own interests. For this reason, participating organisations do not wish, and may not be allowed to know the details of their partner organisations. Therefore, each participating organisation only has a partial and restricted view of the whole collaboration [9-12]. Due to diverse partnerships and authorities, different organisations may view the same collaboration differently.

### 3.1 Workflow Cardinality

Figure 2 shows a possible instance correspondence situation of the collaborative business process in the motivating example. In general, there are four possible cardinality relationships between a pair of interacting business processes, viz., single-to-single, single-to-many, many-to-single and many-to-many. In the organisation-oriented view, we substitute the four bilateral cardinality relationships with the pair of unidirectional cardinality relationships. For example, a single-to-many relationship between business processes $p_B$ and $p_C$ can be represented by a "to-many" relationship from $p_B$ to $p_C$ and a "to-one" relationship from $p_C$ to $p_B$. A many-to-many relationship between $p_A$ and $p_C$ can be represented by a "to-many" relationship from $p_A$ to $p_C$ and a "to-many" relationship from $p_C$ to $p_A$. In this paper, we define these two cardinality relationships with two *workflow cardinality parameters*,

[:1], denotes a *to-one* cardinality relationship;
[:*n*], denotes a *to-many* cardinality relationship.

As process interactions are implemented in the form of messaging behaviours, we incorporate these two cardinality parameters to message modelling. Conceptually, a message type can be defined as follows:

**Definition** *message type***.** A message type *m* is defined as a tuple ( $\rho$, $\alpha$, $\beta$, $f$, $\chi$ ), where

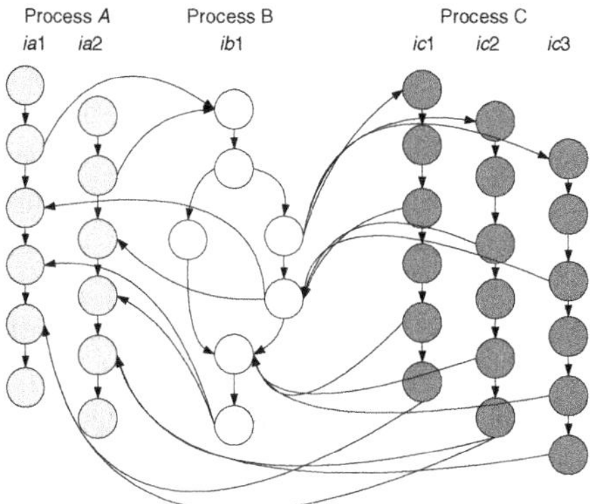


**Fig. 2.** Workflow cardinality of motivating example

- $\rho$ is $m$'s messaging direction, '*in*' or '*out*'. These two values denote that $m$ stands for an incoming message or an outgoing message, respectively.
- $\alpha$ is a task of a business process. $\alpha$ represents $m$'s source task, if $m$ is an outgoing message; or it represents $m$'s target task.
- $\beta$ is a set of tasks. This set of tasks represents $m$'s possible source tasks, if $m$ stands for an incoming message; or it represents $m$'s possible target tasks. Each task in $\beta$ is likely to send or receive an instance of $m$ according to $m$'s direction.
- $f : \beta \rightarrow \{ [:1], [:n] \}$ is a mapping from $\beta$ to the two discussed cardinality parameters.
- $\chi$ denotes the specification of the message body.

Here, $\alpha$ and $\beta$ together represent the cardinality between business processes at type level. Two message types are said to be a pair if they have complementary source / target tasks and the same message body specification. The details of linking internal business processes into a collaborative business process via message types will be discussed in Section 5.

### 3.2 Workflow Correlation

Workflow correlation denotes the semantic relation between business process instances in the same business collaboration. Instances are directly correlated, when they "shake hands" during interactions. In addition, some instances may inherit pre-existing correlations from their counterparts during interactions. This correlation inheritance reflects the extending of business semantic relations.

In the scenario shown in Figure 2, firstly instances $ia_1$ and $ia_2$ are correlated with instance $ib_1$, when $ib_1$ accepts orders from $ia_1$ and $ia_2$; Secondly, $ib_1$ contacts instances $ic_1$, $ic_2$ and $ic_3$ for delivery booking. Here, suppose $ib_1$ assigns $ic_1$ and $ic_2$ to transfer products for $ia_1$, and assigns $ic_2$ and $ic_3$ to transfer products for $ia_2$. Thereby instances $ic_1$, $ic_2$ and $ic_3$ are directly correlated with $ib_1$, and they also inherit previous correlations from $ib_1$. In this example, $ic_1$ and $ic_2$ inherit the correlation between $ia_1$ and $ib_1$ from $ib_1$, while instances $ic_2$ and $ic_3$ inherit the correlation between $ia_2$ and $ib_1$ from $ib_1$. This inheritance implies that shippers require consignees' information to arrange their shipping schedules. Corresponding shipping instances are therefore indirectly correlated with retailers' product-ordering instances. This inheritance is realised by passing retailers' order numbers from the manufacturer to shippers.

Based on these workflow correlations, we can derive a *logical instance* of a participating business process instance in the organisation-oriented view. From a business process instance $\zeta$ of organisation $g$, a so-called logical instance $\xi$ consists of $\zeta$ and all its related instances of business processes belonging to other organisations through the instance correlations at run time. Here, organisation $g$ is called *host organisation* of $\xi$, and $\zeta$ is called *base business process instance* of $\xi$. In terms of workflow correlations, we can define a logical instance as follows,

**Definition** *logical instance*. In the context of a collaborative business process $\Lambda$, the logical instance for a base business process instance $\zeta$ is defined as tuple $(\zeta, \Lambda, \Delta)$, where $\Delta$ is the set of business process instances that are correlated with $\zeta$ in the context of $\Lambda$.

The set of correlated business process instances evolves during the business collaboration. For example, if we start from instance $ia_1$, the set of correlated business process instances for $ia_1$ contains no instances at the beginning; while it includes instance $ib_1$ right after $ib_1$ accepts its order, i.e., $\Delta = \{ ib_1 \}$; afterwards instances $ic_1$ and $ic_2$ may be added after $ib_1$ books delivery with $ic_1$ and $ic_2$, then $\Delta = \{ ib_1, ic_1, ic_2 \}$.

# 4 Correspondence Representation Methodology

Petri nets were invented by Carl Petri in the sixties for modelling concurrent behaviours of a distributed system. A Petri net is a bipartite graph whose nodes can be distinguished in places and transitions, which are graphically represented by circles and rectangles, respectively. A Predicate / Transition or coloured Petri net can differentiate tokens with unique identifications or a set of colours. Each place can contain tokens of different identifications or colours at the same time. Each arc may be assigned with an expression to restrict what tokens and the number of tokens that can transfer through. Therefore, a Petri net can represent multiple process executions within one net. Now, Petri nets are widely applied in concurrency control and process simulation [13].

## 4.1 Extensions to Petri Nets

To support workflow cardinality and correlation, we extend traditional Petri nets with new parameters and functions together with special places and transitions.

1. Cardinality parameters

An auxiliary place is used to denote a message between two business processes, which may be represented by two sub nets. In Figure 3 (a), auxiliary place $p$ is drawn as a shaded circle, while sub nets $A$ and $B$ are differentiated by white and striped circles.

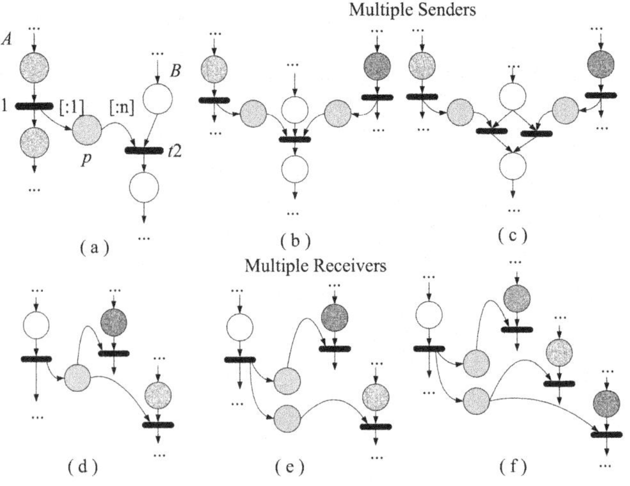


**Fig. 3.** Cardinality parameters

Transition $t_1$ of $A$ is an *interaction requesting transition*, while transition $t_2$ of $B$ is an *interaction responding transition.* Unidirectional cardinality parameter "[:1]" on the arc linking $t_1$ to $p$ denotes that $A$ views this interaction as a "to-one" cardinality, i.e., each token in $A$ interacts with one token in $B$ from $A$'s view. Parameter "[:$n$]" on the arc linking $p$ to $t_2$ denotes a "to-many" cardinality, i.e., each token in $B$ corresponds multiple tokens in $A$ from $B$'s view. Therefore, we see that an auxiliary place separates the cardinality views from different perspectives.

2. Multiple message senders / receivers
Particular structures are used to represent the scenarios where multiple possible senders or receivers are instances of different business processes. In regard to multiple senders, Figure 3 (b) shows an interaction receiving messages from two senders; while Figure 3 (c) shows an interaction receiving one message from two senders. In regard to multiple receivers, Figure 3 (d) shows an interaction in which one of two receivers is expected to receive the message; while Figure 3 (e) shows an interaction that a message is sent to both receivers. With these basic interaction schemes, we can represent more complicated ones. For example, Figure 3 (f) shows a scenario that a task sends a message to three receivers, and one of the three will receive it definitely, while only one of the other two is expected to do so.

3. Special transitions
In some cases, an interaction may result in generating new instances. For example, in the book-delivery interaction between a manufacturer and a shipper, the shipper may generate several new shipping instances to handle it. In Petri net context, this requires the corresponding transition to be capable of generating new tokens. In this paper, we classify such transitions as *token-generating transitions*. In this way, we represent the book-delivery interaction as the Petri net segment shown in Figure 4.

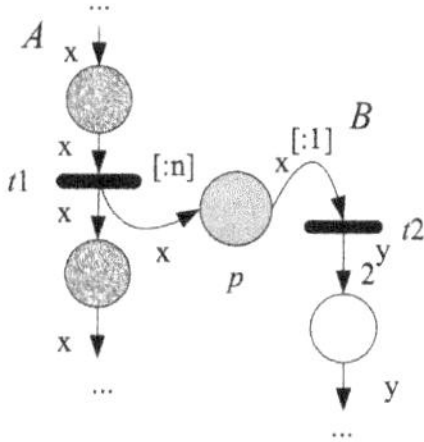


**Fig. 4.** Correlation function attached structures

In Figure 4, variable $x$ or $y$ ( $x \neq y$ ) is labelled along an arc to denote the type of tokens that may go through this arc. For example, the token that flows from transition $t_1$ to place $p$ is different from the token that flows out of transition $t_2$. In addition, expression $2^y$ is labelled along the arc linking $t_2$ to the adjacent place, as $t_2$ is a token-generating transition. Thereby, this arc allows that more than one token representing instances of the same business process to pass through at one time.

4. Correlation structures
To record the run time workflow correlation, we combine a *correlation structure* with each token. A correlation structure is defined as follows:

**Definition** *correlation structure*. In a Petri net, the correlation structure for token $\varsigma$ is defined as $r^{\varsigma} = \{ \varsigma, D_1, D_2, \ldots, D_n, \mathcal{R} \}$, where

- each $D_i$ ( $1 \le i \le n$ ) denotes a set of tokens, which represent correlated instances of a business process. All tokens in $D_1, D_2, \ldots, D_n$ are correlated with $\varsigma$.
- $\mathcal{R}$ is a binary relation defined between tokens in $\bigcup_{i=1}^{n} D_i$. Here, $d_x \mathcal{R} d_y$, ( $d_x, d_y \in \bigcup_i D_i$ ), denotes that tokens $d_x$ and $d_y$ are correlated via token $\varsigma$.

$\varsigma$ is called *base token* of this correlation structure. Token sets $D_1, D_2, \ldots, D_i$ may be dynamically updated during collaboration. For example, Figure 5 shows a part of the collaboration scenario mentioned in the motivating example using a Petri net. Each sub net stands for a business process, and is distinguished with different circles. The tiny circles within places denote tokens, and each token such as $ia_1$, $ib_1$, $ic_1$ stands for a business process instance. Each transition such as $ta_2$, $tb_1$, $tc_1$ stands for a task. When $ia_1$ and $ia_2$ flow to transition $tb_1$ via auxiliary place $ap_1$, it means that the production instance accepts the orders from two retailers. Therefore, correlation structure $r^{ib1}$ at this moment is { $ib_1$, { $ia_1$, $ia_2$ }, ∅ }. Tokens $ia_1$ and $ia_2$ may have correlation structures $r^{ia1}$={ $ia_1$, { $ib_1$ }, ∅ } and $r^{ia2}$={ $ia_2$, { $ib_1$ }, ∅ }, respectively.

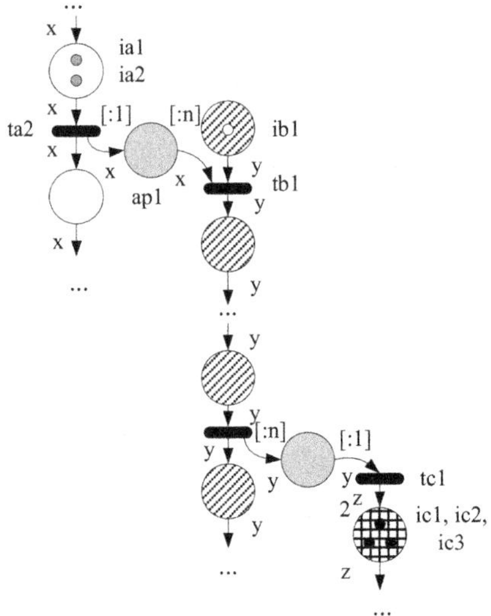


**Fig. 5.** Correlation scenario

This correlation structure accordingly evolves as the base token flows and interacts with other tokens. When $ib_1$ contacts $ic_1$, $ic_2$ and $ic_3$ to arrange the goods delivery for $ia_1$ and $ia_2$, we suppose that $ib_1$ assigns $ic_1$ and $ic_2$ to serve $ia_1$, while assigns $ic_2$ and $ic_3$ to serve $ia_2$. Thus, $r^{ib1}$ will change to { $ib_1$, { $ia_1$, $ia_2$ }, { $ic_1$, $ic_2$, $ic_3$ }, {( $ia_1$, $ic_1$ ), ( $ia_1$, $ic_2$ ), ( $ia_2$, $ic_2$ ), ( $ia_2$, $ic_3$ )}}. Here, the last set denotes the correlated tokens via $ib_1$. As the consignee information, the order numbers from $ia_1$ and $ia_2$ are passed to $ic_1$ and $ic_2$, $ic_2$ and $ic_3$ by $ib_1$, respectively. Therefore, $r^{ic1}$ is set as { $ic_1$, { $ib_1$ }, { $ia_1$ }, ∅ }, $r^{ic2}$ is set as { $ic_2$, { $ib_1$ }, { $ia_1$, $ia_2$ }, ∅ } and $r^{ic3}$ is set as { $ic_3$, { $ib_1$ }, { $ia_2$ }, ∅ }.

### 4.2 Correspondence Petri Net

According to the above discussion, we establish a novel Petri net, called correspondence Petri net (CorPN), by extending the traditional Place / Transition Petri net. The definition of this CorPN is given below.

**Definition** *Correspondence Petri net.* A CorPN is represented as tuple $\Sigma = (\mathcal{P}, \mathcal{T}, \mathcal{F}, \mathcal{P}^{\circ}, \mathcal{F}^{\circ}, \mathcal{D}, \mathcal{V}, \mathcal{G}, \mathcal{E}, \mathcal{C}, \mathcal{Q}, I)$, where

(*i*) $(\mathcal{P}, \mathcal{T}, \mathcal{F})$ is a directed net, called the base net of $\Sigma$. Here, $\mathcal{P}$, $\mathcal{T}$ and $\mathcal{F}$ stand for the sets of places, transitions and arcs, respectively. $\mathcal{P} \cap \mathcal{T} = \varnothing$; $\mathcal{P} \cup \mathcal{T} \neq \varnothing$; $\mathcal{F} \subseteq \mathcal{P} \times \mathcal{T} \cup \mathcal{T} \times \mathcal{P}$;

(*ii*) $\mathcal{P}^{\circ} \subset \mathcal{P}$, is the set of auxiliary places, which represent the messaging relations between business processes of a collaborative business process.

(*iii*) $\mathcal{F}^{\circ} \subset \mathcal{F}$, is the set of arcs that connect auxiliary places, i.e., $\mathcal{F}^{\circ} \subseteq \mathcal{P}^{\circ} \times \mathcal{T} \cup \mathcal{T} \times \mathcal{P}^{\circ}$.

(*iv*) $\mathcal{D}$ is a set of tokens, each of which stands for a possible participating business process instance. Here, $\mathcal{D} = \mathcal{D}_1 \cup \mathcal{D}_2 \cup \ldots \cup \mathcal{D}_n$, $\mathcal{D}_i \cap \mathcal{D}_j = \varnothing$, where $1 \leq i, j \leq n$ and $i \neq j$. Precisely, each $\mathcal{D}_i$ denotes a *token group*, which includes instances of the same business process. $n$ is the number of token groups.

(*v*) $\mathcal{V}$ is a set of variables for token groups, and $\mathcal{V} = \{ v_1, v_2 \ldots, v_n \}$. Actually, each element $v_i$ of $\mathcal{V}$ is defined on a token group, i.e., $v_i \in \mathcal{V}$. $v_i$ is defined on $\mathcal{D}_i$, where $1 \leq i \leq n$ and $n$ is the number of token groups.

(*vi*) $\mathcal{G}: \mathcal{P} \rightarrow \tau$, where each element $\tau_i$ of set $\tau$ is a set of possible tokens, i.e., $\tau_i \in \tau$ and $\tau_i \in 2^{\mathcal{D}}$.

(*vii*) $\mathcal{E}: \mathcal{F} \rightarrow \sigma$, where $\sigma$ is a set of expressions defined on $\mathcal{V}$.

(*viii*) $\mathcal{C}: \mathcal{F}^{\circ} \rightarrow \varepsilon$, where $\varepsilon$ is the set of cardinality parameters, i.e., $\varepsilon = \{ [:1], [:n] \}$.

(*ix*) $\mathcal{Q}: \mathcal{D} \rightarrow \lambda$, where $\lambda$ is a set of correlation structures.

(*x*) $I: \mathcal{P} \rightarrow \theta$, where $\theta$ is a set of possible composition of tokens defined in $\mathcal{D}$.

Explanation:

(1) $(\mathcal{P}, \mathcal{T}, \mathcal{F})$ determines the component net structures of this CorPN.

(2) $\mathcal{P}^{\circ}$ and $\mathcal{F}^{\circ}$ describe the messaging behaviours between the business processes of the underlying collaborative business process.

(3) The variables in $\mathcal{V}$ are defined according to each token group, which represents the instances of a business process. Thus, the variables can be used to differentiate the instances of participating business processes and abstract the common behaviours of each business process.

(4) Mapping $\mathcal{G}$ sets up the capacity of each place defined in $\mathcal{P}$.

(5) Mapping $\mathcal{E}$ sets up the arc expressions to restrict the flowing of tokens.

(6) Mapping $\mathcal{C}$ maps a cardinality parameter onto each arc that connects with an auxiliary place.

(7) Mapping $\mathcal{Q}$ combines a correlation structure to each token, and this evolving correlation structure is responsible for recording tokens that correlated with the combined token. Actually, the combined token is the base token of this correlation structure.

(8) Mapping $I$ denotes the initial distribution of tokens.

## 5 Applying Correspondence Petri Nets

### 5.1 Generating Correspondence Petri Nets

To generate a CorPN, we first need to collect the participating business processes of this collaborative business process, as well as the messages to use. The conversion from a single business process to an individual Petri net encompasses the following steps:

(1) Build up token set $\mathcal{D}$ and variable set $\mathcal{V}$;
(2) Set up place capacity expression set $\mathcal{G}$ and arc expression set $\mathcal{E}$ to designate the flowing range of tokens;
(3) In regard to token producible transitions, we mark a variable symbol $2^{\mathcal{V}}$ to adjacent outgoing arcs to represent the possibility of all available tokens defined for this business process.
(4) Initialise correlation set $\mathcal{C}$.

After the four steps, we can obtain the tuple sets for a business process. By incorporating all the obtained tuple sets of all business processes participating in a business collaboration, we may obtain the tuple of a pre-processed CorPN, Σ, for the corresponding collaborative business process. Due to the page limit, we do not discuss this issue intensively.

Algorithm 1 formalises the procedure of assembling these individual Petri nets into a CorPN via message types for the underlying collaborative business process. As this CorPN is created at process level rather than instance level, messages types are therefore used in this algorithm instead of message instances.

In Algorithm 1, function *transition*( Σ, *t* ) returns the transition that stands for task *t* in CorPN Σ; function *link*( *t* / *p*, *p* / *t* ) creates an arc linking transition *t* to place *p*, or place *p* to transition *t*, and *t* or *p* can also be set null to denote an undetermined transition or place; function *priorP* / *posteriorP* ( Σ, *tr* ) returns the prior / posterior place of transition *tr* in CorPN Σ; function *priorA* / *posteriorA* ( Σ, *tr* ) returns the prior / posterior arc of transition *tr* in CorPN Σ; function *relink*( Σ, *a* / *p*, *p* / *a* ) adjusts a half-determined arc *a* to connect to / from place *p* in CorPN Σ.

In this algorithm, line 4 to line 10 first generates arcs for outgoing message types, and line 12 to line 23 generates arcs for incoming message types. At this stage, these generated arcs are half-determined ones, because we only designate one end of an arc while leave the other end open. To keep the information of multiple receivers or senders of a message, two mapping functions, Π and Ω, are used to record the correspondence between the interaction participating transitions and the generated half-determined arcs. Based on these two mappings, line 26 to line 35 generates auxiliary places and re-links the open ends of those half-determined arcs to proper auxiliary places. In this way, we can connect the individual Petri nets together according to the messaging behaviours between participating business processes.

Following this algorithms, we can generate a CorPN as shown in Figure 6 for the collaborative business process of the motivating example. The sub nets for different

**Algorithm 1.** Assembling Petri nets

**Input**: *MSG*: the set of unidirectional message types used by business processes in *WP*.
$\Sigma$: the tuple of the pre-processed CorPN.

**Output**: $\Sigma'$: the CorPN tuple that is updated with auxiliary places, corresponding arcs etc.

```
set Σ' = Σ; ∏ = null; Ω = null; sendingArcs = ∅;
for each m∈ MSG
   if m. ρ = 'out' then // handling for outgoing message types
      tempT = ∅;  // create a half-determined arc for each outgoing message type
      k = link ( transition( Σ', m.α ), null );
      Σ'.ℱ°← k;
      for each t'∈ m.β
         Σ'.C°← ( k → m.f( t' )); tempT ← transition( Σ', t' ); sendingArcs ← k;
      end for
      ∏←( k→ tempT );
   else                // handling for incoming message types
      tempA = ∅;
      for each task t' ∈ m.β  // decompose the message-receiving transition into a
         create transition tr;  // series of transitions, please refer to Figure 6 (b)
         k = link( priorP( transition( Σ', m.α ), tr );
         Σ'.ℱ°←k;
         b = link( tr, posteriorP( transition( m.α ) );
         c = link( null, tr ); Σ'.ℱ°← c; Σ'.
         C°← ( c→m.f( t' ) );
         /* create a half-determined arc for each potential incoming route of this
   message type */
         Ω ←( transition( Σ', t' ) → c );
      end for
      Σ'.T = Σ'.T - { transition( Σ', m.α )};
      Σ'.ℱ = Σ'.ℱ - { priorA( Σ', transition( m.α )), posteriorA( Σ', transition( m.α ))};
   end if
end for
for each k ∈ sendingArcs  // link half-determined arcs with proper auxiliary places
   create auxiliary place px;
   relink( Σ', k, px );
   for each transition tr∈ ∏( a )
         b = Ω( tr );
         ∏←( k→( ∏( k )-{ k }));
         Ω←( tr→( Ω( tr ) –{ b } ));
      relink( Σ', px, b );
   end for
end for
```

business processes are distinguished with different circles, and the auxiliary places are marked as shaded circles.

Because this CorPN simulates the interaction between multiple business processes, it may own more than one starting place and ending place.

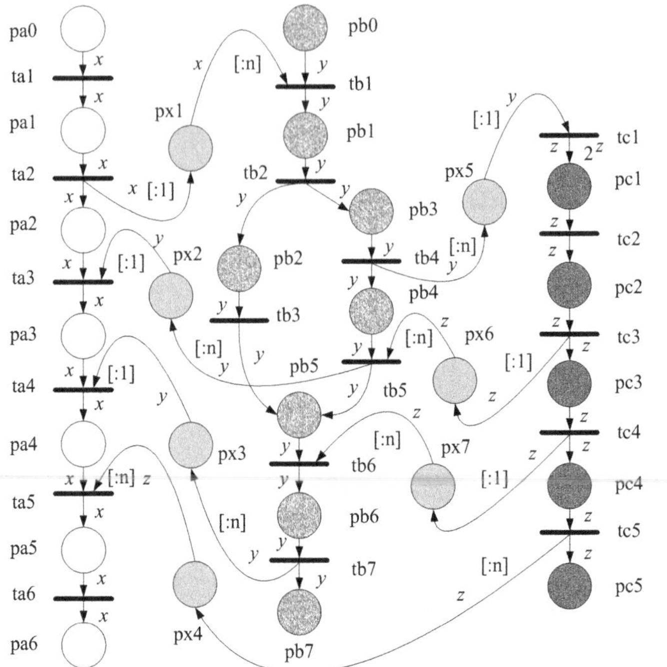


**Fig. 6.** a CorPN for a collaborative business processes

## 5.2 Run Time Execution

As discussed in Section 3, workflow correlations occur when business process instances interact. During interactions, a participating instance may inherit pre-existing workflow correlations from its counterparts in case that this interaction has some relation with previous correlations. To update these correlations, each business process instance needs to modify its correlation structure every time after 'shaking hands' with partner business process instances.

For example, when the manufacturer contacts shippers for delivery, the manufacturer's production instance may update its correlation structure with the correlations between retailers' product-ordering instances and shippers' assigned shipping instances. In the meantime, these shipping instances also update their correlation structures with the production instance and retailers' product-ordering instances that are to be served. As for retailers' product-ordering instances, they may not know these new correlations until the manufacturer notifies them of the delivery date after booking deliveries. Actually, to timely update their correlation strictures, retailers need to proactively trace such potential correlations rather than passively wait for feedbacks. Thus, the correlation handling comprises two procedures, i.e., to generate correlations after interactions and to trace existing correlations through coupled instances.

Algorithm 2 details the procedure of updating correlation structures after collaborating business process instances 'shake hands'. Following the organisation-oriented view, we classify the tokens representing the host organisation's involved instances as local tokens, and the ones representing the involved instances of partner business processes as foreign tokens. In this algorithm, function *TYPE( setTK )* returns which token group that tokens in *setTK* belong to; function *relatedTK( tk, setTK, ψ )* returns the set of tokens correlated with token *tk* from set *setTK* during interaction *ψ*; function *update( tk, setTk )* updates the content of token *tk*'s correlation structure with tokens in *setTk*. The details of function *update* are given at the end of Algorithm 2.

**Algorithm 2.** Updating correlation structures

**Input**: $\Sigma$: a CorPN.
$\psi$: a real interaction.
*localTK*: the set of participating local tokens during interaction $\psi$.
*foreignTK*: the set of participating foreign tokens during interaction $\psi$.
**Output**: $\Sigma'$: the updated CorPN.

**1.** **set** $f$ = null;
**2.** **set** $\Sigma'=\Sigma$;
**3.** **for** each $tk \in localTK$
**4.** $setTK' = relatedTK(\ tk, foreignTK,\ \psi\ )$;
**5.** *update( tk, setTK′ ); // update the correlation structures of local tokens.*
**6.** **for** each $tk^{\circ} \in setTK'$
**7.** $f \leftarrow (\ tk^{\circ},\ tk\ )$;
**8.** **end for**
**9.** **for** each $tk^{\circ} \in foreignTK$
**10.** *update( tk°, f( tk° )); // update the correlation structures of foreign tokens.*

*// function update is given below*
*update( tk, setTK )*
**u-1.** $r^{tk} = \Sigma'.Q(\ tk\ )$;
**u-2.** **if** $\exists D_i, D_i \in r^{tk}$ ( $TYPE(D_i)=TYPE(setTK)$) **then** $r^{tk}.D_i \leftarrow setTK$;
**u-3.** **else** $r^{tk}.D_i \leftarrow \{\ setTK\ \}$;
**u-4.** **for** each $tk_1 \in \bigcup_i r^{tk}.D_i$, $tk_2 \in setTK$
**u-5.** **if** $tk_1$ is coupled with $tk_2$ via $tk$ **then** $r^{tk}.\mathcal{R} \leftarrow (tk_1,\ tk_2)$;

Once an interaction occurs, each participating business process instance needs to run Algorithm 2 to update its correlation structure. For each local token, this algorithm searches all participated tokens for the correlated ones with this local token. This job is done by line 3 to line 8. Line 9 and line 10 call function *update* to update these correlated tokens in the correlation structures of local tokens. In addition, function *update* also generates proper tuples in relation *R* of each participated local token, if there exist tokens that are correlated via this local token.

Algorithm 3 describes the procedure of tracing potentially correlated tokens. An organisation may use this algorithm to proactively detect correlated business process instances for its own business process instance. In this algorithm, function *update( tk, setTk )* is the same with the one in Algorithm 2.

**Algorithm 3.** Tracing correlated tokens

**Input**: $tk^{\circ}$: the original token to update correlation structure.
$\Sigma$: the CorPN.
**Output**: $\Sigma'$: the updated CorPN.

1. **set** $\Sigma' = \Sigma$;
2. $List = \varnothing$;
3. $oldList = \varnothing$;
4. $r^{tk^{\circ}} = \Sigma.Q(tk^{\circ})$;
5. $List \leftarrow \bigcup_i r^{tk^{\circ}}.D_i$; *// List is used to store the tokens to check.*
6. **do while** $List \neq \varnothing$
7.     **select** $tk \in List$; **remove** $tk$ from $List$;
8.     $oldList \leftarrow tk$; *// oldList is used to store the checked tokens.*
9.     **for** each $tk' \in \bigcup_i r^{tk}.D_i$
10.         **if** $\exists(tk^{\circ}, tk') \in r^{tk}.\mathcal{R} \wedge tk' \notin oldList$ **then** $List \leftarrow tk'$;
11. **end while**
12. $update(tk^{\circ}, oldList)$;

This tracing procedure, from line 6 to line 11, follows a depth-first strategy to search for correlated tokens. After finding correlated tokens, the host organisation updates the retrieved tokens to its correlation structure by invoking function *update*. This correlation structure determines the logical instance of the specified business process instance. This procedure may be called upon request by the host organisation, for example, at a point that a retailer wants to know shippers' details while waiting for goods delivered by several shippers. Therefore, we do not have to derive this correlation structure for all instances involved in a collaborative business process.

## 6 Discussion and Conclusion

This paper looked into the problem of instance correspondence in an inter-organisational setting, which is of great importance to business process management yet has not been extensively studied in the literature. By establishing a CorPN model, we proposed a novel method to specify instance correspondences among collaborating business processes. This method captures the dynamics and diversity of business collaboration in terms of workflow cardinality and correlation. With this method, an organisation can clearly track its involvement over a collaborative business process. The detailed contributions of this paper are as follows:

(1) Unidirectional cardinality parameters and correlation structures to characterise instance correspondence at build time and run time, respectively;

(2) A correspondence Petri net model with proposed cardinality parameters and correlation structures etc., for inter-organisational workflow monitoring;

(3) An algorithm for assembling individual business processes into a collaborative process;

(4) Algorithms for specifying workflow correlations and tracing workflow correlations on the fly.

Our future work is to incorporate the proposed method into Business Process Management Notation (BPMN) or BPEL languages, and combine it with our existing relative workflow framework [9]. This future work is expected to provide a comprehensive solution for collaborative business process applications.

## References


1. Bussler, C.: B2B Integration. Springer, Heidelberg (2003)
2. Chen, Q., Hsu, M.: Inter-Enterprise Collaborative Business Process Management. In: Proceedings of the 17th International Conference on Data Engineering. Heidelberg, Germany, pp.253-260 (2001)
3. Dumas, M., ter Hofstede, A.H.M.: UML Activity Diagrams as a Workflow Specification Language. In: Proceedings of 4th International Conference on the Unified Modeling Language, Modeling Languages, Concepts, and Tools. Toronto, Canada, pp. 76-90 (2001)
4. Barros, A.P., Dumas, M., and ter Hofstede, A.H.M.: Service Interaction Patterns. In: Proceedings of the 3rd International Conference on Business Process Management (BPM, Nancy, France pp. 302–318 (2005)
5. van der Aalst, W.M.P., ter Hofstede, A.H.M., Kiepuszewski, B., Barros, A.P.: Workflow Patterns. Distributed and Parallel Databases 14(1), 5–51 (2003)
6. van der Aalst, W.M.P., ter Hofstede, A.H.M.: YAWL: Yet Another Workflow Language. Information Systems 30(4), 245–275 (2005)
7. Guabtni, A., Charoy, F.: Multiple Instantiation in a Dynamic Workflow Environment. In: Proceedings of 16th International Conference on Advanced Information Systems Engineering (CAiSE, Riga, Latvia, pp. 175–188 (2004)
8. Andrews, T., Curbera, F., Dholakia, H., Goland, Y., Klein, J., Leymann, F., Liu, K., Roller, D., Smith, D., Thatte, S., Trickovic, I., Weerawarana, S.: Business Process Execution Language for Web Services (2003)
9. Zhao, X., Liu, C., Yang, Y.: An Organisational Perspective on Collaborative Business Processes. In: Proceedings of the 3rd International Conference on Business Process Management. Nancy, France, pp.17-31 (2005)
10. Schulz, K., Orlowska, M.: Facilitating Cross-organisational Workflows with a Workflow View Approach. Data & Knowledge Engineering 51(1), 109–147 (2004)
11. Chiu, D.K.W., Karlapalem, K., Li, Q., Kafeza, E.: Workflow View Based E-Contracts in a Cross-Organizational E-Services Environment. Distributed and Parallel Databases 12(2-3), 193–216 (2002)
12. Zhao, X., Liu, C.: Tracking over Collaborative Business Processes. In: Proceedings of the 4th International Conference on Business Process Management. Vienna, Austria, pp. 33-48 (2006)
13. Reisig, W.: A Primer in Petri Net Design. Springer, Berlin (1992)

# Assessing Feasibility of IT-Enabled Networked Value Constellations: A Case Study in the Electricity Sector

Zsófia Derzsi[1,3], Jaap Gordijn[1], Koen Kok[1,2], Hans Akkermans[1], and Yao-Hua Tan[3]

[1] Free University, FEW/Business Informatics, De Boelelaan 1083a, 1081 HV Amsterdam, The Netherlands
{derzsi, gordijn, elly}@few.vu.nl

[2] ECN, Intelligent Energy Grids, PO Box 1, Petten, The Netherlands
j.kok@ecn.nl

[3] Free University, FEWEB, De Boelelaan 1083a, 1081 HV Amsterdam, The Netherlands
ytan@feweb.vu.nl

**Abstract.** Innovative networked value constellations, such as Cisco or Dell, are often enabled by Information Technology (IT). The same holds for the Distributed Electricity Balancing Service (DBS), which we present in this case study. To explore *feasibility* of such constellations while designing them, we need at least to develop a *financial* and *technical* understanding of the constellation at hand. In this paper, we take a multiple perspective approach, by taking a *business value* perspective (using $e^3$*-value* ) and an *information system* perspective (using UML-deployment diagrams) on the case at hand. We present a novel, structured approach to relate both perspectives, thus enabling a financial and technical feasibility assessment of the constellation, using a real-life case study in the field of electricity supply and consumption.

## 1 Introduction

Companies increasingly form networked value constellations to jointly satisfy complex needs. Well known examples include the networked business model of Cisco Systems [17] -actually consisting of a series of well integrated companies-, and the virtual integration of Dell Computers [11]. In a networked value constellation, enterprises use each other core-competencies to offer a product or service that each individual enterprise could not offer on its own.

Such a constellation requires more than just a few enterprises alone, co-producing things of economic value. To coordinate their processes properly, information and communication technology is indispensable. Actually, for the distributed electricity balancing service case study, as to be presented in this paper, information technology is a prerequisite.

To our consulting experience [3,7,9], one of the issues in designing a networked value constellation is first to find a constellation (in terms of participating enterprises, cross-organizational processes, *and* interworking information systems)

J. Krogstie, A.L. Opdahl, and G. Sindre (Eds.): CAiSE 2007, LNCS 4495, pp. 66–80, 2007.

that seems to be *feasible*. Such an explorative feasibility assessment track should be done in a light-weight fashion, to be able to develop a comprehensive, yet global, understanding of the constellation at hand, within a reasonable time-frame (time-to-market is typically just a few months). Such an understanding, while shared and agreed upon by the enterprises involved, can then provide further direction for a more detailed and focused requirements engineering and system design track.

In this paper, we consider two types of feasibility: (1) *economic* feasibility and (2) *technical* feasibility. Economic feasibility refers to the question whether *all* enterprises participating in a constellation can be economically sustainable over a reasonable period of time with respect to their participation in the constellation. It is then important to know *substantial* economical effects (in terms of expenses, investments, and revenues). Technical feasibility is about the question whether we can find an acceptable solution to put the value constellation into operation e.g. by deploying information technology (the focus of this paper) and inter-organizational business processes.

In this paper we combine two modeling techniques ($e^3$*-value* and UML deployment diagrams) to reason about feasibility. An $e^3$*-value* model has constructs for reasoning about financial feasibility by definition; however, since important financial effects can come from investments and expenses in IT, we feed the $e^3$-*value* model by financial annotations of a UML deployment diagram. Another contribution of this paper is that we show that a value model and deployment model, if correctly related, can be used to reason about scalability issues, which are of importance while considering technical feasibility.

This paper is structured as follows. In section 2 we introduce how to explore feasibility of networked value constellations and in section 3 we present our case study-based research approach. A first step is to explore the networked constellation from an economic value perspective (section 4); a second step is to understand the information system perspective (section 5) of the case at hand. In section 6, both perspectives are structurally related with each other. Finally, in section 7 we present the lessons learned and conclusions.

## 2 Perspectives to Understand Feasibility of IT-Enabled Networked Value Constellations

To our experience [7], to assess feasibility of networked value constellations, multiple *perspectives* (e.g. strategic goals, value transfers, business processes, and information systems) need to be considered. We consider the following perspectives, amongst others inspired on frameworks such as TOGAF or Zachman [16,1]. The *strategic goal* perspective represents the long-term objectives of enterprises, such as cost leadership or differentiation of products and services (see e.g. [15]). To explore feasible networked value constellations, it is important to know that the individual objectives of participating enterprises are aligned, and that no crucial conflicts exist. For the paper at hand, we do not elaborate on this perspective; instead the reader is referred to [13,10,8,19]. The *economic value transfer* perspective explores *what* enterprises offer of *economic value* to each other, and request *what* in return. The value transfer perspective shows how the strategy is put into operation. For feasibility understanding, this perspective is useful to assess *economic* sustainability, in terms of in-going and out-going money flows. In this paper, we employ the $e^3$*-value* [7] approach for representing the value transfer perspective (see section 3.1); other possibilities are BMO, [12], or REA [6]. The *business process* perspective shows *how* value transfers are carried out (e.g. time ordering, parallelism), by processes, including coordinating processes between enterprises. For feasibility purposes, this perspective is usable to understand *economic* feasibility (since resources such as workers cost money). In this paper, we do not explore this perspective further; the reader is referred to [14,18]. The *information system* perspective presents the software and hardware components, their communication, etc. In fact, this perspective may contain many sub-perspectives, depending on the modeling aim, and contributes to the understanding of *economic* feasibility (e.g. IS-components require investments and expenses for maintenance). Additionally, the perspective may give a clue regarding *technical* feasibility; whether it is possible to design an information system that satisfies the requirements expressed by e.g. the economic value transfer and business process perspectives. In this paper, we use UML [2], and as we will motivate later on, specifically deployment diagrams to capture the information system perspective.

## 3 Research Approach

### 3.1 The $e^3$*-value* Methodology for Economical Feasibility

To evaluate feasibility in this paper, we employ $e^3$*-value* and UML-deployment diagrams. To make this paper self-contained, we briefly introduce the $e^3$*-value* modeling concepts below as well as the $e^3$*-value* way of reasoning about economic feasibility (see for a more detailed explanation [7]). The $e^3$*-value* methodology provides modeling constructs for representing and analyzing a network of enterprises, exchanging things of economic value with each other. The methodology is ontologically well founded and has been expressed as UML classes, Prolog code, RDF/S, and a Java-based graphical $e^3$*-value* ontology editor and an analysis tool, which is available for download (see http://www.e3value.com/) [7]. In the following text, we use an educational example (see Figure 1) to explain the ontological constructs.

An *actor* is perceived by his/her environment as an economically independent entity. The Store and Manufacturer are examples of actors. Actors exchange *value objects* (e.g. Money). A value object is a service, a good, money, or even an experience, which is of economic value for at least one of the actors. An actor uses a *value port* to provide or request value objects to or from other actors. Actors have one or more *value interfaces*, grouping value ports, and showing economic reciprocity. So, in the example, Goods can only be obtained for Money and vice versa. A *value transfer* is used to connect two value ports with each other. In the example, a transfer of Good or Payment are both examples of value transfers.

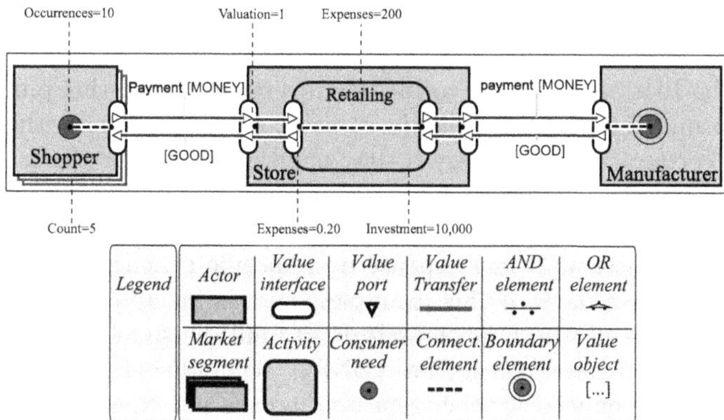


**Fig. 1.** Educational example

A *value transaction* groups value transfers that all should happen, or none at all. A *market segment* composes actors into segments of actors that assign economic value to objects equally. The Shopper is a market segment, consisting of a number of individual shoppers. An actor performs one or more *value activities*. These are assumed to yield a profit. In the example, the value activity of the Store is Retailing. A *dependency path* is used to reason about the number of value transfers as well as their economic value. A path consists of *consumer needs*, *connections*, *dependency elements* and *dependency boundaries*. A consumer need is satisfied by exchanging value objects (via one or more interfaces). A connection relates a consumer need to a value interface, or relates various value interfaces internally, of a same actor. A path can take complex forms, using AND/OR dependency elements taken from UCM scenarios [5]. A dependency boundary represents that we do not consider any more value transfers for the path. In the example, by following the path we can see that, to satisfy the need of the Shopper, the Manufacturer ultimately has to provide Goods.

An $e^3$-*value* model can be attributed with (financial) numbers (e.g. the number of occurrences of consumer needs, the size (count) of a market segment, and the valuation of objects transfered) that are used to generate Net Value Flow Sheets (NVF) (for a free software tool see http://www.e3value.com/). Such sheets show the net cash flow for each actor involved and are a first indication whether the model at hand can be commercially successful for each actor. In the example, the Store has $5 * 10 = 50$ transfers with the Shoppers, so the incoming money stream = €50,-∗1 =€50,-.

It is also possible to add various kinds of *expenses* and *investments*. These are cash-out flows that are significant, but for which it is not important to understand the actor receiving the expenses. Additionally, expenses are the hook to include financials obtained from other modeling perspectives (e.g. UML deployment diagrams) into the financial picture. In the example, the activity Retailing has a *fixed expense* (meaning independent from the number of transfers handled) of €200,-. Moreover, there is a *variable expense* of €0.20,- per outgoing Good transfer (so for this case the expense is 50∗€0.20,-=€10,-. The $e^3$-*value* ontology is capable of *assigning* the expenses related to activities to the *performing* actor automatically. Moreover, fixed and variable expenses can also be attributed to actors directly. It is also possible to include investments. For this purpose, it is important to understand that a *single* $e^3$*-value* model represents the financial effects for a certain *timeframe*, say a day, month, or year. A *series* of potentially different $e^3$*-value* models can be combined into an $e^3$*-timeseries* model to consider a number of timeframes (say many years). An investment is then actually an one-time (often upfront) expense in a specific timeframe that does not occur in other timeframes. In this example, there is an upfront investment of €10,000.-. Summing up the financial effects of multiple timeframes is done via the Discounted Net Present Cash Flow (DNPC) technique [4]. This results in a, hopefully positive, financial number representing the net financial effects for an actor, thereby accounting properly for the time-value of money.

### 3.2 Explorative Case Study: Distributed Balancing Services

We explore a model-based way of assessing feasibility of networked value constellations, by taking a value transfer and information system perspective (the other perspectives are also important but simply not in the scope of this paper due to space restrictions). Our ultimate goal is to arrive at a set of relevant, well integrated models that allows for feasibility studies. In this paper, we use a case-study on electricity supply and consumption (see [9]).

Due to the physical nature of electricity power, the amount of electricity supplied to the network must be *exactly equal* to the amount of electricity consumed, including inevitable transport losses. This balance has to be maintained at every instance otherwise power outages will occur. This requirement is at all time ensured by the Transmission System Operator (TSO). The TSO does so by asking large consumers and generators for their consumption/production *plans* a day ahead, matches these, and returns consumption/production plans that ensure consumption and supply balance. However, at runtime there are always deviations from the plans since it is impossible to *precisely* consume/produce the amount of electricity as planned. Since deviation from the plans causes imbalance, and adequate yet costly counter measures have to be taken, suppliers/consumers have to pay a penalty fee for causing imbalance to the TSO, who by default compensates for system imbalance.

The innovative idea for the case at hand is to create an IT-enabled service that reduces real-time imbalance in a *portfolio* of generators and consumers by allowing near-real time, distributed control over the electricity production and consumption of portfolio's participants: the Distributed Balancing Service (DBS). In case of imbalance, actors are asked to change their level of production and/or consumption. Because the imbalance is reduced for the portfolio, the penalties decrease also, and thus for the suppliers/consumers participating in the portfolio. Obviously, such near real-time control is only possible using advanced information technology, giving the time-scale (minutes) and the number of actors.

The aim of this case-study for us is to explore financial- and technical feasibility assessment of an IT-enabled value constellation by considering an economic value transfer and information system perspective. We have selected this specific case-study because: (1) The constellation and the information technology for imbalance reduction has already been built. We want to focus on the *conceptual constructs* required to understand feasibility issues and not yet on the *process* of assessment itself. (2) We have access to the financial data. (3) We have access to the developers of the DBS case. (4) The DBS case relies heavily on IT. (5) The case-study is of industrial strength. For the DBS case, we first study the available materials and do interviews with the domain experts, and we construct an $e^3$-*value* model of the DBS case (see section 4). Also, we construct a UML deployment diagram of the DBS (see section 5). By annotating the UML deployment diagram, such that financials related to information systems can be represented, *and* by structurally relating these annotations to the elements in the value model, we derive comprehensive (discounted) net value flow sheets for both perspectives. For relating the value and deployment perspectives, we present a sub-ontology (see section 6). Finally, we reason about *technical feasibility* using the presented models with an emphasis on *scalability*.

## 4 An $e^3$-*value* Model for the Distributed Balancing Service

Figure 2 shows an $e^3$-*value* model for the DBS case study. The focus is on the participating enterprises and what they *transfer* of economic value, and not on the required soft- and hardware components yet.

There are different market segments of 'electricity generators' in the form of 'wind turbines', 'Combined Heat Power generators' (CHPs) and 'emergency generators'. All these generators offer 'electricity' and request 'money' in return. Different types of generators exist because, due the nature of the generator (volume of total electricity power, predictability of this volume), the pricing schemes may be different. Additionally, they offer 'operational flexibility', meaning that a portfolio holder (here the 'supplier') may influence the amount of electricity production, in return for 'money'. There are 'consumers' who buy 'electricity' and pay 'money' in return. Also, they offer 'operational flexibility' so that a portfolio holder can influence their amount of electricity consumption, and they request some 'money' in return for that. Normally, the 'generators' and 'consumers' must also pay a fee to the 'Transmission System Operator' (TSO), if their real-life production/consumption deviates from their forecasted production/consumption (which is always the case). This *balance-responsibility* is in the DBS $e^3$-*value* model taken over by a 'supplier' of which we have one. The 'generators' and 'consumers' are all in the portfolio of the 'supplier'. The 'supplier' pays a penalty ('money') to the TSO for the amount of imbalance caused. This amount can be reduced by controlling the 'generators' and 'consumers' near real-time. Finally, there is a 'wholesale market operator'. The role of this operator is to sell electricity to the 'supplier' in case of shortage or to buy electricity from the 'supplier' in case of a surplus.

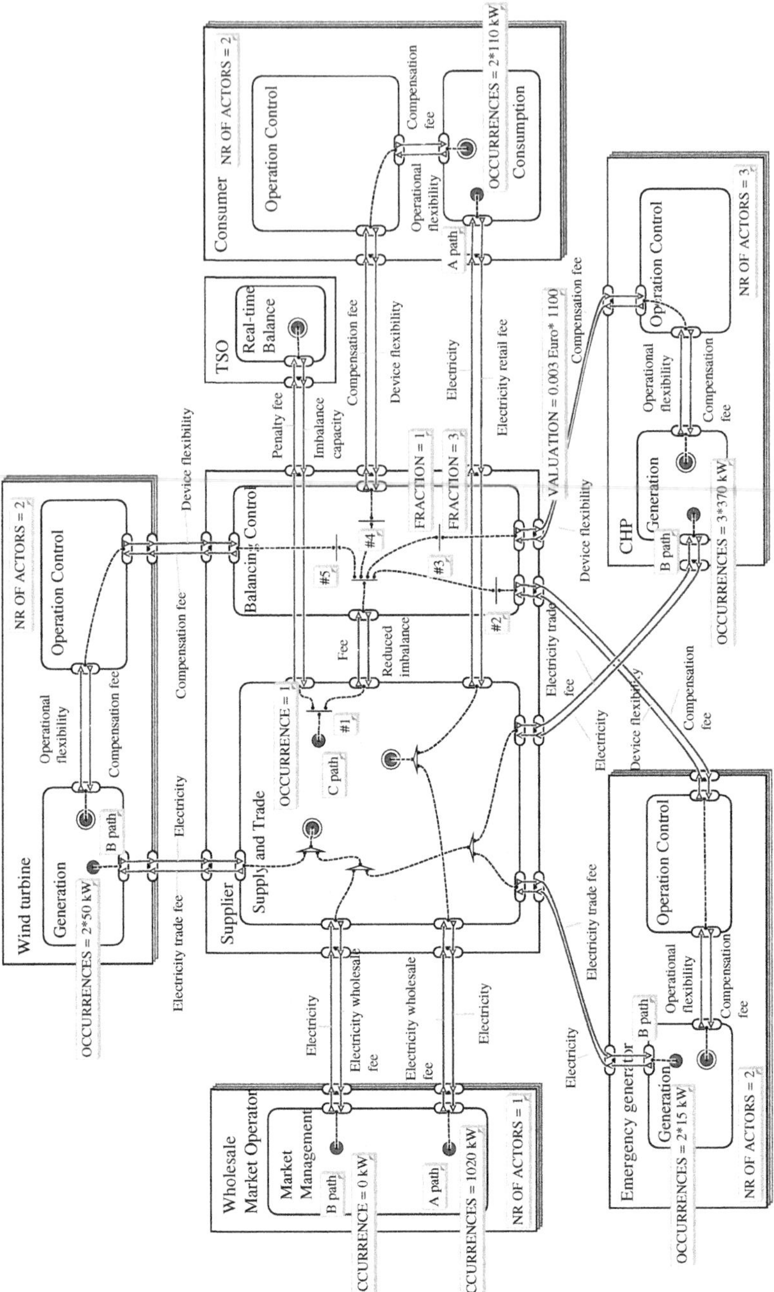


**Fig. 2.** *$e^3$-value* model of the Distributed Balance System

An $e^3$*-value* model provides a snapshot of value transfers for a certain timeframe; here, for 15 minutes, since it is used as a discrete interval to calculate fees, based on the actual production/consumption. All the modeled *consumer needs* occur within this timeframe.

Now, tracing through the 'A' dependency path, the 'consumer' has a need for a certain amount of kilowatt-hours (kWh) (see Figure 2). The 'wholesale market operator' has also a need for electricity. These needs are satisfied by the 'supplier'. He buys electricity from the 'generators' of his portfolio, and from the 'wholesale market operator' in case of a shortage, as can be seen from the 'B' path. From the 'C' path it can be seen that the 'supply & trade' activity requires 'balancing control', and so control of the operation of 'generators' and/or 'consumers' in terms of operational flexibility. 'Balancing control' operates together with the 'operation control' activity, which is executed by consumers and generators. Since such a control moment is needed once per 15 minutes (timeframe of the model), there will be precisely one occurrence, so one 'operational flexibility' transfer between the 'supplier' and the 'generators'/'consumers'. However, due to the fact that market segments aggregate actors, explosion elements are needed (fork (#2)-(#5)) in order to achieve one occurrence per actor in such a market segment. Despite the efforts of the 'supplier', there will always be some imbalance (because the 'supplier' can control *near* real-time). This is modeled by the *AND* fork (#1).

The $e^3$*-value* model calculates, as shown, the occurrences for each dependency path element for the 15-minutes timeframe. We assume that investments in generators and in consumption control equipment were done earlier, so we do not consider these. Investments related to IT are explored in section 5. If we assign pricing schemes (valuation functions) to the model (see Figure 2 for an example), assume an amount electricity power needed, assume a number of generators and consumers, and assume how much required electricity power can be satisfied by the portfolio's participants (and the wholesale market), we can derive for each 15 minute timeframe net value sheets for each enterprise involved. With $e^3$*-timeseries* , it is possible to concatenate a series of $e^3$*-value* model snapshots, capturing many sequential timeframes of each 15 minutes. Then, a Discounted Net Present Cash Flow [4] sheet per actor can be derived to judge the financial attractiveness of the DBS, which we do not discuss in detail due to space restrictions. In table 1, such a sheet is (as an example) given for the 'CHP generator', including both the $e^3$*-value* and UML-deployment perspective.

## 5 IS-Perspective: A UML Deployment Diagram Annotated with Expenses and Investments

### 5.1 Why a Deployment Diagram?

We now take an IS perspective on the DBS case. In this case study, we consider an already up-and-running system, for which the designed IS-models are available in UML. For our purpose, assessing economic and technical feasibility (in terms of scalability), we restrict ourselves to deployment diagrams for a few reasons. (1) Deployment diagrams show, statically, *components* and *nodes* on which these components run. These components and nodes require *investments* (one-time upfront expenses) and regularly occurring *fixed expenses* (e.g. for maintenance). So, for annotating UML with financials, deployment diagrams provide sufficient handles. (2) Components themselves have *interfaces* which offer or request *services* via *ports*. Both offering and requesting services may result in *variable expenses*. 'Variable' means here that the expense relates to the number of service invocations; in case of a 'fixed' expense we have to do the expense always -even if there are zero service invocations. Service invocations are related to value transfers on the economic value transfer perspective (see section 6.1). (3) Ports offering and requesting services are annotated, e.g. with the *maximum* number of service invocations per timeframe. If the previously mentioned value transfers are related to service invocations, we can reason about scalability issues. (4) Deployment diagrams are sufficiently course-grained, so they are of use in a light-weight feasibility assessment approach.

## 5.2 A DBS Deployment Diagram

Figure 3 shows a deployment diagram (with components) for the case at hand. The 'generators' and 'consumers' all have the same, complex components derived from the value activity 'operation control'. They consist of several subcomponents, namely (a) a *computation* component, (b) a *database*, and (c) a *measuring&control* component. The computation component computes for each 15-minute timeframe a *pricing-function* that can be used to calculate, given the amount of electricity supplied/required, the price willing to obtain/pay. For this calculation, historical data from the local database is used. The measuring&control component directly influences the generation/consumption device, e.g. by adjusting the produced/required electricity power.

The 'supplier' managing the portfolio operates the 'balancing control' component, which consist of (a) a *computation* component, and (b) a *database* component. These components are used to collect the forementioned pricing-functions from each 'generator' and 'consumer'. Then, supply and demand is balanced, and the 'generators' and 'consumers' are each reported back the required/consumed electricity power.

There are two services: (1) the generators and consumers offer a service that returns the forementioned pricing-function, (2) the supplier offers a service that tells the generator/consumer how much electricity they must produce/consume the coming 15-minutes timeframe by using the above pricing functions. Effectively, this controls the generator/consumer behavior.

Components are assigned to *devices* and eventually to *nodes*, being physical resources. The nodes, devices and components are *classes*, which have one or more *instances*. So, the deployment diagram tells that *one* supplier node (instance) is associated with six device nodes (instances) (consumer and generator PC's) connected via ADSL, and with three device nodes (instances) (in this specific case, the CHP nodes) connected via wireless-UMTS. The latter

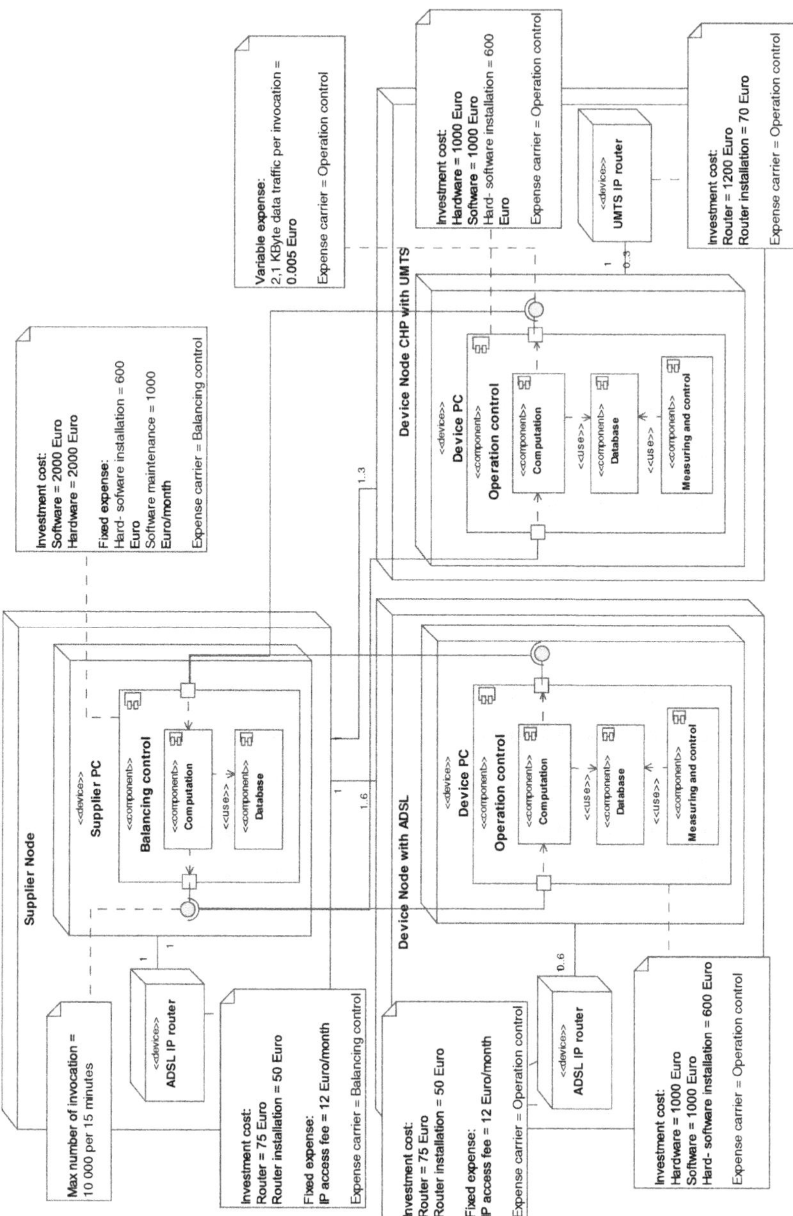


**Fig. 3.** UML deployment model of the DBS case

sub-classification, based on connection technology, is motivated by the very different expense-profiles of these technologies. The number of required instances are *derived* from the $e^3$-*value* model, by counting the number of generators and consumers.

### 5.3 Annotating the Deployment Diagram for Feasibility Reasoning

Figure 3 shows various annotations to different constructs of the deployment diagram. The *financial annotations* are structured along the lines of Figure 4, as an *extension* to the UML 2.0 metamodel [2] and to the $e^3$*-value* ontology [7]. Moreover, Figure 4 shows how the UML *relates* to $e^3$*-value* . Figure 4 distinguishes various kinds of IT-expenses. *Fixed expenses* are expenses that occur once per timeframe in an $e^3$*-timeseries* sequence of value models; *investments* occur only once per timeseries (typically these are upfront investments in equipment, software, etc., to enable future cash flow). A fixed expense is *for* precisely one *asset*, which in UML terminology is a *device* or a software *artifact*; an asset *has* one or more *fixed expenses*. An expense is *assigned-to* to one *expense carrier*. An expense carrier is an $e^3$*-value actor*, *value activity* or *market segment*. This way, expenses can be assigned to business entities that create revenues to pay these expenses.

In a UML deployment diagram *ports* attached to components are used to offer and request services from the environment. Requesting or offering a service via a port may result for each invocation in expenses themselves. The connection between requested and offered services via ports is in UML stated as an *assembly connection*. Such a connection *is caused by* one or more *value transfers* in an $e^3$*-value* model. Conversely, a *value transfer causes* one or more invocations (as represented by an assembly connection showing the invoked and invokeed port). This model-fragment allows for modeling *variable expenses*; the amount of expense is based on the number of service invocations, which in turn depends on the number of value transfers in the $e^3$*-value* business model.

As an example consider *investments*. The financial annotations of Figure 3 show the hard- and software costs related to components that are, as mentioned before, derived from value activities. By executing value activities (i.e. the 'operation control' activity) e.g. hard- and software investments (€1,000.- and €1,000.- respectively) are required. This can be fed into the $e^3$*-value* model for the appropriate value activity (being an *expense carrier*). The use of wireless-UMTS routers results in data-traffic accounted on a per KByte basis and thus

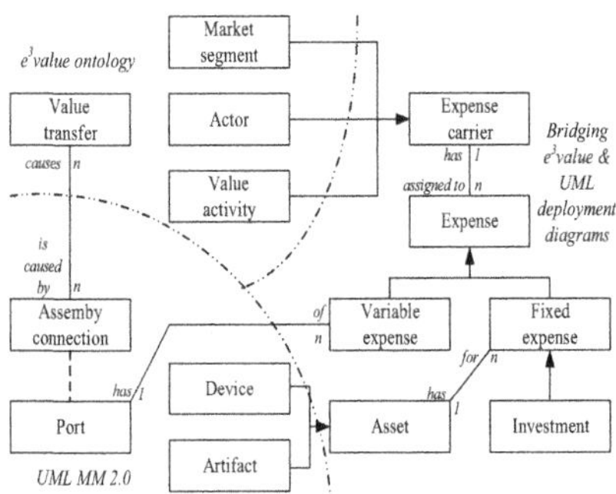


**Fig. 4.** Relating an $e^3$*-value* model and a UML deployment diagram

in extra *variable expenses* each time services (using the UMTS connection) are invoked (see Figure 3). In this example, each invocation results in sending of 2,1 KB, resulting in an expense of €0.005,-. For this expense, an *expense carrier* is identified (here the 'operation control' activity of the 'CHP').

# 6 Relating the $e^3$-*value* and UML Deployment Perspective

## 6.1 Financial Feasibility

Financial feasibility is assessed by summing up the net cash flow (*revenues – expenses – investments*) for each actor involved over a series of timeframes. From a value transfer perspective, we consider a series of $e^3$-*value* models, each describing a (here 15 minutes) *timeframe*, together forming an $e^3$-*timeseries* . Each $e^3$-*value* model, or each timeframe, contributes revenues, expenses, and possibly investments for each actor.

From an information system perspective, each timeframe may result in (fixed or variable) expenses and investments. Fixed expenses and investments of IT are directly assigned to *expense carriers* (along the lines of of Figure 4). Based on the number of value transfers (per timeframe, per actor), the amount of variable expenses is calculated that stem from IT-service invocations.

The result of the above calculation is exemplified in table 1 for the 'Operation control' activity that is executed by a CHP. The table normally lists all considered timeframes (here only period 0 -showing the initial investment-, and period 1 -in which the first value transfers are done- are shown, for brevity reasons). Since many sequential timeframes can be considered as equal, the number of timeframes with *different* financials is often much less. For each timeframe, first the cash transferred (both to- and from an actor) is shown as a result of doing value transfers according to the stated $e^3$-*value* model for that timeframe. Then, expenses and investments are shown for that timeframe that result from the information system perspective. Hereafter, the net cash is calculated for each timeframe, just by subtracting expenses and investments from revenues. Finally, all net cash flows for all timeframes are summed up using the Discounted Net Present Cash Flow method [4], thereby accounting properly for the time value of money, cost of capital, and risk associated with participating in the constellation.

## 6.2 Technical Feasibility

The technical feasibility assessment may contain various perspectives; here we explore *scalability* only. More specifically the question is: what happens if from an economic value transfer perspective things scale up (e.g. a significant increase in consumers, or generators).

To facilitate such reasoning, we have annotated the port of the 'balancing control' component (see Figure 3) with a *maximum* number of *invocations* per timeframe (here 15 minutes). We have already explained that the number of value transfers (for the economic value transfer perspective) indicates the number

**Table 1.** Net value flow sheet for 'Operation control' activity of one CHP

| **Actor/Activity:** | CHP- 'Operation control' | | | | | |
|---|---|---|---|---|---|---|
| **Timeframe:** | period 0 | | | | | |
| | | | | | **Economic Value** | **Total** |
| INVESTMENT | | | | | 3,870 | 3,870 |
| **Timeframe:** | period 1 | | | | | |
| **Value Interface** | **Value Port** | **Value Transfer** | **Occurrences** | **Valuation** | **Economic Value** | **Total** |
| Device flexibility,MONEY | | | 1 | | 0.735 | |
| | out: Device flexibility | (EXPENSES) | 1 | 0.005 | -0.005 | |
| | in: MONEY (Compensation fee) | MONEY | 1 | 0.002 | 0.74 | |
| Operational flexibility,MONEY | | | 1 | | -0.666 | |
| | out: MONEY (Compensation fee) | MONEY | 1 | 0.0018 | -0.666 | |
| | | | | | | |
| Net Cash Flow: | | | | | | 0.069 |
| **Timeframe:** | period 103,680 + 1 | | | | | |
| | ... | | | | | |
| Discounted Net Cash Flow: | | | | | | 1,715.12 |

of port-invocations per timeframe. If this number is larger than the maximum number of invocations, there is a scalability issue. Perhaps it can be solved by using different hardware, but at some point, it can be possible that an entirely different architecture should be selected.

Given the numbers for the case at hand, problems occur if the size of a market segment increases since it effects the number of value transfers. As an example, let the number of consumers increase to 15,000. This results in 15,000 value transfers and so in 15,000 port-invocations of the 'balancing control' component initiated by consumers. As can be seen from the annotation of the component port (see Figure 3), this is already larger than the maximum number of invocations that can be handled by the current IS design. Obviously, this is just an example how we can reason about scalability, but it shows that for addressing a scalability issue, an integrated view on the value transfer and IS perspective is useful.

## 7 Lessons Learned and Conclusions

In this paper, we have shown how an $e^3$*-value* model, taking an *economic* perspective on networked value constellations, can be structurally related to a corresponding UML deployment diagram, representing a *technical* perspective. As a result, a comprehensive discounted net value flow sheet can be produced for each actor involved, for the purpose to assess *economic* feasibility. Additionally, scalability, being an aspect of the larger construct, *technical* feasibility, can be be reasoned about.

While doing this case study, we experienced some learnings, of which we articulate here two (due to space considerations). First, with respect to the notion of time, it is important that the timeframes as considered by the different perspectives are *indeed* about the *same* timeframe. Specifically, if the timeframe is determined by IT (e.g. timeframe of invocations), the $e^3$*-value* model should possess the same timeframe as well. This puts certain requirements on the models constructed. Second, we have seen that -for the purpose of *feasibility* assessment-, the selection of the relevant components and nodes as distinguished by the UML deployment diagrams are influenced by the size of expenses of these components and nodes. We experienced that these deployment diagrams are sufficient for our purposes; obviously, they need to be detailed if stakeholders really decide to develop the case at hand further.

Many continuing research lines are possible. Deployment diagrams are typically constructed if information system requirements and design are already somewhat clear. Consequently, it is important to understand how $e^3$*-value* models influence other UML-type of diagrams and vice versa (i.e. use-cases, activity / state transition diagrams, class diagrams) that are usually built in an earlier stage of requirements engineering and system design. Additionally, other aspects, specifically of *technical* feasibility need to be addressed, in conjunction with the $e^3$*-value* model (think of flexibility, maintainability, etc.). Another line of research is the development of *guidelines* that help to make architectural decisions, which are then expressed using the integrated models we have proposed.

On the short term, we continue our research by developing a DBS to be used in Woking/UK. As the current DBS is developed for The Netherlands, we expect changes in the $e^3$*-value* model and IS architecture for the Woking case, due to specific UK-regulations about electricity supply. We intend to use this case study to develop an integrated view on flexibility, by considering to what extent the Dutch system is usable for the UK. Additionally, we will work on an entirely different case, a ship container tracking system, to validate our proposed model-based feasibility assessment approach for IT-intensive networked value constellations.

**Acknowledgements.** This work has been partly sponsored by the EESD-IST funded project FENIX (518272), the JACQUARD/NWO funded project VITAL (838.003.407), and the VUA funded project VUBIS.

## References

1. Togaf enterprise edition, version 8.1. accessed november (2006) `http://www.opengroup.org/architecture/togaf8-doc/arch/`
2. Unified modeling language: Superstructure. accessed november (2006) `http://www.omg.org/docs/formal/05-07-04.pdf`
3. Akkermans, J.M., Baida, Z., Gordijn, J., Peña, N., Altuna, A., Laresgoiti, I.: Value webs: Using ontologies to bundle real-world services. IEEE Intelligent Systems 19(4), 57–66 (2004)

4. Brealey, R., Myers, S., Allen, F.: Corporate Finance. McGraw Hill Higher Education (2005)
5. Buhr, R.J.A.: Use case maps as architectural entities for complex systems. IEEE Transactions on Software Engineering 24(12), 1131–1155 (1998)
6. Geerts, G., McCarthy, W.E.: An accounting object infrastructure for knowledge-based enterprise models. IEEE Intelligent Systems and Their Applications, pp. 89–94 (1999)
7. Gordijn, J., Akkermans, J.M.: Value-based requirements engineering: Exploring innovative e-commerce ideas. Requirements Engineering Journal 8(2), 114–134 (2003)
8. Gordijn, J., Yu, E., Van der Raadt, B.: e-Service design using i and e3 value modeling. IEEE Software 23(3), 26–33 (2006)
9. Gordijn, J., Akkermans, H.: Business models for distributed energy resources in a liberalized market environment. The Electric Power Systems Research Journal, 2007. Accepted by the Electric Power Systems Research Journal. Preprint available. doi:10.1016/j.epsr.2006.08.008
10. Gordijn, J., Petit, M., Wieringa, R.: Understanding business strategies of networked value constellations using goal- and value modeling. In: Glinz, M., Lutz, R. (eds.) Proceedings of the 14th IEEE International Requirements Engineering Conference, pp. 129–138. IEEE CS, Los Alamitos, CA (2006)
11. Magretta, J.: The power of virtual integration: an interview with dell computers michael dell. Harvard Business Review 76(2), 72–84 (March-April 1998)
12. Osterwalder, A., Pigneur, Y.: An ontology for e-Business models. In: Currie, W.L. (ed.) Value Creation From e-Business Models (ch.4), pp. 65–97. Elsevier Butterworth-Heinemann, Oxford, UK (2004)
13. Pijpers, V., Gordijn, J.: Bridging business value models and business process models in aviation value webs via possession rights. Accepted by HICSS 2007, see (2007), `http://docs.e3value.com/bibtex/pdf/PijpersBridging2007.pdf`
14. Pijpers, V., Gordijn, J.: e3forces: Understanding the environment of networked value constellations for strategic goal and business model analysis. Submitted, see ( 2007), `http://docs.e3value.com/bibtex/pdf/PijpersStrategy2007.pdf`
15. Porter, M.E.: Strategy and the Internet. Harvard Business Review, 63–78 (march 2001)
16. Sowa, J., Zachman, J.: Extending and formalizing the framework for information systems architecture. IBM Systems Journal 31, 590–616 (1998)
17. Tapscott, D., Ticoll, D., Lowy, A.: Digital Capital - Harnessing the Power of Business Webs. Nicholas Brealy Publishing, London, UK (2000)
18. Weigand, H.: On the notion of value object. In: Dubois, E., Pohl, K. (eds.) CAiSE 2006. LNCS, vol. 4001, pp. 321–335. Springer, Heidelberg (2006)
19. Weigand, H., Johannesson, P., Andersson, B., Bergholtz, M., Edirisuriya, A., Ilayperuma, T.: Strategic analysis using value modeling - the c3-value approach. Accepted by HICSS 2007 (2007)

# Behavioral Consistency for B2B Process Integration

Gero Decker and Mathias Weske

Hasso-Plattner-Institute, University of Potsdam, Germany
{gero.decker,mathias.weske}@hpi.uni-potsdam.de

**Abstract.** Interacting services are at the center of attention in business-to-business (B2B) process integration scenarios. Global interaction models specify the interaction behavior of each service and serve as contractual basis for the collaboration. Consequently, service implementations have to be consistent with the specifications. Consistency checking ensures that an implemented service is compatible with other services, i.e. that it can interact successfully with them. This is important in order to avoid deadlocks and guarantee proper termination of a collaboration. Different notions of compatibility between interacting services and consistency between specification and implementation are available but they are typically discussed independently from each other. This paper presents a unifying framework for compatibility and consistency and shows how these two notions relate to one another. Criteria for an optimal consistency relation with respect to a given compatibility relation are presented. Based on these criteria weak bi-simulation is evaluated.

## 1 Introduction

In the case of business-to-business (B2B) process integration different business partners interact with each other to reach a common goal. Each partner exposes its communication behavior as services that exchange business documents with the other partners' services in a certain order. Choreography languages such as WS-CDL ([5]) and Let's Dance ([15]) were put forward for capturing the interaction behavior from a global perspective. The interaction behavior is represented by interaction models, which serve as contractual basis for the collaboration between the partners. Interface processes, i.e. the behavioral specifications for the individual partners, can be generated (cf. [16]). These interface processes are the starting point for implementing new services or for adapting existing ones e.g. using BPEL ([1]). Consequently, the behavior of a service has to be consistent with the specified interface process. Such a consistency relation should ensure that an implemented service is in fact compatible with the partners' services without needing to check compatibility between the actual implementations. The latter is not desired since internal process details should not be revealed. Furthermore, if there is a large number of interacting partners, the number of possible combinations of partners that have to be checked for compatibility is so high that the testing phase becomes complex in itself.

J. Krogstie, A.L. Opdahl, and G. Sindre (Eds.): CAiSE 2007, LNCS 4495, pp. 81–95, 2007.

While different consistency relations have been reported in the literature, overarching criteria for evaluating consistency relations for specific purposes has not been proposed yet. This paper argues that a consistency relation can be checked for suitability with respect to a specific compatibility relation. E.g. the compatibility relation could allow that certain interactions never happen or that messages sent by one service are ignored by another one. Having chosen a suitable compatibility relation for a given context, we provide the criteria to decide whether a given consistency relation is optimal or not.

Existing notions of compatibility and consistency are discussed in the next section. Section 4 provides a formal definition of what an optimal consistency relation is with respect to a given compatibility notion. Section 5 elaborates on possible refinements from a process specification to an implementation. Since weak bi-simulation is a common formal basis for consistency checking, we will investigate in Section 6 whether it is optimal for the selected compatibility notions. Section 7 concludes and gives an outlook to future work.

## 2 Compatibility and Consistency in B2B Scenarios

Compatibility is the ability of a set of interconnected services to interact successfully. Consistency between a service implementation and a service specification is given if the implementation is valid with respect to the specification. In the literature also other names such as process inheritance are used as synonyms for consistency (cf. [4], [2]).

This section will further explain the need for compatibility and behavioral consistency using a B2B scenario. Figure 1 gives an overview over the partners in that scenario: A buyer (e.g., car manufacturer) uses reverse auctioning for procuring specially designed components. In order to get help with selecting the right suppliers and organizing and managing the auction, the buyer outsources these activities to an auctioning service. The auctioning service advertises the auction, before different suppliers can request the permission to participate in it. The suppliers determine the shipper that would deliver the components to the buyer or provide a list of shippers with different transport costs and quality levels, where the buyer can choose from. Once the auction has started, the suppliers can bid for the lowest price. At the end, the buyer selects the supplier according to the lowest bid or according to other criteria. After the auction is over, the auctioning service has to be paid for and shipment details are dealt with. Finally, the components are delivered to the buyer and are paid for.

It is obvious that there can be several suppliers, auctioning services, shippers and buyers. Therefore, different combinations of participants must be able to interact successfully. Unsuccessful interaction behavior could arise e.g., if different message formats are used in the collaboration and one participant does not understand the message content sent by other participants. Another source of incompatibility, which we will mainly focus on, is behavioral incompatibility. Imagine that a participant expects a notification at some point in a process before it can proceed and none of the other participants ever sends such a notification.

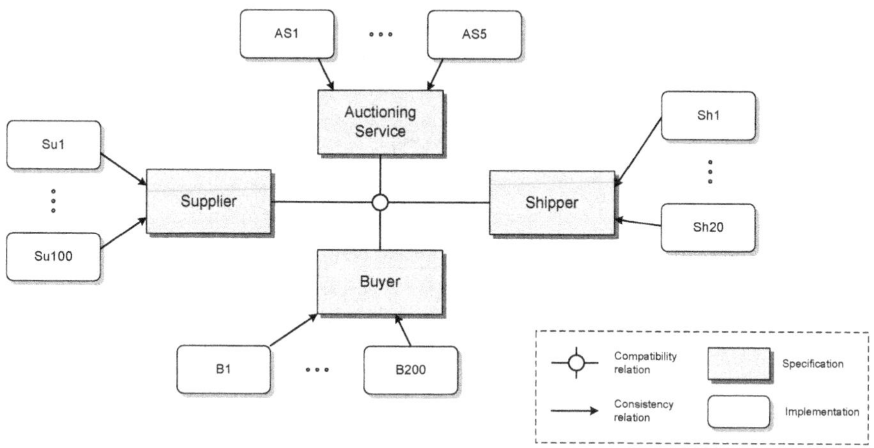


**Fig. 1.** Participants and roles in a reverse auctioning scenario

We call this situation a *deadlock*. In order to avoid deadlock situations and to ensure interoperability, the participants can agree on a certain desired interaction behavior. The behavioral constraints between message exchanges would be captured from the perspective of an ideal observer and the constraints for the communication behavior of every participant (the interface process) could be derived from such an interaction model (the choreography). This specification then serves as contractual basis for the collaboration and violations of the interaction contract could have legal consequences.

Figure 2 depicts a part of the collaboration specification where suppliers can request permission to the auction. We see that the roles supplier, auctioning service and seller take part in this collaboration. For each of these roles an interface process is given in the form of a Petri net. The places on the border of the dashed rectangle depict the structural interface of each role, i.e. the types of messages a role can potentially send or receive. The control flow between the communication actions constrains the execution. A "?" symbolizes a receive action and "!" a send action.

The supplier places a participation request at the auctioning service. The service formulates a recommendation whether to accept this supplier or not. This recommendation is normally based on previous experience with the supplier or legal requirements. The auctioning service sends the recommendation to the buyer. The buyer is not bound to this recommendation and can freely chose whether to accept or reject the supplier. Finally, the auctioning service forwards the decision of the buyer on to the supplier.

This specification does not tell the individual participants how their internal behavior should look like. The auctioning service could, for instance, lookup historical data about the supplier before coming up with a recommendation; also the buyer could have an internal decision making process possibly spanning different organizational units. No matter how the internal processes look like, we are concerned that the different participants successfully collaborate, i.e.

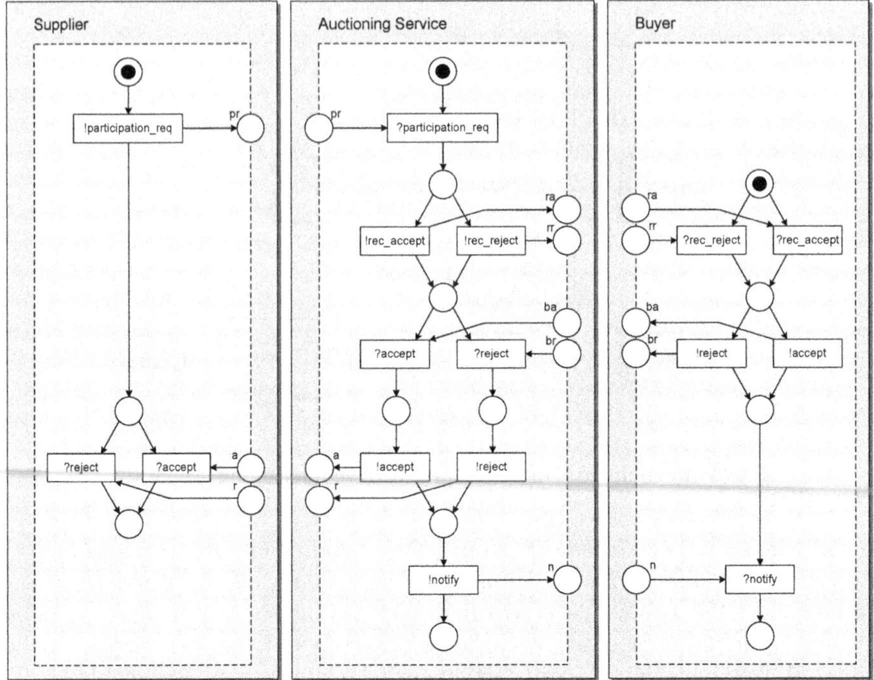


**Fig. 2.** Interface processes: Getting a participation permission

that the implementations are compatible. Ensuring compatibility is a challenging and cumbersome task when dealing with a large number of participants in this auctioning scenario, involving e.g., 100 suppliers, 20 shippers, 5 auctioning services and 200 buyers.

A remedy for this situation is the notion of consistency between interface and executable processes. The interface processes are the specifications for the different roles. Consistency between an interface process and the actual process implementation should ensure that the given implementation can interact with implementations for the other roles (provided that they in turn are consistent with their respective interface process). That way, we can locally check whether or not a participant should be allowed to be involved in the collaboration scenario. Compatibility between different implementations does not need to be checked any more.

## 3 Compatibility and Consistency Notions

In recent years there has been extensive work on different compatibility notions for interacting processes. This section compares four different notions, namely the compatibility notion by Martens [6], the compatibility notion by Canal et al. [4], interaction soundness by Puhlmann et al. [10] and the well-communicating requirement used in the operating guidelines approach by Massuthe et al. [8,9]. Furthermore, this section will present different existing consistency notions.

**Compatibility.** First, we can distinguish between structural compatibility and behavioral compatibility. Structural compatibility like presented as "syntactic compatibility" in [6] demands that for every message that can be sent, the corresponding interaction partner must be able to receive it. Futhermore, for every message that can be received the corresponding partner must be able to send such a message. I.e. in the case of web service architectures the receiving service must have a matching operation for every outgoing SOAP message of the sending service, and for every operation a sending service must be able to send a corresponding message. We call this notion *strong structural compatibility.* In other compatibility notions, e.g. the well-communicating requirement, strong structural compatibility is not required. This acknowledges the fact that if a service provides a certain operation, the partners do not necessarily need to invoke this operation. However, it is still required that for every message sent there must be a corresponding operation. We call this *weak structural compatibility.* One could also think of examples where even weak structural compatibility is too restrictive: a middleware platform might be configurable in such a way that unprocessable messages are simply ignored. E.g. notifications that are not necessarily needed might not be received in some process. For such scenarios we introduce the notion of *minimal structural compatibility.* Minimal structural compatibility requires that there is at least one potential message send with a corresponding message receive by another participant.

Figure 3 presents three alternative service implementations for the buyer in our reverse auctioning example. We see that internal actions were added in the case of $B1$ and $B3$ (e.g. "Add to blacklist" and "Store decision"). $B1$ is structurally equivalent to the buyer in Figure 2 but has a different control flow structure. This alternative has strong structural compatibility with the Supplier and Auctioning Service, since every message sent can be received and for every message that can be received there is a message sent. This applies for both incoming and outgoing messages. $B2$ has only weak structural compatibility, since the Auctioning Service could receive a *reject* message from the buyer but the buyer never sends one. $B3$ does not have weak structural compatibility with the other participants: the notification sent by the Auctioning Service cannot be received by the buyer. However, since there are messages sent that can be received by the buyer, minimal structural compatibility is still given.

In contrast to structural compatibility, behavioral compatibility considers behavioral dependencies, i.e. control flow, between different message exchanges within one conversation. In most approaches the interface processes of the interacting partners are interconnected and reasoning is done on the resulting global process.

Martens bases his compatibility notion on interconnected workflow modules and requires strong structural compatibility. These modules are Petri nets with input, output and internal places (like the examples in Figures 2 and 3). When

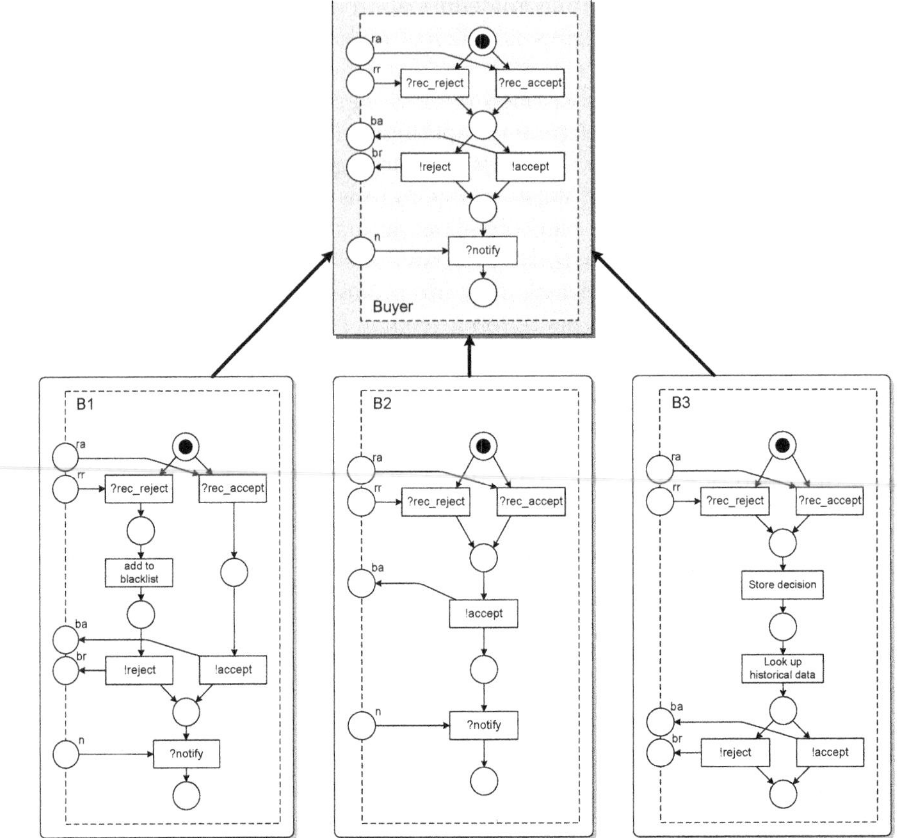


**Fig. 3.** Alternative implementations for the buyer

composing them, corresponding input and output places of interacting processes are merged and a global initial place and a global final place are added. Martens defines "weak soundness" on the global process, requiring that the final marking must always be reachable. This ensures that the global process is free of deadlocks and livelocks.

Canal et al. have also defined a compatibility notion for interacting $\pi$-processes. A main advantage of using $\pi$-calculus is the availability of *link passing mobility*. I.e. communication channels between interacting processes do not need to be statically defined but can be established at runtime. In real world settings this is called dynamic binding. E.g. a service broker passes the reference to a provided service on to a service consumer who can then use the service. $\pi$-interactions are atomic, i.e. sending and receiving of messages happen at the same time. Therefore, it is not possible that one $\pi$-process sends a message which is not consumed by the other. The compatibility notion by Canal et al. requires that both processes complete, i.e. that no more sending or receiving action is left to be performed. A major drawback of the given compatibility notion is that it is defined for bi-lateral settings only.

Interaction soundness is based on "lazy soundness" of the global process. It is required that the process always completes, while some activities are still allowed to run even after completion. Considering these "lazy activities" is essential for coping with advanced control flow constructs such as Discrimators (cf. [12]) but leads to the fact that livelocks cannot be detected in some situations. Interaction soundness is defined for a combination of a service and its environment. We find a mixture of strong and minimal structural compatibility between the service and its environment: The environment must be able to send and receive all those kind of messages that the service can receive or send. Therefore, there must be strong structural compatibility in one direction. However, the service is not required to send and receive all those kind of messages that the environment is able to receive or send. Interaction soundness is defined on $\pi$-calculus. Therefore, it can also deal with link passing mobility.

The operating guidelines approach to checking compatibility [8,9] is different to the three previous approaches in that it does not reason on interconnected interface processes. Rather an "operating guideline" can be generated for an interface process which includes all valid interaction behavior that respects the well-communicating requirement. This requirement includes weak structural compatibility and the absence of deadlocks and livelocks. Operating guidelines are annotated state machines and represent the most permissive interaction behavior for the interaction partners. The interface processes of the partners (also given as state machines) must then be sub state machines of the most permissive behavior. The main motivation behind the operating guidelines approach is to reach a smaller computational complexity for compatibility checking. A current limitation of operating guidelines is that only acyclic processes are allowed.

**Consistency.** There has been quite some research work on consistency relations between specified interface processes and process implementations comparing their observable behavior. Basten et al. introduce different notions of process inheritance in [2], namely protocol inheritance, projection inheritance, protocol/projection inheritance and life-cycle inheritance. In order to determine whether an implementation is a subclass of a given specification, encapsulation and abstraction mechanisms are employed. Encapsulation deletes additional activities from the implementation before comparing it to the specification, while abstraction re-labels certain activities as $\tau$-actions, i.e. they are not considered in the weak bi-simulation relation. Once encapsulation and abstraction is applied, branching bi-simulation is used to compare the two process definitions. Bi-simulation relations were defined for different formalisms. These results are used in the public to private approach reported in [13], where participants agree on a global interaction model and a partitioning of this model to participants. Using inheritance mechanisms, each partner can implement an arbitrary subclass of their public process as a local, private implementation. The inheritance rules make sure that the private implementations satisfy the interaction constraints defined in the public model.

Other examples for (bi-)simulation relations are weak open (bi-)simulation for $\pi$-calculus by Sangiorgi [11] and branching bi-simulation for Petri nets by van Glabbeek and Weijland [14]. Bi-simulation in general will be discussed in section 6. Busi et al. have introduced their own calculi for choreographies and orchestrations in [3]. Consistency between orchestration and the specified behavior, which is given in the choreography, is shown through a bi-simulation-like relation, which is also defined by the authors. Martens presents a consistency relation in [7] where the implementation must accept at least those messages specified and must produce at most those messages specified.

## 4 Optimal Consistency Relations

The previous section recapitulates different notions of compatibility and consistency. Consistency is dependant on compatibility and should go hand in hand with it. Therefore, this section introduces a means to judge whether or not a given consistency relation is suited for a given compatibility notion. The following two requirements for consistency relations can be identified:

1. *A consistency relation should ensure compatibility.* If a set of specified service definitions are compatible and a set of implementations are consistent with these specifications then this set must also be compatible. Figure 4 illustrates this. If $A_{spec}$, $B_{spec}$ and $C_{spec}$ are compatible and $A_{impl}$ is consistent with $A_{spec}$, $B_{impl}$ with $B_{spec}$ and $C_{impl}$ with $C_{spec}$, then $A_{impl}$, $B_{impl}$ and $C_{impl}$ must also be compatible.
2. *A consistency relation should not be too restrictive.* The consistency relation should be maximal, i.e., every possible extension to the consistency relation must result in the violation of the previous requirement.

In the remainder of this section the abovementioned requirements are formalized. To allow for reusing the definitions for arbitrary compatibility and consistency notions, the formalization is independent from a particular formalism (such as Petri nets or $\pi$-calculus). We introduce as follows:

- *$S$ is a set of service definitions,*
- *$C \subseteq \wp(S)$ is the set of all compatible combinations of service definitions* and
- *$\succeq \subseteq S \times S$ is a binary relation on $S$ where $s_{impl} \succeq s_{spec}$ denotes that service definition $s_{impl}$ is consistent with service definition $s_{spec}$.*

In the example shown in Figure 4, all spefications and implementations are service definitions: $A_{spec}$, $B_{spec}$, $C_{spec}$, $A_{impl}$, $B_{impl}$, $C_{impl} \in S$, the set of specifications and the set of implementations are compatible, respectively: $\{A_{spec}$, $B_{spec}$, $C_{spec}\}$, $\{A_{impl}$, $B_{impl}$, $C_{impl}\} \in C$ and the implementations are consistent with their respective specifications: $(A_{spec}, A_{impl})$, $(B_{spec}, B_{impl})$, $(C_{spec}, C_{impl}) \in \succeq$.

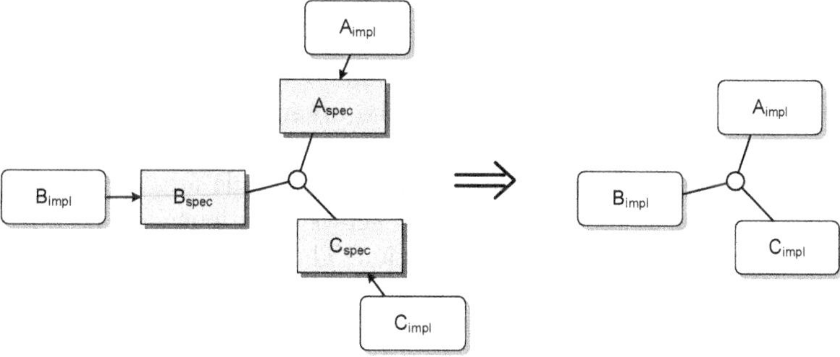


**Fig. 4.** The consistency relation must respect the compatibility notion

An auxiliary relation is introduced, $\succeq' := \{(c_1, c_2) \in \wp(S) \times \wp(S) \mid \forall s_1 \in c_1\ [\exists s_2 \in c_2\ (s_1 \succeq s_2)] \wedge \forall s_2 \in c_2\ [\exists s_1 \in c_1\ (s_1 \succeq s_2)]$: All service definitions $s_1$ in one set are consistent with at least one service definition $s_2$ in the other set and there is no $s_2$ without at least one $s_1$ that is consistent with it.

Based on these definitions, we can formalize the two requirements for optimal consistency relations: A consistency relation $\succeq$ is optimal for a compatibility notion $C$ if and only if

1. $\succeq$ *respects* $C$, i.e. $\forall c_1, c_2 \in \wp(S)\ [(c_2 \in C \wedge c_1 \succeq' c_2) \Rightarrow c_1 \in C]$
2. $\forall (s_1, s_2) \in S \times S\ [\neg(s_1 \succeq s_2) \Rightarrow \neg((\succeq \cup \{(s_1, s_2)\})\ \textit{respects}\ C)]$

The first line states that a combination of service definitions $c_1$ must be compatible, if $c_1$ is consistent with a combination of service definitions $c_2$ that are compatible. We have introduced the relation $\textit{respects} \subseteq (S \times S) \times \wp(S)$ indicating which consistency relation respects which compatibility notion. The second line states that adding a new tuple $(s_1, s_2)$ of service definitions to the consistency relation must result in breaking the first requirement. I.e. we must not disallow any implementation that would successfully interact with the other allowed implementations.

## 5 Process Refinement Categories

The definition of optimal consistency relations in the previous section allows to decide a true/false decision about the suitability of a given consistency relation. In this section we want to describe some typical process refinements, i.e. differences between specified interface processes and process implementations. An optimal consistency relation is expected to support all these process refinements, provided that the compatibility notion is permissive enough. The list of refinement categories allows to compare compatibility and consistency notions with respect to engineering needs. If a consistency relation is not optimal for a given compatibility notion it probably provides less support for at least one of the categories. Unlike the requirements for optimal consistency relations we do not provide formal definitions of the refinements.

1. **Addition of internal actions.** Specified interface processes indicate what interaction behavior other participants can expect from an organization. On the other hand, a process implementation covers all internal activities and dependencies that are present within the organization. In other words, the specified interface process constrains the interaction behavior of an organization while the implementation contains all details for actually executing the process. The implementation also shows interdependencies with other processes running within the organization. Therefore, additional internal activities that are not visible outside the organization can be found in the process implementation.
2. **Addition of communication actions.** The process implementation sometimes needs to be able to comply to the constraints given in different interface processes. E.g. there might be different interaction contracts with upstream and downstream partners in a supply chain scenario. In order to cope with such a situation the process implementation contains more communication actions with partners than specified in one interface process.
3. **Deciding choices at design-time.** If a partner is allowed to do choices, e.g. an organizational unit can decide whether to send an accept or reject message (cf. the buyer in Figure 3), the specification indicates the latest moment where the choice can be made. However, an organization might decide that always the same branch is taken. E.g. the buyer in the example of the previous section might decide to always accept a supplier (cf. B2 in Figure 3). Therefore, the choice is already done at configuration time of the partner's system, i.e. design-time of the process implementation.
4. **Removal of communication actions.** When deciding at design-time that particular branches should be taken then this results in removing communication actions from the process that are part of the other branches. This has an influence on structural compatibility if all communication actions for a particular message type are deleted. Another reason for deleting receive actions could be the knowledge that the sending party has done a design-time decision never to take a certain branch. However, it can be argued that such an implementation should not be allowed by the consistency relation because such an agreement between the two partners is not reflected in the specification.
5. **Deciding choices earlier at runtime.** It is also imaginable that a choice is done sometime earlier in the process. E.g. an organizational unit decides dependending on what message comes in sometimes earlier in the process. B1 in Figure 3 sends an accept or reject message if a corresponding recommendation comes in. Therefore, the choice whether to send an accept or reject message is not done right before such a message is to be sent, as it is the case in the specification, but the choice is rather done as soon as a recommendation is received. In other cases, deciding choices earlier at runtime does not have any visible effect for the outside world.
6. **Sequentialization of communication actions.** Ideally, specified interface processes do not make any restrictions on the ordering of message production and consumption if not absolutely necessary. E.g. once a supplier is selected,

a buyer is required to initiate the payment for the auctioning service and send delivery details to the designated shipper. Although there is no constraint in what order the buyer has to send the two messages, the process implementation might sequentialize it. E.g. the shipper is always notified as soon as possible while the payment is delayed for a while.

7. **Reordering of communication actions.** Consider a similar scenario like the previous one: it is specified that a seller first receives payment details before he receives the delivery details from the buyer. Assuming asynchronous communication, it might be allowed that the seller can reorder the consumption of the two incoming messages for process optimization purposes: He first processes the delivery details and initiates transport before he processes the payment details.

## 6 Assessment of Bi-simulation for Consistency Checking

Section 3 has shown that weak bi-simulation is the basis for several consistency relations. The main idea behind weak bi-simulation is that a process $A$ can simulate the communication behavior of process $B$ and vice-versa, while internal actions are not considered. Therefore, if $A$ is capable of doing some communication action c then $B$ must also be capable of doing c and again vice-versa. In the case of consistency checking we can therefore compare a specified interface process with a process implementation in terms of bi-similarity.

Since two bi-simulation related processes $A$ and $B$ show equivalent communication behavior, it is easy to see that bi-simulation *respects* a wide range of compatibility notions. The first criterion for an optimal consistency relation is therefore given. The remainder of this section investigates whether bi-simulation is too restrictive, i.e. if the second requirement for an optimal relation is met.

Consider an example similar to the B2B scenario introduced in Section 2. In Figure 5 processes $P_1$ and $P_2$ are depicted having the same structural interface to the environment: Messages of type $a$ and $b$ can be received, and messages of type $c$ and $d$ can be produced. Due to this structural equivalence, $a$ and $b$ have the same degree of structural compatibility with any given environment. However, these processes show different behavior. In terms of combinations of communication actions that are performed within one process instance, process $P_1$ allows $?a.!c$, $?a.!d$, $?b.!c!$ and $?b.!d$. On the other hand, $P_2$ only allows $?a.!c$ and $?b.!d$. In this example, $P_1$ can simulate all behavior of $P_2$ but $P_2$ cannot simulate all behavior of $P_1$. Therefore, $P_1$ and $P_2$ are not bi-simulation related.

The second example (Figure 6) leads to a similar situation. In $P_3$ no ordering constraint between $!f$ and $!g$ is given, while in $P_4$ $!f$ always happens before $!g$. $P_3$ therefore allows $?e.!f.!g$ and $?e.!g.!f$, while $P_4$ only allows $?e.!f.!g$. $P_3$ and $P_4$ are not bi-simulation related.

According to the consistency notions in [2] and [3], which are based on bi-simulation, $P_2$ and $P_4$ would not be allowed as implementations for $P_1$ and $P_3$. However, $P_2$ would be compatible with all environments that $P_1$ is compatible with. This is due to the fact that the environment must be able to receive $c$ and $d$.

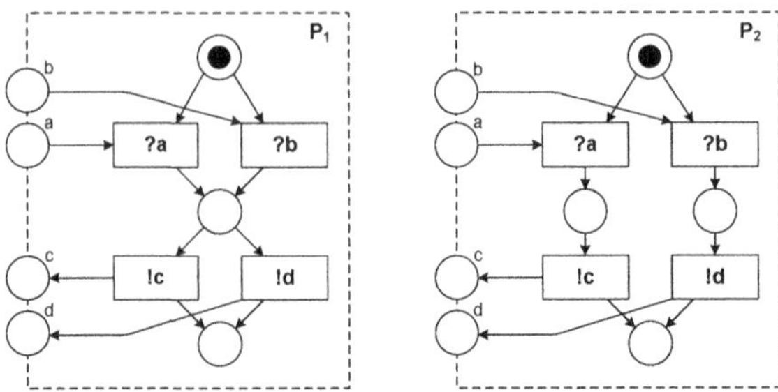


**Fig. 5.** $P_1$ and $P_2$ are not bi-simulation related

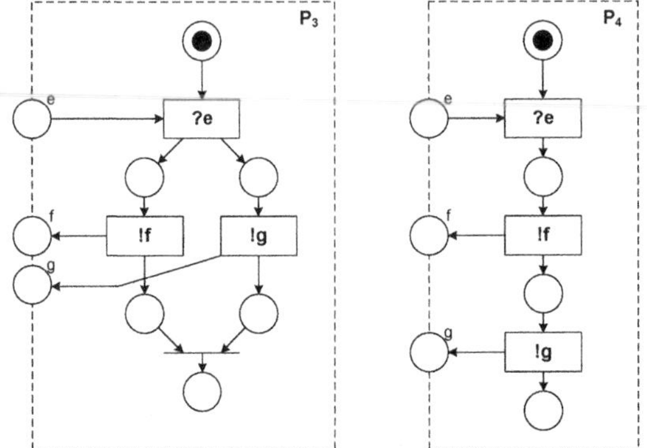


**Fig. 6.** $P_3$ and $P_4$ are not bi-simulation related

The environment is not allowed to do any assumptions about which message is to be received, otherwise it would be incompatible with $P_1$. For this reason, it is just a special case that in $P_2$ the choice for $c$ or $d$ is linked to the previously received message. In analogy to this, we know that $P_4$ would be compatible with all environments that $P_3$ is compatible with.

Therefore we can conclude that bi-simulation is too restrictive as compatibility notion for the compatibility notions presented in section 3, since adding the tuples $(P_1, P_2)$ and $(P_3, P_4)$ to the consistency relation would not result in not respecting any of the compatibility notions presented in section 3 any longer.

These examples also show that consistency relations do not need to be symmetric, i.e. that if a process $A$ is consistent with process $B$, $B$ does not need to be consistent with $A$. Assume a process $X$ allowing the two combinations $!a.?c$ and $!b.?d$ would be compatible with $P_2$ but not with $P_1$: Assume that $P_1$ produces a message of type $d$ after having received a message of type $a$, i.e. $?a.!d$, the conversation would deadlock since $X$ would expect a message of type $c$ which in turn would never be sent in that conversation. Figure 7 depicts this situation. Since all presented compatibility notions detect such simple potential deadlocks, incompatibility holds with respect to all these notions.

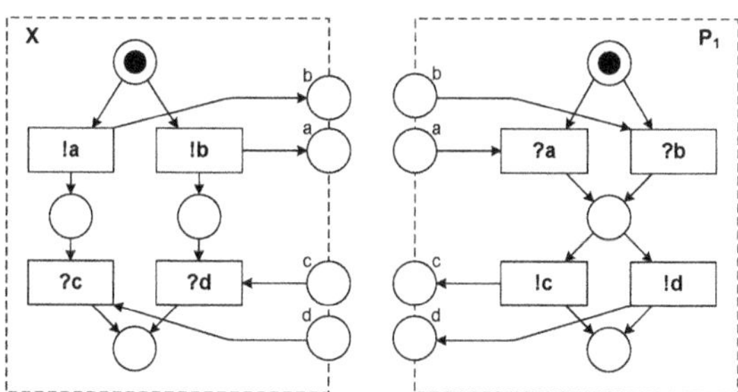


**Fig. 7.** $X$ and $P_1$ are not compatible

The processes in Figures 5 and 6 are examples for the process refinement categories presented in the previous section. In $P_2$ the choice whether a message of type $c$ or $d$ is sent is made earlier than in $P_1$, namely already as soon as a message is consumed. $P_4$ is an example for the sequentialization of actions.

Let us now take a look at the other categories from Section 5. Weak bi-simulation directly supports the category *addition of internal activities.* The added activities are simply treated as $\tau$-actions and are therefore ignored. For *addition of communication actions* it might be possible to re-label the added actions as $\tau$-actions before testing for bi-similarity. This pre-processing step is suggested in [2]. However, it cannot be generally allowed. Especially if message exchanges for existing message types are added, compatibility with other processes might be affected. *Removing communication actions* is not supported through bi-simulation in the case of reachable communication actions. *Deciding choices at design-time* is not supported (assuming that it affects the communication behavior). When *deciding choices earlier at runtime* the communication behavior is not affected in many real-world cases. Therefore, we conclude that there is partial support for this category. *Sequentialization* and *reordering of communication actions* are not supported through bi-simulation.

Table 1 summarizes what categories of process refinements are supported through weak bi-simulation. A "+" denotes that there is full support, "+/–" partial support and "–" indicates no support (including the assumptions made in the previous paragraph). The table highlights that weak-simulation does not

**Table 1.** Common process refinements and support through weak bi-simulation

| Process refinements | Weak bi-simulation |
|---|---|
| 1. Addition of internal actions | + |
| 2. Addition of communication actions | +/– |
| 3. Deciding choices at design-time | – |
| 4. Removal of communication actions | – |
| 5. Deciding choices earlier at runtime | +/– |
| 6. Sequentialization of communication actions | – |
| 7. Reordering of communication actions | – |

fully support a wide range of common process refinements and therefore has limited suitability.

## 7 Conclusion

This paper motivates the need for behavioral consistency checking in B2B process integration scenarios. Especially in choreography-driven settings such consistency is of key importance. We have introduced a unifying framework for behavioral compatibility and consistency of services. Two requirements for consistency relations have been introduced for classifying whether a consistency relation preserves compatibility and if it is too restrictive with respect to the compatibility relation. It was shown that interacting partners only need to agree on a suitable compatibility notion for their purposes and no further discussion about the consistency relation is required, since it can be determined whether or not a consistency relation is optimal for the chosen compatibility notion.

Furthermore, it was shown that classical weak bi-simulation relations do not fulfill the two requirements and refinement categories were highlighted that are fully, partially, or not supported by weak bi-simulation.

In future work, we are going to investigate other consistency relations with respect to corresponding compatibility notions. The consistency relation introduced by Martens ([7]) is promising. It might turn out to be optimal for his weak-soundness-based compatibility notion introduced in [6]. In addition we are going to define optimal consistency relations for selected compatibility notions.

## References


1. Andrews, T., Curbera, F., Dholakia, H., Goland, Y., Klein, J., Leymann, F., Liu, K., Roller, D., Smith, D., Thatte, S., Trickovic, I., Weerawarana, S.: Business Process Execution Language for Web Services, version 1.1. Technical report, OASIS, (May 2003). `http://www-106.ibm.com/developerworks/webservices/library/ws-bpel`
2. Basten, T., van der Aalst, W.M.P.: Inheritance of behavior. JLAP 47(2), 47–145 (2001)
3. Busi, N., Gorrieri, R., Guidi, C., Lucchi, R., Zavattaro, G.: Choreography and Orchestration: A Synergic Approach for System Design. In: Proceedings 3rd International Conference on Service Oriented Computing (ICSOC, Amsterdam, The Netherlands, Springer (December 2005)
4. Canal, C., Pimentel, E., Troya, J.M.: Compatibility and inheritance in software architectures. Sci. Comput. Program. 41(2), 105–138 (2001)
5. Kavantzas, N., Burdett, D., Ritzinger, G., Lafon, Y.: Web Services Choreography Description Language Version 1.0, W3C Candidate Recommendation. Technical report,(November 2005) `http://www.w3.org/TR/ws-cdl-10`
6. Martens, A.: Analyzing Web Service based Business Processes. In: Cerioli, M. (ed.) FASE 2005. LNCS, vol. 3442, Springer, Heidelberg (2005)
7. Martens, A.: Consistency between Executable and Abstract Processes. In: Proceedings IEEE International Conference on e-Technology, e-Commerce, and e-Services (EEE 2005), Hong Kong, China, pp. 60–67. IEEE Computer Society Press, Los Alamitos (March 2005)

8. Massuthe, P., Reisig, W., Schmidt, K.: An Operating Guideline Approach to the SOA. Annals of Mathematics, Computing and Teleinformatics 1(3), 35–43 (2005)
9. Massuthe, P., Schmidt, K.: Operating guidelines - an automata-theoretic foundation for the service-oriented architecture. In: Proceedings Fifth International Conference on Quality Software (QSIC 2005), pp. 452–457. IEEE Computer Society Press, Washington, DC, USA (2005)
10. Puhlmann, F., Weske, M.: Interaction Soundness for Service Orchestrations. In: Dan, A., Lamersdorf, W. (eds.) Proceedings of the 4th International Conference on Service Oriented Computing ICSOC 2006, LNCS, vol. 4294, pp. 302–313. Springer, Heidelberg (2006)
11. Sangiorgi, D.: A Theory of Bisimulation for the pi-Calculus. Acta Informatica 16(33), 69–97 (1996)
12. van der Aalst, W.M.P., ter Hofstede, A.H.M., Kiepuszewski, B., Barros, A.P.: Workflow Patterns. Distributed and Parallel Databases 14(1), 5–51 (2003)
13. van der Aalst, W.M.P., Weske, M.: The P2P Approach to Interorganizational Workflows. In: Dittrich, K.R., Geppert, A., Norrie, M.C. (eds.) CAiSE 2001. LNCS, vol. 2068, pp. 140–156. Springer, Heidelberg (2001)
14. van Glabbeek, R., Weijland, W.: Branching Time and Abstraction in Bisimulation Semantics. Journal of the ACM 43(3), 555–600 (1996)
15. Zaha, J.M., Barros, A., Dumas, M., ter Hofstede, A.: A Language for Service Behavior Modeling. In: Proceedings 14th International Conference on Cooperative Information Systems (CoopIS 2006), Montpellier, France, Springer, Heidelberg (November 2006)
16. Zaha, J.M., Dumas, M., ter Hofstede, A., Barros, A., Decker, G.: Service Interaction Modeling: Bridging Global and Local Views. In: Proceedings 10th IEEE International EDOC Conference (EDOC 2006), Hong Kong (October 2006)

# Declarative XML Data Cleaning with XClean

Melanie Weis[1] and Ioana Manolescu[2]

[1] HPI für Softwaresystemtechnik GmbH
Prof.-Dr.-Helmert-Str. 2-3, D-14482 Potsdam
melanie.weis@hpi.uni-potsdam.de
[2] INRIA Futurs
2-4, rue Jacques Monod, 91893 Orsay Cedex France
Ioana.Manolescu@inria.fr

**Abstract.** Data cleaning is the process of correcting anomalies in a data source, that may for instance be due to typographical errors, or duplicate representations of an entity. It is a crucial task in customer relationship management, data mining, and data integration. With the growing amount of XML data, approaches to effectively and efficiently clean XML are needed, an issue not addressed by existing data cleaning systems that mostly specialize on relational data.

We present XClean, a data cleaning framework specifically geared towards cleaning XML data. XClean's approach is based on a set of cleaning operators, whose semantics is well-defined in terms of XML algebraic operators. Users may specify cleaning programs by combining operators by means of a declarative XClean/PL program, which is then compiled into XQuery. We describe XClean's operators, language, and compilation approach, and validate its effectiveness through a series of case studies.

## 1 Motivation

Data cleaning is the process of correcting anomalies in a data source, that may for instance be due to typographical errors, formatting differences, or duplicate representations of an entity. It is a crucial task in customer relationship management, data mining, and data integration. Relational data cleaning is performed in specialized frameworks [13,20,25], or by specialized modules in modern RDBMSs [7].

With the growing popularity of XML and the large volumes of XML data becoming available, approaches to effectively and efficiently clean XML data are needed. For example, consider DBLP[1] whose data is available in XML format. Fig. 1 shows an excerpt of the DBLP entry of one of this paper's authors, on which we observe several XML data cleaning issues. First, the SIGMOD conference is represented by the conference abbreviation, the string "Conference", and the year of the conference, whereas VLDB is only represented by its abbreviation and year. A second example is the representation of author names. In the bottom publication, the first author is represented by its firstname and lastname, whereas the second author's firstname is abbreviated. The last inconsistency is that the bottom publication has actually not been written by the same author as the other two publications. When looking at the paper, the first author is represented as

[1] http://www.informatik.uni-trier.de/ ley/db/

J. Krogstie, A.L. Opdahl, and G. Sindre (Eds.): CAiSE 2007, LNCS 4495, pp. 96–110, 2007.

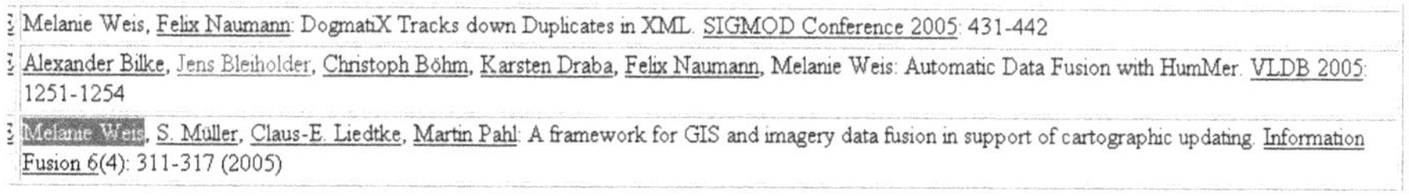

Melanie Weis, Felix Naumann: DogmatiX Tracks down Duplicates in XML. SIGMOD Conference 2005: 431-442

Alexander Bilke, Jens Bleiholder, Christoph Böhm, Karsten Draba, Felix Naumann, Melanie Weis: Automatic Data Fusion with HumMer. VLDB 2005: 1251-1254

Melanie Weis, S. Müller, Claus-E. Liedtke, Martin Pahl: A framework for GIS and imagery data fusion in support of cartographic updating. Information Fusion 6(4): 311-317 (2005)

**Fig. 1.** Excerpt of DBLP entry

M. Weis, and it has been falsely matched to author Melanie Weis. This problem, known as entity resolution [4] is also part of data cleaning. This example shows that XML data cleaning is a problem of practical relevance. Therefore, we develop XClean, a system for declarative XML data cleaning.

In developing such a system, lessons learned from relational data cleaning clearly apply, but have to be rethought due to the significantly different data structure.

**Modularity.** Data cleaning processes should be *modular* in order to allow the composition of such processes from a set of smaller, interchangeable building blocks. Modularity brings several benefits. It facilitates reusing existing cleaning transformations, simplifies the process of debugging and inspecting the data transformation process, and it allows incremental development, maintenance and evolution of the cleaning process. To achieve modularity, relational data cleaning systems such as [13] have defined cleaning operators. For XClean, we also define operators, which distinguish themselves from existing relational cleaning operators because they have to deal with the nested and semi-structured nature of XML data. For example, object properties may be multi-valued (e.g., a publication has several authors) or missing (opposed to empty content). Furthermore, crucial information describing the way XML nodes relate to one another is given by their parent-child relationships, whereas relational data cleaning concentrates on cleaning flat tuples of a single table. Consequently, XML data cleaning operators need to preserve these relationships, but also have the opportunity to exploit them.

**DBMS-backed data cleaning.** Many transformations involved in data cleaning are closely related to those typically applied inside database management systems (DBMSs). Therefore, cleaning data on top of a DBMS allows taking advantage of its functionalities, including persistence, transactions etc. but also query optimization, which may speed up the cleaning process. Relational data cleaning's reliance on RDBMSs was limited by expressive power mismatches between the cleaning primitives and SQL. Features such as user-defined aggregate functions, transitive closure computation, nested tables etc. are either not fully supported by the language, or not well supported by existing systems. In contrast, the standard XML query language, XQuery, is Turing-complete, raising the question whether simply writing XQuery queries may not suffice for data cleaning? While this approach can be made to work, it amounts to writing fresh code for every new cleaning problem, which does not agree with the last requirement applying to relational data cleaning systems, as well as to XClean.

**Declarativity.** By declaratively describing the cleaning process, its logic can be decoupled from the actual processing and its implementation. This makes data cleaning processes easier to write and to debug than alternative approaches, based on imperative code. Declarative cleaning programs allow concentrating on the cleaning tasks, while delegating storage and optimization issues to the underlying data management systems. As XML cleaning operators are significantly different from relational cleaning operators, it is natural that declaring them is also different. In XClean, we provide a declarative programming language, called XClean/PL to specify the cleaning process. This program is then compiled to an XQuery, and executed using any XQuery engine.

In this paper, we present XClean, the first modular, declarative system for native XML data cleaning. Our main contributions are $(i)$ The definition of *cleaning operators*, to be combined in arbitrary complex cleaning processes, viewed as operator graphs. $(ii)$ A high-level *operator definition language*, called XClean/PL, which is *compiled* into XQuery, to be executed on top of any XQuery processor.

We outline the XClean architecture and define its operators, in Sec. 2. XClean/PL and its compilation to XQuery are outlined in Sec. 3. We evaluate XClean's expressive power and ease of use on several case studies in Sec. 4, discuss related works in Sec. 5, then we conclude.

## 2 XClean Overview

The XClean system described in this paper is a data cleaning system that allows declarative and modular specification of a cleaning process. In this section, we first present the overall XClean system, and then introduce the XClean operators enabling modularity.

### 2.1 XClean Architecture

The architecture of the XClean system is depicted in Fig. 2(a). A user specifies an *XClean program* in our proposed declarative XClean/PL language (see Sec. 3). An XClean/PL program specifies a set of XClean operators, and the way their inputs and outputs are connected. We design XClean/PL with the goal of minimizing the cognitive effort for the average XClean user. XClean/PL provides a custom syntax for cleaning-specific operators, increasing the *readability and ease of maintenance* of cleaning programs, while being significantly more *concise* than the XQuery programs resulting from the *compilation of XClean/PL to XQuery*.

XClean provides a *function library* including commonly used functions (e.g., date formatting for scrubbing, edit distance for string similarity) which may be used in XClean/PL programs. These functions can be defined as XQuery functions, implemented either in XQuery or in an external language [27].

XQuery is a feature-rich language widely implemented by major DBMS vendors (such as IBM, Oracle, Microsoft etc.) and free-source projects (e.g. Saxon, BerkeleyDB/XML etc.), so using XQuery allows to execute the compiled XClean/PL programs on top of any XQuery-enabled platform. XQuery execution plans can be *optimized* to make query *execution* more efficient. Executing the XQuery results in the clean XML data. Note that we do not discuss XQuery optimization in this thesis, as it is a separate research area by itself. Instead, we focus our discussion on XClean operators, XClean/PL, and its compilation to XQuery.

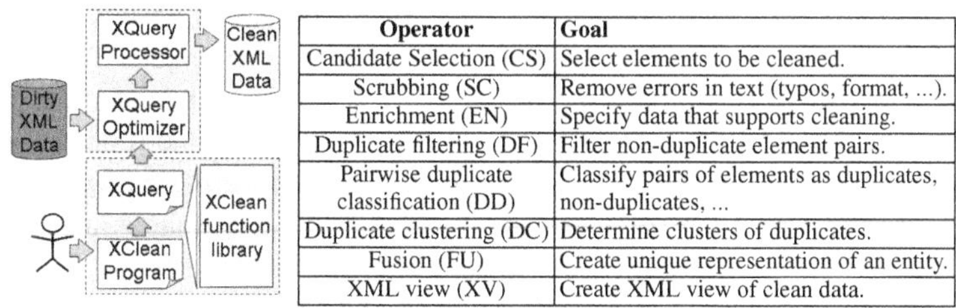


| Operator | Goal |
|---|---|
| Candidate Selection (CS) | Select elements to be cleaned. |
| Scrubbing (SC) | Remove errors in text (typos, format, ...). |
| Enrichment (EN) | Specify data that supports cleaning. |
| Duplicate filtering (DF) | Filter non-duplicate element pairs. |
| Pairwise duplicate classification (DD) | Classify pairs of elements as duplicates, non-duplicates, ... |
| Duplicate clustering (DC) | Determine clusters of duplicates. |
| Fusion (FU) | Create unique representation of an entity. |
| XML view (XV) | Create XML view of clean data. |

(a) XClean architecture (b) XClean operator averview

**Fig. 2.** Overview of the XClean architecture and its cleaning operators

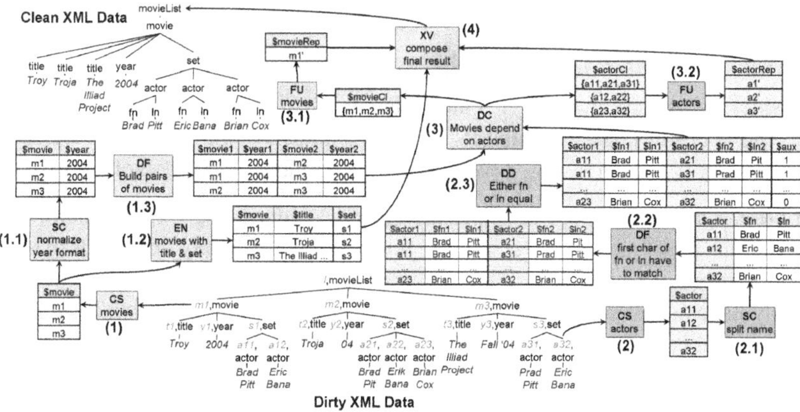


**Fig. 3.** Sample cleaning process overview

## 2.2 Operators

XClean's cleaning operators are summarized in Fig. 2(b), and are defined over a data model described in detail in [15], for which we only summarize features and notations relevant for defining XClean operators' algebra.

Any XClean operator inputs and outputs collections of (nested) tuples, having the structure ($\$a_1 = val_1, \ldots, \$a_k = val_k$), where each $\$a_i = val_i$ is a variable-value pair. Variable names such as $\$a_1$, $\$a_2$ etc. are \$-prefixed, following XQuery conventions, and are unique within a tuple. A value may be (*i*) the special constant $\perp$ (*null*), (*ii*) an XML node or value, or (*iii*) a (nested) set, list, or bag of tuples. Given a tuple ($\$a_1 = val_1, \ldots, \$a_k = val_k$) the list of names $[\$a_1, \ldots, \$a_k]$ is the tuple's *schema*.

We refer to the set of all tuples as $\mathcal{T}$, and denote the set of $n$-ary tuples $\mathcal{T}_n$. We use $\mathcal{P}(\mathcal{T})$ to denote all sets of tuples from $\mathcal{T}$. Given a tuple $t = (\ldots \$x = v \ldots)$ we say that $\$x$ maps to $v$ in the context of $t$. We represent by $t.\$x$ the value that the variable $\$x$ maps to in the tuple $t$. The notation $t' = t + (\$var = v)$ indicates that the tuple $t'$ contains all the variable-value pairs of $t$ and, in addition, the variable-value pair $\$var = v$. The tuple $t'' = t + t'$ contains all the variable-value pairs of both $t$ and $t'$.

Fig. 3 (bottom) presents a sample XML document containing three versions of the same real-world movie, with their respective title, year and actor sets. The labels $m1$, $a1$ etc. uniquely identify an element and are used to reference them in our example. Assume that the goal of the cleaning process is: $(i)$ obtaining one representation for each movie, including *all alternative titles, one year*, and *all actors (but each actor only once)*, and $(ii)$ *restructuring each actor element* into a firstname and a lastname element. A possible result of this process is shown at the top of Fig. 3. Using this example, we introduce the XClean operators that define a cleaning process.

**Candidate Selection.** Candidate selection is used to designate elements that are subject to the cleaning process. Candidates are designated by a set of queries $q_1, q_2, \ldots, q_k$, and the effect of the $CS$ operator is to evaluate all queries and union their results into a flow of 1-tuples. Candidate selection is the first step in the process of cleaning, therefore, the $CS$ operator has no input (child) operator. Formally:

$$CS_{q_1,\ldots,q_k}() = q_1 \cup \ldots \cup q_k$$

Consider the selection of movie candidates. Let $q_m$ = $doc/moviedb/movie. Fig. 3 depicts the operator $CS_{q_m}$ (1) and its output. Similarly, actor candidates are selected by the operator $CS_{q_a}$ (2) in Fig. 3, where $q_a$ = $doc/moviedb/movie/set/actor.

**Scrubbing.** Scrubbing is used for normalizing and standardizing formats and/or values. We model this based on a set of *scrubbing functions*, which apply on (tuples of) atomic values and produce (tuples of) atomic values. For generality, XClean scrubbing functions may have one or several inputs and one or several outputs. We deliberately chose to restrict scrubbing functions to atomic values. We argue that functions which apply more complex object analysis and transformation for cleaning would benefit from being decomposed in elementary steps, which help reasoning and optimization. Our framework does allow to model such transformations by the XML View operator, described later. Formally, let $f : \mathcal{A}^n \rightarrow \mathcal{A}^m$ be a scrubbing function, and $IN$ be a flow of tuples of arity $k$, such that $n \leq k$, and let $i_1, i_2, \ldots, i_n$ be a set of integers such that for any $1 \leq j \leq n$, we have $1 \leq i_j \leq k$. Furthermore, let $q_1, q_2, \ldots, q_n$ be some XML queries, which are used to extract from the (potentially complex) input the atomic inputs of the scrubbing function. Then:

$$\begin{aligned} SC_{f,i_1,i_2,\ldots,i_n,q_1,q_2,\ldots,q_n}(IN) = \{ & t + (\$b_1 = v_1^t, \$b_2 = v_2^t, \ldots, \$b_m = v_m^t) \mid \\ & t \in IN, \exists (u_1^t, \ldots, u_n^t) \ s.t. \ u_1^t \in q_1(t.\$a_{i_1}), u_2^t \in q_2(t.\$a_{i_2}), \ldots, u_n^t \in q_n(t.\$a_{i_n}), \\ & \text{and } f(u_1^t, u_2^t, \ldots, u_n^t) = (v_1^t, v_2^t, \ldots, v_m^t) \} \end{aligned}$$

where $\$b_1, \$b_2, \ldots, \$b_m$ do not appear in $IN$'s schema. The definition accounts for the general case where each XML query $q_i$ may return a sequence of results. $SC$ semantics requires that every combination of atomic inputs be used to call $f$.

To split actor names into a firstname and a lastname, we use a scrubbing function $f_{actor} : \mathcal{A} \rightarrow \mathcal{A}^2$. Let $q_{actor}$ = ./actors/actor/text() be the query extracting the initial actor names, and assume $IN$ contains tuples having just one attribute, namely the actor candidates. We apply $SC_{f_{actor},1,q_{actor}}(IN)$ (2.1) in Fig. 3. The second scrubbing operation is the standardization of years, performed by applying $SC_{f_{year},1,q_{year}}(IN)$ (1.2): $f_{year} : \mathcal{A} \rightarrow \mathcal{A}$ scrubs movie year values (by normalizing them to four digits), and $q_{year}$ = ./year/text(). This time, $IN$ contains the set of movie candidates.

**Enrichment.** Enrichment allows specifying which data to use for comparing two candidate duplicates. Let $IN$ be a flow of tuples, $\$c$ be the name of one attribute in these tuples, and $q_1, q_2, \ldots, q_k$ be a set of XML queries, which may be absolute (i.e., navigate from the root of some given document) or relative (i.e., navigate from $\$c$). We have:

$$EN_{\$c,q_1,\ldots,q_k}(IN) = \{t + (\$a_1 = q_1(t.\$c), \ldots, \$a_k = q_k(t.\$c)) \mid t \in IN\}$$

where $\$a_1, \$a_2, \ldots, \$a_k$ do not appear in $IN$'s schema.

Consider a movie is described by its title and its set. Therefore, we specify $q_{en1}$ = ./title, and $q_{en2}$ = ./set. Assuming as input the result of the movie candidate selection operator $CS_{q_m}$, the operator $EN_{\$movie,q_{en1},q_{en2}}(IN)$ corresponds to step (1.2).

**Duplicate Filtering.** Duplicate filtering constructs (a subset of) the cartesian product of a flow of candidates with itself, to be used in order to identify duplicate objects later. If only a subset of the cartesian product is built, the operator has been used to restrict the space of comparisons by pruning out some pairs of objects about which it can be said with certainty that they are not duplicates. Only on the pairs of objects which may be duplicates, other measures will later be applied to determine whether they are duplicates indeed. The second, separate output of this operator is the set of pairs of input tuples, which are definitely deemed to be non-duplicates. Although they will not be used in the main cleaning process, they may be needed, e.g., for further analysis by the user. Formally, let $m$ be the arity of the tuples in the input $IN$. Let $f_1, f_2, \ldots, f_k$ be $k$ functions such that $f_i : \mathcal{T}_m \times \mathcal{T}_m \rightarrow \{true, false\}$. The duplicate filtering operator $DF$ has two outputs, denoted $D$(duplicates) and $ND$(non-duplicates), and is defined as follows:

$$DF_{f_1,\ldots,f_k}.D(IN) = \{t_1 + t_2 \parallel t_1, t_2 \in IN, f_1(t_1,t_2) = \ldots = f_k(t_1,t_2) = true\}$$
$$DF_{f_1,\ldots,f_k}.ND(IN) = \{t_1 + t_2 \parallel t_1, t_2 \in IN, \exists 1 \leq i \leq k \ s.t. \ f_i(t_1,t_2) = false\}$$

Clearly, several $DF$ operators can be used to apply (conjunctively) several filters on potential duplicates. More complex (not necessarily conjunctive) filtering combinations can be devised by creating a complex function from simple ones, and using a single $DF$ operator based on the complex function.

Let $f_{firstLetter}$ be a filter function that returns true if the string values of either \$firstname or \$lastname of an actor are equal, false otherwise. Further, let $f_{equal}$ return true if the \$actor nodes are equal according to node identity, false otherwise. Finally, let $f_{order}$ return true if the \$actor node of the first tuple appears before the \$actor node of the second tuple in the document. Then, $DF_{f_{firstLetter},f_{equal},f_{order}}$ is the operator labeled (2.2) in Fig. 3, and its $DUP$ output is the table depicted right above it.

**Pairwise Duplicate Detection.** Duplicate detection expects input tuples that include two possibly enriched candidates, and outputs one or more tuple classes, according to a classifier function. If only one class of output is produced, it is understood as containing duplicates. If more classes are produced, their semantics depend on the classifier. For instance, one classifier may identify "certain duplicates", "likely duplicates" on which another duplicate detection classifier on pairs is applied, and "others", for which human user expertise is needed. Moreover, each tuple is annotated with a classifier-produced data structure which may encapsulate auxiliary information about the classification result (such as the confidence in the announced score, similarity, etc.).

Formally, let $f_{class} : \mathcal{T}_{2m} \rightarrow \{1, 2, \ldots, m\} \times \mathcal{N}$ be a classifier function returning for every input tuple, the index of a class, and an auxiliary data structure, modeled as an XML node. The duplicate detection operator $DD$ has $m$ outputs, denoted $OUT_1, OUT_2, \ldots, OUT_m$, defined as follows:

$$DD_{f_{class}}.OUT_i(IN) = \{t + (\$n = f_{class}(t).aux) \mid t \in IN, f_{class}(t).c = i\}$$

where the class output of the classifier is denoted $f_{class}.c$, the auxiliary data structure is denoted $f_{class}.aux$, and $\$n$ does not appear in $IN$'s schema.

We detect duplicates in actors using a classifier *dActor* returning a single $DUP$ class. It classifies a pair of actors as duplicates if either firstname or lastname are equal (this simple function could already have been used for filtering, but we use it here to keep the example simple). The auxiliary information of the classifier returns the edit distance between the names. The operator $DD_{dActor}$ is numbered (2.3) in Fig. 3.

**Duplicate Clustering.** Duplicate clustering takes as input one or several sets of potential duplicates, and outputs as many sets of duplicate clusters. A tuple in every output has one attribute, whose value is a set of tuples from the corresponding input flow, representing a set of candidates which represent the same real-world object. Clustering algorithms need to examine their whole input before producing their output, therefore, this operator is not defined on a per-tuple basis, as the previous ones. Moreover, some clustering algorithms take advantage of candidates from one input to determine how to cluster candidates from another input [21,10], therefore this operator has multiple inputs and outputs. Formally, let $k$ be an integer, and $IN_1, IN_2, \ldots, IN_k$ be some operators such that tuples in the output of $IN_i$, $1 \leq i \leq k$, have arity $2\,n_i$, for some integer $n_i$. $\{IN_i\}$ denote the set of tuples output by $IN_i$. Let

$$f_{clust} : \mathcal{P}(\mathcal{T}_{2\,n_1}) \times \ldots \times \mathcal{P}(\mathcal{T}_{2\,n_k}) \rightarrow \mathcal{P}(\mathcal{P}(\mathcal{T}_{n_1})) \times \ldots \times \mathcal{P}(\mathcal{P}(\mathcal{T}_{n_k}))$$

be a clustering function that takes as input $k$ whole sets of tuples, and outputs $k$ sets of sets of tuples (representing clusters). Let $\{IN_i\}$ denote the set of tuples output by $IN_i$, and $OUT_1, OUT_2, \ldots, OUT_k$ be the outputs of $DC$. Then:

$$DC_{f_{clust}}(IN_1, \ldots, IN_k).OUT_i = f_{clust}(\{IN_1\}, \ldots, \{IN_k\}).i$$

where the $i$-th attribute of $f_{clust}$'s output is denoted $OUT_i$. Note that $DC$ breaks down every two-candidate duplicate in two, folding all duplicate tuples into a single cluster.

Consider the clustering of movies, described by their titles and actor sets. To detect duplicates in movies, information about duplicates among their actor sets is helpful, so we perform duplicate detection using clustering on the input $IN_{actors}$, consisting of the pairs produced by actor duplicate detection (2.3), and $IN_{movies}$, holding movie pairs. In our example, the clustering operator $DC_{f_{clust}}(IN_{actors}, IN_{movies})$ is labeled (3), and it produces two sets of tuples. Each tuple in the first set is a cluster of actors considered duplicates, and each tuple in the second set is a cluster of duplicate movies.

**Fusion.** The fusion operator applies on clustered tuples. Its purpose is to construct a single representative, or cleaned version, from every cluster of tuples in its input. Formally, let $f_{fuse} : \mathcal{P}(\mathcal{T}) \rightarrow \mathcal{T}$ be a function that, for every cluster of $\mathcal{T}$ tuples, returns a cleaned tuple representing the unified cluster. Assume $IN$ contains 1-tuple attributes, such that every attribute value belongs to $\mathcal{P}(\mathcal{T})$. Then:

$$F_{f_{fuse}}(IN) = \{f_{fuse}(t.\$a_1) \mid t \in IN\}$$

This generalizes to $IN$ having several attributes, one of which is a nested table.

We fuse movies by unifying all their descendant sequences, which results in a new element denoted $m'_i$, for the $i$-th tuple in the input (3.1). We fuse actors by choosing the first actor (according to document order) as a cluster representative (3.2).

As for duplicate detection, fusion may involve more complex logic than the simple aggregation above. E.g., when detecting duplicates in movies, and simultaneously in actors, which are descendants of movies, the fused result of movies also depends on the fusion of actors. Hence, information about duplicate movies and actors is required, similarly as for relationship-based duplicate detection. As fusion usually requires a previous clustering, we decide to let fusion be part of the clustering function when necessary.

**XML View.** During cleaning, it may be necessary at several points to apply some "adjustment" transformation to one operator's output prior to sending it into another operator's input. Furthermore, if only parts of the input data have been cleaned, an extra query may be needed to combine the cleaned data with the document it originated from. Such transformations can be accomplished via the $XV$ operator, standing for XML View. Let $IN$ contain some tuples in $\mathcal{T}_n$, and $i_1, i_2, \ldots, i_k$ be some column indices in $IN$. Let $q(\$x_1, \$x_2, \ldots, \$x_k)$ be a parameterized XML query. Then:

$$XV_{q,i_1,i_2,\ldots,i_k}(IN) = \{t + (\$a = q(t.\$a_{i_1}, t.\$a_{i_2}, \ldots, t.\$a_{i_k})) \mid t \in IN\}$$

Actor elements need to be restructured in the final result, with ⟨firstname⟩ and ⟨lastname⟩ children. Names have been split by the scrubbing operator, we now need to create a new representation of every candidate movie, including the complex-structure actor names. This transformation can be specified by an XQuery $q_{xv}$. Fig. 3 shows the $XV_{q_{xv}}(IN)$ operator (4) and its output.

## 3 XClean Programming

Having discussed the XClean architecture and the operators used to define a cleaning process, we present how these can be specified using the programming language XClean/PL, and compiled to an executable XQuery.

### 3.1 Language Rationale and Design

The specification of a cleaning process can be decomposed in ($i$) choosing the specific filters, distance functions, duplicate detection algorithms, etc., and ($ii$) writing the "surrounding" code necessary to implement the operator tree using these functions.

Previous experience in data cleaning [13,20,25] demonstrates that creating or choosing the cleaning functions and algorithms requires a human expert, and cannot be automated. In contrast, the second task is repetitive, and amenable to automation. Based on this observation, we designed the XClean/PL language as follows.

**Table 1.** Sample XClean/PL clauses

| (a) XClean/PL clauses | (b) Compiled XQuery clauses |
|---|---|
| **ENRICH** $m **IN** $scrubbedMovies<br>**INTO** $enrichedMovies<br>**BY** $m.mCand/movie/title/text() **AS** $title,<br>$m.mCand/movie/set/actor **AS** $set; | let $enrichedMovies := for $m in $scrubbedMovies<br>return <tSYS>{$m/element(),<br>element title {$m/mCand/movie/title/text()},<br>element set {$m/mCand/movie/set/actor} </tSYS> |
| **CLUSTER CLASSIFICATION USING**<br>xcl:radc($actorDups,<br>$candMovieDups)<br>**INTO** $movieClusters<br>**SCHEMA** [$movieCluster],<br>$actorClusters<br>**SCHEMA** [$actorCluster]; | CLUSTERSSYS := xcl:radd( $actorDups, $candMovieDups),<br>$movieClusters :=<br>for $VSYS in $CLUSTERSSYS/element()[1]/element()<br>return <tSYS>{element mClust{$VSYS/element()}}</tSYS>,<br>$actorClusters :=<br>for $VSYS in $CLUSTERSSYS/element()[2]/element()<br>return <tSYS>{element aClust{$VSYS/element()}}</tSYS> |
| **PAIR CLASSIFICATION**<br>**PAIR** $p **IN** $candActorDups<br>**WITH** detectActorDups($p),<br>eDist($p) **AS** $aux<br>**INTO** $actorDups **IF CLASS** = 1; | $PAIRSSYS := for $p in $candActorDups return<br><CLASSSYS>{ attribute CLASSSYS{detectActorDups($p)},<br><tSYS>{$p/element(), element aux{eDist($p)}}</tSYS><br>}</CLASSSYS>,<br>$actorDups := $PAIRSSYS[@CLASSSYS = '1']/element() |

An XClean/PL program is a set of clauses, each of which defines a cleaning operator. Operators input and output tuples from shared, global XClean/PL variables. Sample XClean/PL clauses appear in Tab. 1(a), the full description of XClean/PL's syntax being provided in [1]. XClean/PL keywords appear in bold font.

The top enrichment clause defines the operator labeled (1.2) in Fig. 3. The clause refers to two named tuple sets, globally visible in the XClean/PL program: $scrubbed-Movies, the operator's input, and $enrichedMovies, its output. The tuple variable $m iterates over the input. The **BY** clause introduces the two enrichments: the result of each query is added as a new variable, part of the output flow.

The cluster classification clause defines the operator labeled (3) in Fig. 3. The classifier function xcl:radc denotes a relationship-aware duplicate clustering function (e.g.,[21]), which is one among the possible classifiers to be used here. The classification function returns two sets of clusters, one containing movies and another one actors. The **INTO** keyword is used, as previously, to capture the outputs of xcl:radc, and make them visible in the XClean program for further usage. This clause also explicitly renames the attributes in each set of cluster's schema, through the **SCHEMA** clause.

The last clause in Tab. 1(a) defines the pairwise duplicate operator $DD_{dActor}$ used for illustration in Section 2.2 (numbered 2.3 in Fig. 3). Function detectActorDups returns 1 if the two actors are considered duplicates and 2 otherwise, while function eDist is a simple edit distance. The **WITH** keyword is immediately followed by a call to the classifier function, after which the (optional) function producing the auxiliary information is invoked. The difference between the **WITH** keyword and the **USING** keyword is that the first calls a function within an iteration, whereas the second calls a function that takes sets of tuples as input.

### 3.2 Compiling XClean/PL to XQuery

An XClean/PL program is compiled based on a few principles, discussed next. Tab. 1(b) shows XQuery snippets obtained when compiling the XClean/PL clauses of Tab. 1(a).

First, given that XQuery does not support tuples, every tuple manipulated during the cleaning process is translated into a system-generated XML element, named ⟨tSYS⟩. For every variable of an XClean tuple, the system-generated ⟨tSYS⟩ element has a child element named after the variable (without the leading $), its content being variable value. Nested tuples are translated into nested ⟨tSYS⟩ elements. While this element generation is generally computationally expensive, [11] has characterized situations when element construction can be avoided. These are the situations when the *identity* of the constructed node is never used in the remainder of the same XQuery program. Fortunately, all ⟨tSYS⟩ element creation done in XClean are in this situation.

Second, global XClean/PL variables are compiled in XQuery variables introduced by let clauses, bound to the lists of ⟨tSYS⟩ elements.

Third, XClean/PL operators defined by iterating over input tuples ($SC$, $EN$, d$DD$, $FU$, and $XV$) are compiled into for-where-return expressions, with XClean/PL's iteration variables (such as $m) compiled into XQuery for clause variables.

Finally, XClean adds an internal xcl:id identifier attributes to XML elements manipulated during cleaning. One reason for this is related to XQuery semantics: when creating an element having the value of the variable $x as a child, such as, e.g., ⟨a⟩{$x}⟨/a⟩, the nodes associated to $x are *copied*, thus they are no longer equal to the original nodes. However, the cleaning process needs to reason on the relationships between the nodes, e.g., when de-duplicating movie candidates based on related actor duplicates. Moreover, when re-assembling the cleaned elements (last step in Fig. 3), IDs are also needed. A second usage of system-introduced IDs is to enable *lineage tracing*, i.e., discovering the operators (and inputs) that have led to obtaining a given output (clean) element. Lineage issues are often central in data cleaning processes [13,25], to help users understand cleaning results, inspect and refine the process. To keep ID insertion and manipulation overhead low, XClean IDs are added to cleaning candidates only.

## 4 Usage Report

The approach described in this paper has been implemented in our XClean Java-based prototype ([22]) following the architecture shown in Fig. 2(a). XClean/PL programs are compiled using the antlr tool[2], into XQuery programs.

Section 4.1 reports on three use cases. Section 4.2 outlines a quantitative evaluation of XClean performance. Details on use cases, sample data sets, full XClean/PL programs, and their resulting XQueries are available at [1].

### 4.1 Use Cases

**FreeDB Use Case.** This use case concerns CD description FreeDB data[3]. The cleaning process (see Fig. 4(a)) includes correcting errors in artist names (common errors include different capitalization schemes, Various Artists is also represented by V.A., Various ), standardizing dates, correcting titles (e.g., the title element often includes Artist/Title), and track titles (again, capitalization). All these operations correspond

[2] http://www.antlr.org/

[3] http://www.freedb.org

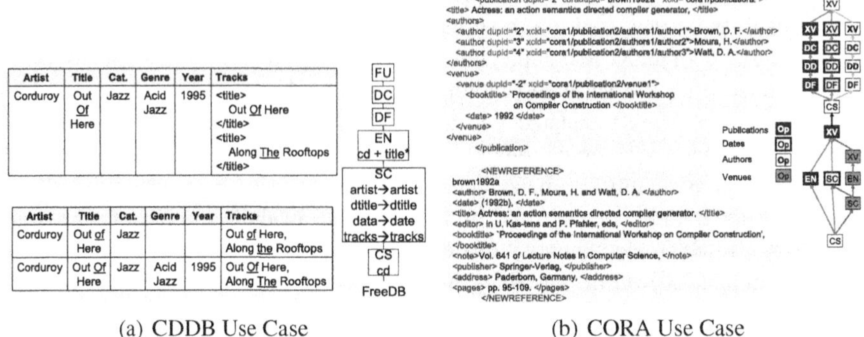


(a) CDDB Use Case (b) CORA Use Case

**Fig. 4.** CDDB and CORA use case description

to scrubbing operators. We further enrich CDs with track ⟨title⟩ elements obtained by splitting the comma-separated list of track titles into individual elements. Using the enriched CDs, the final task is to deduplicate CDs: if both ⟨artist⟩ and ⟨title⟩ are equal, we consider CDs to be duplicates, which is a boolean function without auxiliary information that can directly be used in the DF operator. Clustering performs transitive closure over CD pairs and fusion creates a single representative for every CD. During fusion, conflicts may appear in category, genre, year, and tracks: different categories, genres, and years are concatenated, whereas sets of ⟨title⟩ elements are unified. Note that the table representation has only been used for readability, the actual data is XML.

**MOVIE Use Case.** This is a data integration scenario, in which movies from two sources are first mapped to a common schema, and then de-duplicated. The sources are the Internet Movie Database IMDB and the German Movie Repository FILMDIENST[4]. Fig. 5(a) outlines the two source schemas and the target schema. In IMDB titles, the possible leading "The" or "An" is separated in an ⟨article⟩ element. Non-trivial correspondences between source and target types are rendered by curved arrows, possibly annotated with transformation functions. For instance, IMDB names are split into firstname and lastname, and gender is set to "m" for actors, or to "f" for actresses.

XClean allows specifying this process in several ways, as illustrated in Fig. 5(b). In this figure, the central tree of connected operators represents one possible cleaning process, denoted $P_1$. An alternative XClean operator graph for the same task, which we denote $P_2$, can be obtained by modifying $P_1$, as we will shortly explain.

In $P_1$, the operators in the shaded bottom areas are used to select and scrub the cleaning candidates from IMDB (left), and FILMDIENST (right). From IMDB, we extract: aka-title, movie and name candidates. Then, title scrubbing (upper/lower case normalization) is applied on the text content of aka-title elements, as well as on the text of movie/title; actor name scrubbing separates first names from last names. FILMDIENST

[4] http://www.imdb.com and http://film-dienst.kim-info.de/, respectively.

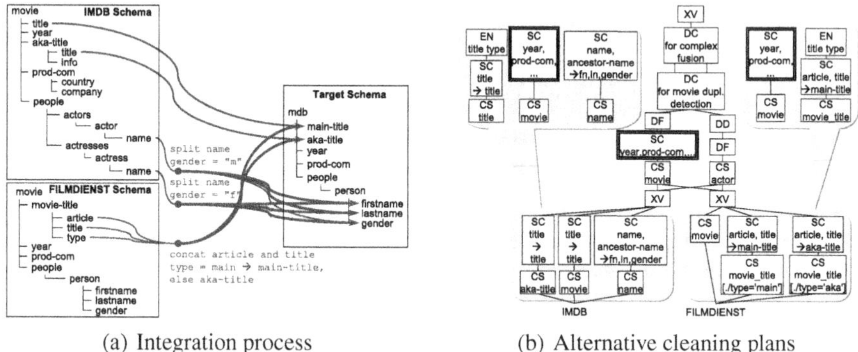


(a) Integration process (b) Alternative cleaning plans

**Fig. 5.** Two equivalent cleaning processes for the MOVIE Use Case

candidates are: movies, movie main titles, and movie aka-titles. Both title types undergo scrubbing. The $XV$ operators align the movies to the target schema as shown in Fig. 5. The $CS$ operators merge candidate movies (respectively, candidate actors) from both data sources, to be used together in the rest of the process: year and prod-com scrubbing, duplicate filtering and clustering. Note that the second duplicate clustering operator performs fusion, illustrating a case of complex fusion that requires several inputs (movies and actors). The final result is composed using an XV operator.

The second plan $P_2$ differs from $P_1$ in the following two ways: First, the scrubbing operator on year and prod-com (thick line) is pushed down below the intermediate XV operators. Because integration has not been performed below this point, the scrubbing operator is split into two individual scrubbing operators, one for each source. Second, scrubbing in $P_1$ has been performed separately on main-titles and aka-titles in both sources. In $P_2$, titles of each source are scrubbed in one operation, are then enriched by their type and split in the corresponding title elements in the intermediate XV operator.

This scenario illustrates the benefits of *modular* data cleaning: the same scrubbing function can be used for all titles. It also demonstrates the interest of *declarativity*: a given process can be specified in multiple ways, and separating its specification from its actual implementation allows for its automatic optimization.

**CORA Use Case.** The CORA bibliographic data set is frequently used to evaluate duplicate detection algorithms [10,21]. Fig. 4 outlines an XClean operator graph (right) as well as a sample input reference (bottom left) and its corresponding clean version (bottom right). Colors of operator boxes indicate the bibliographic data components on which they apply, e.g. authors, dates etc. The sample XClean process scrubs, enriches and restructures the dirty data (from the bottom to the upper $CS$). We assume the restructured data is then fed into three parallel cleaning chains, with the purpose of comparing their respective clean outputs.

In this example, we again scrub dates, reusing a standard function available in the XClean function library already used in the MOVIE scenario. Also, detecting duplicates in author names is similar to detecting duplicates in actor names, so we can reuse the

**Table 2.** Use Case Statistics

| Use Case | #Nodes | #Chars | Time (s) | Word Savings |
|---|---|---|---|---|
| FreeDB | 2001 | 36103 | 4.7 | 61% |
| MOVIE ($P_1$) | 3108 | 9666 | 20.4 | 59% |
| MOVIE ($P_2$) | 3108 | 9666 | 1.8 | 57% |
| CORA | 1116 | 9705 | 6.1 | 45% |

same pairwise duplicate detection function as in the MOVIE scenario, showing the advantage of modularity.

### 4.2 Quantitative Aspects

We have run the XQuery programs on several freely available XML engines: XML Spy, QizX/Open, and Saxon B[5]. We found the latter to be the most efficient for our generated XQuery set. Tab. 2 provides an overview of the size of the data sets of the three use cases, actual runtimes (averaged on 10 runs), and savings in word counts when using XClean/PL, relative to the word count of the respective generated XQueries.

On the MOVIE use case, we observe a difference on runtime, depending on the cleaning plan. Indeed, the difference between the two plans for the MOVIE scenario is of an order of magnitude. This is mainly due to the higher number of function calls and the multiple joins performed in the intermediate $XV$ operator of $P_1$. Some of these joins are avoided in $P_2$, because candidates have already been enriched with the information via the $EN$ operators. This recalls the classical optimization consisting of pushing function calls under joins, if the function call results are not cached [9]. As XQuery optimizers grow more efficient and include such mainstream techniques, the performance of XClean programs translated to XQuery is likely to improve. (Admittedly, more efficient XQuery processors are there today - on the Saxon website, the commercial version of Saxon is said to be two orders of magnitude faster, on some queries, than the free version we used.)

In the CORA use case, although the data set is quite small, the runtime is worse than for FreeDB or MOVIE ($P_2$), because it applies more expensive operators on pairs ($DF$, $DD$, $DC$).

Finally, we observe significant savings in the size of a XClean/PL program over the generated XQuery, as shown in the last column of Tab. 2. This indicates that a specialized language like XClean/PL makes the specification of cleaning tasks more concise, thus, we believe, more convienent for the user.

## 5 Related Work

Due to the lack of space, we only briefly discuss selected related work. A survey on relational data cleaning is made in [19], and more recent approaches include AJAX [13] and Potter's Wheel [20]. XClean is conceptually close to AJAX by its operator-based approach. However, our operators consider the existence of more than one candidate type, which can be related to each other. Relationships between candidate types are

[5] http://www.altova.com, http://www.xfra.net/qizxopen/, and http://saxon.sourceforge.net/

maintained throughout an XClean process, and can be used by various algorithms, e.g., for duplicate detection or fusion. Another difference to AJAX is that the XML context lifts the expressive power barriers that confronted AJAX. In our context, advantages of a declarative, modular approach are: ease of specification and maintenance, and opportunities for optimization. AJAX moreover provided an exception handling mechanism, which we plan to consider as well in the future.

XClean is not meant to replace existing algorithms for specific cleaning tasks, such as clustering, distance computation etc. Instead, these approaches can be plugged in as physical implementations of specific operators, thus re-using existing results and running code. For duplicate detection, numerous algorithms have been developed, for relational data [14,16], XML/hierarchical data [2,18,23], and more complex graph data [10,21,24]; a survey is provided in [26]. For similarity joins, the computationally expensive part of duplicate detection, a relational operator has been proposed in [8]. Fusion has received less attention, and all work focuses on relational data. The authors of [6] propose an operator that extends SQL to support declarative fusion and implemented in the HumMer system [5], and we plan to develop a similar technique for XML data. Other solutions include TSIMMIS [17] relying on source preference in the context of data integration, and ConQuer[12] that filters inconsistencies out of query results.

XClean's internal model includes tuples [1], which have made it easy to model associations between objects. Existing works suggested a controlled inclusion of tuples in XQuery to facilitate analytic queries rich in group-by [3]. The difference is that we include tuples as XClean internals and compile in standard XQuery, whereas [3] add new syntactic constructs.

## 6 Conclusion

We presented XClean, a system for declarative XML data cleaning. Users of the system write an XClean/PL program that reflects the desired cleaning process and which is automatically compiled into an XQuery, that can be optimized and executed by an XQuery engine. The result of this query is a clean version of the data. We defined several operators that can be combined in a modular way to form a cleaning process, and for each of which an XClean/PL clause exists. Use case based studies show that using XClean/PL to define a cleaning process is more convenient than writing a custom XQuery, and operators can be easily reused.

However, efficiency for a given cleaning task depends on the actual cleaning plan. The performance attained by the XQuery processors used in our evaluation could clearly be improved; as part of our future work, we intend to investigate which (intra-engine, external to XClean) XQuery optimizations would most help for such queries. XClean extensions we envision in the short term are: a GUI to support the design of the cleaning process, and exception handling (also absent from XQuery !), which is very important since exceptions may arise from a variety of sources in a cleaning context, and they include valuable information for the user seeking to refine the process.

**Acknowldgements.** This research was partly funded by a "DAAD Doktorandenstipendium" scholarship.

## References

1. XClean, A.: system for declarative XML data cleaning. http://www.hpi.uni-potsdam.de/~naumann/xclean/
2. Ananthakrishna, R., Chaudhuri, S., Ganti, V.: Eliminating fuzzy duplicates in data warehouses. In: VLDB (2002)
3. Beyer, K., Chamberlin, D.D., Colby, L., Ozcan, F., Pirahesh, H., Xu, Y.: Extending xquery for analytics. In: SIGMOD (2005)
4. Bhattacharya, I., Getoor, L.: A latent dirichlet model for unsupervised entity resolution. In: SIAM Conference on Data Mining (SDM), Bethesda, MD (2006)
5. Bilke, A., Bleiholder, J., Böhm, C., Draba, K., Naumann, F., Weis, M.: Automatic data fusion with HumMer. In: VLDB (2005)
6. Bleiholder, J., Naumann, F.: Declarative data fusion - syntax, semantics, and implementation. In: ADBIS (2005)
7. Chaudhuri, S., Ganjam, K., Ganti, V., Kapoor, R., Narasayya, V., Vassilakis, T.: Data cleaning in Microsoft SQL server 2005. In: SIGMOD (2005)
8. Chaudhuri, S., Ganti, V., Kaushik, R.: A primitive operator for similarity joins in data cleaning. In: ICDE (2006)
9. Chaudhuri, S., Shim, K.: Query optimization in the presence of foreign functions. In: VLDB (1993)
10. Dong, X., Halevy, A., Madhavan, J.: Reference reconciliation in complex information spaces. In: SIGMOD (2005)
11. Florescu, D., Kossmann, D.: XML query processing. In: ICDE (2004)
12. Fuxman, A., Fazli, E., Miller, R.J.: ConQuer: Efficient management of inconsistent databases. In: SIGMOD (2005)
13. Galhardas, H., Florescu, D., Shasha, D., Simon, E., Saita, C.: Declarative data cleaning: Language, model, and algorithms. In: VLDB (2001)
14. Hernández, M.A., Stolfo, S.J.: The merge/purge problem for large databases. In: SIGMOD (May 1995)
15. Manolescu, I., Papakonstantinou, Y.: An unified tuple-based algebra for XQuery. Technical report (2005)
16. Monge, A.E., Elkan, C.P.: An efficient domain-independent algorithm for detecting approximately duplicate database records. In: SIGMOD-1997 DMKD Workshop (May 1997)
17. Papakonstantinou, Y., Abiteboul, S., Garcia-Molina, H.: Object fusion in mediator systems. In: VLDB (1996)
18. Puhlmann, S., Weis, M., Naumann, F.: XML duplicate detection using sorted neigborhoods. In: EDBT (2006)
19. Rahm, E., Do, H. H.: Data cleaning: Problems and current approaches. IEEE Data Engineering Bulletin, Vol. 23 (2000)
20. Raman, V., Hellerstein, J.: Potter's wheel: An interactive data cleaning system. In: VLDB (2001)
21. Singla, P., Domingos, P.: Object identification with attribute-mediated dependences. In: Conference on Principles and Practice of Knowledge Discovery in Databases (PKDD), Porto, Portugal (2005)
22. Weis, M., Manolescu, I.: Xclean in action (4 page demo, to appear). In: CIDR (2007)
23. Weis, M., Naumann, F.: DogmatiX tracks down duplicates in XML. In: SIGMOD (2005)
24. Weis, M., Naumann, F.: Detecting duplicates in complex XML data(poster). In: ICDE (2006)
25. Widom, J.: Trio: A system for integrated management of data, accuracy, and lineage. In: CIDR (2005)
26. Winkler, W.E.: Overview of record linkage and current research directions. Technical report, U. S. Bureau of the Census (2006)
27. XQuery 1.0. (2006) http://www.w3.org/TR/XQuery

# Personalizing PageRank-Based Ranking over Distributed Collections

Stefania Costache, Wolfgang Nejdl, and Raluca Paiu

L3S Research Center / University of Hanover
Deutscher Pavillon, Expo Plaza 1
30539 Hanover, Germany
{costache,nejdl,paiu}@l3s.de

**Abstract.** In distributed work environments, where users are sharing and searching resources, ensuring an appropriate ranking at remote peers is a key problem. While this issue has been investigated for federated libraries, where the exchange of collection specific information suffices to enable homogeneous TFxIDF rankings across the participating collections, no solutions are known for PageRank-based ranking schemes, important for personalized retrieval on the desktop.

Connected users share fulltext resources and metadata expressing information about them and connecting them. Based on which information is shared or private, we propose several algorithms for computing personalized PageRank-based rankings for these connected peers. We discuss which information is needed for the ranking computation and how PageRank values can be estimated in case of incomplete information. We analyze the performance of our algorithms through a set of experiments, and conclude with suggestions for choosing among these algorithms.

**Keywords:** PageRank, distributed search, personalization, privacy.

## 1 Introduction

Collaborative work has become a key factor on the way to success in every company - people do not work isolated, but rather interact with each other by exchanging information, using tools like email clients, IM, blogs, wikis or shared repositories. Every personal desktop thus becomes the sum of all other desktops it interacts with. Accessing these connected information sources in such a collaborative work environment becomes a crucial functionality, which so far has only been partially tackled.

Personal information management [9,10] is a subject of growing interest to the database community, and (distributed and heterogeneous) dataspaces will extend databases beyond centralized and structured information repositories [11]. The *Social Semantic Desktop* paradigm integrates data annotation, organization and search on the desktop, and promises to provide collaborative work environments through connecting all shared data resources in a work group. The

J. Krogstie, A.L. Opdahl, and G. Sindre (Eds.): CAiSE 2007, LNCS 4495, pp. 111–126, 2007.

NEPOMUK[1] project [2] aims to create such an infrastructure, which improves the state of the art in online collaboration and personal data management, by providing seamless access to all information created by single or group efforts.

Peers in the NEPOMUK context share fulltext and semi-structured information, referring to publications, reports and other desktop documents, emails, browsed web pages, address books, etc. These metadata represent additional information about these resources and connect them through semantic relations, such as authorship of papers and reports, sender and recipient information for emails or email attachments. Based on this infrastructure, advanced searching and ranking capabilities can utilize both conventional Information Retrieval (IR)-based information like term frequency in documents and collections, as well as link-related information, the basis of PageRank-like algorithms, e.g., ObjectRank [7,6].

Extending these ranking schemes to a distributed setup is not trivial, because it involves (partial) sharing of possibly private information. Solutions for distributed collections in federated libraries exist, but they provide just traditional IR-based rankings based on TFxIDF metrics through the exchange of collection specific information. We will investigate which resources and information need to be shared to enable personalized PageRank-based ranking among peers, and how algorithms can take privacy constraints for these resources into account. Specifically, we propose and evaluate new algorithms for consistently computing ObjectRank, a PageRank variant appropriate for ranking these connected resources on the desktop.

In Section 2 we will start with the discussion of a search scenario in a distributed work group, and then discuss in detail which information needs to be exchanged in order to achieve appropriate rankings of results. In section 3 we propose and discuss several new algorithms for computing ObjectRank over a set of distributed collections / semantically enabled desktops. In section 4 we describe the experimental setup to evaluate our algorithms, present the experiments we performed and analyze the results. Finally, we discuss related work in section 5 and conclude (section 6).

## 2 Which Information Should We Exchange?

### 2.1 A Motivating Scenario

Let's imagine Alice, working in a team with five other students for a research project. Alice's team uses the NEPOMUK-enabled desktop to interact and share information. The team members share papers, project documents and group emails, among others. Papers are annotated with bibliographic information, and connected to the emails they have been attached to. Alice participates in other teams as well, where she shares some of the same documents as well as other information specific only to these other projects. The NEPOMUK infrastructure

[1] This work was supported by the NEPOMUK project funded by the European Commission under the 6th Framework Programme (IST Contract No. 027705).

allows her to search resources on her own desktop as well as on the desktops of her team members, to which Alice's queries are propagated.

The importance of documents (important for the ranking of search results) is influenced by the importance of their authors and conferences, or by the importance of team members sending the document as attachment. These factors are not necessarily the same on each desktop, but are rather based on the conferences relevant to each team member, the number of documents authored by a given person stored on a specific desktop, or the emails connected to these documents. Part of this information (importance of conferences, papers stored on a desktop) can be exchanged easily. Other information such as private emails, or reports from other projects referencing specific papers, should not be exchanged among all participants.

In general, there will be resources that Alice can make public and thus share with everyone, there will be other resources which she will make available only to her trusted friends or to her work mates and there are of course some resources she will never want to share with anybody. This is also true for her contextual metadata generated and stored on her computer, which connects all her resources. Keeping (parts of) her metadata graph private, however, also means that search result rankings at other peers will not be comparable to her own. This unfortunately collides with Alice's desire to get the best ranked matching resources from all her team members connected in her NEPOMUK network (remember that best ranked in this case means "according to Alice's interests / set of resources").

What do we need to exchange in order to provide an appropriate ranking over all document collections Alice asks for results? Clearly, given that the metadata graph determines Alice's ObjectRank scores for all resources (details are described in [7]), we have to exchange PageRank/ObjectRank-related information in addition to the usual IR statistics. We will discuss in the next sections, what can and should be exchanged, in order to rank results for Alice's query on her team members desktops in a way compatible with Alice's ranking. We will take into account the constraint that Alice and her team members do not want to exchange their complete data graphs, which would provide information about all resources they have on their machines.

## 2.2 Exchanging IR Related Information

Let us first look at a typical scenario in which a user is doing a full-text search over several distributed collections, and wants to rank results according to the usual TFxIDF measures ([4,12]). A query $q$ will consist of several keywords, say $q_1$ and $q_2$, and is posed to a broker, which forwards it to a set of $m$ search engines / peers, $P_i'$, which will then send back to the broker their document rankings $R_i'$. In practice the user is only interested in the best "top-$k$" results, where $k$ is usually between 5 and 20. For this, all rankings $R_i'$ have to be merged into one ranked list $Rm$ and the top-$k$ results are presented to the user. Our goal is to achieve the same ranking in the distributed case as produced by the same search on a single collection $C$ containing all documents.

The ranking of the documents in a collection is based on TFxIDF weights, which measure the significance of a word with respect to a document in a collection. The significance of a term increases proportionally to the number of times the term appears in the document, but decreases with the frequency of the term in the whole collection. So, Term Frequency (*TF*) in the given document gives a measure of importance of the term $t_i$ within that particular document, whereas the Inverted Document Frequency (*IDF*) is a measure of the general importance of the term. A high weight in TFxIDF is reached by a high *TF* (in the given document) and a low Document Frequency (*DF*) of the term in the whole collection of documents.

For distributed retrieval, we want to make the distributed similarity score equal to the similarity scores computed on a single collection *C*. Therefore, the collection specific values, number of documents (*N*) and *DF*, need to be computed before query time (see for example [4]), and recomputed when changes in the collections occur (such as document additions, deletions and updates). To exchange and aggregate them over all collections, we need to send them to the query broker, which can compute the overall Global Inverted Document Frequency (*GIDF*) value, which is then sent back to all search engines. During query execution, all peers will rank results with comparable scores, since they use the common *GIDF*, propagated together with the query. A globally ranked list is achieved by merging the sub-result list entries in descending order of global similarity score. Figure 1 illustrates this process in detail.

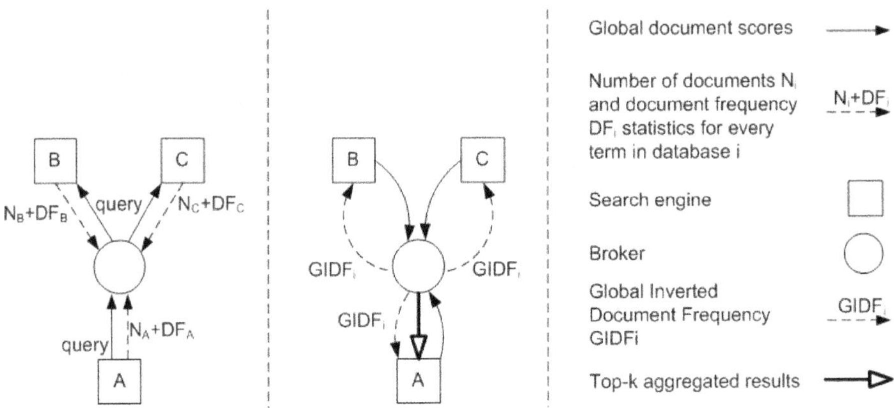


**Fig. 1.** Statistics propagation for results merging

1. *A*, *B*, *C* send to the *Broker* the total number of documents in the collections ($N_A$, $N_B$, $N_C$) and the DF values. [2]
2. Peer *A* sends a query to the *Broker* and the *Broker* forwards it to *B* and *C*.
3. The *Broker* computes the $GIDF_i$ for each keyword $q_i$ and sends them back to all peers.
4. *A*, *B*, *C* find the matching results for the query and send the top-k results to the *Broker* sorted by the Global Document Scores.
5. The *Broker* merges the results from all peers and sends back to peer *A* the top-k results.

[2] *TF* values need not be exchanged since they are document-dependent and therefore do not influence the order of the aggregated result list entries.

### 2.3 Exchanging ObjectRank Related Information

Let us now look at PageRank / ObjectRank based ranking and which information has to be exchanged to make such rankings on distributed peers compatible with each other. Recall that the computation of PageRank is based on the random surfer model, with the surfer traversing links through the graph of resources, and sometimes jumping randomly to another resource. Then the PageRank value of a resource represents the probability that the random surfer stays on this resource at a given time. If we represent the link structure between all resources through the adjacency matrix $A$ and the random jump through the $e$ vector and the dampening factor $d$ (usually 0.85), PageRank values $R$ are computed through the following eigenvector computation:

$$R = d \cdot A \cdot R + (1 - d) \cdot e \tag{1}$$

For ObjectRank computation, we do not assume the same weight for each link, but rather define link weights based on the type of the connected nodes, through an authority transfer schema [8]. Such a schema specifies how much importance (represented as a real number between 0 and 1) is transferred between connected nodes. The weights of the links between the instances correspond to the weights specified in the authority transfer schema divided by the number of links of the same type. For example, 70% of the importance of a conference node is distributed evenly to each of the publications which are presented at this conference (see [6] for a more detailed description of the algorithm). Let us

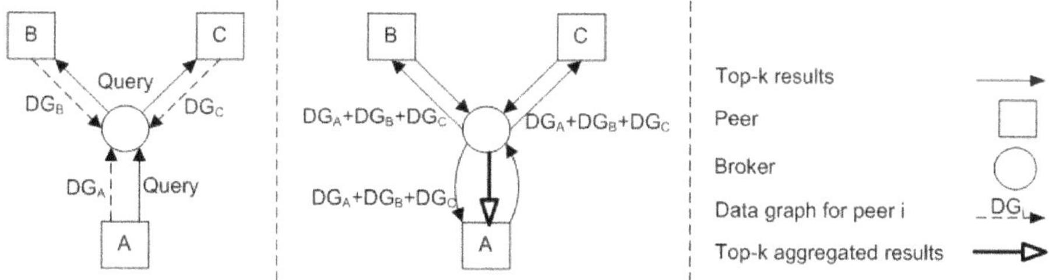


**Fig. 2.** Aggregated ObjectRank computation

assume, without loss of generality, that all peers use the same authority transfer schema as basis for the ranking computation. Each peer computes ObjectRank scores for its collection. Since this ObjectRank computation is based on the data graph, the adjacency matrix of each peer needs to be updated so that it reflects the new structure created by the integration of the other peers' resources into its own data graph. Therefore, peers need to exchange the URIs of the resources they are sharing, together with the links connecting them. External URIs are integrated into each peer's own data graph of resources. The more resources are shared among peers, the more accurate the aggregated ranked results will be.

Figure 2 presents the necessary steps for computing the aggregated ObjectRank scores in the ideal case, where peers share all resources they own:

1. Peer $A$ sends a query to the *Broker*[3] and the *Broker* forwards it to $B$ and $C$.
2. The data graph, $DG_i$ is sent to the *Broker* by each peer.
3. The *Broker* merges $DG_A+DG_B+DG_C$ and sends the results to the peers.
4. Peers compute ObjectRank on $DG_A+DG_B+DG_C$ and send top-k results to the *Broker*.
5. The *Broker* merges the results from all peers and sends back to peer $A$ the top-k results.

# 3 Information Exchange and Rank Computation

## 3.1 Privacy vs. Information Exchange

The discussion in the previous section assumed the ideal case, where peers share everything they have on their machines. This is usually not the case, instead peers will decide to share only parts of their data graphs and protect the rest. Moreover, peers usually do not want to involve third parties in the exchange process, because this would imply additional privacy and security issues, so they do not want to send data through a broker. We therefore need to develop strategies which do not involve a broker and which allow sending only specific parts of the data graph to the other peers.

As we have already seen, to be able to appropriately rank resources for their neighbors, peers need to know their corresponding data graphs, or at least parts of them. For exchanging this information, peers have the following alternatives:

1. send all nodes in the graph
2. send some of the nodes in the graph
3. send all nodes in the graph, part of them anonymized (the items they want to keep private have hidden URIs, e.g. "hidden_41323")
4. send all nodes in the graph, part of them hashed - which keeps the nodes secret if the other peer does not have them and makes them identifiable if the other peer has them too and uses the same hashing function
5. send all nodes summarized into a world node [5] (which appropriately aggregates node and link information of the graph)

Ranking computation can be based on: a) simple ObjectRank; or b) ObjectRank with biasing [6] on the resources coming from the other peers. We will discuss appropriate combinations of these alternatives in the following.

To describe the graphs used by the different algorithms, we will use the following notations: let $G_i = (V_i, E_i)$ be the data graph of peer i, where $V_i$ and $E_i$ are the corresponding sets of nodes and weighted edges, respectively. In this context, the nodes model the desktop resources (files, emails, visited web pages, etc.), while the edges represent the semantic relationships between them [7]. $G_i' = (V_i', E_i')$ represents the data graph corresponding only to the shared resources, where $G_i' \subset G_i$, $V_i' \subset V_i$, $E_i' \subset E_i$ and $E_i' = \{e_{jk} | j, k \in V_i', j \neq k\}$.

[3] We assume that the peers have already agreed on the authority transfer schema to be used for the ObjectRank computation.

$G_i^{anon} = (V_i^{anon}, E_i^{anon})$ denotes the anonymized data graph of peer i, where $G_i^{anon} = G'_i \cup anonymized(G_i^{unshared})$, $V_i^{anon} = V'_i \cup anonymized(V_i^{unshared})$, $G_i^{unshared} = G_i \; G'_i$ and $E_i^{anon} = E_i$. With $G_i^h = (V_i^h, E_i^h)$ we refer to the hashed data graph, where $G_i^h = hash(G_i)$, $V_i^h = hash(V_i)$ and $E_i^h = E_i$. An example covering all these graphs is presented in figure 3[4].

### 3.2 Aggregating Graphs into World Nodes

One especially interesting possibility of keeping a graph private, yet provide some information about its connections to the graphs of other peers, is to aggregate all nodes in the graph into a world node and aggregate his connections to the other graphs as well. An example is presented in figure 4, where P2 creates a world node out of its nodes and connects it to the data graph of P1. Using a similar notation as in section 3.1 we define $G_i^{WN} = (V_i^{WN}, E_i^{WN})$, where $V_i^{WN} = WN$ and $E_i^{WN}$ is formed as follows:

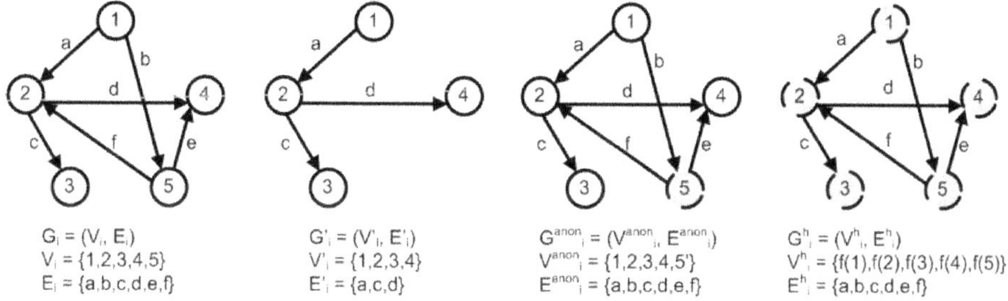


**Fig. 3.** Example of weighted data graphs - different setups

1. All links from nodes in the other peers' graphs pointing to the nodes in the graph of the peer aggregated into the world node become inlinks of the world node.
2. All links from the nodes of the peer creating the world node pointing to nodes of other peers become outlinks of the world node.
   For a better approximation of the total authority score mass that is received from nodes aggregated in the world node, we weigh every outlink from the world node based on the sum of the weights aggregated into it (the links from the world node to a node of other peers), divided by the number of nodes summarized into the world node.
3. To represent internal links between nodes aggregated into the world node, we create a self-loop link at the world node.
   The weight of this self-loop link is given by the sum of all weights corresponding to the internal links inside the world node, divided by the number of nodes in the world node. The self-loop link represents the probability that a random surfer remains inside the graph that was aggregated into the world node, when following links.

[4] $a$ to $f$ are real numbers, representing the weights of the edges.

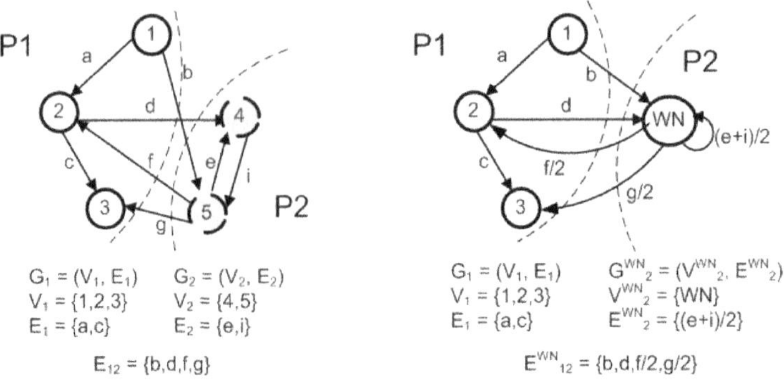


**Fig. 4.** Example of world node creation

In figure 4 we defined $E_{12}$ as the edges between peers 1 and 2 and $E_{12}^{WN}$ as the edges between P1 and the world node representing P2. An important observation is that for being able to consistently create the world node, a peer needs to know at least a partial structure of the graph of the other peers, otherwise it cannot connect the world node to the other peers' graphs. This means for our setup in figure 4 that P1, who is sending the query, also needs to send its data graph (either the original graph or a hashed version), or at least a part of its graph (original / hashed), such that P2 can correctly put the corresponding inlinks/outlinks to/from its world node.

The big advantage of aggregating everything into a world node is that this protects all internal information about resources and their connections from the receiving peers, while still disclosing (most) information related to external connections and overall weights / scores of the aggregated graph.

### 3.3 Query Processing and Ranking

Using these notations, we can now distinguish between 8 different query processing and ranking algorithms. These 8 algorithms result as appropriate combinations of the 5 possibilities of exchanging information with the 2 modalities of ranking computation (section 3.1). We eliminated several cases as they proved to be equivalent to the remaining 8 ones. We will describe our algorithms in the following, using 3 peers $P_1$, $P_2$ and $P_3$, with $P_1$ always sending the query to P2 and P3. In each case $P_1$ will eventually have a ranked list of results from all peers, including himself.

**Algorithm 1** represents the ideal setup, where everything is shared among the three peers, so that each of them can access the aggregated data graph (all peers' graphs merged into one). **Algorithm 2** describes the situation when P1 shares all its resources, but P2 and P3 share only some parts of their data items and anonymize the rest. So P2 and P3 will have complete information regarding P1's graph, but P1 will not know the exact data structures of P2 and P3.

**Algorithm 1.**

1: $P_1$ sends $G_1$ to $P_2$ and $P_3$
2: $P_2$ sends $G_2$ to $P_1$ and $P_3$
3: $P_3$ sends $G_3$ to $P_1$ and $P_2$
4: Peers aggregate $G_a = G_1 \cup G_2 \cup G_3$
5: Peers compute ObjectRank on $G_a$

**Algorithm 2.**

1: $P_1$ sends $G_1$ to $P_2$ and $P_3$
2: $P_2$ computes ObjectRank on $G2 = G_1 \cup G_2$
$P_3$ computes ObjectRank on $G3 = G_1 \cup G_3$
3: $P_2$ sends $G_2^{anon}$ to $P_1$
$P_3$ sends $G_3^{anon}$ to $P_1$
4: $P_1$ aggregates $G_a = G_1 \cup G_2^{anon} \cup G_3^{anon}$
5: $P_1$ computes ObjectRank on $G_a$

**Algorithm 3.**

1: $P_1$ sends $G_1^{anon}$ to $P_2$ and $P_3$
2: $P_2$ computes ObjectRank on
$G2 = G_1^{anon} \cup G_2$
$P_3$ computes ObjectRank on
$G3 = G_1^{anon} \cup G_3$
3: $P_2$ sends $G_2^{anon}$ and $R_2 = rank(G2)$ to $P_1$
$P_3$ sends $G_3^{anon}$ and $R_3 = rank(G3)$ to $P_1$
4: $P_1$ aggregates $G_a = G_1 \cup G_2^{anon} \cup G_3^{anon}$
5: $P_1$ computes ObjectRank on $G_a$, biasing on
$R_2$ and $R_3$

**Algorithm 4.**

1: $P_1$ sends $G_1'$ to $P_2$ and $P_3$
2: $P_2$ computes ObjectRank on $G2 = G_1' \cup G_2$
$P_3$ computes ObjectRank on $G3 = G_1' \cup G_3$
3: $P_2$ sends $G_2^{anon}$ and $R_2 = rank(G2)$ to $P_1$
$P_3$ sends $G_3^{anon}$ and $R_3 = rank(G3)$ to $P_1$
4: $P_1$ aggregates $G_a = G_1 \cup G_2^{anon} \cup G_3^{anon}$
5: $P_1$ computes ObjectRank on $G_a$, biasing on
$R_2$ and $R_3$

**Algorithm 5.**

1: $P_1$ sends $G_1'$ to $P_2$ and $P_3$
2: $P_2$ computes ObjectRank on $G2 = G_1' \cup G_2$
$P_3$ computes ObjectRank on $G3 = G_1' \cup G_3$
$P_2$ and $P_3$ bias on resources from $P_1$
3: $P_2$ sends $G_2^{anon}$ and $R_2 = rank(G2)$ to $P_1$
$P_3$ sends $G_3^{anon}$ and $R_3 = rank(G3)$ to $P_1$
4: $P_1$ aggregates $G_a = G_1 \cup G_2^{anon} \cup G_3^{anon}$
5: $P_1$ computes ObjectRank on $G_a$, biasing on
$R_2$ and $R_3$

**Algorithm 6.**

1: $P_1$ sends $G_1'$ to $P_2$ and $P_3$
2: $P_2$ computes ObjectRank on $G2 = G_1' \cup G_2$
$P_3$ computes ObjectRank on $G3 = G_1' \cup G_3$
$P_2$ and $P_3$ bias on resources from $P_1$
3: $P_2$ sends $G_2'$ and $R_2 = rank(G2)$ to $P_1$
$P_3$ sends $G_3'$ and $R_3 = rank(G3)$ to $P_1$
4: $P_1$ aggregates $G_a = G_1 \cup G_2' \cup G_3'$
5: $P_1$ computes ObjectRank on $G_a$, biasing on
$R_2$ and $R_3$

**Algorithm 7.**

1: $P_1$ sends $G_1$ to $P_2$ and $P_3$
2: $P_2$ computes ObjectRank on $G2 = G_1 \cup G_2$
$P_3$ computes ObjectRank on $G3 = G_1 \cup G_3$
3: $P_2$ sends $G_2^{WN}$ and $E_{12}^{WN}$ to $P_1$
$P_2$ sends ranked results matching the query
$P_3$ sends $G_3^{WN}$ and $E_{13}^{WN}$ to $P_1$
$P_3$ sends ranked results matching the query
4: $P_1$ aggregates $G_a = G_1 \cup G_2^{WN} \cup G_3^{WN}$
5: $P_1$ adds to $G_a$ the edges from $E_{12}^{WN} \cup E_{13}^{WN}$
6: $P_1$ computes ObjectRank on $G_a$
$P_1$ merges P2 and P3 results into final list

**Algorithm 8.**

1: $P_1$ sends $G_1'$ to $P_2$ and $P_3$
2: $P_2$ computes ObjectRank on $G2 = G_1' \cup G_2$
$P_3$ computes ObjectRank on $G3 = G_1' \cup G_3$
3: $P_2$ sends $G_2^{WN}$ and $E_{12}^{WN}$ to $P_1$
$P_2$ sends ranked results matching the query
$P_3$ sends $G_3^{WN}$ and $E_{13}^{WN}$ to $P_1$
$P_3$ sends ranked results matching the query
4: $P_1$ aggregates $G_a = G_1 \cup G_2^{WN} \cup G_3^{WN}$
5: $P_1$ adds to $G_a$ the edges from $E_{12}^{WN} \cup E_{13}^{WN}$
6: $P_1$ computes ObjectRank on $G_a$
$P_1$ merges P2 and P3 results into final list

We can also bias ranking computation at P1 on the graphs sent by P2 and P3. In **Algorithm 3**, P1, P2 and P3 share only parts of their resources and anonymize their corresponding data graphs for the items they want to keep private. P2 and P3 compute ObjectRank on the data graph resulting from merging the anonymized data graph of P1 and their own data graph. Results are sent back to P1, which computes ObjectRank on the graph including its own data graph and the anonymized graphs of P2 and P3, biasing the computation on the results coming from P2 and P3. **Algorithm 4**, with P1 sending a subgraph containing only the resources it wants to share, is similar to Algorithm 3.

We can also bias ranking computation at P2 and P3 on the resources received from P1, and then get **Algorithm 5**, based on Algorithm 3, and **Algorithm 6**, based on Algorithm 4. Note that when peers send hashed data graphs, the results will not differ from the case where they anonymize nodes in the private part of their graph. This is because for hashed resources, the receiving peers can identify all resources they share with the sending peers if they use the same hashing function. For the resources they do not share, they will get all information about the link structure, but with the node names unknown / anonymized.

**Algorithms 7** and **8** represent the situations where P2 and P3 protect their resources as much as possible, while still providing useful information to P1 using world node aggregation. **Algorithm 7** is a special case of Algorithm 2: P1 shares all its resources but P2 and P3 aggregate their graphs into a world node, keeping the connections to and from P1's graph. **Algorithm 8** is similar to Algorithm 7, only that P1 sends only part of his graph to P2 and P3. In both algorithms, P1 will have to merge results received from P2 and P3 with its own resources, and still keep the relative importance of the items it received, which it can estimate through the information transmitted from P2 and P3 in form of their world nodes, connected to the graph of P1.

All the algorithms we presented can be obviously extended to the general case where a peer is querying in a larger network with more than 2 neighbours.

# 4 Experiments

## 4.1 Experimental Setup

To evaluate our algorithms, we gathered metadata from 9 different users (a total of 46500 RDF triples) and partitioned them into 3 sets, the 3 peers. Metadata were produced by a number of metadata generators integrated in Beagle++ [1], and correspond to several types of resources: files, web pages, emails, attachments, publications, persons and conferences. The data set from a single user did not get partitioned into different peers, since we wanted to simulate real peers, with their own profile, but metadata from some of the physical users was copied to more than one peer to simulate different sizes of overlap between the peers. In all considered scenarios, our peers have a common set of data, as we are dealing with peers collaborating with each other. Figure 5 gives an overview: a) resources residing in $X$ are common to all peers; b) slice R contains resources appearing only at peer 1; c) slice O contains resources only from peer 2 and d) slice T contains private resources of peer 3. Based on the amount and type of resources the three peers are sharing, we have three different setups:

1. P1, P2 and P3 share everything, except of some items they want to protect from the uncommon parts, T, O and R;
2. P1, P2, P3 protect resources which can be located both in the common part X, as well as in the uncommon parts of the graph, T, O and R;
3. We experimented with different sizes of the common part X, i.e. the overlap among the peers: a) small; b) medium; and c) large.

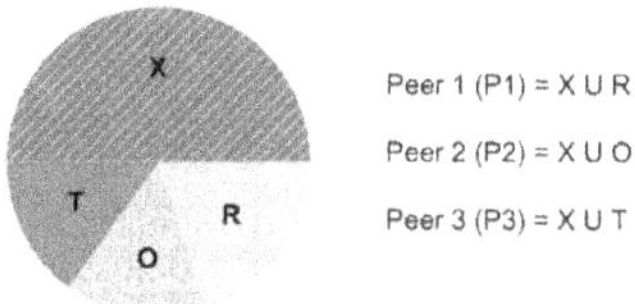


**Fig. 5.** Peers' resource distribution

For SETUPs 1 and 2 we used **Partitioning 1**, having P1 with 40264 triples, P2 with 7700, P3 with 1786 and a size of the overlap of 1624 triples. For SETUP 3 (**Partitioning 2**) we used a different partitioning: for the big overlap case we divided the set into 45512, 45434, 45584 triples for P1, P2 and P3 respectively and 45015 triples the size of the overlap; for medium overlap 6815 (P1), 44715 (P2), 7120 (P3) and 6075 triples the overlap. The small overlap was simulated with a partitioning of 1215 (P1), 6785 (P2), 38780 (P3) and 140 common triples.

In all our algorithms P1 initiates the query, thus we observe the rank evolution for P1. For all three setups and each algorithm described in section 3, we investigated how the scores of the resources evolve. We compared the ObjectRank scores using 2 similarity metrics between the ObjectRank scores obtained in different algorithms and the ideal case for P1, defined as follows (see also [13]):

1. **OSim** indicates the degree of overlap between the top $n$ elements of two ranked lists $\tau_1$ and $\tau_2$. It is defined as

$$\frac{|Top_n(\tau_1) \cap Top_n(\tau_2)|}{n} \tag{2}$$

2. **KSim** is a variant of Kendall's $\tau$ distance measure. Unlike OSim, it measures the *degree of agreement* between the two ranked lists. If $U$ is the union of items in $\tau_1$ and $\tau_2$ and $\delta_1$ is $U \setminus \tau_1$, then let $\tau_1'$ be the extension of $\tau_1$ containing $\delta_1$ apearing after all items in $\tau_1$. Similarly, $\tau_2'$ is defined as an extension of $\tau_2$. Using these notations, KSim is defined as follows:

$$KSim(\tau_1, \tau_2) = \frac{\left|(u,v) : \begin{array}{c}\tau_1' \text{ and } \tau_2' \text{ agree on order} \\ \text{(u,v), and } u \neq v\end{array}\right|}{|U| \cdot |U-1|} \tag{3}$$

### 4.2 Results and Analysis

For all three setups we computed KSim and OSim measures (tables 1-5), comparing the ObjectRank results we obtained for algorithms 2-6/2-8 (column 2) against algorithm 1 (column 1), representing the ideal situation, where all peers share everything they have. We analyzed the top 5, 10, 20, 50 and 100[5] ranked results for each algorithm.

[5] ObjectRank is not query dependent, which means that the rankings for specific queries will be a combination between the ObjectRank values and TFxIDF and therefore the matching results can be located beyond top-20.

**Table 1.** SETUP 1 - OSim, KSim

| SETUP 1 | | | | | | | | | | | |
|---|---|---|---|---|---|---|---|---|---|---|---|
| Vs. | | Top 5 | | Top 10 | | Top 20 | | Top 50 | | Top 100 | |
| Algorithm | Algorithm | OSim | KSim | OSim | KSim | OSim | KSim | OSim | KSim | OSim | KSim |
| 1 | 2 | **1.0** | **1.0** | **0.9** | **0.927** | **1.0** | **0.926** | **1.0** | **0.977** | **1.0** | **0.991** |
| 1 | 3 | 0.4 | 0.607 | 0.6 | 0.582 | 0.9 | 0.670 | 1.0 | 0.909 | 0.96 | 0.936 |
| 1 | 4 | 0.4 | 0.607 | 0.6 | 0.582 | 0.9 | 0.670 | 1.0 | 0.909 | 0.96 | 0.936 |
| 1 | 5 | 0.4 | 0.607 | 0.4 | 0.5 | 0.55 | 0.586 | 0.98 | 0.805 | 0.94 | 0.897 |
| 1 | 6 | 0.2 | 0.472 | 0.3 | 0.448 | 0.55 | 0.534 | 0.98 | 0.755 | 0.91 | 0.871 |

**Partitioning 1.** In SETUP 1 (Table 1), the peers protect resources located only in the non-shared parts, *R*, *O*, or *T*. Given this restriction and the way the world node is constructed we do not need to perform simulations for algorithms 7 and 8, since they yield the same results as in setup 2[6]. In terms of both KSim and OSim, the second algorithm performs best: P1 integrates into its own data graph the anonymized data graphs of P2 and P3, but since P1 is dominating from the point of the number of triples in the graph, this does not have any significant impact on the final scores of P1. Algorithm 6, when every peer biases on the resources received from the others and when only the subgraphs containing the shared resources are sent through the network, performs worst. The reason is that P1 is dominant and the final result will be too much biased on the shared resources of P1. Algorithms 3 and 4 perform the same, as P1 receives the same data graphs in both algorithms.

SETUP 2 (Table 2) differs from SETUP 1 by the fact that the peers can keep private resources from any parts of the graph, *X*, *R*, *O*, or *T*. When looking at the top-5 ranked results, algorithm 2 still performs good, but as we increase top-k, algorithm 6 gets considerably better. If we consider a small value for k, then for P1 it is better to send part of its data graph containing only the shared resources rather than anonymizing the graph, because anonymization introduces errors (peers are not able to identify what the anonymized resources represent and therefore can introduce duplicates - the resource itself and its anonymized copy). For algorithm 6 with increasing k, biasing on both P2/P3's and P1's side significantly improves the results. Algorithms 7 and 8, using the world node-based approach, perform best, both in terms of OSim and KSim. Evaluating these last two algorithms is done as follows (remember that the list of results contains all nodes of P1 plus the world nodes representing P2 and P3): We merged into the list of P1 (without the world nodes) the lists that P2 and P3 computed after integrating the resources of P1. The way we construct the world node and determine the weights of its outlinks and of the self-loop link models with high fidelity the internal structure of the original graph. Even if the receiving peers do not know the graph structure residing at the other peers - that is the

[6] In algorithm 7 P1 sends all his graph, so that no anonymization is involved which makes SETUP 1 and SETUP 2 exactly the same. For algorithm 8 in SETUP 2, the resources that P1 does not share from X (common part) will still appear in the graphs of P2 and P3, therefore this setup is the same as SETUP 1.

**Table 2.** SETUP 2 - OSim, KSim

| **SETUP 2** | | | | | | | | | | | |
|---|---|---|---|---|---|---|---|---|---|---|---|
| Vs. | | Top 5 | | Top 10 | | Top 20 | | Top 50 | | Top 100 | |
| Algorithm | Algorithm | OSim | KSim | OSim | KSim | OSim | KSim | OSim | KSim | OSim | KSim |
| 1 | 2 | 0.8 | 0.6 | 0.7 | 0.705 | 0.9 | 0.757 | 0.88 | 0.873 | 0.75 | 0.827 |
| 1 | 3 | 0.4 | 0.607 | 0.6 | 0.626 | 0.9 | 0.701 | 0.86 | 0.855 | 0.81 | 0.836 |
| 1 | 4 | 0.6 | 0.666 | 0.6 | 0.648 | 0.95 | 0.647 | 0.8 | 0.853 | 0.74 | 0.806 |
| 1 | 5 | 0.4 | 0.607 | 0.3 | 0.573 | 0.65 | 0.581 | 0.86 | 0.8 | 0.86 | 0.835 |
| 1 | 6 | 0.4 | 0.607 | 0.4 | 0.558 | 0.65 | 0.581 | 0.92 | 0.796 | 0.89 | 0.853 |
| 1 | 7 | **1.0** | **0.9** | **0.8** | **0.893** | **1.0** | **0.815** | 0.96 | **0.923** | **0.94** | **0.929** |
| 1 | 8 | **1.0** | **0.9** | **0.8** | **0.893** | **1.0** | **0.815** | **0.98** | **0.923** | 0.93 | 0.912 |

**Table 3.** SETUP 3 - Small Overlap

| **SETUP 3 - Small Overlap** | | | | | | | | | | | |
|---|---|---|---|---|---|---|---|---|---|---|---|
| Vs. | | Top 5 | | Top 10 | | Top 20 | | Top 50 | | Top 100 | |
| Algorithm | Algorithm | OSim | KSim | OSim | KSim | OSim | KSim | OSim | KSim | OSim | KSim |
| 1 | 2 | **1.0** | 0.9 | **1.0** | **0.977** | **1.0** | **0.989** | 0.84 | **0.934** | 0.87 | 0.834 |
| 1 | 3 | 0.6 | 0.761 | 0.7 | 0.666 | 0.9 | 0.744 | 0.88 | 0.906 | 0.8 | 0.835 |
| 1 | 4 | 0.4 | 0.607 | 0.6 | 0.582 | 0.85 | 0.683 | 0.88 | 0.883 | 0.87 | **0.869** |
| 1 | 5 | 0.6 | 0.761 | 0.7 | 0.666 | 0.9 | 0.740 | 0.82 | 0.879 | 0.86 | 0.846 |
| 1 | 6 | 0.6 | 0.666 | 0.4 | 0.616 | 0.6 | 0.658 | 0.86 | 0.780 | **0.9** | 0.822 |
| 1 | 7 | **1.0** | **1.0** | 0.6 | 0.824 | **1.0** | 0.7 | **0.9** | 0.888 | 0.88 | 0.841 |
| 1 | 8 | **1.0** | **1.0** | 0.6 | 0.824 | **1.0** | 0.7 | **0.9** | 0.878 | 0.85 | 0.817 |

peer does not disclose any sensible information - the authority transfer among the peers is captured within this model.

**Partitioning 2.** In SETUP 3 (Tables 3-5) we experimented with 3 different sizes of the overlap.

If the overlap is small or medium, algorithm 2 still performs best for the top-10 and 20 results. If the overlap is big, algorithm 7 performs best for all top-k we consider, followed by algorithm 8 with really small differences. In this case, world nodes (algorithms 7, 8) are strongly connected to the rest of the graph and can therefore very accurately model the influence of the hidden parts of the graph. When looking at top-5 in all variants, algorithms 7 and 8 are the best ones. Algorithms 3 and 4 now perform differently, the biggest difference being for the top-5 ranked results.

**Table 4.** SETUP 3 - Medium Overlap

| **SETUP 3 - Medium Overlap** | | | | | | | | | | | |
|---|---|---|---|---|---|---|---|---|---|---|---|
| Vs. | | Top 5 | | Top 10 | | Top 20 | | Top 50 | | Top 100 | |
| Algorithm | Algorithm | OSim | KSim | OSim | KSim | OSim | KSim | OSim | KSim | OSim | KSim |
| 1 | 2 | **1.0** | 0.8 | **1.0** | **0.955** | **1.0** | **0.984** | 0.88 | **0.944** | 0.77 | 0.828 |
| 1 | 3 | 0.6 | 0.714 | 0.3 | 0.625 | 0.75 | 0.623 | 0.86 | 0.818 | 0.82 | 0.797 |
| 1 | 4 | 0.6 | 0.714 | 0.5 | 0.628 | 0.7 | 0.68 | 0.84 | 0.829 | 0.76 | 0.814 |
| 1 | 5 | 0.4 | 0.642 | 0.5 | 0.590 | 0.75 | 0.68 | 0.88 | 0.801 | 0.82 | 0.805 |
| 1 | 6 | 0.4 | 0.678 | 0.5 | 0.638 | 0.75 | 0.686 | **0.96** | 0.811 | **0.89** | 0.846 |
| 1 | 7 | **1.0** | **1.0** | 0.6 | 0.824 | **1.0** | 0.736 | 0.9 | 0.881 | 0.88 | **0.847** |
| 1 | 8 | **1.0** | **1.0** | 0.6 | 0.824 | **1.0** | 0.7 | 0.9 | 0.878 | 0.86 | 0.832 |

**Table 5.** SETUP 3 - Big Overlap

| **SETUP 3 - Big Overlap** | | | | | | | | | | | |
|---|---|---|---|---|---|---|---|---|---|---|---|
| Vs. | | Top 5 | | Top 10 | | Top 20 | | Top 50 | | Top 100 | |
| Algorithm | Algorithm | OSim | KSim | OSim | KSim | OSim | KSim | OSim | KSim | OSim | KSim |
| 1 | 2 | 0.8 | 0.6 | 0.6 | 0.692 | 0.85 | 0.664 | 0.86 | 0.864 | 0.8 | 0.818 |
| 1 | 3 | 0.8 | 0.866 | 0.6 | 0.703 | 0.95 | 0.661 | 0.86 | 0.865 | 0.8 | 0.830 |
| 1 | 4 | 0.6 | 0.761 | 0.5 | 0.619 | 0.95 | 0.628 | 0.86 | 0.874 | 0.81 | 0.845 |
| 1 | 5 | 0.8 | 0.866 | 0.5 | 0.704 | 0.8 | 0.673 | 0.94 | 0.828 | 0.86 | 0.881 |
| 1 | 6 | 0.4 | 0.678 | 0.5 | 0.561 | 0.6 | 0.648 | 0.96 | 0.779 | 0.89 | 0.844 |
| 1 | 7 | **1.0** | **1.0** | **0.7** | **0.884** | **1.0** | **0.784** | **1.0** | **0.935** | **0.98** | **0.981** |
| 1 | 8 | **1.0** | 0.9 | **0.7** | 0.846 | **1.0** | 0.684 | **1.0** | 0.902 | 0.92 | 0.931 |

## 5 Related Work

In the last two years researchers have investigated how to compute PageRank in a distributed manner. [15] proposes a distributed search engine framework, in which every web server answers queries over its data, and results from multiple web servers are merged into one ranked list. Each web server constructs a web link graph based on its own pages to compute a Local PageRank vector, then they exchange their inter-server link information and compute a ServerRank vector, which is used to refine their Local PageRank vectors. Similarly, [16] computes SiteRank, based on applying PageRank to the graph of Web sites, i.e., the Web graph at the granularity of Web sites instead of Web pages. Aggregating the rankings from multiple sites produces results similar to the true PageRank scores. Both approaches aim to distribute the PageRank computation using several servers and iterations, such that the computational load is reduced, but still the final scores are similar enough with the ones obtained from a global computation. Our goal is to ensure a personalized view over heterogeneous collections, distributed over several desktops, using exchange of appropriate collection/link information before the computation.

[5] was the first paper to introduce the concept of "world node", to incrementally compute a good approximation of PageRank as links evolve. They identify a small portion of the web graph in the vicinity of changes and model the rest of the Web as a single node in this small graph, onto which they compute a version of PageRank and suitably transfer back the results to the original graph. Building on this work, [14] describes a P2P search engine architecture where peers are autonomous, crawl Web fragments and index them locally, but collaborate for query routing and execution. Each peer computes the PageRank scores for the pages it has in its local index. Peers meet and exchange information, and then recompute their PageRank scores. Their original local graph G is extended by adding a special node W, *world node*, representing all pages in the network that do not belong to *G*. Their algorithm assumes that URLs of pages in the world node are known, only their content is not known (not yet crawled). In our scenario, peers do not know the URIs of the external resources and therefore need to send at least part of their data graph to the other peers so that these can create the world node for them. As our world node is used to keep link and

node information private, no inner structure is known. Moreover, all other approaches perform ranking computation on graphs containing only web pages and hyperlinks, while in our case we have different types of links among the nodes, based on their type and on the desktop ontology.

The idea of how communities influence each other is investigated in [3]. They introduce the interesting notion of "energy" of communities, which they define for subsets of the global graph. A community can be viewed as a set of pages on a given topic and the corresponding energy is a measure of the community's authority. The "energy" concept is also applicable in our case, since we are investigating how peers influence each other through the data they are sharing. However, their formulas assume all information about the graph at one location is known, which is not the case in our scenario. It will be interesting to find suitable formulas for approximating energy level and flow for our scenarios, where we have only partial information about the whole graph.

## 6 Conclusions

An important functionality in distributed work environments is to provide searching and ranking capabilities over collections distributed over the desktops of a work group. In this paper we introduced several algorithms for retrieving resources over a network of such desktops, which rely on the exchange of collection specific information between the participating peers in order to achieve appropriate ranking using PageRank-based algorithms. All our algorithms take privacy into account, i.e. peers want to exchange only certain parts of their desktop content, a constraint which has been neglected so far in all previous work on distributed PageRank computation.

We analyzed in detail how our algorithms perform in several setups of resource sharing. In particular, we experimented with different sizes of data sets residing on the peers' desktops and with different dimensions of the overlapping information. Our experiments show that we can compute appropriate ObjectRank values even if the peers do not share everything they have. Specifically, algorithms aggregating node and link information into one "world node" proved to be the best tradeoff between privacy and quality. They offer the best way of protecting resources, since peers do not reveal any of their nodes or the way they are interconnected, approximate ObjectRank values very well, and guarantee the smallest network load. In future work we will extend these algorithms with methods to estimate the potential of peers to influence results of other peers, and come up with incremental update schemes when peer content changes.

## References

1. Beagle++. (2006) `http://beagle.kbs.uni-hannover.de/`
2. NEPOMUK - The Social Semantic Desktop. (2006) `http://nepomuk.semanticdesktop.org`
3. Bianchini, M., Gori, M., Scarselli, F.: Inside pagerank. ACM Trans. Inter. Tech. 5(1), 92–128 (2005)

4. Callan, J.P., Lu, Z., Croft, W.B.: Searching distributed collections with inference networks. In: Proc. of the Intl. Conf. on Research and Development in Information Retrieval (SIGIR) (1995)
5. Chien, S., Dwork, C., Kumar, S., Sivakumar, D.: Towards exploiting link evolution. In: Unpublished manuscript (2001)
6. Chirita, P.A., Costache, S., Nejdl, W., Paiu, R.: Beagle++: Semantically enhanced searching and ranking on the desktop. In: Proc. of the European Semantic Web Conf. (ESWC) (2006)
7. Chirita, P.A., Ghita, S., Nejdl, W., Paiu, R.: Semantically enhanced searching and ranking on the desktop. In: Proc. of the Semantic Desktop Workshop held at the Intl. Semantic Web Conf (2005)
8. Damian, A., Nejdl, W., Paiu, R.: Peer-sensitive objectrank: Valuing contextual information in social networks. In: Proc. of the Intl. Conf. on Web Information Systems Engineering (2005)
9. Dong, X., Halevy, A.Y.: A platform for personal information management and integration. In: Proc. of Conf. on Innovative Data Systems Research (CIDR) (2005)
10. Dong, X., Halevy, A.Y., Nemes, E., Sigundsson, S.B., Domingos, P.: Semex: Toward on-the-fly personal information integration. In: Proc. of the Workshop on Information Integration on the Web (2004)
11. Franklin, M., Halevy, A.Y., Maier, D.: From databases to dataspaces: a new abstraction for information management. SIGMOD Rec. 34(4), 27–33 (2005)
12. Green, N., Ipeirotis, P.G., Gravano, L.: SDLIP + STARTS = SDARTS a protocol and toolkit for metasearching. In: ACM/IEEE Joint Conference on Digital Libraries, pp. 207–214 (2001)
13. Haveliwala, T.: Topic-sensitive pagerank. In: Proc. of the Intl. WWW Conf (2002)
14. Parreira, J.X., Donato, D., Michel, S., Weikum, G.: Efficient and decentralized pagerank approximation in a peer-to-peer web search network. In: Proc. of the Intl. Conf. on Very Large Data Bases (VLDB) (2006)
15. Wang, Y., DeWitt, D.: Computing pagerank in a distributed internet search system. In: Proc. of the Intl. Conf. on Very Large Databases (VLDB) (2004)
16. Wu, J., Aberer, K.: Using siterank for decentralized computation of web document ranking. In: Proc. of Intl. Conf. on Adaptive Hypermedia and Adaptive WebBased Systems (2004)

# Generic Schema Merging

Christoph Quix, David Kensche, and Xiang Li

RWTH Aachen University, Informatik V (Information Systems), 52056 Aachen, Germany
{quix,kensche,lixiang}@i5.informatik.rwth-aachen.de

**Abstract.** Schema merging is the process of integrating several schemas into a common, unified schema. There have been various approaches to schema merging, focusing on particular modeling languages, or using a lightweight, abstract metamodel. Having a semantically rich representation of models and mappings is particularly important for merging as semantic information is required to resolve the conflicts encountered. Therefore, our approach to schema merging is based on the generic role-based metamodel *GeRoMe* and intensional mappings based on the real world states of model elements. We give a formal definition of the merged schema and present an algorithm implementing these formalizations.

## 1 Introduction

Management of models is an important activity in the design of complex information systems. The availability of data sources and the need to analyze the existing data in an integrated way, has led to applications which are able to integrate and present data from various sources in a uniform way. The integration of the models of data sources into a unified schema of the integrated information system is a prerequisite to build such applications. *Schema integration* (or *schema merging*) is the process of integrating several schemas into a common, unified schema. This problem is also addressed in *Model Management* [3], which aims at defining an algebra for models and mappings. Merge is one of the proposed operators in this algebra and addresses the problem of generating a merged model given two input models and a mapping between them. The merged model should contain all the information contained in the input models and the mapping; it should *dominate* the inputs in terms of information capacity [8,15].

A mapping is not just a simple set of 1:1 correspondences between model elements; it might have itself a complex structure and is therefore often regarded also as a *mapping model*. A mapping model is necessary because the models to be merged also have complex structures, which usually do not correspond to each other [16] (e.g. the address of a person is represented in one ER model as a complex attribute, in another model as a separate entity type with a relationship type to person). These structural heterogeneities are one class of conflicts which occur in Merge. Other types of conflicts are semantic conflicts (model elements describe overlapping sets of objects), descriptive conflicts (the same elements are described by different sets of properties; this includes also name conflicts), and heterogeneity conflicts (models are described in different modeling languages) [19]. The resolution of these conflicts is the main problem in Merge.

Schema integration has been addressed for various metamodels, such as variants of the ER metamodel [19], relational and conceptual models in the context of data warehouses [5], graph-based models [18], or a simple generic metamodel [16]. In these

J. Krogstie, A.L. Opdahl, and G. Sindre (Eds.): CAiSE 2007, LNCS 4495, pp. 127–141, 2007.

works, it has also been argued, that a semantically rich representation of the models and mappings simplifies schema integration, as semantic information is required to resolve the conflicts. Our work is therefore based on the semantically rich generic metamodel *GeRoMe* [10]. It provides a generic, but yet detailed representation of models originally represented in different metamodels. Therefore, an implementation of the Merge operator based on *GeRoMe* can merge models from different metamodels (e.g. XML Schema and Relational). It is not always possible to represent the result of Merge in one of the input metamodels. For instance, merging a column with a table yields a model with a composite attribute that is not allowed in the relational model. Therefore, *GeRoMe* enables the representation of such results. The transformation of the merge result into a specific metamodel is the task of other model management operators: ModelGen transforms the modeling constructs which are not allowed in the target model, and Export translates the *GeRoMe* representation into the representation in a native metamodel.

Another important aspect is the representation of mappings between models. As argued before, the mapping is itself a model and contains information required for Merge. In contrast to recent approaches to mapping composition and executable mappings [7,13], which focus on the *extensional* relationships between models for data translation, the mappings here represent the *intensional* relationships of model elements. These mapping definitions often overlap, but are not necessarily identical. This distinction between intensional and extensional relationships has also been made in [5].

In this work, we present a merge algorithm which is based on the intensional relationships between models. The contributions of this paper are (1) a definition of the intensional semantics of models, (2) a mapping model representing intensional relationships, (3) a formal characterization of the desired merged model, and (4) an algorithm implementing these formalizations using *GeRoMe*, thereby enabling the integration of models represented in different modeling languages.

The paper is structured as follows. The next section discusses existing work on model and mapping representation. Section 3 introduces a real world semantics, that we use to define the semantics of merging, and the mapping representation. In section 4 we describe our algorithm for merging *GeRoMe* models given a mapping between them. Section 5 compares our approach to some existing works on schema merging. The last section concludes the paper and gives an outlook.

## 2 Background

Considerable research has already been done in the fields of model management and data integration. The Merge operator in model management receives two models and a mapping as inputs. Hence, besides existing algorithms, the representation of models and mappings are particularly important prerequisites in the context of schema merging.

**Mapping Representations.** Depending on the application area, such as data translation, query translation or model merging, schema mappings come in different flavors. The first step in specifying a mapping between two schemas is usually the automatic derivation of informal correspondences between elements of the two respective schemas, called schema matching [17]. These simple binary correspondences, often called *morphisms* [14], state only informally that the respective model elements are similar. Thus, relationships between model elements must be represented more accurately, but morphisms are usually the starting point for specifying more formal mappings.

Formalization of mappings is done in form of view definitions or as correspondence assertions. These will be called in the following *extensional* and *intensional* mappings, respectively. Extensional mappings are defined as local-as-view (LAV), global-as-view (GAV), source-to-target tuple generating dependencies (tgds) [12,13], second order tuple generating dependencies (SO tgds) [7], or similar formalisms. These are pairs of queries with an implication or equivalence operator in between. Each of these classes has certain advantages and disadvantages when it comes to properties such as composability, invertibility or execution of the mappings.

Extensional mappings are used for tasks such as data translation or query rewriting but they are inappropriate for ontology alignment and schema merging. Schema merging is about integrating models according to their intensional semantics. It has the goal to construct a duplicate free union of two input models and a mapping in between. The union is "duplicate free" with respect to the real world concepts described by the model elements. So, the mappings interrelate the *intensions* of model elements. The integrated model describes each real world concept only once. On the other hand, if mappings are to be executed by using them for query rewriting or data translation, intensional mappings are not useful. Thus, these are two options of mapping representation, each of which has certain advantages with respect to the goal of the mapping.

**Model Representations.** Any system that allows the usage of different native metamodels should employ some generic schema representation. Most model management systems either refrain from providing a generic representation and instead require some operators to be implemented for different combinations of native metamodels or employ very lightweight metamodels [14,16]. However, some model manipulation operations require much more information about the schemas than is expressable in such lightweight languages. Resolving conflicts in model integration is eased by additional information about the schemas to be merged [16,19]. This particular challenge of providing a generic model representation [3] has been adressed in particular in [1,2,10].

The metamodel that we use for our implementation of Merge is the generic role based metamodel *GeRoMe* [10,9]. In *GeRoMe* each model element of a native model (e.g. an XML Schema or a relational schema) is represented as an object that plays a set of roles which decorate it with features and act as interfaces to the model element.

Figure 1 shows in the left part a simple EER model for a "Customers ordering Products" scenario. It contains a relationship type with two attributes and two entity types, the attributes of which we omitted from the example. The right part of the figure presents the equivalent representation in our generic metamodel (we omitted some details of the model for clarity of the presentation). Each entity type, attribute, and relationship type is represented by an individual model element (shown as grey rectangles). Every *GeRoMe* model element plays a number of roles depending on the features of the source element that it represents. The date element plays an *Attribute* role (Att) as it represents an EER attribute, the entity types play ObjectSet roles (OS) since their instances have object identity (as opposed to the instances of a relational table) and *Aggregate* roles (Ag) as they are aggregates of attributes (not shown here). The orders

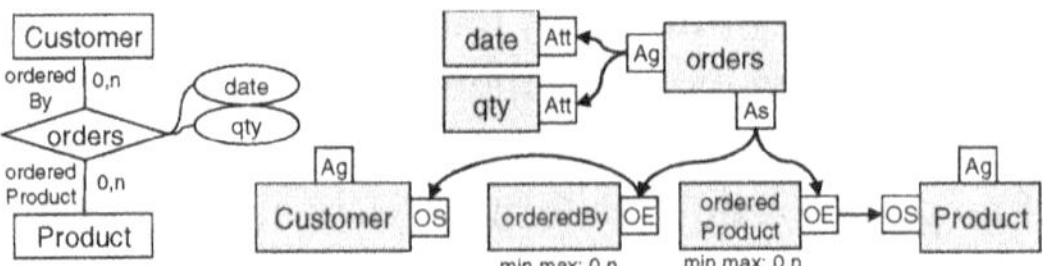


**Fig. 1.** A small EER model and its representation in *GeRoMe*

element is represented similarly. It plays the role of an *Association* which connects to two *ObjectAssociationEnds* (OE) (orderedBy and orderedProduct). The *ObjectAssociationEnds* specify also the cardinality constraints. Other features of native models can be represented in a similar fashion in *GeRoMe*. The same role classes are used to describe models in different metamodels, thereby providing a common datastructure for the polymorphic implementation of model management operators.

It is important to emphasize that this representation is not to be used by end users. Instead, it is a representation employed internally by model management applications, with the goal to generically provide more information to model management operators than the usual graph based model.

## 3 Semantics of Models and Mappings

Mappings between the models to be merged are an important source of information for Merge. In order to define the semantics of the mappings in a clear and formal way, we first need to specify the semantics of the models and their elements which are going to be related. As discussed in the previous section, mappings in the form of view definitions are not useful for schema integration, as they usually specify many-to-many (or at least many-to-one) relationships between model elements. Such relationships cannot be used to detect elements in two models which have the same semantics. Relationships must be based on the intended semantics of the model elements rather than extensional relationships. For example, consider the schemas of two universities, each representing the students of that university. The extensions of the databases are disjoint, but the concept "student" should be merged if the schemas are going to be integrated.

Therefore, we need relationships between model elements which are based on their real world semantics, i.e. the set of real world objects a model element represents. Only with respect to this semantics, we can decide whether two elements should be merged. Such an approach based on real world semantics has also been taken in [11,19]. In contrast to these previous approaches, our definition of real world objects is more detailed, i.e. it aims at making this abstract concept more concrete so that it is possible to use it in an implementation of the Merge operator.

In the following, we will first define model elements with respect to their real world semantics, then explain how this representation is related to our generic metamodel *GeRoMe*, and finally define the mapping model which we will use to express intensional relationships between models.

**Model Elements.** To define the semantics of a model element, we first define the real world objects it should represent.

**Definition 1 (Real World Object).** *A* real world object (RWO) *is defined as a vector in the feature space with some arity. Each dimension is called an* axis. *The* universe $\mathcal{U}$ *is defined as the set of all RWOs. A* projection *of a real world object o wrt. one axis* $\alpha$, $\pi_\alpha(o)$, *is* $\epsilon$, *a literal or RWO, a set of literals or RWOs, or a tuple of literals or RWOs. A projection wrt. a tuple of axes is a tuple of which each component is the projection wrt. the corresponding axis:* $\pi_{\alpha_1,\ldots,\alpha_n}(o) =< \pi_{\alpha_1}(o), \ldots, \pi_{\alpha_n}(o) >$.

When a projection wrt. one axis is $\epsilon$, this means that the RWO is not defined over this axis. The empty set denotes that the RWO is defined for that axis, but it has no value. For each axis $\alpha$, we denote all the RWOs over which the axis $\alpha$ is defined as $\mathcal{U}_\alpha$.

**Definition 2 (Model Element and Real World Set).** *A* model element *m consists of a tuple of axes, denoted by* $axes(m) = \{\alpha_1, \ldots, \alpha_n\}$. *The* real world set (RWS) *of a model element* $m$ *is a subset of the RWOs for which all the axes of* $m$ *are defined:* $RWS(m) \subseteq \mathcal{U}_{\alpha_1} \cap \ldots \cap \mathcal{U}_{\alpha_n}$, *if* $axes(m) = \{\alpha_1, \ldots, \alpha_n\}$.

**Relationship to *GeRoMe*.** These definitions characterize the real world semantics of a model element. In [10], we defined also a formal semantics for *GeRoMe* models which characterizes the structure of their instances. In *GeRoMe*, there are four different roles for which the corresponding model elements can have instances. These roles are *Domain*, *ObjectSet*, *Aggregate*, and *Association*. Elements playing a *Domain* role are a special case; these model elements cannot play any of the other roles. Their instances are just sets of literal values. However, *ObjectSet*, *Aggregate*, and *Association* roles can be combined. The instances of model elements playing at least one of these roles are specified by a triple which has the following components: (i) an object identifier, if it plays an *ObjectSet* role, (ii) a tuple of object identifiers, denoting the participating objects in an association (one for each association end), if it plays an *Association* role, and (iii) a tuple of literal value sets (one for each attribute), if it plays an *Aggregate* role.

Thus, a RWO can be easily mapped to a *GeRoMe* instance. The axes having a literal value or a set of literal values correspond to the third component of a *GeRoMe* instance which specifies the attribute values. The axes referring to RWOs are mapped to the second component which expresses associations to other objects.

Based on this relationship between the real world semantics and *GeRoMe*, we can later use the mapping model defined in the following to specify mappings between *GeRoMe* models. The transition from the real world semantics to *GeRoMe* makes this definition useful for the implementation of a model management system.

**Mappings.** A mapping specifies how the two models will be merged. In [16], the mapping model is a nested structure consisting of mapping elements; each mapping element is related to at least one model element. The mapping elements can specify, that the related model elements are either equivalent or similar. A similarity mapping states that the elements are related by a complex expression which is not further specified.

A richer set of relationships between model elements is defined in [19], which is also based on the real world semantics of model elements. Possible relationships are equivalence, inclusion, intersection and exclusion. However, only equivalence relationships are used in the integration rules. A simple form of nesting of elements can be specified using the "with corresponding attributes" clause for two related model elements.

Our approach is a combination of the ideas of both approaches: a nested mapping model with rich semantic relationships based on the RWS of model elements. In addition, all this information will be used in the Merge algorithm as we will see in section 4. We will first define how model elements can be related at the top-level.

**Definition 3 (Element Mapping).** *An element mapping $\phi$ between two model elements $m$ and $m'$ is an expression $m\theta m'$ with $\theta \in \{=, \subseteq, \cap, \neq\}$. The semantics of the mapping is defined by the RWS of the model elements:*

1. *$m\theta m'$ with $\theta \in \{=, \subseteq\}$ implies that $RWS(m)\theta RWS(m')$.*
2. *$m \cap m'$ states that $RWS(m)$ and $RWS(m')$ have a non-empty intersection.*
3. *$m \neq m'$ specifies that $RWS(m)$ and $RWS(m')$ are disjoint, but there is some $m''$ with $RWS(m) \subseteq RWS(m'')$ and $RWS(m') \subseteq RWS(m'')$.*

The disjointness of elements is useful for a case where two model elements have a common super-type (cf. the example of fig. 4 in sec. 4). In *GeRoMe*, element mappings are only allowed between model elements playing either the *ObjectSet* or the *Aggregate* role, as they have instances. Associations may also have instances, but this information is already covered by the axes representing *AssociationEnds*. In addition to this simple type of mapping, a complex mapping represents 1:N relationships between elements.

**Definition 4 (Complex Mapping).** *A complex mapping $\phi$ between a set of model elements is an expression $m\theta f(m_1, \ldots, m_n)$ with $\theta \in \{=, \subseteq\}$ for some function $f$. The semantics of this mapping is defined by applying the corresponding operations to the RWS of the model elements.*

This enables us to represent that a model element in one model is represented by a combination of model elements in the other model, e.g. $Parent = Mother \cup Father$. This also subsumes the definition of *paths* as defined in [19] as functions can specify arbitrary relationships between model elements.

A nested mapping is a mapping between axes of model elements. It must be nested into an element or complex mapping, so that a context for this mapping is given, i.e. we need to know the model elements of which the axes are mapped.

**Definition 5 (Nested Mapping).** *A nested mapping under a mapping $\phi$ between model elements $m$ and $m'$ is an expression (i) $\alpha\theta\beta$ with $\theta \in \{=, \subseteq\}$ and $RWS(m) \cup RWS(m') \subseteq \mathcal{U}_\alpha \cap \mathcal{U}_\beta$. The semantics is $\forall o \in RWS(m) \cup RWS(m') : \pi_\alpha(o)\theta\pi_\beta(o)$; (ii) $\alpha = f(\alpha_1, \ldots, \alpha_n)$ with $RWS(m) \cup RWS(m') \subseteq \mathcal{U}_\alpha \cap \mathcal{U}_{\alpha_1} \cap \cdots \cap \mathcal{U}_{\alpha_n}$. The semantics is $\forall o \in RWS(m) \cup RWS(m') : \pi_\alpha(o) = f(\pi_{\alpha_1}(o), \ldots, \pi_{\alpha_n}(o))$.*

The functional relationships between axes are necessary to represent complex relationships between axes (as before between model elements). Examples are amounts represented in different currencies or aggregations (e.g. salary=base salary+bonus).

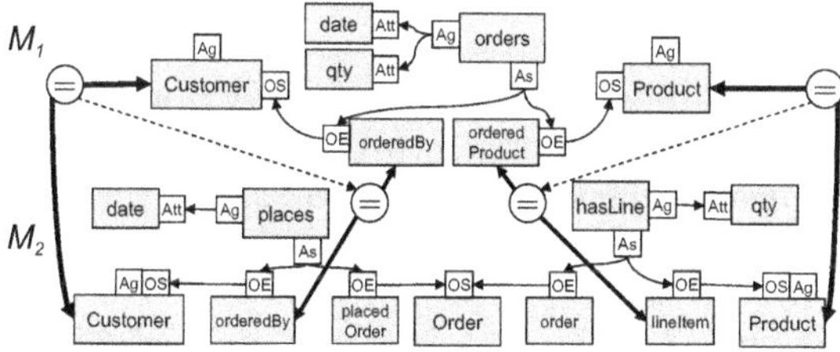


**Fig. 2.** Mapping including complex and nested mappings

In *GeRoMe*, nested mappings can be applied to model elements playing the *Attribute* or *AssociationEnd* roles. Fig. 2 shows an example of a mapping including a nested mapping. The upper model $M_1$ is as in section 2, the lower model $M_2$ reifies the orders relationship as entity type and has therefore two relationship types places and hasLine. The mapping relates the corresponding Customer and Product entity types. The mappings of the *ObjectAssociationEnds* are nested within these mappings. The example contains also a complex mapping (not shown in the figure), as the attributes of the orders relationship are distributed over two relationships in $M_2$. Therefore, the complex mapping orders $\subseteq f$(places, Order, hasLine) needs to be defined in which $f$ performs a join over these elements. In addition, mappings for the attributes date and qty will be nested into this mapping.

As a consequence of this mapping, the Merge algorithm will produce $M_2$ as result, as it contains "more detailed" information than $M_1$: the entity types are represented by the corresponding entity types in $M_2$; the same holds for the *ObjectAssociationEnds*; the orders association is represented by a combination of elements in $M_2$ and its attributes are also present in $M_2$. The difficult question in the definition of Merge is, what does it exactly mean when we say that a model contains "more detailed" information than another model, and how can we verify that $M_2$ is a correct result of the Merge operator in this example. These issues will be addressed in the following section.

# 4 Model Merging

In this section we first provide a definition of the concept of a merged model using our previously defined real world semantics. Then, we explain our solution to the problem of merging of schemas represented in *GeRoMe*.

**Definition of Merged Model.** Our definition of schema merging is closest related to that of [19] as it also defines the problem of schema integration with respect to the real world sets described by the input schemas. The difference is that we use our real world sets to define the notion of a merged model. In addition, we relate the real world sets to information capacity [15] and, in doing so, enrich this notion with meaning.

The following definition introduces successively more general concepts of subsumption ending with a definition of when a model element is subsumed by a set of other elements. This means that all the information represented by the model element is also represented by combinations of the properties of the other model elements.

**Definition 6 (Subsumption ($\sqsubseteq$)).**

a. *Given two axes $\alpha$ and $\beta$, we say $\alpha$ is subsumed by $\beta$ wrt. a RWO $o$ ($\alpha \sqsubseteq_o \beta$) if these two axes are defined over $o$ and $\pi_\alpha(o) \subseteq \pi_\beta(o)$.*
b. *an axis $\alpha$ is subsumed by a set of axes $\mathcal{A} = \{\beta_1, \ldots, \beta_n\}$ wrt. a RWO $o$ ($\alpha \sqsubseteq_o \mathcal{A}$) if the axes $\mathcal{A} \cup \alpha$ are defined over $o$ and $\pi_\alpha(o) \subseteq \pi_{\beta_1}(o) \cup \ldots \cup \pi_{\beta_n}(o)$ or more general $\exists f : \pi_\alpha(o) \subseteq f(\pi_{\beta_1}(o), \ldots, \pi_{\beta_n}(o))$.*
c. *an axis $\alpha$ is subsumed by a set of axes $\mathcal{A}$ wrt. a set of RWOs $R$ ($\alpha \sqsubseteq_R \mathcal{A}$) if $\forall o \in R : \alpha \sqsubseteq_o \mathcal{A}$.*
d. *an axis $\alpha$ of model element $m$ is subsumed by a set of model elements $\mathcal{M}$ ($\alpha \sqsubseteq_{RWS(m)} \mathcal{M}$) if $\forall o \in RWS(m), \exists m_1, \ldots, m_n \in \mathcal{M} : o \in RWS(m_1) \cap \ldots \cap RWS(m_n) \land \alpha \sqsubseteq_o axes(m_1) \cup \ldots \cup axes(m_n)$.*
e. *a model element $m$ is subsumed by a set of model elements $\mathcal{M}$ ($m \sqsubseteq \mathcal{M}$) if $\forall \alpha \in axes(m) : \alpha \sqsubseteq_{RWS(m)} \mathcal{M}$.*

By defining $RWS(\mathcal{M}) = \bigcup_{m \in \mathcal{M}} RWS(m)$, the definition is extended to a definition for subsumption of models, which is used to define the *upper bound* of two models.

**Definition 7 (Subsumption and Upper Bound Models).** *Let $\mathcal{M}$ and $\mathcal{M}'$ be two models. We say $\mathcal{M} \sqsubseteq \mathcal{M}'$ if $RWS(\mathcal{M}) \subseteq RWS(\mathcal{M}') \land \forall m \in \mathcal{M} : m \sqsubseteq_{RWS(\mathcal{M})} \mathcal{M}'$. A model $\mathcal{M}_{UB}$ is an* upper bound model *of two models $\mathcal{M}_1$ and $\mathcal{M}_2$ if $\mathcal{M}_1 \sqsubseteq \mathcal{M}_{UB} \land \mathcal{M}_2 \sqsubseteq \mathcal{M}_{UB}$.*

An upper bound model is a model that represents the same (or a larger) set of real world objects and that does not lose any properties of the two models. However, this definition allows that subclasses that add no axes are removed from the models. Therefore, we need the following definition of granularity.

**Definition 8.** *A model element $m$ is* retained in granularity *by a model $\mathcal{M}$ if (i) $\exists m' \in \mathcal{M}$ such that $RWS(m) = RWS(m')$, and (ii) $\forall \alpha \in axes(m)\ \exists m_1, \ldots, m_n \in \mathcal{M} : \alpha \sqsubseteq_{RWS(m)} \{m', m_1, \ldots, m_n\}$.*

The first condition requires that $\mathcal{M}$ must include a model element $m'$ that represents exactly that same set of real world objects that $m$ represents. The second condition states that each axis of each object in the set represented by $m$, is either explicitly represented in $\mathcal{M}$ or can be computed from some of its axes. This also includes any notion of inheritance between the model elements in $\mathcal{M}$. This is because it is not required that the axes representing $\alpha$ are axes of $m'$ but they may be inherited as well, whereas the object itself must be in $m'$ and all the other model elements.

A model $\mathcal{M}'$ subsumes a model $\mathcal{M}$ while retaining all its model elements in granularity, if it represents the same (or a larger) set of RWOs and of all objects represented by $\mathcal{M}$ it represents the same (or more) information either explicitly or by means of some functional relationship. Also, we do not want a model to be redundant. The last definition before we can define the notion of a *merged model* is that of duplicate-freeness:

**Definition 9 (duplicate-free).** *A model $\mathcal{M}$ is* duplicate-free *if (i) for each model element $m \in \mathcal{M}$ there is no other model element $m' \in \mathcal{M}$ that represents the same set of real world objects, and (ii) for each axis $\alpha$ of any model element in $\mathcal{M}$ there exists no other axis $\beta$ that represents the same property of the model element.*

**Definition 10 (Merged Model).** *Let $\mathcal{M}_1$ and $\mathcal{M}_2$ be two input models and let $M$ be a mapping between $\mathcal{M}_1$ and $\mathcal{M}_2$. A model $\mathcal{G}$ is the result of $Merge < \mathcal{M}_1, \mathcal{M}_2, M >$ (a merged model), if it satisfies the following properties:*

- *$\mathcal{G}$ is an upper bound model of $\mathcal{M}_1$ and $\mathcal{M}_2$.*
- *$\mathcal{G}$ is duplicate-free.*
- *$\mathcal{G}$ retains granularity of all model elements in $\mathcal{M}_1$ and $\mathcal{M}_2$.*
- *$\mathcal{G}$ contains all constraints of the input models and the mapping, and in case of conflicting constraints, the least restrictive constraint.*
- *There is no other model $\mathcal{G}'$ with $\mathcal{G}' \sqsubset \mathcal{G}$, which fulfills these conditions.*

Informally, given two models and a mapping, merging these models means to create a model that contains no duplicate model elements or axes and only structural elements from any of the two input models. However, derivations or constraints may be added to the integrated model in order to relate these model elements to each other. The information necessary for adding such elements stems from the mapping model.

Constraints of the input models and the mapping should be retained in $\mathcal{G}$. If there is a conflict between constraints, the least restrictive constraint is represented in $\mathcal{G}$. For instance, given two cardinality constraints $(0, 1)$ and $(1, n)$, $\mathcal{G}$ contains $(0, n)$.

This definition may be relaxed such that the resulting model is allowed to represent axes that can be derived from other represented axes. These are, for instance, derived attributes in an EER model or methods in an object oriented model that compute values from member variables. Such an extension would necessitate to partition the axes of the models into two kinds of axes and define subsumption only with respect to one of these classes. This extension is straightforward and adds only little information.

The definitions of subsumption and merged models can also be related to the notion of relative information capacity of schemas [15].

**Definition 11.** *An* information capacity preserving mapping *between the instances of two schemas $S1$ and $S2$ is a total, injective function $f : I(S1) \rightarrow I(S2)$. If such a mapping exists $S2$* dominates *$S1$ via $f$, denoted $S1 \preceq S2$.*

A common criticism about this definition is that it allows the models and the function to be arbitrary, given only that such a function can be constructed. By using our relationship to real world sets, we add some meaning to this definition.

**Theorem 1.** *If $\mathcal{M} \sqsubseteq \mathcal{M}'$ then $\mathcal{M} \preceq \mathcal{M}'$* and *the domain of discourse of $\mathcal{M}'$ encompasses the domain of discourse of $\mathcal{M}$ ($RWS(\mathcal{M}) \subseteq RWS(\mathcal{M}')$).*

*Proof.* The last part of the theorem ($RWS(\mathcal{M}) \subseteq RWS(\mathcal{M}')$) follows directly from the definition of subsumption between models. To show that $\mathcal{M}'$ dominates $\mathcal{M}$, we have to construct an information capacity preserving mapping. Instances of the models in our context are sets of RWOs. According to the definition of subsumption, the axes of $\mathcal{M}$ are represented in $\mathcal{M}'$, either directly or they can be computed by some function $f$. Thus, each RWO in $RWS(\mathcal{M}')$ is represented in more detail than in $RWS(\mathcal{M})$. This means that there are no RWOs $o_1, o_2 \in RWS(\mathcal{M})$, such that both correspond to the same RWO in $RWS(\mathcal{M}')$. As the RWS of $\mathcal{M}'$ is a superset of the RWS of $\mathcal{M}$, the information capacity preserving mapping is the identity function on the RWOs.

**Input:** Two models $\mathcal{M}_1$, $\mathcal{M}_2$, and a mapping $M$
**Output:** Merged model $\mathcal{G}$ according definition 10

1. Group equivalent model elements transitively. For each group, introduce a corresponding model element in $\mathcal{G}$. Each of the new model element is created as follows:
   (a) Singletons are copied with no linking axes to other model elements.
   (b) Collapse groups into one model element with a union of all properties and roles.
   (c) Conflicts are resolved according to the strategy described below.
2. Introduce all the classification relationships (IsA) and constraints (Disjoint) from both the input models and the mapping. Remove redundant or cyclic IsA links.
3. Insert in $\mathcal{G}$ a most specific supertype for model elements connected by $\cap$ or $\neq$.
4. The singleton side of complex mappings are removed from the merged model if all of its axes are related in the nesting mappings.
5. For all axes nested under model elements connected by element mappings
   (a) Equally related axes are collapsed and pulled up to their most specific supertype.
   (b) $\subseteq$ related axes will lead to a partition of the subsumer axes.
   (c) Axes related by functions are handled depending on configuration: leaving only the inputs, the output or both.
   (d) The remaining axes are retained at the corresponding element in $\mathcal{G}$.
6. All links between model elements are handled as follows:
   (a) duplicate links are only introduced once;
   (b) conflicting links such as multiple types for one attribute are resolved according to the strategy described below.
7. Check "local" constraints and resolve possible conflicts as described below.
8. Check "global" conflicts that can not detected locally, e.g. recursion of complex attribute.

**Fig. 3.** Merge Algorithm

Our definition of a merged model has the following consequences with respect to information capacity preserving mappings: (i) $\mathcal{G} \succeq \mathcal{M}_1$ and $\mathcal{G} \succeq \mathcal{M}_2$ because it retains each element $m \in \mathcal{M}_1 \cup \mathcal{M}_2$ in granularity, and (ii) $RWS(\mathcal{G}) = RWS(\mathcal{M}_1) \cup RWS(\mathcal{M}_2)$. This adds some meaning to the notion of information capacity as now outright absurd functions are not possible. Please note that it is not necessarily the case that $\mathcal{M}_1 \cup \mathcal{M}_2 \succeq \mathcal{G}$ because the mapping may add relationships to the models.

**Merge Algorithm.** Most previous algorithms are rule-like [11,19] operational procedures or semi-automatic procedures [16]. The first type usually goes through a continuous pattern-transformation procedure. The latter type first collapses all equivalent elements and then introduce links between the grouped elements.

Our merging algorithm consists of several steps described in fig. 3. In the first step, all model elements equally related in the mapping are grouped transitively. Groups of equivalent model elements are collapsed into one element. This might cause conflicts which must be resolved as described below. The second and third steps deal with *IsA*, *Overlapping*, and *Disjoint* relationships and introduce helper elements as necessary.

Step 4 deals with complex mappings. For the orders-example in section 3 the orders element $\mathcal{M}_1$ is removed as it can be represented by combination of other elements. The next step addresses the axes of elements. If we state that $\alpha \subseteq \beta$, then we will keep $\alpha$ and introduce a new element "$\beta - \alpha$" for the remaining part of $\beta$ (e.g. $father \subseteq parent$ is replaced by $father$ and $mother$). The handling of axes connected by functions depends on the setting; it may be useful to keep the inputs, the output, or both. In step 6, redundant and conflicting links introduced by Merge are removed.

The remaining steps deal with conflicts and constraints; these must be handled by specific procedures as described below. Due to space constraints, we can only sketch them briefly. The general rule is, first to check whether it is solvable by an existing automatic resolution strategy depending on the given configuration. Please note that our merge algorithm can be configured with several parameters which allow to fine tune the algorithm for a given situation. The general context can be specified (database vs. view integration), or the handling of the inputs and outputs of functions. For instance, given two equivalent attributes with a simple type and a complex type, respectively, this is a structural conflict which can be resolved by chosing the more detailed representation (e.g. the complex type). Then, try to resolve the conflict using the mapping or preferred model. If the conflict cannot be resolved with the given information, the user has to be involved. The handling of conflicts is often done on a case-by-case basis; for each specific problem an individual conflict resolution strategy must be found. Often, different strategies have to be applied to different scenarios (e.g. view integration vs. database integration). However, as we based our implementation on *GeRoMe*, these resolution strategies have to be implemented only once for *GeRoMe* and can then be applied to several merge scenarios involving different modeling languages.

**Constraints Integration.** Several types of constraints are represented in *GeRoMe* explicitly (such as keys, references, derivation, and type constraints) and can therefore be addressed in Merge. Each type of constraint requires individual methods for merging and conflict resolution. Even worse, conflict resolution might depend on the scenario. A non-null constraint on an axis not represented in one input model might be removed if we insist that the integrated schema should host all data instances of the input models for database integration, while in view integration we would better retain it.

Key constraints are handled in Merge in the following way. A key is a set of axes of an element. Assume there are two elements with different keys. As the key of one model element might not be unique within the other model element, we can only introduce a uniqueness constraint over the union of the keys by default. If a new key is required, different strategies are possible: using the union as a new key (if it is the smallest key), introducing an artificial key or asking the user for a key. All foreign keys are rechecked after integration of keys, and the key components are updated to match the new keys. Other constraints such as default values for attributes and sequences (ordered attributes/associations) cannot be handled by a reasonable integration strategy. They can only be merged by preferring one input model, or asking the user.

**Conflicts in Merging.** In general, the types of conflicts caused by merging schemas are determined by the nature of the target metamodel. For example, we have several options of names for one model element in the merged model. In the relational model, this leads to a name conflict, while in OWL multiple labels for one model element are allowed and thus there is no conflict at all. As we represent our models in *GeRoMe*, such metamodel conflicts have to be addressed by the ModelGen operator, as this operator translates constructs not supported by a target metamodel. Please note that in general the task of the ModelGen operator is to translate a model from one metamodel to another metamodel (e.g. EER to Relational). The use of a generic metamodel allows us to reuse the functionality of this operator in the context of schema merging.

However, there are still conflicts which might also occur in *GeRoMe*, e.g. multiple roles of the same type for one model element, an attribute with more than one type, or recursion in the types of complex attributes. Resolution of conflicts is ad-hoc, there is usually no universal way to solve all of them [16].

We handle conflicts in a multi-level procedure. Firstly, we take an automatic resolution strategy if possible. The information to resolve the conflicts might be given by the parameters of the merge algorithm (e.g. view vs. database integration), the input models and mapping, or the metamodel. For example, if we have two conflicting roles for one model element, we keep the more general one (e.g. if a key reference is in conflict with an association, we keep the association). Secondly, metamodel heterogeneity conflicts are resolved by taking the most flexible construct. For example, conflicts among foreign keys, complex attributes and associations lead to a representation as an association. Thirdly, explicitly encoded choices (e.g. prefer one input model) are taken for mutually excluded properties such as name of the element or default values.

The final fallback is to ask the user for a resolution. It has to be noted that the problem of schema merging will remain an activity which requires human intervention as schema merging is a *design* activity. Some of the conflicts addressed in [16] are solved by having a complex mapping model as input. However, this input needs to be defined by some user. Our current on-going implementation of the mapping editor and merge algorithm integrates the process of mapping definition and merging in an interactive way, i.e. while defining the mapping, the user will be notified about conflicts during merging the schemas.

**An Application of the Merge Algorithm.** Fig. 4 depicts an example of two models to be merged with our algorithm. The models are the partial *GeRoMe* representations of two models showing courses with the students assisting in these courses. Schema $M_1$ is the representation of an XML Schema. It shows a Course complex type with a nested element hasAssistant of type CSStudent. The XML Schema element is represented in *GeRoMe* as an *Association* with an anonymous *ObjectAssociationEnd* (OE) and a *CompositionEnd* (CE) due to the fact that in an XML document students must be nested into courses. For the same reason, the association end linking to the CSStudent type has a cardinality constraint of $min = max = 1$; whenever a CSStudent occurs in a document it *must* be nested into a Course via this element link. Schema $M_2$ is the representation of an object oriented model, e.g. a UML object model. Again, we have two classes Course and GradStudent with the same relationship in between. The CourseStud element represents an association class which has an attribute hours. Also, the association ends in this relationship are both named and do have more relaxed cardinality constraints as the used metamodel is not constrained to define tree structures. Students' names are represented as an attribute with a composite type instead of two simple attributes and the class adds an attribute program.

The mapping between the models equates the Course types, the associations, their association ends, and the the firstname and lastname attributes (not shown in the figure). The GradStudent and CSStudent are declared overlapping, as each graduate student that studies computer science is also a computer science student.

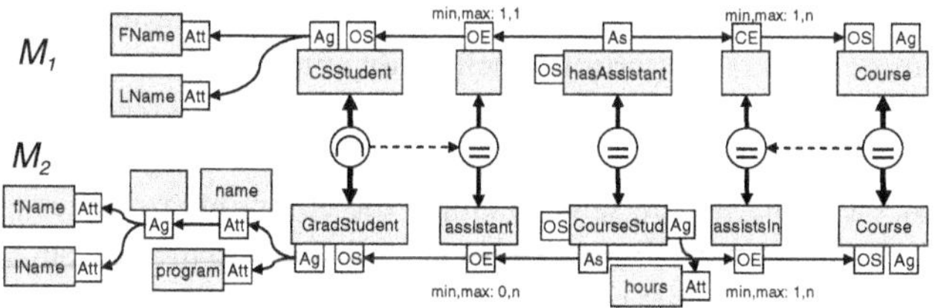


**Fig. 4.** Merging *GeRoMe* representations of an XML Schema and an OO model

The merging procedure first copies all elements that are not mapped to the merged model. This includes the hours, program, and name attributes and the anonymous type. Then, equally related model elements that are not attributes are collapsed. This involves collapsing the Course types, the associations and the association ends. When collapsing the association ends, conflicts between constraints will be resolved by using the least restrictive constraints. That is, $\mathcal{G}$ will contain two association ends with cardinality constraints (0, n) and (1, n) respectively. The corresponding element in $\mathcal{G}$ to assistsIn will play an *ObjectAssociationEnd* as in $M_2$ which is less restrictive than the *CompositionEnd* in the XML Schema. During collapsing these elements, unmapped axes of the elements will be linked to their types in $\mathcal{G}$. The hours attribute will be an attribute of the merged association. The same applies to the program attribute and GradStudent. As CSStudent and GradStudent are overlapping, a most specific supertype of the two types will be introduced. According to step 5 of the merge algorithm the new type will have a composite attribute name that results from merging the original attributes.

It must be emphasized that, although one of the input schemas was an XML Schema, the merge result is not an XML Schema. This is because merging has destroyed the nesting structure of the XML element hasAssistant and the result contains an attribute with composite type, which is not allowed in XML Schema. The only requirement in our approach is that the result is a valid model in the generic metamodel *GeRoMe*.

## 5 Comparison with Other Approaches

Most existing work on schema merging deals with integration of models in one particular metamodel and rarely considers integration across various metamodels. In Rondo, the Merge operator is implemented using morphisms and simple graph representations of models [14]. In [16] a nested mapping model is utilized for merging of simple object oriented models. However, as the generic metamodel is relatively simple, some constraints cannot be described and hence cannot be used by the merging algorithm.

One result of [16] is a list of *generic merge requirements*. Our Merge solution satisfies all but two of these requirements which can be adapted to our mapping representation: *extraneous item prohibition* and *property preservation* demand that no new model elements are added and that a model element in the merged model has a property if and only if one of the corresponding source elements had that property. In [16] there are only two types of mappings allowed, namely similarity and equality. Therefore, the main operations are collapsing of elements declared equal and nesting of similar elements under a helper element that is given in the mapping model. Our mapping model allows more kinds of assertions such as disjointness, overlap, and subset relationships. In the example in the last section a new element *Student* has been added as a common superclass of the original elements *GradStudent* and *CSStudent*, due to the overlap of these elements. In the same example the common properties of these overlapping, (but unequal) elements have also been "pulled up" to the new supertype (representing their union). Because the two requirements demand all elements and properties in the merged model to be given in either of the input models *or* the mapping model, an element such as the superclass and its relationship to the original elements must be defined in the mapping. This amounts to giving the result of merging in the input of merging. In our case it suffices to declare the elements as overlapping to achieve the same result.

Another approach to merging is that of [19] which also presents a comprehensive taxonomy of schema integration conflicts. Like ours, their work is based on the real world sets of model elements. We have used our extended notion of real world sets to define mappings and the merged model formally. Our solution does not only allow a metamodel independent specification of mappings but also the merging algorithm itself is independent of native modeling languages. We provide just one solution for merging models represented in our generic metamodel [10]. The Merge algorithm can therefore be applied polymorphically. Other approaches [4,19] use also generic metamodels, but these are not as detailed as our generic metamodel *GeRoMe*.

While a variety of integration approaches exist in database practices [6,16,19], theoretical aspects of merging are first covered in Buneman et al [4]. The authors introduce the notion of a least upper bound for merging. We extend their work to accommodate more complex mappings using real world semantics and allow configuration options for different scenarios or requirements instead of one single solution.

## 6 Conclusion and Outlook

By giving formal definitions of models, mappings, and merging based on their intensional semantics and relating this to the notion of information capacity we formalized the term "duplicate-free union" that is usually used informally to describe the merge result. We also gave a Merge algorithm that uses accurate intensional mappings. Strategies for solving conflicts in schema merging are highly case based. This problem is aggravated by the number of metamodels. Our merge solution contributes to solving such heterogeneity conflicts [19] as it is based on the rich generic metamodel *GeRoMe* which makes it possible to apply resolutions polymorphically for different metamodels.

In the future we will develop generic conflict resolution strategies for a representative set of structural conflicts, and we will investigate the question how intensional mappings can be derived from extensional mappings (cf. section 2) and vice versa.

**Acknowledgements.** The work is supported by the Research Cluster on Ultra High-Speed Mobile Information and Communcation UMIC (`www.umic.rwth-aachen.de`).

## References


1. Atzeni, P., Cappellari, P., Bernstein, P.A.: A Multilevel Dictionary for Model Management. In: Proc. Conf. Conceptual Modeling(ER 2005), LNCS, vol. 3716, pp. 160–175. Springer, Heidelberg (2005)
2. Atzeni, P., Torlone, R.: Management of Multiple Models in an Extensible Database Design Tool. In: Apers, P.M.G., Bouzeghoub, M., Gardarin, G. (eds.) EDBT 1996. LNCS, vol. 1057, pp. 79–95. Springer, Heidelberg (1996)
3. Bernstein, P.A., Halevy, A.Y., Pottinger, R.: A Vision for Management of Complex Models. SIGMOD Record 29(4), 55–63 (2000)
4. Buneman, P., Davidson, S., Kosky, A.: Theoretical Aspects of Schema Merging. In: Pirotte, A., Delobel, C., Gottlob, G. (eds.) EDBT 1992. LNCS, vol. 580, pp. 152–167. Springer, Heidelberg (1992)
5. Calvanese, D., Giacomo, G.D., Lenzerini, M., Nardi, D., Rosati, R.: Description Logic Framework for Information Integration. Proc. KR, pp. 2–13 (1998)
6. Euzenat, J.: State of the art on ontology alignment. Deliv. D2.2.3, KnowledgeWeb (2004)
7. Fagin, R., Kolaitis, P.G., Popa, L., Tan, W.C.: Composing schema mappings: Second-order dependencies to the rescue. ACM Trans. Database Syst. 30(4), 994–1055 (2005)
8. Hull, R.: Relative Information Capacity of Simple Relational Database Schemata. SIAM Journal of Computing 15(3), 856–886 (August 1986)
9. Kensche, D., Quix, C.: Transformation of Models in(to) a Generic Metamodel. Proc. BTW Workshop on Model and Metadata Management, pp. 4–15 (2007)
10. Kensche, D., Quix, C., Chatti, M.A., Jarke, M.: GeRoMe: A Generic Role Based Metamodel for Model Management. Journal on Data Semantics VIII, 82–117 (2007)
11. Larson, J.A., Navathe, S.B., Elmasri, R.: A Theory of Attribute Equivalence in Databases with Application to Schema Integration. IEEE Trans. Software Eng. 15(4), 449–463 (1989)
12. Lenzerini, M.: Data Integration: A Theoretical Perspective. Proc. PODS, pp. 233–246 (2002)
13. Melnik, S., Bernstein, P.A., Halevy, A.Y., Rahm, E.: Supporting Executable Mappings in Model Management. In: Proc. SIGMOD Conf, pp. 167–178. ACM Press, New York (2005)
14. Melnik, S., Rahm, E., Bernstein, P.A.: Rondo: A Programming Platform for Generic Model Management. In: Proc. SIGMOD, pp. 193–204. ACM, New York (2003)
15. Miller, R.J., Ioannidis, Y.E., Ramakrishnan, R.: The Use of Information Capacity in Schema Integration and Translation. In: Proc. VLDB, pp. 120–133. Morgan Kaufmann, Washington (1993)
16. Pottinger, R., Bernstein, P.A.: Merging Models Based on Given Correspondences. In: Proc. VLDB, pp. 826–873. Morgan Kaufmann, Washington (2003)
17. Rahm, E., Bernstein, P.A.: A Survey of Approaches to Automatic Schema Matching. VLDB Journal 10(4), 334–350 (2001)
18. Sabetzadeh, M., Easterbrook, S.: View merging in the presence of incompleteness and inconsistency. Requirements Engineering 11(3), 174–193 (2006)
19. Spaccapietra, S., Parent, C., Dupont, Y.: Model Independent Assertions for Integration of Heterogeneous Schemas. VLDB Journal 1(1), 81–126 (1992)

# Discovering Web Services to Specify More Complete System Requirements

Konstantinos Zachos, Neil Maiden, Xiaohong Zhu, and Sara Jones

Centre for HCI Design, City University, London
kzachos@soi.city.ac.uk, n.a.m.maiden@city.ac.uk,
x.zhu@soi.city.ac.uk, s.v.jones@city.ac.uk

**Abstract.** Service-centric systems pose new challenges and opportunities for requirements processes and techniques. This paper reports new techniques developed by the EU-funded SeCSE Integrated Project that enable service discovery during early requirements processes and exploit discovered services to enhance requirements specifications. The paper describes the algorithm for discovering services from requirements expressed using structured natural language, and demonstrates it using an automotive example. The paper also reports a first evaluation of the utility of the environment that implements this algorithm when improving the specification of requirements with retrieved services.

## 1 Developing with Web Services

Web and software services are operations that users access via the internet through a well-defined interface independent of where the service is executed [15]. Service-centric systems integrate software services from different providers seamlessly into applications that discover, compose and monitor these services. Developments in service-centric computing have been rapid. Leavitt [3] reports that worldwide spending on web services-based software projects will reach $11 billion by 2008. However, there has been little reported software engineering research to address how to engineer service-centric systems.

One consequence of service-centric systems is that requirements processes might change due to the availability of services. Discovering candidate services can enable analysts to increase the completeness of system requirements based on available service features. However, for this to happen, new tools and techniques are needed to form service queries from incomplete requirements specifications – tools and techniques developed in the EU-funded SeCSE Integrated Project.

SeCSE's mission statement is to create new methods, tools and techniques for requirements analysts, system integrators and service providers that support the cost-effective development and use of dependable services and service-centric applications in the European automotive and telecommunication sectors [12]. The four-year research project covers four main activity areas – service engineering, service discovery, service-centric systems engineering, and service delivery. In this technical research paper we describe new tools and algorithms for discovering services to use to make requirements specifications more complete.

J. Krogstie, A.L. Opdahl, and G. Sindre (Eds.): CAiSE 2007, LNCS 4495, pp. 142–157, 2007.

Sections 2 and 3 describe SeCSE's service-centric requirements process and two research challenges that it generates. Sections 4 and 5 describe our response to these challenges – SeCSE's environment and service discovery algorithm. The algorithm extends information retrieval techniques to overcome these two challenges. Section 6 reports a first evaluation of the usefulness of the environment and algorithm when specifying requirements for service-centric systems. The paper ends with a discussion of future work to extend the environment and its algorithms.

## 2 Discovering Services in SeCSE

In previous SeCSE work we report an iterative and incremental requirements process for service-centric systems [2]. Requirements analysts form queries from a requirements specification to discover services that are related to the requirements in some form. Descriptions of these discovered services are retrieved and explained to stakeholders, then used to refine and complete the requirements specification to enable more accurate service discovery, and so on.

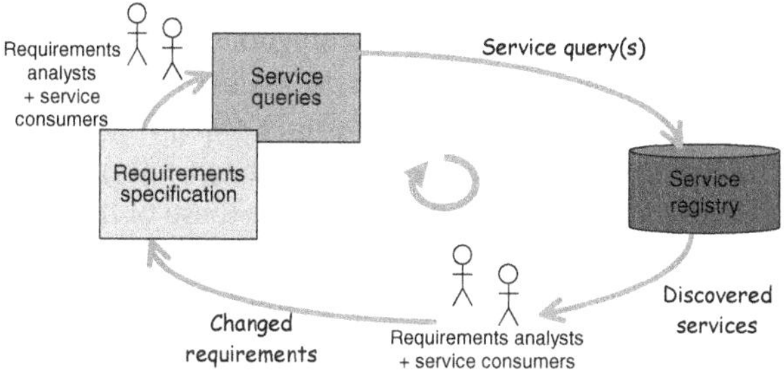


**Fig. 1.** SeCSE's Requirements Process

Relevance feedback, as it is known, has important advantages for the requirements process. Stakeholders such as service consumers will rarely express complete requirements at the correct levels of abstraction and granularity to match to the descriptions of available services. Relevance feedback enables service consumers and analysts to specify new requirements and re-express current ones to increase the likelihood of discovering compliant services. Furthermore, accurate relevance feedback provides information about whether requirements can be satisfied by available services, to guide the analysts to consider alternative build, buy or lease alternatives or trade-off whether requirements can be met by the available services.

The process has 2 important features. Firstly, to ensure its widespread industrial uptake, the process uses established specification techniques based on structured natural language. For example, to specify system behaviour the process uses UML use case specifications. To specify the required properties in a testable form for generating service monitoring policies it uses the VOLERE requirements shell [6]. As such the process extends the Rational Unified Process (RUP) without enforcing its use or mandating unnecessary specification or service querying activities.

Secondly, the process uses services that are discovered from service registries to support requirements processes in different ways [2]. During early requirements processes it uses services to challenge system boundaries and discover new requirements. For example, if no services are found with an initial query, SeCSE provides advice on how to broaden the query to find services that, though not exactly matching the needs of the future system, might provide a useful basis for further specification. During late requirements processes it uses services to decompose and refine requirements and restructure them to enable more effective service monitoring. Service descriptions provide the requirements team with important quality-of-service information, for example about likely system performance and reliability, used to specify measurable fit criteria for requirements [6].

## 3 Two Research Challenges for SeCSE

To deliver the SeCSE requirements process we need two new capabilities that overcome common characteristics of natural language requirement specifications – ambiguity and incompleteness. These capabilities are designed to generate queries that will discover services using requirements that are ambiguous and incomplete. Consider the requirement for a car's route planning system: *the system shall provide the driver with directions to a chosen destination by the most direct route*. It is incomplete because it does not state what the directions are and what direction information is needed. It is also ambiguous because it does not define what is the sense of the "most direct" route. There are several possible meanings of *direct*. Does the analyst mean *direct in spatial dimensions; proceeding without deviation or* interruption; straight and short, or does s/he mean having *no intervening persons or agents*?

To handle incompleteness and ambiguity when discovering services we have designed and implemented the two capabilities listed in Table 1. We extend query expansion techniques previously only applied to WSDL service specifications [16] to incomplete statements of requirement to generate more complete service queries. And we apply term disambiguation techniques from information retrieval [9] to ambiguous statements of requirement to generate unambiguous service queries. The claimed innovation is to import research from related disciplines and extend it to handle problems specific to requirements engineering and service discovery.

**Table 1.** SeCSE's two new querying capabilities

| Requirements | SeCSE querying capabilities |
|---|---|
| Incompleteness | Expansion of service query with terms that have similar meanings |
| Ambiguity | Disambiguation of query terms using pre-defined term senses |

In SeCSE we adopted an engineering paradigm and prototyped a requirements-based service discovery environment, reported in the next section, that implemented these new capabilities as proof of concept and to enable evaluation of their usefulness with our industrial partners.

# 4 SeCSE's Service Discovery Environment

The environment has three main components: (i) UCaRE, a module to document requirements and generate service queries; (ii) EDDiE, the service discovery engine; (iii) the service registries. We describe these 3 components in turn.

## 4.1 SeCSE's Service Registries

The environment discovers services from federated SeCSE service registries that store both the service implementation that applications invoke and one or more facets that specify different aspects of each service. Current service registries such as UDDI are inadequate for discovering services using criteria such as cost, quality of service and exception handling. Therefore SeCSE has defined 6 facets of a service – signature, description, operational semantics, exception, quality-of-service, cost/commerce, and testing [8] – that specify features that are important when discovering services. Each facet is described using an XML data structure. The requirements-based service discovery reported in this paper uses the description and quality-of-service facets. The quality-of-service facet is used to refine selection once services are discovered.

Figure 2 presents an example of part of the description facet for a service called *YNavigation*, which finds the location of a reseller of some commodity. Again, to facilitate industrial take-up, SeCSE assumes that service providers describe each facet of a service using structured natural language, due to the excessive effort needed to document services more formally. For example the *ServiceGoal* attribute describes the purpose of the service as an end-state expressed in structured natural language. The *ShortServiceDescription* atribute describes the service's behaviour using a short paragraph similar to a use case précis, whilst the *LongServiceDescription* attribute describes this behaviour in more detail using structured English similar to a use case normal course. Service providers specify and publish services in SeCSE registries using SeCSE's service specification tool reported at length in [8].

**Owner**: FIAT
**Service goal**: A reseller for a commodity to purchase is found
**Short service description**: This service helps you to find the nearest location where you could talk with our reseller. Calculates the arrival time on the basis of the current car position, road preferences and car features.
**Service rationale**: Car drivers this service to find the commodity.

**Fig. 2.** Example of part of one service with SeCSE's description facet

SeCSE's service registries are implemented using eXist, an Open Source native XML database featuring index-based XQuery processing, automatic indexing. The EDDiE service discovery engine queries these registries using XQuery, a query language designed for processing XML data and data whose structure is similar to XML. Generated queries are transformed into XQueries that are fired at the service description facets of services in the SeCSE service registries.

## 4.2 The UCaRE Requirements Module

Analysts express requirements for new applications using UCaRE, a web-based .NET application. UCaRE supports tight integration of use case and requirements specifications – a requirement expressed using VOLERE can describe a system-wide requirement, a requirement on the behavior specified in one use case, or a requirement on the behavior expressed in one use case action. UCaRE allows analysts to create service queries from use case and requirements specifications.

At the start of the requirements process, analysts work with future service consumers to develop simple use case précis that describe the required behaviour of a new system. Figure 3(a) shows a use case précis expressed in UCaRE, taken from our industrial automotive partners, to specify what a driver might want from an in-car route planner. The précis is repeated in a readable form in Figure 4. Figure 3(b) shows a simple requirement, also from these partners, associated with the précis expressed using the UCaRE VOLERE shell. An analyst can specify functional and qualities requirements such as the two also shown in Figure 4.

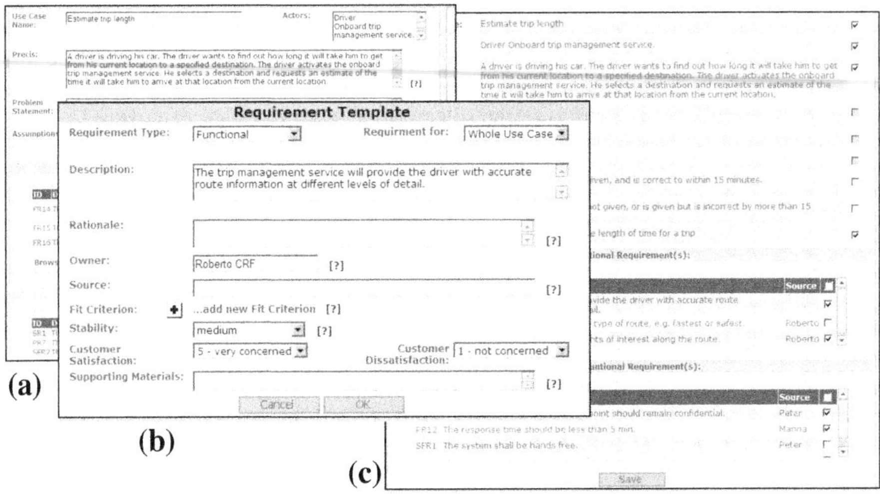


**Fig. 3.** Specification of a use case (a) and requirement (b) in UCaRE, and selection of use case and requirements attributes to generate service queries (c)

| | |
|---|---|
| **Precis:** | A driver is driving his car. The driver wants to find out how long it will take him to get from his current location to a specified destination. The driver activates the onboard trip management service. He selects a destination and requests an estimate of the time it will take him to arrive at that location from the current location. |
| **FR1:** | The trip management service will provide the driver with route information at different levels of detail. |
| **PR1:** | The trip management service will provide the driver with route information within 10 seconds. |

**Fig. 4.** A simple use case précis and requirements for an in-car route planner application, which are used to formulate queries with which to discover services

The analyst then uses the simple tick-box feature shown in Figure 3(c) to select attributes of use cases and requirements to include in a service query. Each service query is formed of one or more elements of a pre-defined type such as a requirement description or rationale, or a use case précis, pre-condition or action. UCaRE maps these element types to service query elements to deliver the seamless integration of service querying with requirements specification that we believe is important for industrial uptake of UCaRE. These integration features are described at length in [18]. The analyst then uses additional UCaRE features described in the next section to refine each generated service query.

An analyst using UCaRE can generate one or more service queries from the specification of a system. Each query is a structured XML file containing structured natural language statements. Because these statements are derived from requirements and use cases, each is potentially ambiguous and incomplete. Each query is then passed to EDDiE, the service discovery engine.

# 5 SeCSE's Service Discovery Algorithm

The main function of the service discovery algorithm is to discover descriptions of candidate services expressed using the service description facet shown in Figure 1 with service queries composed on the structured natural language statements. Non-functional requirement types fulfil important roles during service selection once discovered, but their use is not described further in this paper.

The algorithm implements SeCSE's two new capabilities:

1. Query expansion – the addition of terms in the service query that have the same or similar meaning to existing query terms, to make the query more complete;
2. Term disambiguation – selecting the meaning, or sense of each term in the query to enable query expansion, thus making the query unambiguous.

The algorithm has the 4 key components shown in Figure 5. In the first the service query is divided into sentences, then tokenized and part-of-speech tagged and modified to include each term's morphological root (e.g. *driving* to *drive*, and *drivers* to *driver*). Secondly, the algorithm applies procedures to disambiguate each term by defining its correct sense and tagging it with that sense (e.g. defining a *driver* to be a *vehicle* rather than a *type of golf club*). Thirdly, the algorithm expands each term with

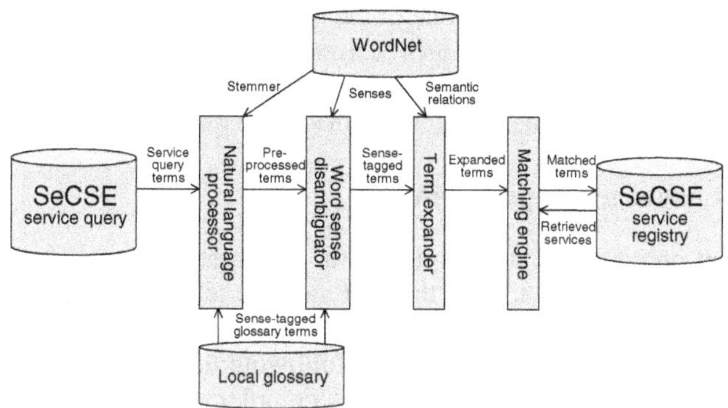


**Fig. 5.** SeCSE service discovery algorithm

other terms that have similar meaning according to the tagged sense, to make it more complete and increase the likelihood of a match with a service description (e.g. the term *driver* is synonymous with the term *motorist* which is also included in the query). In the fourth component the algorithm matches all expanded and sense-tagged query terms to a similar set of terms that describe each candidate service, expressed using the service description facet, in the SeCSE service registry. Query matching is in 2 steps: (i) XQuery text-searching functions to discover an initial set of services descriptions that satisfy global search constraints; (ii) traditional vector-space model information retrieval, enhanced with WordNet, to further refine and assess the quality of the candidate service set. This two-step approach overcomes XQuery's limited text-based search capabilities.

The WordNet on-line lexicon fulfils an important role for three of the algorithm's components. WordNet is a lexical database inspired by current psycholinguistic theories of human lexical memory [5]. It has two important features. Firstly it divides the lexicon into four categories: nouns, verbs, adjectives and adverbs. Word meanings, called senses, for each category are organized into synonym sets (synsets) that represent concepts, and each synset is followed by its definition or gloss that contains a defining phrase, an optional comment and one or more examples. Secondly WordNet is structured using semantic relations between word meanings that link concepts. Relationships between concepts such as hypernym and hyponym relations are represented as semantic pointers between related concepts [5]. A hypernym is a generic term used to designate a whole class of specific instances. For example, *vehicle* denotes all the things that are separately denoted by the words *train*, *chariot*, *dogsled*, *airplane*, and *automobile*, and is therefore a hypernym of each of those words. On the other hand, a hyponym is a specific term used to designate a member of a class, e.g. *chauffeur*, *taxidriver* and *motorist* are all hyponyms of *driver*. A semantic relation between word meanings, such as a hypernymy, links concepts.

WordNet is an essential component of SeCSE's two new capabilities. WordNet's word senses and definitions provide the data with which to disambiguate terms in service queries. WordNet's semantic relations link terms to other terms with similar meanings with which to make service queries more complete.

EDDiE implements the WordNet.Net library, the .Net Framework library for WordNet [10]. The library provides public classes that can be accessed through public interfaces. For example, to look up a word to see if it is in the dictionary, the following code sample achieves this using one of the public classes:

`if(WNDB.is_defined(word,pos).NonEmpty)` where *word* is a string, *pos* is a part-of-speech. The next sections describe in more detail how the algorithm exploits WordNet to discover service descriptions from service queries.

## 5.1 Natural Language Pre-processing

This component prepares the structured natural language service query for sense disambiguation and term expansion. In the first step the text is split into sentences and word tokens. For example, when using white space as the delimiter for splitting the sentence *the engine is misfiring*, we get the following tokens: *the*, *engine*, *is*, *misfiring*. In the second step the algorithm identifies complex nominals (e.g. the term *automotive highway*) based on domain-specific terms defined within a glossary (see 5.2.1) and term definitions in WordNet. In the third step the algorithm identifies and removes all terms defined in a list of stop words (e.g. prepositions and pronouns). Next, all terms are tagged with their corresponding part-of-speech (e.g. singular common noun, comparative adjective, etc.) and classified accordingly using an improved version of the Brill Tagger [1]. In the fifth step each term is converted to its morphological root (e.g. *driving* to *drive*). Finally, all duplicate occurrences of a term are removed so that each term is stored only once with its cardinality, as reported in [16].

Returning to our example of the service query generated for requirements shown in Figure 3, the algorithm produces the first version of the XML service query, showing only the noun terms (e.g. *driver, car, location, destination, service*, etc) processed from the use case précis element in Figure 4. The next section describes how EDDiE determines the correct sense of each term.

### 5.2 Word Sense Disambiguation

Assigning the correct sense to a word in context requires syntactic, semantic and pragmatic knowledge about the word itself, its part of speech, and its context [13]. The pre-processing described in section 5.1 adds syntactic and semantic information to query terms through part-of-speech tagging and WordNet. With this component the algorithm completes disambiguation by iteratively using context knowledge from the project glossary, requirements analyst and other terms in the service query through 7 procedures. Word sense ambiguity is problematic in information retrieval with small queries [14]. However requirements-based service discovery uses larger queries that offer more terms with which to disambiguate. Each procedure is increasingly costly to apply. The first, the cheapest, exploits prior analysis work that the analyst normally undertakes with UCaRE. The next 5 are applied automatically. The seventh, the most expensive, demands analyst input. The 7 procedures are: (i) defining the glossary; (ii) defining single term senses; (iii) defining synonyms; (iv) defining hypernyms; (v) frequency-based senses; (vi) context-based senses, and; (vii) user selection.

For each term T the algorithm determines its sense S using one of 7 procedures that are applied in order. If Procedure *i* does not provide any positive result, then Procedure *i+1* will be applied. In a generic iteration of the algorithm the input is a list of pre-processing terms $T = [t_i,\ldots,t_n]$, and a list of associated senses $S = [St_i,\ldots,St_n]$. $T$ represents all terms to be disambiguated and $S$ represents the semantic meaning of $T$, where $St_i$ is either the chosen sense for $t_i$ or the empty set, i.e. the term is not yet disambiguated. A set of ambiguous terms $A = \{t_i | St_i = \varnothing\}$ is also maintained. $T$ is initialized with the empty set $T = \{\}$ and $A$ with the list formed by all terms parsed from the Natural Language Processor. The output is the updated list $S$ of senses associated with the input terms $T$.

The disambiguation procedures are described below.

**Procedure 1: Defining the Glossary.** This procedure minimizes ambiguity in the original service query. During SeCSE's requirements process UCaRE maintains a project glossary. Terms in the requirement and use case specification are defined in an interactive glossary that accesses WordNet directly to offer one or more pre-defined senses that the analyst selects and assigns to the term. Figure 6 shows a UCaRE screenshot in which the analyst selects the correct sense #1 for *driver* and tags it in the project glossary, i.e. *a vehicle carrying many passengers; used for public transport.*

Hence, the term *driver* is stored in the glossary with the sense #1. If the term *driver* appears in the list *T*, this procedure finds the term in the glossary and marks it as having sense #1 in *S*.

**Procedure 2: Defining Single-Sense Terms.** This procedure exploits the existence of terms with only one sense in WordNet, called monosenous terms, and tags them automatically with that sense. For example, the compound noun *motor vehicle* has one sense defined in WordNet and is tagged with that sense #1.

**Procedures 3&4: Defining Synonyms and Hypernyms.** Procedure 3 finds query terms that are semantically connected to already-disambiguated terms (i.e. terms with a tagged sense) and for which the connection distance is 0 as computed using WordNet hierarchies. A semantic distance of 0 between two terms defines that both belong to the same synset, and therefore the new term is tagged with the same meaning as the connected term. For example consider the terms *passenger* and *rider* in *T*. The noun *passenger* is a monosemous word disambiguated with Procedure 2. One of the senses of the noun *rider*, sense #4 (*a traveler riding in a vehicle (a boat or bus or car or plane or train etc) who is not operating it*), appears in the same synset with *passenger* #1, so the procedure tags *rider* with sense #4.

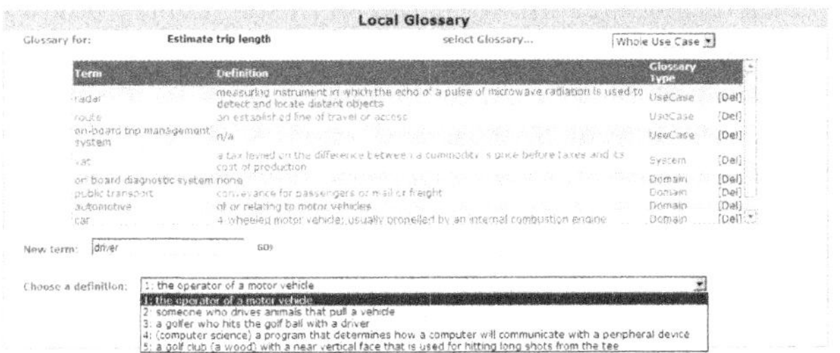


**Fig. 6.** Sense definition during requirements specification with UCaRE

Procedure 4, defining hypernyms, works in a similar way. It finds query terms that are semantically connected to already disambiguated terms but for which the connection distance is the maximum 1 as computed using WordNet hierarchies. A semantic distance of 1 between two words indicates that both belong to the same hypernymy/hyponymy relation and therefore the new term is tagged with the same meaning as the connected term.

**Procedure 5: Frequency-Based Senses.** This procedure assigns the most frequent sense to a term irrespective of its context [17]. This heuristic has been used to baseline supervised word sense disambiguation systems [4]. Its high performance is due to the skewed frequency distribution of word senses. WordNet has a 200,000-word sample of hand-tagged senses through the SemCor project [5]. However, infrequent words can lead to sense bias. Therefore our solution is to constrain the use of this procedure to terms that achieve both a threshold on the frequency of the predominant sense, and a threshold on the ratio between the first sense and the next.

If both are satisfied, the term is tagged with sense #1 from WordNet. Consider the noun *location* appearing in *T*. The term has 3 senses and all senses have appeared in the semantically tagged corpora. The first sense has 992 semantic tags, the second and third senses have both 2 tags. As this scenario satisfies the condition described earlier, sense #1 (*a point or extent in space*) is selected for *location.*

**Procedure 6: Context-based Senses.** The SemCor bigrams method forms two pairs, one with the previous word, the other with the next word, and searches for these pairs in SemCor corpus [5]. If in all of the occurrences of these pairs, the given word has the same sense, and the number of occurrences is bigger than a preferred threshold, then we assign that sense to the word. For example the term *approval* in *T* has the query context *committee approval of.* The pairs formed are *committee approval* and *approval of.* There are no occurrences of the first pair but four occurrences of the second, more than the set threshold 3, and in all these occurrences the sense of approval is sense #1 (*the formal act of approving*), hence this sense is tagged.

**Procedure 7: User Selection.** In this most costly procedure the analyst selects the sense for any term that could not be disambiguated using the 6 previous procedures. All pre-defined senses (represented through the gloss) for the term are presented to the analyst to select and assign. Reliance on the other 6 procedures means that user selection should only be needed for a small number of terms. Consider the ambiguous noun *direction* in *T*. The analyst can select the correct sense #1 for *direction* and tags it in the definition window, i.e. *a line leading to a place or point.*

If a term $t_i$ could not be disambiguated through any procedure, then $St_i$ becomes 0, i.e. the term is still ambiguous. A term $t_j$ which is not included in WordNet and defined in the project glossary, $St_j$ becomes -1.

Now let us return to our automotive service query after all 7 procedures have been applied. The partial XML service query in Figure 7 shows the noun terms processed from the use case précis. For example, the term *car* is tagged with sense #1, i.e. *4-wheeled motor vehicle; usually propelled by an internal combustion engine*, whilst the term destination is tagged with sense #3 i.e. *written directions for finding some location; written on letters or packages that are to be delivered to that location.*


```
<SingleTerm>
  <Term termID="1" occur="1" pos="NN" wnsn="-1">onboard trip management </Term>
  <Term termID="2" occur="1" pos="NN" wnsn="1">car</Term>
  <Term termID="3" occur="3" pos="NN" wnsn="1">driver</Term>
  <Term termID="4" occur="1" pos="NN" wnsn="1">estimate</Term>
  <Term termID="5" occur="1" pos="NNS" wnsn="1">request</Term>
  <Term termID="6" occur="2" pos="NN" wnsn="3">destination </Term>
  <Term termID="8" occur="3" pos="NN" wnsn="-1">location</Term>
  <Term termID="11" occur="1" pos="NN" wnsn="2">time</Term>
</SingleTerm>
```


**Fig. 7.** An extract of the XML service query after word sense disambiguation, showing the sense number (wnsn) of different nouns in use case précis elements

## 5.3 Query Expansion

Word mismatches are a fundamental problem to overcome in service discovery. Simply stated, it means that service consumers and providers use different words to express their requirements and service descriptions [11]. The severity of the problem decreases as queries get longer and the likelihood of words co-occurring in the query and service descriptions increases. Through query expansion in EDDiE, the query is expanded using words or phrases with similar meaning to those in the query so that the chance of matching words in relevant service descriptions is increased. Query expansion techniques from information retrieval are essential for effective requirements-based service queries. More formal ontologies for most requirements domains are not available, so synonym-based query expansion using ontological information in WordNet is one of the few viable options.

Elsewhere Wang & Stroulia [16] report a web service discovery technique that combines WordNet with matching on the structure of the WSDL service specification to expand queries with semantically similar words. However, it is limited to formal representations in WSDL. Our innovation is to expand service queries to handle requirements expressed in natural language and compatible with established processes.

EDDiE uses ontological information from WordNet to extract semantically related terms for query terms. As such prior disambiguation is essential to ensure that term expansion uses the correct sense, otherwise queries are expanded incorrectly. Hence only disambiguated terms are considered. EDDiE uses 3 expansion methods:

- *Synset expansion*: terms are replaced by their synsets, for example the term *car* is replaced with its synset for sense #2 [*car, auto, automobile, machine, motorcar*].
- *Hypernym expansion*: terms are augmented by their WordNet direct hypernyms, for example the hypernym of *car* is *motor vehicle*.
- *Gloss words expansion*: terms are augmented with the terms in their glosses, for example the sense #1 definition of the term *garage* is a *4-wheeled motor vehicle; usually propelled by an internal combustion engine*. Hence *motor vehicle* and *engine* are extracted.

Continuing with our example, Figure 8 shows an extract of the XML service query after query expansion. Two terms – *car* and *estimate* – have been expanded with synonyms, hypernyms and gloss terms. For example the query now also contains terms that include *auto*, *automobile*, *motorcar* and *vehicle* as well as *car*, and *approximation*, thus increasing the likelihood of discovering relevant service descriptions. All other terms in Figure 5 are expanded in the same manner, creating a larger query composed of more terms with similar meanings.

```
<SingleTerm>
  <Term termID="2" occur ="1" pos="NN" car</Term>
  <Term termID="4" occur ="1" pos="NN" wnsn="1">estimate</Term>
  ...
  <Term termID="13" pos="NN" refTerm="2" expType="synonym">auto</Term>
  <Term termID="14" pos="NN" refTerm="2" expType="synonym">automobile</Term>
  <Term termID="16" pos="NN" refTerm="2" expType="synonym">motorcar</Term>
  <Term termID="15" pos="NN" refTerm="4" expType="synonym">estimation</Term>
  <Term termID="15" pos="NN" refTerm="4" expType="synonym">approximation</Term>
  ...
  <Term termID="31" pos="NN" refTerm="2" expType="hypernym"> motor vehicle</Term>
   <Term termID="32" pos="NNS" refTerm="4" expType="hypernym">calculation</Term>
  ...
  <Term termID="32" pos="NNS" refTerm="2" expType="gloss">vehicle</Term>
</SingleTerm>
```

**Fig. 8.** The same extract of XML service query after query expansion, showing synonyms, hypernyms and gloss terms for original terms *car* and *estimate*

### 5.4 Query Matching

The expanded query is transformed into one or more XQueries that are fired at the service description facets of services in SeCSE service registries. Once an initial set of service descriptions has been retrieved using XQueries, a traditional vector-space [7] model information-retrieval step, enhanced with WordNet, is applied to refine and extract the most similar services from the set. As reported earlier, SeCSE's service description facet in Figure 2 is structured using attributes that facilitate matching with the elements of the service query. For example, expanded terms describing use case actors in the query are matched to terms that describe service consumers, expanded terms from the use case précis are matched to terms in the short service description, and expanded terms in normal course actions of the use case specification are matched to terms describing atomic service operations.

In the traditional vector-space model, documents and queries are represented as T-dimensional vectors, where T is the total number of distinct words in the document collection after pre-processing. Each term in the vector is assigned a weight that reflects the importance of a word in the document. This value is proportional to the frequency a word appears in a document and inversely proportional to number of documents in which this word appears [7]. The WordNet vector-space model involves the maintenance of vectors for each service description property and corresponding expanded query elements, atomic and compound terms, where compound terms consist of multiple atomic terms, for example *nearest location*. We employ the WordNet vector space model to retrieve services that are most similar to the input description on the respective vectors. Corresponding vectors from service description properties and expanded query elements are matched to provide similarity scores. Matching scores of original terms are assigned twice the weight as matching scores of expanded terms (synonyms, hypernyms, gloss terms). A higher overall score indicates a closer similarity between the source and target specifications.

Figure 9 describes part of XML service match between part of the example expanded query and an extract of the XML service description facet for a service that calculates the arrival time on the basis of the current car position shown in Figure 8. It shows that expanded terms – for example *calculation* (a hypernym of *estimate*) and *journey* (a gloss term of *trip*) are needed to match the service because these terms were missing from the original query. Without such expansion of disambiguated terms, retrieval of this service would not be possible. The match values shown in Figure 7 represent the computed semantic distance between the terms *calculation* and *journey*. These match values are used to compute an overall score for the match between the service query and description. Use of these overall scores is demonstrated in the next section.

```
<SingleTerm>
<QueryTerm QId="11" QueryTerm="trip">
   <MatchTerm MId="12" MatchValue="0.543" Expansion Type="gloss">journey</MatchTerm>
  </QueryTerm>
  <QueryTerm QId="16" QueryTerm="estimate">
  <MatchTerm MId="17" MatchValue="0.368" ExpansionType="hypernym">calculation</MatchTerm>
</QueryTerm> ...
</SingleTerm>...
```

**Fig. 9.** An extract of the XML service match for the service *YNavigation* with the expanded service query

## 6 Evaluating UCaRE and EDDiE

We evaluated UCaRE and EDDiE with SeCSE's industrial partners. Rather than investigate traditional measures of precision and recall properties of the EDDiE algorithm itself we investigated the utility of algorithm in a requirements workshop. More specifically we explored whether discovered services were sufficient to trigger specification of requirements that had not been specified prior to service discovery. Our assumption underlying this strategy was that high levels of precision and recall were not essential for service discovery – service specifications with lower similarity scores might still enable analysts to discover new requirements.

Four experienced practitioners from SeCSE industrial partners – 2 from Fiat, one from DaimlerChrysler and one from CA, discovered and documented requirements for the in-car route planner system reported throughout the paper. Three of them had extensive experience with such automotive applications. Two of the authors ran the workshop – one facilitated requirements discovery whilst the scribe operated UCaRE.

The workshop was in 2 stages. In the first the facilitator walked the practitioners through the use case précis then normal course to discover requirements for the in-car route planner application that the scribe documented in UCaRE. This process continued until the practitioners were unable to discover more requirements. The scribe then generated a service query from the use case précis and searched a SeCSE-enabled registry containing 112 services for applications that included weather reporting and flight booking as well route planning taken from existing public UDDI registries. The query expanded all term types with possible synonyms, hypernyms and gloss terms.

In the second stage UCaRE displayed each discovered service description as shown on the left-hand side of Figure 10. The practitioners selected which service descriptions to retain as pertinent to the route planner application. The facilitator then walked the practitioners through each service to discover additional route planner requirements from the retained services that the scribe also documented in UCaRE. The facilitator took care to avoid bias and use the same prompts and guidelines before and after service discovery. After the workshop each practitioner completed a questionnaire that ranked each requirement documented during the workshop for its importance and novelty on a simple scale of 1 to 3.

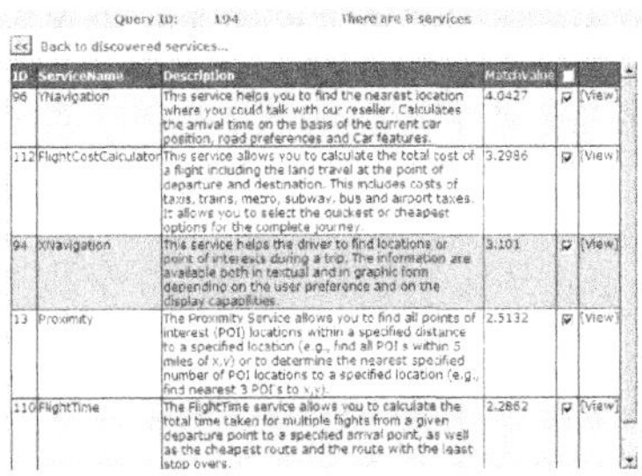

Query ID: 194 There are 8 services

<< Back to discovered services...

| ID | ServiceName | Description | MatchValue | | |
|---|---|---|---|---|---|
| 96 | YNavigation | This service helps you to find the nearest location where you could talk with our reseller. Calculates the arrival time on the basis of the current car position, road preferences and Car features. | 4.0427 | ☑ | [View] |
| 112 | FlightCostCalculator | This service allows you to calculate the total cost of a flight including the land travel at the point of departure and destination. This includes costs of taxis, trains, metro, subway, bus and airport taxes. It allows you to select the quickest or cheapest options for the complete journey. | 3.2986 | ☑ | [View] |
| 94 | XNavigation | This service helps the driver to find locations or point of interests during a trip. The information are available both in textual and in graphic form depending on the user preference and on the display capabilities. | 3.101 | ☑ | [View] |
| 13 | Proximity | The Proximity Service allows you to find all points of interest (POI) locations within a specified distance to a specified location (e.g., find all POI s within 5 miles of x,y) or to determine the nearest specified number of POI locations to a specified location (e.g., find nearest 3 POI's to x,y). | 2.5132 | ☑ | [View] |
| 110 | FlightTime | The FlightTime service allows you to calculate the total time taken for multiple flights from a given departure point to a specified arrival point, as well as the cheapest route and the route with the least stop overs. | 2.2862 | ☑ | [View] |

| | Practi - tioner | R | M | S | P |
|---|---|---|---|---|---|
| Importance rating | Stage1 | 2.7 | 2.5 | 2.0 | 2.6 |
| | Stage2 | 2.3 | 2.1 | 1.9 | 2.2 |
| Novelty rating | Stage1 | 1.3 | 2.0 | 1.6 | 2.4 |
| | Stage2 | 1.7 | 2.2 | 2.3 | 2.9 |

**Fig. 10.** Discovered service descriptions shown in the SeCSE environment (on left), and average importance and novelty ratings for requirements discovered before and after service retrieval (on right)

The workshop took place as planned. The first stage lasted 60 minutes, during which the practitioners discovered 27 requirements that were documented in UCaRE. The service query then retrieved 11 services, 8 of which the practitioners retained as relevant. The second stage lasted 50 minutes and led to a further 20 requirements from services that were ranked with high and low similarity scores.

At the end of the workshop each practitioner completed the questionnaire (the ordering of the requirements was pseudo-randomized), and their average ratings of the importance and novelty of the 47 requirements are shown on the right-hand side of Figure 8. Results revealed a clear pattern – each practitioner ranked requirements specified prior to service discovery as more important than requirements documented after service discovery, but requirements specified from discovered services were more novel, suggesting that the retrieved services complemented walkthroughs by discovering requirements unlikely to be considered during the requirements process.

To explore how requirements were generated from retrieved services, we undertook a post-workshop analysis to reveal the generation patterns reported in Table 2.

**Table 2.** Identified patterns of requirements discovery from services

| Service-requirement pattern | Number occurrences |
|---|---|
| Requirements expressed new system features that were a consequence of an application that implemented the retrieved service | 7 |
| Requirements were expression of a refinement of the features of the discovered service applied to the system under analysis | 4 |
| Requirements expressed required inputs to an application that invoked the discovered service | 2 |
| Requirements expressed a function that has the potential to satisfy service qualities described in the service description | 1 |
| Requirements were associated with the preceding requirement generated from a service description | 3 |
| Requirements and services shared concepts that were deeper than the above input, output and consequence relations | 2 |
| Requirements and services had no discernible similarities | 1 |

Several patterns, such as the first, suggest that the practitioners would implement a service-centric application. Other patterns represent good practice that we will validate and reinforce in SeCSE's requirements process.

To conclude the workshop, although limited by the coverage of services in the registry, revealed that EDDiE is capable of discovering services deemed relevant from first-cut requirements and a use case specification specified in a one-hour workshop. Furthermore 8 of the discovered services with different similarity scores enabled the experienced practitioners to specify new requirements that they deemed more novel than the earlier requirements, thus providing evidence that EDDiE and UCaRE can deliver SeCSE's iterative and incremental requirements process.

## 7 Discussion and Future Work

This paper reports a research-based software environment for constructing service queries from natural language requirements specifications, disambiguating query terms using 7 procedures then expanding them with defined senses, and retrieving discovered services from service registries. An evaluation of the environment revealed that experienced practitioners used retrieved services with a range of similarity scores to generate new requirements that were later ranked as more novel than requirements discovered using traditional use case walkthrough techniques. This positive outcome supports our fundamental claim – that candidate services can enable analysts to increase the completeness of requirements. It suggests the potential to use services that the final application might not invoke to inform later architecture design, service composition and implementation tasks. We encourage researchers to think more innovatively about how to use web services in information systems engineering.

## Acknowledgements

SeCSE is funded by the EU 511680 Integrated Project.

## References

[1] Brill, E.: A simple rule-based part of speech tagger. In: Proc. Third Conference on Applied Natural Language Processing, ACL, Trento, Italy (1992)

[2] Jones, S.V., Maiden, N.A.M., Zachos, K., Zhu, X.: How Service-Centric Systems Change the Requirements Process. In: Proceedings REFSQ'2005 Workshop, CaiSE'2005, pp.13–14, Porto (2005)

[3] Leavitt, N.: Are Web Services Finally Ready to Deliver? IEEE Computer 37(11), 14–18 (2004)

[4] McCarthy, D., Koeling, R., Weeds, J., Carroll, J.: Using Automatically Acquired Predominant Senses for Word Sense Disambiguation. In: Proceedings of the ACL 2004 Senseval-3 Workshop Barcelona, Spain (2004)

[5] Miller, K.: Introduction to WordNet: an On-line Lexical Database Distributed with WordNet software (1993)

[6] Robertson, S., Robertson, J.: Mastering the Requirements Process. Addison-Wesley-Longman, Redwood City (1999)

[7] Salton, G., Wong, A., Yang, C.S.: A vector-space model for information retrieval. In: Journal of the American Society for Information Science. vol.18, pp. 13–620. ACM Press, New York (1975)

[8] Sawyer, P., Hutchinson, J., Walkerdine, J., Sommerville, I.: Faceted Service Specification. In: Proceedings SOCCER (Service-Oriented Computing: Consequences for Engineering Requirements) Workshop, at RE'05 Conference, Paris (August 2005)

[9] Schutze, H., Pedersen, J.: Information retrieval based on word senses. Proceedings of the Symposium on Document Analysis and Information Retrieval 4, 161–175 (1995)

[10] Simpson, T.: (2005) opensource.ebswift.com/WordNet.Net

[11] Singhal, A., Pereira, F.: Document expansion for speech retrieval. In: Proceedings of ACM SIGIR, pp. 34–41, Berkeley, CA, USA (1999)

[12] SeCSE 2005, secse.eng.it

[13] Stevenson, M., Wilks, Y.: The Interaction of Knowledge Sources in Word Sense Disambiguation. Computational Linguistics 27(3), 321–349 (2001)

[14] Stokoe, C.M, Oakes, M.J, Tait, J.I: Word Sense Disambiguation in Information Retrieval Revisited. In: Proceedings of the 26th ACM SIGIR. pp. 159 – 166 Toronto, Canada (2003)

[15] Tetlow, P., Pan, J., Oberle, D., Wallace, E., Uschold, M., Kendall, E.: Ontology Driven Architectures and Potential Uses of the Semantic Web in Software Engineering, W3C (2005)

[16] Wang, Y., Stroulia, E.: Semantic Structure Matching for Assessing Web-Service Similarity, First International Conference on Service Oriented Computing, Trento, Italy (December 15-18 (2003)

[17] Wilks, Y., Stevenson, M.: The grammar of sense: Is word-sense tagging much more than part-of-speech tagging? Sheffield Department of Computer Science, Research Memoranda (1996)

[18] Zachos, K., Zhu, X., Maiden, N., Jones, S.: Seamlessly Integrating Service Discovery into UML Requirements Processes. In: Proceedings 2006 International Workshop on Service-Oriented Software Engineering (SoSE'2006), Shanghai, China, ACM Press, New York (2006)

[19] Zhu, H., Maiden, N.A.M., Jones, S.V., Zachos, K.: Applying Patterns in Service Discovery.In: Proceedings SOCCER (Service-Oriented Computing: Consequences for Engineering Requirements) Workshop, at RE'05 Conference, Paris (August 2005)

# On ISOA: Intentional Services Oriented Architecture

Colette Rolland[1], Rim Samia Kaabi[1], and Naoufel Kraiem[2]

[1] Université Paris1 Panthéon Sorbonne, 90, rue de Tolbiac, 75013 Paris, France
[2] Ecole Nationale des Sciences de l'Informatique, 2035 Manouba, Tunis, Tunisia
rolland@univ-paris1.fr, rim-samia.kaabi@malix.univ-paris1.fr, Naoufel.Kraiem@ensi.rnu.tn

**Abstract.** Despite the growing acceptance of SOA, service-oriented computing remains a computing mechanism to speed-up the design of software applications by assembling ready-made services. We argue that it is difficult for business people to fully benefit of the SOA if it remains at the software level. The paper proposes a move towards a description of services in business terms, i.e. intentions and strategies to achieve them and to organize their publication, search and composition on the basis of these descriptions. In this way, it leverages on the SOA to an intentional level, the ISOA. We present *ISM*, the model to describe intentional services, and to populate the service registry. We highlight its intention driven perspective for service description, retrieval and composition. Thereafter, we propose a methodology to determine intentional services that meet business goals. Finally, we introduce agent architecture to support model driven execution of intentional services.

**Keywords:** Service-oriented computing, service-oriented architecture, intentional service-oriented architecture, intentional service modelling, intention-driven service composition.

## 1 Introduction

Service-Oriented Computing (SOC) is the computing paradigm that utilizes services as fundamental elements for developing software applications [1][2]. SOC relies on the Service-Oriented Architecture (SOA) [3] that is a way of reorganizing a portfolio of previously developed applications into services that are self-describing, platform agnostic computational elements performing functions, accessible through standard interfaces and that can be assembled in complex compositions based on standard messaging protocols. As shown in Fig.1, the basic SOA defines an interaction between three kinds of software agents [4], namely, the *service provider*, the *service client* and the *service registry* involving the *publish, find* and *bind* operations. Services are offered by *service providers* that procure the service implementations and supply their descriptions to a service registry. The service registry *publishes* services by making their descriptions. The service client uses the *find* operation to retrieve the service description matching his functional needs and uses it to *bind* with the service provider and invoke the service.

J. Krogstie, A.L. Opdahl, and G. Sindre (Eds.): CAiSE 2007, LNCS 4495, pp. 158–172, 2007.

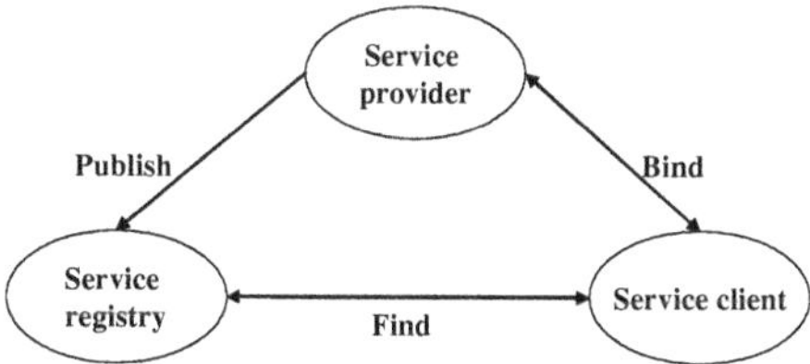


**Fig. 1.** Function-driven SOA

SOA is a way of designing a software system that is *function-driven*. Services perform functions implemented in software, wrapped with formal documented interfaces which provide the mechanism by which services can communicate with one another in compositions to perform higher level functions. The service interface (that provides the signatures of the available operations) is central to the SOA view as it is the only thing, which is exposed to the client to invoke the function.

However, it shall be noticed that interface descriptions are low level, technical statements (cf. WSDL statements [5]) that are understandable by software professionals but far to be comprehensible by business people. At the same time, the notion of a service is familiar to the management world [6] and with the growing acceptance and popularity of SOA, computing systems now aim to extend far beyond the firewall to automate enterprise-wide business processes, covering sales, supply chain, manufacturing, delivery, payment, human resources, and more. To attain this, it is necessary to adapt SOA to a mainstream practitioners' level and bridge the gap between high level business services and low level software services [7], [8].

The position adopted in this paper is to suggest a move from the *function-driven* SOA to *intention-driven* SOA. Whereas the former lies on a functional view of services, the latter proposes to spell out the purpose, the intention behind a service. As a consequence, interfaces of these services will bring out the business goal that the service allows to fulfil instead of defining the signatures of basic operations that can be invoked on class objects. This will avoid the current mismatch of languages between low level services expressions such as WSDL statements and business perceived services. We refer to these services as *intentional services* and present in this paper *ISM*, a model for intentional service modelling.

While complying with the SOA model, our model, the intentional SOA, ISOA is a proposal for leveraging the 3-SOA tuple <Publish, Find, Bind> to an intentional level matching the business mainstream needs. In adapting the roles and operations of the SOA model, the ISOA (Fig. 2) introduces two main departures:

(i) in the interaction, business agents replace software agents,
(ii) intentional service descriptions replace functional software service descriptions.

The ISOA implies that business centric organizations offering e-business services shall describe their services in an intentional manner, and publish them to an e-business service registry that makes these descriptions available in an intentional service registry. Business agents who are searching for services use an intention matching mechanism to retrieve service descriptions fitting their needs and use them to bind to the e-business provider.

**Fig. 2.** Intention-driven SOA

In this paper, we use the three roles mentioned in the ISOA architecture to structure the discussion on ISOA. For the registry, we introduce the notion of *intentional service*, highlight its relationship with software services and present *ISM*, a model for intentional service modelling. We show that an intentional service description shall include variability, i.e. propose alternative variations of a given component service. The model gives to us the capability to populate the intentional service registry. This is the subject of section 2.

It is for the e-business provider to define the services that are to be provided in the business. We propose to represent business intentions in a graphical representation called *Map*. This Map takes the form of an intention/strategy graph with intentions as nodes and strategies to achieve them as edges. The e-business provider derives the services that can be published from a map following a set of guidelines. This role of the provider is considered in section 3.

Finally, in carrying out his role, the business agent must be provided with the appropriate execution architecture particularly to handle variability. In section 4, we outline an agent architecture for service execution.

## 2 Populating the Registry with Intentional Services

In this section we consider the Intentional Service Registry. The aim is to develop a model that defines the contents of the Registry. Towards this end, we clarify the notion of an intentional service and present the Intentional Service Model, *ISM*, to model different types of intentional services.

### 2.1 Intentional Service Model

An *intentional service* is a service captured at the business level, in business comprehensible terms and described in an intentional perspective, i.e. focusing on the intention it allows to achieve rather than on the functionality it performs. Fig. 3 presents *ISM* using UML notations. As shown by the colors used in the Figure, there are three different aspects in the description of an intentional service, namely the service interface, the service behavior and the service composition. We describe the three in turn.

**First,** central to the Figure is the fact that a service permits the fulfillment of an *intention,* given an *initial situation* and terminating in a *final situation.* These three elements constitute the interface of an intentional service; the intention replaces the

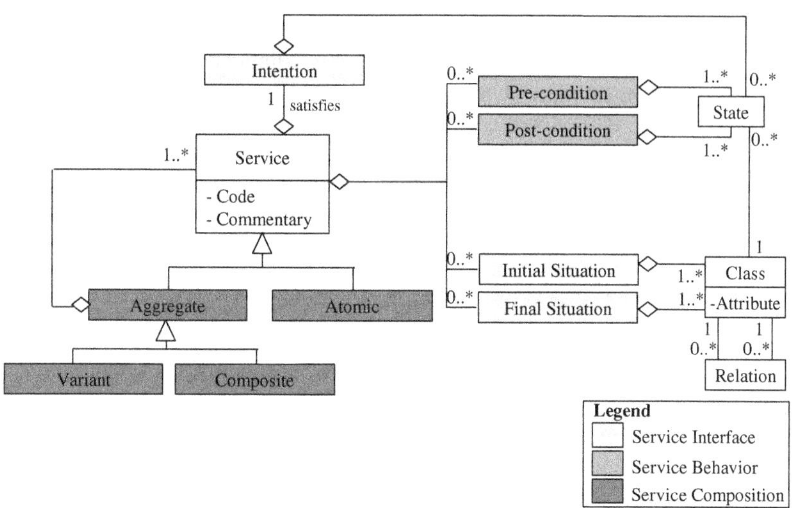


**Fig. 3.** Intentional Service Model

operations that are part of a typical software interface whereas the initial and final situations are the input and output parameters structured as business object classes.

We view an intention in the same sense as a goal. A goal is 'an optative' statement [9], that expresses what is wanted i.e. a state that is expected to be reached or maintained. Thus, *Make Room Booking* is the intention to make a reservation for rooms in a hotel. The achievement of this intention leaves the system in the state, *Booking made.* If *Accept Payment* is the intention of a service then the initial situation refers to the *booking* and *customer* classes whereas the final situation comprises the *payment* class in addition.

**Second**, the behavior of the service is specified through its *pre* and *post conditions* that are the initial and final sets of states characterizing the initial and the final situation respectively. In the *Accept Payment* service example, <booking.state ='OK' ∧ customer.status='registered'> and <booking.state='paid' ∧ payment.status = 'done' > are the pre and post-conditions respectively.

**Finally,** services are classified as *aggregate* or *atomic*. The former are composed of other services whereas the latter are not. *Atomic services* have intentions that are fulfilled by SOA level functional services. In contrast *aggregate services* have high-level intentions that need to be decomposed in lower level ISOA services till atomic intentional services are found. Therefore, it can be understood that aggregate intentional services lie on an intention-driven composition that is necessary to bridge the gap between the actual functionality (captured in the atomic service) and the high level perception of business executives for a service fulfilling their strategic/tactical intentions.

Fig. 3 shows that aggregate services are further refined. Aggregation of services can involve *variants*, i.e. services which are alternative to the others or result from simple composition, leading to *composite services.*

*Composite services* reflect the precedence/succession relationship between their intentions. For example, in the room booking case, *Make Room Booking* must precede

*Accept Payment.* The composition of this two services leads to the satisfaction of the intention *Make Confirmed Booking*. This form of composition is grounded on the AND goal decomposition as used in goal modelling [10].

The composition is denoted "•" when there is a sequential order between component services and "//" when they can run in parallel. Every service in a composition can be executed repeatedly, this is denoted by the "*" symbol. Thus, the composite service to fulfil the *Make Confirmed Booking* intention is defined as follows:

$$S_{\mathit{Make\ Confirmed\ Booking}} = \bullet(S_{\mathit{Make\ Room\ Booking}}, S_{\mathit{Accept\ Payment}})$$

Introduction of *variability* in intentional service modelling is justified by the need to introduce flexibility in intention achievement and adaptability in intentional service execution. There are three types of variants in *ISM*, namely *alternative, choice* and *multi-path*.

An *alternative* variation corresponds to an XOR relationship between the service intentions involved. For example, assume that *Accept Payment* can be achieved in exclusively one of the following ways, *By electronic transfer* or *By credit card* or *By cash.* This leads to define the service $S_{\mathit{Accept\ Payment}}$ as a variant aggregate with three alternative components. We use the symbol "⊗" to denote alternative and therefore:

$$S_{\mathit{Accept\ Payment}} = \otimes(S_{\mathit{Accept\ Payment\ by\ electronic\ transfer}}, S_{\mathit{Accept\ Payment\ by\ credit\ card}}, S_{\mathit{Accept\ Payment\ by\ cash}})$$

A *choice* variation corresponds to an OR relationship between the service intentions involved. For example, assume that *Investigate Candidate Booking* can be achieved either *On the Internet* or *By visiting a travel agent* or by both. The aggregate service $S_{\mathit{Investigate\ Candidate\ Booking}}$ is therefore defined as variant with two components $S_{\mathit{Investigate\ Candidate\ Booking\ on\ the\ Internet}}$ and $S_{\mathit{Investigate\ Candidate\ Booking\ by\ visiting\ a\ travel\ agent}}$. We use the symbol "**v**" to denote the choice variation and therefore:

$$S_{\mathit{Investigate\ Candidate\ Booking}} = \mathbf{v}(S_{\mathit{Investigate\ Candidate\ Booking\ on\ the\ Internet}}, S_{\mathit{Investigate\ Candidate\ Booking\ by\ visiting\ a\ travel\ agent}})$$

Finally a *multi-path* variation occurs when several compositions of an intentional service allow to achieve the same intentional service. Let us assume in our example that it is possible that the customer gets a booking as a reward for loyalty to the hotel chain. Thus, there are two paths to providing the intentional service *Make a Confirmed Booking:* one by achieving the sequence of intentional services *Make a Booking*, *Accept payment* and the other one *Get a Rewarded Booking*. The multi-path is denoted "∪" and the multi-path service $S_{\mathit{Make\ Confirmed\ Booking}}$ is defined as follows:

$$S_{\mathit{Make\ Confirmed\ Booking}} = \cup(\bullet(S_{\mathit{Make\ Room\ Booking}}, S_{\mathit{Accept\ Payment}}), S_{\mathit{Get\ a\ Rewarded\ Booking}})$$

The foregoing demonstrates that services are defined recursively; an aggregate service being possibly composed of other aggregate services; besides, components of an aggregate service can be related directly through composition links (•, //, *) or in a more complex manner through relationships (∪,⊗,v). Relationships between intentional services introduce variability in the composition. Overall, services are defined in an intention-driven manner focusing on the '*whys*' of the functionality provided by the underlying SOA level software service. Moreover, composition is itself intention-driven and grounded in XOR, OR, AND relationships among intentional services. Thus, whereas the service interface exhibits the '*whys*' of the service, its actual implemented functionality is embedded in the related atomic services.

Now, consider the issue of populating the Intentional Service Registry. Evidently, every service must be available in the Registry. That is, every atomic and aggregate service is kept here. For an aggregate, information about composition links and relationships is kept. This enables (Fig. 2) retrieval of complete aggregate services, their binding and adaptation to conform to the task at hand. Retrieval is based on intention matching and thereafter on situation and condition matching. That is, given the need to find a service with intention *I*, the registry is searched to retrieve a service with the same or similar intention. Once such a service is found, one drills down to assure oneself that the pre and post conditions match. Finally, the initial and final situations yield the input and output parameters.

# 3 Discovering Services for Publication

We believe that the services that populate the Registry arise in the business of organizations. Services to be provided relate to business objectives and, indeed, help to achieve these. This requires that a model of the business can be developed using which the E-business provider (Fig. 2) discovers services for publication. In this section, we propose the use of the Map formalism [11] to represent businesses in intentional terms and provide guidelines to determine services from this representation. We use Materials Management (MM) to illustrate service publication (see [11] for full details of the MM map).

## 3.1 Capturing Business Intentionality in Maps

*Map* is a representation system that was originally developed to represent a process model expressed in intentional terms. It provides a representation mechanism based on a non-deterministic ordering of *intentions* and *strategies*. We will use it here as a means for modelling intention-driven composition of services.

A *map* is a labelled directed graph with *intentions* as nodes and *strategies* as edges. An edge enters a node if its strategy can be used to achieve the intention of the node. There can be multiple edges entering a node.

An *intention* is a goal that can be achieved by the performance of a process. For example, the MM map in Fig. 4 has *Purchase Material* and *Monitor Stock* as intentions. Furthermore, each map has two special intentions, *Start* and *Stop*, to respectively start and end the process.

A *strategy* is an approach, a manner to achieve an intention. In Fig.4, *By reorder point planning* is a manner to place an order to *Purchase Material,* any time the stock of this material falls under the reorder point.

A *section* is the key element of a map. It is a triplet as for instance <*Start, Purchase Material, Manual Strategy*> which couples a source intention (*Start*) to a target intention (*Purchase Material*) through a strategy (*Manual strategy*) and represents a way to achieve the target intention *Purchase Material* from the source intention *Start* following the *Manual Strategy*.

Sections in a map are related to each other by four kinds of relationships namely *multi-thread, bundle, path* and *multi-path* relationships.

*Bundle relationship*: Several sections having the same pair of source and target intention, which are mutually exclusive are in a *bundle relationship*. For example in Fig.4, the *Planning strategy* is a bundle consisting of the *Reorder point strategy* and *Forecast based strategy*. Similarly, the *Inventory balance strategy* is a bundle of *periodic, continuous* and *sampling* strategies.

*Multi-thread relationship*: It is possible for a target intention to be achieved from a source intention in many different ways. Each of these ways is expressed as a section in the map and these sections are in a *multi-thread relationship* with one another. In Fig.4 the *Planning strategy* and the *Manual strategy* are in a *multi-thread* relationship. The difference between a *multi-thread* and a bundle relationship is that of an exclusive OR of sections in the latter versus an OR in the former.

*Path relationship*: This establishes a precedence/succession relationship between sections. For a section to succeed another, its source intention must be the target intention of the preceding one. For example the two sections, *<Start, Purchase Material, Manual strategy >, <Purchase Material, Monitor Stock, Out-In strategy >* constitutes a *path*.

*Multi-path*: Given the three previous relationships, an intention can be achieved by several combinations of sections. Such a topology is called a *multi-path*. In general, a map from its *Start* to its *Stop* intentions is a multi-path and contains multi-threads. For

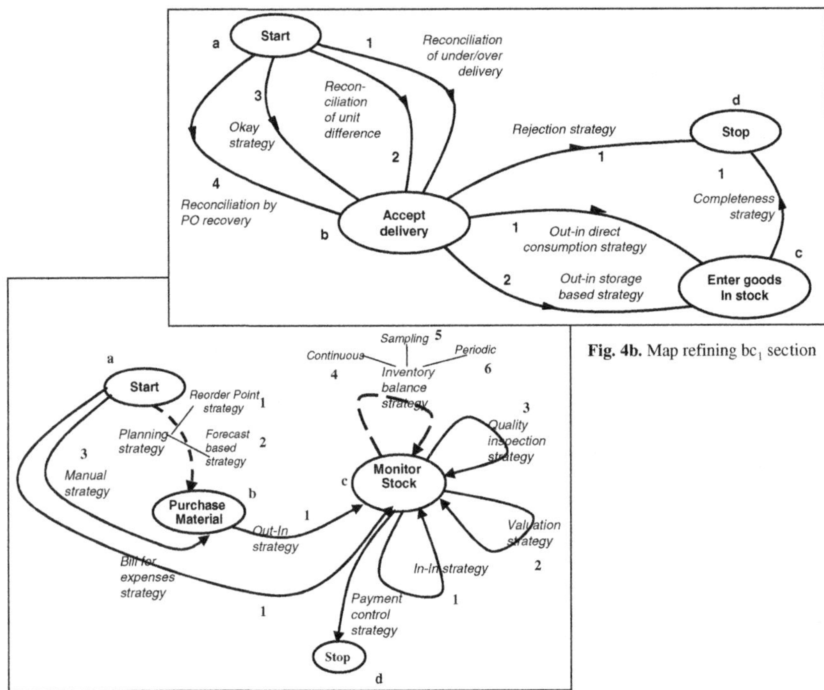


Fig. 4b. Map refining $bc_1$ section

Fig. 4a. The Material Management Map

**Fig. 4.** The Material Management Map (Fig. 4a) and the Map refining $bc_1$ section (Fig. 4b)

example, there is a *multi-path* to achieve the intention *Purchase Material*; either the path from *Start* to *Monitor Stock* via *Purchase Material* can be followed or the direct path from *Start* to *Monitor Stock* can be used.

Finally, it is possible to *refine* a section of a map into an entire map at a lower level of abstraction. For example, Fig. 4 shows the refinement of the section *<Purchase Material, Monitor Stock, Out-in strategy>* as a map (Fig. 4a). This refinement mechanism leads to model business intentionality as a hierarchy of maps.

## 3.2 Deriving Intentional Services from Maps

Having represented business intentionality as maps, we now proceed to determine services and their composition according to the *ISM*. We propose three key guidelines[1] to do this:

1- associate every section to an *atomic service*,
2- calculate all the *paths* of a map using an adaptation of the MacNaughton and Yamada's algorithm [12],
3- determine the aggregate services using the following *correspondences* between sections relationships in maps and service composition operators in *ISM* <path – composite>, <bundle – alternative>, <multi-thread – choice>, <multi-path – multi-path>. Since the entire map is, in general, a multi-path, it corresponds to an aggregate service.

We consider the three steps in turn and illustrate them with the MM map.

### 3.2.1 Associating Map Sections to Atomic Services

The first step consists of associating every section of a map to an atomic service. This correspondence leads in the case of the MM example, to services shown in Table1 in

**Table 1.** Services of the MM map

| MM map sections | Intentional Services |
|---|---|
| $ab_1$ | $S$ *Purchase Material with reorder point strategy* |
| $ab_2$ | $S$ *Purchase Material with forecast strategy* |
| $ab_3$ | $S$ *Purchase Material Manually* |
| $bc_1$ | $S$ *Receive stock of purchased material* |
| $ac_1$ | $S$ *Receive stock by bill for expenses* |
| $cc_1$ | $S$ *Move stock* |
| $cc_2$ | $S$ *Evaluate value of stock* |
| $cc_3$ | $S$ *Inspect stock* |
| $cc_4$ | $S$ *Conduct Physical Inventory continuously* |
| $cc_5$ | $S$ *Conduct Physical Inventory by sampling* |
| $cc_6$ | $S$ *Conduct Physical Inventory periodically* |
| $cd_1$ | $S$ *Verify invoice against delivery* |

[1] For sake of clarity, we deal here with guidelines for one single map whereas the entire process must deal with a hierarchy of maps. Rule 1 above needs then to be adapted (non refined sections are associated to atomic services) and an iteration step for every refined map shall be added.

correspondence with each of the 12 sections of the MM map. For sake of conciseness we use an abbreviated notation to refer to a section. We refer to each intention by a letter and to each strategy between a pair of intentions by a digit starting from 1 (see Fig. 4). Therefore, *ab3* is the reference of section *<Start, Monitor Stock, Manual strategy>* between the source intention *Start*, the target intention *Monitor Stock*, with the *Manual strategy* coded 3.

It can be seen that the name of each service reflects the business intention that can be achieved as well as the strategy to achieve it.

### 3.2.2 Calculating all Paths

We sketch an algorithm that automatically generates paths in the map and therefore, allow us to determine aggregate services as well as their nature, composite or variant. This algorithm is an adaptation of the MacNaughton and Yamada's algorithm [12] to calculate paths in a graph. This algorithm uses the different types of relationships between sections in a map that we introduced earlier.

The MacNaughton's algorithm is based on the two following formula:

Let *s* and *t* be the source and target intentions, *Q* the set of intermediary intentions including *s* and *t* and *P* the set of intermediate intentions excluding *s* and *t*.

The initial formula $Y_{s,Q,t}$ used to discover the set of all possible paths using the three operators that are the union ("∪"), the composition operator (".") and the iteration operator ("*") is:

$$Y_{s,Q,t} = \bullet\,(X^*_{s,Q\setminus\{s\},s},\; X_{s,Q\setminus\{s,t\},t},\; X^*_{t,Q\setminus\{s,t\},t})$$

And given a particular intention *q* of *P*, the formula $X_{s,P,t}$ applied to discover the set of possible paths is:

$$X_{s,P,t} = \cup\,(X_{s,P\setminus\{q\},t},\; \bullet(X_{s,P\setminus\{q\},q},\; X^*_{q,P\setminus\{q\},q},\; X_{q,P\setminus\{q\},t}))$$

We specialize the $X_{s,P,t}$ into paths, multi-paths, multi-threads and bundle relationships that we note as follows:

**Bundle relationship** between two intentions *k* and *l* is denoted $B_{kl} = \otimes(kl_1,\; kl_2 \ldots kl_n)$ where the $kl_i$ are the exclusive sections related by the bundle relationship. In Fig. 4, the bundle of planning strategies is $B_{ab} = \otimes(ab_1, ab_2)$.

**Multi-thread relationship** between two intentions *k* and *l* is denoted $MT_{kl} = \vee(kl_1\; kl_2, kl_n)$ where the $kl_i$ are the sections related by the multi-thread relationship. Thus, the multi-thread between *Start* and *Purchase Material* in Fig. 4 is $MT_{ab} = \vee(B_{ab}, ab_3)$.

**Path relationship** between two intentions *k* and *l* is denoted $P_{k,Q,l}$ where *Q* designates the set of intermediary intentions used to achieve the target intention *l* from the source intention *k*. A path relationship is based on the sequential composition operator "." between sections and relationships of any kind. As an example, the path relationship in Fig. 4 between *Start* and *Monitor Stock* is denoted $P_{a,\{b\},c} = \bullet(MT_{ab}, bc_1)$.

**Multi-path relationship** between two intentions *k* and *l* is denoted $MP_{k,Q,,l}$ where *Q* designates the set of intermediary intentions used to achieve the target intention *l* from the source one *k*. A multi-path relationship is based on the union operator "∪" between alternative paths. Thus, the multi-path in Fig. 4 between *Start* and *Stop* is denoted $MP_{a,\{b\},c} = \cup(ac_1, P_{a,\{b\},c})$.

The initial formula generating all the paths between the intentions a and d of Fig. 4 map, is:

$$Y_{a,\{a,b,c,d\},d} = \bullet(X^*_{a,\{b,c,d\},a}, X_{a,\{b,c\},d}, X^*_{d,\{b,c\},d})$$

The identified paths are summarized in Table 2.

**Table 2.** List of sections relationships

| Type of relationship | Identified relationships |
|---|---|
| Path | $P_{a,\{b\},c} = \bullet\ (MT_{ab}, bc_1)$ |
| | $P_{a,\{b,c\},d} = \bullet\ (MP_{a,\{b\},c}, MT_{cc}^*, cd_1)$ |
| Multi-Path | $MP_{a,\{b\},c} = \cup\ (ac_1, P_{a,\{b\},c})$ |
| Bundle | $B_{ab} = \otimes(ab_1, ab_2)$ |
| | $B_{cc} = \otimes(cc_4, cc_5, cc_6)$ |
| Multi-Thread | $MT_{ab} = \vee(B_{ab}, ab_3)$ |
| | $MT_{cc} = \vee(cc_1, cc_2, cc_3, B_{cc})$ |

### 3.2.3 Determine Aggregate Services

Now, we establish a correspondence between section relationships in the map and aggregate service types. This correspondence is as follows: <path – *composite*>, <bundle – *alternative*>, <multi-thread- *choice*>, <multi-path- *multi-path*>. Table 3 presents the variant and composite services associated to the MM map. These are expressed with the set of variant and composite operators, namely $\vee$, $\otimes$, $\cup$, $\bullet$, * introduced earlier in section 2.

**Table 3.** Components of the aggregate service $S_{\textit{Satisfy Material Need Efficiently}}$

| Aggregate Types | Services |
|---|---|
| *Variant services* | $S_{\textit{Purchase Material Planning strategy}} = \otimes\ (S_{\textit{PM reorder point strategy}}, S_{\textit{PM with forecast strategy}})$ |
| | $S_{\textit{Conduct Physical Inventory}} = \otimes\ (S_{\textit{CPI continuously}}, S_{\textit{CPI by periodically}}, S_{\textit{CPI by sampling}})$ |
| | $S_{\textit{Purchase Material}} = \vee\ (S_{\textit{Purchase Material Manually}}, S_{\textit{Purchase Material Planning Strategy}})$ |
| | $S_{\textit{Monitor Stock}} = \vee\ (S_{\textit{Conduct physical inventory}}, S_{\textit{Inspect stock}}, S_{\textit{Move stock}}, S_{\textit{Evaluate value of stock}})$ |
| | $S_{\textit{Receive stock}} = \cup\ (S_{\textit{Receive stock by bill for expenses}}, S_{\textit{Receive stock normally}})$ |
| *Composite services* | $S_{\textit{Satisfy Material Need Efficiently}} = \bullet\ (S_{\textit{Receive stock}}, S_{\textit{Monitor Stock}}^*, S_{\textit{Verify invoice against delivery}})$ |
| | $S_{\textit{Receive stock normally}} = \bullet\ (S_{\textit{Purchase material}}, S_{\textit{Receive stock of purchased material}})$ |

It is to be noted that the entire MM map is associated to a composite service $S_{\textit{Satisfy Material Need Efficiently}}$ having the intention *Satisfy Material Need Efficiently.* This aggregate service is a composition of three services, $S_{\textit{Receive stock}}$, $S^*_{\textit{Monitor Stock}}$, and $S_{\textit{Verify invoice against delivery}}$. The first one of these is a multi-path with the intention to *Receive Material in stock.* The second is a set of variant services to achieve the intention *Monitor Stock.* The third one is an atomic service intended to *Verify invoices against delivery.*

# 4 Adapting Services

Since an aggregate service captures a full range of variants to achieve the root service intention, when the business agent (Fig. 2) desires to use the service he has selected there is an adaptation issue. The issue of adaptation is that of determining which variant services and which combination of variant services are relevant to the situation at hand.

We again believe that adaptation must be driven by business intentions and identified two different ways in which adaptation can be done:

- *Design time adaptation* permits a selection of a combination of variants that might result in only one composite service; i.e. one path from Start to Stop in the map.
- *Run time adaptation* allows to leave a large degree of variability in the adapted aggregate service and the desired variant services can then be selected *dynamically* at enactment time.

This section describes how the different combinations of services in an $\mathcal{ISM}$ aggregate service can be mapped to an agent architecture that monitors the navigation through service relationships and thus allows dynamic service selection at run time.

## 4.1 The Agent Architecture

In order to monitor the navigation among the composition of services and offer to the business agent the choice of variants he/she wants to execute, we build a hierarchy of agents to managing service relationships and handing over the execution of atomic services. The hierarchy is composed of two kinds of agents: control and executor agents.

- *An executor agent* is a self-contained unit that implements an atomic service; this can be done by handing over the control to a traditional service composition engine such as the BPEL4WS engine [13].
- *A control agent* controls the selection and execution of a given composition, i.e. a path in a map (executors or/and other control agents). We distinguish four kinds of control agents for each of the four operators "∪" *(multi-path),* "**.**" *(path),* "∨" *(thread)* and "⊗" *(bundle)* control agents. They respectively control the selection and execution of the paths related by multi-path, path, multi-thread and bundle relationships.

In order to build the hierarchy, we defined mapping rules that are briefly sketched in the following. We first introduce one executor for each atomic service. As can be seen in Fig. 5, there is a one to one correspondence between atomic services and executors. For example, the service $ab_1$ is mapped to an executor having the same name. Executor agents are the leaves of the hierarchy.

Higher levels correspond to control agents. There is a kind of control agent for each kind of service relationship. For example, in Fig. 5, the multi-thread relationship $MT_{ab}$ (see Table 2) is associated to a multi-thread control agent having the same name. We first identify control agents using a one-to-one correspondence and then, make some simplifications, for example, a path relationship composed of one atomic service in not mapped to a control agent.

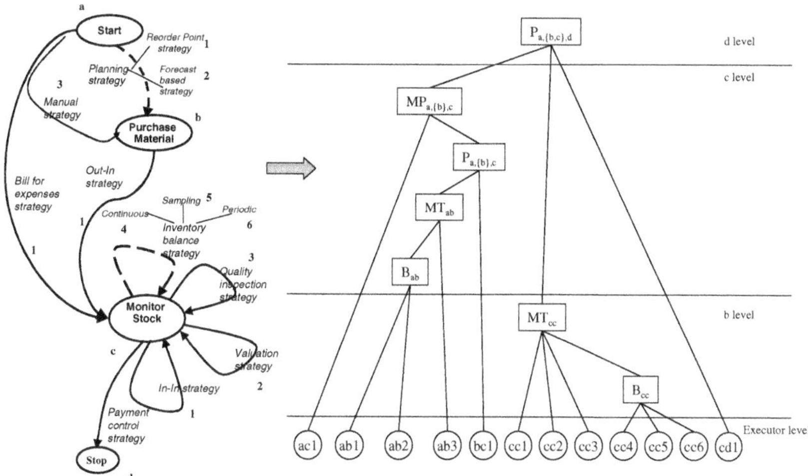


**Fig. 5.** Agent architecture

Control agents are organized at different levels. Each level is responsible for the achievement of a service intention. The top level of the hierarchy corresponds to the *Stop* intention and is responsible of the achievement of the goal of the whole aggregate service (map). The next level is related to the intention preceding the *Stop* intention. The bottom level of the hierarchy is composed of the executor agents. The hierarchy of Fig. 5 is composed of four levels related to the intentions of the MM map. The control agents of each level control children agents belonging to the same level or to the levels below.

### 4.2 Service Agent Support

Clearly, the ISOA departs from the usual SOA binding mechanism in providing an enactment mechanism that permits dynamic selection of services at run time. This is compatible with the business-oriented view of the *ISM* and the need for business agents to adapt decision making 'on the fly'. We believe that it is possible for business people to perform this adaptation. This is because knowledge of the business characteristics and an analysis based on these is enough to make the adaptation decision.

## 5 Related Work

Generally speaking, research on service description, composition and adaptation is relevant for our work [3].

Typical descriptions of services are based on finite state formalisms, e.g., in [14] [15] services are represented as state charts, in [16] services are modeled as Mealy machines and in [17], services are represented as finite state machines. The *ISM* shares with these approaches the need to describe service to ease their retrieval but departs from their *function driven* perspective to propose an *intention drive* of service description. As a consequence, *ISM* service descriptions will bring out the business intention that the service allows to fulfill and pre and post conditions instead of defining the signatures of operations that can be invoked on class objects. We believe that this contribute to avoid the current mismatch of languages between low level services descriptions such as WSDL statements and business perceived services.

Our description of intentional services has some similarities with semantic descriptions as found in [18][19][20]. Annotations which provide these semantic descriptions are compared to ontology elements in order to enrich usual retrieval mechanism. However, none of these semantic descriptions seem to be based on goal matching.

Our approach borrows from goal driven approaches in Requirements Engineering [21][22] the idea of goal decomposition and goal refinement through AND/OR graphs. This leads to an intention driven service composition: an *ISM* aggregate service has a high level, strategic intention as its key characteristic and its composition is reflecting the intention decomposition into sub-intentions that can be themselves fulfilled thanks to a composition of lower level sub-intentions etc. till operational intentions related to atomic services are found. By contrast most proposals are based on the idea of flow-composed services in which services are black boxes exchanging input/output parameters [23][17][24].

A large body of research work [25][26][14][15][16] deals with service execution: (i) the peer-to-peer architecture in which the individual service interact among themselves and with the client directly, and (ii) the mediated architecture in which the control over the available services is centralized. Our approach fits best to the peer-to-peer perspective but needs specific mechanism to cope with the adaptation issue.

From a methodological viewpoint, our proposal is close to [27] as both share the idea to capture service needs from exploring business goals. In [27] a revised *Tropos* design process is used to support service discovery and composition by offering a roadmap that relates stakeholder goals to collections of services available in different directories.

## 6 Conclusion

In this paper we introduced the notion of intentional service as one described in terms of the business goal it allows to fulfill. We also showed that ISOA service composition is intention driven and reflects business needs. This is in accordance with our view that business executives must be provided with a description of services available in a service portfolio that is adapted to their own perceived needs.

The paper considered in some detail the three roles of our ISOA architecture:

- E-business provider, who looks at a business, identifies its intentions, derives and publishes services in the intentional service registry.
- Intentional service registry where services are available. The descriptors of services and the typology of services being kept are modeled in the *ISM*.
- Business agent who retrieves services from the registry and dynamically navigates through aggregate services composition graphs using the agent architecture. The appropriate aggregate variant is thus available for execution.

Whereas the three roles of ISOA correspond to the service provider, registry and client roles of the SOA, it is to be noted that ISOA services, aside from supporting business intentions, are also more complex than SOA ones. This is because of aggregate variants that provide flexibility to the business agent in performing the task at hand. In contrast SOA services are fixed and are available on a 'take it or leave it' basis.

The proposed approach is still work in progress. Current research aims at developing (a) an intention driven search mechanism for the selection of services on the basis of the business goal they allow to fulfil and (b) a software tool to guide the discovery of aggregate service through maps.

## References


1. Papazoglou, M-P., Giunchiglia, F., Kraemer, B., Traverso, P.: Service Oriented Computing Network, The new computing paradigm for the network world (2003)
2. Papazoglou, M-P.: Service-Oriented Computing: Concepts, Characteristics and Directions, WISE'03, Rome, Italy (2003)
3. Alonso, G., Casati, F., Kuno, H., Machiraju, V.: Web Services. In: Concepts, Architectures and Applications, Springer, Heidelberg (2004)
4. Papazoglou, M-P., Georgakopoulos, D.: Service-Oriented Computing. Communication of the ACM, 46(10) (2003)
5. W3C Web Service Description Language (WSDL) Version 1.2. W3C Working Draft 3, (2003) http://www.w3.org/TR/wsdl12/
6. Piccinelli, G., Emmerich, W., Williams, S-L., Stearns, M.: A Model-Driven Architecture for Electronic Service Management Systems. In: Orlowska, M.E., Weerawarana, S., Papazoglou, M.M.P., Yang, J. (eds.) ICSOC 2003. LNCS, vol. 2910, pp. 241–255. Springer, Heidelberg (2003)
7. Arsanjani, A.: Service-oriented modelling and architecture. (November 2004) http://www-128.ibm.com/developerworks/webservices/library/ws-soa-design1/
8. Zimmermann, O., Krogdahl, P., Gee, C.: Elements of Service-Oriented Analysis and Design (2004) http://www-128.ibm.com/developmentworks/webservices/library/ws-soad1/
9. Jackson, M.: Software Requirements and Specifications. In: A lexicon of practice, principles and prejudices, p. 256. Addison-Wesley, New York (August 1995)
10. Rolland, C., Souveyet, C., Ben Achour, C.: Guiding Goal Modelling using Scenarios. IEEE Transactions on Software Engineering, Special Issue on Scenario Management 24(12), 1055–1071 (1998)
11. Rolland, C., Prakash, N.: Bridging the gap between Organizational needs and ERP functionality. Requirement Engineering Journal (2000)
12. MacNaughton, R.: Yamada: Regular expressions and state graphs for automata. IEEE transactions on electronic computers EC-9, 39–47 (1960)
13. Andrews, T., Curbera, F., Dholakia, H.: Microsoft, IBM, and SAP. BPEL4WS version 1.1, (2003) http://www-106.ibm.com/developerworks/webservices/library/ws-bpel/
14. Mecella, M., Pernici, B., Craca, P.: Compatibility of e-Services in a Cooperative Multi-Platform Environment. In: Proc. VLDB-TES (2001)
15. Fauvet, M-C., Dumas, M., Benatallah, B., Paik, H.: Peer-to-Peer Traced Execution of Composite Services.In: Proc. of VLDB-TES (2001)

16. Bultan, T., Fu, X., Hull, R., Su, J.: Conversation Specification : A New Approach to design and analysis of E-Service Composition. In: Proc. of the WWW'03 Conference (2003)
17. Berardi, D., Calvanese, D., De Giacomo, G., Lenzerini, M., Mecella, M.: Automatic Composition of e-Services that Export their Behavior. In: Proc. of WES 2003 (2003)
18. DAML-S Defense Advanced Research Projects Agency: DARPA agents markup language- Services (DAML-S). http://www. Daml.org/services/
19. Sirin, E., Parsia, B.: Planning for semantic web services in Semantic web services workshop at ISWC'04. (2004) http://www.mindswap.org/papers/SWS-ISWC04.pdf
20. Horrocks, I., van Harmelen, F., Patel-Schneider, P., Berners-Lee, T., Brickley, D., Connoly, D., Dean, M., Decker, S., Fensel, D., Hayes, P., Heflin, J., Hendler, J., Lassila, O., McGuinness, D., Stein, L.A.: DAML+OIL (2001) http://www. Daml.org/2001/03/daml+oil-index.html
21. Van Lamsweerde, A., Dairmont, R., Massonet, P.: Goal Directed Elaboration of Requirements for a Meeting Scheduler: Problems and Lessons Learnt. In: Proc. Of RE'95, pp. 194–204. IEEE, New York (1995)
22. Yu, E.: Towards Modelling and Reasoning Support for Early-Phase Requirements Engineering. In: Proceedings of the 3rd IEEE Int. Symp. On RE'97. pp. 226-235 Washington D.C., USA (January 6-8 1997)
23. Yang, J., Papazoglou, M-P.: Service Components for Managing the Life-Cycle of Service Compositions. Information Systems Journal (2003)
24. Dijkman, R-M.: A Basic Design Model for Service-Oriented Design, ArCo/WP1/T1/D2/V1.00, (2003)
25. Casati, F., Shan, M.: Dynamic and Adaptive Composition of e-Services, Information Systems, 6(3) (2001)
26. McIltraith, S., Son, T., Zeng, H.: Semantic web services, IEEE Intelligent Systems, 16(2) (2001)
27. Perini, A., Susi, A., Mylopoulos, J.: Tropos Design Process for Web Services, 1st Int. Workshop on SOC: Consequences for Engineering Requirements, Paris (2005)

# WSXplorer: Searching for Desired Web Services

Yanan Hao[1], Yanchun Zhang[1], and Jinli Cao[2]

[1] School of Computer Science and Mathematics, Victoria University
PO Box 14428, Melbourne, VIC 8001, Australia
{haoyn, yzhang}@csm.vu.edu.au

[2] Department of Computer Science and Computer Engineering, La Trobe University
Bundoora, VIC 3086, Australia
j.cao@latrobe.edu.au

**Abstract.** With the rapid development of e-commerce over Internet, web services have attracted much attention in recent years. Nowadays, enterprises are able to outsource their internal business processes as services and make them accessible via the Web. Then they can dynamically combine individual services to provide new value-added services. A main problem that remains is how to discover desired web services. In this paper, we propose WSXplorer, a novel scheme for identifying potentially relevant web services given a textual description of services. In particular, we propose a new schema matching algorithm for supporting web-service operations matching. The matching algorithm catches not only structures, but even better semantic information of schemas. Based on service operations matching, the concept of *attribute closure* is introduced to identify associations between web-service operations. We also propose a ranking strategy to satisfy a user's top-$k$ requirements. Experimental evaluation shows that our approach can achieve high precision and recall ratio.

## 1 Introduction

A web service is programmatically available application logic exposed over Internet. It has a set of operations and data types. The current set of web service specifications defines how to specify reusable operations through the Web-Service Description Language(WSDL), how these operations can be discovered and reused through the Universal Description, Discovery, and Integration(UDDI) API, and how the requests to and responses from web-service operations can be transmitted through the Simple Object Access Protocol API(SOAP).

With the rapid development of e-commerce over Internet, web services have attracted much attention in recent years. Nowadays, enterprises are able to outsource their internal business processes as services and make them accessible via the Web (see, e.g.,[1,2,3,4,5]). Then they can combine individual services into more complex, orchestrated services. A main problem that remains is how to discover desired web services. To find a service in UDDI, a user needs to input some keywords about the required service and then to browse the relevant UDDI category to locate relevant web services. Considering a large amount

J. Krogstie, A.L. Opdahl, and G. Sindre (Eds.): CAiSE 2007, LNCS 4495, pp. 173–187, 2007.

of service entries, this process is time consuming and frustrating. Furthermore, this method does not provide a mechanism assisting users in selecting relevant services and composing with them. Since a web service is usually used as part of an application, users often would like to know relevant services as much as possible. For example, consider the examples shown in Fig. 1. A user searching for a *CreateOrder* service may also be interested in a *TransportOrder* service. There is an association between these two services, in which the output of *CreateOrderService*, *Order*, is also the input of *TransportOrderService*. This form of association potentially involves more web services. It is particularly useful and challenging in service composition.

**Fig. 1.** Sample web-service operations

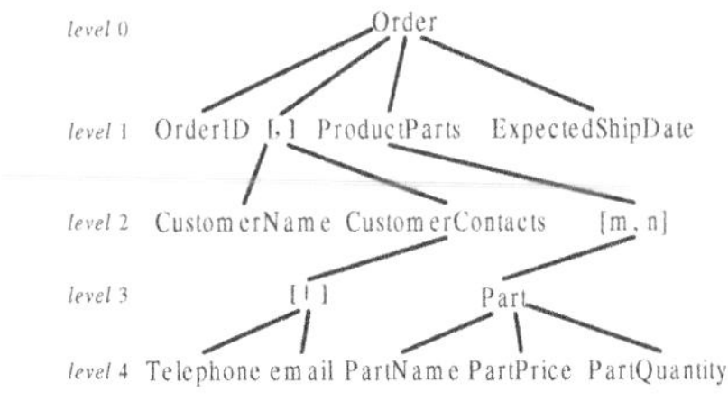


**Fig. 2.** XML schema tree of *Order* type

To address the problems above in searching for web services, we propose WSXplorer (Web Services eXplorer), a novel scheme for identifying potentially relevant web services given a textual description of services. The contribution of the work reported here is summarized as follows:

1. We propose algorithms for supporting web-service operations matching. The key part of our algorithms is a schema tree matching algorithm, which employs a new cost model to compute tree edit distances. Our new schema tree matching algorithm can not only catch structures, but also the semantic information of schemas.
2. Based on service operations matching, an approach to identify associations between web-service operations is presented. This approach uses the concept of *attribute closure* to obtain sets of operations. Each set is composed of associated web-service operations.
3. We also introduce a ranking strategy to satisfy a user's top-$k$ requirements. Experimental evaluation shows that WSXplorer can achieve acceptable result with high performance.

The rest of this paper is organized as follows. Section 2 reviews the related work. Section 3 gives an overview of WSXplorer. Section 4 describes a web-service operation matching algorithm, in which a new cost model and some XML schema transformation rules are defined. In section 5 we present how to cluster web-service operations and how to find associations between them. Section 6 describes our experimental evaluation. Section 7 gives some concluding remarks.

## 2 Related Work

Finding similar web-service is closely related to software components matching. In [6], signatures are used to describe a component's type information (which is usually statically checkable), and formal specifications are defined to describe the component's dynamic behaviour. Two components match if their signatures and specifications match. However, the formal specifications used there are function's post conditions, which are not available in web services.

Several approaches use text or structural matching to find similar web services for a given web service. The earlier technique tModel presents an abstract interface to enhance service matching process. But the tModel needs to be defined while authors publishing in UDDI [7]. In [8], the authors propose a SVD-Based algorithm to locate matched services for a given service. This algorithm uses characteristics of singular value decomposition to find relationships among services. But it only considers textual descriptions and can not reveal the semantic relationship between web services. Wang etc. [9] discover similar web services based on structure matching of data types in WSDL. The drawback is that simple structural matching may be invalid when two web-service operations have many similar substructures on data types.

Recently, some methods have been proposed to annotate web services with additional semantic information. These annotations are used to match and compose services. For example, in [10] the authors extended DAML-S to support service specifications, including behavior specifications of operations; The Web Service Modeling Ontology (WSMO) [11] is a conceptual model for describing Web services semantically, and defines the four main aspects of semantic Web service, namely Ontologies, Web services, Goals and Mediators. However, currently, most of existing web services use WSDL specifications, which do not contain semantics. Annotating the collection of services requires much effort, and it is infeasible in our case. [12] formally defines a behaviour model for web service by automata and logic formalisms. However, the behaviour signature and query statements need to be constructed manually, which can be very hard for common users.

Woogle [13] develops a clustering algorithm to group names of parameters of web-service operations into semantically meaningful concepts. Then these concepts are used to measure similarity of web-service operations. It relies too much on names of parameters and does not deal with composition problem however. In our previous work [14] we use schema to find web services, but the associations between services are not considered. In [15] the authors propose a syntactic approach to web service composition, given only the input-output types of web services available in their WSDL descriptions. Discover [16,17] and DBXplore [18] operate on relational databases and facilitates information discovery on them by allowing users to issue keyword queries without any knowledge of the database schema. They return sets of tuples that are associated by joining on their primary and foreign keys. Inspired by these methods, we model each web-service operation as a dependency (schema) according to its data types (attributes), and then find associations between web-service operations.

## 3 Overview of WSXplorer

The goal of WSXplorer is to find relevant web-service operations given a natural language description of desired web services and WSDL specifications of all available services published through UDDI. The WSDL files consist of textual description of web-service operations. Thus, firstly we use traditional IR technique TF (*term frequency*) and IDF (*inverse document frequency*) [19] to find service operations that are most similar to the given description. We call these operations *candidate operations*. To do this, we extract words from web-service operation descriptions in WSDL. These words are pre-processed and assigned weight based on IDF. According to these weights, the similarity between the given description and a web-service operation description can be measured. A higher score indicates a closer similarity. For more details on measuring similarity among documents interested readers are referred to see [20]. After obtaining candidate operations, we employ a schema-match based method to measure similarity among them. Then based on the matching result the candidate operations are clustered into some *operation sets*. Each operation set contains a group of similar operations. Finally, all associations between operation sets are generated using the concept of *type closure*. Operations involved in one association are considered as a search result. Since each candidate operation has a score, we can rank search results simply by accumulating the score of operations. In the following sections we describe the models and algorithms underlying WSXplorer, in particular we show how to measure similarity between web-service operations based on schema matching.

## 4 Web-Service Operation Matching

### 4.1 Web-Service Operation Modelling

**Definition 1.** *A web service is a triple* $ws = (TpSet, MsgSet, OpSet)$*, where TpSet is a set of data types; MsgSet is a set of messages(parameters) conforming to the data types defined in TpSet;* $OpSet = \{op_i(input_i, output_i) | i = 1, 2, ..., n\}$ *is a set of operations, where* $input_i$ *and* $output_i$ *are parameters(messages) for exchanging data between web-service operations.*

Figure 1 gives three web-service operations used as examples in this paper. According to definition 1, a web service can be briefly described as a set of operations.

**Definition 2.** *Each web-service operation is a multi-input-multi-output function of the form* $f : s_1, s_2, ..., s_n \rightarrow t_1, t_2, ..., t_m$*, where* $s_i$ *and* $t_j$ *are data types in according with XML schema specification. We call* $f$ *a* ***dependency*** *and* $s_i/t_j$ *a* ***dependency attribute****.*

A dependency attribute can be a complex data type or a primitive data type. Complex data types, such as *Order* and *PurchaseOrder* in Fig. 1, define the structure, content, and semantics of parameters, whereas primitive data types, like *int* and *string*, are typically too coarse to reflect semantic information. Since parameters usually can be regarded as data types, we can convert primitive data types to complex data types by replacing them with their corresponding parameters. For example, in Fig. 1 *string* is converted into *UserName* type while *int* is converted into *UserID* type. Both *UserName* and *UserID* are considered as complex data types with semantics. Thus, each data type defined in a web-service operation carries semantic meaning. An XML schema can be modelled as a tree of labelled nodes. We categorize a node $n$ by its label:

1. **Tag node**: Each tag node $n$ is associated with an element type $T$. $T$ is also the tag name of node $n$.
2. **Constraint node**:
   - **Sequence node**: A sequence node indicates its children are an ordered set of element types. We use [ , ] to denote a sequence node.
   - **Union node**: A union node represents a choice complex-type, that is, the instance of which can only be one of the children types in accordance with the XML Schema specification. We use [ | ] to denote a union node.
   - **Multiplicity node**: Each node may optionally have a multiplicity modifier $[m, n]$ indicating that in the instance, its occurrence is between $m$ and $n$. This corresponds to the *minOccurs* and *maxOccurs* constraints in XML Schema. We use $[m, n]$ to denote a multiplicity node.

As an example, the schema tree of data type *Order* is shown in Fig. 2.

As we can see, data types defined in web-service operations carry semantic information. Intuitively, we consider two web-service operations similar if they have similar input/output data types. Thus the problem of web-service operation matching is converted to the problem of schema tree matching.

### 4.2 Tree Edit Distance

Many works have been done on the similarity computation on trees. Among them *tree edit distance* is one of the efficient approaches to describe difference between two trees. We introduce tree edit operations first. Generally, the tree edit distance operations include: (a) *node removal*, (b) *node insertion*, and (c) *node relabelling*. Such a set of operations can be represented by a mapping with minimum cost between the two trees. The concept of mapping is formally defined as follows [21]:

**Definition 3.** *Let $T_x$ be a tree and let $T_x[i]$ be the* i*th node of tree $T_x$ in a pre-order traverse of the tree. A mapping between a tree $T_1$ and a tree $T_2$ is a set* M *of ordered pairs $(i, j)$, satisfying the following conditions for all $(i_1, j_1)$,$(i_2, j_2) \in M$*

*1. $i_1 = i_2$ iff $j_1 = j_2$;*
*2. $T_1[i_1]$ is on the left of $T_1[i_2]$ iff $T_2[j_1]$ is on the left of $T_2[j_2]$;*
*3. $T_1[i_1]$ is an ancestor of $T_1[i_2]$ iff $T_2[j_1]$ is an ancestor of $T_2[j_2]$.*

Figure 3 gives an example of tree mapping. This mapping also shows the way of transforming the left tree to the right one. A dotted line from a node of $T_1$ to a

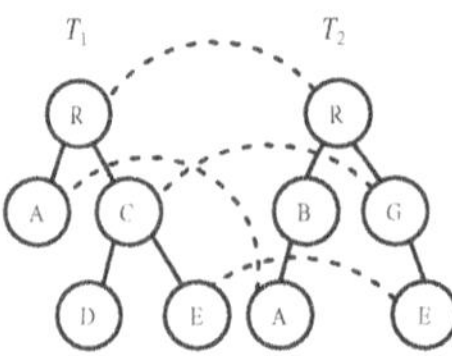


**Fig. 3.** Example of tree mapping

node of $T_2$ indicates that the node of $T_1$ should be changed if the corresponding nodes are different, remaining unchanged otherwise. Nodes of $T_1$ not connected by dotted lines are deleted, and nodes of $T_2$ not connected are inserted. Each of these operations is assigned a cost. The tree edit distance between two trees is defined as the minimal set of operations to transform one tree into the other.

Our schema matching algorithm is based on tree edit distance. However, the problem in our case is more complex than the traditional tree edit distance for the following reasons:

1. The labels of an XML Schema tree can carry complex type information (e.g., union, multiplicity) which makes simple relabelling operations inapplicable. For instance, let $T_1$ and $T_2$ be the schema trees of *Order* and *PurchaseOrder* respectively. Let us imagine there exits a mapping $M$ between $T_1$ and $T_2$, and there are two node-mapping pairs $(i_1, j_1)$,$(i_2, j_2) \in M$, where $T_1[i_1]$ =[*telephone*|*email*], $T_2[j_1]$ =*email*, $T_1[i_2]$=*price*, and $T_2[j_2]$=*quantity*. The edit operation of $(i_1, j_1)$ should have less cost than that of $(i_2, j_2)$. But the existing work consider all tree edit operations to have same unit distance.
2. The labels of nodes carry semantic information. So a relabelling from one node to another unrelated node will have more cost than to a semantic related node. For example, relabelling *part* to *item* is less costing than relabelling *price* to *email.*
3. We argue that tree edit operations on low-level nodes of a tree should have more influence than operations on high-level nodes. For example, in Fig. 2, node *Order* is more important than node *PartPrice*, because *Order* denotes broader semantics information than *PartPrice*. So, if a *PartPrice* node of the first tree is mapped into an *Order* node of the second tree, the edit operation cost should not be zero. But the traditional works on tree edit distance do not consider the difference and assign each edit operation unit cost.

In the next section, we present a new cost model to compute the cost of tree edit operation, as a consequence, the tree edit distance of two schema trees.

### 4.3 Cost Model

Measuring similarity between two XML schema trees equals to finding a mapping with minimum cost. So, the cost of each edit operation involved in the mapping needs to be computed first. [22] proposed a algorithm for fast computing tree edit distance, but it assigned the same cost for each unit edit operations on all nodes and overlooked nodes difference. Authors in [23] introduced a summary structure for computing structural distance and took weight information into account for nodes in distance computation, but it did not consider the semantic difference or similarity. In this section we introduce a new cost mode based on tree edit distance presented in [22,23]. The new cost model integrates weights of nodes and semantic connections between nodes. Let $T_1, T_2$ be two schema trees and let $n$, $node_1$ and $node_2$ be tree nodes. Formally, the cost model is defined as

$$cost(\rho) = \begin{cases} weight(n)/W(T_1,T_2), & \text{if} \rho = insert(n) \\ weight(n)/W(T_1,T_2), & \text{if} \rho = delete(n) \\ \alpha \times wd(node_1, node_2) & \text{if} \rho \text{ relabels} \\ +\beta \times sd(node_1, node_2) & node_1 \text{to } node_2 \end{cases} \quad (1)$$

where $\rho$ indicates a tree edit operation. $weight(n)$ shows the weight of node $n$, which is defined in definition 6. $wd(node_1, node_2)$ and $sd(node_1, node_2)$ give the weight and semantic difference of $node_1$ and $node_2$, respectively. $\alpha$ and $\beta$ are weights of $wd$ and $sd$, satisfying $\alpha + \beta = 1$. $W(T_1, T_2)$ is defined as $W(T_1, T_2) = weight(T_1) + weight(T_2)$, where $weight(T_i)$ is the sum of all node weights of tree $T_i (i = 1, 2)$. $wd(node_1, node_2)$ is defined as

$$wd(node_1, node_2) = \frac{\|weight(node_1) - weight(node_2)\|}{W(T_1, T_2)} \quad (2)$$

where $node_1 \in T_1$ and $node_2 \in T_2$ .

In equation 1, $weight(n)/W(T_1, T_2)$ explains the cost of inserting or deleting node $n$. For the relabel operation, both weight and semantics of $node_1$ and $node_2$ can be different, so we use the combination of weight and semantic difference as the relabel cost. All the costs are normalized by $W(T_1, T_2)$, i.e. the sum of all nodes weights of tree $T_1$ and $T_2$.

In the next two sections, we propose a set of schema-tree transformation rules and a semantic similarity measure to compute $wd$ and $sd$, i.e. the weight and semantic difference of nodes.

### 4.4 XML Schema Tree Transformation

**Definition 4.** *The tag name of a node is typically a sequence of concatenated words, with the first letter of every word capitalized (e.g., ExpectedShipDate). Such a set of words is referred to as a* **word bag***. We use $\pi(n)$ to denote the word bag of node $n$.*

**Definition 5.** *Two word bags $\pi(n_1)$ and $\pi(n_2)$ are said to be equal, only if they have same words.*

Two nodes are considered different if they have different word bags. The word bag reflects semantic meaning of a node. As we shall see later, using word bags we can measure the semantic similarity between two schema-tree nodes.

**Definition 6.** *Let level(n) denote the level of node n in schema tree T. The weight of node n is defined by a weight function:*

$$weight(n) = 2^{depth(T)-level(n)} (\forall n \in T) \tag{3}$$

The weights of all nodes fall in the range of $[2, 2^{depth(T)}]$. Each weight reflects the importance of a node in schema tree $T$.

From section 4.2, it can be seen that traditional tree-edit-distance algorithm is not suitable for XML schema trees. It does not deal with constraint nodes. We propose three transformation rules to solve this problem. These rules are used to transform constraint nodes, specifically, sequence nodes, union nodes and multiplicity nodes to tag nodes. At the same time, the weights of nodes are reassigned.

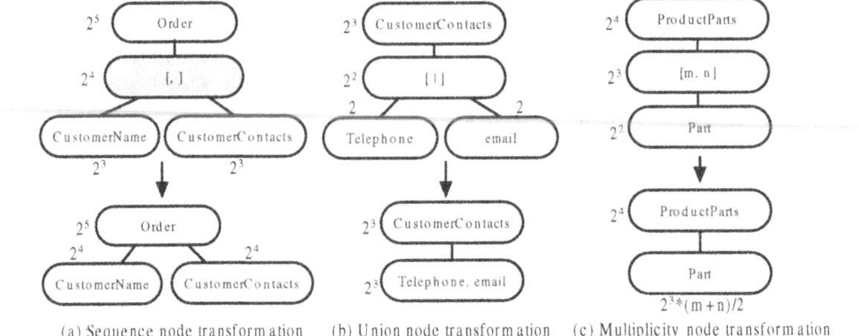


**Fig. 4.** XML Schema tree transformation

1. *split*: This rule is applied to sequence nodes. A sequence node $l = [l_1, l_2, ..., l_s]$ is split into an ordered list of nodes $l_1, l_2, ..., l_s$, where $l_i (i = 1, 2, ..., s)$ is a child node of the sequence node $l$. After the split process, each sequence node is replaced by its child nodes. Each child node $l_i$ inherits the weight of its parent node $l$ as a new weight. Figure 4(a) gives an example of the split rule.
2. *merge*: This rule is applied to union nodes. After the merge process, each union node is replaced by all its option nodes, i.e. all its child nodes. All child nodes of the union node $l = [l_1|l_2|...|l_s]$ are merged into a new node $l^*$, while the union node $l$ is deleted. The weight of node $l^*$ is $s$ times the weight of $l$. Each $l_i$'s$(i = 1, 2, ..., s)$ word bag is also merged into a new word bag. Formally, we have $weight(l^*) = weight(l) \times s$. Figure 4(b) gives an example of the merge rule.
3. *delete*: This rule is applied to multiplicity nodes. We delete a multiplicity node $l = [m, n](m, n \in N)$ and scale up the weight of each of its child nodes $l_i$. After the deletion process, each multiplicity node is replaced by its child nodes. We have $weight(l_i) = weight(l) \times (m + n)/2$. Figure 4(c) gives an example of the delete rule.

Note that the definition of complex types can be nested according to XML schema specification. Thus, given a schema tree, we apply the three transformation rules to its nodes level by level, from bottom to top. This process is formally

```
input  : schema tree T
output: transformed schema tree T*
d = GetDepth(T);
for i ← d to 0 do
    foreach node p ∈ level_i do
        if p is a sequence node then
            weight(each of p's child nodes)=weight(p);
            add p's child nodes to p's parent's child list;
            delete p;
        end
        if p is a union node with s options {l_i|i = 1, ..s} then
            merge p's child nodes into a new node q;
            add q to p's parent's child list;
            weight(q) = weight(p) × s;
            π(q) = ⋃_{i=1}^{s} π(l_i) ;
            delete p;
        end
        if p is a multiplicity node [m, n] then
            add p's child node to p's parent's child list;
            weight(p's child node)=weight(p) × (m + n)/2;
            delete p;
        end
    end
end
```

**Algorithm 1.** Bottom-up-transformation

described as the *bottom-up-transformation* algorithm (see Algorithm 1). The time complexity of Bottom-up-transformation is $O(n)$, where $n$ is the number of nodes in the XML schema tree.

## 4.5 Semantic Measurement Between Schema-Tree Nodes

After the bottom-up transformation, schema tree $T$ is converted into a new schema tree $T^*$. Each node $n$ of $T^*$ is a tag node, whose word bag may come from two or more word tags because of nodes merge by the merge rule. Formally, node $n$ can be regarded as a vector $(W, B)$, where $W$ is the weight of node $n$ and $B$ is the word bag of node $n$. As we can see, after transformation the weight difference between two nodes can be computed by the new cost model. In this section, we present a strategy to determine the semantic similarity of two schema-tree nodes, i.e. the semantic distance between two word bags.

WSXplorer relies on a hypothesis that two co-occurrence words in a WSDL description tend to have same semantics. We exploit the co-occurrence of words in word bags to cluster them into meaningful concepts. To improve accuracy of semantic measurement, a pre-processing step is carried out first before words clustering. Pre-processing includes word stemming, removing stop words and expanding abbreviations and acronyms into the original forms.

Let $I = \{w_1, w_2, ..., w_m\}$ be a set of words. These words come from word bags of all schema-tree nodes to which similarity measurement is applied. Let $D$ be a set of candidate web-service operation descriptions available in WSDL files. We introduce association rules to reflect the notion of word co-occurrence. An *association rule* is an implication of the form $w_i \rightarrow w_j$, where $w_i, w_j \in I$. The rule $w_i \rightarrow w_j$ holds in the descriptions set $D$ with *support* $s$ and *confidence* $c$, where $s$ is the probability that $w_i$ occurs in an web-service operation description; $c$ is the probability that $w_j$ occurs in an operation description, given $w_i$ is known to occur in it. All association rules can be found by the A-Priori algorithm [24]. We are only interested in rules that have confidence above a certain threshold $t$.

We use the agglomeration algorithm [24] to cluster words set $I = \{w_1, w_2, ..., w_m\}$ into concept set $C = \{C_1, C_2, ...\}$. There are three steps in the clustering process. It begins with each word forming its own cluster and gradually merges similar clusters.

1. Set up a confidence matrix $M_{m \times m}$. $M_{ij}$ is a two-dimensional vector $(s_{ij}, c_{ij})$, where $s_{ij}$ and $c_{ij}$ are the support and confidence of association rule $w_i \rightarrow w_j$, respectively.
2. Find the two-dimensional vector $M_{ij} = (s_{ij}, c_{ij})$ with the largest $c_{ij}$ in the confidence matrix $M$. If, for both of them, $c_{ij} > t$ and $s_{ij} > t$ then merge these two clusters and update $M$ by replacing the two rows with a new row that describes the association between the merged cluster and the remaining clusters. The distance between two clusters is given by the distance between their closest members. There are now $m - 1$ clusters and $m - 1$ rows in $M$.
3. Repeat the merge step until no more clusters can be merged.

Finally, we get a set of concepts $C$. Each concept $C_i$ consists a set of words $\{w_1, w_2, ...\}$. To compute semantic similarity between schema-tree nodes, we replace each word in word bags with its corresponding concept, and then use the TF/IDF measure. After schema-tree transformation and semantic similarity measure, the tree edit distance can be applied to match two XML schema trees by the new cost model.

### 4.6 Web-Service Operations Matching

As it has been mentioned before, we use tree edit distance to match two schema trees. It is equivalent to finding the minimum cost mapping. Let $M$ be a mapping between schema tree $T_1$ and $T_2$, let $S$ be a subset of pairs $(i, j) \in M$ with distinct word bags. Let $D$ be the set of nodes in $T_1$ that are not mapped by $M$, and $I$ be the set of nodes in $T_2$ that are not mapped by $M$. The mapping cost is given by $C = Sp + Iq + Dr$, where $p$, $q$ and $r$ are the costs assigned to the relabel, insertion, and removal operations according to the cost model proposed in section 4.3. We call $C$ the *match distance* between $T_1$ and $T_2$, denoted as $C = ED(T_1, T_2)$. Match distance reflects semantic similarity of two schema trees.

Now let us see how to match web-service operations. Given two web-service operations $op_1 : s_1, s_2, ..., s_n \rightarrow t_1, t_2, ..., t_m$ and $op_2 : x_1, x_2, ..., x_l \rightarrow y_1, y_2, ..., y_k$, for each schema tree of $op_1$, we find its corresponding schema tree of $op_2$ with the minimum match distance. We simply identify all possible matches between two lists of schema trees, and return the source-target correspondence that minimizes the overall match distance between the two lists. It does not depends on whether the number of parameters in the same or not between the two operations. We omit the algorithm details because of space limit.

## 5 Finding Associated Web-Service Operations

### 5.1 Clustering Web-Service Operations

Suppose $OP = \{op_1, op_2, ..., op_q\}$ is a set of web-service operations and each pair of operations $op_i$ and $op_j$ $(i, j = 1, 2, ..., q)$ match with the distance of $z_{ij}$. We classify $OP$ into a set of clusters $\{op_{c1}, op_{c2}, ...\}$. The clustering algorithm is described as below. It begins with each operation forming its own cluster and gradually merges similar clusters.

1. Set up a match matrix $M_{q \times q}$. $M_{ij}$ is the match distance of operation $op_i$ and $op_j$.
2. Find the smallest $M_{ij}$ in the match matrix $M$. If $M_{ij} <$ threshold $\delta$ then merge these two clusters and update $M$ by replacing the two rows with a new row that describes the association between the merged cluster and the remaining clusters. The distance between two clusters is given by the distance between their closest members. There are now $q - 1$ clusters and $q - 1$ rows in $M$.
3. Repeat the merge step until no more clusters can be merged.

Finally, a set of clusters $\{OPC_1, OPC_2, ...\}$ is obtained. Given a cluster $OPC_i$ and an operation $OPC_{ik} \in OPC_i$, $OPC_{ik}$ is called a *pivot* of $OPC_i$ if i t minimizes the sum of match distances to all the other operations in $OPC_i$. We consider all operations in $OPC_i$ as *instances* of $OPC_{ik}$.

For example, in Fig. 1 we give a clustering result. There are two clusters of web-service operations. One is $\{WS1, WS2\}$, and the other is $\{WS3\}$. In cluster $\{WS1, WS2\}$ the pivot is *GetOrder* and the instrances of *GetOrder* are *GetOrder* and *OrderBuilder*. In cluster $\{WS3\}$ the pivot is *ShippingOrder*, which is also an instance of itself.

### 5.2 Identifying Associations

A set of web-service operations is said to be *associated* if they potentially contribute to a user's web-service composition. Clearly, given two web-service operations $op_1$ and $op_2$, if the output attributes of $op_1$ are similar to the input attributes of $op_2$ then $op_1$ and $op_2$ may participate in a user's service composition together. The objective of this step is to find all associations between web-service operations. To do this, we first find associations among clusters $\{OPC_1, OPC_2, ...\}$. Let $OPC_{ik}$, say $x_1, x_2, ..., x_k \rightarrow y_1, y_2, ..., y_j$ be a pivot of

$OPC_i$. Let $X = \{x_1, x_2, ..., x_k\}$ and $Y = \{y_1, y_2, ..., y_j\}$.We first compute the *attribute closure* $X^+$ with respect to $X$, which is the set of attributes $A$ such that $X \to A$ can be inferred by transitivity. At the same time, a pivot set $PS$ associated with $OPC_{ik}$ is computed. The overall process is shown as algorithm 2. We perform a worst case time analysis of algorithm 2. The repeat loop is executed as most $|S|$ times, where $|S|$ is the total number of pivots corresponding to all clusters. The calculation of $q$ takes time $|S| - |T|$, where $T$ is the number of pivots in the pivot set $PS$. Hence the total execution time takes in the worst case time $O(S^2)$.

**input** : A pivot $p : x_1, x_2, ..., x_k \to y_1, y_2, ..., y_j$
**output**: A pivot set $PS$ containing associated pivots

$X = \{x_1, x_2, ..., x_k\}$; $Y = \{y_1, y_2, ..., y_j\}$;
*Closure* $= X$;
$PS = \{X \to Y\}$;
**repeat**
    **if** *there is a pivot* $q : U \to V$ *such that the match distance of* $U$ *and Closure is less than threshold* $\delta$ **then**
        set *Closure* $=$ *Closure* $\bigcup V$;
        set $PS = PS \bigcup q$;
    **end**
**until** *there is no change* ;

**Algorithm 2.** Algorithm for computing attribute closure and pivot set

We first choose a pivot $OPC_{ik}$ for each cluster $OPC_i$. For each pivot, we compute a pivot set. We eliminate duplicate pivot sets. If two pivots are in a same pivot set, then their corresponding instances are associated.

Each pivot set $PS = \{p_1, p_2, ..., p_k, ...\}$ can generate a set of operation groups in the form of $\{p'_1, p'_2, ..., p'_k, ...\}$, where $p'_i$ is an instance of $p_i$. Operations in a same group are associated. To obtain an operation group, we simply replace each pivot $p_i$ in $PS$ with one of its corresponding instances. All possible operation groups are outputted as search results.

For example, a pivot set for the clusters given in Fig. 1 is {*GetOrder*, *ShippingOrder*}. It can generate two search results, one is {*GetOrder*, *ShippingOrder*} and the other is {*OrderBuilder*, *ShippingOrder*}.

Recall that each candidate web-service operation is assigned a score indicating similarity to the given description. Thus, each operation group acquires a *group score* by counting the sum of operation scores in it. A higher group score indicates a more desirable search result, so the user's top-$k$ requirements can be satisfied.

## 6 Experiments and Evaluations

We have implemented a prototype system, called WSXplorer, and conducted some experiments to evaluate the effectiveness and efficiency. The data set used in our tests is a group of web services collected from [25,26,27]. Their WSDL specifications are available so we can obtain the textual descriptions and XML schemas of input/output data types. The data contains 223 web services including 930 web-service operations. We chose 7 web-service operations from three domains: *order*(3), *travel*(2) and *finance*(2). Each operation description was used as the basis for desired operations.

We used *recall* and *precision* ratio to evaluate the effectiveness of our approach. The precision($p$) and recall($r$) are defined as $p = \frac{A}{A+B}, r = \frac{A}{A+C}$ where $A$ stands for the number of returned relevant operations, $B$ stands for the number of returned irrelevant operations, $C$ stands for the number of missing relevant operations, $A+C$ stands for the total number of relevant operations, and $A+B$ stands for the total number of returned operations. Specially, the top 100 search results were considered in our experiments for each web-service operation search.

We first evaluated the efficiency of WSXplorer by comparing the recall and precision of operation search with three other methods: keyword searching method, structure matching [9] and Woogle [13]. We computed the recall/precision ratio manully and plotted them in Fig. 5(a) and Fig. 5(b), respectively. As can be seen, the precisions of WSXplorer are 92%, 87% and 78% respectively, almost always outperforming that of keyword, structrure and Woogle. The precision is higher on *order* operations but lower in *finance* operations because *order* operations have more complex structures and richer semantics in input/output data types. This indicates that, by combining structural and semantic information, the precision of WSXplorer improves significantly, compared to the results obtained with structural or semantic information only. It is also can be seen that by keyword method the precision is rather low whereas the recall is rather high. This demonstrates textual description of operations contain much useful information but also much noise at the same time.

Then, we labeled the associated web-service operations in data set manually. The average recall/precision curve is used in Fig. 5(c) to evaluate the performance of WSXplorer on identifying associated operations. This figure illustrates that WSXplorer can achieve good recall and precision by integrating structural and semantic measurements.

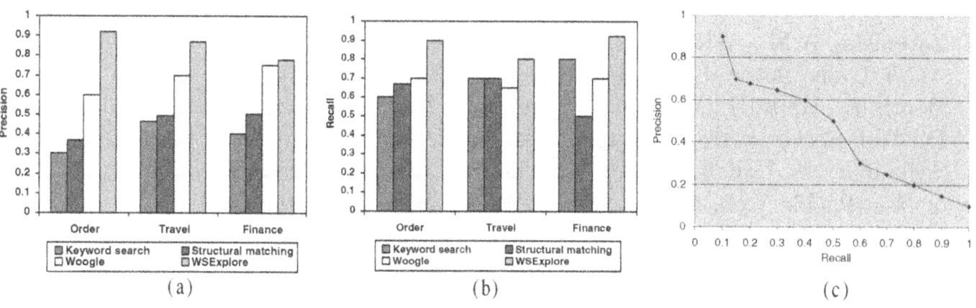


**Fig. 5.** Performance of WSXplorer

## 7 Conclusions

In this paper we have presented WSXplorer, a novel method to retrieve desired web-service operations of a given textual description. The concept of tree edit distance is employed to match web-service operations. Meanwhile, some algorithms are proposed for measuring and grouping similar operations. The proposed matching algorithm catches not only structures, but even better semantic information of schemas. We also introduced attribute closure for identifying associations between web-service operations. Our approach can be used for web-service searching tasks with top-$k$ requirements.

As part of on-going work, we are interested in improving efficiency of the web-service operation matching algorithm in terms of running time, since the computation of extended tree edit distance is costly. Our proposed technique assumes structures of XML schema are trees. However, their structures may also be graphs and contain cycles. In the future, we plan to extend our algorithm to support graph matching. In order to further understand the semantics of web services descriptions and integrate more semantic information to our system, we also plan to use WordNet to handle word stems and synonyms to improve the precision of our algorithm.

## References

1. Wang, H., Zhang, Y., Cao, J., Varadharajan, V.: Achieving secure and flexible M-services through tickets. IEEE Transactions on Systems, Man, and Cybernetics, Part A. 33(6), 697–708 (2003)
2. Bhiri, S., Perrin, O., Godart, C.: Ensuring required failure atomicity of composite Web services. In: Proc. of WWW Conference. pp.138–147 (2005)
3. Wang, H., Cao, J., Zhang, Y.: A Flexible Payment Scheme and Its Role-Based Access Control. IEEE Transactions on Knowledge and Data Engineering 17(3), 425–436 (2005)
4. Limthanmaphon, B., Zhang, Y.: Web Service Composition Transaction Management. In: Proceedings of Australasian Database Conference (ADC). pp.171–179 (2004)
5. Limthanmaphon, B., Zhang, Y.: Web Service Composition with Case-Based Reasoning. In: Proceedings of Australasian Database Conference (ADC). pp.201–208 (2003)
6. Zaremski, A.M., Wing, J.M.: Specification Matching of Software Components. ACM Trans. Softw. Eng. Methodol. 6(4), 333–369 (1997)
7. Booth, D., Haas, H., McCab, F., Newcomer, E., Champion, M., Ferris, C., Orchard, D.: Web Services Architecture. (2004) http://www.w3.org/TR/ws-arch/
8. Sajjanhar, A., Hou, J., Zhang, Y.: Algorithm for Web Services Matching. In: Proc. of Asia-Pacific Web. Conference (APWeb) 3007, 665–670 (2004)
9. Wang, Y., Stroulia, E.: Flexible Interface Matching for Web-Service Discovery. In: Proc. of International Conference on Web Information Systems Engineering (WISE) (2003)
10. Sycara, K.P., Widoff, S., Klusch, M., Lu, J.: Larks: Dynamic Matchmaking Among Heterogeneous Software Agents in Cyberspace. Autonomous Agents and Multi-Agent Systems 5(2), 173–203 (2002)

11. Roman, D., Lausen, H., Keller, U.: Web Service Modeling Ontology (WSMO). WSMO Final Draft 10 (2005)
12. Shen, Z., Su, J.: Web service discovery based on behavior signatures. In: Proc. of International Conference on Services Computing (SCC) 1, 279–286 (2005)
13. Dong, X., Halevy, A.Y., Madhavan, J., Nemes, E., Zhang, J.: Simlarity Search for Web Services. In: Proc. of International Conference on Very Large Data Bases (VLDB). pp. 372–383 (2004)
14. Hao, Y., Zhang, Y.: Web Services Discovery Based on Schema Matching. In: Proceedings of Australasian Computer Science Conference (ACSC) (2007)
15. Pu, K., Hristidis, V., Koudas, N.: Syntactic Rule Based Approach to Web Service Composition. In: Proc. of International Conference on Data Engineering (ICDE). vol.31 (2006)
16. Hristidis, V., Gravano, L., Papakonstantinou, Y.: Efficient IR-Style Keyword Search over Relational Databases. In: Proc. of VLDB. pp.850–861 (2003)
17. Hristidis, V., Papakonstantinou, Y.: DISCOVER: Keyword Search in Relational Databases. In: Proc. of VLDB). pp.670–681 (2002)
18. Agrawal, S., Chaudhuri, S., Das, G.: DBXplorer: A System for Keyword-Based Search over Relational Databases. In: Proc. of International Conference on Data Engineering (ICDE) (2002)
19. Salton, G.: The SMART Retrieval System - Experiments in Automatic Document Processing. Prentice-Hall, Inc, Upper Saddle River, NJ, USA (1971)
20. Salton, G., Wong, A., Yang, C.S.: A Vector Space Model for Automatic Indexing. Communications of the ACM (CACM) 18(11), 613–620 (1975)
21. Reis, D.D.C., Golgher, P.B., d.Silva, A.S., Laender, A.H.F.: Automatic web news extraction using tree edit distance. In: Proc. of WWW Conference. pp.502–511 (2004)
22. Zhang, K., Shasha, D.: Simple Fast Algorithms for the Editing Distance Between Trees and Related Problems. SIAM Journal on Computing 18(6), 1245–1262 (1989)
23. Xie, T., Sha, C., Wang, X., Zhou, A.: Approximate Top-k Structural Similarity Search over XML Documents. In: Proc. of Asia-Pacific Web. Conference (AP-Web) 3841, 319–330 (2006)
24. Kaufman, L., Rousseeuw, P.J.: Finding Groups in Data: An Introduction to Cluster Analysis. John Wiley, New York (1990)
25. http://www.xmethods.org. (XMethod)
26. http://www.bindingpoint.com. (BindingPoint)
27. http://www.webservicelist.com. (WebServiceList)

# $e^3$*forces* : Understanding Strategies of Networked $e^3$*value* Constellations by Analyzing Environmental Forces

Vincent Pijpers and Jaap Gordijn

Free University, FEW/Business Informatics,
De Boelelaan 1083a, 1081 HV Amsterdam, The Netherlands
(v.pijpers, gordijn)@few.vu.nl

**Abstract.** Enterprises increasingly form networked value constellations; networks of enterprises that can jointly satisfy complex consumer needs, while still focusing on core competencies. Information technology and information systems play an important role for such constellations, for instance to coordinate inter-organizational business processes and/or to offer an IT-intensive product, such as music or games. To do successful requirements engineering for these information systems it is important to understand its context; being here the constellation itself. To this end, business value modeling approaches for networked constellations, such as $e^3$*value* , BMO, or REA, can be used. In this paper, we extend these business value modeling approaches to understand the *strategic rationale* of business value models. We introduce two dominant schools on strategic thinking: (1) the "environment" school and (2) the "core competences" school, and present the $e^3$*forces* ontology that considers business strategy as a positioning problem in a complex environment. We illustrate the practical use and reasoning capabilities of the $e^3$*forces* ontology by using a case study in the Dutch aviation industry.

## 1 Introduction

With the rise of the world wide web, enterprises are migrating from participation in linear value chains [15] to participation in *networked value constellations*, which are sets of organizations who *together* create value for their environment [17]. Various ontologically founded modeling techniques have been developed to analyze and reason about business models of networked value constellations. Worth mentioning are: $e^3$*value* , developed by Gordijn and Akkermans, showing how objects of value are produced, transferred, and consumed in a networked constellation [8,9]; *BMO*, developed by Osterwalder and Pigneur, expressing the business logic of firms [14]; and finally, *REA*, developed by Geerts and McCarthy, taking an accounting view on the economic relationship between various economic entities [7].

All three techniques are able to analyze the business model of a networked value constellation and are able to link the business model to the constellations IT infrastructure (eg. [6]). But, although the importance of *strategy* on

J. Krogstie, A.L. Opdahl, and G. Sindre (Eds.): CAiSE 2007, LNCS 4495, pp. 188–202, 2007.

business models and IT *and* IT on strategy has been stressed by multiple authors (eg. [2, 11]), these techniques do not consider *strategic* motivations of organizations underpinning the networked value constellation [18]. The mentioned techniques mainly provide a (graphical) representation of *how* a constellation looks like in terms of participating enterprises and *what* these enterprises exchange of economic value with each other, but do not show *why* a business model is as it is. By looking at strategic dependencies and strategic rationales of actors in a constellation, *i* (eye-star)*, developed by Yu and Mylopoulos, *does* take the "why" into consideration [19,21]. The *i** concepts of "strategic dependency" and "strategic rationale" are however grounded in quite general *agent*-based theories and not in specific *business strategy* theories. To put it differently, well known basic business strategy concepts such as "core competences", "competitive advantage" and "environment" are not considered in *i** explicitly.

Our contribution is to add to the existing *business model* ontologies (which formalize theory on networked value constellations, thereby enabling computer-supported reasoning about these) a *business strategy* ontology. This business strategy ontology is based on accepted business strategy theories. An important requirement for such an ontology is that it represents a *shared* understanding [4]. By using *accepted* theories we conceptualize a shared understanding of "business strategy" as such. In a multi-enterprise setting, as a networked value constellation is, a shared understanding is obviously essential to arrive at a sustainable constellation. Shared and better understanding of strategic motivations underpinning a networked value constellation is not only important from a business perspective, but also from an IT perspective (see eg. [2, 11]).

There are at least two distinctive, yet complementary, schools on "business strategy". One school considers the *environment* of an organization as an important strategic motivator; the other school focuses on *internal competences* of an organization. The first school originated from the work of Porter [15, 16], and successors [17]. It believes that *forces* in the *environment* of an organization determine the strategy the organization should chose. An organization should position itself such that competitive advantage is achieved over the competition and threats from the environment are limited. The second school considers the *inside* of an organization to determine the best strategy. This school is rooted in the belief that an organization should focus on its *unique resources* [3] and *core competences* [12]. Core competences are those activities which with an organization is capable of making solid profits [12]. According to this school, the best path to ensure the continuity of the organization is to focus on the unique resources and core competences the organization posses.

In this paper Porter's five-forces model [15, 16] will be used to create an ontology, named $e^3$*forces* , which provides a graphical and semi-formal model of environmental forces that influence actors in a networked value constellation. The $e^3$*forces* ontology will provide a means to reason about strategic considerations (the "why") of a business model in general, and specifically an $e^3$*value* model [8,9]. So, the $e^3$*forces* ontology bridges Porter's five forces framework and

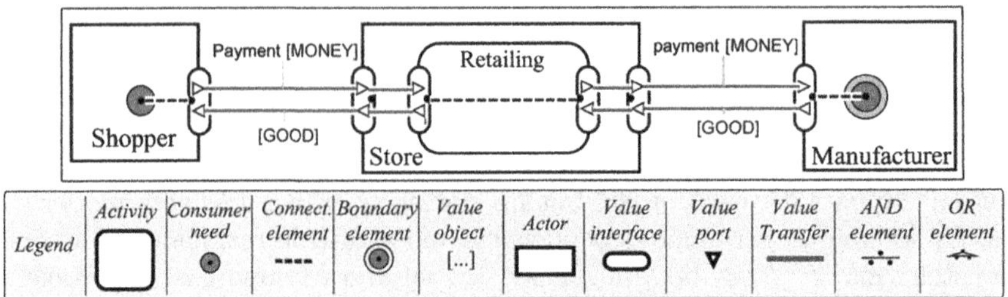


**Fig. 1.** Educational example

the $e^3$*value* ontology by representing how *environmental forces* influence a *business value model.*

The paper is structured as follows. First, to make the paper self-contained, we briefly present the $e^3$*value* ontology. Second, an industrial strength case study will be introduced, which is used to develop and exemplify the $e^3$*forces* ontology. Then we present the conceptual foundation of the $e^3$*forces* ontology. Subsequently, we show, using the ontological construct, how the environment of a constellation may influence actors in this constellation for the case at hand, and we show how to reason with the $e^3$*forces* ontology. Finally, we present our conclusions.

## 2 The $e^3$*value* Ontology

The aim of this paper is to provide an ontologically well founded motivation for business value models of networked value constellations in terms of business strategies. Since we use $e^3$*value* to model such constellations, we summarize $e^3$*value* below (for more information, see [9]). The $e^3$*value* methodology provides modeling constructs for representing and analyzing a network of enterprises, exchanging things of economic value with each other. The methodology is ontologically well founded and has been expressed as UML classes, Prolog code, RDF/S, and a Java-based graphical $e^3$*value* ontology editor as well as analysis tool is available for download (see http://www.e3value.com) [9]. We use an educational example (see Fig. 1) to explain the ontological constructs.

*Actors* (often enterprises or final customers) are perceived by their environment as economically independent entities, meaning that actors can take economic decisions on their own. The Store and Manufacturer are examples of actors. *Value objects* are services, goods, money, or even experiences, which are of economic value for at least one of the actors. Value objects are exchanged by actors. *Value ports* are used by actors to provide or request value objects to or from other actors. *Value interfaces*, owned by actors, group value ports and show economic reciprocity. Actors are only willing to offer objects to someone else, if they receive adequate compensation in return. Either all ports in a value interface each precisely exchange one value object, or none at all. So, in the example,

Goods can only be obtained for Money and vice versa. *Value transfers* are used to connect two value ports with each other. It represents one or more potential trades of value objects. In the example, the transfer of a Good or a Payment are both examples of value transfers. *Value transactions* group all value transfers that should happen, or none should happen at all. In most cases, value transactions can be derived from how value transfers connect ports in interfaces. *Value activities* are performed by actors. These activities are assumed to yield profits. In the example, the value activity of the Store is Retailing. *Dependency paths* are used to reason about the number of value transfers as well as their economic values. A path consists of *consumer needs*, *connections*, *dependency elements* and *dependency boundaries*. A consumer need is satisfied by exchanging value objects (via one or more interfaces). A connection relates a consumer need to a value interface, or relates various value interfaces internally, of a same actor. A path can take complex forms, using AND/OR dependency elements taken from UCM scenarios [5]. A dependency boundary represents that we do not consider any more value transfers for the path. In the example, by following the path we can see that, to satisfy the need of the Shopper, the Manufacturer ultimately has to provide Goods.

## 3 Case Study: Dutch Aviation Constellation

To develop and test the $e^3$*forces* ontology we conducted a case study at the Dutch aviation industry, in which multiple organizations cooperate to offer flights to, from, and via the Netherlands. From the large number of actors in the Dutch Aviation constellation we have chosen only key players for further analysis. The key players were identified with the help of a "power/interest matrix" [12]. *Power* is defined as the capability to influence the strategic decision making of other actors [12]. An actor can do so when s/he is able to influence the capacity or quality of the products/services offered by others to the environment. *Interest* is defined as the active attitude and amount of activities taken to influence the strategic choices of other actors. The matrix axis' have the value high and low. Actors with high interest and high power are considered key players [12]. As a result, we identified the following key actors: (1)*Amsterdam Airport Schiphol*, hereafter referred to as "AAS", is the common name for the organization NV Schiphol Group, who owns and is responsible for the operations of the actual airport Schiphol. "AAS" 's core business activity is to provide infrastructural services, in the form of a physical airport and other necessary services, to various other actors who exploit these facilities. (2)*AirFrance-KLM*, hereafter referred to as "KLM", This hub carrier is a recent merger between "AirFrance" and "KLM". Because one of the home bases of "KLM" is Amsterdam, they are part of the Dutch aviation industry. "KLM" is responsible for the largest share of flights to, from and via "AAS". The core business of "KLM" is to provide (hubbed) air transportation to customers such as passengers and freight transporters. (3) *Air Traffic Control*, hereafter referred to as "ATC", is responsible for guiding planes through Dutch airspace, which includes the landing and take-off of planes at

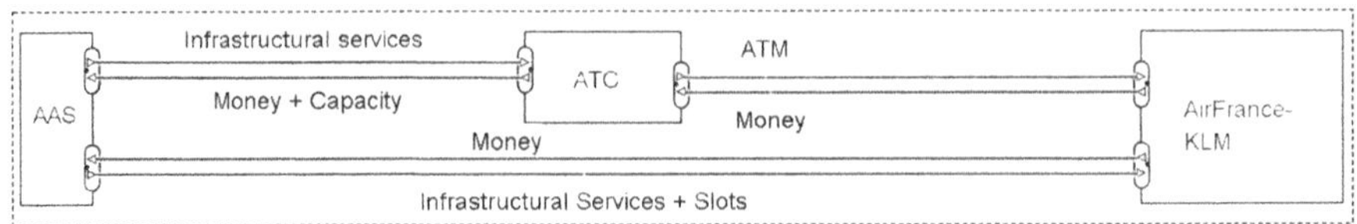


**Fig. 2.** The Dutch Aviation Constellation

"AAS". This service is called "Air Traffic Management", which is the core business activity of "ATC".

Fig. 2 shows an introductionary $e^3$*value* model for the Dutch aviation constellation. "AAS" offers infrastructural services (e.g. baggage handling) plus landing and starting slots to "KLM", who pays money for this. In addition, "AAS" offers to "ATC" infrastructural services (e.g. control tower), and gets paid for in return (and also gets landing and starting capacity). Finally, "ATC" provides "KLM" with "Air-Traffic Management", and gets paid in return. We will use this baseline value model to develop and demonstrate $e^3$*forces* , motivating the value model at hand. A more comprehensive model, *with* the environmental forces, can be found in Fig. 6.

## 4 The $e^3$*forces* Ontology

The $e^3$*forces* ontology extends existing business value ontologies by modeling their strategic motivations that stem from environmental forces. Because an ontology is a formal specification of a shared conceptualization, with the purpose of creating shared understanding between various actors [4], most concepts are based on broadly *accepted* knowledge from either business literature (eg. [15, 13, 12]) or other networked value constellation ontologies (eg. [8, 21]).

Although the $e^3$*forces* ontology is closely related to the $e^3$*value* ontology, with the advantage that consistency is easily achieved and both models could be partly derived from one another, they *significantly differ*. The focus of $e^3$*value* is on *value transfers* between actors in a constellation and their *profitability*. Factors, other then value transfers, that influence the relationship between actors are *not* considered in the $e^3$*value* ontology. In contrast the $e^3$*forces* ontology *does* consider factors in the *environment* which *influence* the constellation. Instead of focusing on value transfers, $e^3$*forces* focuses on the *strategic position* of a constellation in its environment. Below, we introduce $e^3$*forces* 's constructs (due to lack of space, we do not show the ontology in a more formal way, such as in RDF/S or OWL):

*Constellation.* A constellation is a *coherent* set of two or more actors who *cooperate* to create value to their environment [17]. As in $e^3$*value* , actors are independent economic (and often also legal) entities [13, 12]. Obviously, we need a criterion to decide whether an actor should be in a constellation or not. For each of the actors in the constellation it holds that if the actor would seize its core business, then all other actors would not be able to execute a certain share (roughly 50% or more) of their core business or a certain share would no longer be valuable. The required share expresses the supposed coherence in the constellation. For example, "AAS", "KLM" and "ATC" form a constellation because if one of the actors would seize its activities the other actors would not be able to perform their core business, or their core business would loose its value. In an $e^3$*forces* model the constellation itself shows up as a dashed box that surrounds the actors it consists of. The actors are related using value transfers, cf. $e^3$*value* [8,9].

*Market.* A constellation operates in an *environment* [12,15] consisting of *markets*. Markets are sets of actors in the environment of the constellation (modeled as a layered rectangle). The actors in a market 1) are *not part* of the constellation 2) operate in the *same industry* as the constellation 3) are considered as *peers*; they offer similar or even equal value objects to the world 4) are in terms of $e^3$*value* value transfers cf. [8] (in)directly *related* to actors in the constellation [15]. For instance carriers form a market, because they include all carriers not part of the Dutch aviation constellation, have economic relationships with actors in the constellation, are in the same industry and, carriers offer similar value objects to their environment. Note that although "KLM" is a carrier they are not part of the "Carrier" market, because they are already part of the constellation. The organizations are grouped in a market because by considering sets of organizations, we abstract away from the individual and limited [15] influence on actors in the constellation of many single organizations. Therefore, the notion of "market" is motivated by the need to reduce modeling and analysis complexity. By doing so, we consider forces between *actors in the constellation* and specific *markets in the environment*, rather than the many forces between actors in the constellation and each *individual* actor in the environment.

*Dominant Actor.* A market may contain *dominant actors*. Such actors have a power to influence the market and thus actors in the constellation. If a market is constructed out of a single large organization and a few small organizations, then it is the large organization who determines the strength of a market and is it less relevant to consider the small organizations. Usually dominant actors posses a considerable large share of the market. What is "considerable large" depends on the industry in which the analysis is performed. For instance in the market of operation systems Microsoft (over 70% market share) is a dominant actor, while Toyota can be considered a dominant actor in the automotive industry with only 13% market. Dominant actors are modeled as a rectangle *within* an market.

*Submarket.* It is possible to model *submarkets* of a market. A submarket is a market, but has a *special type* of value object that is offered or requested from the constellation. For instance, *low cost* carriers are a submarket of the *carrier* market. A submarket is shown in the interior of a market.

*Industry.* An industry unites all actors shown in an $e^3$*forces* model. So, the actors of the constellation, and actors in a (sub)market are all in an *industry*.

*Force.* Markets in the environment of a constellation influence actors in the constellation, by exercising a *force*, this is expressed by a "strength" arrow. Such an arrow is shown near an $e^3$*value* value transfer. In the following sections, we illustrate specific forces, as derived from Porter's five forces model [15].

# 5 Modeling Porter's Five Forces Using $e^3$*forces*

Using the $e^3$*forces* ontology, we model various forces between actors and markets. Porter distinguishes five kinds of forces [12,15,16]: *bargaining power of suppliers, bargaining power of buyers, competitive rivalry among competitors, threat of new entrants* and *threat of substitutions.*

## 5.1 Bargaining Power of Suppliers

Suppliers are those organizations which are part of the environment of a constellation (because they do not satisfy the previously discussed "coherence" criterion) and *provide* value objects to actors in the constellation [12]. For the case at hand, suppliers are e.g. "Airplane Manufacturers". Suppliers influence actors in a constellation by threatening to alter the configuration of goods/services, to increase the price or to limit availability of products [12, 15]. These are changes related to the value objects and/or their transfers between actors and their environment. So, a first step is to elicit (important) suppliers for each actor part of the constellation. Suppliers are identified by finding organization which *provide* value objects *to* the constellation, but who are *not* part of the constellation.

Next the strength of the bargaining power of the suppliers in relationship to the actors in the constellation must be analyzed. According to [15], five factors determine the strength of a supplier market: 1) *The concentration of (dominant) suppliers.* Suppliers are able to exert more influence if they are with few and when buyers are fragmented. 2) *The necessity of the object provided by the suppliers.* If the value object is essential then the actors in the constellation can make less demands. 3) *The importance of actors in the constellation to the suppliers.* If actors in the constellation are not the supplier market's main buyer, then the supplier is stronger. 4) *The costs of changing suppliers.* If the costs are high, then actors in the constellation are less likely to choose another supplier, which give the supplier more strength. 5) *Threat of taking over an actor in the constellation.* The supplier might plan to take over an actor in the constellation to strengthen its position in the environment.

Using these questions, the relative strength of the power of a supplier market is determined for each transfer (connected to an actor in the constellation), and is shown as a *strength arrow* along the lines of the connected value transfers (which are the transfer of the value object provided by the supplier market to the actor in the constellation *and* the transfer of the value object provided as a compensation (e.g. money)). Note that since we model the power the supplier market exercises over an actor in the constellation, the strength arrow always points from the supplier's interface of the market *toward* the buyer interface of

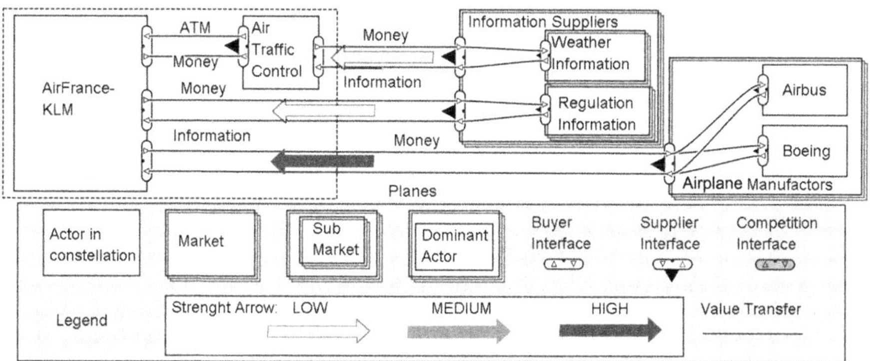


**Fig. 3.** $e^3$*forces* :Suppliers

the actor in the constellation. The relative strength of the arrow is based on the analysis of the supplier market given above. Also note that a market can be a *supplier* market, a *buyer* market, a *competition* market or any combination, since markets can have *supplier interface(s)* and/or *buyer interface(s)*, depending on the role. A supplier interface is, via value transfers, connected to a buyer interface of an actor in the constellation.

Fig. 3 demonstrates some supplier forces for the case at hand. For example "Airplane Manufacturers" is a supplier market to "KLM", having two dominant actors: "Boeing" and "Airbus". This market exercises a power of *high* strength because: a) there is a concentration of dominant suppliers, b) the value object is essential to "KLM", and c) "KLM" is only one of many buyers. Due to lack of space, we can not explain each power relation in a more detailed way.

## 5.2 Bargaining Power of Buyers

Buyers are environmental actors that *acquire* value objects from actors in the constellation [12]. Buyers can exercise a force because they negotiate down prices, bargain for higher quality, desire more goods/services and, try to play competitors against each other [15, 16]. All this is at the expense of the profitability of the actors in the constellation [15, 16]. Buyer markets have value transfers with actors in the constellation similar to supplier markets.

After eliciting possible buyer markets, the strength of the power they exercise is analyzed. According to [15], seven factors determine the strength of buyer markets: 1) *The concentration of (dominant) buyers.* If a few large buyers acquire a vast amount of sales, then they are very important to actors in the constellation, which gives them more strength. 2) *The number of similar value objects available.* A buyer market is stronger, if there is a wide range of suppliers from which the buyer market can chose. 3) *Alternative resources of supply.* If the buyer market can chose between many alternative value objects then the buyer market is powerful. 4) *Costs of changing supplier.* If costs are low, then buyers can easily choose another supplier, which gives the buyer market strength. 5)

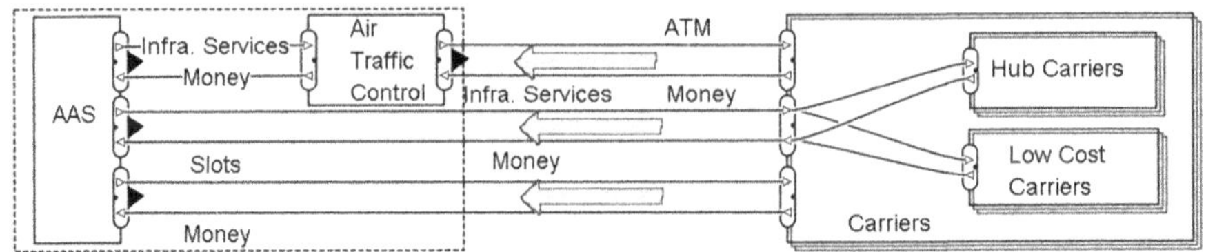


**Fig. 4.** $e^3$*forces* : Buyers

*The importance of the value object.* If the value object is not important to the buyer market, it is harder for actors in the constellation to maintain an economic feasible relationship. 6) *Low profits.* The actors in the constellation have to sell large volumes to make profits, giving the buyer market more bargaining power. 7) *Threat of taking over an actor in the constellation.* A buyer is willing and capable to purchase an actor in the constellation, which the purpose to strengthen its own position.

Similar to supplier markers, by using these questions, the relative strength of the power of a buyer market is determined for each transfer (connected to an actor in the constellation), and is shown as a *strength arrow* along the lines of the connected value transfer.

In Fig. 4, two actors of the constellation are given: "AAS "and "ATC". One buyer market (carriers) is modeled, in which two submarkets are present ("Hub Carriers" and "Low Cost Carriers"). "ATC" provides a service to the entire carrier market, resulting in a low strength. "AAS" provides "Infrastructural Service" to "Carriers", but these services slightly differ for "Hub Carriers" and "Low Cost Carriers". Consequently, both submarkets are connected to the buyer interface of the entire market. This buyer market is in turn connected to the supplier interface of the "AAS".

### 5.3 Competitive Rivalry Among Competitors

An additional force is exercised by *competitors*; actors that operate in the same industry as the constellation and try to satisfy the same needs of buyers by offering the same value objects to buyer markets as the constellation does [12]. Competitors are a threat for actors because they try to increase their own market share, influence prices and profits and influence customer needs; in short: they create competitive rivalry [15,16].

So far, forces exercised by markets on actors in the constellations have been expressed along the lines of *direct* value transfers between markets and actors. Such a representation can not be used anymore for modeling competitive rivalry. In case of competitive rivalry, (competitive) markets aim to transfer same value objects to the same buyer markets as the actors in the constellation do. Consequently, competitive rivalry is represented as: a) value transfers of a constellation's actor to a *buyer* value interface of a (buyer) market, *and* b) *competing*

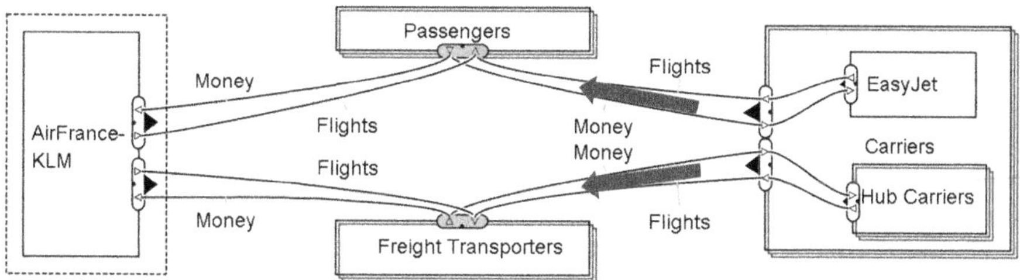


**Fig. 5.** $e^3$*forces* : Competitors

transfers of a competition market to the *same* buyer interface of the market. The extent of competitive rivalry is expressed by incorporating a *strength arrow* that points from the competition market toward the *buyer market.* This is because competitive rivalry, as expressed by the strength arrow, is located at the *buyer market*, and *not* at the actor in the constellation [15]. The buyer interface of a market for which competition occurs is called the "competition" interface, and is explicitly stated. Also, it is worthwhile to show dominant actors for a competitive market; these are considered the most important competitors.

To decide upon the strength of the competitive force, seven factors are used [15]: 1) *The balance between competitors.* If competitors are equal in size, strength and market share, then it is harder to become a dominant actor, which leads to more rivalry. 2) *Low growth rates.* If industry growth rates are low then competitors have to make more effort to increase their own growth rates, which leads to higher competitive rivalry. 3) *High fixed costs for competitors.* This can result in price-wars and low profit margins, which increase competitive rivalry. 4) *High exit barriers.* In this case competitors cannot easily leave the market. To remain profitable they will increase their effort to increase or maintain their market share. 5) *Differentiation between competitors.* If there is no difference between value objects offered by competitors, then it is harder to sell value objects to customers. 6) *Capacity augmented in large increments.* This can lead to recurring overcapacity and price cutting. 7) *Sacrificing profitability.* If actors are willing to sacrificing profitability to increase market share and achieve strategic goals, other organization have to follow; leading to more competition. [15].

Fig. 5 shows that the constellation "KLM", has two buyer markets; "Freight Transport" and "Passengers". In the competition market "Carriers" a *submarket* is modeled and a *dominant actor.* The submarket "Hub Carriers" is connected with its own supplier interface, and via an interface of the total market, to the buyer market "Freight Transport". This indicates that this *submarket* is responsible for the competitive rivalry at the buyer market and *not* the entire carrier market. Furthermore, the dominant actor modeled, "EasyJet", is connect to the "Passengers" buyer market. This indicates that this particular actor is responsible for a large amount of the competitive rivalry at the "Passengers" buyer market.

### 5.4 Threat of New Entrants

Potential *entrants* are actors who *can become* competitors, but who are currently *not*, or who do not exist yet [12, 15]. Consequently, we consider new entrants as a *future* competitive market. To determine the threat of a potential entrant, the following aspects need to be analyzed [15]: 1) The *economics of scale* needed to become profitable. 2) The *capital* required to facilitate the entry in an industry. 3) The extent of *access to distribution channels* are accessible. 4) The *experience and understanding* of the market of the new entrant. 5) The *possibility of retaliation* by existing organizations in an industry, with the goal to force new entrants out of the industry. 6) *Legal restraints* which place boundaries on potential entrants. 7) The difficulty of *differentiating* from existing organizations.

Potential entrants are modeled (as rounded squares) *within* a competitive market and labeled after the potential entrant. Furthermore, the potential entrant has a supplier interface which is connected to the relevant supplier interface of the competition market. The threat of a potential entrant is expressed by a strength arrow, which originates at the potential entrant and point toward the supplier interface of the entire competition market. The strength of the arrow is based on the analysis of potential entrants given above.

### 5.5 Threat of Substitutions

Actors may offer *substitutions*, so different value objects, to a buyer market, yet satisfy the same need of the buyers [12,15]. Substitution markets are seen as competitive markets who offer different value objects, as an alternatives to objects offered by actors in the constellation, to the *same* buyer markets. Substitution markets are modeled in the same way as competition markets, but value objects of actors in the constellation and of the substitution markets differ. In brief, the strength of the arrow is determined by the likelihood that the substitution will reduce the market share of the constellation for this buyer market [15, 16].

## 6 An $e^3$*forces* Model for the Dutch Aviation Industry

Fig. 6 shows an $e^3$*forces* model for the Dutch aviation constellation. It first shows how the key actors are internally and externally connected in terms of $e^3$*value* value transfers. Furthermore, the strengths of the forces that influence the (actors in the) constellation are shown. A number of small suppliers, who have low strength, are grouped into "supplier" markets for space purposes.

At a first glance, the model shows that environmental forces have the least impact on "ATC". Moreover, "ATC" does not have any competitors. Second, the model shows that "AAS" mostly acts as a provider and that environmental forces have a low impact on "AAS": most forces have low strength. The third actor, "KLM", has to deal with the strongest forces. This is due to the competitive rivalry at the buyer markets of "KLM".

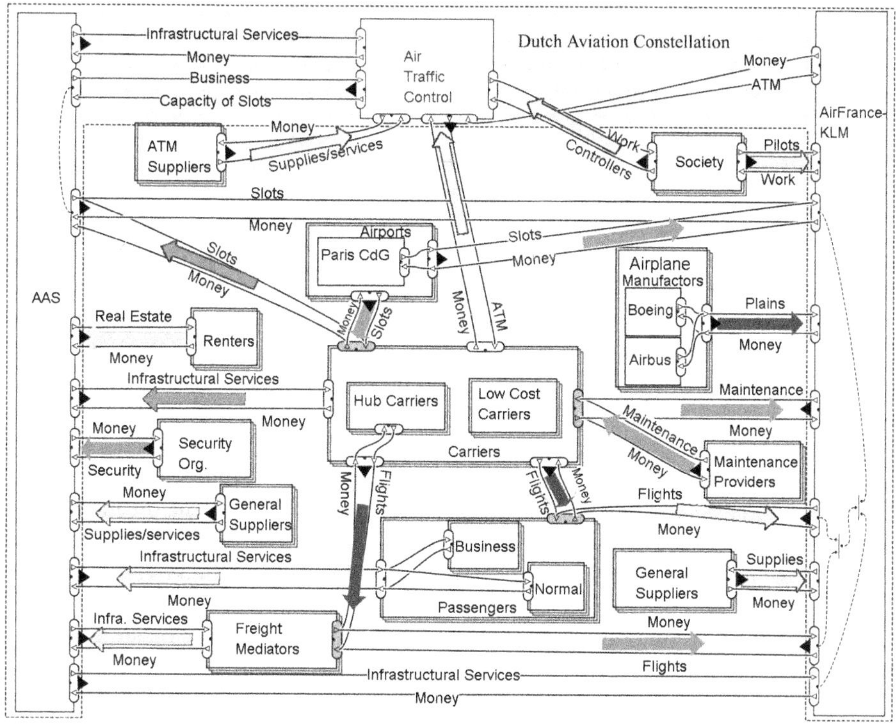


Fig. 6. $e^3$*forces* : Complete

## 6.1 Reasoning with $e^3$*forces* and Practical Use for Information Systems

The aim of the $e^3$*forces* ontology is to understand strategic considerations of actors in a constellation in terms of environmental forces. Is this possible? With the aid of the $e^3$*forces* model we are able to understand that: (1) As a result of the high competitive rivalry at "KLM" 's buyer markets (See Fig. 6), "KLM" needs to reduce costs per unit through economics of scale (eg. increase capacity) to remain profitable [15]. For achieving this goal "KLM" partly depends on services provided by "AAS" and "ATC", as seen by the *dependency relations* between the actors, which we have introduced in the model to facilitate dependency-tracing reasoning (see e.g. $i*$ [21,19] and $e^3$*value* [9] for examples of such reasoning). This motivates "KLM" desire for improved inter-organizational operations. (2) "AAS", although in a constellation with "KLM", provides value objects to competitors of "KLM"; possibly leading to conflicts. Furthermore, due to the high rivalry between carriers and their medium strength, there is pressure on the profits margins of the value objects offered by "AAS" to the carriers (See Fig. 6). Therefore "AAS" is also exploiting other buyer markets (eg. "Renters") to generate additional profits. Finally, "AAS" partly depends on "ATC", which motivates their desire for better inter-organizational operations. (3) “ATC” is dependent on by “AAS” and “KLM” , but is in a luxury position due to the monopoly it possesses. “ATC” however only has one buyer: “AAS” (See Fig. 6). Therefore “ATC” is willing to cooperate with “AAS” and “KLM” to improve operations and increase profits.

In addition, ontologies such as $e^3$*value* , $i*$ and $e^3$*forces* are most relevant for the early phases of requirements engineering [20]. Information system analysts can use such analysis methods for better understanding their organization and designing processes and IT accordingly [18]. For instance, it is understood that “electronic marketplaces” can be exploited for strategic purposes [1], but $e^3$*forces* aids in understanding *where* (eg. which markets) and *how* (eg. limitations enforced by forces) electronic marketplaces can be exploited. It is also possible to use $e^3$*forces* model to analyze changes in the environment of the constellation when for instance an electronic marketplace is introduced. To illustrate we use the well known e-ticket system. Introducing the e-ticket system has enabled carriers to sell tickets directly to passengers, meaning that mediators are no longer necessary. In this new situation carriers are no longer dependent on mediators. Furthermore the relationship between carriers and passengers is now direct. The application of IT has thus changed the environment of the constellation. For information system developers it is important to understand that users of the e-ticket system are primarily passengers and secondary mediators (assuming here that passengers have different needs for the e-ticket system than mediators).

An $e^3$*forces* model can also be used for reasoning about the sustainability of competitive advantage achieved by exploiting IT. If for instance “KLM” would introduce an electronic marketplace for the freight market they would create *competitive advantage* over the competition (assuming lower costs for “KLM”). Due to the *high competitive rivalry* at this market, as seen in the model, it is important for organizations to maintain a profitable market share [15]. Therefore competitors will also invest in electronic marketplaces, thereby reducing the competitive advantage of “KLM”. IS developers can use this information to understand that the IS will only generate additional profits in the early phase of its life cycle and that additional or new innovations need to be developed to sustain competitive advantage.

## 7 Related Work

Closely related to this research is the work performed by Weigand, Johannesson, Andersson, Bergholtz, Edirisuriya and Ilayperuma [18]. They propose the c3-value approach in which the $e^3$*value* ontology [8,9] is extended to do competition analysis, customer analysis and to do capabilities analysis. They, however, do not provide a complete set of constructs or methodologies for the three models. Therefore the models are currently quite abstract and give rise to both modeling and conceptual questions. Furthermore, the authors seem to focus more on the composition of value objects (in terms of second order value transfers), than on the *strategic motivation* for a business value model.

Also related to this research is the work done by Gordijn, Yu and Van der Raadt [10]. In this research, the authors try to combine $e^3$*value* and *i**, with the purpose to better understand the strategic motivations for e-service business models. The $e^3$*value* model is used to analyze the profitability of the e-services; *i** is used to analyze the (strategic) goals of the participants offering/requesting the e-services. The $e^3$*forces* ontology adds a specific *vocabulary* on business strategy, which is lacking in both $e^3$*value* and *i**.

## 8 Conclusion

With the aid of an industrial strength case study we were able to create an ontology for modeling and analyzing the forces that influence a networked value constellation. By using the $e^3$*value* ontology and Porter's Five Forces framework as a basis, we used existing and accepted knowledge on networked value constellations and environmental influences on business strategies to create a solid theoretic base for the $e^3$*forces* ontology. This solid theoretic base enabled us to reason about the configuration of networked value constellations; as demonstrated by the case study. In this study we presented a clear model of 1) the value transfers *within* the constellation, but more important: 2) the value transfers *between* actors in the constellation and *markets* in the *environment* of the constellation and, 3) the *strength* of forces, created by the markets, which influence actors in the constellation. Via this model and strategy theories we were able to use semi-formal reasoning to explain dependencies between actors. In addition we were able to analyze the position and roles of the actors in the constellation. This enabled use to reason about the configuration of the networked value constellation by considering the question of "Why".

The $e^3$*forces* ontology is a step to arrive at a more comprehensive $e^3$*strategy* ontology which can be used to capture the business strategy goals of organizations in networked value constellation. In future research, we complement $e^3$*strategy* with a more *internal competencies*-oriented view on the notion of business strategy.

**Acknowledgments.** The authors wish to thank Paul Riemens, Hans Wrekenhorst and Jasper Daams from Air-Traffic Control The Netherlands for providing case study material and for having many fruitful discussions. This work has been partly sponsored by NWO project COOP 600.065.120.24N16.

## References

1. Bakos, J.Y.: A strategic analysis of electronic marketplaces. MIS Quarterly 15(3), 295–310 (September 1991)
2. Bakos, J.Y., Tracy, M.E.: Information technology and corporate strategy: A research perspective. MIS Quarterly 10(2), 107–119 (June 1986)
3. Barney, J.B.: The resource-based theory of the firm. Organization Science 7(5), 131–136 (1994)

4. Borst, W.N., Akkermans, J.M., Top, J.L.: Engineering ontologies. International Journal of Human-Computer Studies 46, 365–406 (1997)
5. Buhr, R.J.A.: Use case maps as architectural entities for complex systems. Software Engineering 24(12), 1131–1155 (1998)
6. Derzsi, Z., Gordijn, J., Kok, K., Akkermans, H., Tan, Y.H.: Feasibility of it-enabled networked value constellations: A case study in the electricity sector.(2007) (Accepted at CAISE (2007))
7. Geerts, G., McCarthy, W.E.: An accounting object infrastructure for knowledge-based enterprise models. IEEE Intelligent Systems and Their Applications, pp. 89–94 (July- August 1999)
8. Gordijn, J., Akkermans, H.: E3-value: Design and evaluation of e-business models. IEEE Intelligent Systems 16(4), 11–17 (2001)
9. Gordijn, J., Akkermans, H.: Value based requirements engineering: Exploring innovative e-commerce idea. Requirements Engineering Journal 8(2), 114–134 (2003)
10. Gordijn, J., Yu, E., Van Der Raadt, B.: E-service design using i* and e3value modeling. IEEE Software 23(3), 26–33 (2006)
11. Hidding, G.J.: Sustaining strategic advantage in the information age. In: Proceedings of the 32nd Hawaii International Conference on System Sciences, IEEE, Orlando (1999)
12. Johnson, G., Scholes, K.: Exploring Corporate Strategy. Pearson Education Limited, Edinburgh, UK (2002)
13. Mintzberg, H.: The Structur of Organizations. Prentice-Hall, New York (1979)
14. Osterwalder, A.: The Business Model Ontology - a proposition in a design science approach. PhD thesis, University of Lausanne, Lausanne, Switzerland (2004)
15. Porter, M.E. (ed.): Competetive Strategy. Techniques for analyzing industries and competitors. The Free Press, New York (1980)
16. Porter, M.E. (ed.): Competitive advantage. Creating and sustaining superior performance. The Free Press, New York (1985)
17. Tapscott, D., Ticoll, D., Lowy, A.: Digital Capital - Harnessing the Power of Business Webs. Harvard Business School Press, Boston, MA (2000)
18. Weigand, H., Johannesson, P., Andersson, B., Bergholtz, M., Edirisuriya, A., Ilayperuma, T.: Strategic analysis using value modeling - the c3-value approach. In: Proceedings of the 32nd Hawaii International Conference on System Sciences, IEEE, New York (2007)
19. Yu, E.: Models for supporting the redesign of organizational work. In: COCS '95: Proceedings of conference on Organizational computing systems, pp. 226–236. ACM Press, New York (1995)
20. Yu, E.: Towards modelling and reasoning support for early-phase requirements engineering. In: Proceedings of the 3rd IEEE Int. Symp. on Requirements Engineering (RE'97), pp. 226–235 (1997)
21. Yu, E., Mylopoulos, J.: An actor dependency model of organizational work - with application to business process reengineering. In: COCS '93: Proceedings of the conference on Organizational computing systems, pp. 258–268. ACM Press, New York (1993)

# Aligning IS to Organization's Strategy: The INSTAL Method

Laure-Hélène Thevenet[1,2] and Camille Salinesi[1]

[1] Université Paris 1 - Panthéon Sorbonne, CRI, 90 rue de Tolbiac
75013 Paris, France
[2] BNP Paribas, Système d'Information Groupe, 41 rue de Valmy
93100 Montreuil Sous Bois, France
Laure-Helene.Thevenet@malix.univ-paris1.fr,
Camille.Salinesi@univ-paris1.fr

**Abstract.** Aligning Information Systems (IS) to organization's strategic business objectives is one of organizations' top preoccupations. Misalignment is considered as a reason of IT's failure to improve organizational performance. If strategic alignment is relatively simple to understand, it is not so easy to implement. Our experience showed us that organizations are not really able to systematically evaluate whether there is alignment, mainly because of the lack of documentation on strategic alignment. This paper intends to deal with this issue by proposing an approach to describe organizations' strategic objectives and its IS, in order to document and analyze strategic alignment, i.e. how the IS contributes to strategic objectives satisfaction. The proposed method, called INSTAL (Intentional Strategic Alignment), reuses organization documents as a basis to formalize strategic alignment. INSTAL was created following the principles of an action research approach, which consists in developing the approach while exploring issues raised by the case study.

**Keywords:** Alignment, Strategic Alignment Documentation, Organization strategy, Requirements Engineering.

## 1 Introduction

As [1] [2] [3] [4] or [5] already showed, aligning Information Strategy (IS) and business objectives has been considered as a top priority by CIOs and IT executives since several years.

There is a large corpus of empirical and theoretical evidence that alignment improves organizational performance (e.g. [6], [7], [8], [9]). Indeed this latter depends on structures and capabilities that support the successful realizations of strategic decisions. Furthermore, studies highlight the lack of alignment as a major cause for business processes failure in providing return on investments.

Although it is admitted that alignment is impacted by the changing environment, it is still unclear how to achieve and sustain strategic alignment over time. IS and business processes must support the strategy, i.e. the organization's strategic business objectives. Therefore, IS/IT must be deployed to help meeting those objectives.

J. Krogstie, A.L. Opdahl, and G. Sindre (Eds.): CAiSE 2007, LNCS 4495, pp. 203–217, 2007.

However, strategic alignment, i.e. the synergy between the strategic level (organization strategy) and the operational level (BP/IS), must also be controlled and maintained over time despite possible evolutions of IS/IT, of the organization strategy and of the environment.

According to [10], a crucial issue when dealing with strategic alignment lies in the lack of common understanding and communication between the strategy and IS worlds. In fact, actors who define organization strategies, like enterprise executives, (a) do not speak the same language as operational actors (e.g. IS engineers) and (b) do not have the same vision of the organization. The consequence is that IT does not provide the expected value to the organization.

Our method, named INSTAL (INtentional STrategic ALignment), proposes to consider organizations at two levels: (i) the strategic level, which includes the decision-makers' strategy and high level requirements, and (ii) the operational level, which comprises the IS/IT. Based on the observation that documentation usually exists at the two levels while correspondence is seldom systematically documented with the degree of formality needed to support systematic analysis, INSTAL was designed to reuse the two levels' documentation in order to create a third kind of document that describes the synergy between the two levels (i.e. strategic alignment) and the existing links with organization elements such as IT applications, business processes, strategic documentation, etc.

For the organization, the outcomes expected from considering and documenting strategic alignment as a unified view of the strategic and operational levels are: (i) to improve the enterprise agility – its capability to respond to unexpected environmental changes, (ii) to reduce resistance to change, as IS users get a better view on design rationales and on the IS contribution to their own performance, (iii) to improve the visibility of top managers on the IS ROI, in terms of cost savings but also in terms of added value, and on its capacity to answer organization needs, and (iv) to help improve performance evaluation.

On the IS management side, the goals are: (i) to improve the IS flexibility, (ii) to better trace IS evolutions, (iii) to better manage project portfolio, by identifying which IS components are obsolete and redundant and which ones deliver value (needed for arbitration), and (iv) to facilitate impact analysis of evolution requirements.

With these goals in mind, our requirements for a good strategic alignment documentation were to: (i) be formalized using modeling rules, (ii) reflect the complexity of strategic alignment while being able to represent the strategic alignment in a simple manner (using a black box/white box approach), (iii) show alignment as well as mis-alignment (iv) be scalable to real-world organization sizes, (v) deal with strategy and IS on different levels of granularity, and (vi) ensure interoperability with the tools that are already used in organizations.

INSTAL was created using the principles of an action research approach. Basic principles of the method were identified based on bibliography research, then the Seven Eleven Japan (SEJ) case study was explored and the method was constructed each time a new kind of strategic alignment-related issue was encountered.

This paper is structured as follows. Section 2 gives an overview of the INSTAL method, and reports its development with the SEJ case study. Section 3 describes the process model into more detail. Section 4 compares our approach with related works. Our conclusions on this research are given in section 5.

## 2 Overview of the INSTAL Method: Principles, and Alignment Meta-Model Developed with the Case Study

The INSTAL method is presented in the next section, followed by its development using the SEJ case study and the resulting meta model.

### 2.1 The INSTAL Method

Usually, documents about strategy definition and IT components are already present in the organization. For instance business plans, annual reports, strategic reports, performance indicators and scorecards are part of the strategic documentation. IT components, legacy system, IT functionalities, business process are most often documented by specifications. These documents can easily be reused and referenced. However, whereas organization strategies are defined in terms of goals, actors and performance indicators, systems are specified using concepts such as objects, events and functions. This conceptual mismatch results in a difficulty to draw links between them.

One way to obviate this issue is to express both the strategic and operational levels in terms of requirements and with the same language. This approach, recommended by the TOGAF in the context of Enterprise Architecture, allows expressing alignment in a straightforward manner. A review of the RE literature shows that goal-centred languages seem to be the most adequate for this purpose, as they explicitly capture the why and how of both system functionality and organization businesses [11] [12] [13]. Goal modeling has several advantages: goals subsume different concepts such as systems functionalities, business processes or organizational objectives, goals can be refined and therefore be defined at different levels, they can handle the scalability by abstraction mechanisms, and they can be considered as ambivalent as they can integrate different perspectives.

As Fig. 1 shows, the basic principle of the INSTAL method stands in documenting strategic alignment by (i) using a goal-oriented model representing both strategic and operational levels and (ii) defining *links* between this model and the existing organization's strategic and operational components. Indeed documenting strategic alignment can lead to a network of links between multiple artifacts. For this reason, we use MAPs as an intermediate formalism rather than linking elements directly: they provide a unified and purposeful view on strategic alignment. MAPs sections can be used as a starting point to focus on one particular aspect of alignment based on the purpose we want to deal with. The resulting maps, called *Strategic Alignment Maps* (SAMs), capture the organization strategy through formalization and the operational level through abstraction. *Contribution links* are defined between organization's elements and the SAMs and between organization's operational elements and the SAMs. This approach was chosen because we observed in previous experiences that using the MAP formalism to represent both the IS level and the Business Process level is an efficient way to deal with complex problems at multiple levels of granularity and under multiple perspectives ([13]).

Several questions were addressed while exploring the case study: (i) is the MAP technique adapted? Are extensions needed?, (ii) is it well-suited to model both

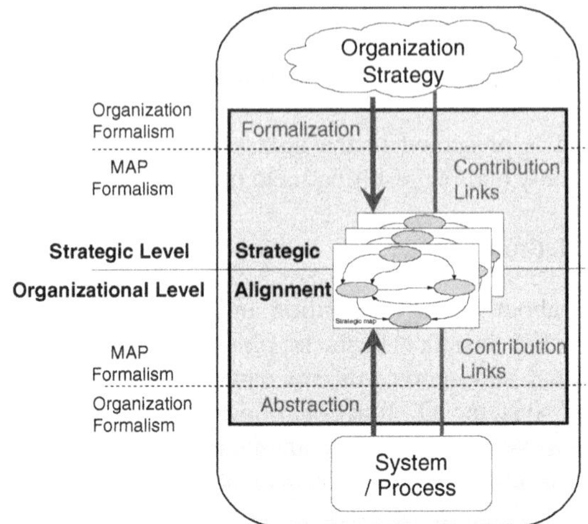


**Fig. 1.** Overview of the INSTAL method

organization strategy (non functional items) and functional items in the same model?, (iii) what are the types of links needed between SAMs and existing organization's elements?, and (iv) what is the process to apply the INSTAL method?

## 2.2 Description of the SEJ Case Study

The SEJ case study was already well described in the literature (e.g. [14]). This particular case study was chosen to develop INSTAL because it is a real case study (and not a toy example), a large amount of data is available, and more importantly, it had already been used to evaluate another strategic alignment approach [15] [16].

SEJ is the largest chain in the Japanese convenience retailing industry. The enterprise has franchise contracts with local shops all over Japan and supplies exclusive products and services to franchisees. The distribution centers, that distribute these products and services to shops, are also independent proprietors.

The SEJ supply chain is complex and implies several partners like suppliers, distributors, logistics providers, and franchise stores.

SEJ's major asset is information rather than physical properties. Indeed, SEJ's strategy is to use information to meet customer's demands, so that they can always find "what they need when they need it" in SEJ franchised stores. Coupled with an effective delivery service, this strategy helps in increasing sales, lowering the number of unsold items, and reducing the need for storage space, which is important in Japan where space is rare.

Having the right product at the right time calls for gathering very diverse information: purchasing habits, the store's neighborhood from both social and environmental perspectives, weather, local events, etc. All this data is analyzed in real-time in order to forecast what the customer might need at the exact time they shall need it.

Several information sources are available for the strategy and the IS. We particularly used annual reports[1]. The analysis of these documents allowed identifying the following strategic objectives: (i) Get better value of SEJ stores by answering to any client's needs ("To meet customer needs, products must be delivered just in time and only when needed"), (ii) Live in harmony with the local communities and (iii) Respect the environment.

As defined earlier, a primary concern was to create a unified model of these strategic objectives that would also provide a view on SEJ's IS. This task was undertaken using the MAP formalism, which was adapted at the same time to achieve the actual goal of documenting strategic alignment.

## 2.3 The MAP Formalism

A map is an oriented graph where nodes are *goals* (or intentions) and edges are *strategies*. A map is composed of sections that contribute to achieving a high level goal. A *section* is a triplet $<G_i, G_j, S_{ij}>$ and represents a way to achieve the target goal $G_j$ following the strategy $S_i$ taking into account that the goal $G_i$ should have already been achieved. Maps organize goals and strategies to represent a flow of decisions.

Indeed, a section has for source a goal when its achievement is a precondition to undertake the strategy. The other way round, as soon as a goal is achieved, any section that starts from it can be undertaken at anytime.

Each map has two special goals, *Start* and *Stop*. The *Start* goal corresponds to the entry point of the process. Sections containing this goal can always be undertaken. The *Stop* intention allows to specify sections that aim at completing the goal described by the map. A more detailed definition of MAP can be found in [17].

## 2.4 Documenting Strategic Alignment Using MAP

Fig. 2 shows an example of two SAMs that were described to document the strategic objective "Get better value of SEJ stores by answering to any client's needs" and part of the IS that supports it. The study of SEJ values and aspirations (visibility, availability towards customers, innovation, anticipation etc), made emerge two main goals in the high-level SAM: *Ensure the Control of resources* (such as time, space, stores, products) and *Increase sources of value* (such as customers, products quality and organization efficiency). These two goals are ambivalent, they can represent the organization strategy, but also tackle the operational level. Besides the strategies attached to these goals can be refined. Therefore, goals can be defined at different levels of abstraction and a collection of goals can be structured in a goal graph.

These goals can be attained in different ways (strategies in MAP formalism). The $M_1$ map describes sections composed of goals and strategies. The latter details how SEJ can attain its strategic goals in accordance with its organization strategy. For example, one way to increase sources of value is by having the lowest prices, but as we can see with strategies between start and (c) *Increase the sources of value,* this option has not been retained by SEJ. SEJ chose to develop quality products/services and to have the right products rather than having only basic products at a lower price.

[1] SEJ website: http://www.sej.co.jp/english/

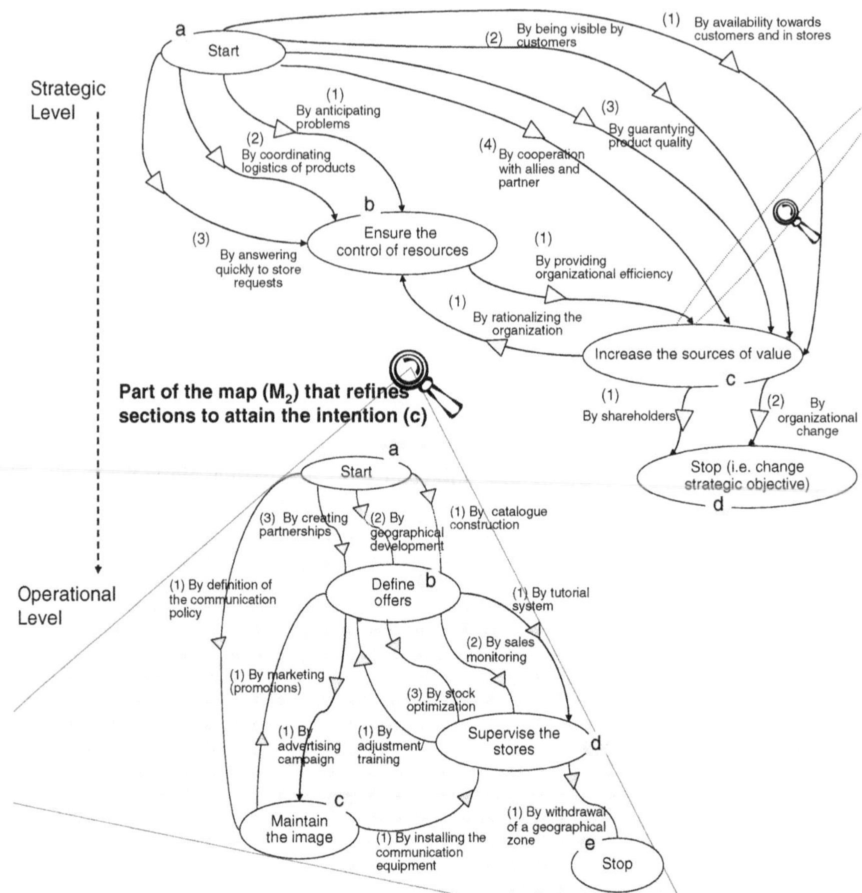


**Fig. 2.** Part of Strategic alignment maps representing the strategic objective: "Get better value of SEJ stores by answering any client's needs"

The second map $M_2$, describes, using a black box / white box approach, details on how SEJ addresses the strategic goal (c) *Increase the sources of value*. Goals present in $M_2$ are: "Define Offers", "Supervise the stores" and "Maintain the image". The goal "Define Offers" can be attained by three strategies that detail what SEJ must do to offer product and services in accordance with the organization strategy (have the right product at the right time, have products/services of quality etc.) which includes: (i) to define the catalogue of products and services (adapted to the customers' needs), (ii) to develop partnerships with suppliers, organizations (e.g. to allow clients to pay electricity bills), transport companies etc., and (iii) to develop its sites: stores and warehouses. These strategies contribute to increase the sources of value (customers, sales etc.) by availability (in catalogue, in stores), by visibility (network of stores), and by developing quality (products in catalogue and logistics for fresh products).

We designed the $M_2$ map by focusing on goals that are important to SEJ without dealing with their operational details. For example, the goal "Define offers" presents an important goal for SEJ that relies on the operational goal "define catalogue", which can appear in the MAP refining the strategy ab1.

Developing the maps presented in Fig. 2 revealed that it is possible to represent both the strategic and operational level in a single map with an abstract view, and that the black box/white box approach could be used to deal with complex problems.

However, while the map was developed using strategic documentation and specifications of IS components, the links between all these documents were still not explicit and even sometimes unclear. For example, sections of the high-level map $M_1$ can be linked to some strategic documents like business plans, annual reports and internal documents. The set of SAMs describe the AS IS based on strategic documentation and IS components and the AS Wished that can for example highlights manual processes or processes to redesign. The refined map $M_2$ is, by construction, mostly related to IS components. For example the section *<Start, Define offers, By catalogue construction>* was identified by analyzing IS features. If we refine this section, we can see systems (i) that analyze customers' profiles from questionnaires about their marital status, activities when visiting stores, their proximity with the store in POS (Point of sale) and others (ii) to analyze purchases hour by hour and product by product. These clearly correspond to items to include in strategic alignment documentation as they are needed to anticipate sales, to optimize store storage and so to contribute to having the right product at the right time.

While creating the SAMs, we were lead to reason about the links between strategy and the IS, and discovered that there are a number of different links and that these links can be combined in complex constructions. The different types of link discovered during this analysis are presented in the next section.

### 2.5 Resulting Alignment Meta Model

Fig. 3 presents the product meta model associated to the INSTAL method using the UML formalism. As the meta model shows, strategic alignment is documented by contribution links defined between the *strategic alignment map* (SAM) and the *strategic components* on the one hand, and *operational components* on the other hand.

The SAM meta model reuses the basic concepts of the MAP formalism except for the refinement mechanism which is handled differently. Indeed, in SAMs, several sections can be refined in one SAM, while in the traditional MAP meta model, refinement is a one-to-one relationship between a section and a finer grain map.

As shown in Fig. 3, contribution links are defined between an element and one or several *strategic* or *operational components*. An element is either a SAM section or a *formula*, a formula being defined as a set of SAM sections separated by logic *AND/OR operators*. For example, *(ab1 AND ab2)* is a formula composed of two members (in this case two SAM sections) separated by the AND operator. Components refer to any organization documents: strategic components are for example business plans or annual reports. Operational components can be application specifications, models (e.g. UML use case diagrams) or business processes. Links are oriented from the components to the SAM. They can also be defined in the other direction. For example, specifying that a component "is necessary to" achieve a section of a SAM is equivalent to specifying that the SAM section "requires" the component.

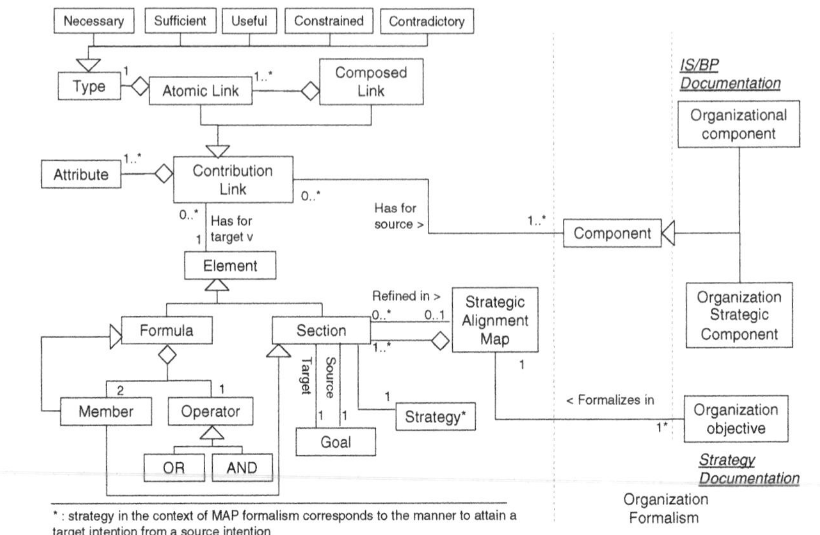


**Fig. 3.** Strategic Alignment Meta model

Five types of atomic links were identified while exploring the SEJ case study: "is necessary to", "is sufficient to", "is useful to", "is constrained by" and "is contradictory with". Contribution links can also be composed. The valid composition of contribution link types are: "necessary and sufficient", "useful and sufficient", "useful and constrained by". As the definition of contribution links is not an easy task, quite subjective, attributes related to contribution links have been defined in the meta model as a date of last revision; percentage of trust and associated metrics/measures.

Let C be a component (strategic or operational) and M an element of a SAM, the atomic links defined between C and M are defined as follows.

***Is Necessary To*:** C is necessary to M (equivalent to M requires C) means that the fulfillment of a map element M cannot occur if that of the strategic or operational component C is not performed.

If evolution impacts C or M, it is inevitable to verify that the link is preserved, and if not, that the choice is intentional. If M or C evolves, either the necessary link is preserved or it is modified, the latter implying the update of the link type and the verification that there is still one or more links of *necessary type* going to M. In the SEJ case study, the strategic document that describes the SEJ's objective to answer any clients' needs is necessary to the section ab1: <Start, Define offers, by catalogue construction>. In the same way, the three IT applications that allow managing catalogue (OC1), having the suppliers catalogue (OC2) and having the result of sales by different criteria (OC3) (e.g. by products, by stores, by geographical zone) in real time are necessary and sufficient to the section ab1. It means that OC1 is necessary to M, OC2 is necessary to M, OC3 is necessary to M and (OC1 and OC2 and OC3) are necessary and sufficient to M. If OC3 evolves and is no more able to provide sales report in real-time, the system regresses for this functionality. So either it is a choice to be less reactive but to have others gains (e.g. more cross information, reports), or it is not and so the link between OC3 and M is no more of type necessary, and the set of OC1, OC2 and OC3 is no more necessary and sufficient to M.

***Is Useful To*:** C is useful to M (equivalent to M draws part of C) indicates a dependency of a weaker nature than the "is necessary to" link. This link specifies that the strategic or operational component C helps in realizing/satisfying the element M, but is not in any way mandatory.

In the case study, the applications that allow managing the best practices catalogue, reporting past problems and giving an overview of sales and purchases by annual period are useful to the section ab1: <Start, Control the resources, By anticipating problems> because they help supervising the shops, but they are not sufficient. For example, the analysis of some data (e.g. sales by shop, sales by product, and sales by geographical zone) is necessary to better help and advice stores.

If several components are useful to an element, it means that these components can be complimentary or alternative. It is interesting to study them to examine if there is some redundancy in the IS. For example, the best practice management and the past problem report might be in the same system, in particular if these two systems are used by the same users and if there is some possible consolidation. It might have an application that proposes best practices with some of them based on real problems that occurred in the past.

***Is Sufficient To*:** C is sufficient to M (equivalent to M satisfies by C) means that realizing the strategic or operational component C is enough to satisfy the fulfillment of the element M. The study of the other types of links (necessary and useful) allows to highlight some redundant and obsolete systems which don't provide added value and so that should be replaced or deleted.

***Is Constrained By*:** C is constrained by M (equivalent to M constraints C) (e.g. the realization of M is influenced or limited by those of C). This type of link means that the strategic or operational components have more capabilities than what they are referenced for in the SAM. It is interesting to analyze these types of link to either update the SAM to integrate these opportunities, or to document this link with a justification. For example, an ERP can have more possible functionalities than the ones implemented in the organization.

***Is Contradictory With*:** C is contradictory with M (equivalent to M excludes the realization of C in a certain context) indicates that a strategic or operational component C can be contradictory with an element. This link puts in perspective the cases where alignment is not assured or outlines conflicting decisions.

## 3 The INSTAL Process Model

Fig. 4 provides an overview of the process that was used to apply the INSTAL method in the SEJ case study. The process model is formalized with the traditional MAP formalism so as to focus on methodological goals and supports to achieve these goals rather than on process details such as sequences of activities.

Three methodological goals are addressed by the method: (b) *Identify Strategic/ Operational Items*, (c) *Construct a SAM* and (d) *Define link between section and Strategic/Operational components.*

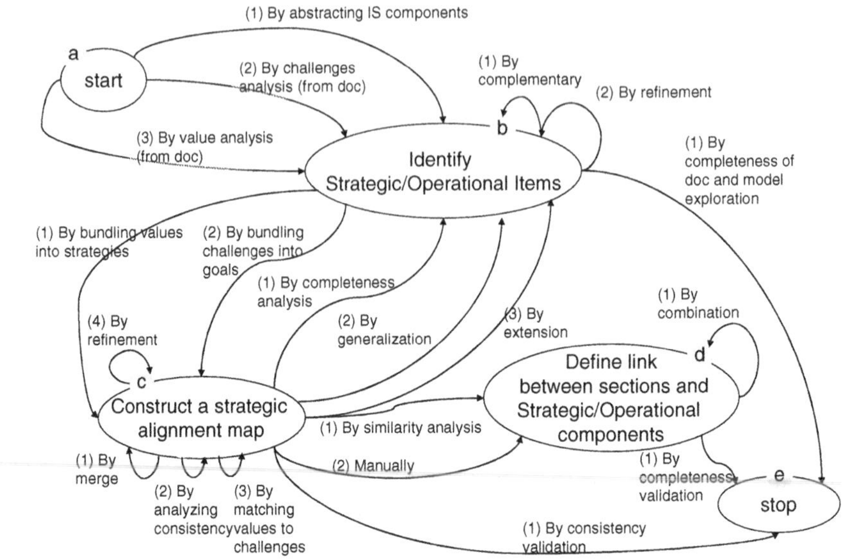


**Fig. 4.** Process model to document strategic alignment

"Items" is the general term used to address strategic and operational items. They are identified from documents and interviews and contribute to define the sections of SAMs. Once items are identified, they are used to define the SAMs.

Once the SAMs are specified, it is possible to link elements (i.e. SAM sections or formula) with strategic or operational components already present in the organization. The SEJ experience revealed that several strategies could be used to identify items, construct a SAM, define contribution links and terminate the process.

### 3.1 INSTAL Strategies to Achieve the Goal (b) *Identify Strategic/Operational Items*

The goal (b) *Identify Strategic/Operational Items* can be achieved through three strategies originating from (a) *Start*: (1) *by abstracting IS components*, (2) *by challenges analysis from strategic documents* and (3) *by value analysis from strategic documents,* and from two reflexive strategies *(1) by complementarity and (2) by refinement.* As the SAM is ambivalent, the map definition is based on both the operational and strategic elements. In the strategy (1) the IS components are abstracted to have high level functionalities, in strategies (2) and (3) strategic documents are analyzed to find organization's challenges and values. Challenges are finally expressed as elements that should not be lost to increase benefits. In the SEJ case study, examples of challenges are: clients, space, stores, market share, time, organization's quality, suppliers, products quality. Values are high level qualities chosen by the organization to address the challenges. Examples of values in the SEJ case study are: availability, visibility, products' quality, coordination, speed, rationality, and anticipation.

The two reflexive strategies allow finding complementary and more detailed items.

### 3.2 INSTAL Strategies to Achieve the Goal (c) *Construct a Strategic Alignment Map*

The strategic goal (c) *Construct a strategic alignment map* can be attained from the source goal (b) *Identify Strategic/Operational Items* by two strategies (1) *By bundling values into strategies* and (2) *By bundling challenges into goals.* Two categories of challenges were defined in the SEJ case study: (i) challenges related to resources, and (ii) challenges related to the sources of values that should be increased. The former leads to the strategic goal "ensure the control of the resources" (time, space, stores, products and services). The latter leads to the goal "increase the sources of values" (market share, client, products' and services' quality, and organization's quality). Several strategies to reach these goals were specified by studying SEJ values. The strategic goal "*Increase the sources of value*" can be attained by four strategies. These are based on different values such as availability, visibility, quality, coordination and cooperation (see sections ac1, ac2, ac3 and ac4 in Fig. 2).

Four recursive strategies on the goal (c) *Construct a strategic alignment map,* are proposed: (1) *by merge*, (2) *by analyzing consistency*, (3) *by matching values to challenges* and (4) *by refinement.* These strategies allow to improve a SAM or to define other SAM by refinement of section(s).

*From a SAM* it is possible to identify strategic and operational items (1) *by completeness analysis* and (2) *by refinement.* For example, the section ab1 (Fig. 2) in $M_2$: <*Start, Define offers, By construction catalogue*> can be refined in an other map to describe how SEJ defines offers, for example by analyzing the customers' needs from sales, local events, weather forecasting etc. New items might be identified to construct the refined map.

### 3.3 INSTAL Strategies to Achieve the Goal (d) *Define Link Between Sections and Strategic/Operational Components* and to Stop the Process

Once all items have been identified and used to define the set of SAMs, it is possible to stop the process (1) *by completeness of documentation and model exploration.* Besides, further exploration of the case study revealed several different kinds of links between strategic or operational components, and SAM's section which have been defined previously. The strategies used to achieve this task are (1) *by similarity analysis* and (2) *manually.* Searching in components' documentation about a part of a goal or a strategy can help finding the automatic link between a section and a component. Manual link means interviewing the person in charge of the component (e.g. top-managers, functional architect) or using documentation to define the appropriate type(s) of link and define a new link.

The recursive strategy (1) *by combination* allows to define formula (several sections with OR/AND links) and a link between this set of sections and components.

It is possible to stop the process from the goals: (b) *Identify Strategic/operational Items* (1) *by completeness of documentation and model exploration*, (c) *Construct a strategic alignment map* (1) *by consistency validation* and (d) *Link items with Strategic/Operational components* (1) *by completeness validation.*

# 4 Related Works

## 4.1 Goal Modeling in RE

Different approaches have been developed in the Requirements Engineering (RE) community to express high level requirements for IS. The goal modeling approaches allow defining the purpose of the system from an external point of view, in particular from a user perspective. Among the best known methods, there are i* [18], CREWS-*L'Ecritoire* [19], and KAOS [20].

I* is a method that aims at modeling the relationships between actors and their goals. Bleistein *et al.* [15] [16] have adapted i* to model both strategic and operational goals. The authors uses i* representation (adapted using the BRG-Model conceptual framework) to represent the strategy then the context diagram from Jackson's Problem Frames to model the IT context of strategy achievement. The approach represents the business strategies and the IS description in the same model, with the same formalism and relates them through simple and mono-typed contribution links. One issue with this approach relates to the i* notation, which does not use the black box/white box strategy and produces complex models when the number of goals increases. Indeed the main difficulties with i* is that it lacks (i) systematic goal refinement mechanisms (all goals are in the same model), and has no (ii) goal-strategy couple to help clarifying the multiple ways in which a goal can be achieved. Besides, our experience with the MAP formalism showed us that MAPs are particularly adapted to dealing with multi purpose systems, which helps managing complex situations, whereas it has been clearly demonstrated that i* models find their limit when the situation gets complex.

CREWS-L'Ecritoire combines goal modeling and scenario analysis to guide the elicitation of user goals. Scenario are not used in the INSTAL method, nevertheless the CREWS-L'Ecritoire uses pre-defined levels that could be reused to better guide the application of the black box / white box paradigm.

In KAOS goal models, goals are linked through AND/OR decomposition links. This allows refining high-level goals into finer grain goals down to concrete system requirements. Refinement cannot be separated from OR decomposition and AND decomposition which introduces artificial complexity in the goal hierarchy, while the main issue is to sort goals according to their level of abstraction and relate goals when they belong to the same level of abstraction.

As Bleistein *et al* show [15][16], although some aspects of the organization strategy can be taken into account with these goal models, none of them (except for Bleistein's adaptation of i*) really addresses the issue of strategic alignment.

## 4.2 Research on IS Alignment

Different kinds of concepts are involved in strategic alignment depending on authors. A well known model is Henderson's and Venkatraman's Strategic Alignment Model [21] that inspired numerous models such as [22] [23] [24]. The Strategic Alignment Model defines the interrelationship between business strategies and IT strategies in a clear way. However, it does not provide a practical framework to document strategic alignment. Luftman [22] and Maes [23] define alignment between organization strategy and IT people. However, they do not propose a method for designing the strategy. Other authors focus on alignment of software architecture and business process architecture [25], [26], or on alignment between system and business processes [27] [28] [29] [30]. These approaches concern the improvement of alignment. However, alignment is not considered as a concept by itself that can be documented and therefore used to support systematic reasoning.

Salinesi [31] showed that a systematic method should use a notation to document and reason on IS alignment. This view is also supported by EA approaches [32] [33].

Notations for documenting alignment are considered in [31] [34] [35] and [36], but not at the strategic level, which is different by nature from the operational level.

## 5 Conclusions

This paper has described the INSTAL method for IS strategic alignment, which was developed using the principles of an action research approach that consists in exploring the SEJ case study. The paper has shown that (i) MAP formalism is adapted to document models both at the strategic and operational levels, and that (ii) complex links can be defined to document organization's strategic and operational components involvement in strategic alignment. A process model was also presented.

The experience revealed some limitations in our approach:

- Only research documents, articles and SEJ's annual reports were used in the case study. No SEJ internal documents or interviews were used, which could be considered as a bias of the case study with respect to real world situation.
- Our exploration of the SEJ case study did not consider the entire company. This is a bias towards validating the scalability of the approach. Besides, different kinds of problems might have arisen if the study had been complete.
- No quantitative evaluation is proposed. Our only validation stands in the SEJ case study which was addressed as an empirical and subjective experiment.
- We did not use a tool to systematically check the degree of completeness and consistency of the documentation of strategic alignment.

Our current works concern the development of a technique to analyze SAMs and contribution links with the aim of improving strategic alignment. We also believe that a more complete validation should be undertaken using empirical evaluations, and interviews of experts to explore the usability of the SAMs and the effectiveness of the INSTAL method. Last, our experience at BNP-Paribas and our participation in industrial workgroups showed us that organizations face different alignment-related problems such as: difficulty to maintain legacy systems that would be replaced if sufficient benefit had justified the high cost of change, difficulty for top managers to acquire visibility on the IS results and on strategic alignment to determine priorities for investment etc. To meet these needs, we think that it is essential to explore further extension of the INSTAL method in the following directions:

– Facilitate traceability between high-level requirements and IS components in order to support impact analysis and to systematically examine how new projects contribute to the organization strategy and to the strategic alignment.

- Support IS redundancy analysis: IS can be huge and lack good documentation, so it happens that two systems that have similar functionalities are developed in two different departments, which generates double costs. We believe the analysis of strategic alignment documentation could highlight this kind of redundancy and either avoid new developments or at least trace its rationale.
- Measures are essential to analyze strategic alignment and answer the concrete preoccupations of top managers, it would be interesting to study the possibility of associating measures and metrics to contribution links.
- Document To-Be SAMs and support comparative analysis between the AS-IS and To-Be SAMs so as to develop a better vision of what must be addressed in the target and the path to reach it (using contribution links). Any project could show how it contributes to the target strategic alignment.

## References


1. Luftman, J., Maclean, E.R.: Key issues for IT executives. MIS Quarterly Executive 4(2), 89–104 (2004)
2. Reich, B.H., Nelson, K.M.: In Their Own Words: CIO Visions About the Future of In-House IT Organizations. The DATA BASE for Advances in ISs 34, 28–44 (2003)
3. Tallon, P.P., Kraemer, K.L.: Executives' Perspectives on IT: Unraveling the Link between Business Strategy, Management Practices and IT Business Value. ACIS2002, USA (2002)
4. Watson, R.T., Kelly, G.G., Galliers, R.D., Brancheau, J.: Key issues in information systems management: an international perspective. Journal of Management ISs 13, 91–115 (1997)
5. Brancheau, J.C, Janz, B., Wetherbe, J.C.: Key issues in information systems management. 1994-95. SIM Delphi results. MIS Quarterly 20, n°(2), 225–242 (1996)
6. Chan, Y.E., Huff, S.L., Copeland, D.G., Barclay, D.W.: Business Strategic Orientation, Information Systems Strategic Orientation and Strategic Alignment. Information Systems Research 8(2), 125–150 (1997)
7. Kefi, H., Kalika, M.: Survey of Strategic Alignment Impacts on Organizational Performance in Int. European Companies, Hawaii Int. Conf. on System Sciences (2005)
8. Chan, Y.E.: Why haven't we mastered alignment? The importance of the informal organization structure, MIS Quarterly Executive 1, 97–112 (2002)
9. Croteau, A.M., Bergeron, F.: An information technology trilogy: business strategy, technological deployment and organizational performance. Journal of Strategic Information Systems 10, 77–99 (2001)
10. Luftman, J.: Assessing Business-IT Alignment Maturity. Communications of the Association for Information Systems 4, n°(14), 1–50 (2000)
11. Yu, E.: Agent Orientation as a Modelling Paradigm. Wirtschaftsinformatik. Vol.43 (2) (2001)
12. Van Lamsweerde, A.: Goal-Oriented Requirements Engineering: A Guided Tour, Invited Paper, Int. Symposium on Requirements Engineering (RE), pp.249-263 Toronto, Canada (2001)
13. Rolland, C., Salinesi, C.: Modeling Goals and Reasoning with Them, Engineering and Managing Software Requirements (EMSR), Aurum, A., Wohlin, C. Springer Verlag (2005)
14. Bensaou, M.: Seven-Eleven Japan: Managing a Networked Organization, INSEAD EuroAsia Centre, Case Study (1997)

15. Bleistein, S., Aurum, A., Cox, K., Ray, P.: Strategy-Oriented Alignment in Requirements Engineering: Linking Business Strategy to Requirements of e-Business Systems using the SOARE approach. Journal of Research and Practice in IT 36, 259–276 (2004)
16. Bleistein, S., Cox, K., Verner, J., Phalp, K.: B-SCP: a requirements analysis framework for validating strategic alignment of organisational IT based on strategy, context and process. Information and Software Technology 46, 846–868 (2006)
17. Rolland, C., Prakash, N.: Matching ERP System Functionality to Customer Requirements. RE01, Canada pp. 66-75 (2001)
18. Yu, E.: Towards Modelling and Reasoning Support for Early-Phase Requirements Engineering. RE97 Washington D.C., USA pp. 226-235 (1997)
19. Rolland, C., Souveyet, C., Ben Achour, C.: Guiding goal modelling using scenarios. IEEE Transactions on Software Engineering vol. 24(12) (1998)
20. Dardenne, A., Lamsweerde, A., Fickas, S.: Goal-directed Requirements Acquisition. Science of Computer Programming 20, 3–50 (1993)
21. Henderson, J.C., Venkatraman, N.: Strategic alignment: Leveraging information technology for transforming organizations. IBM Systems Journal 32(1), 4–16 (1993)
22. Luftman, J.N.: Competing in the Information Age. Oxford University Press, Oxford (1996)
23. Maes, R.: A Generic Framework for Information Management. Prime Vera Working Paper, Universiteit Van Amsterdam (1999)
24. Goedvolk, H., van Schijndel, A., van Swede, V., Tolido, R.: The Design, Development and Deployment of ICT Systems in the 21st Century: Integrated Architecture Framework (IAF). Cap Gemini Ernst and Young (2000)
25. Aerts, A.T.M., Goossenaerts, J.B.M., Hammer, D.K., Wortmann, J.C.: Architectures in context: On the evolution of business, application software, and ICT platform architectures. Information and Management 41, N°(6), 781–794 (2004)
26. Wieringa, R.J., Blanken, H.M., Fokkinga, M.M., Grefen, P.W.P.J.: Aligning application architecture to the business context, CAiSE03, Austria, pp.209–225 (2003)
27. Bodhuin, T., Esposito, R., Pacelli, C., Tortorella, M.: Impact Analysis for Supporting the Co-Evolution of Business Processes and Supporting Software Systems. BPMDS04, Latvia
28. Arsanjani, A., Alpigini, J.: Using Grammar-oriented Object Design to Seamlessly Map Business Models to Component-based Software Architectures. Int. Symposium of Modelling and Simulation, USA, pp.186–191 (2001)
29. Giaglis, G.M.: On the Integrated Design and Evaluation of Business Processes and Information Systems. Communications of the AIS, Vol 2, N°5, July (1999)
30. Kardasis, P., Loucopoulos, P.: Aligning Legacy Information Systems to Business Processes, CAiSE98, Italy, pp. 25-40 (1998)
31. Salinesi, C., Rolland, C.: Fitting Business Models to Systems Functionality Exploring the Fitness Relationship. CAiSE03, Austria (2003)
32. Zachman, J.A.: A framework for Information Systems architecture. IBM Systems Journal 26, N°(3), 276–292 (1987)
33. Longépé, C.: The Enterprise Architecture IT Project: The Urbanisation Paradigm. Elsevier Health Sciences, New York (2005)
34. Etien, A., Rolland, C.: A Process for Generating Fitness Measures, CAISE05, pp. 277–292. Springer, Portugal (2005)
35. Wegmann, A., Regev, R., Loison, B.: Business and IT Alignment with SEAM, REBNITA, RE05, France (2005)
36. Soffer, P.: Fit Measurement: How to Distinguish Between Fit and Misfit, note for BPMDS'04, Latvia (2004

# Towards a Framework for Tracking Legal Compliance in Healthcare

Sepideh Ghanavati, Daniel Amyot, and Liam Peyton

SITE, University of Ottawa, Canada
{sghanava,damyot,lpeyton}@site.uottawa.ca

**Abstract.** Hospitals strive to improve the quality of the healthcare they provide. To achieve this, they require access to health data. These data are sensitive since they contain personal information. Governments have legislation to ensure that privacy is respected and hospitals must comply with it. Unfortunately, most of the procedures meant to control access to health information remain paper-based, making it difficult to trace. In this paper, we introduce a framework based on the User Requirements Notation that models the business processes of a hospital and links them with legislation such as the Ontario Personal Health Information Privacy Act (PHIPA). We analyze different types of links, their functionality, and usefulness in complying with privacy law. This framework will help health information custodians track compliance and indicate how their business processes can be improved.

**Keywords:** Business Process, Compliance, Health Information Custodian, Privacy Legislation, Requirements Engineering.

## 1 Introduction

Hospitals strive to improve the quality of the healthcare they provide. To achieve this, they require access to health data. These data are sensitive since they contain personal information. Disclosing this information accidentally, may affect negatively the individual's life. To prevent such situations, governments have established legislation to ensure that patient privacy is respected and hospitals, as health information custodians, must comply with it. For example, the Personal Health Information Privacy Act (PHIPA) protects electronic patient information from being disclosed to unauthorized third-parties in the Canadian province of Ontario [1]. Our objective is to provide health information custodians with tools that will allow them to protect patient data and track their compliance to legislation like PHIPA.

This paper describes a requirement management framework which connects privacy laws to business processes and helps health information custodians to ensure their business processes comply with these laws. This framework has been developed iteratively based on a case study. This framework uses the User Requirements Notation (URN) [2,3] to model both the business processes of a health information custodian and the applicable privacy legislation. Links are

J. Krogstie, A.L. Opdahl, and G. Sindre (Eds.): CAiSE 2007, LNCS 4495, pp. 218–232, 2007.

created between the two models to track the custodian's compliance to the law. To provide this traceability, we use a commercial requirements management system (Telelogic DOORS) [4] combined with an Eclipse-based URN modeling tool (jUCMNav) [5]. With these tools we are able to specify a variety of link types that connect the two models, each providing a different function. For instance, *traceability* links are used to handle compliance with the non-functional requirements defined in legal documents while *compliance* links are used to handle exceptions and constraints. Using these link types, we are able to find missing goals, special constraints, and discrepancies in the responsibilities of the various entities involved in a case study of The Ottawa Hospital (TOH).

Our technical framework will enable health information custodians to evaluate their business processes in terms of their compliance with privacy legislation. It will also allow them to make decisions about how they will remain compliant as business processes and legislation evolve over time.

## 2 Background and Related Work

### 2.1 Personal Health Information Privacy Act (PHIPA)

PHIPA [1] is legislation specific to healthcare in the Canadian province of Ontario within the framework of the federal Personal Information Protection and Electronic Documents (PIPEDA) act [6]. PIPEDA has been recognized by the European Commission as being compliant with the European Union's Data Protection [7]. In the United States, there is similar legislation for healthcare in the form of the Health Insurance Portability and Accountability Act (HIPAA) [8].

PHIPA is divided into seven parts with a total of 75 sections. It establishes a set of rules pertaining to the collection, use, and disclosure of personal health information with the goal of protecting the privacy of the individual (e.g., the patient). These rules specify that health information custodians (e.g., hospitals) obtain data with consent; that they use it only for the purposes stated; and that they do not disclose the data without the consent of the individual. The health information custodian must also provide the individual with access to his/her data with the capability to amend it if desired. Individuals must also be allowed an avenue for an independent review with respect to the handling of their personal information and remedies must be provided if it is deemed that the information was handled inappropriately.

### 2.2 User Requirements Notation (URN)

The User Requirements Notation is a draft ITU-T standard that combines goals and scenarios in order to help capture, model, and analyze user requirements in the early stages of development [2]. It can be applied to describe most kinds of reactive and distributed systems as well as business processes.

URN is composed of two complementary notations: *Goal-oriented Requirement Language* (GRL) and *Use Case Maps* (UCM) [9]. These notations together connect goals and business processes. GRL models business objectives, rationales, tradeoffs, and non-functional aspects (the "why" aspects) while UCM focuses more on architectures and functional or operational aspects of business processes (the "who", "what", "where", and "when" aspects).

GRL combines a subset of the Non-Functional Requirements (NFR) [10] and the $i^*$ [11] frameworks. The main concepts (e.g., actors, intentional elements, and links) are borrowed from $i^*$ supplemented with the NFR framework's evaluation mechanism (i.e., qualitative labels associated to lower-level intentional elements used to compute the satisfaction degree of high-level intentional elements.) GRL's intentional elements include goals, softgoals (which can never be fully satisfied), and tasks (solutions). Such elements can contribute positively or negatively to each other and be decomposed in an AND/OR graph. In addition, they can be allocated to actors, who as a result may have conflicting concerns. See Figure 1 (left side) for an overview of the main notation elements and Figure 3 for an example.

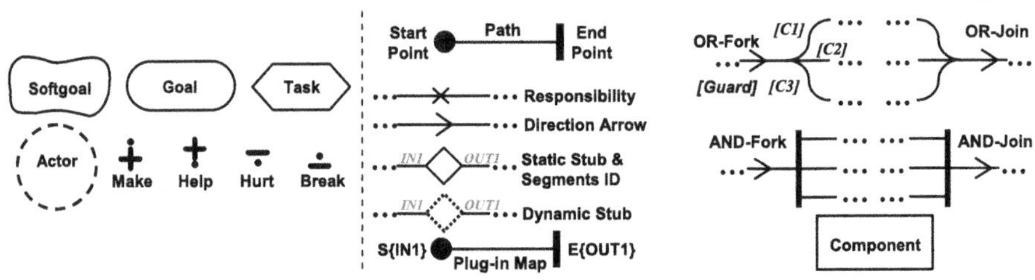


**Fig. 1.** Main elements of the GRL and UCM notations

The UCM notation is used to model related scenarios and use cases. As illustrated in Figure 1 (right side) and in Figure 4, scenario paths connect start points (preconditions and triggering events), end points (post-conditions and resulting events), and responsibilities. Responsibilities indicate where actions, transformations, or processing is required. They can be performed in sequence, concurrently, or as alternatives.

Complex scenario maps can be decomposed using path elements called stubs. Sub-maps in stubs are called plug-in maps. Stubs have identified input and output segments that can be connected to the start points and end points in the plug-in, hence ensuring scenario continuity across various levels of details. Dynamic stubs are used to specify alternative maps in the same location. The path elements (and especially responsibilities) can be allocated to components, which can represent actors, roles, software modules, sub-systems, etc. Components can also be decomposed recursively with sub-components.

### 2.3 URN for BPM and Requirements Management

Business Process Modeling (BPM) is used by an organization to represent its current and planned business processes as a basis for improving the mechanisms used to achieve business goals while taking into consideration the interests of the various stakeholders [12,13]. In [3], the authors illustrate how URN can be effective in modeling business processes and goals while including stakeholders in the modeling process. GRL helps to model the risks and benefits for different alternative business processes as well as the dependencies between participants, and allow refinements of business goals into high-level tasks and/or low-level UCM responsibilities, scenarios, and plug-ins.

In [5,14], the authors have introduced a metamodel which defines URN models and combines them with external requirements documents in a Requirement Management System (RMS), namely Telelogic DOORS. We reuse and extend this approach to implement a generic framework to track compliance between two URN document-based models.

### 2.4 Related Work

Darimont *et al.* describe an approach where one of the main goal-oriented requirements engineering methodologies (called KAOS) is used to model regulations [15]. They explain how to incrementally transform regulation documents into three models for goals, objects, and threats while maintaining a level of traceability from the source document to the models. This method, however, does not combine the three models into one integrated model. The integration of the models would help exploit traceability in a more effective manner. A modeling language such as URN has the capacity to represent high-level goals, actors, and tasks (activities) in one model. It employs different strategies to illustrate conflicting intentions and their impact on the main high-level objectives and scenarios of the system.

He *et al.* introduce the Requirement-based Access Control Analysis and Policy Specification (ReCAPS) method [16], which integrates components of access control analysis, improves software quality, and ensures policy- and requirements-compliant systems. It emphasizes traceability and compliance between different policy levels, requirements, and system designs. ReCAPS includes a set of process descriptions and heuristics to help analysts derive and specify access control policies (ACPs) and establish traceability from source documents to these ACPs. This approach is presented in the context of the software development process and thus applies less generally than what we propose in this paper. Our method provides traceability for a compliance mechanism between business processes and legal documents, with consideration for how they evolve.

In [17], the authors apply goal-based modelling on the implementation of a financial system to ensure that it complies with Basel II regulations. In this method, the organization and its business processes are divided with respect to different organizational layers. The objectives, strategies, policies, and indicators (based on the definition of a goal model) are defined for each layer and provide a structure for the design of a regulation-compliant financial system. However, this method does not provide a traceability mechanism that highlights situations of non-compliance for the goals and business processes of the organization.

## 3 Compliance Framework

The framework we introduce here demonstrates how compliance can be tracked by defining and managing external links between two models: a model of the health information custodian's policies and business processes in terms of GRL and UCM notations, and a model of privacy legislation in terms of GRL notation.

As shown on the left-hand side of Figure 2, we use GRL to capture the policies of a health information custodian and UCM to represent the business processes that implement them. The figure further serves to illustrate the types of links that connect the different levels of the health information custodian model. We identify two types of links, namely:

- *Source Links*: These are the links between actual policies and procedure definitions in the original textual documents and the hospital GRL or UCM model elements.
- *Responsibility Links*: Each GRL element can be linked to one or more UCM elements. Softgoals and goals can be linked to maps of the UCM business processes that realize them while tasks and actors can be linked to UCM elements like responsibilities and agents.

On the right-hand side of Figure 2, we show how GRL is used to model privacy legislation in terms of softgoals, goals, tasks and actors. Since privacy legislation usually includes few or no operational procedures, it is usually not worth investing in UCM models for such legislation. The only link type here is:

- *Source Links*: Similar to source links for the health information custodian, these are the links between the actual legislative documents and the privacy legislation GRL elements.

After developing the health information custodian model and the privacy legislation model, we can establish links between them. Since we have two different ways of representing legal documents (textual document format, and

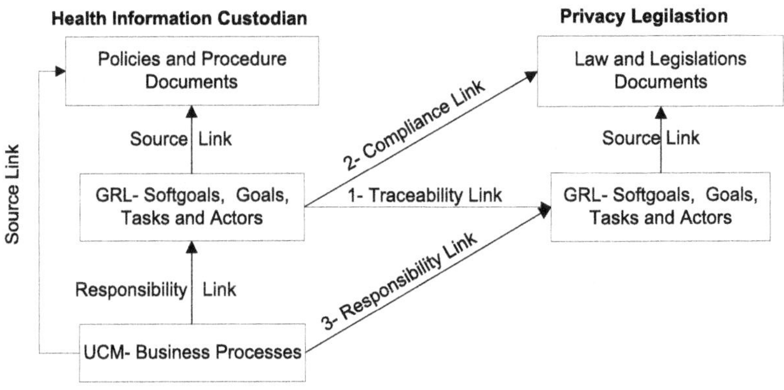


**Fig. 2.** Modeling compliance of health information custodian to privacy legislation

GRL model), we can construct different sets of links from these representations between the health information custodian and privacy legislation. These links can be added depending on the functionality desired. The links defined in our framework are shown in Figure 2:

1. *Traceability Links*: between health information custodian GRL elements and privacy legislation GRL elements (softgoals, goals, tasks, and actors).
2. *Compliance Links*: between health information custodian GRL elements and the actual text of the law and legislative documents.
3. *Responsibility Links*: between health information custodian UCM elements and privacy legislation GRL elements.

These links can be used to highlight the difference between what is implemented in business processes and what is required by privacy legislation. Missing and unnecessary elements in the business processes can be addressed and compliance can be tracked and managed.

## 4 Application to a Teaching Hospital and PHIPA

In our example, we study the business process that is in place to control access to a major teaching hospital's data warehouse in Ontario. This hospital is interested in improving the effectiveness and the efficiency of its healthcare and its support of health services research. Its plan for achieving these goals includes making its data more readily accessible to its stakeholders, including doctors, researchers, other hospitals, and patients. However, due to the existence of legislation protecting the use of health information, the hospital has established policies and heavy procedures to control the access to the data warehouse. Anyone requesting access to the data warehouse must follow this process.

### 4.1 Hospital Model

The hospital GRL model was derived from the hospital's data warehouse policies and guidelines document [18]. The process that controls access to the data

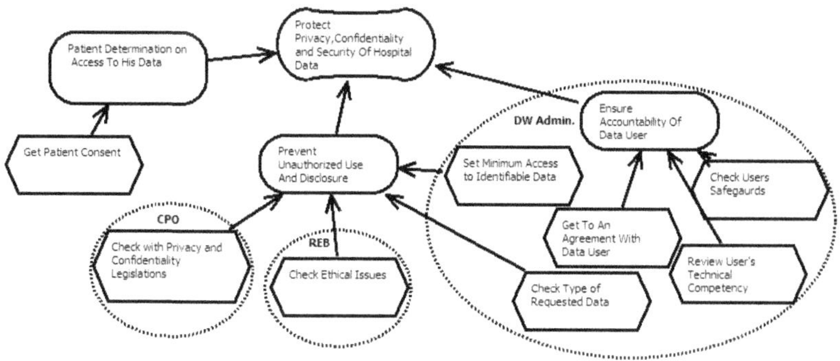


**Fig. 3.** Partial hospital's GRL model

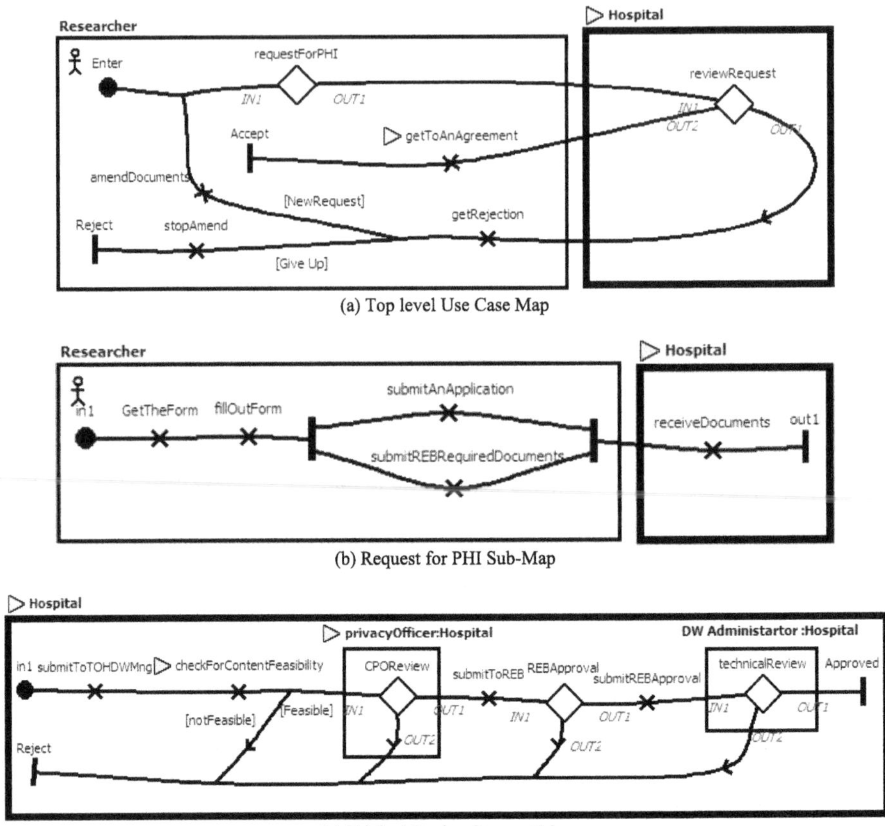


(a) Top level Use Case Map

(b) Request for PHI Sub-Map

(c) Original Review Request Map

**Fig. 4.** Partial hospital's UCM model

warehouse is modeled with the UCM notation. We then used jUCMNav to create the GRL and UCM elements as well as the links between them. jUCMNav is an open source Eclipse-based graphical editor and analysis tool for the User Requirements Notation [5].

The hospital must ensure that stakeholders get the required information while protecting the privacy, confidentiality, and security of health data. Therefore, the partial GRL diagram of Figure 3 contains a softgoal called Protect Privacy, Confidentiality and Security of Hospital Data. Other goals contribute positively to this objective, such as Ensure Accountability of Data User, Prevent Unauthorized Use and Disclosure and Patient Determination on Access to His Data. In addition, this GRL diagram allocates the general goals and concerns to their respective actors: Research Ethic Board (REB), Data Warehouse Administrator (DW Admin), and Privacy Officer (CPO). GRL tasks are used to operationalize the parent softgoals or goals and they can correspond to responsibilities in the UCM model.

This GRL diagram illustrates some of the necessary business process scenarios and some of the activities required in the corresponding UCMs. To model how the goal Prevent Unauthorized Use and Disclosure would be operationalized, we built a top-level UCM diagram (Figure 4(a)) and six sub-maps. This diagram shows how a researcher who needs personal health information (PHI) interacts with the hospital. This map contains Request for PHI and Review Request sub-maps, also shown in the figure. The Review Request sub-map also includes three stubs containing a Privacy Officer Review sub-map (CPO Review), a Research Ethics Board Approval sub-map (REBApproval), and a Review Request Technically sub-map (technicalReview).

Each part of these UCM diagrams potentially corresponds to a GRL element. Therefore, there are some links between them, i.e., *responsibility* links. Some of the links can be created manually inside jUCMNav (indicated by a ▷ triangle next to the label). In this example, we created links between GRL actors and UCM components, and between GRL tasks and UCM responsibilities. In Figure 4(c), the link labeled privacyOfficer:Hospital is a link from a UCM component to the GRL actor CPO in Figure 3. Also, UCM responsibility checkForContentFeasibility is linked to GRL task Check Type of Requested Data in Figure 3.

### 4.2 Privacy Legislation Model

The relevant sections of the PHIPA legislation were also modeled with URN. Figure 5 shows a partial GRL diagram that highlights PHIPA's major softgoal: Satisfy Privacy Regulations and Protect Confidentiality. This softgoal has many other softgoals, not shown here, that contribute to its satisfaction. Such softgoals can be broken down into goals such as Limiting the Collection of Data, Limiting the Use of Data, Secure Transfer, and Limiting the Disclosure of Data.

This GRL diagram also contains tasks that operationalize several goals. For example, for the goal Limiting the Disclosure of Data, four tasks have to be performed. One of them (Ask for REB Approval) is also decomposed into Check for Adequate Safeguards and Check Ethical Issues in Research Plan subtasks ([1], Chapter 3, Schedule A, s.44).

### 4.3 Model Linking

The dependencies and links that exist between PHIPA documentation, hospital documentation, GRL elements, and UCM elements were managed using Telelogic DOORS [4]. DOORS is used to collect, organize, and link requirements in a database as well as to trace, analyze, and manage changes to information in order to ensure compliance to the specified requirements and standards. jUCMNav has a filter that can be used to export GRL and UCM elements to DOORS (including internal links) so that they can be maintained [5,14]. In DOORS, we establish links between the PHIPA and hospital models and look for situations of non-compliance or any areas that require modification. In addition, we test the different types of links (described in the previous section) and determine which ones are best in terms of functionality, precision, quantity of manual links, difficulty, and completeness. A portion of the framework, along with its defined links, is illustrated in Figure 6. This figure provides a high-level overview of what

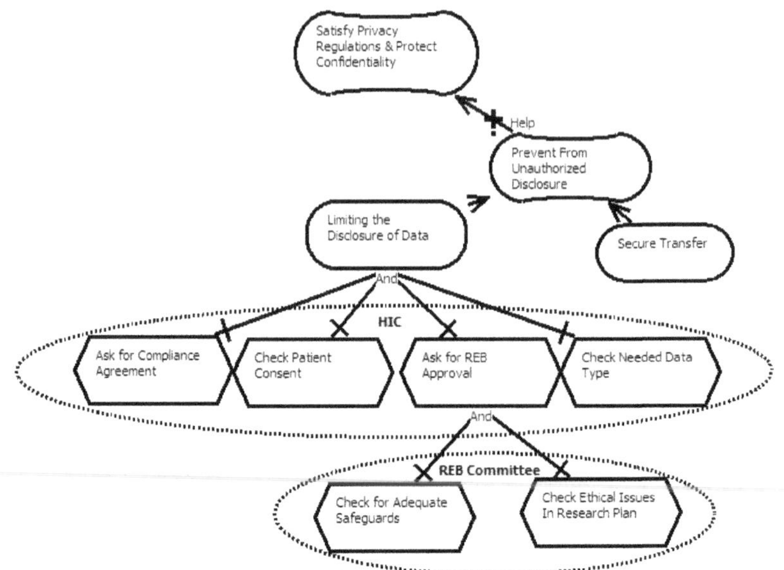


**Fig. 5.** Partial PHIPA GRL model

exists in DOORS and describes the different types of links that exist between elements of the hospital and PHIPA models.

After establishing manual and automatic links in DOORS, we analyze each type of link to find potential non-compliance issues. Figure 7 shows a partial overview of traceability links as they exist between the hospital GRL elements and the PHIPA GRL elements. For example, there is a link between softgoals Protect Privacy and Confidentiality of Hospital Data and Satisfy Privacy Regulations and Protect Confidentiality. We also find links between tasks Get to an Agreement with Data User and Ask for Compliance Agreement as well as between actors REB and REB Committee. These links illustrate that the hospital is trying to be compliant with PHIPA.

On the other hand, by studying these *traceability* links, it is obvious that there are some elements in PHIPA which do not have any corresponding element in the hospital model. For example, the PHIPA goal Secure Transfer is not linked to any task or goal at the hospital. This is however of critical importance to the hospital. It shows that the hospital may not comply with PHIPA thoroughly. As a result, the hospital may need to add this goal or a task to its model and ensure that processes are implemented to support it.

Moreover, a more detailed analysis of these links reveals further areas of potential non-compliance. From Figure 7, we can identify that there are some tasks, which are currently performed by a specific actor in the hospital model which have to be done by a different actor in the PHIPA model. For example, the task Check for Adequate Safeguards is handled by the REB committee but at the hospital the Data Warehouse Administrator is in charge of it. These discrepancies

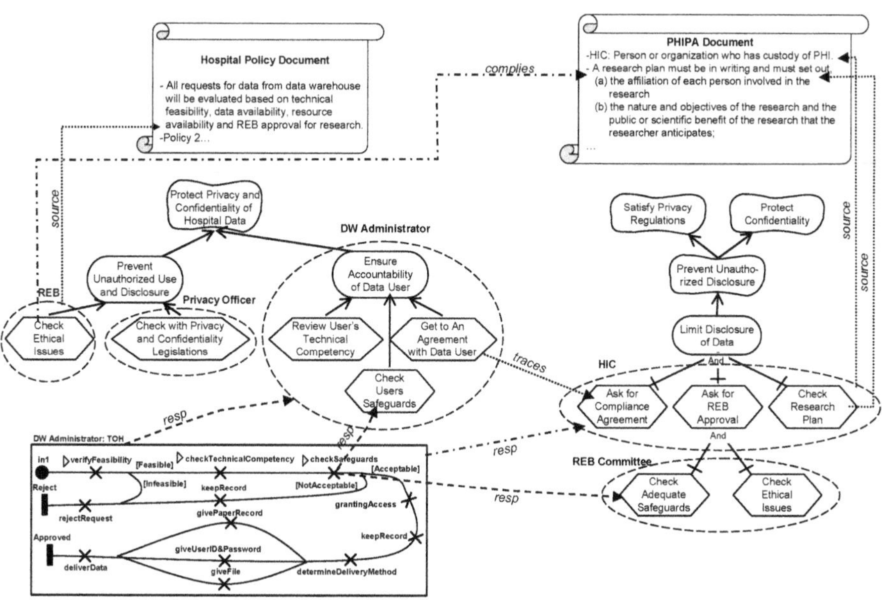


**Fig. 6.** Example of privacy compliance links in the hospital model

may lead to changes in the hospital model and clarification of the processes that implement the tasks.

The next links established are *compliance* links as they exist between the hospital GRL elements and the PHIPA document. This link set illustrates the details of PHIPA, the exceptions, and certain definitions that cannot be modeled using URN. An example is the goal Prevent Unauthorized Disclosure, for which the REB needs to check the ethical issues of the request. In PHIPA such a request is

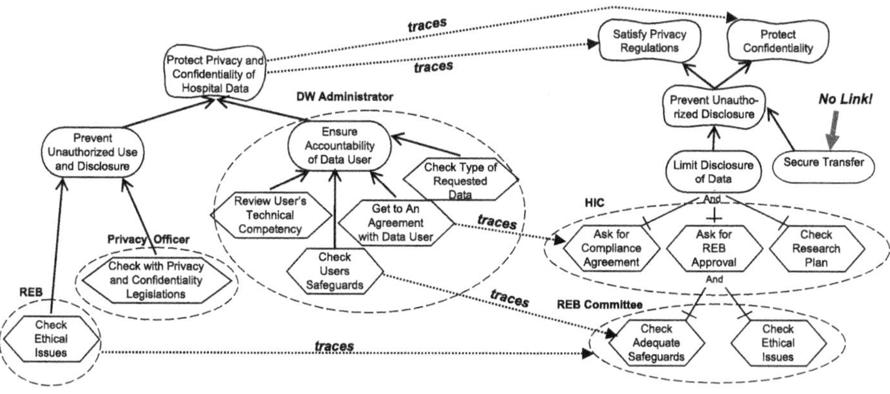


**Fig. 7.** Link set between hospital GRL and PHIPA GRL models

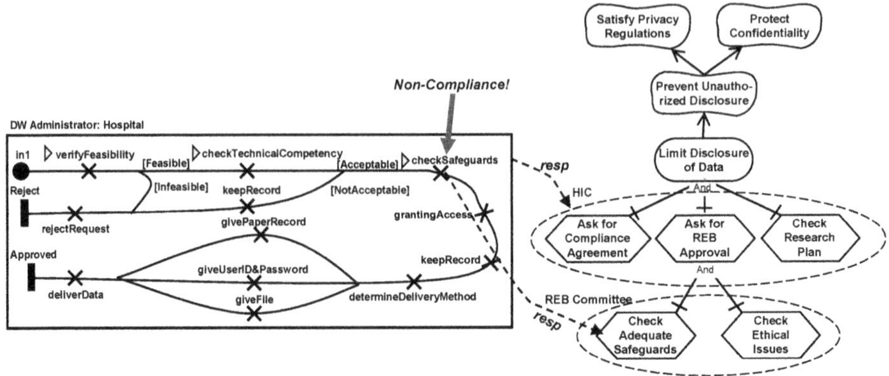


**Fig. 8.** Link set between hospital UCM and PHIPA GRL models

called the "Research Plan" and it has some requirements that cannot be defined with softgoals, goals, or tasks. In PHIPA, Chapter 3, Schedule A, s.44 (2), it is written that "*A research plan must be in writing and must set out, (a) the affiliation of each person involved in the research, (b) the nature and objectives of the research and the public or scientific benefit of the research that the researcher anticipates; and (c) all other prescribed matters related to the research.*" As a result the task Check Ethical Issues in the hospital model is linked to this text to ensure that the research plan satisfies the PHIPA requirements (Figure 6).

The last link set is concerned with *responsibility* links. These links are created between responsibilities, components, and maps in the hospital UCM model and tasks, actors, goals, and softgoals in the PHIPA GRL model. Figure 8 shows some *responsibility* links (represent as "resp") between a UCM (ReviewRequest Technically map) element and the partial PHIPA GRL model. This link type is similar to the *traceability* type in terms of utility.

As explained before, the task Check for Adequate Safeguards should be performed by the Research Ethics Board (REB) according to PHIPA. However, as seen in Figure 8, the corresponding responsibility checkSafeguards in the map ReviewRequestTechnically indicates that it is the Data Warehouse Administrator who is responsible for it. In order to address this example of non-compliance, the UCM model has to be revised and the checkSafeguards responsibility needs to be moved to a different part of the process.

## 5 Analysis

In this section we analyze the four types of links based on the following criteria: functionality, precision, number of manual links, difficulty, and importance of completeness.

*Traceability Links*: This link type is found between the HIC GRL elements and the privacy legislation GRL elements. It shows what is missing or unnecessary in terms of the hospitals' goals and tasks (and consequently in their processes) and who is in charge of what activity. A missing softgoal, goal, or task can be a strong indication that the hospital does not completely comply with the law. Therefore, this link set is quite precise and it can help hospitals to measure their compliance very accurately. Traceability links are created manually. However, establishing this link set is not very difficult since both models are expressed at the same level of abstraction.

*Compliance Links*: This set differs from the first one in that instead of using GRL elements to model the privacy legislation document, we use the document itself. In practice, this set only contains those links between HIC GRL elements and the special constraints and exceptions in the text documents that cannot be modeled in the privacy GRL model. Therefore, this set is very precise and provides hospitals with additional information in order to define or improve their processes in terms of legal compliance. Creating this link set manually needs much effort but the number of manual links is fairly small and most of the links can be created through jUCMNav's auto-completion mechanism.

*Responsibility Links*: The main difference between these links and *traceability* links is that the hospital UCM model is linked directly to the privacy legislation GRL model. This link set is very precise since it includes fine-grained details of the business processes, so the traceability between processes and privacy legislation GRL is much easier than with the other links. However, its functionality is similar to the *traceability* links. Thus, it is often only necessary to create one of these two alternatives. In addition, as with *traceability* links, this link set needs to be complete and the number of links involved is high. However, most of these links can be created automatically by transitivity.

We evaluated each of these link sets based on the criteria mentioned above. Table 1 shows the summary of our analysis. As seen in this table, *traceability*

**Table 1.** Evaluation of Different Link Types

| Criteria \ Links | Traceability Link | Compliance Link | Responsibility Link |
|---|---|---|---|
| Granularity | Softgoals, Goals, Tasks, and Actors | Legislative Text | Responsibilities, Components (Actors), Maps (Operational Processes) |
| Functionality | Handles Traceability of Non-Functional Requirements and Tasks | Handles Exceptions and Constraints | Handles Traceability of Business Processes |
| Quantity of Manual Links | Many | Few | Few |
| Precision | Precise | Very Precise | Very Precise |
| Difficulty | Moderate | Difficult | Moderate |
| Importance of Completeness | Very Important | Not Important | Very Important |

and *responsibility* links are very similar in what they achieve and the amount of effort required. In particular, they both require complete coverage in order to be useful. *Responsibility* links are a bit more specific and precise but there is much overlap in the content they communicate, namely the mapping of roles and tasks or actors and processes at the HIC to the GRL elements in the privacy legislation model. It would only make sense for one or the other of these two types of links to be used in order to track the legal compliance. *Responsibility* links are a bit more specific but either set is adequate for the job.

Finally, if the HIC wants to ensure that their processes comply thoroughly with the legislation and laws, it would be necessary to use *compliance* links as well. These links can be used to highlight exceptions and specific constraints that are not captured in GRL or UCM models but which are critical for ensuring compliance. They can be difficult and time consuming to define, since they require direct reference to legal text, but it is only necessary for specific critical parts of the privacy legislation documents. It is likely that a privacy expert or a lawyer would highlight relevant HIC document passages that should be linked to the privacy model.

## 6 Conclusions

In this paper we have presented a framework that helps health information custodians analyze and improve their business processes in order to comply with relevant privacy legislation. A case study involving a major teaching hospital in Ontario and PHIPA privacy legislation was used to illustrate the framework. The User Requirement Notation (URN) was used to model the goals and business processes related to the access of confidential information stored in a data warehouse. Links were then made between this model and a model of the PHIPA privacy legislation in a requirements management system (DOORS). Different types of links were used at different levels and their functionality and accuracy were analyzed. In doing so, discrepancies were discovered that indicated possible instances of non-compliance with PHIPA legislation that would need to be addressed.

Both privacy legislation and the business processes of health information custodians are continually evolving in the face of changing technology and greater public awareness. Therefore, we will study how changes to a section of legislation will affect the goals and processes of the organization (and vice-versa) and how our framework can help guarantee that the processes will still comply with the legislation.

Finally, modeling legislation is not a new problem and our approach could benefit from recent work in that domain. For instance, Breaux *et al.* describe how to apply semantic parameterization to HIPAA privacy rules to extract rights and obligations from HIPAA text [19]. This approach could facilitate the extraction of our privacy GRL goal model. We also expect to extract generic goal and scenario models for privacy laws that could be used as patterns to kick start this process in multiple environments (PHIPA, HIPPA, PIPEDA, U.S. Sarbanes-Oxley Act, etc.), as was done for the software architecture domain [20]. In addition, in the privacy domain, GRL models could benefit from the privacy goal catalogues and patterns suggested in [21], which also focus on the Canadian healthcare sector. This work could accelerate the creation of the models and help determine suitable operationalizations that must be found in the related business processes. Moreover, in terms of transforming privacy policies into business process, Antón *et al.* provide a taxonomy for classifying privacy goals, and examining privacy policies in order to extract system requirements using goal-mining techniques [22]. In other words, they introduce a set of guidelines for requirement engineers and policy makers to follow when they analyze and evaluate privacy policies.

**Acknowledgments.** This work was supported by the Ontario Research Network for Electronic Commerce. We thank Jason Kealey and Jean-François Roy for their help with jUCMNav and Alan Forster for his insights into hospital processes. Telelogic provided us with the latest release of DOORS.

## References

1. Government of Ontario: Personal health information protection act (2004) (Accessed March 2007) `http://www.e-laws.gov.on.ca/DBLaws/Statutes/English/04p03_e.htm`
2. ITU-T: User Requirements Notation (URN) – language requirements and framework. ITU-T Recommendation Z.150. Geneva (February 2003)
3. Weiss, M., Amyot, D.: Business process modeling with URN. International Journal of E-Business Research 1(3), 63–90 (2005)
4. Telelogic AB: Doors. (Accessed March 2007) `http://www.telelogic.com/products/doors/doors/`
5. Roy, J.F., Kealey, J., Amyot, D.: Towards integrated tool support for the User Requirements Notation. In: Gotzhein, R., Reed, R. (eds.) SAM 2006. LNCS, vol. 4320, pp. 198–215. Springer, Heidelberg (2006)
6. Government of Canada: Health information custodians in the province of Ontario exemption order. (Accessed March 2007) `http://canadagazette.gc.ca/partII/2005/20051214/html/sor399-e.html`
7. European Union: Directive on privacy and electronic communication (2002) (Accessed March 2007) `http://eur-lex.europa.eu/LexUriServ/site/en/oj/2002/l_201/l_20120020731 en00370047.pdf`
8. US Dept. of Health and Human Services: Medical privacy - national standards to protect the privacy of personal health information. (Accessed March 2007) `http://www.hhs.gov/ocr/hipaa/`
9. Amyot, D.: Introduction to the User Requirements Notation: learning by example. Computer Networks 42(3), 285–301 (2003)
10. Chung, L., Nixon, B.A., Yu, E., Mylopoulos, J.: Formalizing Functional Requirements in Software Engineering. Kluwer Academic, Dordrecht, USA (2000)
11. Yu, E.: Towards modelling and reasoning support for early-phase requirements engineering. In: RE'97. Proc. 3rd IEEE Int. Symp. on Requirements Engineering, pp. 226–235. IEEE Computer Society Press, Los Alamitos (1997)

12. Caetano, A., Silva, A.R., Tribolet, J.: Using roles and business objects to model and understand business processes. In: SAC 2005. LNCS, ACM Press, New York, USA (2005)
13. Staccini, P., Joubert, M., Quaranta, J.F., Fieschi, D., Fieschi, M.: Modelling healthcare processes for eliciting user requirements: a way to link a quality paradigm and clinical information system design. International Journal of Medical Informatics 64(2-3), 129–142 (2001)
14. Kealey, J., Kim, Y., Amyot, D., Mussbacher, G.: Integrating an Eclipse-based scenario modeling environment with a requirements management system. In: CCECE06: IEEE Canadian Conf. on Electrical and Computer Engineering, Ottawa, Canada, pp.2432–2435 (2006)
15. Darimont, R., Lemoine, M.: Goal-oriented analysis of regulations. In: REMO2V06: Int. Workshop on Regulations Modelling and their Verification & Validation, Luxemburg June 2006 (2006)
16. He, Q., Otto, P., Antón, A.I., Jones, L.: Ensuring compliance between policies, requirements and software design: A case study. In: WIA 2006. Fourth IEEE Int. Workshop on Information Assurance, pp. 79–92. IEEE Computer Society Press, Washington, USA (2006)
17. Rifaut, A., Feltus, C.: Improving operational risk management systems by formalizing the Basel II regulation with goal models and the ISO/IEC 15504 approach. In: REMO2V06: Int. Workshop on Regulations Modelling and their Verification & Validation, Luxemburg (2006)
18. Fairfield, D.: The Ottawa Hospital data warehouse - governance and operation procedures - phase 1 research. Technical report, The Ottawa Hospital (2004)
19. Breaux, T.D., Vail, M.W., Antón, A.I.: Towards regulatory compliance: Extracting rights and obligations to align requirements with regulations. In: RE'06: Proc. 14th Int. Conf. on Requirements Engineering, pp. 46–55. IEEE Computer Society Press, Washington, USA (2006)
20. Amyot, D., Mussbacher, G., Weiss, M.: Formalizing patterns with the User Requirements Notation. In: Taibi, T. (ed.) Design Pattern Formalization Techniques, Idea Group Publishing, Hershey, USA (2007)
21. Webster, I., Ivanova, V., Cysneiros, L.M.: Reusable knowledge for achieving privacy: A canadian health information technologies perspective. In: WER'05: Workshop em Engenharia de Requisitos, pp. 112–122 (2005)
22. Antón, A.I., Earp, J.B., Reese, A.: Analyzing website privacy requirements using a privacy goal taxonomy. In: RE'02: Proc. 10th Int. Conf. on Requirements Engineering, pp. 23–31. IEEE Computer Society Press, Washington, USA (2002)

# Conceptual Modeling of Privacy-Aware Web Service Protocols

Rachid Hamadi, Hye-Young Paik, and Boualem Benatallah

School of Computer Science and Engineering
The University of New South Wales, Sydney NSW 2052, Australia
{rhamadi,hpaik,boualem}@cse.unsw.edu.au

**Abstract.** Internet users are becoming increasingly concerned about their personal information being collected and used by Web service providers. Since the privacy policies are mainly developed and maintained separately from the business process that collects and manipulates data, it is hard to perform analysis and management of the processes in terms of privacy policies. We propose a formal technique with which Web service providers describe the use and storage of requesters' personal data. The description is integrated with a Web service protocol using an extended state machine model. Having such a conceptual model will enable model-driven development and management of Web service protocols with respect to their privacy aspects such as collection, disclosure, and obligation.

**Keywords:** Web services, privacy policies, conceptual modeling.

## 1 Introduction

More and more Internet users are concerned about their personal information. The fact that modern business applications are extremely complex and often involve interactions with many other autonomous and heterogeneous partner systems makes the task of preserving privacy more complicated.

Web services are emerging as a promising technology for the effective *automation of inter-organizational interactions* [1]. Despite their growing popularity, the development of technologies addressing privacy issues in Web services has not kept the same pace. For example, the customers of Amazon.com use a uniform interface provided by the company, however, the actual processing of an order, delivery, analysis of sales data, and personalised services may involve passing the customer's personal data to third parties (e.g., the individual second-hand booksellers). In the midst of these business operations, it is not clear where and how the statements in the privacy policies apply to the activities and whether they will be enforced or not.

One of the problems is that there is no proper modelling technique for capturing the privacy aspects for a Web service. That is, no current Web service modelling technologies offer a simple way to state a privacy requirement (e.g., "The

J. Krogstie, A.L. Opdahl, and G. Sindre (Eds.): CAiSE 2007, LNCS 4495, pp. 233–248, 2007.

intended recipient of this message is a delivery service and the data should be removed after the delivery is completed") in a Web service model.

So far, online companies have dealt with privacy issues largely by publishing privacy policies. Privacy policies describe the organisation's general business practices based on the criteria set by government rules and regulations. However, they do not discuss the behaviour of individual business applications within the organisation that actually collect, analyse, and distribute personal data. This makes the enforcement of the policies difficult. We argue that a model-driven approach, where privacy policies are modelled explicitly as part of the Web service behaviour, can contribute to making the privacy policies explicit and enforceable. Having such a conceptual model will enable model-driven development and management of Web service protocols with respect to their privacy policies.

In this paper, we propose a Web service modeling technique purposely designed to capture privacy abstractions while describing the behaviour of a Web service. We use *Web service protocols* [2,3] to represent the way Web services interact with others. Our contributions are as follows:

- We identify common privacy abstractions in Web service protocols by studying publicly available privacy policies in Web portals.
- We propose an extended state machine as a conceptual model that incorporates the privacy abstractions into a Web service protocol model. In the model, we introduce the concept of *states with multiple privacy properties* and reflect the consequence of a state transition in terms of privacy properties such as access, disclosure and retention.

This paper is organized as follows. Section 2 introduces privacy and the related terminology. Section 3 discusses the privacy policies in Web services and gives some observations. Section 4 introduces the proposed conceptual model. Section 5 describes the tool supporting the model proposed as well as the application of privacy-aware Web service protocols. Finally, Sect. 6 reviews related work and concludes the paper.

## 2 Overview of Privacy Policies

Before we discuss the modelling of privacy in Web services, we first introduce the main issues and terminology in privacy policies in general.

What kind of privacy aspects are addressed or declared in a privacy policy may be different depending on the rules and regulations. However, studying privacy policies of the online companies tells us that there are some standard elements that commonly appear in all privacy policies. In the following, we summarise the gist of privacy policies.

**Personal data.** This identifies personal data collected by a Web site. The statements may differentiate information collected expectedly (e.g., user account registration and payment account information) and automatically such as cookies. It may also declare information collected from other sources (e.g., credit history from credit bureaus).

**Purpose.** This states the purpose of using the personal data. The statement may not refer to the specific data and we may only see generic purposes such as "to fulfill your request" or "to customise advertising".

**Recipient.** This identifies the recipient(s) of the personal data. Unless stated otherwise, the intended recipient of the data is the organisation itself.

**Disclosure.** This declares any business partners that may share the data.

**Data retention.** This states how long the collected data are retained by the organisation. More than often, data may be retained indefinitely, unless stated otherwise.

**Access to data.** Some of the retained personal data can be accessed by the owner for correction or update purposes. It is noted that, as explained, the term "Access to data" refers to the right of the data owner to access her/his personal data after it is collected. It does not refer to access by third parties.

**Opt in/out.** A company may provide services that users can choose to receive or not to receive. This feature of privacy policy is often referred to by the research community as *user consent management.*

Although majority of the publicly accessible privacy policies are written in plain English, there are standard languages designed for encoding the various aspects of privacy policies such as P3P (Platform for Privacy Preferences) [4] which is designed for Web site operators and EPAL (Enterprise Privacy Authorization Language) [5] which is designed for inter-organisational privacy policies.

Our work does not make any assumption about the choice of the language in which the policies are written. They could be in plain English or in one of the standard languages. What we would like to focus on is to encode such policies into Web service modelling process. If the policies are written in a standard language (e.g., P3P), of course, some of the encoding process can be done automatically by "reading" in the policies into the tool we provide.

## 3 Web Services and Privacy Policies

The model we use to describe the behaviour of a Web service is based on the concept of "Web service protocol" [2,3]. A Web service protocol model is understood as the set of acceptable message exchanges and the order in which they should occur when interacting with the service. In the model, a Web service protocol is described by a set of states and transitions. States are labeled with a logical name, such as `Logged` or `Ordered`. Transitions are labeled by either input or output messages (or service operations), i.e., messages sent by the requester (input) or by the provider (output). We use the notation `m(+)` (respectively, `m(-)`) to denote an *input* (respectively, an *output*) message.

In our previous work, we observed that there are cases in which transitions occur without an explicit invocation by requesters [2,3]. We refer to them as *implicit transitions.* The majority of implicit transitions are due to timing issues (i.e., deadline expirations). To characterise this observation, we use timed transitions. We will use the term *timed Web service protocol* (or *protocol* for short) to denote a Web service protocol whose definition contains timed transitions. We also use the term *protocol schema* to denote the specification of a protocol, while the term *protocol instance* will refer to an individual execution of a protocol between a service provider and a service requester.

### 3.1 A Running Example: Snowy.com

We have constructed a running example which is largely based on the behaviour of the Amazon.com Web site and its privacy policy [6]. We have simplified and modified the actual behaviour and policy to make it easier to illustrate our approach throughout the paper. We named the fictitious book-selling company `Snowy.com`.

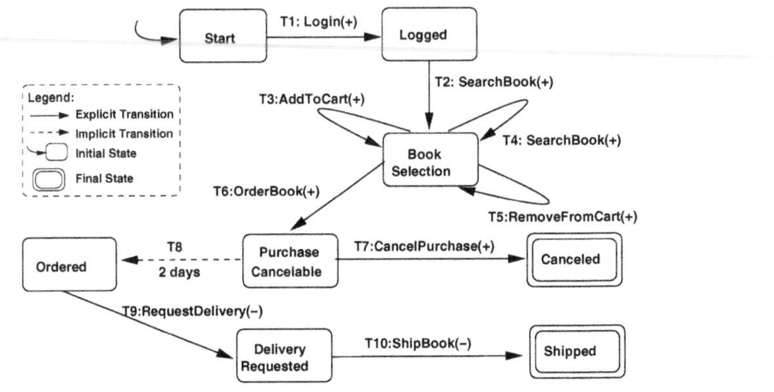


**Fig. 1.** The behaviour of `Snowy.com` as a timed Web service protocol

**The Scenario.** Figure 1 presents an example of a protocol schema. According to the model, once the user is logged in, s/he can search books, add/remove books to/from a shopping cart and order the books. Once the user has ordered the books, s/he will have two days to cancel the order. Otherwise, the ordered books are shipped, which completes the scenario. Note that `T8` is a timed (i.e., implicit) transition.

Let us assume that Snowy.com has the following key elements in its privacy policy. For ease of understanding, we use plain English to specify the policies instead of P3P or other language:

*As a result of actions such as* `OrderBook`*, you supply us with name, address, phone numbers, and credit card number which* ***we expectedly collect****. Some data may be* ***collected automatically*** *such as login id, password, your order history, and products you added/removed to/from your shopping cart.*

***We share*** *your name, address, and phone numbers with our delivery service partners. This enables us to update your delivery status in a prompt manner.*

***We use*** *your personnel information for purposes such as responding to your requests, customizing future shopping for you, and improving our stores. We also use your order history data in our market analysis.*

***We retain the collected data for the following periods:*** *the shopping cart history data for 6 months, login data for 2 months, and order detail for 3 months. However, it is assured that in case you cancel your order, we delete the order detail from your order history immediately. As long as the data are retained with us, you will be able to* ***access the following data through our user management system:*** *login id, password, payment information, and order history.*

**Applying the Policies to the Scenario.** Although studies suggest that people feel more comfortable with the Web sites that have privacy policies, only a small number of people actually read them [7]. Even if one fully understands the policies, it will not be easy to be on alert for every step of the way while s/he interacts with a Web site.

According to the policies above, when a customer visits the Web site, quite a lot of data about her/him are collected through automatic means. For example, let us consider when and how the policies should apply to Joe (the customer) when he interacts with Snowy.com. We will examine some of the transitions and see whether any part of the privacy policies are relevant to them.

The first transition is via `T1:Login` which carries a message that is likely to contain login id and password of Joe. A statement in the policies says login id and password are collected. It is reasonable to think that the purpose of collecting the data is to process Joe's login request. Once Joe is logged in, he may try to search some books (`T2:SearchBook`). The message should contain the search terms. We do not see any claim in the policy about collecting search terms, so transition `T2` seems to bear no privacy concern.

However, data associated with activities such as adding/removing an item to/from the shopping cart are collected. This means the data involved in `T3` and `T5` will be retained by the company and, according to the policy, they will be used for "possibly" purposes such as *responding to your requests, customizing future shopping for you, and improving our stores.*

The statement also claims that in the case of cancellation, the order details will be deleted from Joe's order history. That is, an obligation for the company is to make sure that the data are not retained after cancellation. If Joe places any order and receives the goods, he should be able to access the details of the order through the user management system, as the policy states. Also, the collected data are retained for three months.

### 3.2 Discussions and Observations

In this section, we discuss the main lessons learned during the analysis of privacy policies and Web portals.

**Privacy statements need to have clear semantics.** Although policies include the standard elements that are required by the rules and regulations, it is often difficult to extract information about a particular action (i.e., a transition in a protocol schema) due to informality of the language used. For the policies to be enforceable, we need to be explicit about the identification of data and purpose of using them. This information should be explicitly expressed in the model without ambiguity.

**Explicit transitions and their privacy implication.** Most transitions between states occur due to explicit operation invocations (i.e., message exchanges). We refer to these transitions as "explicit" transitions. We showed in the running example that some transitions in a protocol schema may have associated privacy aspects which are identified from the privacy policies. We argue that for such transitions, one should consider any *privacy implications* generated by them. For example, a privacy implication of the transitions `T2:SearchBook` or `T3:AddToCart` is that after they are fired, some personal data are collected; or a consequence of firing `T9:RequestDelivery` is that some personal data will be disclosed to a third-party service (i.e., the delivery service). The proposed model should be able to express these implications.

**Obligations.** Some of the privacy implications could mean more than collection or disclosure of personal data. They may lead to an action that the organisation is obligated to implement. Consider two of the privacy elements discussed in Sect. 2: data retention and access to data. First, the organisation must implement an action that will remove the data from their system to honour the data retention obligation. For example, when `T7:CancelPurchase` is fired, the privacy implication is (according to the policy) that the data collected during `T6:OrderBook` (i.e., order history) should be deleted immediately. Second, access to data ensures that the personal data owner can check and verify that the data are up-to-date. This means that the organisation is obligated to provide a user interface and operations for the users to access/update their personal data. For example, `T6:OrderBook` will collect payment information and order history. The privacy implication is that the data should be viewable by the owner. The proposed model should be able to express these obligations.

## 4 Conceptual Modelling of Privacy-Aware Web Service Protocols

In this section, we introduce the proposed conceptual model for privacy-aware Web service protocols. The proposed model will allow service providers to describe the use and storage of requesters' personal data. The description is integrated with a Web service protocol using an extended state machine model. We use state machines although other analogous models are possible.

To cater for privacy policies, states (and consequently transitions) are extended beyond the traditional timed state machine model [3]. We generalize the

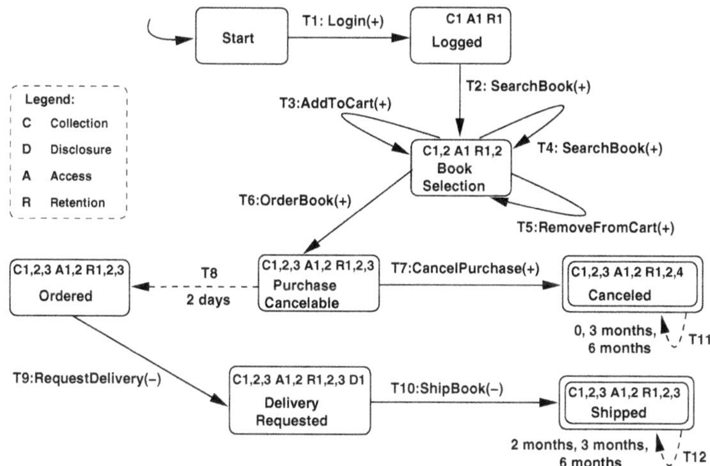


**Fig. 2.** The augmented protocol of Snowy.com with privacy properties

approach by enabling the association of several *privacy properties* with states to characterize when a privacy property enforcement should occur and what are its implications (e.g., destroy information when retention period expires). Having such a conceptual model will enable analysis of privacy aspects such as data collection, disclosure, access, and retention.

Figure 2 represents Snowy.com of Fig. 1 as an extended state machine augmented with privacy properties. Some extra implicit transitions, as they are not caused by explicit operation invocations, are added. They represent the implication of data retention, that is, deleting the collected data when their retention period is expired. The symbol Ci (respectively, Di, Ai, and Ri), $i \in \mathbb{N}^+$, within states means *Collection* (respectively, *Disclosure*, *Access*, and *Retention*) privacy property.

## 4.1 States with Multiple Privacy Properties

In this section, we will describe the states of our privacy-aware protocol model. We discuss a list of privacy properties that can be used to capture privacy policies described in Sect. 2. These privacy properties consist of an initial set of privacy abstractions that are commonly needed in practical situations, namely *collection*, *disclosure*, and *obligation* privacy properties. The model is extensible in the sense that other privacy properties may be defined and used.

The conceptual model shown in Fig. 3 represents a UML static model for the different components that constitute the privacy properties of a state. Each privacy property is described using a set of attributes. The model is also open to the extension of the definitions of privacy properties by adding new domain-specific attributes. The remainder of this section gives details about the identified state privacy properties, namely *collection*, *disclosure*, and *obligation*.

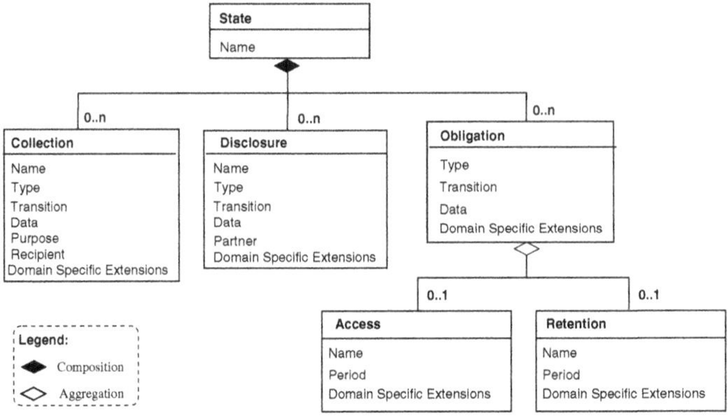


**Fig. 3.** UML conceptual model for privacy properties

**Collection Privacy Property.** In the extended state machine model, besides the fact that a state has a name, a *collection* privacy property expresses that data (or group of data) have been collected by the service provider when invoking an operation such as `T6:OrderBook` (see Fig. 2). More precisely, it specifies that the data are either automatically or expectedly collected. Hence the collection type is either *automatic* (`Type="automatic"`) or *expected* (`Type="expected"`). The attribute `Transition` is the name of the triggered transition. The attribute `Data` specifies the data or group of data collected, `Purpose` specifies the purpose of the data, and `Recipient` expresses the recipient of the data. It is important to note that the collection privacy property will be carried over by the subsequent states until the collected data are deleted (see *obligation* privacy property below). In this case, if there are `n` collection privacy properties `C1,C2,..,Cn` within a state, they will be represented as `C1,2,..,n`.

We use XPath [8] to express queries and conditions since privacy-aware protocol objects and requester profiles are represented using XML.

Let us consider the description of the collection privacy property `C1` (respectively, `C3`) of the state `Logged` (respectively, `PurchaseCancelable`) (see Fig. 2). The following XML codes represent the description of the collection privacy properties `C1` and `C3`:

```
<state name="Logged">
  <collection name="C1", type="automatic", transition="T1",
        data="/user[@login_data]", Purpose="", Recipient=""/>
</state>
<state name="PurchaseCancelable">
  <collection name="C3, "type="expected", transition="T6",
        data="/user[@order_data]", Purpose="", Recipient=""/>
</state>
```

**Disclosure Privacy Property.** The disclosure privacy property of a state `S` declares that data (or group of data) are shared with service partners when invoking an operation that leads to S. Similarly to collection privacy property, this privacy property specifies that the data are either automatically or expectedly disclosed. The other attributes of this property are the disclosure privacy property name (Name), the name of the triggered transition (Transition), the data disclosed (Data), and the service partner to which the data have been disclosed (Partner). Since the disclosure privacy property has no implications, it will not be carried over by the subsequent states. The only purpose is to annotate the states in which data have been disclosed.

The following XML code represents the description of the disclosure privacy property D1 of the state DeliveryRequested (see Fig. 2):

```
<state name="DeliveryRequested">
  <disclosure name="D1", type="expected", transition="T9",
     data="/user[@delivery_data]", partner="Delivery Service Partner"/>
</state>
```

**Obligation Privacy Property.** This privacy property models data retention and data access. We distinguish the following types of the obligation privacy property:

- *Access* to denote that some collected personal data are accessible by its owner for a specific period of time or indefinitely. We use Ai, $i \in \mathbb{N}^+$, to annotate this type of obligation privacy property.
- *Retention* to denote that certain requester's collected data are retained for a specific period of time or indefinitely. We use Ri, $i \in \mathbb{N}^+$, to annotate this type of obligation privacy property.

The obligation attribute Type indicates whether some collected data are retained (Type="R"), can be accessed (Type="A"), or both (type="mixed"). The other attributes of this property are data accessed and/or retained (Data), the name of the triggered transition (Transition), the obligation privacy property name (Name), and the period of time the data are accessed or retained (Period). The implications of this property are the implicit transitions Access and/or Retention.

When an obligation privacy property contains an Access transition as implication, that is when Type="A" or Type="mixed", the privacy property will be carried over by the subsequent states until the expiration of the access period. A state S carrying an Ai, $i \in \mathbb{N}^+$, annotation expresses that an implicit transition, for which S is both source and target state, is created. This means certain personal data are accessible by the requester from S. The time associated with this implicit transition is equal to Period the first time Ai appears and will decrease in subsequent states as the execution of the business process progresses. For clarity of presentation, we omitted the representation of these implicit transitions in Fig. 2.

When an obligation privacy property contains a Retention transition as implication, that is when Type="R" or Type="mixed", the privacy property will be carried over by the subsequent states until the expiration of the retention period.

A state S carrying an Ri, $i \in \mathbb{N}^+$, annotation expresses that an implicit transition is created for each final state F of the extended state machine for which F is both source and target state. The time associated with these implicit transitions is equal to Period the first time Ri appears and will decrease as the execution of the business process progresses. When the retention period expires, the corresponding retained data will be deleted (and its annotation removed from the final state).

The definition of temporal constraints uses XPath[8] time functions (e.g., current-time()). The following XML code represents the description of the obligation privacy property R2 of the state BookSelection (see Fig. 2):

```
<state name="BookSelection">
  <obligation type="R", data="/user[@cart_data]", transition="T3">
    <retention name="R2", period="6 months"/>
  </obligation>
</state>
```

The obligation privacy property R2 specifies that the data will be removed after a period of six months according to the associated Retention transition. But there is no Access transition associated with it.

The following XML code represents the description of the obligation privacy properties A2 and R3 of the state PurchaseCancelable (see Fig. 2):

```
<state name="PurchaseCancelable">
  <obligation type="mixed", data="/user[@order_data]", transition="T6">
    <access name="A2", period="3 months"/>
    <retention name="R3", period="3 months"/>
  </obligation>
</state>
```

This obligation privacy property contains an Access transition A2 as implication which states that requesters are allowed to access their personal data. This access data privacy property will be carried over by the subsequent states.

Finally, the following XML code represents the description of the obligation privacy properties R4 of the state Canceled (see Fig. 2):

```
<state name="Canceled">
  <obligation type="R", data="/user[@order_data]", transition="T7">
    <retention name="R4", period="0"/>
  </obligation>
</state>
```

The obligation privacy property R4 expresses that the order data collected by the service provider must be deleted immediately if the requester cancels her/his purchase of the books. This will override the obligation privacy property R3 which states that the same order data will be deleted after three months. Hence, R3 will not be carried over from the state PurchaseCancelable to the state Canceled but instead will be replaced by R4.

### 4.2 Privacy-Aware Protocol Formal Model

Formally, a timed Web service protocol can be modelled as a finite state machine as follows:

**Definition 1 (Timed Web Service Protocol).**
*A Timed Web Service Protocol is a tuple* $\mathcal{P} = (S, \mathcal{O}, T, s^0, \ell)$ *where:*

- $S$ *is a finite set of states,*
- $\mathcal{O}$ *is a set of operation names,*
- $T \subseteq S \times (\mathcal{O} \cup \{(\varepsilon, t) \mid t \in \mathbb{R}^+\}) \times S$ *is a finite set of transitions. Implicit transitions will be given the empty operation name* $\varepsilon$ *and a time* $t \in \mathbb{R}^+$*,*
- $s^0 \in S$ *is the initial state of* $\mathcal{P}$*,*
- $\ell : S \rightarrow \mathcal{SN}$ *is a naming function where* $\mathcal{SN}$ *is a set of state names.* □

The state machine of `Snowy.com` timed Web service protocol (see Fig. 1) is defined as $\mathcal{P} = (S, \mathcal{O}, T, s^0, \ell)$ where:

- $S = \{s_1, ..., s_8\}$,
- $T = \{t_1, ..., t_{10}\}$,
- $t_1 = (s_1, Login, s_2)$, $t_2 = (s_2, SearchBook, s_3)$, $t_3 = (s_3, AddToCart, s_3)$, $t_4 = (s_3, SearchBook, s_3)$, $t_5 = (s_3, RemoveFromCart, s_3)$, $t_6 = (s_3, OrderBook, s_4)$, $t_7 = (s_4, CancelPurchase, s_5)$, $t_8 = (s_4, (\varepsilon, 2\ days), s_6)$, $t_9 = (s_6, RequestDelivery, s_7)$, and $t_{10} = (s_7, ShipBook, s_8)$,
- $s^0 = s_1$,
- $\ell = \{(s_1, Start), (s_2, Logged), (s_3, BookSelection), (s_4, PurchaseCancelable), (s_5, Canceled), (s_6, Ordered), (s_7, DeliveryRequested), (s_8, Shipped)\}$.

The multiple privacy properties of a state are formally defined as follows:

**Definition 2 (State Privacy Properties).**
*Let Col (respectively, Dis and Obl) be a set of all Collection (respectively, Disclosure and Obligation) privacy properties. Given a Web Service Protocol* $\mathcal{P} = (S, \mathcal{O}, T, s^0, \ell)$*. The privacy properties of a state* $s \in S$ *are defined as a triple* $(\mathcal{C}, \mathcal{D}, \mathcal{O})$ *where:*

- $\mathcal{C} \in \mathcal{P}(Col)$ *denotes the set of collection privacy properties, each with:*
  - *Name is the name of the collection privacy property,*
  - *Type* $\in \{expected, automatic\}$ *denotes the collection type,*
  - *Transition is the triggered transition,*
  - *Data are the data or group of data collected,*
  - *Purpose specifies the purpose of the data, and*
  - *Recipient is the recipient of the data.*
- $\mathcal{D} \in \mathcal{P}(Dis)$ *denotes the set of disclosure privacy properties, each with:*
  - *Name is the name of the disclosure privacy property,*
  - *Type* $\in \{expected, automatic\}$ *denotes the disclosure type,*
  - *Transition is the triggered transition,*
  - *Data are the data or group of data disclosed, and*
  - *Partner is the service partner to which the data have been disclosed.*

- $\mathcal{O} \in \mathcal{P}(Obl)$ *denotes the set of obligation privacy properties, each with:*
  - $Type \in \{A, R, mixed\}$ *denotes the disclosure type which can be A (Access), R (Retention), or mixed meaning both Access and Retention,*
  - $Transition$ *is the triggered transition,*
  - $Data$ *are the data accessed and/or retained,*
  - $Access = (Name, Period)$ *specifies the implication of the access data obligation privacy property where:*
    - $Name$ *is the name of the access obligation privacy property,*
    - $Period$ *is the period of time the collected data are accessed,*
  - $Retained = (Name, Period)$ *specifies the implication of the retention obligation privacy property where:*
    - $Name$ *is the name of the retention obligation privacy property, and*
    - $Period$ *is the period of time the collected data are retained.* □

The extended state machine that models the privacy-aware Web service protocol is defined as follows:

**Definition 3 (Privacy-Aware Service Protocol).**
*A Privacy-Aware Service Protocol is a tuple* $\mathcal{PP} = (S, \mathcal{O}, T, P, s^0, \ell)$ *where:*

- $S$ *is a finite set of states,*
- $\mathcal{O}$ *is a set of operation names,*
- $T \subseteq S \times (\mathcal{O} \cup \{(\varepsilon, t) \mid t \in \mathbb{R}^+\}) \times S$ *is a finite set of transitions. Implicit transitions will be given the empty operation name* $\varepsilon$ *and a time* $t \in \mathbb{R}^+$*,*
- $P : S \longrightarrow \mathcal{P}(Col) \times \mathcal{P}(Dis) \times \mathcal{P}(Obl)$ *is the state privacy property function,*
- $s^0 \in S$ *is the initial state of* $\mathcal{PP}$*,*
- $\ell : S \longrightarrow \mathcal{SN}$ *is a naming function where* $\mathcal{SN}$ *is a set of state names.* □

Having such a formal model will help, for instance, in *replaceability* analysis which is concerned with verifying whether two privacy-aware protocols, e.g., of two different service providers, are equivalent. That is, if they can support the same set of privacy policies. Replaceability analysis also involves finding the subset of privacy policies that both Web services can support if they are not equivalent.

## 5 Tool Support and Application of Privacy-Aware Web Service Protocols

In this section, we present the implementation of the tool supporting the model proposed as well as possible applications of privacy-aware Web service protocols.

To simplify the entire service development and management lifecycle, we need to consider the following [9]:

(1) *Models and languages.* Users should have at their disposal protocol models that are easy to understand and use. The key is to include frequently needed privacy aspects, but avoid overloading the model with too many features. Another

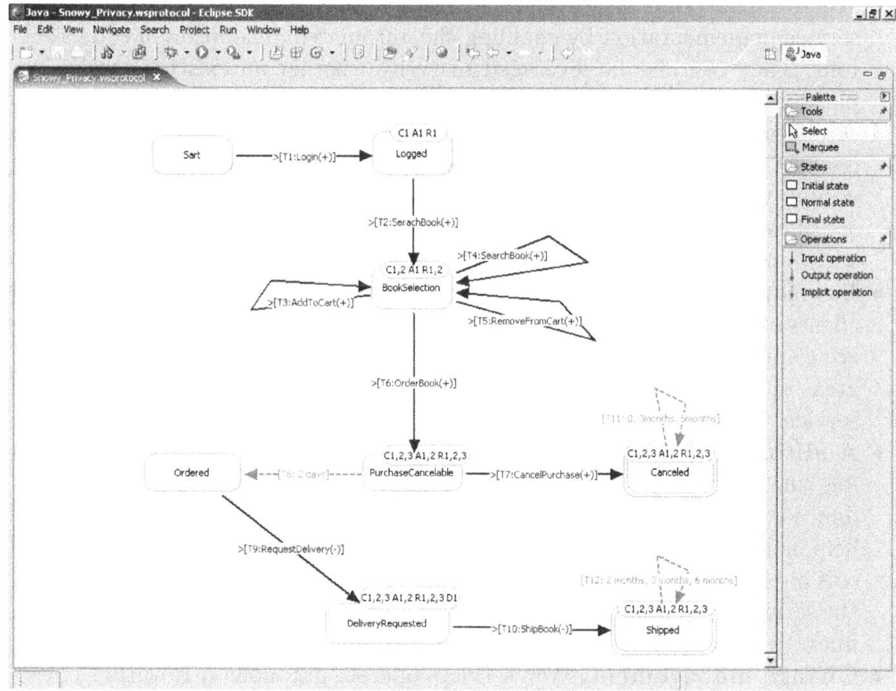


**Fig. 4.** Privacy-aware service protocol editor as part of ServiceMosaic

important aspect is that the privacy-aware protocol model should be formal enough to allow automated analysis and code generation.

(2) *Tools.* In the end what users really need and work with are tools. Hence, models and languages need also to be developed by considering how tools can leverage the concepts to provide concrete benefits to developers.

We have developed a privacy-aware Web service protocol tool to facilitate the creation, management, and analysis of privacy-aware Web service protocols. It is implemented as part of the `ServiceMosaic` model-driven framework for Web services lifecycle management. A description of `ServiceMosaic` framework can be found in [10]. The privacy-aware Web service protocol tool assists designers creating privacy-aware protocol definitions. A privacy-aware protocol definition is edited through a visual interface (see Fig. 4), and translated into an XML document. The visual interface offers an editor for describing an extended state machine diagram of a privacy-aware protocol. It also provides means to describe the privacy properties of states.

Our future objectives of this research is to provide developers with a privacy-aware Web service development environment. We are currently considering the following applications of our privacy-aware Web service protocol model:

- **Automated code generation.** Privacy-aware protocols can support Web service implementation by enabling the automated generation of code skeletons. They can also be leveraged to verify whether an existing service implementation can support the privacy-aware protocol as declared.
- **Automated exception handling.** Privacy-aware protocol specifications enable the development of generic tools that read protocol specifications and verify at runtime that the interaction is occurring in compliance with the specification, raising an exception otherwise (e.g., returning a fault message when an "illegal" use of personal data is detected).
- **Development-time analysis.** During service development, protocols of clients and providers can be analyzed to identify which part of the protocols are compatible with respect to privacy policies, therefore suggesting possible areas of modifications to increase the level of compatibility with a desired service.
- **Auditing and compliance.** A privacy audit has the objective of discovering whether records of personal information are being maintained in accordance with the service provider's privacy policies. Having a formal model will help in developing an auditing framework. Furthermore, privacy-aware protocols analysis and management provide opportunities to understand whether the service is compliant with certain organisations' privacy policies or guidelines.
- **Change management.** Web services operate autonomously within potentially dynamic environments. As a result, their privacy policies may evolve, e.g., because of changes to laws and regulations and changing business strategies. Consequently, services may fail to invoke required operations when needed.

## 6 Related Work and Conclusions

Although many research projects are looking into various aspects of privacy and computing, to the best of our knowledge, there is no other work done in terms of embedding privacy concerns during the modelling of Web services. However, there are some alternative approaches which are discussed here.

For example, [11] suggested the *Integrated Privacy View* system, in which privacy policy "anchors" are attached to each element of the HTML Forms. The anchors are used to link each Form element to the relevant part of P3P statements and to visualise them. This helps users easily identify how and where privacy policies apply to the Web site. However, the approach is ad-hoc in nature and involves augmenting the HTML code. Also, it is limited to HTML pages which do not represent business processes. Our approach deals with Web service protocols that depict business processes. Therefore, it is possible to formally analyse any conflicts and violation that might occur during the execution of the process.

There are database-centric approaches. The concept of Hippocratic databases is introduced in [12]. The core idea is to protect access to the personal data inside databases using SQL rewriting to include privacy policy evaluation. [13] presents another approach to authorised privacy data access in Web services. The actual prevention of inappropriate disclosure happens at the database component of the architecture via query rewriting. Similar to the database-centric approach, [14] proposed a framework that adds privacy enforcement to existing applications. The work considers the entire lifecycle of personal data (i.e., collection, use, disclosure and deletion) by attaching a policy statement, that needs to be tracked and enforced, to the data. The drawback of these approaches is that they focus on the database level (relational databases in particular). We believe that a higher-level view (i.e., business process and orchestration) consideration is required to cater for the privacy concerns that may span across different systems and different types of data repositories.

A model-driven approach described in [15] is developed for security applications. From UML models, suitable access control infrastructures for server-based applications are derived. Although the privacy aspects are not discussed in their work, similar principles can apply to build privacy enforcement mechanisms with proper consideration for the privacy concerns we discussed.

To conclude, we proposed a conceptual model for privacy-aware Web service protocols. The description of the use and storage of personal data is integrated with a Web service protocol using an extended state machine model. This conceptual model enables model-driven development and management of Web service protocols with respect to their privacy aspects such as data collection, disclosure, access, and retention. A tool support has been implemented, as part of `ServiceMosaic`, to let designers model privacy aspects within the Web service protocol. Currently, we are further extending our prototype to include support for automatic BPEL [16] skeleton code generation. As future work, we will add a provision for analysis of conflicts between Web service protocols in terms of privacy policies and storage and analysis of personal data usage logs for auditing purpose.

## References


1. Curbera, F., Duftler, M., Khalaf, R., Nagy, W., Mukhi, N., Weerawarana, S.: Unraveling the Web Services Web: An Introduction to SOAP, WSDL, and UDDI. IEEE Internet Computing 6(2), 86–93 (2002)
2. Benatallah, B., Casati, F., Toumani, F., Hamadi, R.: Conceptual Modeling of Web Service Conversations. In: Proc. of the 15th Int. Conf. on Advanced Information Systems Engineering, CAiSE 2003, LNCS, vol. 2681, pp. 449–467. Springer, Heidelberg (2003)
3. Benatallah, B., Casati, F., Ponge, J., Toumani, F.: On Temporal Abstractions of Web Service Protocols. In: CAiSE'05 Short Paper Proceedings, Porto, Portugal (2005)
4. Cranor, L., Langheinrich, M., Marchiori, M., Presler-Marshall, M., Reagle, J.: The Platform for Privacy Preferences 1.0 (P3P1.0) Specification. W3C Recommendation (2002)

5. Ashley, P., Hada, S., Karjoth, G., Powers, C., Schunter, M.: Enterprise Privacy Authorization Language (EPAL 1.1) Specification. IBM Research Report. (2003) http://www.zurich.ibm.com/security/enterprise-privacy/epal
6. Amazon.com: Amazon.com Privacy Notice. (2006) http://www.amazon.com/gp/help/customer/display.html?nodeId=468496
7. Cranor, L.F.: Web Privacy with P3P. O'Reilly (2002)
8. Clark, J., DeRose, S.: XML Path Language (XPath) Version 1.0. (1999) http://www.w3.org/TR/xpath
9. Benatallah, B., Casati, F., Toumani, F.: Representing, Analysing and Managing Web Service Protocols. Data and Knowledge Engineering 58(3), 327–357 (2006)
10. Benatallah, B., Casati, F., Toumani, F., Ponge, J., Motahari Nezhad, H.: Service Mosaic: A Model-Driven Framework for Web Services Life-Cycle Management. IEEE Internet Computing 10(4), 55–63 (2006)
11. Levy, S., Gutwin, C.: Improving Understanding of Website Privacy Policies with Fine-Grained Policy Anchors. In: Proc. of the 14th Int. World Wide Web Conference, Chiba, Japan, pp. 480–488. ACM, New York (2005)
12. Agrawal, R., Kiernan, J., Srikant, R., Xu, Y.: Hippocratic Databases. In: Proc. of the 28th Int. Conf. on Very Large Data Bases, pp. 143–154. Morgan Kaufmann, Washington (2002)
13. Rezgui, A., Ouzzani, M., Bouguettaya, A., Medjahed, B.: Preserving Privacy in Web Services. In: Proc. of the 4th Int. Workshop on Web Information and Data Management, Virginia, USA, pp. 56–62. ACM, New York (2002)
14. Berghe, C.V., Schunter, M.: Privacy Injector - Automated Privacy Enforcement Through Aspects. In: Proc. of 6th Workshop on Privacy Enhancing Technologies. pp. 99–117 (2006)
15. Basin, D., Doser, J., Lodderstedt, T.: Model driven security: From UML to access control infrastructures. ACM Trans. Soft. Eng. Methodol. 15(1), 39–91 (2006)
16. Curbera, F., Goland, Y., Klein, J., Leymann, F., Roller, D., Thatte, S., Weerawarana, S.: Business Process Execution Language for Web Services (BPEL4WS). (2002) http://dev2dev.bea.com/techtrack/BPEL4WS.jsp

# Policies for Context-Driven Transactional Web Services

Zakaria Maamar[1], Nanjangud C. Narendra[2], Djamal Benslimane[3], and Sattanathan Subramanian[4]

[1] Zayed University, U.A.E
zakaria.maamar@zu.ac.ae
[2] IBM India Research Lab, India
narendra@in.ibm.com
[3] Claude Bernard University, Lyon, France
djamal.benslimane@liris.cnrs.fr
[4] IMRU-FUNDP, University of Namur, Belgium
subramanian.sattanathan@fundp.ac.be

**Abstract.** This paper presents an approach that uses policies to manage context-driven transactional Web services. Context feeds policies with details on Web services like current status, which permits aligning the behavior of these Web services to the transactional properties they need to satisfy. Context refers here to any information on the interactions a Web service initiates with peers and external environment. Three types of transactional properties are used namely pivot, compensatable, and retriable. Each property satisfaction calls for a set of policies that are specified with a policy language like WSPL. This paper also presents the adaptation strategy that supports developing context-driven transactional Web services. A prototype that implements this strategy is discussed in the paper, too.

**Keywords:** Adaptation, Context, Policy, Transaction, Web service.

## 1 Introduction

For the W3C, a Web service *"is a software application identified by a URI, whose interfaces and binding are capable of being defined, described, and discovered by XML artifacts and supports direct interactions with other software applications using XML-based messages via Internet-based applications"*. Though this definition highlights the potential and multiple uses of Web services, it does not stress the obstacles that hinder Web services execution and the way these obstacles could be first, identified prior to execution and second, overcome as part of the exception handling strategy. Guidelines backing the correct execution of a Web service need to be stated and checked prior execution. To this end we suggest mapping these guidelines onto transactional properties to be associated with a Web service. The role of a transactional property is to define the acceptable behavior of a Web service. For example the failure of a Web service could be tolerated in one scenario but not in another one. Different transactional properties are reported in literature and different specifications exist (e.g., Web Services Transaction[1], Web Services Transaction Management[2]). In this paper the focus is on pivot, retriable,

[1] dev2dev.bea.com/pub/a/2004/01/ws-transaction.html
[2] developers.sun.com/techtopics/webservices/wscaf/wstxm.pdf

J. Krogstie, A.L. Opdahl, and G. Sindre (Eds.): CAiSE 2007, LNCS 4495, pp. 249–263, 2007.

and compensatable transactional properties. A Web service is defined as *retriable* if it can be retried one or more times after failure. A Web service is defined as *compensatable* if it offers mechanisms to undo its effects. Finally, a Web service is defined as *pivot* if once it successfully completes, its effects remain unchanged for ever and cannot be semantically undone. Additionally, a pivot Web service cannot be retried following failure, and thus, will need to be aborted.

Exceptions altering a Web service's behavior need to be monitored so, appropriate corrective actions for satisfying the transactional properties of this Web service are taken. We propose to run the monitoring operation upon a structure, which receives, refines, and stores the necessary information for this operation. We refer to this structure as context. Context "... *is not simply the state of a predefined environment with a fixed set of interaction resources. It is part of a process of interacting with an ever-changing environment composed of reconfigurable, migratory, distributed, and multiscale resources*" [5]. In this paper, context not only supports the operation of monitoring a Web service execution, but supports also a Web service in making decisions based on the status of the surrounding environment [8]. The environment could be related to users (e.g., stationary user, mobile user), computing resources (e.g., fixed device, handheld device), time of day (e.g., in the afternoon, in the morning), physical locations (e.g., shopping center, movie theater), etc.

Satisfying the transactional properties of a Web service happens through mechanisms, which we specify and implement as policies. In [7], we used policies to support the behavior flexibility of a Web service, so this latter can align its capabilities to users' requirements and resources' constraints. In this paper we motivate behavior flexibility because of the multiple execution situations a Web service encounters. Indeed a Web service has to consider its internal execution status, has to know how to perform exception handling in case it gets disrupted, etc. In this paper as well, policies not only permit checking the satisfaction of the transactional properties of a Web service, but permit also a clear separation between the functionality of a Web service and the different cases that make up the acceptable behavior of a Web service.

In this paper we discuss our approach for using policies to develop context-driven transactional Web services. Context feeds policies with details required for their execution prior to claiming the satisfaction of the transactional property of a Web service in that specific context. Section 2 presents an illustrative scenario and some related works. Section 3 discusses the approach to develop context-driven transactional Web services using policies. Section 4 presents the adaptation strategy that accommodates these Web services' features and requirements. Prior to concluding and highlighting future work in Section 6, prototype of the approach is presented in Section 5.

## 2 Background

**Illustrative scenario.** It is about Amin who travels to Trondheim in Norway to meet his friend Melissa. One day they agree to meet in a coffee shop, not far from Melissa's office. Amin has two options to reach the meeting place: by taxi or by bus. A specification of Amin scenario using state chart diagrams and service chart diagrams [9] is illustrated with Fig. 1. The component Web services of this specification are: trip (`TP`),

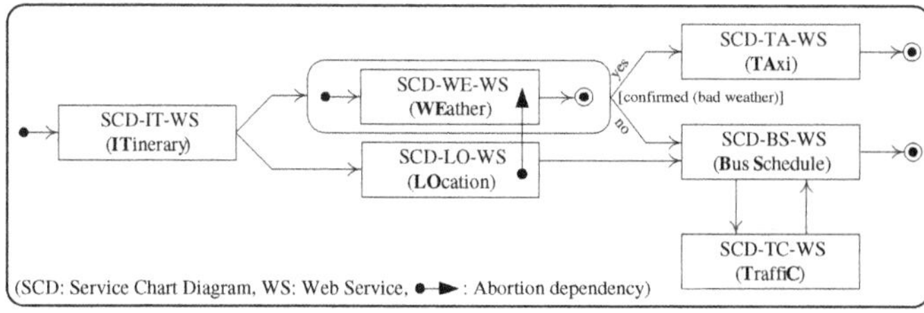


**Fig. 1.** Specification of Amin scenario

weather (WE), location (LO), taxi (TA), bus schedule (BS), and traffic (TC). Amin scenario specification could be done with BPEL for example, without any changes in the various policies and strategies that will be defined later.

At his hotel, Amin browses some Web sites about transportation in Trondheim. A site has *Itinerary WS* that proposes routes between two specific places like Amin's hotel and the coffee shop. The proposed routes are subject to weather forecasts: cold weather results in recommending taxis, otherwise public transportation like tramways and buses are recommended. Parallel to checking weather forecasts with *Weather WS*, *Itinerary WS* requests details about the origin and destination places using *Location WS*. Amin appreciates using *Location WS* as he is not familiar with the city.

In case *Weather WS* forecasts bad weather, a taxi booking is made using *Taxi WS* upon Amin's approval. Otherwise, i.e., pleasant day, Amin uses public transportation. The location of both Amin's hotel and coffee shop are submitted to *Bus Schedule WS*, which returns for example the bus numbers Amin has to take. Potential traffic jams force *Bus Schedule WS* to regularly interact with *Traffic WS* that monitors the status of the traffic network. This status is fed into *Bus Schedule WS* so adjustments to bus numbers and correspondences between buses can occur.

From a transactional perspective the designer of Amin scenario needs to pay attention among other things to (i) the Web services that are critical to the successful completeness of this scenario, (ii) the failure details that hinder Web services execution, and (iii) how much these failures impact Web services' and composite Web service's completeness. Hereafter, we list some cases the designer will look into: (i) ensure that either *Taxi WS* or *Bus Schedule WS* completes their execution; (ii) ensure that *Weather WS* successfully completes its execution; and (iii) compensate *Taxi WS* in case the meeting is canceled so the taxi booking is canceled, too.

**Related Work.** Compared to traditional transactions that comply with the Atomicity, Consistency, Isolation, and Durability (ACID) model, Verma and Deswal discuss the non-suitability of this model for Web services because of the following reasons [14]: transactions may be of a long duration (sometimes lasting hours, days, or more), participants may not allow their resources to be locked for long durations, some of the ACID properties are not mandatory, a transaction may succeed even if only some of the participants choose to confirm and others choose to cancel, transactions that have to be rolled back have the concept of compensation, etc. Interesting to emphasize here the overall success of a transaction despite the failure of some of this transaction's portions.

Bhiri et al. propose a transactional approach to guarantee the failure atomicity of a composite Web service [4]. They use the accepted termination states property as a means for guaranteeing this atomicity. The correctness criterion associated with a composite Web service execution varies from one designer to another. Bhiri et al. claim that this criterion defines the transactional behavior of a composite Web service. This behavior needs to be consistent with the transactional properties that are associated with the component Web services of this composite Web service.

For Younas et al., specifications and protocols developed for Web services transactions such as WS-Transactions, OASIS Business Transaction Protocol (BTP), and Business Transaction Framework are mainly based on the database transaction models such as ACID and extended/advanced transaction models [15]. Although these specifications and protocols have been useful in various domains, they are inappropriate for long running business activities like the ones involving Web services. Younas et al. suggest a new set of transactional properties that are specifically devoted to Web services namely Semantic Atomicity, Consistency, Resiliency, and Durability (SACReD) and are extensively explained in [16]. For instance, semantic atomicity allows the unilateral commit of component service transactions regardless of the commits of their sibling component service transactions.

Pires et al. discuss how to build reliable Web services compositions [10]. Unlike components in traditional business processes, the building task of these compositions is much more difficult due to Web services heterogeneity and autonomy. To face both obstacles, Pires et al. suggest `WebTransact` framework, which is implemented with a multi-layered architecture associated with an XML-based language named *Web Services Transaction Language* (`WSTL`) and a transaction model. Some components that populate this architecture include composite mediator services and remote services.

An interesting perspective on exception handling during process activity failures is built upon forward recovery strategies. This is reported in [3] where Bassil et al. claim that not all failures can be dealt with using roll-back mechanisms such as undoing or compensating activities. Examples of such failures include an already accomplished surgery or a vehicle transporting containers that breaks down. Bassil et al.'s solution suggests a set of factors that may influence the right choice of a forward recovery solution. Two of these factors include knowing the current data context of a failed activity and knowing how far process execution has progressed.

## 3 Context and Transactional Web Services

### 3.1 Design and Operation

Achieving transactional Web services using the information that context provides led us to identify the following four levels: composition, component, instance, and state ([9] explains how these levels get deployed). The composition level shows the composite Web services that are developed according to users' needs. The component level shows the Web services that providers develop and advertise so, users' needs are satisfied. The participation of Web services in composite Web services occurs thanks to the instance level [9]. This level shows the Web service instances that are created upon composition participation acceptance. Finally the state level shows the behavior of a

Web service instance using an UML state chart diagram. Each level is associated with a specific type of context: $\mathcal{C}$-context for $\mathcal{C}$omposite Web service, $\mathcal{W}$-context for $\mathcal{W}$eb service, $\mathcal{I}$-context for Web service $\mathcal{I}$nstance, and $\mathcal{S}$-context for $\mathcal{S}$tate chart diagram of a Web service instance. The $\mathcal{W}$-context of a Web service returns information on the participations of this Web service in different compositions. These participations happen according to the Web services instantiation principle [9]. The $\mathcal{C}$-context of a composite Web service is built upon the $\mathcal{W}$-contexts of its component Web services and permits overseeing the progress of a composition. The $\mathcal{I}$-context of a Web service instance records the progress of the execution of this instance, including the states it takes on during execution. Details on the state information of a Web service instance are later recorded in its $\mathcal{S}$-context. Fig. 2 illustrates our proposed context-driven three-level approach for transactional Web services. Not represented in this figure are the composition level and its respective $\mathcal{C}$-context. Interesting to note the $\mathcal{S}$-context of a state chart diagram. $\mathcal{S}$-context tracks the states that permit claiming the satisfaction of the transactional properties of a Web service instance. We recall that composite Web services are made up of Web service instances and not of Web services. In Fig. 2, active means the state that a Web service instance takes now on. Passive means the opposite.

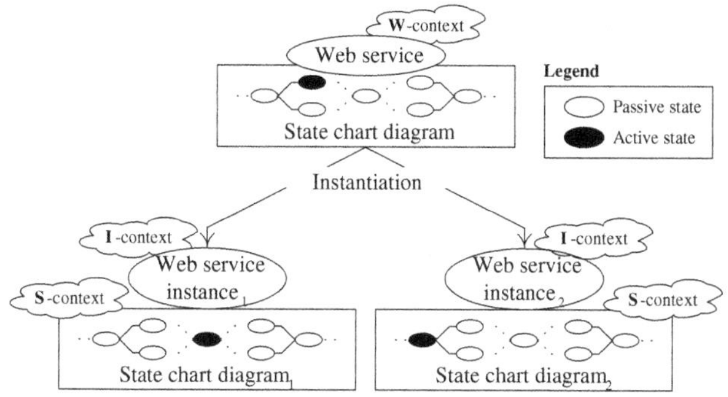


**Fig. 2.** Context-driven approach for transactional Web services

The operation of this approach concerns Web services of type instance. This operation is about first, context assessment and policy triggering and second, the way context and policy permit meeting the transactional properties of a Web service instance. In this paper context assessment is excluded. Initially the designer associates a Web service, to be deployed later as a Web service instance, with a set of transactional properties like pivot and compensatable. This association per Web service depends on the business logic that underpins the composition scenario. As discussed earlier, the failure of a Web service can be tolerated in one scenario but not in another one.

At run-time, the Web service instance gets triggered according to the specification of the composite Web service. The Web service instance takes on various states like activated, failed, and suspended, which form its state chart diagram. This diagram is context-aware since it has an $\mathcal{S}$-context. As a result tracking the various states that a Web service instance binds to, is now possible. Fig. 3 shows that the monitoring of a

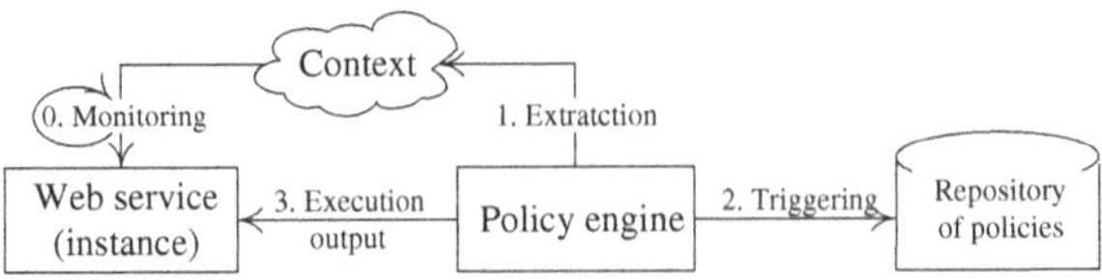


**Fig. 3.** Operation of the approach

Web service instance is continuous so, relevant details are collected and fed into the context. Additional details are collected as well from the respective contexts of the Web service and composite Web service. All these details are submitted to the policy engine that next, consults the repository of policies. Currently we only assume that one policy executes so conflicts between policies' outcomes are avoided. Execution means making the Web service instance transitions to a new state (active as in Fig. 2), which could allow this Web service instance to satisfy its transactional property. For example if a Web service instance is declared as pivot, then the various policies have to guarantee that this Web service instance only gets aborted in case of execution failure, i.e., no compensation actions are tolerated. We assume in this paper that a Web service does not take on any state that is not included in the acceptable states of its state chart diagram.

We recall that three types of context were defined: $\mathcal{S}$-context, $\mathcal{I}$-context, and $\mathcal{W}$-context. For this paper's requirements the emphasis is on the contexts of state and Web service instance. Each context type has a set of arguments that permit feeding the policy engine with the necessary details for triggering the appropriate policies as depicted in Fig. 3. $\mathcal{S}$**-context**'s arguments include: *StateIdentifier*: Identifier of current state; *StateLabel*: not-activated, activated, suspended, done, compensated, aborted; *PreviousState*: name of previous state from which the Web service instance has transitioned to current state; *NextEffectiveState*: name of next state that the Web service instance has effectively transitioned to; *TransitionIn*: name of transition that permitted transiting the Web service instance to current state; and *TransitionOut*: name of transition that permitted transiting the Web service instance to next effective state; $\mathcal{I}$**-context's arguments** include: *WSIdentifier*: name of Web service instance; *CurrentState*: not-activated, activated, suspended, done, compensated, aborted; *TransactionalProperty*: null, pivot, retriable, compensatable; *MaximumNumberOfRetries*: maximum number of times that the failed execution is authorized to be retried; and *CurrentNumberOfRetries*: current number of times that the Web service instance execution has been retried.

### 3.2 Transactional Properties and Web Services Modeling

As per Bhiri et al.'s transactional properties namely pivot, compensatable, and retriable (that could be combined as well) [4], we bind to the same properties. We show in the rest of this section how a Web service's behavior is continuously aligned, by using policies, in order to meet the requirements of its associated transactional property. To represent a Web service's behavior we use UML state chart diagram. Since we selected three transactional properties we developed three separate state chart diagrams for clarity reasons. In addition for each state chart diagram we provide a discussion on the role of context

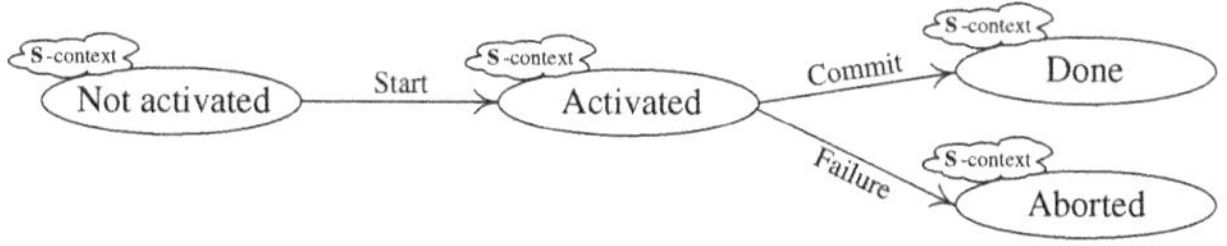


**Fig. 4.** State chart diagram for a pivot Web service

in feeding the policies with the information that permits achieving the associated transactional property. We recall that the following description applies to Web services of type instance. For illustration purposes we show how a rule is mapped onto WSPL (its syntax is based on the OASIS eXtensible Access Control Markup Language) [2]. The selection of this policy specification language is based on our previous research [7]. In addition, we only detail the pivot transactional property. Fig. 4 shows the acceptable state chart diagram of a pivot Web-service. The key state in this diagram is *activated* from which the Web service could transition to either *done* or *aborted*. We present hereafter the policies that describe the acceptable behavior of a pivot Web service. All the necessary details for policy specification exist in $\mathcal{S}/\mathcal{I}$-contexts.

`WS-Pivot.Policy`$_{done}$ states that a pivot Web service transitions from *activated* state to *done* state if-and-only-if the transactional property is *pivot*, the current state is *activated*, the previous state is *not activated*, and the transition name that was successfully fired is *commit*. This policy is shown below in WSPL.

```
Policy(Aspect="PivotPolicyDone"){
<Rule xmlns="urn:oasis:names:tc:xacml:3.0:generalization:policy:schema:wd:01"
RuleId="PivotPolicyDoneWS">
  <Condition>
    <Apply FunctionId="and">
      <Apply FunctionId="equal" DataType="boolean">
      <SubjectAttributeDesignator AttributeId="TransactionalProperty" DataType="string"/>
      <AttributeValue DataType="string"/> "pivot" </AttributeValue> </Apply>
      <Apply FunctionId="equal" DataType="boolean">
      <SubjectAttributeDesignator AttributeId="CurrentState" DataType="string"/>
      <AttributeValue DataType="string"/> "activated" </AttributeValue> </Apply>
      <Apply FunctionId="equal" DataType="boolean">
      <SubjectAttributeDesignator AttributeId="PreviousState" DataType="string"/>
      <AttributeValue DataType="string"/> "notactivated" </AttributeValue></Apply>
      <Apply FunctionId="equal" DataType="boolean">
      <SubjectAttributeDesignator AttributeId="TransitionOut" DataType="string"/>
      <AttributeValue DataType="string"/> "commit" </AttributeValue></Apply>
    </Apply>
  </Condition>
  <Conclusions> <TrueConclusion PivotPolicyDone = "Permit"/> </Conclusions>
</Rule>}
```

`WS-Pivot.Policy`$_{aborted}$ states that a Web service transitions from *activated* state to *aborted* state if-and-only-if the transactional property is *pivot*, the current state is *activated*, the previous state is *not activated*, and the transition name that was successfully fired is *failure*.

Similar state chart diagrams and their related policies are defined for the retriable and compensatable cases. The retriable case will contain an additional *Suspended* state between *Activated* and *Aborted* states. The compensatable case will extend the retriable case with an additional *Compensated* state between *Done* and *Not-Activated* states.

In addition, the compensatable case will not contain the *Aborted* state; rather, it will contain a transition from *Suspended* to the *Compensated* state.

We discuss hereafter the dependencies among transactional Web services. In particular, we comply with the dependencies suggested by Bhiri et al. in [4], namely activation, abortion, and compensation. We recall that dependencies become effective at the Web service instance level. It is expected that the multiple policies that implement the dependencies will feed also the repository of policies of Fig. 3. Later we will show how these dependencies are used during adaptation (Section 4). In this paper, we expose the transactional properties of the peers of a Web service using two arguments available in $\mathcal{I}$-context: *TransactionalPropertyPerPreviousWebServiceInstance(s)* and *TransactionalPropertyPerNextWebServiceInstance(s)*. For illustration purposes, we present only the compensation dependency. There is a compensation dependency from $\text{WS}_x$ to $\text{WS}_y$ if the compensation of $\text{WS}_x$ fires the compensation of $\text{WS}_y$ (or abortion of $\text{WS}_y$, in case $\text{WS}_y$ is retriable or pivot). This dependency is reported using $\texttt{WS}_x\texttt{.Policy}_{Compensation(WS_y)}$ and is defined as follows:

```
If    WSx.I-context.CurrentState(Aborted⊕Compensated) &
      WSy.I-Context.CurrentState(Done⊕Suspended) &
      WSy.S-context.TransitionOut(CompensateAfterCommit⊕AbortAfterFailedRetries)
Then  WSy.S-context.NextEffectiveState=Aborted⊕Compensated &
      WSy.I-context.CurrentState=WSy.S-Context.NextEffectiveState &
      WSy.I-context.CurrentPolicyForNextState=WSx.Policy_Compensation(WSy)
```

Note: ⊕ stands for exclusive or.

## 4 Context-Driven Transactional Web Services Adaptation

In this section, we discuss the adaptation of context-driven transactional Web services during exception handling. Our strategy is to modify the composition specification with minimal disruption to the previously run or already running Web service instances. An exception occurs if a Web service instance execution fails due for example to lack of resources [12]. Exception handling for the Web service instance, called $\text{WS.I}_{failed}$, tightly depends on its transactional property. If it is pivot, then the entire composite Web service will fail, since the effects of the Web service instance cannot be undone. If it is retriable or compensatable, its failure needs to be propagated to the *affected Web service instances* because of the failure's side effects. These affected Web service instances are defined later, but are classified into two types according to their execution order to $\text{WS.I}_{failed}$:

1. Post-affected Web service instances are yet to be performed. This requires a *forward adaptation* strategy.
2. Pre-affected Web service instances are either concurrently executing (perhaps there exists an abortion or compensation dependency from $\text{WS.I}_{failed}$ to some of these Web service instance) or have already executed. This requires a *backward adaptation* strategy. This strategy is not discussed in this paper.

### 4.1 Some Definitions

From now on, we assume that the failed Web service instance $\text{WS.I}_{failed}$ is either retriable or compensatable. Before we describe our forward adaptation strategy, some basic definitions are needed. We first, assume that the composition specification like the one in Fig. 1 is mapped onto a graph. The graph's nodes and edges correspond to the Web service instances and the dependencies between these Web service instances, respectively. The dependencies, derived from workflow models [11], are modeled as follows:

- We define the graph of the composition specification as $G = (V,E)$, where $V$ is the set of nodes representing Web service instances, and $E$ is the set of edges depicting dependencies between the Web service instances. Each edge is a tuple of the form ¡$WS_i$,$WS_j$¿, where the edge is directed from $WS_i$ to $WS_j$. The graph also has two unique nodes: *START* (has no predecessors) and *END* (has no successors). The graph is supposed to meet two basic conditions: (a) every node in the graph is directly or indirectly reachable from *START* node, and (b) *END* node is reachable from every node in the graph.
- *Forward edge* ¡$\mathrm{WS}_i \rightarrow \mathrm{WS}_j$¿ depicts the activation dependency between a Web service instance and one of its direct, successor Web service instances in the graph.
- *Backward edge* ¡$\mathrm{WS}_i \leftarrow \mathrm{WS}_j$¿ depicts an edge from a Web service instance to one of its direct, predecessor Web service instances in the graph. This edge depicts repeated execution within a loop, and represents a backward activation dependency.
- Abortion/compensation dependency edges - as described earlier in Section 3.2.

## 4.2 Forward Adaptation Strategy

In any exception situation, forward adaptation is always preferable, due to its minimal impact on the already executed or currently running Web service instances. The forward adaptation strategy consists of two main steps: (1) determination of the set of the affected Web service instances, and (2) forward adaptation itself.

**Determination of the affected Web service instances.** Two types of Web service instances are affected by the failed Web service instance $\mathrm{WS.I}_{failed}$: currently executing Web service instances that have an abortion or compensation dependency starting from $\mathrm{WS.I}_{failed}$, and yet to start executing Web service instances that are connected to $\mathrm{WS.I}_{failed}$ with forward edges. We extend the former category to include those Web service instances that have abortion/compensation dependencies pointing to the currently executing Web service instances, and so on, in a recursive manner. In other words, the former category of Web service instances includes now all those Web service instances that are directly or indirectly dependent on $\mathrm{WS.I}_{failed}$. The determination algorithm of the set of affected Web service instances is given in Fig. 5 and can be described as follows:

1. Mark all Web service instances that are either currently executing or are yet to execute, in the graph of the composition specification, as "not-visited".
2. For each abortion/compensation dependency pointing from $\mathrm{WS.I}_{failed}$, perform a backward traversal marking all visited Web service instances as "visited", until a Web service instance is reached to which no abortion/compensation dependency edge points.

```
PROC FS(wsif,G)
Input: wsif WebServiceInstanceFailed, G graph
Output: fs SetOfWebServiceInstances
Auxiliary: S, T, K, L
Begin
  ▷ fs represent the Web service instances that are directly or indirectly dependent on wsif
  fs← ∅
  WSI← WEBSERVINST(G)
  ▷ WebServInst(G) is a function which returns the whole Web service instances of the graph G
  for wsi∈WSI do
    if CURRENTSTATE(wsi) ∈ {activated, not − activated} then
      wsi.tag← not_visited
    end if
  end for
  ▷ abortion(x) and completion(x) are two functions which return the set of Web service instances
  ▷ that have an abortion (respectively a completion) dependency with a Web service instance x.
  S← ABORTION(wsif) ⋃ COMPLETION(wsif)
  T← ∅
  while S≠ ∅ do
    for wsx∈S do
      wsx.tag← visited
      T←T⋃ ABORTION(wsx) ⋃ COMPLETION(wsx)
      S←S−{wsx}
      fs←fs⋃{wsx}
    end for
    S←T
    T← ∅
  end while
  K← FORWARD_ALL(wsif)
  ▷ forward_all is a function which returns the set of all Web service instances directly connected to
  ▷ wsif by a forward edge leading out of wsif and its successors.
  L← ∅
  while K≠ ∅ do
    for wsx∈K do
      wsx.tag← visited
      L←L⋃ ABORTION(wsx) ⋃ COMPLETION(wsx)
      K←K−{wsx}
      fs←fs⋃{wsx}
    end for
    K←L
    L← ∅
    M← IN-LOOP(wsif)
    ▷ if wsif belongs to a loop, this returns the set of instances in the loop. Otherwise, it returns an empty set.
    if M≠ ∅ then
      wsx←wsif
      while M≠ ∅andwsx≠M.first do
        ▷ M.first is the beginning instance in the loop
        wsx← PREDECESSOR(wsx)
        wsx.tag← visited
        fs←fs⋃{wsx}
        M←M−{wsx}
      end while
      wsx.tag← visited
    end if
  end while
  return fs
End
```

**Fig. 5.** Forward sphere calculation

3. Starting at $\text{WS.I}_{failed}$, move to each Web service instance along individual forward edges from $\text{WS.I}_{failed}$ by marking it as "visited". Continue doing this until the end of the composition specification graph is reached, i.e., *END* node. In case of multiple forward edges leading out of $\text{WS.I}_{failed}$, this step should be implemented in parallel for each forward edge.

4. If WS.I$_{failed}$ belongs to a loop, then traverse the composition specification graph backward from WS.I$_{failed}$, until the beginning of the loop is reached, marking all the visited Web service instances as "visited". The semantics of the loop dictates that such a case needs to be considered, since control flow in a loop could flow via backward edges also.
5. The collection of Web service instances labeled "visited" constitute now the *forward sphere* for WS.I$_{failed}$.

```
PROC FA(wsif,G)
Input: wsif WebServiceInstanceFailed, G graph
Output: -
Begin
  for wsx∈ fs(wsif,G) do
  | ▷ suspend all the Web service instances that are in the forward sphere
  | if TransactionalProperty(wsx) = retriable then
  | | executeWS − Retriable.Policy_suspended(wsx)
  | else executeWS − Compensatable.Policy_suspended(wsx)
  | end if
  end for
  while CurrentNumberOfRetries(wsif) < MaximumNumberOfRetries(wsif)orCurrentState(wsfi) ≠ done do
  | Retryexecutionofwsif
  | IncrementationofCurrentNumberOfRetriesofwsif
  | ▷ The CurrentState of wsif is automatically updated after each execution
  end while
  if CurrentState(wsif) = done then
  | resumetheexecutionofeachwsxfs(wsif,G)
  else if TransactionalProperty(wsif) = Retriable then
  | | executeWS − Retriable.Policy_aborted(wsif)
  | else if TransactionalProperty(wsif) = Compensatable then
  | | | executeWS − Compensatable.Policy_compensated(wsif)
  | | end if
  | | for wsx∈ fs(wsif,G) do
  | | | if TransactionalProperty(wsx) = retriable then
  | | | | executeWS − Retriable.Policy_aborted(wsx)
  | | | end if
  | | | if TransactionalProperty(wsx) = compensatable then
  | | | | executeWS − Compensated.Policy_compensated(wsx)
  | | | end if
  | | end for
  | end if
  end if
End
```

**Fig. 6.** Forward strategy algorithm

**Forward adaptation.** Once the forward sphere for WS.I$_{failed}$ is calculated, the forward adaptation algorithm given in Fig. 6 is executed. It consists of the following:

1. While WS.I$_{failed}$ is being retried, all currently, running Web service instances in the forward sphere are to be suspended using either WS-Retriable.Policy$_{suspended}$ or WS-Compensatable.Policy$_{suspended}$, as the case maybe.
2. Retry executing WS.I$_{failed}$ until one of the following happens: (i) maximum number of retries is reached without success, or (ii) one of the retries succeeds.
3. If one of the retries of WS.I$_{failed}$ succeeds, then WS.I$_{failed}$ execution will be resumed as well as the execution of the currently running Web service instances in the forward sphere.
4. In case the retry of WS.I$_{failed}$ fails, or the maximum number of retries without success is reached, WS.I$_{failed}$ will be either aborted via WS-Retriable.Policy$_{aborted}$, or compensated via WS-Compensatable.Policy$_{compensated}$ followed by WS-Compensatable.Policy$_{not-activated}$, as per its transactional property.

4.(a) If WS.$I_{failed}$ is either aborted or compensated, the Web service instances in the forward sphere are then aborted (respectively, compensated) if they are retriable (respectively, compensatable) in the reverse order in which they were executed. This is implemented via the pairwise abortion (respectively, compensation) dependency between consecutively aborted (respectively, compensated) Web service instances, as listed in Section 3.2.

4.(b) Execution control now returns to the state before the occurrence of the exception. The composite Web service designer can redesign the rest of the specification composition by taking into account the changed situation after all the aborts and compensations.

### 4.3 Illustration of the Forward Strategy Using Amin Scenario

Let us assume that *Location WS* has failed, so its forward sphere consists of {*Bus Schedule WS, Traffic WS, Taxi WS, Weather WS*}.

If *Location WS* can be retried successfully, the execution will then proceed normally. If not, *Location WS* needs to be aborted. While it is being retried, *Weather WS* is kept suspended, until *Location WS* either succeeds or fails. In case *Location WS* fails, *Weather WS* should also be aborted, as per the abortion dependency between *Location WS* and *Weather WS* (Section. 3.2). This will lead to a redesign of the composition specification starting from *Itinerary WS*. Perhaps, during redesign, *Location WS* is associated with another Web service to be offered from within the hotel itself. This extra Web service will be triggered as per an alternate dependency policy.

## 5 Implementation

Our prototype is developed with the use of *JDK1.4.2* as a high level language, *W3C DOM* for processing XML information, *XACML* for transactional policies, *SWT* for GUI, and *Eclipse 3.2* as a development environment. Fig. 7 shows the initial $\mathcal{I}$-context and $\mathcal{S}$-context values of *Weather-Instance*$_1$ and *Location-Instance*$_1$ when *Itinerary WS* gets requested. In Fig. 7 (a), it can be seen that *Weather-Instance*$_1$ has got *retriable* as transactional property with *2* as *MaximumNumberOfRetries*, and its successor Web services are *Taxi WS* or *Bus Schedule WS* whose transactional properties are *compensatable* and *retriable*, respectively. In Fig. 7-(c) details on the state chart diagram of *Weather-Instance*$_1$ are given. For instance, the current state is *activated* while the transition in is *start*. Finally, Fig. 7-(b,d) show the initial $\mathcal{I}$-context and $\mathcal{S}$-context values of *Location-Instance*$_1$. In Fig. 7-(c,d) it is shown the same $\mathcal{S}$-context for both *Weather-Instance*$_1$ and *Location-Instance*$_1$. This is due to the following reasons: both have got the same transactional property namely retriable; both are getting activated at the same time since they are to be executed in parallel.

For prototyping purposes, we assume that *Location-Instance*$_1$ gets failed in the middle of execution. As a result, $\mathcal{S}/\mathcal{I}$-context's arguments get updated as per WS-Retriable.Policy$_{suspended}$. i.e., $\mathcal{S}$-context.*NextEffectiveState* gets changed to *suspended* and $\mathcal{I}$-context.*CurrentState* gets changed to *suspended*, too. Following the failure of *Location-Instance*$_1$, *Weather-Instance*$_1$ gets also suspended as per the adaptation strategy of Section 4.2. Since *Location-Instance*$_1$ has got *retriable* as transactional

(a)

| I-Context Parameters | Values |
|---|---|
| WSIdentifier | WEather-Instance-1 |
| CurrentState | Activated |
| TransactionalProperty | Retriable |
| PreviousWebServiceInstance | Nil |
| TransactionalPropertyPerPreviousWebSe... | Nil |
| NextWebServiceInstance | Taxi-Instance-1/BusSchedule-Instan |
| TransactionalPropertyPerNextWebServic... | Compensatable/Retriable |
| PreviousPolicyForCurrentState | Nil |
| CurrentPolicyForNextState | WS-Retriable.Policy.done |
| MaximumNumberOfRetries | 2 |
| CurrentNumberOfRetries | 0 |

(b)

| I-Context Parameters | Values |
|---|---|
| WSIdentifier | Location-Instance-1 |
| CurrentState | Activated |
| TransactionalProperty | Retriable |
| PreviousWebServiceInstance | Nil |
| TransactionalPropertyPerPreviousWebSe... | Nil |
| NextWebServiceInstance | BusSchedule-Instance-1 |
| TransactionalPropertyPerNextWebServic... | Retriable |
| PreviousPolicyForCurrentState | Nil |
| CurrentPolicyForNextState | WS-Retriable.Policy.done |
| MaximumNumberOfRetries | 2 |
| CurrentNumberOfRetries | 0 |

(c)

| S-Context Parameters | Values |
|---|---|
| StateIdentifier | StateId-1 |
| StateLabel | Activated |
| PreviousState | Not-Activated |
| NextExpectedState | Done or Suspended or Aborted |
| NextEffectiveState | Nil |
| TransitionIn | Start |
| TransitionOut | Nil |

(d)

| S-Context Parameters | Values |
|---|---|
| StateIdentifier | StateId-1 |
| StateLabel | Activated |
| PreviousState | Not-Activated |
| NextExpectedState | Done or Suspended or Aborted |
| NextEffectiveState | Nil |
| TransitionIn | Start |
| TransitionOut | Nil |

**Fig. 7.** $\mathcal{I}/\mathcal{S}$-contexts of *Weather-Instance*$_1$ and *Location-Instance*$_1$

(a)

| S-Context Parameters | Values |
|---|---|
| StateIdentifier | StateId-4 |
| StateLabel | Suspended |
| PreviousState | Activated |
| NextExpectedState | Activated or Aborted |
| NextEffectiveState | Activated |
| TransitionIn | Failure |
| TransitionOut | Retry |

(b)

| I-Context Parameters | Values |
|---|---|
| WSIdentifier | Location-Instance-1 |
| CurrentState | Activated |
| TransactionalProperty | Retriable |
| PreviousWebServiceInstance | Nil |
| TransactionalPropertyPerPreviousWebSe... | Nil |
| NextWebServiceInstance | BusSchedule-Instance-1 |
| TransactionalPropertyPerNextWebServic... | Retriable |
| PreviousPolicyForCurrentState | WS-Retriable.Policy.suspended |
| CurrentPolicyForNextState | WS-Retriable.Policy.activated |
| MaximumNumberOfRetries | 2 |
| CurrentNumberOfRetries | 1 |

**Fig. 8.** Updated $\mathcal{I}/\mathcal{S}$-contexts of *Location-Instance*$_1$ after successful retry

property, it gets retried with *WS-Retriable.Policy*$_{activated}$ transactional policy. Luckily the first attempt of retry itself is successful. Fig. 8 shows *Location-Instance*$_1$'s $\mathcal{I}/\mathcal{S}$-contexts with focus on *NextEffectiveState*, *CurrentState*, and *CurrentNumberOfRetries* arguments.

Next, Fig. 9 shows the further updated $\mathcal{I}/\mathcal{S}$-contexts of *Location-Instance*$_1$. For $\mathcal{S}$-context (Fig. 9-(a)), the state identifier, state label, previous state, next expected state, and transition out values are modified with respect to the current state, i.e., Activated. Same comment is made for $\mathcal{I}$-context's arguments (Fig. 9-(b)).

Once *Location-Instance*$_1$ gets activated, *Weather-Instance*$_1$ is also retried and successfully activated. Its respective $\mathcal{I}/\mathcal{S}$-contexts are updated with respect to its *WS-Retriable.Policy*$_{activated}$ policy. Eventually, both *Weather-Instance*$_1$ and *Location-Instance*$_1$ successfully complete execution. Finally, after the completion of *Weather-Instance*$_1$, *Bus-Schedule-Instance*$_1$ invoked, which in turn invokes *Traffic-Instance*$_1$ for obtaining traffic information. Eventually, both *Bus-Schedule-Instance*$_1$ and *Traffic-Instance*$_1$ successfully complete execution.

(a)

| S-Context Parameters | Values |
|---|---|
| StateIdentifier | StateId-1 |
| StateLabel | Activated |
| PreviousState | Suspended |
| NextExpectedState | Done or Suspended or Aborted |
| NextEffectiveState | Nil |
| TransitionIn | Retry |
| TransitionOut | Commit |

(b)

| I-Context Parameters | Values |
|---|---|
| WSIdentifier | Location-Instance-1 |
| CurrentState | Activated |
| TransactionalProperty | Retriable |
| PreviousWebServiceInstance | Nil |
| TransactionalPropertyPerPreviousWebSe... | Nil |
| NextWebServiceInstance | BusSchedule-Instance-1 |
| TransactionalPropertyPerNextWebServic... | Retriable |
| PreviousPolicyForCurrentState | WS-Retriable.Policy.activated |
| CurrentPolicyForNextState | WS-Retriable.Policy.done |
| MaximumNumberOfRetries | 2 |
| CurrentNumberOfRetries | 1 |

**Fig. 9.** Further updated (as per Activated state) $\mathcal{I}/\mathcal{S}$-contexts of *Location-Instance*$_1$

## 6 Conclusion

In this paper, we presented an approach to develop context-driven transactional Web services. We defined transactional properties on Web services that permit them to be composed together, and their joint execution managed, via policies. We also discussed how this approach helps handle exceptions, via the application of an adaptation strategy.

Our future work concerns a more thorough experimentation and evaluation activity to compare our approach against some other approaches presented in the literature [1,3,4,6,14,15]. In particular, we plan to demonstrate the feasibility of our approach on larger examples. Another interesting future work concerns the definition of a composite Web services framework that can manage transactional properties and ensure substitution mechanisms. This substitution can play an important role, mainly for pivot Web services. Some preliminary results are already reported in [13]. Finally, we plan to study how some fault tolerance concepts of distributed systems can be adapted to the requirements of transactional Web services.

## References


1. Alrifai, M., Dolog, P., Nejdl, W.: Transactions Concurrency Control in Web Service Environment. In: Proceedings of The 4th IEEE European Conference on Web Services (ECOWS'2006), Zurich, Switzerland (2006)
2. Anderson, A.H.: An Introduction to The Web Services Policy Language (WSPL). In: Proceedings of The 5th IEEE International Workshop on Policies for Distributed Systems and Networks (POLICY'2004), New-York, USA (2004)
3. Bassil, S., Rinderle, S., Keller, R., Kropf, P., Reichert, M.: Preserving the Context of Interrupted Business Process. In: Proceedings of The 7th International Conference on Enterprise Information Systems (ICEIS'2005), Miami, USA (2005)
4. Bhiri, S., Perrin, O., Godart, C.: Ensuring Required Failure Atomicity of Composite Web Services. In: Proceedings of The Fourteenth International World Wide Web Conference (WWW'2005), Chiba, Japan (2005)
5. Coutaz, J., Crowley, J.L., Dobson, S., Garlan, D.: Context is Key. Communications of the ACM, 48(3) (March 2005)
6. Fauvet, M.-C., Duarte, H., Dumas, M., Benatallah, B.: Handling Transactional Properties in Web Service Composition. In: Proceedings of The 6th International Conference on Web Information Systems Engineering, (WISE'2005), New-York, USA (2005)
7. Maamar, Z., Benslimane, D., Anderson, A.: Using Policies to Manage Composite Web Services. IEEE IT Professional, 8(5) (September/October 2006)

8. Maamar, Z., Benslimane, D., Narendra, N.C.: What Can Context do for Web Services? Communications of the ACM, 49(12) (December 2006)
9. Maamar, Z., Mostéfaoui, S.K., Yahyaoui, H.: Towards an Agent-based and Context-oriented Approach for Web Services Composition. IEEE Transactions on Knowledge and Data Engineering, 17(5) (May 2005)
10. Pires, P.F., Benevides, M.R.F., Mattoso, M.: Building Reliable Web Services Compositions. In: Proceedings of The International Workshop on Web Services Research, Standardization, and Deployment (WS-RSD'2002), Erfurt, Germany (2002)
11. Reichert, M., Dadam, P.: ADEPTflex - Supporting Dynamic Changes of Workflows without Losing Control. Journal of Intelligent Information Systems, 10(2) (1998)
12. Russell, N., van der Aalst, W.M.P., ter Hofstede, A.H.M.: Exception Handling Patterns. In: Process-Aware Information Systems. Technical report, BPM Center Report BPM-06-04, BPMcenter.org. (2006)
13. Taher, Y., Benslimane, D., Fauvet, M.-C., Maamar, Z.: Towards an Approach for Web Services Substitution. In: Proceedings of The 10th International Database Engineering & Applications Symposium (IDEAS'2006), Delhi, India (2006)
14. Verma, M., Deswal, P.: Approaching Web Services Transactions. Technical report, Second Foundation Inc., February 2003. Visited ( February 2005) http://www-128.ibm.com/developerworks/webservices/library/ws-tranart
15. Younas, M., Chao, K.M., Lo, C.C., Li, Y.: An Efficient Transaction Commit Protocol for Composite Web Services. In: Proceedings of The IEEE 20th International Conference on Advanced Information Networking and Applications (AINA'2006), Vienna, Austria (2006)
16. Younas, M., Eaglestone, B., Chao, K.M.: A Low Latency Resilient Protocol for E-Business Transactions. International Journal of Web Engineering and Technology, 1(3) (2004)

# On Automated Generation of Web Service Level Agreements

Cinzia Cappiello, Marco Comuzzi, and Pierluigi Plebani

Dipartimento di Elettronica e Informazione – Politecnico di Milano
Piazza Leonardo da Vinci 32, 20133 Milano (Italy)
{cappiello,comuzzi,plebani}@elet.polimi.it

**Abstract.** Before a service invocation takes place, an agreement between the service provider and the service user might be required. Such an agreement is the result of a negotiation process between the two parties and defines how the service invocation has to occur. Considering the Service Oriented Computing paradigm, the relationship among providers and users is extremely loose. Traditional agreements are likely to concern long term relationships and to be manually performed. In this paper, we propose a model to generate service level agreement on-the-fly. Just before the invocation commences, the quality of the service is negotiated in order to generate a service level agreement tied to that specific invocation. Such an approach relies on a quality model that supports both users requirements and providers capabilities definition.

## 1 Introduction

Organizations are increasingly exporting their services as Web services [1]. Such a proliferation increases the likelihood that users may find several services satisfying their functional requirements [2,3,4]. When users can choose among a set of functionally equivalent services, non-functional requirements become the driver for Web service selection. As a consequence, we need to define and manage Service Level Agreements (SLAs) between service providers and users [5].

In Service Oriented Computing paradigm, an SLA is defined as a binding contract which formally specifies user expectations about the solution and tolerances. SLA is a collection of service level requirements that have been negotiated and mutually agreed upon by the information providers and the information consumers. Usually, providers define some service levels as a fixed combination of their specific capabilities on a set of quality dimensions, and users must choose one these levels. Reasonable service levels that meet user requirements can be achieved by increasing the flexibility of the SLA definition. We argue that this could be obtained by allowing parties, i.e., users and providers, to re-examine and to negotiate defined levels. It is worth noting that identifying attainable service levels is a time consuming activity for the providers. Adding negotiation features creates further overhead during SLA definition activity. For these reasons, our approach does not identify service levels in advance. Providers only clarify their capabilities

J. Krogstie, A.L. Opdahl, and G. Sindre (Eds.): CAiSE 2007, LNCS 4495, pp. 264–278, 2007.

and service levels will be identified on-the-fly considering the users expectations. Service levels negotiation is also performed on-the-fly to reduce its overhead.

The discussion of mechanisms for on-the-fly generation of the SLA will be tied to a running example. We focus on a *TrafficMonitoring* service example. The *TrafficMonitoring* Web service provides up-to-date information about local traffic to business and retail customers across the US. The quality of such a service is defined by two classes of quality dimensions: *technical* and *domain dependent*.

Technical quality dimensions refer to technical aspects of service provisioning. Quality dimensions belonging to this class can be associated with any Web service, and do not explicitly depend on a characterization of the domain in which a Web service operates. For the sake of simplicity, we consider three quality dimensions, that is, *availability*, *data encryption*, and *response time*. Readers may refer to [6,7] for an extensive review of Web service technical quality. Availability refers to the expected percentage of time the system is up and accessible. Data encryption refers to the algorithms adopted for protecting data from malicious accesses. Eventually, response time refers to the expected delay between the moment in which a request is sent and the moment in which results are received [6].

Domain dependent quality dimensions strongly rely on the type of Web service that is under consideration. For the *TrafficMonitoring* example, we consider the *covered area*, *routes set*, and *detail level* dimensions. The *covered area* dimension characterizes the extensiveness of the area over which the service is able to provide traffic information. A service, for instance, may provide information only on national highways, while other ones may also consider interstate or local routes and downtown traffic conditions. Similarly, the detail level of traffic information provided by a service may also vary. A service may provide information on accidents and traffic jams, while other ones may also provide information about closed routes, detours, and predictions about future conditions of local traffic.

The paper is organized as follows. Section 2 presents a model to describe Web service quality, provider capabilities, and user requirements. Section 3 describes the negotiation model by which SLAs can be obtained on-the-fly. Section 4 discusses related work, while conclusions are finally drawn in Section 5.

## 2 Quality Model

A negotiation process occurs whenever both a user and a provider are able to define the documents specifying the requirements and the capabilities, namely. In a Web service environment, where users and providers might not know each others in advance, these documents must rely on the same language. In [8], a model able to express the quality of a Web service is discussed. The same model, discusses in the following, will be adopted in this work as well.

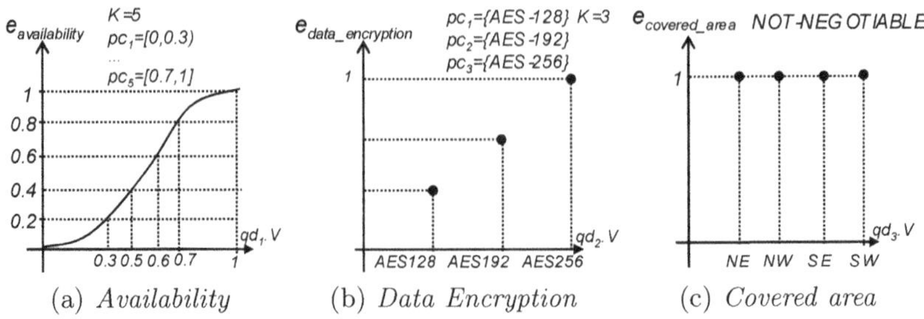


(a) *Availability* (b) *Data Encryption* (c) *Covered area*

**Fig. 1.** Evaluation functions and primitive service classes for *availability*, *data encryption*, and *covered area*

The quality of a Web service is defined by a set of quality dimensions[1] each of them associated to a given quality aspect. More formally, we define a quality dimension $qd_i$ as:

$$qd_i = \langle name, V, ef(V), PC \rangle \;\; i = 1, \ldots, I. \tag{1}$$

The *name* uniquely identifies the quality dimension. The element $V$ corresponds to either categorical or interval admissible values. In the former case, the admissible values will be included in a specific vector $V = \{v_h\}$ $(h = 1, \ldots, H)$, while, in the latter case $V$ will be defined by its extremes, i.e., $V = [v_{min}, v_{max}]$. The function $ef : V \rightarrow [0..1]$ represents the *quality evaluation function*, i.e., how the quality increases or decreases with respect to the admissible values: 0 means lowest quality, 1 highest quality. The trend of $ef$ is usually defined by an utility function, e.g., linear, logarithmic, exponential, sigmoidal. The admissible value set $V$ is organized in disjoint primitive service classes $PC = \{pc_k\}$ $(k = 1, \ldots, K)$ and are obtained as follows:

- In case of categorical values, the primitive service classes coincide with the values that the dimension may assume: i.e, $qd_i.PC \equiv qd_i.V \;\;, H = K$.
- In case of interval values, primitive service classes are obtained by splitting $V = [v_{min}, v_{max}]$ into $K$ intervals, so $PC = \{pc_k = [pc_{k_{min}}; pc_{k_{max}}]\}$ where $pc_{k_{max}} = pc_{(k+1)_{min}}$, $pc_{1_{min}} = v_{min}$, $pc_{K_{max}} = v_{max}$. $pc_k$ ranges are obtained as follows: let divide $qd_i.ef(V)$ in $K$ ranges $\{[e_{k_{min}}; e_{k_{max}}]\}$, then $p_{k_{min}} = qd_i.ef^{-1}(e_{k_{min}})$ and $pd_{k_{max}} = qd_i.ef^{-1}(e_{k_{max}})$.

Figure 1(a) and 1(b) show, respectively, this methodology applied to the *availability* and *data encryption* dimensions in the running example. The definition of primitive service classes is exploited by the negotiation algorithms described in Section 3. We assume that additional elements, such as measurement units or metrics, are also defined. We do not explicitly include them in $qd_i$ since they are not relevant for our approach.

[1] In the literature, quality dimensions are also named quality attributes or quality parameters.

Given a Web service, its quality is defined by the set $QD = \{qd_i\}$. As mentioned break above, negotiation takes place only if both requirements and capabilities are expressed on the same quality dimensions set. For this reason we assume that a third party, called *community*, is in charge of identifying the set of relevant quality dimensions. In this way, the quality dimensions included in $QD$ will be used (i) by the provider to express the offered quality, i.e., *capabilities* $C$ and (ii) by the user to define the required quality, i.e., *user requirements* $UR$.

As defined in [9], a community is a group of people which aims at proposing a specification for a group of objects with some relevant common characteristics. More generally, given an application domain, we suppose that a community exists and produces the set of relevant quality dimensions. Sometimes, the community can be easily identified since it is explicitly constituted (e.g., tourism community, financial community). Most of the times the community associated with an application domain does not explicitly exist. For example, if we want to buy a laptop then everyone can list the set of relevant quality dimensions which the evaluation of the laptop quality relies on, e.g., CPU, memory, HD capacity, screen resolution, and so on. Roughly speaking, the agreement on $QD$ between providers and users definitely exists but it is implicit. In some way, introducing the actor community means to make explicit this implicit common understatement.

Table 1 shows the quality dimensions included in $QD$ for the *TrafficMonitoring* example. Once the community decides to include a $qd_i$ in $QD$, the community also defines the range of admissible values, the related evaluation function $qd_i.ef$, and the primitive service classes $qd_i.PC$. In Table 1, all the $qd_i \in QD$ are described. In some case (e.g., covered area), the community cannot state which are the best and worst values, since they depend on the user preferences. So, the evaluation function always returns 1. This kind of dimensions, as explained in Section 3, are non-negotiable.

It is worth noting that the range of admissible values has been identified regardless of a specific Web service implementation. So, we assume that all the existing Web services, given a quality dimension, can only offer a subset of the

**Table 1.** Quality parameters for Traffic Monitoring example

| name | V | ef | P |
|---|---|---|---|
| *availability* | [0,1] | sigmoidal (see Figure 1(a)) | $\{[0, 0.3); [0.3, 0.5); \ldots; [0.7, 1]\}$ |
| *data encryption* | [AES-128;AES-192;...] | linear | [AES-128;AES-192;...] |
| *response time* | [0,10] | inverse linear | $\{[0.2, 1], \ldots, [9, 10]\}$ |
| *covered area* | [SouthEast;SouthWest; NorthEast;NorthWest] | $1\ \forall v_h \in V$ (see Figure 1(c)) | [SouthEast;...] |
| *routes set* | [Highways;interstate; local;...] | $1\ \forall v_h \in V$ | [Highways;interstate; local;...] |
| *detail level* | [jams; detours; toll;...] | $1\ \forall v_h \in V$ | [jams; detours; toll;...] |

admissible values defined by the community. In addition, users will customize the quality dimensions accordingly to their preferences.

Starting from the $QD$ defined by the community, Sections 2.1 and 2.2 describe, respectively, how the capabilities and the requirements can be defined.

## 2.1 Capabilities

Capabilities reflect the quality offered by a Web service provider. Focusing on the service description, the provider before publishing its Web service will define a document expressing the functional aspects. About this, WSDL represents the *de-facto* standard that identifies the set of available operations and exchanged messages. Along with the functional aspects, the service provider also needs to attach a document in which the offered quality is described. At this stage, the literature does not include a language for quality description with the same consensus as WSDL does for the functional aspects. Anyway, we think that the capabilities as introduced in the following can be simply expressed according to languages such as WSOL [10] or WS-Policy [11].

We define a capability $c(qd_i)$ as a restriction on the range of admissible values of the quality dimension $qd_i$. More precisely:

$$c(qd_i) = \langle qd_i.name, offering, qdprice(offering)\rangle, \tag{2}$$

where $offering \subseteq qd_i.V$ represents the restriction on the range of admissible values. In this way, the provider defines, given a quality dimension, which are the actual values the provider is able to support. In addition, the provider also defines *qdprice* function which maps the dependency between the offered values and the price per user associated with such a provisioning.

According to this model, the provider during the publication process of a Web service, will attach a document $C$ collecting all the supported capabilities. In particular:

$$C = \{c(qd_i)\} \quad \forall qd_i \in QD. \tag{3}$$

In other words, a capability document must include all the quality dimensions previously identified by the community. Table 2 lists the capabilities of a hypothetical *TrafficMonitoring* service provider. For instance, the offered *availability* is included in the range $[0.5, 1.0]$ and the price for such a provisioning is given by a fixed amount (e.g., 30\$) and a variable one that varies according to the actual value of the availability (e.g., availability*5\$). Similarly, different prices will be associated to different *covered area.* Since US NorthEast is more populated than US NorthWest then the price varies accordingly (e.g., 5\$ rather than 3\$).

## 2.2 Requirement Model

Similarly to the capabilities, the user requirements are expressed on the basis of the quality dimensions identified by the community. In particular, for each $qd_i \in QD$ users operate a restriction on the admissible range of values. With this

**Table 2.** Capabilities for TrafficMonitoring service

| qd | offering | qdprice |
|---|---|---|
| availability | [0.5,1.0] | 30$+(availability*5$) |
| data_encryption | [AES-128] | 500$ |
| response time | [1,2] | 3$*(5$/timeliness) |
| covered_area | [NorthEast;NorthWest] | 5$-NE;3$-NW |
| route_set | [interstate;local] | 5$-interstate;10$-local |
| detail_level | [detours] | 10$ |

operation, the users state which is the required quality. Hence, a user requirement $R$ that will be compared to the capabilities $C$ during to the negotiation process is defined as:

$$UR = \langle \{ur(qd_i)\}, budget \rangle, \tag{4}$$

where the $\{ur(qd_i)\}$ represents the user requirements of a specific $qd_i$ and $budget$ is the amount of money that the user is willing to pay for the service. In detail:

$$ur(qd_i) = \langle qd_i.name, request, w \rangle. \tag{5}$$

Here $request \in qd_i.V$ represents the restriction on the range of admissible values. This restriction corresponds to the values required by the user for the given quality dimension.

The element $w$ in $ur$ represents the weight that identifies how much the related quality dimension $qd_i$ influences the overall quality of the service. It is worth noting that the weight assignment activity is a crucial point of the method. It can be performed in different ways. The simplest way could be to let users associate with each quality dimension a weight to express the importance that the dimension has for the specific user class. In this case the only constraint is that the sum of the weights associated with all the dimensions has to be equal to 1. This method is difficult to apply, since the absolute relevance of a dimension on the total quality is hardly identifiable. For this reason, in this model we assume that the weight assignment is driven by the AHP (Analytic Hierarchy Process) approach, a decision making technique developed by T.L. Saaty [12]. This is a qualitative approach in which the user only states if a sub-dimension is more influent than another one on the overall quality. We assume that all the quality dimensions are independent. AHP is a decision-making technique that assigns to each sub-dimension a score that represents the overall performance with respect to the different parameters. AHP is suitable for hierarchical structures as the quality model described previously and proposes to user pairwise comparisons between sub-dimensions.

Considering the difficulty that some users have in the requirements specification, we assume that the community supports them by preliminarily identifying their profile. We borrow the *profiling* concept from the Web Information Systems

(WIS) literature in which it is used for the personalization of content to user expectations. Profiling is the technique through which data are collected and manipulated with the goal of identifying and describing the profile of an entity, such as a user, an object, a product, or a process [13]. A *profile* is a source of user requirements, in fact it is a structured representation of the information that describes users and their preferences along the services that they require. This information can be obtained by suitable architectures and modules operating along with the Web service infrastructure.

In the requirement model proposed in this paper, users are characterized by a profile and assigned to users classes. Each class contains users with similar characteristics. Formally, our model considers a set $U = \{u_y\}$ of user and a set $UC = \{uc_z\}$ of users classes. In particular, we assume that:

$$\forall u_y \in U\ \exists! uc_z \in UC \mid u_y \in uc_z. \tag{6}$$

A class of user $uc_z$ corresponds to the requirements suitable for the users belonging to the class. According to our model:

$$uc_z = \{ur_z(qd_i)\}. \tag{7}$$

We assume that the community is in charge of defining such requirements, therefore, of identifying the users classes. In this way, users of a class $UC$ have a sort of template of requirements that can be customized with respect to specific requirements to produce a specific $UR$.

Given a class of users, the user class requirements $ur_z$ represents the quality of service usually required by the user belonging to that given class. Users take inspiration from these requirements defined by the community to express their specific user requirements. User requirements can be more or less selective than class requirements. Table 3 shows possible user requirements given by a user for the *TrafficMonitoring* service. For example, if class requirements for the availability dimensions are the values included in the range $[0.5, 1.0]$, the user can be more selective by specifying a quality limit greater than 0.7 or, alternatively, decrease the relevance of the data quality dimension by accepting a range such as $[0.7, 0.99]$.

**Table 3.** User requirements for TrafficMonitoring service

| qd | request | w |
|---|---|---|
| availability | [0.5,1.0] | 0.4 |
| data_encryption | [AES-128] | 0.025 |
| response time | [0.5,1] | 0.3 |
| covered_area | [SouthEast;NorthEast] | 0.1 |
| route_set | [highways;local] | 0.15 |
| detail_level | [jams;detours] | 0.025 |

## 3 Negotiation Model

Before negotiation taking place, we need to state if the offerings satisfy the user requirements. So, we have to verify the following statement:

$$\forall qd_i \in QD \;\; isec\,(c(qd_i), ur(qd_i)) = c(qd_i).offerings \cap ur(qd_i).request \neq \emptyset. \quad (8)$$

The service level negotiation occurs within the quality values identified by $isec\,(c(qd_i), ur(qd_i))$.

Automated negotiation is usually defined by three elements: the *negotiation protocol*, the participants *decision models* [14], and the *negotiation objects*. We adopt a very simple negotiation protocol where, the user for each $qd_i$ starts considering the primitive class in $qd_i.PC$ which also belongs to the calculated intersection and which corresponds to the lowest quality. Then, as long as the budget is not fully exploited, the user will consider the primitive class with higher quality. In this mechanism, the decision model controls the way in which the budget is split across the quality dimensions. Finally, negotiation objects refer to the elements over which negotiation is performed. We argue that only the QoS dimension associated to a non-constant evaluation function $qd_i.ef$ are negotiable. For dimensions characterized by a constant evaluation function, e.g., *covered area* in our running example, we hypothesize that the user's requirements are *non-negotiable*. If, for instance, the user identifies $NE$ and $NW$ as required values for the *covered area* dimensions, the user requests can be fulfilled only when the service provides traffic information on $NE$ and $NW$. We hypothesize that it is not possible to negotiate on this kind of dimensions, since the community is not able to define an evaluation function that orders their values. Therefore, the set $QD$ is split in two sets: the set $NQD$ of negotiable quality dimensions, and the set $NNQD$ of non-negotiable quality dimension. More formally:

$$\begin{aligned} QD = NQD \cup NNQD,\;\; NQD \cap NNQD = \emptyset \\ NQD = nqd_l \quad l = 1, \ldots, L \\ NNQD = nnqd_m \quad m = 1, \ldots, M \end{aligned}$$

For each negotiable quality dimension $nqd_l$, we formally define the negotiation objects as:

$$negobj_l\,(c(nqd_l), ur(nqd_l)) = \langle nqd_l.name, NPC, ur(nqd_l).w \rangle \quad l = 1, \ldots, L. \quad (9)$$

As mentioned above, for each quality dimensions we calculate the intersection of related capabilities and user requirements. Since $qd_i$ is divided by definition into $K$ primitive classes, then only a subset of them will be included in the intersection as well. Such a subset is named $NPC$ (negotiation primitive classes) and defines, for each quality dimension $nqd_l$, the set of negotiation service classes $nsc_j$, $j = 1, \ldots, J$ included in the intersection ($J \leq K$). The set $NPC$ includes also the price $price(npc_j)$ associated by the service provider to each negotiation service class:

$$NPC(nqd_l) = \{\langle npc_j, price(npc_j) \rangle\} \;\; j = 1, \ldots, J \;\; l = 1, \ldots, L. \quad (10)$$

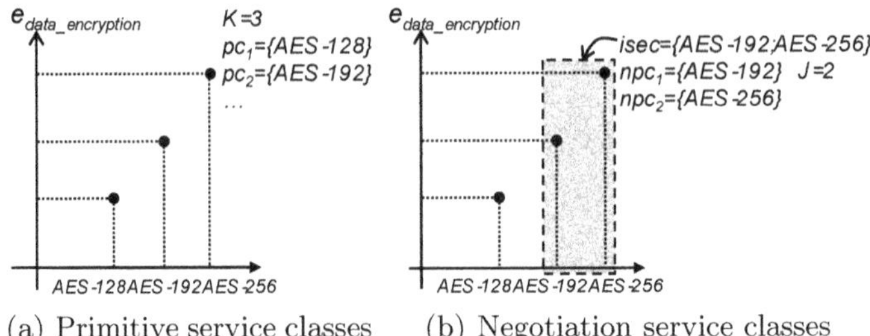


(a) Primitive service classes (b) Negotiation service classes

**Fig. 2.** Defining negotiation service classes for *data encryption*

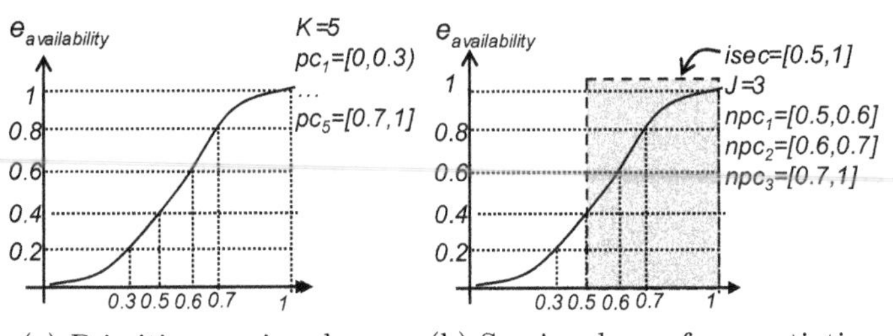


(a) Primitive service classes (b) Service classes for negotiation

**Fig. 3.** Defining negotiation service classes for *availability*

The methodology for defining service classes $npc_j$ and their price differs with respect to the nature of the negotiable quality dimension $nqd_l$.

As reported in Section 2, when considering a dimension $nqd_l$ that assumes categorical values, the primitive service classes $nqd_l.PC$ coincide with the values $nqd_l.V$ identified by the community. Figure 2 shows the methodology to obtain negotiation service classes for the *data encryption* dimension. In this case, the price associated with a service class $npc_j$ is directly obtained from the price information in the provider capabilities. A negotiation service class $npc_j$ includes, in fact, one single value $v_{\bar{h}} \in V$, hence:

$$price(npc_j) = c(nqd_l).qdprice(nqd_l.v_{\bar{h}}). \tag{11}$$

The definition of negotiation service classes $npc_j$ for continuous $nqd_l$ derives from the restriction operated on primitive service classes $nqd_l.PC$ over $isec\,(c(nqd_l), ur(nqd_l))$. How to obtain service classes for the *availability* dimension is graphically reported in Figure 3. Let us refer to $min(npc_j)$ and $max(npc_j)$ as, respectively, the left and right boundaries of the negotiation service class $npc_j$. The price $price(npc_j)$ associated with a service class $npc_j$ is the average between the price associated with its left and right boundaries, that is:

$$price(npc_j) = \frac{c(nqd_l).qdprice[min(npc_j)]) + c(nqd_l).qdprice[max(npc_j)])}{2}. \tag{12}$$

The algorithm adopted to assign a price to a service class can be more general and it is usually defined by the community.

Once having defined $negobj_l\,(c(nqd_l), ur(nqd_l))$, $\forall l,\ l = 1, \ldots, L$, we define the basic quality level $QL_{base}$ of the Web service, which is constituted, for each negotiable quality dimension, by the lowest quality negotiation service class $negobj_l.npc_1$. Then, it will be:

$$QL_{base} = \{negobj_1.npc_1, \ldots, negobj_L.npc_1\}\,. \tag{13}$$

The objective of the negotiation is to obtain a negotiated quality level $QL_{neg}$ which improves the quality of the basic level. The user exploits the declared budget $ur(nqd_l).budget$ to configure the basic quality level and increase the expected quality of the Web service. The price $P(QL_{base})$ associated to the basic quality level is:

$$P(QL_{base}) = \sum_{l=1}^{L} price(npc_1(nqd_l)). \tag{14}$$

Let us define $P_{nn}$ as the price associated with the quality values assumed by non-negotiable dimensions in $isec\,(c(nnqd_m), ur(nnqd_m))$. In the running example, the community may assume that *covered area*, *routes set*, and *detail level* are non-negotiable ($M = 3$). Let us consider a user requirement that specifies *highways* and *local* as required values for the *routes set* dimension. A SLA between a service provider and the user can be generated only if the provided Web service gives traffic information on highways and local routes. If we assume that the user has also required *jams* and *NE* for, respectively, *detail level* and *covered area*, it will be:

$$\begin{aligned} P_{nn} = {} & c(nnqd_1).qdprice_{covered\_area}(NE) + \\ & + c(nnqd_2).qdprice_{detail\_level}(jams) + \\ & + c(nnqd_3).qdprice_{routes\_set}(highways) + \\ & + c(nnqd_3).qdprice_{routes\_set}(local). \end{aligned} \tag{15}$$

We can now define the extra budget $EB$ of the user as:

$$EB = budget - [P(QL_{base}) + P_{nn}]\,. \tag{16}$$

If $EB < 0$, then the service is not going to be provisioned because the user is not able to cover with the budget the total price of the service, that is, the sum of the price associated with the basic quality level for negotiable dimensions and the price of non-negotiable dimensions. In case $EB = 0$, then the service will be provisioned with the basic quality level $QL_{base}$ for negotiable quality dimensions. The negotiation does not take place. The negotiation is executed only if $EB > 0$. Two strategies are available to the user to decide how to split $EB$ across the different negotiable quality dimensions, that we name the *vertical* and the *horizontal* strategies.

```
define ΔEB_l = 0, ∀l //Fraction of EB allocated to the
                          //improvement of nqd_l
define ΔEB = 0 //Exploited fraction of the extra budget
while(END==FALSE)
  select l:max(nqd_l.w) = w_l //select the current nqd with
                              //highest priority
  w_l = w_l − 0.01 //decrease the priority of the selected nqd
  ΔEB_l = price(npc_{j+1}) − price(npc_j) //update EB allocation
                                          //on nqd_l
  npc_j(nqd_l) = npc_{j+1}(nqd_l) //update the nqd_l level
  ΔEB = ΔEB + ΔEB_l //update the EB allocation
  if (ΔEB > EB) //Cannot price increase be covered by EB?
    npc_j(nqd_l) = npc_{j−1}(nqd_l) //restore old nqd_l value
    ΔEB = ΔEB − ΔEB_l //restore EB allocation
    END=TRUE //Exit condition, negotiation stops
  endif
  if (w_l == 0) //Exit condition, negotiation stops
    END=TRUE
endwhile
```

**Fig. 4.** Horizontal negotiation strategy

When adopting the vertical strategy, the user has the objective to maximize the quality associated to the highest priority dimension $nqd_{\bar{l}}$. When the quality of this dimension is maximized, that is, when the remaining extra budget exceeds the price of the negotiation service class $npc_J(nqd_{\bar{l}})$, then the algorithm switches to the maximization of the quality of the second highest priority dimension. The horizontal strategy is adopted when the user wants to split the extra budget on the negotiable quality dimensions proportionally to the priorities $ur(nqd_l).w$, $\forall l \in [1, \ldots, L]$. The horizontal and vertical strategies follow respectively, the algorithms reported in Figure 4 and 5.

Let us refer to $P$ as the total price of a service after quality negotiation:

$$P = P(QL_{neg}) + P_{nn}. \tag{17}$$

The result of the negotiation is a service level agreement $SLA$, generated on-the-fly for Web service, that has the following structure:

$$SLA = \langle QL_{neg}, P, isec\,(c(nnqd_m).ur(nnqd_m)) \rangle\,, \tag{18}$$

where $QL_{neg}$ reports the service class for negotiable quality dimensions obtained from the execution of negotiation, $P$ is the total price associated with the Web service with negotiated quality. Last term refers to the values of the non negotiable quality dimensions.

```
define ΔEB_l = 0, ∀l //Fraction of EB allocated to the
                     //improvement of qd_i
define ΔEB = 0 //Exploited fraction of the extra budget
while(END==FALSE)
  select l:max(nqd_l.w) = w_l //select the current nqd with highest
                              //priority
  w_l = w_l - 0.01 //decrease the priority of the selected nqd
  STOP=FALSE //starting configuration of nqd_l
  while (STOP==FALSE)
    ΔEB_l = price(npc_{j+1}) - price(npc_j) //update EB allocation on
                                            //nqd_l
    npc_j(nqd_l) = npc_{j+1}(nqd_l) //update the nqd_l value
    ΔEB = ΔEB + ΔEB_l //update the EB allocation
    if (ΔEB > EB) //Cannot price increase be covered by EB?
      npc_j(nqd_l) = npc_{j-1}(nqd_l) //restore old nqd_l value
      ΔEB = ΔEB - ΔEB_l //restore EB allocation
      STOP=TRUE //end nqd_l negotiation
      END=TRUE //exit condition, negotiation stops
    endif
    if((j = J)OR(w_l == 0))
      STOP=TRUE //end nqd_l configuration
  endwhile
endwhile
```

**Fig. 5.** Vertical negotiation strategy

# 4 Related Work

This paper presents a model to support the automatic generation of a service level agreement by considering user requirements and provider capabilities. To mediate between these two standpoints, we introduce the community as the actor able to provide a shared knowledge about the quality of a service in a specific application domain. The community defines which relevant aspects of a service can be used as search discriminants in service discovery. In the paper, the community organizes dimensions by using a tree-based structure. This approach for defining service quality has been inspired by [15] and [16], which recognize the correlation among several dimensions. In particular, [15] also refers dimensions to different layers (i.e. system level, resource level, and application level).

The set of dimensions identified by the community is also used as a guideline by the providers to describe the capabilities of the offered service. In fact, a complete service description is an important requirement for users who aim at searching the most suitable Web service. Besides the functional description, for which WSDL represents the most adopted specification, non functional specifications have to be modeled. In [17] a complete comparison of the current quality description languages is presented. Among all the identified contributions, for our work it is important to consider proposed languages for offers and contracts and languages for policies. As regards the former category, WSOL [10], WSLA [18], and WS-Agreement [19] provide some description models that our work can exploit to express quality dimensions. These contributions are particularly relevant, since they also address the definition and monitoring of quality levels. WSOL is suitable for the definition of quality dimensions, their metrics and quality constraints. The language does not formalize the contract terms between user and provider defining service levels but it contains constructs to define simple quality constraints on each quality dimension. A support for the definition and monitoring of Service Level Agreements is, instead, provided by the WSLA language. It allows providers to define quality dimensions and to describe evaluation functions. Furthermore, it provides monitoring of the parameters during operations and invocation of recovery actions when contract violations occur. Similarly, WS-Agreement provides constructs for advertising the capabilities of providers and for creating agreements based on creational offers, and for monitoring agreement compliance at runtime. The latter category includes WS-Policy [11] that can be adopted as a language for defining capabilities and requirements. WS-Policy definitions are independent of any specific quality descriptions. Using this language, users may describe services by using self-defined quality attributes.

Once that the service capabilities description is provided, the selection of the most suitable service is enabled by the definition of the user requirements. In this area, notations and languages to express users requirements have been defined in NoFun [20] and QML [21]. There are also contributions in which quality requirements are expressed by means of standard sentences or linguistic patterns in natural language [22].

In this paper, the automatic generation of a service level agreement is enabled by the use of negotiation mechanisms. In the literature, the only examples that propose policies for automated quality negotiation of Web services can be found in [23,24]. In general, research on SLA management has been carried out in the past couple of years and it has been mainly focused on the SLA specification and on the definition of languages for SLA creation, operation, monitoring, and termination. Examples of SLA management frameworks are WS-agreement [19], WS-negotiation [25], and the Service Negotiation and Acquisition Protocol (SNAP) [26]. However, while these standards are still evolving, they present some limitations. Generally, frameworks for SLA management only define the format and types of messages that can be used in the negotiation, but they do not provide the strategies through which negotiation is performed. In this paper, besides a characterization of negotiation messages built on the underlying Web service quality model, we also define the users' strategies to be adopted in the negotiation.

## 5 Conclusions and Future Work

This paper proposed a framework for the on-the-fly generation of Web service SLAs. The contribution of the paper is twofold. First, we introduced a quality model for Web services that is exploited by providers and users to define, respectively, their capabilities and requirements. Secondly, we provided users with a mechanism to negotiate among the set of service classes at the intersection between capabilities and requirements.

From the quality model definition perspective, future work should deal with an extended multi-level hierarchical model that considers composite dimensions, such as, for instance, security defined as a combination of data encryption, authentication, non-repudiation, and data integrity. Concerning negotiation, this paper focused on SLA generation involving only one provider and one user. Future work should also investigate how negotiation of quality aspects can be used to select a service among a set of functionally equivalent services. In this way, we will be able to add on-the-fly SLA generation capabilities to the common frameworks dealing with service discovery.

## Acknowledgment

The work has been partially supported by the Italian MIUR-FIRB TEKNE Project and by the European WS-DIAMOND Project.

## References

1. Papazoglou, M.P., Georgakopolous, G.: Service Oriented Computing: Introduction. Communications of the ACM 46(10), 1–5 (2003)
2. Bianchini, D., De Antonellis, V., Pernici, B., Plebani, P.: Ontology-based methodology for e-service discovery. Information Systems 31(4-5), 361–380 (2006)
3. Bernstein, A., Klein, M.: Towards High-Precision service retrieval. In: Proc. Int. Semantic Web Conference, ISWC'02 (2002)
4. Stroulia, E., Wang, Y.: Structural and semantic matching for assessing web-service similarity. Int. J. Cooperative Inf. Syst. 14(4), 407–438 (2005)
5. Keller, A., Ludwig, H.: The WSLA framework: Specifying and monitoring service level agreements for Web services. Journal of Network and Systems Management 11(1), 57–81 (2003)
6. Ran, S.: A model for Web services discovery with QoS. ACM SIGCOM Exchange 4(1), 1–10 (2003)
7. Mani, A., Nagarajan, A.: Understanding quality of service for Web services. Technical report, IBM, (2002) `http://www-128.ibm.com/developerworks/library/ws-quality.html`
8. Fugini, M., Plebani, P., Ramoni, F.: A user driven policy selection model. In: ICSOC '06: Proceedings of the 4th international conference on Service oriented computing. To appear (2006)
9. Marchetti, C., Pernici, B., Plebani, P.: A quality model for multichannel adaptive information. In: WWW Alt. '04: Proceedings of the 13th international World Wide Web conference on Alternate track papers & posters, ACM Press, New York (2004)
10. Tosic, V., Ma, W., Pagurek, B., Esfandiari, B.: Web Service Offerings Infrastructure (WSOI) - a management infrastructure for XML Web services. In: Network Operations and Management Symposium, 2004. NOMS 2004. IEEE/IFIP 1, 817–830 (2004)

11. Vedamuthu, A., Orchard, D., Hondo, M., Boubez, T., Yendluri, P.: Web Services Policy 1.5 - Primer. (2006) http://www.w3.org/TR/2006/WD-ws-policy-primer-20061018
12. Saaty, T.L.: The Analytic Hierarchy Process. Mc Graw Hill, New York (1980)
13. Olson, J.: Data Quality: The Accuracy Dimension. Morgan Kaufmann, San Francisco (2002)
14. Jennings, N., Faratin, P., Lomuscio, A., Parsons, S., Wooldridge, M., Sierra, C.: Automated negotiation: Prospects, methods and challenges. Group Decision and Negotiation 10(2), 199–215 (2001)
15. Sabata, B., Chatterjee, S., Davis, M., Sydir, J., Lawrence, T.: Taxonomy for QoS Specifications. In: Object-Oriented Real-Time Dependable Systems, 1997. Proceedings. Third International Workshop on. pp.100–107 (1997)
16. Chung, L., Nixon, B., Yu, E., Mylopoulos, J.: Non-Functional Requirements in Software Engineering. Kluwer Academic Publishers, Boston (2000)
17. Ruckert, J., Paech, B.: Web Service Quality Descriptions for Web Service consumers. In: CONQUEST2006. Proceedings (2006)
18. Keller, A., Ludwig, H.: The WSLA Framework: Specifying and Monitoring Service Level Agreements for Web Services. Technical Report RC22456(W0205-171), IBM Research Division, T.J. Watson Research Center (2002)
19. GRAAP Working Group: WS-Agreement Framework. (2003) https://forge.gridforum.org/projects/graap-wg
20. Franch, X.: Systematic formulation of non-functional characteristics of software. In: 3rd International Conference on Requirements Engineering (ICRE '98). pp.174–181 (1998)
21. Frølund, S., Koistinen, J.: Quality-of-service specification in distributed object systems. Distributed Systems Engineering Journal 5(4) (1998)
22. Duran, A., Bernardez, B., Toro, M., Corchuelo, E., Ruiz, A., Perez, J.: Expressing customer requirements using natural language requirements templates and patterns. In: Proceedings of the third Conference on Circuits, Systems, Communications and Computers (CSCC '99) (1999)
23. Lamparter, S., Agarwal, S.: Specification of policies for Web service negotiations. In: Proc. Semantic Web and Policy Workshop (2005)
24. Gimpel, H., Ludwig, H., Dan, A., Kearney, R.: PANDA: Specifying policies for automated negotiations of service contracts. In: Proc. 1st Int. Conf. Service Oriented Computing, ICSOC'03, pp. 287–302 ( 2003)
25. Rahwan, I., Kowalczyk, R., Pham, H.H.: Intelligent agents for automated one-to-many e-commerce negotiation. In: Computer Science 2002, Twenty-Fifth Australasian Computer Science Conference (ACSC2002). pp.197–203 (2002)
26. Czajkowski, K., Foster, I.T., Kesselman, C., Sander, V., Tuecke, S.: Snap: A protocol for negotiating service level agreements and coordinating resource management in distributed systems. In: Job Scheduling Strategies for Parallel Processing, 8th International Workshop, JSSPP 2002, pp. 153–183 ( 2002)

# RED-PL, a Method for Deriving Product Requirements from a Product Line Requirements Model

Olfa Djebbi[1,2] and Camille Salinesi[1]

[1] CRI, Université Paris 1 – Sorbonne, 90, rue de Tolbiac, 75013 Paris, France
[2] Stago Instruments, 136 avenue Louis Roche, 92341 Gennevilliers, France
olfa.djebbi@malix.univ-paris1.fr,
Camille.Salinesi@univ-paris1.fr, odjebbi@stago.fr

**Abstract.** Software product lines (SPL) modeling has proven to be an effective approach to reuse in software development. Several variability approaches were developed to plan requirements reuse, but only little of them actually address the issue of deriving product requirements. Indeed, while the modeling approaches sell on requirements reuse, the associated derivation techniques actually focus on deriving and reusing technical product data.

This paper presents a method that intends to support requirements derivation. Its underlying principle is to take advantage of approaches made for reuse PL requirements and to complete them by a requirements development process by reuse for single products. The proposed approach matches users' product requirements with PL requirements models and derives a collection of requirements that is (i) consistent, and (ii) optimal with respect to users' priorities and company's constraints. The proposed methodological process was validated in an industrial setting by considering the requirement engineering phase of a product line of blood analyzers.

**Keywords:** Requirements, Derivation, Product Line.

## 1 Introduction

As defined by the Software Engineering Institute (SEI), "*a software product line (SPL) is a set of software-intensive systems that share a common, managed set of features satisfying the specific needs of a particular market segment or mission and that are developed from a common set of core assets in a prescribed way*".

Software Product Line Engineering is rapidly emerging as a viable and important software development paradigm allowing companies to realize order-of-magnitude improvements in time to market, cost, productivity, quality and flexibility.

These new outcomes can be attributed to *strategic software reuse*. Software product line techniques explicitly capitalize on commonality and formally manage the variations among products in the product line. As a result, the main effort to design a product from the product line is due to the variations and the impact of the choices made for the required product.

Compared with conventional techniques, companies that manage a software product line report success stories in which they decreased their time-to-market for new

J. Krogstie, A.L. Opdahl, and G. Sindre (Eds.): CAiSE 2007, LNCS 4495, pp. 279–293, 2007.

products by factors of 2 to 50, reduced defect rates as high as 96% and multiplied productivity by a factor of 2 to 3 [1].

As Fig. 1 shows it, software products are developed, in the context of product line engineering, according to a two-stage process: the domain engineering stage and the application engineering stage [2]. Domain engineering involves implementing commonalities between product family members through a set of shared software artifacts, while preserving at the same time the ability to vary the products. During application engineering, individual products are derived from the product family, i.e. constructed using a subset of the shared software artifacts.

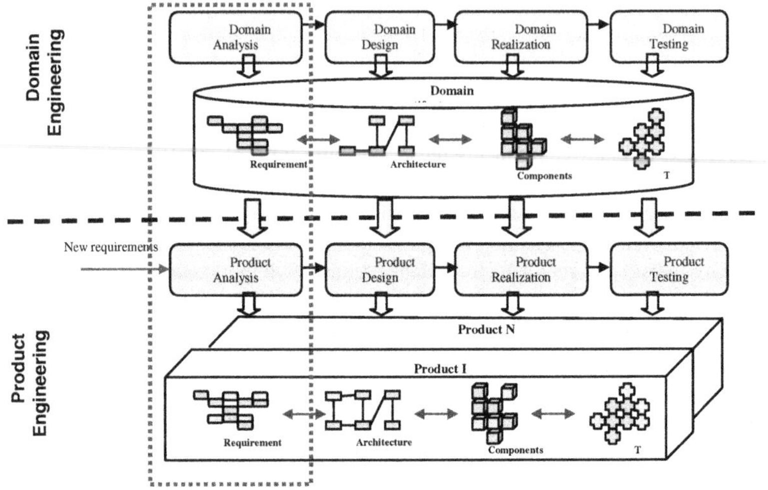


**Fig. 1.** Requirements Engineering challenges in a Software Product Line context (SEI)

In this particular context, Requirements Engineering (RE) processes have two goals: to define and manage requirements within the product line and to coordinate requirements for the single products. To achieve the latter goals, product requirements must be elicited by matching the product line requirements with customers' initial requirements (fig.1).

Some recommendations can be found to manage requirements in the context of SPL, but they always need to be customized [3] [4] [5] [6] [7]. Existing approaches rely on a requirements variability modeling process followed by a requirements selection process to retrieve a requirements collection specifying the single product to build.

Our experience showed us that, as stated by [8] [9], this way of working has several limits:

- *Requirements are solution-driven*: the selection among pre-defined product line requirements models that most often correspond to features already implemented in existing products, can influence stakeholders and skew their choices. They will naturally establish links between their problem and the existing solutions, adopt features with marginal value, and naturally forget about important requirements that are not present in the PL requirements model. As a result, the focus is on model elements that implement the solution rather than on the expression of actual needs.

- *Customer dissatisfaction*: the customer requirements can be different from the ones identified in the PL requirements models. Selecting among existing requirements can lead to miss out important requirements.
- *Innovation damping*: the RE process is inherently characterized by insight-driven evolution episodes. It fosters opportunistic exploration of the conceptual space and promotes creative thinking within the system requirements. On the opposite, selecting among predefined requirements restricts considerably creativity and search for innovative ways to deal with problems, hence reducing the added value of the new products to be developed.
- *Lack of guidance*: customers and marketing people are most often on their own to elicit the requirements for new products. Existing approaches provide little guidance (notation, process, rules …) to assist them in eliciting consistent product requirements, neither are developers guided in adding new requirements to the PL requirements model.
- *Customer training*: interactions between customers and variable requirements models imply that users should make an additional effort to understand the PL models and to seek their requirements in these models.
- *Customer overwhelm*: customers should not have to consider the complete collection of PL requirements as they are only interested in the requirements for a current product. Overwhelmed by a huge amount of data, customers lose track of the initial mission and are naturally lead to inquire about, comment, and even ponder over "requirements" that do not correspond to real needs.

These limits engendered by the requirements selection processing have many impacts on the project processes and artifacts, namely:

- *Quality of the requirements documents*: when stakeholders select requirements from the PL models, the resulting documents consist in a copy of a PL requirements model extract. When, on the contrary, stakeholders come up with new requirements, specifying these independently from the PL requirements model is inefficient. We believe, there is a need for guiding the merge between variability requirements specifications with requirements documentation for single products. Furthermore, product requirements specifications can be inconsistent since PL RE methods do not propose processes to verify the consistency and the compatibility of the new requirements with the older ones.
- *Quality of the resulting product*: it is quite well documented that the outcomes of projects with poor requirements management drive to poor product quality. This applies to products developed in the context of PL as for any other kinds of products even though reuse is facilitated.
- *Project management*: training customers to understand the PL requirements models and to discuss about them is a waste of time and creates an ambiguity between the roles of analysts and customers that inevitably leads to conflicts. This, associated with poor requirements definition in early project phase, generates rework in later phases of the project, extra costs, deadlines overrun, and difficult project management.

- *Strategic objectives of the company*: stopping innovation and market anticipation with new products may harm the company strategic objectives. Besides, applied methods leading to customer dissatisfaction may even threaten the survival of the company.

To overcome these shortcomings of existing methods, we believe there is a need for a product requirements derivation approach that satisfies the following characteristics:

- *Requirements oriented*: customers should be able to express their real needs, and the built product should answer to these needs.
- *Product line based*: the developed product should take advantage of the PL platform and reuse elaborated requirements that are already linked, traced and validated.
- *Unified into the whole product line development cycle*: it should provide means to ensure traceability with the remainder development phases for both the product line and the single product being developed.
- *Easy to apply*
- *Supported by a CASE tool that is integrated into existing toolkits*: appropriate tool support is mandatory to facilitate automate handling of the method processes and artifacts, and hence his large adoption by developers' community.
- *Scalable*: the method should allow modeling large-scale systems.

This paper presents a method that intends to support the requirements listed above. The study was undertaken with the collaboration of the AFIS[1] association and the method was developed by application to a product line of a French company named Stago -a medical company that produces blood analyzers [10]. The experience consisted in gradually introducing basic PL management principles while meeting practical issues in the RE phases of a new product creation project. The selection of these basic principles resulted from extensive bibliography research. Based on this experience, we developed a method, named RED-PL (Requirements Elicitation & Derivation for Product Lines) that guides the elicitation of product requirements by derivation from the PL requirements specification. The approach takes into account both the company's environmental and technical constraints and the specific product requirements as expressed by customers.

RED-PL is based on already existing PL requirements notations. The originality of the RED-PL method is that (i) it is user-oriented and (ii) it guides product requirements elicitation as a decision making activity. Indeed, RED-PL makes it possible to users to express their needs using classic requirements engineering techniques. Then, mechanisms are used to convert these needs and match them with the PL requirements specification. Negotiation and arbitration are finally supported in RED-PL to elicit optimal product requirements while maximizing reuse.

The paper is organized as follows. Section 2 outlines the challenges faced by Stago and the problems encountered while performing RE activities within its SPL management context. Section 3 presents the RED-PL method which was developed to

[1] French Association on Systems Engineering, affiliated to INCOSE (International Council on Systems Engineering) http://www.afis.fr/

meet these challenges. The methodological process is illustrated using the Stago data that were initially used to develop it. Section 4 provides an overview of existing methods and discusses how they deal with these challenges. Finally, conclusions are given in section 5.

## 2 Problem Statement in Stago's Context

Stago Instruments [10] is a company that produces analytical instruments for the haemostasis diagnosis. These instruments are embedded and real-time systems. They are used in hospitals and laboratories in the context of routine analysis or biologic researches.

The automatons produced by the company fit into a product line: all of them share the same core part with the main blood analysis functionalities. Each automaton has also its own characteristics and differs from the others. These variable parts can be as simple as color, weight or user interface of the machine; or more advanced such as biological processes, capacity in term of number of tubes handled, or mechanical and electronic technologies.

In general, instruments make tests on patients' products (total blood, plasma) and return results that are then interpreted by doctors.

In order to make tests, biologists load tubes of patients' products as well as reagent tubes in the instrument. While loading, tubes have to be identified. The biologist must then choose an analysis methodology and launch the tests. A methodology is a series of steps that simulate corpus reactions. Methodologies differ following test types (TP, TCA, etc.), but comprise necessarily a mix step and an incubation step. They may also use mixing and heating steps. Researchers can compose their own methodologies.

The instruments treat tubes, accomplish analyses according to specified methodologies, make measurements, and return the results to the biologist.

Products are loaded by batch. Nevertheless, the instrument is able to interrupt current tests in order to load and treat urgent tubes. Before launching tests, tubes must be treated to separate their constituents. Two processes of separation exist: centrifugation and micro-filtration. All instruments are able to implement theses processes however only one of them is implemented at a time in a given instrument.

There are three kinds of measurements: chronometric, colorimetric and immunologic. Instruments can implement several measure techniques, but an instrument that implements micro-filtration should not implement the photometric measure.

Test results are provided to the biologists in gross unit (Sec, D.O/min, Δ D.O), as well as in calculated unit (INR, µg/ml, UI/ml). To establish correspondences between units, the instrument must support calibrations. Besides, the instrument can view results on the screen, print them, and/or transfer them to the hospital or laboratory' host and put them into the patient case historic.

During projects, Stago teams manage in parallel the requirements documentation for the product line (common requirements) and for the single products (variable requirements).

Fig. 2 presents a model that was developed to document the most important requirements of the Stago instruments product line. The PL requirements are modeled using a Feature-oriented notation.

The figure shows a tree in which nodes are the features that correspond to PL requirements and links describe feature decomposition. There are three types of requirements: mandatory (e.g 'Load products'), optional (e.g. 'Separate constituents') and alternative (e.g. 'Centrifuge' and 'Micro-filter'). A mandatory requirement is common to the PL and must be included in every product of the PL. An optional requirement may, or not, be chosen for the considered product. Alternative requirements are collections of requirements from which some can be selected and others not. A UML-cardinality is associated to the collection to indicate the minimum and maximum number of requirements to be chosen. Additional dependency links between requirements, namely the 'requires' and the 'mutex' relationships, can be defined to specify additional constraints in requirements selection.

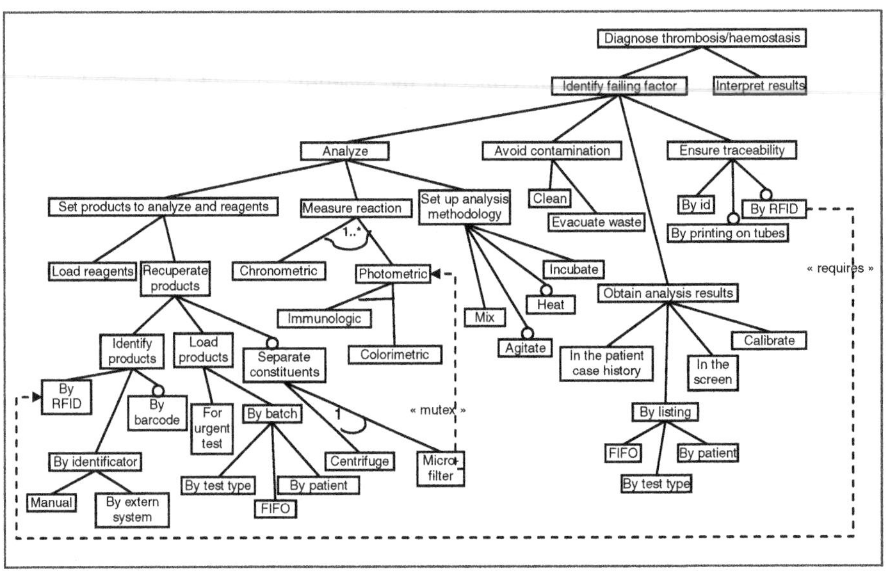


**Fig. 2.** Requirements model of Stago's product line

Since users are free in their way to express requirements, it happens that some requirements already exist in the PL requirements documentation, but with a different form. Users also insist on some requirements and ignore their impacts on other ones, or on the project progress itself. Users also often forget about important requirements and ignore opportunities offered by the product line.

In this context, Stago raised priority questions namely: (i) how to ensure the satisfaction of the real user's needs? and (ii) how to derive an optimal and consistent collection of product requirements that meet users needs and that cost little to the company? The RED-PL approach was developed and tried out on a Stago project to answer these questions.

# 3 The RED-PL Approach

In contrast to the traditional 'Selection' approach, requirements derivation for PLs must take into account stakeholders' original needs. As depicted in Fig. 3, RED-PL consists of:

- eliciting user requirements,
- matching users' requirements with PL requirements. This activity leads to establish the set of requirements that the PL subsumes and that satisfy users' needs. They correspond to a set of possible products to build.
- deriving the optimal set of product requirements, taking into account users' and company's constraints.

These processes are respectively described in the three following sub-sections.

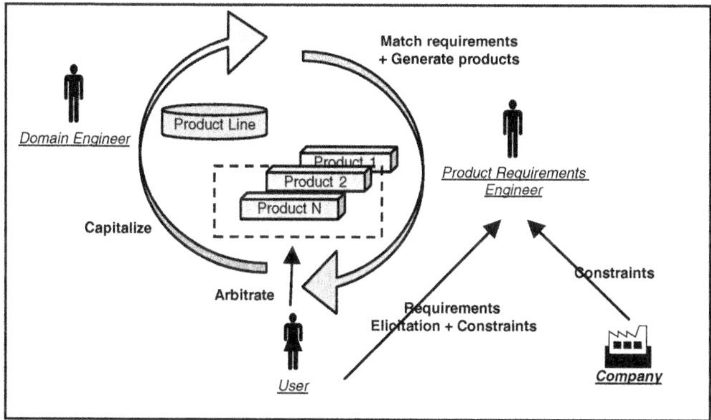


**Fig. 3.** Processes of the RED-PL approach

## 3.1 The Matching Process

The matching process is an iterative process that consists in interpreting users' requirements in terms of the PL requirements. It results in a collection of requirements that shall be implemented in the product (named 'product requirements'). The matching process aims at: (i) eliciting new users' requirements, (ii) avoid missing possible requirements, (iii) refining progressively the final product requirements, and (iv) updating the PL assets.

In the matching process, users' needs can be elicited using classical methods. Then, rules must be applied to construct a valid (i.e. unambiguous, consistent, traceable and verifiable) collection of product requirements. Once this is achieved, users' requirements can be fetched and marked in the PL model.

If users' requirements can not be found in the PL requirements model, then either (i) they are new requirements and they should be added to the PL model as well as links among them and in relation with old requirements, or (ii) they are the same requirements expressed differently, and then consensus should be made on the requirement formulation.

Requirements' matching is guided by using similarity analysis techniques. Two kinds of similarity analysis techniques can be used: surface level and deep level. First techniques are based on lexical similarity where two requirements are considered similar when they use the same term or the same linguistic structures. Conversely, deep level techniques use a structural and a semantic proximity. These techniques need more sophisticated tools such as dictionaries and linguistic parsers. Our similarity analysis approach also uses refinement, as suggested by goal modeling, to progressively improve the quality of the matching and to focus on requirements that are considered more important [11].

Our approach exploits the 30 generic similarity metrics adapted to Dice, Jaccard and Cosine's ratios. As shown below, similarity can be automatically computed by applying a weighted ratio between a number of similarities found between two requirements and the number of elements that define these requirements.

$$S_D^m(A,B) = \frac{\sum_A \underset{B}{MAX}[SIM(Termes_A, Termes_B)] + \sum_B \underset{A}{MAX}[SIM(Termes_A, Termes_B)]}{|\{Termes_A\}| + |\{Termes_B\}|}$$

**(Formula 1)** Adapted Dice ratio

After similarity study, marked requirements and all the associated requirements can then be retrieved from the PL model. This collection of requirements should correspond to a fragment of the PL requirements model, i.e. a sub-tree of requirements that satisfy users' requirements. However, the PL requirements model also contains requirements that are not yet marked. These requirements may be either (i) undesired, they must then be explicitly marked as such, (ii) mandatory then they must be considered in the collection of product requirements, or (iii) variable (optional/ alternative). As long as the tree contains unmarked optional and alternative requirements, a decision must be made on which additional PL requirements to select for the product. Arbitrations must therefore be investigated and discussed with users, as explained in the next sub-section.

### 3.2 The Arbitration Process

The output of the matching process consists in a PL requirements model composed of wanted/unwanted mandatory, optional and alternative requirements. The model fragment composed of desired requirements represents a set of possible releases as it can also contain optional and alternative requirements.

Only wanted optional and alternative requirements are considered in the following to express preferences since mandatory requirements must anyway be included in the collection of product requirements.

Preferences can be expressed by users under the form of weights associated to optional and alternative requirements. A 0 weight means that the requirements should not be selected, a 1 weight means that it should be included in the product requirements collection. The sum of weights of a bunch of alternative requirements must be equal to 1. Implicitly, each mandatory requirement has a 1 value weight.

Users can indicate their constraints on each requirement in terms of costs and benefits. Likewise, managers can state their development constraints on each requirement in terms of human resources, revenues, costs, and implementation/integration time. Although we knew they are important, other constraints such as skills of development teams, team transfers, deadline extension, external resources, were voluntarily ignored because they were too difficult to evaluate and we didn't know if they would really influence arbitration significantly.

Once requirements, priorities and constraints are completely defined, they are formalized using an Integer Linear Programming (ILP) notation. The Akkar approach [12] was selected and adapted to solve the problem at hand. The adapted version allows to define the subset of requirements that composes the optimal release while doing a what-if analysis on a dashboard. The ILP approach generates a collection of requirements that satisfies the constraints values, and is optimal with respect to the optimization criterion.

The following presents our proposal for modeling PL requirements dependencies using Akkar's approach. In Akkar's approach, a requirement $x_a \in \{0,1\}$ with $x_a=1$ if $x_a$ is selected, and $x_a=0$ otherwise. Five kinds of dependencies can be considered: composition, requires, optional composition, exclusion, and alternative. While the four former dependencies were already considered in Akkar's approach under the names 'combination', 'implication' and 'exclusion', the fifth kind had to be created to deal with the specific semantics of PL requirements modeling notations.

*Requires.* If requirement $x_b$ is selected, then requirement $x_a$ must be selected too. In the ILP model, it must be ensured that: $x_b=1 \Rightarrow x_a=1$

The ILP model is extended by the linear inequality $x_b \leq x_a$ ($x_a$ cannot be implemented without implementing $x_b$). In Akkar's initial approach, the corresponding dependency was 'implication'. In terms of PL requirements modeling, "requires" dependencies can be found from the alternative, the optional and the requires relationships.

$$x_b \leq x_a \tag{1}$$

*Composition.* If two requirements $x_a$ and $x_b$ cannot be implemented separately, then it must be ensured that $x_a = x_b$. Composition dependencies can be found in the PL requirements models from composition relationships. In Akkar's terms, it corresponds to the combination dependency.

$$x_a = x_b \tag{2}$$

*Exclusion.* If $R_a$ and $R_b$ cannot both be selected, in the ILP model then the inequality: $x_a + x_b \leq 1$ must be verified. In the PL modeling, exclusion dependencies can be found from "mutex" relationships.

$$x_a + x_b \leq 1 \tag{3}$$

*Alternative.* In PL engineering, a requirement can be realized by one or more requirements among a set. It is partly ensured by the implication relationship from a requirement $x_a$ to its sub-requirements $x_b$.. $x_k$, but needs to be more detailed to model the relationship between sub-requirements themselves. So, the *alternative* dependency (which does not exist in Akkar's model) is defined in ILP model by the following inequality:

$$x_a{*}Card_{min} \leq x_b+..+x_k \leq Card_{max} \tag{4}$$

The following table summarizes the mathematical formulae used to develop the ILP model.

**Table 1.** Recapitulation of requirements dependencies and their representation in the ILP

| Dependency relationship | Explication | Mathematical formula |
|---|---|---|
| 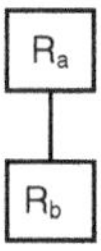 $R_a$ — $R_b$ (composition) | If a requirement is selected then all mandatory requirements composing it must be selected<br>$R_a = 1 \Rightarrow R_b = 1$ \| $R_a = 0 \Rightarrow R_b = 0$<br>$R_b = 1 \Rightarrow R_a = 1$ \| $R_b = 0 \Rightarrow R_a = 0$ | $R_a = R_b$<br>(Combination) |
| 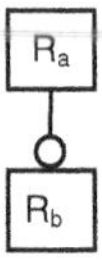 $R_a$ —o $R_b$ (option) | If a requirement is selected then its optional sub-requirements may be selected<br>$R_a = 1 \Rightarrow R_b \in \{0,1\}$ \| $R_b = 1 \Rightarrow R_a = 1$<br>$R_a = 0 \Rightarrow R_b = 0$ \| $R_b = 0 \Rightarrow R_a \in \{0,1\}$ | $R_b \leq R_a$<br>(implication) |
| 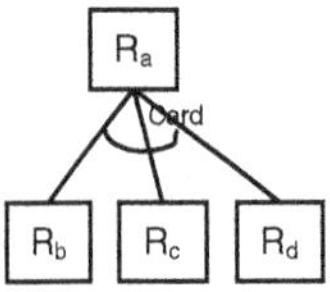 $R_a$ Card $R_b$ $R_c$ $R_d$ (alternative) | If a requirement is selected then alternative sub-requirements must be selected respecting the specified cardinality<br>$R_a = 1 \Rightarrow \begin{cases} R_{b..d} \in \{0,1\} \text{ and} \\ R_b + R_c + R_d \leq Card_{max} \text{ and} \\ R_b + R_c + R_d \geq Card_{min} \end{cases}$<br>$R_a = 0 \Rightarrow R_{b..d} = 0$<br>$R_{b..d} = 1 \Rightarrow R_a = 1$<br>$R_{b..d} = 0 \Rightarrow R_a \in \{0,1\}$ | $R_{b..d} \leq R_a$<br>(implication)<br>$R_a{*}Card_{min} \leq R_b{+}..{+}R_d \leq Card_{max}$<br>(alternative) |
| 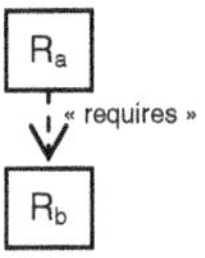 $R_a$ « requires » $R_b$ (requires) | If a requirement is selected then all required requirements must be selected<br>$R_a = 1 \Rightarrow R_b = 1$ \| $R_b = 1 \Rightarrow R_a \in \{0,1\}$<br>$R_a = 0 \Rightarrow R_b \in \{0,1\}$ \| $R_b = 0 \Rightarrow R_a = 0$ | $R_a \leq R_b$<br>(implication) |
| 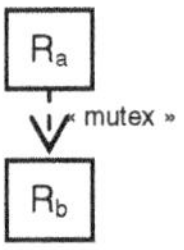 $R_a$ « mutex » $R_b$ (mutex) | If a requirement is selected then all requirements that are mutually exclusive with it must not be selected<br>$R_a = 1 \Rightarrow R_b = 0$ \| $R_b = 1 \Rightarrow R_a = 0$<br>$R_a = 0 \Rightarrow R_b \in \{0,1\}$ \| $R_b = 0 \Rightarrow R_a \in \{0,1\}$ | $R_a + R_b \leq 1$<br>(exclusion) |

The ILP modeling approach presented in the former section was tested in a Stago project with satisfying results. The experience is reported in the next section.

### 3.3 The Case Study

Once user requirements elicited, they were matched with PL requirements as recommended in the RED-PL matching process. The resulting requirements collection is a subset of the PL requirements model. The matching process revealed that users were decided neither on the measurement technique nor on whether the instrument to build should enable indoor constituents separation. Decisions had to be made to generate the optimal collection of requirements for a complete product. The arbitration process presented in section 3.2 was thus used to solve this problem.

First, the PL requirements model was analyzed and a ILP model was developed as defined in section 3.1. All the constraints were recorded in a Microsoft Excel spreadsheet, and analyzed with the Microsoft Excel solver (Fig. 4).

Two criteria were used to guide arbitration, namely cost and revenue. Revenue was evaluated by enquiring salespeople about the perceived value of the functionalities implementing the requirements. Cost evaluations were made by the engineering team who was asked to consider development and integration costs, need for resources (material and human), management costs, test costs, maintenance cost, and installation costs. These evaluations are an ordinary activity of salespeople and engineers, e.g in the context of risk analysis while elaborating the feasibility of the project. Several methods can be used to do this. Our approach does not focus on a particular one as it considers these evaluations as an input.

For confidentiality reasons, revenue and cost are defined in the next figure as relative values rather than under the form of the absolute values that were actually defined.

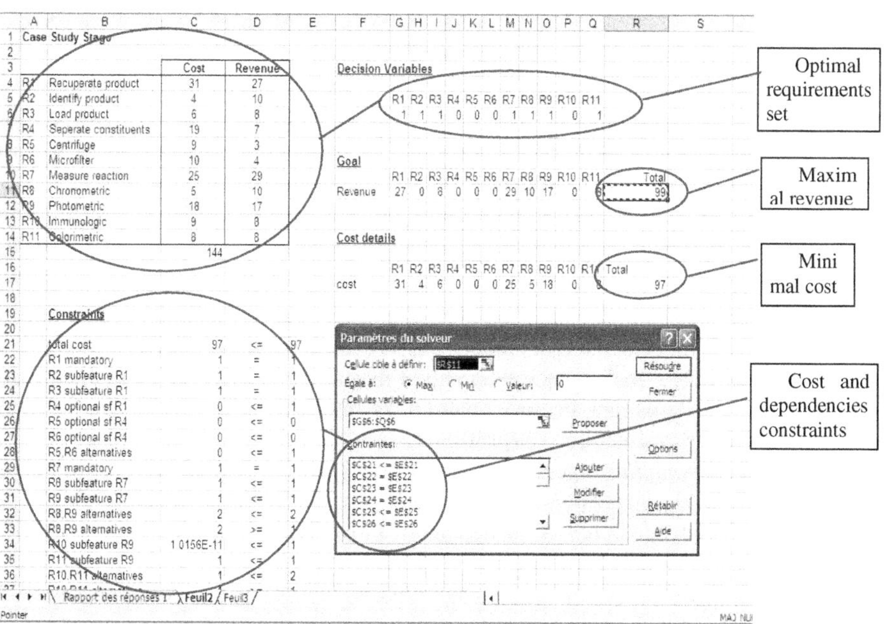


**Fig. 4.** Screenshot of Stago ILP problem after solving

Two goals were considered for optimization: either minimize cost while considering minimal revenue, or maximize revenue taking into account a global cost limitation. Sales and engineer teams agreed to focus on the second goal which is closer to their daily concerns. The collection of requirements generated by the solver using these parameters was found realistic in the sense that the resulting products did correspond to products already developed at Stago. Besides, the product did respond to the users' expressed at the beginning of the project and did correspond to products already identified as being of low cost. It was however difficult to assess if the generated product did really correspond to an optimal product or not.

Some difficulties were observed too while applying the method. First, the matching process was difficult to handle due to a lack of precision in the formulation of users' requirements. The difficulty was due not only to terminology, but also to a conceptual mismatch between users' requirements and the PL requirements (different levels of abstractions, different views). Besides, the ILP technique seemed to be not scalable to large systems, and is limited to optimization requests. We believe that this approach can be replaced by more adequate, flexible and scalable technique such as Constraint Programming. Further applications to other industrial projects are planed to enhance the method and favor its repeatability.

## 4 Related Works

Many different methods interested in constructing SPL assets are available in literature [13] [14] [15] [16] [17] [18]. Product derivation methodologies are on the contrary rather scarce [4] [19] [20]. Besides, while derivation affects the whole product line artifacts, from requirements to code, the derivation issues are mainly addressed in terms of design and implementation [4] [6].

At the requirement engineering level, how to create the right requirements assets of the PL and dependencies among them to develop the right products have been extensively studied [7] [21] [22] [23] [24] [25], but understanding the derivation process itself has received little attention.

In existing approaches, the derivation of the product architecture, code or test artifacts from the product and the PL specifications is performed using the following techniques:

- *Model transformation*: static and dynamic models are instantiated for products from the PL models, using a model transformation language [4] [26] [27] [28].
- *Design patterns:* for instance the method introduced by Jezequel which consists in using the 'Abstract Factory' pattern as interface to create objects of each product in the product line [29] [30].
- *Variability bounding*: generative approaches (e.g. Generative Programming approach [19]) suggest automatic derivation by code generation. Selecting desired product features is sufficient to allow assembling correspondent SPL elementary reusable components and generate the application code. Other approaches introduce aspect programming techniques to assemble components by waving features [31] [32].

Mostly, derivation methods consider as input a collection of PL requirements selected from the SPL requirements model. However, industry experience suggests that simply having the right assets is not sufficient to facilitate its selection and assembly.

So some works tried to propose guidelines to select the appropriate set of assets, but they are still reduced to technical levels.

Namely, the specific assets needed could be specified in a production plan which describes how the core assets are used to develop products [33]. Hunt considers software components and studies the optimal organization proceedings to facilitate finding and selecting them [34]. [35] discusses automating component selection using artificial intelligence techniques. [3] [5] provide a framework of terminology and concepts regarding product derivation as well as a generic software derivation process. It is organized on iterative phases in order to determine the final configuration of the derived product. Once again, the derivation process has by default as input a subset of requirements that originate from customers, legislation, hardware and product family organization. Details about how these requirements are aggregated are not given. [4] also establishes a derivation framework. It indicates that the product requirements derivation is made through a decision process. But, it does not include more details about this process.

Nevertheless, a necessary step in product derivation is to determine the set of requirements to use in order to build the particular product out of the possible products in the product line. This requires some description of the customer needs that allows it to be distinguished from others in the SPL. This description provides a set of product requirements. Someone must then find and select the assets that are needed to meet the product requirements. As presented in the existing approaches, it is often the product developer that makes these decisions as the product is assembled. The role of the user is dumped and mistreated.

While the focus provided by scoping develops mechanisms handling technical derivation, we are interested in instructing requirements derivation processes that originate from users needs, and involve users choices while tacking decisions; which is not typically available in general derivation approaches.

## 5 Conclusions and Future Work

A major addition to existing reuse approaches since the 1990s are software product lines that have been the long standing notion to solve the cost, quality and time-to-market issues associated with development of related software applications.

Over the past few years, domain engineering has received substantial attention from the software engineering community. Most of the researches, however, fail to provide detailed derivation processes namely for deriving requirements, which has been restricted to the selection of a requirements subset.

The idea behind the proposed approach is that the user, the main stakeholder to whom the final product is intended, should be involved in specifying product requirements, in a way that efforts expended in constructing the reusable requirements in domain engineering are outweighed by the benefits in deriving the right individual products that satisfy their mission.

RED-PL includes two processes that are the matching and the arbitration processes. The first establishes the set of possible requirements that meet users' needs. The latter, arbitrate on these requirements in order to derive a consistent requirements set that is optimal for a defined set of users and company constraints (e.g. revenue, cost, resources, time, etc.).

We have thought these processes (namely the mathematical model) based on feature models. But, it is obvious that it may be applied for the different PL modeling languages (Use cases, goals, UML, aspects). That is because these types of dependencies represent fundamental concepts that are implemented by all existing variability languages. Only visual representation is different depending on the language constructs (use cases, classes, etc.) and stereotypes. Besides, the approach viability was tested on real projects developing blood analyzers within a French company named Stago. Obtained results were verified and appreciated.

Further research will focus on the refinement of the approach processes. We aim at defining matching and arbitration processes of variable requirements in correlation with variable PL physical architecture. It is worthwhile in Stago context since it produces instruments where technical requirements impact heavily the decision on functional requirements depending on technology costs and revenues.

We intend next to implement a tool support that interfaces with existing modeling tools and enables such a matching and arbitration processes.

Moreover, the repeatability of the approach will be studied. The purpose is to define a systematic process allowing modeling PLs and deriving products suitable to different companies' contexts. We are confident that if the Integer Linear Programming is not scalable to large systems, it can be replaced by another more adequate Multi Criteria Decision Making method.

## References

1. SEI Product Line Hall of Fame web page, http://www.sei.cmu.edu/productlines/plp_hof.html.
2. Linden, F.: Software Product Families in Europe: The Esaps & Café Projects (2002)
3. Deelstra, S., Sinnema, M., Bosch, J.: Product derivation in software product families: a case study. The Journal of Systems and Software, pp.183–204 (2004)
4. Haugen Ø., Møller-Pedersen B., Oldevik J., Solberg A.: An MDA®-based framework for model-driven product derivation. Software Engineering and Applications, USA (2004)
5. Sinnema, M., Deelstra, S., Nijhuis, J., Bosch, J., COVAMOF,: A Framework for Modeling Variability in Software Product Families. The 3rd Software Product Line Conference (2004)
6. Lee, J., Kang, K.C.: A Feature-Oriented Approach to Developing Dynamically Reconfigurable Products in Product Line Engineering. SPLC (2006)
7. Halmans, G., Pohl, K.: Communicating the variability of a software-product family to customers. In: Proceedings of the Software and Systems Modeling. vol. 2, Springer, Heidelberg (2003)
8. Maiden, N., Gizikis, A., Robertson, S.: Provoking Creativity: Imagine What Your Requirements Could Be Like. IEEE Software 22(5), 68–75 (2004)
9. Michael, G., Kang, K.C.: Issues in Requirements Elicitation. Technical Report (1992)
10. www.stago.fr. Diagnostica Stago Web page
11. Salinesi, C., Etien, A., Zoukar, I.: A Systematic Approach to Express IS Evolution Requirements Using Gap Modelling and Similarity Modelling Techniques. CAiSE Conference, Riga, Latvia. Springer Verlag, Heidelberg (2004)
12. van den Akker, M., Brinkkemper, S., Diepen, G., Versendaal, J.: Flexible Release Planning Using Integer Linear Programming. In: Proceedings of REFSQ, pp.257-272 (2005)
13. Gomaa, H.: Designing Software Product Lines with UML: From Use Cases to Pattern-based Software Architectures. Addison Wesley Object Technology Series (2004)

14. Bayer, J., Flege, O., Knauber, P., Laqua, R., Muthig, D., Schmid, K., Widen, T., DeBaud, J.-M.: Pulse: a methodology to develop software product lines. In: Proceedings of the SSR (1999)
15. Clements, P., Northrop, L.M.: Software Product Lines: Practices and Patterns. Addison Wesley Professional (2001)
16. Bosch, J., Florijn, G., Greefhorst, D., Kuusela, J., Obbink, H., Pohl, K.: Variability Issues in Software Product Lines. The International Workshop on Product Family Engineering (2001)
17. Dobrica, L., Niemelä, E.: UML Notation Extensions for Product Line Architectures Modeling. Australasian Workshop on Software and System Architectures, Australia (2004)
18. Robak, S., Franczyk, B., Politowicz, K.: Extending the UML for modelling variability for system families. International Conference on Algorithmic Mathematics and Computer Science, pp. 295–308 (2002)
19. Czarnecki, K., Eisenecker, U.W.: Generative Programming: Methods, Tools, and Applications. Addison Wesley, New York (2000)
20. Sinnema, M., Deelstra, S., Hoekstra, P.: The COVAMOF Derivation Process. In: Proceedings of the 9th International Conference on Software Reuse (2006)
21. Thompson, J., Heimdahl, M.: Structuring Product Family Requirements for n-Dimensional and Hierarchical Product Lines. Requirements Engineering Journal, vol-8(1) (2002)
22. Streitferdt, D.: Family-Oriented Requirements Engineering. PhD Thesis, Technical University Ilmenau (2003)
23. Kang, K., Lee, K., Lee, J.: Concepts and Guidelines of Feature Modeling for Product Line Software Engineering. In: Proceedings of the 7th International Conference on Software Reuse: Methods, Techniques, and Tools, pp. 62 - 77 (2002)
24. Gibson, J.P.: Feature Requirements Models: Understanding Interactions. In: Feature Interactions, in Telecommunications IV, Montreal, Canada, IOS Press, Amsterdam (1997)
25. Buhne, S., Lauenroth, K., Pohl, K.: Modelling requirements variability across product lines. In 14th IEEE International Conference on Requirements Engineering (2005)
26. Perez Garcia, J., A. Laguna, M., Gonzalez-Carvajal, Y. C., Gonzalez-Baixauli, B.: Requirements variability support through MDD and graph transformation. International Workshop on Graph and Model Transformation, Tallinn, Estonia, pp.171-183 (2006)
27. Ziadi, T.: Manipulation de Lignes de Produits en UML. PhD thesis, Université de Rennes 1, équipe IRISA-TRISKELL, directeur Jean-Marc Jézéquel (2004)
28. Ziadi, T., Hélouët, L., Jézéquel, J-M.: Towards a uml profile for software product Lines. In: the Fifth Internationl Workshop on Product Familly Engineering, Springer Verlag, Heidelberg (2003)
29. Jézéquel, J-M.: Reifying configuration management for object-oriented software. In: Proceedings of the 21th international conference on Software engineering, pp.250–259 (1998)
30. Jézéquel, J-M.: Reifying variants in configuration management. ACM Transaction on Software Engineering and Methodology, pp.294–305 (1999)
31. Jansen, A., Smedinga, R., van Gurp, J., Bosch, J.: First class feature abstractions for product derivation. Special issue on Early Aspects: Aspect-oriented Requirements Engineering and Architecture Design, IEE Proceedings Software, pp.197-207 (2004)
32. Mezini, M., Ostermann, K.: Variability Management with Feature Oriented Programming and Aspects. Foundations of Software Engineering, ACM SIGSOFT (2004)
33. Chastek, G., McGregor, J. D.: Guidelines for developing a product line production plan. Software Engineering Institute, Technical Report CMU/SEI-2102-TR-006 (2002)
34. Hunt, J.M.: Organizing the asset base for product derivation. In 10th SPLC (2006)
35. Asikainen, T., Mnnist, T., Soininen, T.: Using a configurator for modelling and configuring software product lines based on feature models. Software Variability Management for Product Derivation - Towards Tool Support at International Workshop of SPLC (2004)

# Deciding to Adopt Requirements Traceability in Practice

Floris Blaauboer[1], Klaas Sikkel[2], and Mehmet N. Aydin[3]

[1] Accenture, System Integration & Technology, The Netherlands
[2] University of Twente, Faculty of Electrical Engineering, Mathematics and Computer Science, PO Box 217, 7500 AE Enschede, The Netherlands
[3] University of Twente, School of Management and Governance, PO Box 217, 7500 AE Enschede, The Netherlands
floris.blaauboer@accenture.com, {k.sikkel,m.n.aydin}@utwente.nl

**Abstract.** The use of requirements traceability for information systems development (ISD) projects is not very common in practice despite its often mentioned advantages in the literature. We conducted a case study in a large IT company to identify the factors that are relevant for the decision whether or not to adopt traceability in an ISD project. Five dominant factors emerged: development organization awareness, customer awareness, return on investment, stakeholder preferences, and process flow. It turned out that the majority of the software development project leaders we interviewed were not aware of the concept of traceability – with the obvious result that using traceability in software project is not even considered. This fact has possibly been underestimated in the present literature of requirements engineering.

**Keywords:** requirements traceability, decision-making, requirements engineering.

## 1 Introduction

Requirements are a measurable statement of intent about something that a product must do; or a property that a product must have; or a constraint on a system [1]. They are the formal basis for software development. Requirements traceability refers to the ability to describe and follow the life of a requirement, in both a forward and backward direction, ideally through the whole systems life cycle [2].

Despite the fact that many scholars have studied requirements traceability from various perspectives, there is a lack of empirical studies showing if and how it is actually practiced in information systems development (ISD) projects. Most studies, as discussed later on, focus on the execution of requirements traceability and its advantages. Value added aspects of traceability are widely recognized throughout literature, and quality standards and techniques in relation to requirements traceability have been studied. However, some scholars including [3] mention that it is not a concept which is applied in every project. This was also the situation in the organization where we conducted a case study. That is, the claim about advantages and practicing of requirements traceability has been present in the case organization, yet few projects in the research environment have actually adopted traceability explicitly into their

J. Krogstie, A.L. Opdahl, and G. Sindre (Eds.): CAiSE 2007, LNCS 4495, pp. 294–308, 2007.

development process. The exact reason for this has been unclear, which is the motivation of this study.

In line with this motivation, the goal of this study is to understand how practitioners go about deciding to adopt requirements traceability. More specifically, this study is aimed to identify the dominant factors influencing the decision making on adoption of requirements traceability in development projects. In contrast to other studies looking into the adoption of traceability, such as [4], [5], which look at the adoption by a development organization after this decision has been made, this study looks at the preceding stage, where the decision to apply the concept of traceability is to be explicitly made. Whereas other literature focuses on the implementation aspects and the technical aspects of the concept, this study focuses on the factors that influence the decision of applying traceability in a project from a management point of view.

For this study we adopted an explorative research approach. As an empirical analysis we have conducted a case study in a large IT development and management consulting company in the Netherlands that we'll refer to as "ITCC". Since the present literature on requirements engineering and information systems development does not provide theoretical underpinnings, accounts, models or alike, the adopted research approach is found to be appropriate to the nature of subject matter and research goal.

## 2 Traceability

There are many different definitions of traceability [5]. We follow Gotel and Finkelstein [2]: "*Traceability refers to the ability to describe and follow the life of a requirement, in both a forwards and backwards direction (i.e., from its origins, through its development and specification, to its subsequent deployment and use, and through all periods of on-going refinement and iteration in any of these phases).*" In our empirical study we speak of traceability if a conscious effort has been made to record traceability links between different products in the software development life cycle. This implies that some record has been made. It doesn't have to be with a special tool, it could be, say, in MS-Word and Excel. However, if such links exist only in the minds of software developers (conceivable on a small project), they still may be able to describe them, satisfying the definition of Gotel and Finkelstein, but we do not call that traceability.

The main value added by requirements traceability is twofold. First of all, there is the aspect of change management. Through requirements traceability, changes in the context of an application (changing requirements) can easily be analyzed for their impact on the code and test cases and vice versa, which heavily shortens the time required for software maintenance. On the other hand, increased accountability simplifies the verification of a system to its requirements and allows better monitoring of the process. However, establishing and maintaining requirements traceability is an expensive and politically sensitive endeavor. Various techniques [6] and tools [7] have been proposed to support the realization of requirements traceability for systems development projects.

Despite these advantages, traceability is often still just an advocated desirable property of a software development process [8]. Several problems were identified in literature with regard to the implementation of traceability. A central problem is the fact that many developers see traceability as an optional activity, for which there are too few resources available and of which they see too little direct benefits [9], [1]. The reason of this limited amount of resources lies not with the developers, but should be sought one level higher, with project management. They have to release the resources required in the form of time, tooling and training [10].

Though traceability is involved with every effort in development from requirement elicitation to testing, as advocated by most ISD methods, the decision to adopt it should take place at an early stage. According to RUP and PRINCE 2, during both the startup and initiation processes of a project, the project team determines the way in which the work should be performed. The third process involved is the directive process. It is this process in which decisions on accepting or rejecting a project plan are made by a project board, whereas the contents of this plan is determined by others.

All these approaches, methods, techniques and tools proposed for requirements traceability are useful as long as its adoption decision is present preferably at the early stages of a project. However, this subject is undertheorized and calls for studying what underpins the execution of requirements traceability. That is, we need to understand how the decision on requirements traceability is made and which factors influence adoption of traceability. In the following, we present the conceptual treatment of these questions, which eventually provide us with a theoretical lens to examine this adoption in a systematic manner.

## 3 Research Framework and Approach

In this study we adopt the viewpoint that adopting traceability is essentially a matter of choice. One can either make a choice to trace or choose not to trace during development. A decision is *a choice made from available alternatives*, which is exactly what is happening when adopting traceability or not [11]. This definition is somewhat lacking depth, however. Matheson and Howard [12] define a decision as *an irrevocable allocation of resources that is revocable only at a cost in some resource, such as time or money.*

Several different schools of decision making have arisen in the recent years. The classical decision making theory is a normative theory, which stipulates how decisions should be made on a normative basis and what the best outcome would have been [13]. This study adopts Classical Decision Making (CDM), and specifically the Stanford school of decision making of which Howard is one of the most prominent authors [14]. The theory as discussed by Howard identifies a process by which a decision is structured and should be structured for that matter. The goal of this study, however, is not to define what decision *should* be made, but *how* this decision is reached. The model is therefore not used in a normative way, but in a way of describing the process. It is not intended to be followed to see how a decision should be made, but to identify the aspects involved with making this decision.

Howard [14] approaches the decision making process by modeling it as a shift from a real decision problem to a real action, including all the actions required to create this shift. This is a process of elicitation and analysis, leading to clarification of the actual problem in a way that it can be acted on in a logical way. Three phases are distinguished: formulate, evaluate, and appraise a decision problem, along with any actions appropriate for dealing with the problem. A decision problem goes through one or more iterations of the three phases. When the decision maker is content with the outcome of the process, meaning that he sees no need for another iteration and no further information available, he will decide how to tackle a problem and take real action.

Central in the decision making process is the decision basis. It is established during the "formulate" phase, where the required information is elicited and structured as needed. With this basis, the different solutions can be evaluated and appraised, after which, if required, another iteration of the cycle is performed. This entire cycle is based on the decision basis. The decision basis consists of three parts:

- *Choice.* There are different alternative solutions for the decision problem.
- *Information.* In order to judge the alternative solutions, information is needed. In Howard's theory information consists of models and probability assignments.
- *Preferences.* Personal preferences do play a role in decision making. These comprise personal values, time preferences and risk preferences.

Before a problem is subject to the formulate-evaluate-appraise cycle, however, a problem must have been identified as a decision problem – otherwise there is no cycle. When the problem has been identified, an elicitation process yields input to the decision basis.

Howard's theory was used as a structuring principle to identify possible factors influencing the decision to adopt traceability. All relevant factors for the decision to adopt traceability that were found in the literature could be subsumed by categories form Howard's theory. The theoretical framework itself suggested some other factors. For example, the fact that awareness of traceability is needed in order to make any decision. This is self-evident, and for that reason not mentioned in the technical literature. Yet it proved to be a most relevant.

## 3.1 Relevant Factors Identified

To identify existing factors that are considered as relevant, we have reviewed those studies concerned with project management, software engineering process, information systems development, and requirements engineering and structured them using Howard's framework, as elaborated above. Here we elaborate these factors with their descriptions, rationale and the way to examine in empirical setting.

**Problem Identification**

*Development organization awareness.* The choice whether or not to adopt traceability is made in the development organization. If this organization is not aware of the concept, it will never adopt traceability. Several roles within the development organization can create this awareness with the decision maker (the project manager). This factor is measured by checking whether or not it was discussed with the project manager during the project's initial phases. Precondition is that those that were aware did take this effort.

*Customer Awareness.* Traceability can also be a demand of the customer. The customer should be seen in a broader sense here, not just as the manager. Acceptance criteria [15] are examples of concrete demands of the customer on traceability. This factor is measured by checking whether or not any demands with regard to traceability from the customer were known with the project manager.

**Elicitation**

*Sources of Influence.* Literature offers little insight into the roles of members of a project team involved with the creation of the project plan. From the project board the senior user, who represents both the end-user and the IT-management department, is an important source of information required [16].

**Decision Basis – Choice**

*Technological Possibilities.* To make sure that the alternatives are realistic, the technical support has to be adequate for the concept to be applied. For this decision, the decision maker has to be under the impression that the technical support is sufficient. This factor is measured by discovering whether or not the project manager feels the option is realistic.

**Decision Basis – Information**

*Return on Investment.* In business, every decision can be approached as an investment decision. The costs of adopting traceability are often apparent, lying in the resources required for each registered traceability relation. Many researchers have found that one of the main problems, if not the main problem, with regard to adopting traceability in practice is the lack of perceived benefit with management, leading to a diminished support from management [3], [4]. This factor is measured by examining whether or not the project manager is aware of the advantages and drawbacks of traceability in the economic sense

*Quality Standards.* As discussed before, both CMM and ISO 9000 demand the use of traceability in development projects [17], [18]. When an organization wishes to adhere to these standards, in the case of CMM at the level where traceability is required (2 and beyond), adopting traceability is required. The organization can be both the development organization and the customer organization. This factor is measured by looking at certification of the organization.

*Compliance.* Besides quality standards such as ISO and CMM, there are also legal standards which companies have to adhere to. One of these standards is the Sarbanes-Oxley Act of 2002 which applies to all companies tradable on the NYSE [19]. This standard demands traceability from all systems in use in order to ensure transparency. This factor is measured by looking at the required adherence to legal standards in the customer organization.

*Project Complexity.* When the complexity of a project increases, it also becomes harder to comprehend the system for developers. This complexity has to do with the familiarity of the development organization with the technology used. Traceability links between code and other deliverables aid in both the bottom-up and top-down comprehension of code by programmers, increasing their productivity through being able to both derive code from the preceding products and to place chunks of code in perspective [20]. Project complexity can be judged from high to low by the project manager.

*Development Method.* The characteristics of a method influence, amongst other things, the amount of changes that occur during the development project itself. Incremental and iterative development methods lead to more changes on already developed parts of the system during the development project, therefore increasing the value of adopting traceability. Besides simply adding to the total increase of value, the fact that these benefits lie in the development stage also improve the likeliness of the adoption.

*Product Life Expectancy.* The main financial benefit lies in the heavy reduction of the time required for impact analysis when changes occur. The longer that a system is in use, the higher the number of changes that occur during its lifespan, therefore again leading to a higher return on investment. The expected lifespan of the system is often stated at the start of a project.

*Dynamics of the Environment.* The frequency of risks is determined by the amount of changes in its environment. These changes in the environment can consist of changing interfaces, changes in business process layouts, changes in organizational structures, and many other matters. The more dynamic the environment of a system, the higher the amount of changes which is to be expected and the higher the expected return on investment for adopting traceability. This factor is measured based on the project manager's perception of it.

**Decision Basis – Preferences**

*Stakeholder Preferences.* The outcome of a project influences many different parties. Each of these parties might have an influence on the outcome of the decision, which they would want to turn in their own favor. The manager of an IT department might be bound by a tight budget, which means that the solution has to be within that budget, whereas adopting traceability might mean exceeding the budget. This way, what is best for the customer organization is not necessarily what is best for individuals involved. Daft defines this as politics, where in this context power is exercised to influence the decision to strive for a self-serving purpose [11].

*Process Flow.* Adopting traceability increases the workload for those involved with it. It creates additional tasks which are to be performed during the work in development, which are often seen as extra and optional [21]. This factor is measured by discovering the attitude of the project manager towards the influence of traceability on the development process.

*Benefits Outside Project Scope.* The profit of traceability comes along when the links are used in change processes that occur later on in the systems life. The benefits lie not with the development project. Possibly a main driver for using traceability is the customer, in this case the IT management department. First of all, this information has to be made available. Second, the development organization has to be willing to supply what the other party wants, when the investment is of limited benefit to the development organization [3, 4].

## 3.2 The Conduct of Case Study

The case study took place in 2006 at ITCC, an IT consulting company in the Netherlands with about 4600 employees and a yearly turnover of around M€ 450. Within the requirements business unit of ITCC, one of the current topics is the professionalization of the service provided and especially the professionalization of the requirements process. One of the issues that has arisen, and not just in the requirements unit, is the notion that changes within a system can be performed a lot faster, and therefore less

**Table 1.** The profiles of respondents

| Respondent | I | II | III | IV | V | VI |
|---|---|---|---|---|---|---|
| Experience as PM | 10 Years | 8 Years | 10 Years | 8 Years | 15 Years | 2 Years |
| Experience with ISD | 24 Years | 21 Years | 17 Years | 8 Years | 23 Years | 12 Years |
| Work Experience | technical automation | ISD projects | 7 years IT-consultant | IT strategy | Developer, Designer, Analyst | Functional Designer, Team Leader, Architect |
| Product | Forum Application | Call Centre | MIS | ERP system | 1st Line Support System | Batch Conversion System |
| Project Length | 6 Months | 4 Months | 9 Months | 18 months | 24 Months | 8 Months |
| Team Size | 3 | 12 | 17 | 6 | 24 | 14 |
| Traceability Applied? | Yes | No | Yes | Yes | Yes | Yes |

**Table 2.** The opinions of respondents and informants on the relevance of factors

| Factor | Respondent | | | | | | Informant | | | | |
|---|---|---|---|---|---|---|---|---|---|---|---|
| | I | II | III | IV | V | VI | I | II | III | IV | VI |
| Development Organization Awareness | E | E | E | E | E | E | D | D | D | D | D |
| Customer Demand | + | – | – | – | – | – | + | N/A | + | N/A | + |
| Technological Possibilities | E | E | E | 0 | E | E | + | N/A | N/A | N/A | N/A |
| Return on Investment | + | – | 0 | + | + | + | + | + | + | + | N/A |
| Quality Standards | E | – | 0 | 0 | 0 | 0 | N/A | N/A | N/A | + | N/A |
| Compliance | 0 | 0 | 0 | 0 | 0 | 0 | N/A | N/A | N/A | N/A | N/A |
| Project Complexity | 0 | 0 | 0 | 0 | 0 | 0 | N/A | N/A | N/A | N/A | 0 |
| Development Method | 0 | – | 0 | 0 | 0 | 0 | 0 | N/A | + | N/A | N/A |
| Product Life Expectancy | 0 | 0 | 0 | 0 | + | 0 | N/A | N/A | N/A | N/A | + |
| Dynamics of the Environment | 0 | 0 | 0 | 0 | + | 0 | N/A | N/A | N/A | N/A | + |
| Stakeholder Preferences | 0 | – | + | – | 0 | 0 | N/A | – | N/A | 0 | – |
| Process Flow | E | – | E | 0 | E | E | 0 | N/A | + | N/A | – |
| Benefits Outside Project Scope | – | – | 0 | – | – | – | – | – | 0 | – | – |
| Time Pressure | N/A | – | 0 | – | – | 0 | N/A | – | – | + | – |
| Contract Type | N/A | – | + | – | 0 | 0 | N/A | – | – | N/A | – |

Respondents: those who are aware of traceability; Informants: those who are not aware
+ : this factor positively influenced the decision to adopt traceability
– : this factor negatively influenced the decision to adopt traceability
0 : this factor did not weigh in on the decision to adopt traceability
E : this factor was seen as an enabler for the adoption of traceability
D : this factor was a disabler for the adaptation of traceability
N/A this factor is not applicable to the discussed project

expensive, when the entire process is performed in a repeatable, but more specifically, traceable, way. So, the need for the conduct of this study was immediate for the organization, and the researchers got access to whatever data sources they needed.

The unit of analysis in this case study is the person who is responsible for decision making for requirements traceability. At ITCC this is the project manager (we did verify this during the case study). There are project managers with different ranks and with different levels of experience. To get a good sample of the project managers at ITCC, first we have selected eight managers with experience with different kinds of projects. It turned out that only three of them knew about requirements traceability. In some cases traceability techniques were used, but the interviewee was not familiar with the term "traceability". So we operationalized "being aware of traceability" as recognizing it after our concept of traceability (as formulated in the first paragraph of section 2) had been explained.

Awareness about traceability is a precondition for making decision about it. Hence, we expanded the selection and conducted in-depth interviews with in total six managers who are already aware of traceability. Five of them adopted traceability in their projects, but one of them had never adopted it in any project he had been in charge of. Table 1 summarizes their profiles. In addition, interviews were conducted with the five managers in the first sample who had no knowledge of traceability and therefore did not make any decision with regard to its adoption. The interviews with these professionals were a valuable addition, due to their estimated impact of traceability on their work.

The data collection started by performing a pilot interview, as suggested by [22]. A random project manager was selected (with the known prerequisites) and the interview was performed with a focus that lied more on the process than the product. Only minor details were adapted in the interview protocol after the pilot, hence we considered it OK to keep the pilot in the data set. The interviews took place at various locations; half of them were at customer locations and the other half at ITCC headquarters. The data from the interview protocols are summarized in Table 2. The 6 managers with awareness about traceability are referred to as "respondents"; they were interviewed about a project they were currently involved in. The 5 other managers were asked for their opinion after the concept was explained. They are referred to as "informants".

Two factors, contract type and project size, were not on our list in Section 3.2 but came up in the interviews as being relevant.

## 4 Findings and Discussion

Based on the conducted in-depth interviews we have reorganized the factors as shown in Table 3. Obviously there are dependencies between the different factors. We identified five dominant factors and clustered the other ones with the factor they are most related to, yielding the 5 categories shown in Table 3. The five dominant factors are: awareness of the development organization; awareness of the customer organization; the perceived return on investment by the project manager; personal preferences of stakeholders; the way traceability influences the process during the development project. We will discuss each factor and its empirical justification.

**Table 3.** Classification of factors for adopting requirements traceability (dominant factors in italics)

| Development Organization | |
|---|---|
| | *Development Organization Awareness* |
| | Sources of Influence |
| | Quality Standards |
| **Customer Organization** | |
| | *Customer Awareness* |
| | Quality Standards |
| | Compliance |
| **Financial factors** | |
| | *Return on Investment* |
| | Dynamics of the Environment |
| | Project Size |
| | Product Life Expectancy |
| **Political factors** | |
| | *Stakeholder preferences* |
| | Contract Type |
| | Benefits Outside Project Scope |
| **Operational consequences** | |
| | *Process flow* |
| | Technological Possibilities |
| | Time Pressure |

**Development Organization**

*Development Organization Awareness.* Awareness of traceability in the development organization was identified as a prerequisite for the existence of the decision to adopt traceability. Where this factor might seem trivial, its validation showed not only that it was definitely an influencing factor but also that it was a quite important factor. From the first eight interviews that were planned, five interviewees had no knowledge about traceability as a concept that could be applied in systems development. Although it is not possible to draw any quantitative conclusions from this study on how many project managers do and do not know what traceability is, awareness of project managers is quite an important factor because without this awareness it is impossible to make this decision. In the five interviews with informants (i.e. those interviewees not aware of traceability) we could see that they all show development organization awareness as a disabling factor.

In only one of the cases did anybody ever discuss traceability with the project manager before the development started. This was an architect. The fact that the informants did not know anything of the concept also shows that nobody ever bothered them with it, although this might seem like a logical move. Parts of the development organization might be aware of the existence of traceability and its advantages or drawbacks, but these were never communicated to the relevant roles in the project except for the case of one respondent.

*Sources of Influence within the development organization.* In half of the projects mentioned did anyone other then the project manager have an influence on the decision to adopt traceability. Informants only mentioned the program manager as a potential influence once, whereas he was an influence in one of the cases, but the architect was seen as a large potential influence.

*Quality Standards at the development organization.* It was often found that (unofficial) quality standards did not demand traceability which meant that this negatively influenced the decision; it did not have to be adopted. Organizations which do work through certification such as CMM at level 2 or higher would have this influencing the decision, since traceability is a must in this case. This was not found in any of the cases, but is indispensable for organizations which have regular audits on their CMM level.

**Customer Organization**

*Customer Awareness.* Besides the influence and knowledge of the development organization, the customer also has quite an important influence on the decision to adopt traceability. Almost all project managers stated that if a customer demanded traceability, he would adopt it. However, very few customers actually demanded traceability. Whereas the IT management organization of the customer, be this an in-house department or with outsourced IT management, would most likely be the party demanding traceability, this party is often not involved in the initiating phases of a project. Few of the interviewees had contact with this party, and if acceptance criteria were given, they almost never included traceability. The interviewees regarded customer awareness as a causal factor. If the customer demanded traceability, he would perform this (as long as any required budget for this was created). On the other hand, if the customer did not demand it, it made things easier not to adopt it, enforcing the negative decision on this subject. The latter was mostly the case in this study, as can be seen in Table 2.

*Quality Standards at the customer organization.* Only one case was found where the respondent was subjected to an ISO 9000 certified process, but this did not demand traceability. If traceability was performed however, there were strict guidelines with regard to review processes which influenced traceability in a negative way.

*Compliance.* What applies to quality standards is somewhat similar to what applies to compliance. Regulations and laws such as the Sarbanes-Oxley Act and Basel II demand from organizations that they develop their systems in a traceable way. In the empirical study, no cases were found where this was demanded yet, although this would be expected in projects in the financial sector. In this case, it would become part of the customer demand: not as a demand from the IT management department, but from the risk management department as one interviewee remarked.

**Financial factors**

*Return on Investment.* As stated in the CDM theories, the decision to adopt traceability could in practice be approached as an investment decision, and this is exactly what was found in practice. Although not all respondents performed a quantitative analysis of the costs and benefits involved with adopting traceability, they did weigh the pros and cons in a financial sense. This happened mainly because, in the end, they were responsible for finishing the project within budget. This led to the question whether or not adopting traceability would aid in reaching this goal in several of the cases, in which it was answered as true in most cases. What some project managers, in this case two informants, showed as missing in their toolkit was being able to determine whether or not traceability was worth performing at all, especially in their project. The perceived financial benefits of traceability were different for each case. Often, the verification of traceability was seen as a lot more important then the main benefit mentioned in theory, being the simplified impact analysis. Steering the project with better management information, both with respect to the development progress and the correctness of the product was seen as the main financial advantage by one respondent, whereas one person also saw the reduction of tests required as an advantage and only one person performed traceability mainly for the impact analysis advantages.

*Dynamics of the Environment.* As stated, in only one case did the project manager identify impact analysis as an advantage of traceability. The others did not take the dynamics of the environment, and the correlated amount of changes, into account. One respondent considered the environment not very dynamic, hence it didn't have much weight; another was involved in a project where the overriding concern was to get out a system, irrespective of quality, in a very short time frame. This factor is related to 'benefits outside the project scope' discussed below.

*Project Size.* A factor that was not on our list, but did come forward in several interviews, is that of project size. Many of the project managers who were acquainted with the concept of traceability stated that they applied it mainly in larger projects. The longer a project lasts, the larger the benefits. Where the factor "product life expectancy" looks at the life of the product, the life expectancy of the project also matters with regard to the amount of changes found. For a project manager, larger projects are harder to keep track of, to keep a good overview. Traceable development offers more management information. When looking at the adoption of traceability by the respondents, there was also a trend which could be found with regard to the size of projects. Traceability was seldom adopted in small projects, but in larger projects this percentage got a lot higher. This study looked into the factor of project complexity, which focused on the technology used and the familiarity with this. Project size also contributes to complexity. The interviews indicated that size, rather than complexity, is recognized as a factor.

*Product Life Expectancy.* Only one respondent was aware of taking the expected period of use of the system under development into account with regard to the decision on adopting traceability. He felt this weighed in the cost/benefit analysis, which is also what the theory prescribed. Whereas the other respondents did perform a cost/benefit analysis, they often did not see the advantage in the form of the improved impact analysis, so this factor did not apply to them. When one were to recognize this as an advantage of traceability the informants felt that life expectancy of the product would matter. On the other hand, the life expectancy has to do with matters which occur outside of the scope of the project, which is limited to just the development of the system.

**Political factors**

*Stakeholder Preferences.* The different stakeholders involved in the decision could have personal motives with respect to the outcome of the decision which weighed in, and several of these were identified. In many of the interviews, budgets came forward as very important issues. The form of contract available was something discussed very often. If a project manager remains further within budget, this provides him a larger bonus after the project is finished. For many project managers, this was an influence on the decision to adopt traceability, especially in the case where traceability was not adopted. This factor turned out to be quite an important influence in the different interviews. It is closely related to the view the respondent has on the return on investment of traceability however, since for project managers it is often all about the money, the final number is what counts. If he feels traceability will not lead to many extra costs, this factor will not necessarily be of influence, whereas it becomes quite important if he feels there are additional costs involved.

*Contract Type.* The type of contract, to our knowledge, isn't mentioned anywhere in the technical literature as a relevant factor. Yet this was identified by several interviewees as quite important. Following Lauesen [23], we identify two types of contract for software development projects: fixed price and time-material based. The first type is based on a quotation provided by the supplier. This can be a response to a tender, where a request for proposal was sent out by the customer, or because the customer directly approached the supplier. A fixed price is determined for the project, any additional costs are for the supplier, but any savings with respect to the agreed price are also for the supplier. In the second type of contract, a registration of hours is kept and the price paid by the customer is determined by the actual time and materials used in the development project. What is important here, however, is the decision maker's impression of the return on investment of traceability in the development project itself. If he feels traceability will add costs to a fixed price development project, this will negatively influence his decision.

*Benefits Outside Project Scope.* A factor which was not necessarily quite clear in theory became all the more apparent in practice. The scope of a project, as defined by PRINCE2, is limited to the development and does not involve the use of the product [16]. The satisfaction of the customer is only checked at the completion of the project, the project manager is generally not responsible for any aftercare. As one of the interviewees remarked, this is quite important commercially however. A customer who remains satisfied, even after the delivery of the product, is a customer who returns to the company. Many of the project managers did not automatically think of traceability as something that had an improved impact analysis as the main advantage. For them, the fact that most of the benefits generally lie in the IT maintenance phase was irrelevant, they were very clear that all that mattered to them was the project itself and not anything beyond this. Four out of five informants also felt that any benefits obtained beyond from the development project itself were irrelevant and would not be taken along in the decision, therefore negatively influencing this decision.

**Operational consequences**

*Process Flow.* The flow of the process was definitely regarded as an influencing factor by the different project managers. In one case the project manager perceived the adoption of traceability as a negative influence on the process, traceability making the operational processes more complicated. In most other cases, however, the contrary was found. The project managers did not perceive the adoption of traceability as a problem with respect to the flow of the process. In the end, this factor again has to do with the perception of the decision maker of its impact and can only be influenced by experience and demonstration.

*Technological Possibilities.* Technology that supports traceability is something that is quite important to the project managers. If there is insufficient support for traceability in the opinion of the decision maker, this would influence the decision in a negative way. Nearly all of the respondents felt, however, that the technical support of traceability is quite adequate. For this reason, this positively influenced their decision to perform traceability, although it was not really a decisive argument. Technological possibilities were more seen as an enabler of the process. The use of text processors and spreadsheets was always seen as enough support, although tools were deemed as interesting by several respondents. In the case of the adoption of tools the issue of return on investment would become a lot bigger though according to them, a more thorough cost/benefit analysis would have to be performed.

*Time Pressure.* A very clear issue which came forward during one of the first interviews was time pressure. In this specific case, the project had a very strict deadline and the project manager had the feeling that implementing traceability would increase the time required to develop the system. This last thing is quite important; again it has to do with the perception of the project manager on how traceability would influence the project. If time pressure is an issue in a project, it therefore leads to a negative influence on the adoption of traceability. Many informants also stated that traceability would probably be one of the first things that were dropped as soon as time pressure became an issue and a small form of panic appeared. "Quality issues" such as documentation and traceability are seen as a lot less important then functional development, which is critical to the project.

**Discarded Factors**

Two factors identified from the literature were not found to be relevant by the interviewees and therefore discarded from the above list.

*Project Complexity.* The interviewed project managers could not relate to this factor. Note, however, that they did mention project size, as discussed above. An explanation for not perceiving technical complexity as a factor could be that the project manager operates at a higher level of abstraction. According to Antoniol [20], it does play a role for those performing the operational development work.

*Development Method.* This factor has not played a significant role in the practice examined. Although it did influence the decision once, and one informant felt it might influence his decision. This factor was seen as hardly relevant.

## 5 Conclusions

In this study we have investigated how practitioners (project managers) go about making a decision on adopting requirements traceability. From a literature review in requirements engineering, software engineering and information systems development and Howard's theory of Classical Decision Making we identified factors relevant for making a decision about traceability. We validated these factors in a case study conducted in a large software development and management consulting company. As most important factors, in the view of project managers, we found development organization awareness, customer awareness, return on investment, stakeholder preferences, and process flow. Some factors from the literature (development method, project complexity) were not considered important, while some factors not mentioned in the literature (contract type, time pressure) were important for project leaders for making a decision about requirements traceability.

In this company, the majority of software development project leaders are not aware of the concept of traceability – hence using traceability is not even considered. This raises the question whether this company (which has not a bad reputation) is exceptional or whether this awareness is too easily taken for granted.

What stands out in this company as the most prominent reason not to adopt traceability, apart from awareness, is the organizational separation of development projects from the later phases of the software life cycle. Project managers are motivated by and rewarded for achieving the goal for which they are held accountable, which is delivering the right project on time and within budget. There is little incentive to use traceability when most of the benefits are outside the project.

Tender projects in which the supplier is to quote a fixed price are less likely to adopt traceability from a financial perspective – unless the client explicitly requires it.

The technical means for traceability exist. In our case study, the two big obstacles are a lack of awareness and the way software development projects contracted and organized. The latter is more difficult to change than the former. The expected implementation of Sarbanes-Oxley is a chance to change something for the better and to get traceability accepted at a larger scale.

This study adopted an explorative research approach. The findings resulting from one case study should be considered as our contribution to the understanding of decision making on requirements traceability by means of the conceptual articulations and practical insights for the subject matter. To enhance this basic understanding, follow-up research is suggested to test the proposed categories, factors and their relationships in other organizations. We recommend to including different roles, such as software architect, in follow-up studies. Too little is known about their influence on the process. Another interesting topic might be the influence of a development method, but then in the sense of it already proposing the use of traceability, therefore contributing to development organisation awareness. Once the foundation of decision making for requirements traceability is established, one can suggest the appropriate approaches, tools, techniques for adopting and implementing requirements traceability.

## References


1. Robertson, J., Robertson, S.: Mastering the Requirements Process. Addison-Wesley, New York (1999)
2. Gotel, O., Finkelstein, A.: An Analysis of the Requirements Traceability Problem. Int. Conf. on Requirements Engineering (ICRE'94) pp. 94–101 (1994)
3. Arkley, P., Riddle, S.: Overcoming the Traceability Benefit Problem. 13th IEEE Int. Conf. on Requirements Engineering (RE'05) 385–389 (2005)

4. Ramesh, B.: Factors influencing Requirements Traceability Practice. Communications of the ACM 41(12), 37–44 (1998)
5. Ramesh, B., Jarke, M.: Towards Reference Models for Requirements Traceability. IEEE Transactions on Software Engineering 27(1), 58–93 (2001)
6. Hull, E. et al.: Requirements Engineering, 2nd edn. Springer, Heidelberg (2005)
7. INCOSE. INCOSE Requirements Management Tools Survey. International Council on Software Engineering, Retrieved March 6th 2006, from http://www.paper-review.com/tools/rms/read.php. (2006)
8. Lindvall, M., Sandahl, K.: Practical Implications of Traceability. Software – Practice and Experience 26(10), 1161–1180 (1996)
9. Stout, G.A.: Requirements Traceability and the Effect on the Systems Development Lifecycle. Revere Group whitepaper (2001)
10. Dorfman, M., Chardon, R.: Early Experience with Requirements Traceability in an Industrial Environment. Industrial presentation, 5th IEEE International Symposium on Requirements Engineering (ISRE'01) (2001)
11. Daft, R.L.: Management. 5th Edition, Dryden Press, Fort Worth, TX, (2000)
12. Matheson, J.E., Howard, R.A.: An Introduction to Decision Analysis. In: Howard, R. A., Matheson, J. E. (eds.) Readings on the principles and applications of decision analysis I. Strategic Decisions Group, Menlo Park, CA, pp. 17–55 (1983)
13. Aydin, M.N.: Decision-Making and Support for Method Adaptation, Ph.D. Thesis, University of Twente, Enschede, the Netherlands (2006)
14. Howard, R.A.: The Evolution of Decision Analysis. In: Howard, R.A., Matheson, J.E. (eds.) Readings on the principles and applications of decision analysis. Strategic Decisions Group, Menlo Park, CA, pp. 5–16 (1983)
15. Rational, Rational Unified Process, version 2003.06.15, IBM (2006)
16. OGC Managing Successful Projects with PRINCE2. The Stationary Office, London, 4th Edition (2005)
17. ISO 9000-1 Quality systems – Model for Quality Assurance in Design, Development, Production, Installation and Servicing. International Organization for Standardization (1994)
18. Carnegie Mellon SEI, The Capability Maturity Model, Guidelines for Improving the Software Process. Addison Wesley, Reading, Massachusetts. (1999)
19. US Congress Sarbanes-Oxley Act of 2002. Washington, USA, Congress of the United States of America (2002)
20. Antoniol, G.: Recovering Traceability Links between Code and Documentation. IEEE Transactions on Software Engineering 28(10), 970–983 (2002)
21. Jarke, M.: Requirements Traceability. Comm. of the ACM 41(12), 32–36 (1998)
22. Yin, R.K.: Case Study Research; Design and Methods, 2nd edn. Sage Publciations, Thousand Oaks (1994)
23. Lauesen, S.: Software Requirements: Styles and Technique. Pearson Education Ltd (2002)

# Designing Social Patterns Using Advanced Separation of Concerns

Carla Silva[1], João Araújo[2], Ana Moreira[2], and Jaelson Castro[1]

[1] Centro de Informática, Universidade Federal de Pernambuco, 50732-970, Recife, Brazil
{ctlls, jbc}@cin.ufpe.br
[2] Dept. Informática, FCT, Universidade Nova de Lisboa, 2829-516 Caparica, Portugal
{ja, amm}@di.fct.unl.pt

**Abstract.** This paper proposes an approach to support separation and modularization of crosscutting concerns in multi-agent systems (MAS). Crosscutting concerns are properties that do not align well with the decomposition criteria of the chosen approach and, therefore, cannot be modularized. Aspect-Oriented Software Development offers mechanisms to encapsulate such properties in separate modules, the *aspects*. Aspects are used as abstractions to capture social patterns concerns that cut across functional modules in MAS. To achieve this, we propose a technique to describe social patterns in an aspect-oriented context and a systematic way for using them in MAS design.

## 1 Introduction

Tropos [1] has defined a set of design patterns, called social patterns [2], focusing on social and intentional features which are recurrent in cooperative and multi-agent systems (MAS). Although both application core and design patterns concerns are independent from each other [3], the current use of patterns can lead to the scattering and tangling of their concerns with the application functional modules. As a consequence, application core becomes dependent on patterns [3], which do not fully meet quality requirements, such as understandability, maintainability, evolvability and reusability. Thus, design patterns concerns can be called crosscutting concerns, since they cut across the functional modules concerns, decreasing the system reusability and maintainability. These crosscutting concerns can be better addressed by adopting aspect-oriented software development (AOSD) techniques [4]. The purpose of this new technology is to localize crosscutting concerns, avoiding their scattering or tangling through software artifacts [5].

Although many languages for multi-agent systems modeling have been proposed in the last few years, such as AUML [6] and MAS-ML [7], they do not consider the separation of crosscutting concerns. On the other hand, some languages for modeling separation of crosscutting concerns have been proposed, such as aSideML [8] and AODM [9], but they do not address MAS directly. In this context, we propose a notation to describe social patterns using abstractions and mechanisms provided by aspect orientation to support the separation of crosscutting concerns in MAS. To achieve this, we need to: (i) specialize the agency metamodel [10] by using the concept of

J. Krogstie, A.L. Opdahl, and G. Sindre (Eds.): CAiSE 2007, LNCS 4495, pp. 309–323, 2007.

model roles [11] used in Pattern Specifications (PSs) technique [12]; (ii) attach notes to the structural model of the social pattern to capture some abstractions of aspect orientation; and (iii) suggest guidelines to map the proposed notation to the constructs of AspectJ [4] and JADE [13] environments. We have used an integration of JADE and AspectJ to implement our approach. This integration is not difficult as both implementation environments are based on Java, facilitating the combination of JADE and AspectJ code in the same program.

This paper is organised as follows: Section 2 presents some background on Social Patterns, UML-based MAS notation and PSs technique. Section 3 presents the cross-cutting nature of social patterns through a motivation example. Section 4 introduces our notation to describe social patterns in an aspect-oriented context. Section 5 exemplifies the use of our approach. Section 6 presents some related work. Finally, section 7 summarises our work and points out to future work.

## 2 Background

The purpose of this work is to propose a description of social patterns using advanced separation of concerns. This description is achieved using model roles [11] to specialize the agency metamodel introduced in [10]. The concept of model roles has been used to specialize the UML metamodel [14], in the PSs technique [12], to specify what model elements must participate in the pattern. Thus, in this section we introduce the social patterns, the agent oriented notation and the PSs technique.

### 2.1 Social Patterns

Tropos is a requirements-driven framework aimed at building software that operates within a dynamic environment. To promote an efficient development of MAS, Tropos supports five phases of software development: Early Requirements, Late Requirements, Architectural Design, Detailed Design and Implementation. In this work we will concentrate on the late phases.

The Detailed Design phase is intended to introduce additional detail for each architectural component of a system. Designers can be guided by a catalogue of multi-agent patterns which offer a set of standard solutions. In particular, Tropos has defined a set of design patterns, called *social patterns* [2], focusing on social and intentional aspects that are recurrent in multi-agent and cooperative systems. They are inspired by the federated patterns introduced in [15] [16]. Examples of social patterns are booking, subscription, monitor, broker, mediator, wrapper and matchmaker (to be used to illustrate the approach). The Matchmaker pattern involves an intermediary agent (matchmaker) that receives requests from service Providers to subscribe/unsubscribe its services into the Yellow Pages maintained by it. A Client may need a specific service provided by an unknown Provider. The Matchmaker also receives requests from Clients to locate some Providers which offer a specific service. If there is some Provider for the requested service, the Matchmaker informs that Provider's identification to the Client which, in turn, can directly interact with it [17].

## 2.2 UML-Based MAS Notation

In this section, we present the MAS architectural diagram specified according to the agency metamodel introduced in [10] and reflecting the client-server pattern [18] that we have tailored for MAS. We define the MAS architectural diagram (Fig. 1) in terms of AgentRoles and organizational architectural features which include: Goal, MacroPlan, ComplexAction, OrganizationalPort, AgentConnector, Dependum, Dependee and Depender. A Dependum defines an "agreement" of service offer between two agent roles that also play the roles of Depender and Dependee. Thus, the AgentRole responsible for providing the service is the Dependee. The AgentRole that requests the service provided is the Depender. A Dependum can be of four types: goals, softgoals, tasks and resources [19]. AgentRoles need to exchange signals through an AgentConnector to accomplish the contractual agreement of service providing between them. An OrganizationalPort specifies a distinct interaction point between the AgentRole and its environment (depicted as a white square attached to the «AgentRole» class). A Goal is a condition or state of affairs in the world that the actor (Agent or AgentRole) would like to achieve. How the goal is to be achieved is not specified, allowing alternatives to be considered [20]. A MacroPlan encapsulates the recipe for achieving some goal. A ComplexAction determines the steps to perform a MacroPlan.

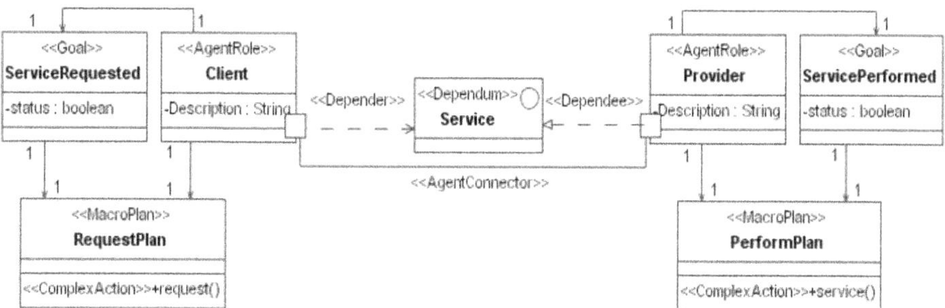


**Fig. 1.** MAS Architectural Diagram

For example, in Fig. 1 we have the Provider AgentRole which is responsible for performing the service defined in the Dependum. This AgentRole aims at achieving the ServicePerformed goal by executing the PerformPlan MacroPlan, which, in turn, consists of performing the service() ComplexAction. The Client AgentRole aims at achieving the ServiceRequested goal by executing the RequestPlan MacroPlan, which, in turn, consists of performing the request() ComplexAction. Therefore, the Client AgentRole is responsible for requesting the service defined in the Dependum. Both the message for requesting the service execution and the message for confirming whether the service was successfully concluded are sent through the AgentConnector.

## 2.3 Pattern Specifications

Pattern Specifications are introduced in [12] as a way of formalizing the structural and behavioral features of a pattern. The notation for PSs is based on the Unified

Modeling Language (UML). The abstract syntax of UML is defined by a UML metamodel [14]. PSs specialize this metamodel by specifying what model elements must participate in the pattern and is defined in terms of roles. Each element in the specification of the patterns is a role, that is, a metaclass specialized by additional properties that any element fulfilling the role must possess. A PS can be instantiated by assigning UML model elements to the roles in the PS. A model conforms to a PS if its model elements that play the roles of the PS satisfy the properties defined by the roles. An example of a PS for class diagram and a conforming UML class diagram is given in Fig. 2. Roles are identified by preceding them with a vertical bar, "|".

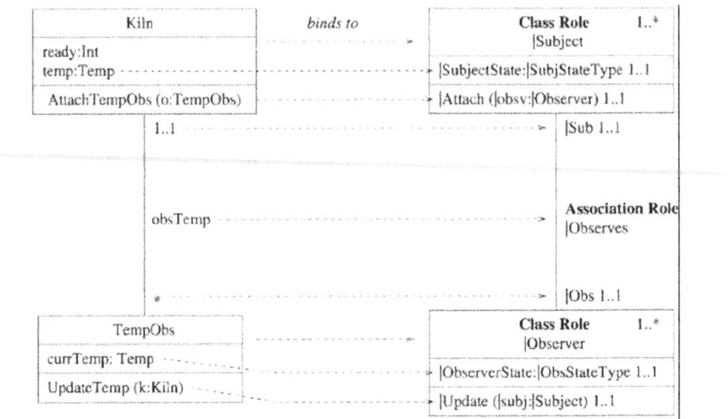


**Fig. 2.** A PS class diagram (right) and a conforming UML class diagram (left). (From [12])

The PS model is a specialization of the metamodel for UML class diagrams but is presented graphically so that it resembles a class diagram not the metamodel. The right-hand side of Fig. 2 is a PS model defining the structure of a general Observer pattern [21]. The left-hand side shows a UML model that conforms to the PS model – *Kiln* plays the role of *|Subject*, *TempObs* plays the role of *|Observer*, *currTemp* plays the role of *|ObserverState*, *AttachTempObs* plays the role of *|Attach*, and so on. Note that the definition of conformance allows other model elements to appear in the conforming model as long as the original role constraints remain satisfied.

In the next section we present a motivation example for the separation of the social patterns concerns in MAS.

## 3 The Crosscutting Nature of the Matchmaker Pattern

Several design patterns for multi-agent systems have been proposed [2, 15, 16] to describe recurring problems and solutions in software design and, therefore, improving software reusability and maintainability. However, current software development paradigms cannot avoid that application core becomes dependent on patterns, decreasing possibilities of reuse [3]. In order to show the crosscutting nature of the Matchmaker pattern, let us consider the implementation of each pattern participant concern (i.e., Client, Matchmaker and Provider concerns) provided by the JADE's [13] API (Application Programming Interface). JADE is a suitable agent platform to support the implementation of MAS. In JADE, a behavior represents a task that an agent can carry out.

Important features that JADE provides are the ability of agents to communicate with each other and the DF (Directory Facilitator) agent, which implements the behavior of the main participant of the Matchmaker pattern [2] i.e., the Matchmaker agent. It provides a Yellow Pages service by means of which an agent (Client) can find other agents providing the services it requires in order to achieve its goals. An agent wishing to publish one or more services (Provider) must provide the DF with a description including its Agent Id (AID) and the list of its services. The services provided by the DF agent (Matchmaker) are usually used by all agents of a MAS implemented in JADE. Thus, all agents who need to register or unregister its services in the Yellow Pages of the DF agent will implement the Provider's concern (Fig. 3 and Fig. 4).

```
DFAgentDescription dfd = new DFAgentDescription();
dfd.setName(genericAgent.getAID());
ServiceDescription sd = new ServiceDescription();
sd.setType(genericAgent.getServiceType());
sd.setName(genericAgent.getServiceName());
dfd.addServices(sd);
try {          DFService.register(genericAgent, dfd);     }
```

**Fig. 3.** Code to register the service in the yellow pages

```
try {          DFService.deregister(genericAgent);     }
```

**Fig. 4.** Code to unregister the service in the yellow pages

Similarly, all agents who need to find a specific provider agent in the Yellow Pages will implement the Client's concern (Fig. 5).

```
DFAgentDescription template = new DFAgentDescription();
ServiceDescription sd = new ServiceDescription();
sd.setType(genericAgent.getServiceType());
template.addServices(sd);
try {       DFAgentDescription[] result =
DFService.search(genericAgent, template);
      for (int i = 0; i < result.length; ++i) {
        genericAgent.getProviders()[i] = result[i].getName();     }
```

**Fig. 5.** Code to search in the yellow pages the agent which provides a service

Generally, the Yellow Pages Provider and Yellow Pages Client concerns (henceforth YPProvider and YPClient concerns, respectively) are going to be implemented by several agents involved in the Matchmaker pattern. Consequently, the YPProvider and YPClient concerns become tangled with the concerns of these agents (i.e., application core functionality) and scattered in the system. To solve this issue, we can use aspect-oriented abstractions both in the design and implementation phases.

Observe that the concern of the Matchmaker participant, i.e. the *Yellow Pages service* concern, is implemented inside the DF agent and its implementation is hidden in the JADE's API. Thus, the Matchmaker participant concern will not crosscut the application functional modules.

Next section presents a notation to describe social patterns using advanced separation of concerns.

# 4 A Standard Technique for Social Pattern Specification

To describe social patterns we propose an approach that specializes the agency metamodel for MAS architectural diagram [10] by using the concept of model roles [11]. In fact, the concept of model roles was used in the Pattern Specifications technique [12] to specialize the UML metamodel for specifying which model elements must participate in a pattern. The work presented in [22] uses the Pattern Specifications (PS) technique [12] to address aspect modeling. Analogously, we can use PSs to promote the separation of social patterns concerns in MAS design.

The following sub-sections describe the template used to specify a social pattern (section 4.1) and the respective structural specification (section 4.2). These are illustrated by the Matchmaker pattern. The communication specification describes a pattern of communications and can be found in [23].

## 4.1 Pattern Template

The template used to describe social patterns (with a subset of GoF's template [21]) was introduced in [17] to illustrate the Matchmaker pattern. Its main attributes are Name, Intent, Applicability, Motivation example, Problem, Solution and Participants.

## 4.2 Structural Agent Pattern Specification

A structural agent pattern specification (SAPS) defines the part of the pattern metamodel that characterizes MAS architectural diagram views of pattern solutions. It defines subtypes of agency metaclasses describing MAS architectural diagram elements (e.g., agency metaclasses AgentRole, AgentConnector) and specifies semantic pattern properties using constraint templates (see Fig. 6). A SAPS consists of a structure of pattern roles [12], where a role specifies properties that a MAS model element must have if it is to be part of a pattern solution model. Formally, a role defines a subtype of an agency metaclass. The metaclass is called the base of the role. For example, a role that has the metaclass AgentConnector as its base specifies a subset of MAS agent connectors. A MAS model element conforms to (or plays) a role if it satisfies the properties defined in the role, that is, the element is an instance of the subtype defined by the role. A role in a SAPS can be a classifier or a relationship role. A role that has the base Classifier or a base that is a subtype of Classifier (e.g., AgentRole, Dependum) is a classifier role. A relationship role is any role that has the base Relationship or a base that is a subtype of Relationship (e.g., AgentConnector).

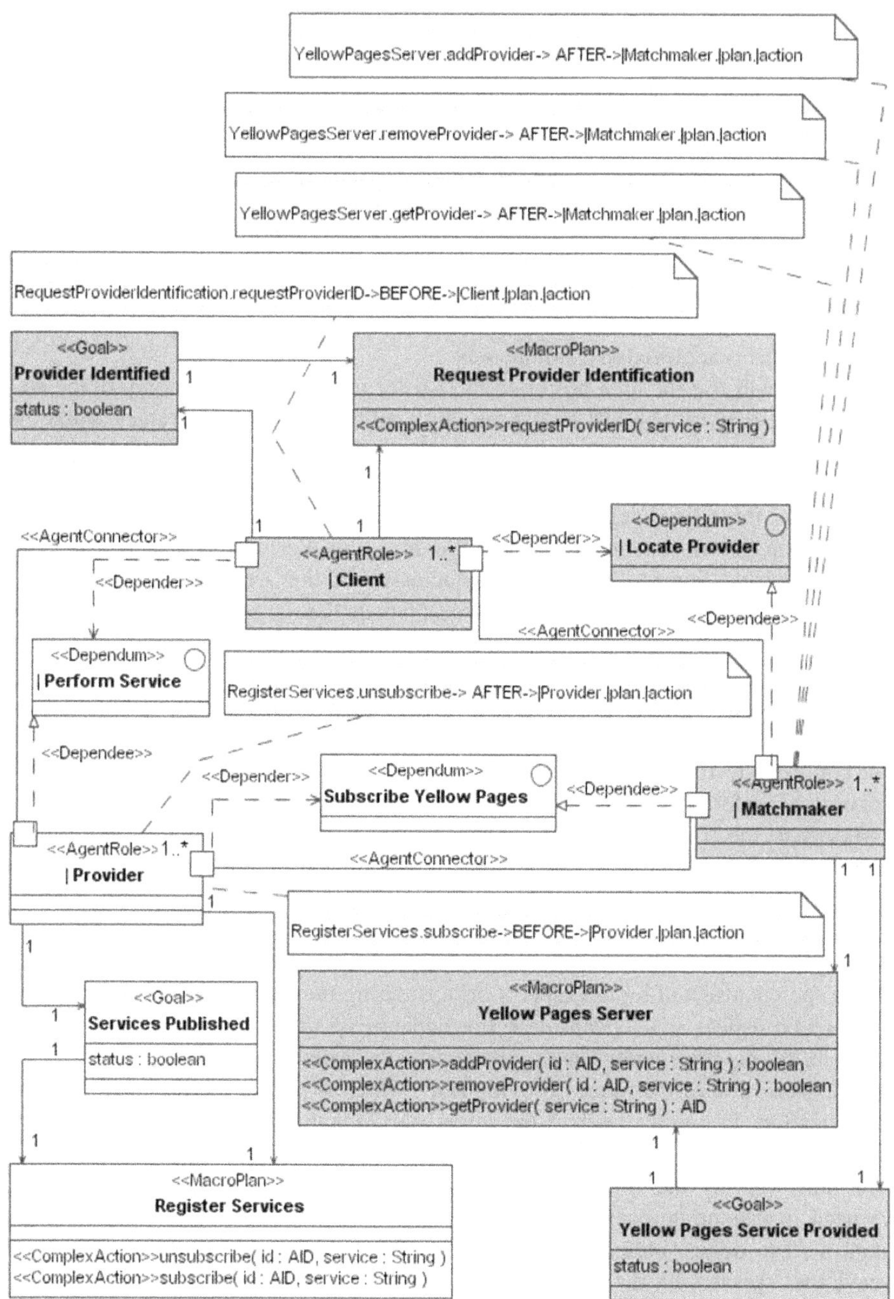


**Fig. 6.** Matchmaker Structural Agent Pattern Specification

Fig. 6 shows a SAPS that specifies solutions for the Matchmaker pattern [2]. This SAPS is described using the client-server architectural pattern (Fig. 1). The Matchmaker pattern has to provide three services (i.e. dependums): *Locate Provider*,

*Perform Service* and *Subscribe Yellow Pages*. For example, the shaded area of Fig. 6 represents the classes involved in the I*Locate Provider* service defined in a Dependum class. When the IClient AgentRole executes the *Request Provider Identification* MacroPlan by performing the requestProviderID ComplexAction to achieve the *Provider Identified* Goal, it triggers a request to the I*Matchmaker* AgentRole to perform the I*Locate Provider* Service. The I*Matchmaker* AgentRole performs the requested service because it does not conflict with the achievement of the *Yellow Pages Service Provided* Goal. A conflict is detected when the service requested to an agent playing a specific agent role, can cause the failure of some of its goals. So, both the requested service and the goal achievement are accomplished by means of the *Yellow Pages Server* MacroPlan. The description of the I*Perform Service* and *Subscribe Yellow Pages* services is achieved in a similar way.

Here we will use the idea proposed in [22] for using model roles only in the elements of the pattern that will vary from one application to another, since when the pattern is applied to a specific application, these roles will be instantiated to model elements of the application. For example, the SAPS in Fig. 6 consists of only five model roles: three for AgentRole metaclass (Client, Provider and Matchmaker) and two for Dependum metaclass (ILocate Provider and IPerform Service). The roles define subtypes (specializations) of metaclasses in the Agency metamodel. For example, the Client role defines a subtype of metaclass AgentRole in the Agency metamodel. The AgentRole roles indicate that conforming architectural diagrams must have at least one AgentRole that conforms to the Client role (as indicated by the 1..* realization multiplicity in the first compartment of the AgentRole role), at least one AgentRole that conforms to the Provider role, and at least one AgentRole that conforms to the Matchmaker role. An AgentRole that conforms to the Matchmaker role (referred to as a Matchmaker class) must have exactly one goal that conforms to the Yellow Pages Service Provided role and exactly one MacroPlan that conforms to the Yellow Pages Server role. The same rationale can be applied to the other roles which are subtypes of metaclass AgentRole in the Agency metamodel.

Moreover, we have the model roles present in the notes which describe the AgentRoles' points affected by the aspects encapsulating the Matchmaker pattern's concerns (each aspect will encapsulate the concern of one pattern participant). For example, let us consider the note attached to the IClient AgentRole. It represents that the execution of the requestProviderID(), which is part of the RequestProviderIdentification, is going to affect the client participant of the Matchmaker pattern before the execution of some of its actions, which is part of some of its plans. To determine who are the client participant in the application and its affected action, we need to instantiate each model role present in the SAPS (e.g., IClient, Iaction, Iplan, etc.) to design elements present in the MAS architectural design. Thus, we weave the aspects with the AgentRoles resulting in the application of the pattern to the system.

In the sequel we propose some mapping guidelines to enable the implementation of social patterns concerns separately from agents concerns through the integration of JADE and AspectJ implementation environments.

### 4.3 Towards Implementation in JADE and AspectJ

To implement our approach we use the integration of JADE and AspectJ environments. AspectJ [4] is a practical aspect-oriented extension to the Java programming language. This section provides some guidelines to map our notation to represent MAS (Fig.1) and social patterns (Fig. 6) in constructs of JADE and AspectJ, respectively. In the affected Architectural Diagram: (i) each «MacroPlan» becomes a Behavior in JADE; (ii) each «ComplexAction» becomes a simple method; (iii) each «AgentRole» becomes an Agent in JADE, since JADE does not support the concept of agent roles played by agents; (iv) the goal element is not mapped to any JADE construct, as JADE does not support the implementation of cognitive agents.

In the SAPS: (i) each «MacroPlan» presented in Fig. 6 becomes an Aspect in AspectJ; (ii) each «ComplexAction» becomes an Advice in the aspect it belongs to; (iii) each «AgentRole» is a module affected by an aspect; (iv) each note attached to an AgentRole becomes a Pointcut and also defines the rule to compose the aspect with the agent; (v) the goal is not mapped to any AspectJ construct.

## 5 An Example

To illustrate our approach, we consider the domain of Newspaper Office introduced in [24]. The e-News system (Fig. 7) enables a user to read news by accessing the newspaper website maintained by a Webmaster AgentRole which is responsible for updating the published information. The information to be published is provided by the Chief Editor AgentRole. The Chief Editor AgentRole depends on the Editor AgentRole to have the news of a specific category. For example, an Editor may be responsible for political news, while another one may be responsible for sports news. Each Editor contacts one or many Photographers-Reporters which can find the news of specific categories (e.g., about sport news). The Chief Editor then edits the Editor' news and forwards them to the Webmaster to publish them.

The following sub-sections describe the architectural diagram (section 5.1), the choice and application of social patterns (section 5.2) and the partial AspectJ code (section 5.3) for the e-News system.

### 5.1 Architectural Diagram

We start by proposing the architectural solution for the e-News problem (Fig. 7) by using the MAS architectural pattern (Fig. 1). The e-News system is composed of four AgentRoles: Editor, Webmaster, Chief Editor and Photographer-Reporter. For example, in Fig. 7 the shaded area corresponds to the interaction between the Editor and Photographer-Reporter to achieve the service *Produce News Article of Specific Subject*. The Editor intends to achieve the *News of Specific Category Edited* goal by means of the *Edit News of Specific Category* MacroPlan. However, to edit the news the Editor has to request the Photographer-Reporter to perform the *Produce News Article of Specific Subject* service. The Photographer-Reporter performs the requested service because it does not conflict with the achievement of the *News Article Produced* goal. Hence, both the requested service and the goal achievement are accomplished by means of the *Contact News Agencies* MacroPlan.

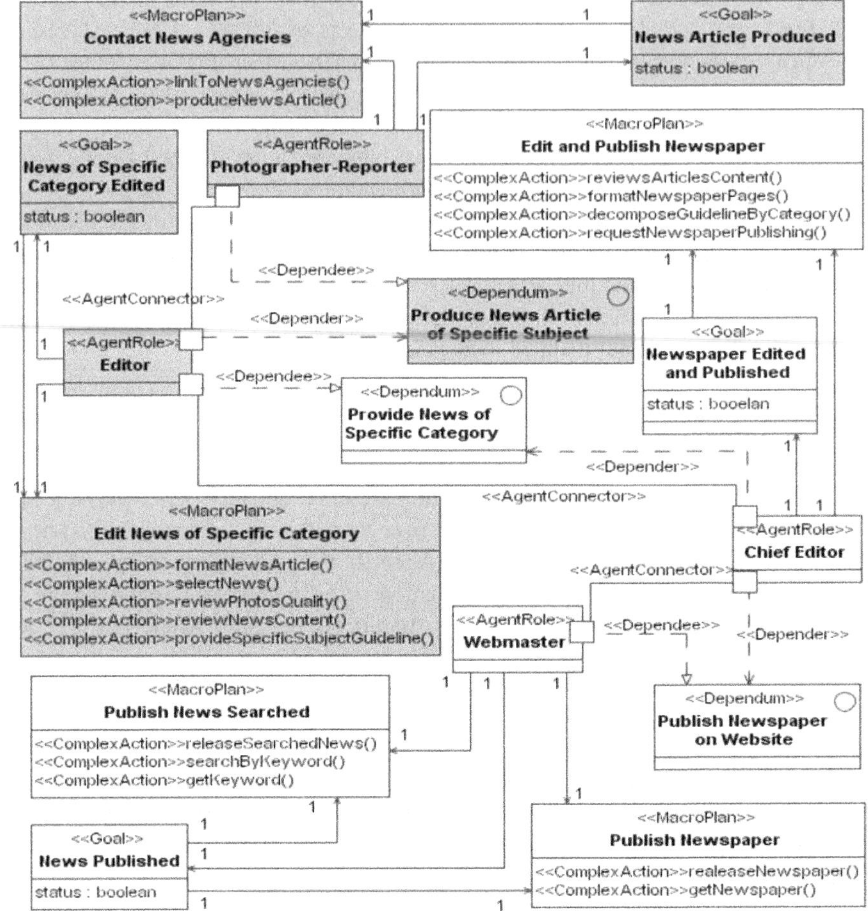


**Fig. 7.** MAS Architectural Diagram for the e-News System

## 5.2 Social Pattern Selection and Application

The last step in the detailed design phase is to select and apply social patterns to refine the architectural design of the e-News system (Fig. 7). One of the key challenges is to choose the proper social pattern to be applied to a system architectural design. One can analyze the template (e.g. [17]) that describes several features of each social pattern to address a specific requirement. For example, the e-News System has the Availability requirement, i.e., the system has to ensure easier recovery of the system if some agent in the system stops running. This requirement could not be shown in the case study because we did not present the requirements models of the e-News System. However, the interested reader can find it in [23].

Analyzing the template of several patterns, we have concluded that the most suitable pattern to address the Availability requirement is the Matchmaker, since it enables the search for another agent to replace the one that has stopped. To apply a social pattern, we need to weave the Matchmaker pattern with the architectural design of e-News system. To achieve this we need to instantiate each model role present in the structure of the pattern (e.g., |Client, |Provider). For example, the following bindings represent the instantiations of the model roles present in the Matchmaker SAPS (Fig. 6) for the e-News system:

1. Bind *|Client* to *Editor*
2. Bind *|Provider* to *Photographer-Reporter*
3. Bind *|Perform Service* to *Produce News Article of Specific Subject*
4. Bind *|Provider|plan.|action* to *Photographer-Reporter.ContactNewsAgencies .getGuideline*
5. Bind *|Client.|plan.|action* to *Editor.EditNewsSpecificCategory. provideSpecific-SubjectGuideline*
6. Bind *|Provider.|plan.|action* to *Photographer-Reporter.ContactNewsAgencies. produceNewsArticle*
7. Bind *|Locate Provider* to *Locate Photographer-Reporter*
8. Bind *|Matchmaker* to *Directory Facilitator*

Observe that the Matchmaker role has been instantiated to Directory Facilitator because we have chosen JADE as the target agent implementation environment.

Instantiating the model roles of the pattern (Fig. 6) for the e-News model elements (Fig. 7), we obtain the pattern applied to the problem (Fig. 8). Thus, we applied the structure of the pattern to the architectural diagram of the e-News System through model roles instantiation (the bindings shown above), which can be easily automated by a CASE tool. Note that the Matchmaker pattern is refining the Editor and Photographer-Reporter AgentRoles by adding the pattern-specific concerns (ports, connectors, dependums, goals, plans). Thus, an agent playing the Editor AgentRole can search for a Photographer-Reporter at run time, as well as an agent playing the Photographer-Reporter AgentRole can publish its services in the DF's Yellow Pages. It ensures the decoupling among agents, because if a Photographer-Reporter contacted by an Editor stops running, for example, the Editor may replace that Photographer-Reporter by requesting to the DF to locate another one in his yellow pages.

### 5.3 The Partial AspectJ Code for the e-News System

In this section we show how the Matchmaker Pattern introduced in Fig. 6 can be codified using the AspectJ environment. To achieve this, we use the mapping guidelines presented in section 4.3. According to these guidelines, the MacroPlans Register Services and Request Provider Identification become aspects in AspectJ. The ComplexActions to subscribe, unsubscribe and requestProviderID become advices in its corresponding aspects. The AgentRoles |Provider and |Client are the agents affected by the aspects, i.e., the Photographer-Reporter and Editor agents. The notes become pointcuts and composition rules in their corresponding aspects.

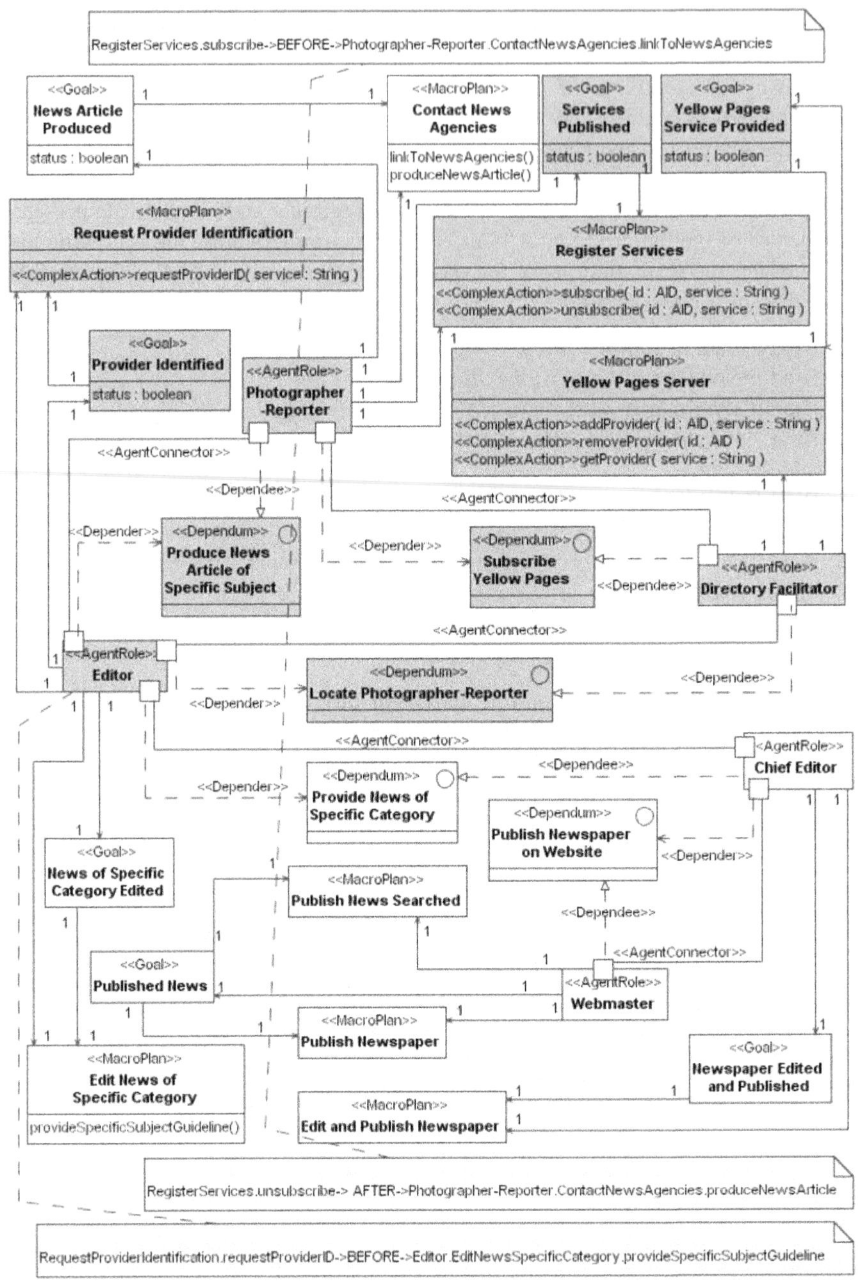


**Fig. 8.** MAS Architectural Diagram weaved with the SAPS

For example, in Fig. 9 we present the RegisterServices aspect weaved with the PhotographerReporter agent in the e-News System. The code after the comments "// Register…" and "// Unregister…" are the ones presented in Fig. 3 and Fig. 4, respectively.

```
public aspect RegisterServices {
     pointcut subscribe() : call (void Photographer-
Reporter.linkToNewsAgencies());
     pointcut unsubscribe() : call (void Photographer-
Reporter.produceNewsArticle ());
     before() : subscribe() {
        Photographer-Reporter genericAgent = (Photographer-
Reporter) thisJoinPoint.getThis();
//Register the PhotographerReporter service in the yellow pages...}
     after() : unsubscribe() {
        Photographer-Reporter genericAgent = (Photographer-
Reporter) thisJoinPoint.getThis();
//Unregister from the yellow pages...}
```

**Fig. 9.** AspectJ Code for RegisterServices Aspect

Fig. 10 presents the RequestProviderIdentification aspect weaved with the Editor agent in the e-News System. The code after the comments "// Update..." is the one presented in Fig. 5. Due to lack of space we could not illustrate the use of the mapping guidelines for the JADE environment, which can be found in [23].

```
public aspect RequestProviderIdentification {
     pointcut requestProviderID() : call (void
Editor.provSpecSubjGuide(..));
     before() : requestProviderID() {
     Editor genericAgent = (Editor) thisJoinPoint.getArgs()[0];
// Update the list of PhotographerReporter agents...)
```

**Fig. 10.** AspectJ Code for RequestProviderIdentification Aspect

# 6 Related Work

Aspect-oriented modeling requires the use of a higher-level aspect model that addresses the aspect-oriented programming concepts at a preliminary design stage (avoiding language specific details), and allows the designer to work at a more abstract level during software construction and evolution. Aspect-Oriented Design Model (AODM) [9] enhances the existing UML specification with aspect-oriented concepts that reproduces the crosscutting characteristics of the AspectJ language. The aSideML [8] is an aspect-oriented modeling language based on the UML that provides notation, semantics and rules for specifying aspects and crosscutting at the design level of OO systems. Composition patterns [25] is an approach that handles crosscutting concerns at design level by means of templates. However, these approaches do not address MAS directly.

It is argued in [26] that the design and implementation of agent internal architecture concerns (e.g. interaction, adaptation, autonomy, knowledge, collaboration, roles, learning and mobility) tend to affect or crosscut many classes of the system design and code, including those representing the basic agent functionality. To address this issue, [26] proposes an aspect-oriented approach to support the separate handling and modularization of MAS specific concerns. The proposed approach encourages the separate handling of agent properties, and provides a disciplined scheme for their composition. In our approach, we are concerned with the separation of social patterns concerns since we consider the agents already built-in (with the agent properties, such as autonomy, interaction, etc.) and supported by some agent implementation platform.

In [17], we have extended the aSideML [8] to incorporate agency features to be used in both the separation of crosscutting concerns in MAS and its latter weaving with the system agents. However, this approach becomes a little complex because we extended the aSideML by using agency features, architectural features and model roles. On the other hand, our current work is much simpler since to model crosscutting concerns of social patterns in MAS, we have created an agency metamodel and specialize it by using model roles.

## 7 Conclusion and Future Work

This work presents an approach to separate crosscutting concerns in MAS. The contribution is threefold: (i) a notation to describe the social patterns concerns separately from MAS functional modules (i.e. agent roles); (ii) a systematic way for weaving the social patterns concerns with the agent roles concerns; and (iii) guidelines to implement a system specified according to our approach using the integration of JADE and AspectJ. In doing so, we envisage several benefits for MAS design and implementation, such as improving concerns modularity, since social patterns concerns are localized into aspects while the application concerns are localized into agent roles. Consequently, this improves software reusability and maintainability.

Future work includes performing an empirical evaluation of the benefits of using aspects in the implementation of social patterns. Firstly, we need to implement MAS using only JADE, then we use the integration of JADE and AspectJ. Finally, we assess the degree of modularity of MAS in each case and compare the results. Moreover, we intent to develop a tool to support both the proposed notation and the generation code for JADE and AspectJ, as well as to apply our approach in other case studies.

## Acknowledgements

This work was supported by several research grants (CNPq Proc. 142248/2004-5 & CAPES/GRICES Proc. 129/05).

## References

1. Giorgini, P., Kolp, M., Mylopoulos, J., Castro, J.: Tropos: A Requirements-Driven Methodology for Agent-Oriented Software. In: Agent-Oriented Methodologies pp. 20-45 (2005)
2. Kolp, M., Do, T., Faulkner, S., Hoang, H.: Introspecting Agent Oriented Design Patterns. In: Advances in Soft. Eng. and Knowledge Engineering, Vol. III. World Publishing (2005)
3. Noda, N., Kishi, T.: Implementing Design Patterns Using Advanced Separation of Concerns. In: OOPSLA'01, Workshop on ASoC in OO Systems, USA (2001)
4. Kiczales, G., Lamping, J., Mendhekar, A., Maeda, C., Lopes, C., Loingtier, J., Irwin, J.: Aspect-Oriented Programming. In: ECOOP'97, Finland, Springer, Heidelberg (1997)
5. Elrad, T., Filman, R., Bader, A.: Aspect-Oriented Programming: Introduction. In: Communications of the ACM 44(10), 29–32 (2001)

6. Odell, J., Parunak, H., Bauer, B.: Extending UML for agents. In: AOIS'00 at the 17th National Conference on AI, Austin, TX, USA, pp. 3 – 17, iCue Publishing (2000)
7. Silva, V., Lucena, C.: From a Conceptual Framework for Agents and Objects to a Multi-Agent System Modeling Language. In: JAAMAS'04, vol. 9(1-2), pp. 145–189. Kluwer, Dordrecht (2004)
8. Chavez, C., Garcia, A., Kulesza, U., Sant'Anna, C., Lucena, C.: Taming Heterogeneous Aspects with Crosscutting Interfaces. In: J. of the Brazilian Computer Society 12, 1 (2006)
9. Stein, D.: An Aspect-Oriented Design Model Based on AspectJ and UML. Master Thesis. University of Essen (2002)
10. Silva, C., Araújo, J., Moreira, A., Castro, J., Tedesco, P., Alencar, F., Ramos, R.: Modeling Multi-Agent Systems using UML. In: SBES'06, Brazil, pp. 81 – 96 (2006)
11. Kim, D., France, R., Ghosh, S., Song, E.: Using Role-Based Modeling. Language (RBML) as Precise Characterizations of Model Families. In: IEEE ICECCS'02, USA (2002)
12. France, F., Kim, D., Ghosh, S., Song, E.: A UML-Based Pattern Specification Technique. In: IEEE Transactions on Software Engineering 30(3), 193–206 (2004)
13. Bellifemine, F., Caire, G., Poggi, A., Rimassa, G.: JADE - A White Paper. In: Special issue on JADE of the TILAB Journal EXP (2003)
14. OMG: Unified Modeling Language (UML): Superstructure. Version 2.0, Available: (2005) www.omg.org/docs/formal/05-07-04.pdf
15. Hayden, S., Carrick, C., Yang, Q.: Architectural design patterns for multiagent coordination. In: Agents'99, Seattle, USA (1999)
16. Woods, S., Barbacci, M.: Architectural Evaluation of Collaborative Agent-Based Systems. Technical Report, CMU/SEI-99-TR-025, Carnegie Mellon University, USA (1999)
17. Silva, C., Castro, J., Araújo, J., Moreira, A., Alencar, F., Ramos, R.: Separation and Modularization of Crosscutting Social Patterns in Detailed Architectural Design. In: CAiSE'06 Forum, Tudor, Luxembourg (2006)
18. Shaw, M., Garlan, D.: Software Architecture: Perspectives on an Emerging Discipline. Prentice-Hall, Englewood Cliffs (1996)
19. Yu, E.: Modelling Strategic Relationships for Process Reengineering. Ph.D Thesis, Department of Computer Science, University of Toronto, Canada (1995)
20. Mylopoulos, J., Kolp, M., Castro, J.: UML for agent-oriented software development: The Tropos proposal. In: UML'01, Toronto, Canada (2001)
21. Gamma, E., Helm, R., Johnson, R., Vlissides, J.: Design Patterns: Elements of Reusable Object-Oriented Software. Addison-Wesley, London (1995)
22. Araújo, J., Whittle, J., Kim, D.: Modeling and Composing Scenario-Based Requirements with Aspects. In: RE'04, Kyoto, Japan (2004)
23. Silva, C.: Agent Pattern Specifications. Technical Report, Available: cin.ufpe.br/~ctlls/APS.pdf
24. Silva, C., Castro, J., Tedesco, P., Silva, I.: Describing Agent-Oriented Design Patterns in Tropos. In: Brazilian Symposium of Software Engineering (SBES'05), Brazil. pp. 10-25 (2005)
25. Clarke, S., Walker, R.: Composition Patterns: An Approach to Designing Reusable Aspects. In: ICSE'01, Toronto, Canada. pp. 12 – 19 (2001)
26. Garcia, A.: From Objects to Agents: An Aspect-Oriented Approach. PhD Thesis, Computer Science Department, PUC-Rio, Rio de Janeiro, Brazil (2004)

# Modeling Business Contexture and Behavior Using Business Artifacts

Rong Liu, Kamal Bhattacharya, and Frederick Y. Wu

IBM T.J. Watson Research Center
19 Skyline Dr. Hawthorne, NY 10532, USA
{rliu, kamalb, fywu}@us.ibm.com

**Abstract.** Traditional process modeling approaches focus on the activities needed to achieve a business goal. However, these approaches often pose obstacles in consolidating processes across an organization because they fail to capture the informational structure pertinent to the business context or *contexture*. In this paper, we discuss business artifact-centered operational modeling. Artifacts capture the contexture of a business and operational models describe how a business goal is achieved by acting upon the business artifact. Business artifacts, such as Purchase Order or Insurance Claim, provide business analysts an additional dimension to model their business. With operational models, they can describe how a business operates by processing business artifacts and adding business value to the artifacts. This approach has been successfully employed in a variety of customer engagements. We summarize our best practices by describing nine operational patterns. Furthermore, we develop a computational model for operational models based on Petri Nets to enable formal analysis and verification thereof.

**Keywords:** Business contexture, business behavior, business artifacts, operational modeling, operational patterns.

## 1 Introduction

Business process modeling is an essential tool for organizations to formalize and reason about how to reach business objectives. A business process model describes actions taken by business (human or system) actors using the resources of an organization to achieve a strategic or operational goal. Business process models convey business intent and serve as the basis of communication amongst a variety of stakeholders in a business, from business management, analysts, process owners, down to system developers. Enterprises of today have often grown through mergers and acquisitions which frequently lead to process redundancies and inconsistencies. Process consolidation efforts when implemented successfully can lead to significant operational improvements and cost savings. In reality, business process consolidation across a large organization is arduous. Part of the problem is cultural, i.e. disagreement over the unified process itself, as the same business process is often implemented in different ways in different organizations. This typically leads to complications in measuring efficiency of business processes and also in setting balanced incentive targets for the process owners.

J. Krogstie, A.L. Opdahl, and G. Sindre (Eds.): CAiSE 2007, LNCS 4495, pp. 324–339, 2007.

One may argue that process consolidation is difficult because different stakeholders employ different process modeling languages, and that transforming from one to another is prone to semantic ambiguities and leads to skewed representations of business intent. We believe, however, that the actual problem of communicating intent using business process models is much more fundamental and independent of modeling semantics. In a variety of client engagements across various industry verticals we have noticed that standardizing the representation of business process models does not necessarily facilitate stakeholder agreement on processes. We find that traditional process models often inhibit consolidation of business operations. Business stakeholders face problems in agreeing on a unified process simply because a business process can be executed in different ways and still achieve the same goal. Agreeing on one process versus another is often a matter of taste.

We propose a different approach to understanding and representing business intent using what we call *business artifacts* (or simply *artifacts*) [6,11,15]. The idea behind business artifacts is the following. Traditional process models like workflows [8] focus on the actions taken to achieve a certain goal (often referred to as "*verb-centric*"). Hence, business stakeholders describe their business by stating "first we do A, then B, then C, and while doing C we also do D." We propose to focus on what is acted upon, thus describing business operations by first identifying business artifacts, the *things* that matter to their business (e.g. Purchase Order, Insurance Claim), and second how these *artifacts* are processed to achieve a certain goal. Business artifacts are so vital to a business that it stops functioning without processing them. Modeling business operations using artifacts is thus a *"noun-centric"* approach. In our engagements we found it relatively easy for stakeholders to agree on business artifacts. This agreement on the artifacts leads more naturally to consolidating business operations across organizational boundaries.

Business processes describe how work is coordinated to achieve operational and strategic business goals. In Hammer's framework of the Seven Dimensions of Work [9] Hammer requires that business process design respect all seven dimensions to successfully drive operational innovation. Based on our noun-centric modeling approach we re-examined Hammer's framework and separated the classification into two parts, dimensions related to information in the work context and dimensions related to the behavioral aspects of work. We refer to these two different sub-spaces as the *contextural space* and *behavioral space*. The contexture of a business is manifested in the business artifacts themselves; the behavior of a business is manifested in all the activities the business performs. In traditional process modeling, the emphasis is on the behavioral space; the contextural aspects are defined as the data attribute inputs and outputs of the work activities. In operational modeling, contextural and behavioral aspects are given equal emphasis; each work task is defined with respect to the business artifact(s) on which the task operates.

Over the past few years we have conducted over a dozen case studies with internal and external clients. A recent case study illustrates the operational approach and the value it demonstrated in consolidating business operations. A major health insurance company was struggling to keep the database of physicians in its provider networks up-to-date. The company has to process large volumes of data coming from physicians, such as requests to become an approved physician, requests to update physician information (e.g. a new address or phone number), and requests to be terminated from networks (e.g. retiring or relocating). Processing these requests frequently requires contacting physicians to ensure the completeness and accuracy of data, and in some cases requires verification of physicians' credentials. These requests are processed at numerous offices across its geographic service areas, and eventually update a centralized provider database.

**Table 1.** Dimensions of Work and Process Modeling Approaches

| Dimensions of Work | Scope |
|---|---|
| What results the work delivers | *Contexture* of business |
| What information the work employs | |
| How thoroughly the work is performed | |
| Who performs the work | *Behavior* of business |
| Where the work is performed (i.e., by which tasks) | |
| When the work is performed (e.g., before or after which tasks) | |
| Whether the work is performed | |

The company had grown by acquisition, and each office had a different method and set of ad hoc tools for handling these requests. The main problem faced by this company was that some requests were taking many months to complete, delaying the processing of claims filed by those physicians. Although operations management had attempted to institute monitoring systems to identify problems, the lack of process consistency led to unintelligible measurements from the various regional offices. Management saw an opportunity to consolidate the processes into one standard set, and thus had asked each regional office to model their provider management processes. The result was a set of drawings that appeared to be very different, although all the representatives agreed that they were doing essentially the same thing. However, none of these models could be accepted as a standard one. For example, offices may have different credentialing requirements. One office requires site visits, but another may need other types of credentials. In addition, these models lacked consistent activity granularity, which complicated matters with respect to identifying fundamental business activities.

We approached the problem by asking the business stakeholders to describe the key business documents used to manage their operations. The stakeholders quickly agreed on four such documents (or artifacts): *Enrollment Request*, *Credentialing Request*, *Update Request*, and *Termination Request*. These four business artifacts are request types that capture all information to manage the on-boarding, updating and terminating of providers. The second step was to identify how these requests are processed. The resulting business operational models describe the lifecycles of these artifacts. These operational models formed an agreeable basis for all stakeholders in all geographies and were further used to implement a business process management system to monitor the performance of request handling.

The remainder of the paper is organized as follows. Section 2 reviews the graphical notation used for operational modeling. Section 3 gives an overview of operational patterns that have been identified by analyzing a wide range of operational models. Operational patterns are a means for modelers to quickly identify a suitable modeling construct for a given business scenario. Section 4 introduces a computational model for business operational modeling based on Petri Nets. Applying Petri net analysis ensures model correctness against several unique correctness criteria. We will describe the automatic transformation from operational models to their Petri net representations. Section 5 compares the operational modeling with other process modeling approaches. Section 6 concludes with a brief description of future work.

## 2 Business Operational Modeling Using Business Artifacts

The goal of operational modeling is to identify business artifacts and describe the lifecycle of artifacts from creation to archiving. A business operational modeling engagement typically consists of two main steps, first, business artifact discovery and second, modeling the lifecycle of the discovered artifacts.

### 2.1 Business Artifact Discovery

As in any business consulting engagements, operational modeling starts with discussions with different business stakeholders to understand the overall business problem and define the modeling scope. Two types of questions are typically asked:

(1) *Scoping:* What are you in the business of producing? What is the outcome of your process?
(2) *Evaluation:* How do you measure that you are doing what you want to do?

The scoping questions aid in understanding the boundaries of the business operations in terms of the actual product produced by the business actors and the input required for successful production. All of this needs to be captured in information terms.

Once the scope is established the evaluation phase will reveal how business stakeholders keep track of their business, i.e. information shared amongst different roles and information recorded within the established scope. Note that the evaluation phase is not about the activities taken to achieve the business goal but about the information managed and maintained to produce the end product. Frequently one can identify physical documents used in the business as candidate business artifacts, such as Purchase Orders, Insurance Claims, Invoices. Sometimes one encounters business stakeholders who have developed spreadsheets, templates, or paper forms as their means of recording relevant information. For example, in a study conducted with Bayer Pharmaceuticals, each lab head uses a specific type of document, which contains a protocol that encodes operational specifications to execute experiments, a placeholder for results and a list that shows the efficacy of a chemical applied against the biological target. We used this document as the basis for designing a business artifact called *experiment record* (or EXP for short). The details of artifact discovery can be found in [6, 15]. Next, we briefly review some fundamental properties of business artifacts and define the semantics for operational modeling.

A business artifact is an *identifiable, self-describing unit-of-information through which business stakeholders add value to the business* [15]. An artifact has an id which identifies itself uniquely within a given enterprise. This uniqueness property has the most important consequence that an artifact cannot be split in two. For example, the experiment record artifact can be worked on by only one role at a time, meaning the unique artifact cannot be split in two.

A business artifact is *self-describing* in the sense that its attributes are so named that its use in a given business domain is apparent. Information contained in the artifact can be listed as name-value pairs. Therefore, artifacts are not business objects, a technical notion from object oriented techniques [15], as artifacts are, in principle, self-describing pure instances without predefined classes. However, during implementation, the information model of an artifact can be modeled using any suitable information modeling approach such as an ER diagram or an XML Schema.

### 2.2 Modeling the Lifecycle of the Discovered Artifacts

Next, we explain the various modeling primitives used in operational models. Their graphical notations are shown in Fig. 1. These primitives mainly mimic physical locations storing artifacts during their lifecycles, as operational models are understood and used by users at the business level.

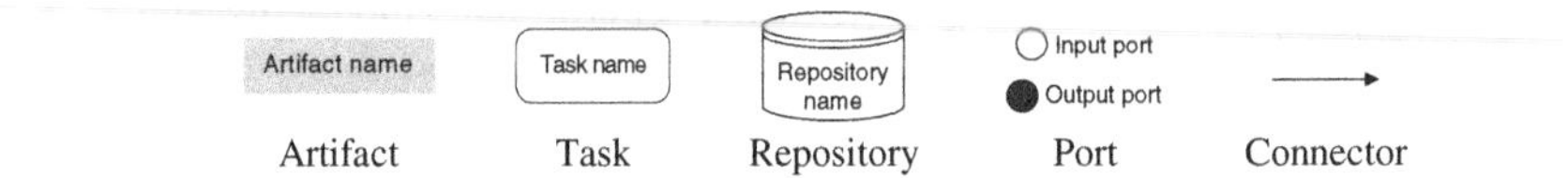


**Fig. 1.** Graphical Notations of Modeling Primitives

A *business task* (or simply *task*) describes the work acting upon an artifact by which *a business role adds measurable business value to this artifact.* We require a task to generate business value and hence, require an update of an artifact. This condition is necessary in our modeling approach and helps in defining the granularity of a task or the task boundaries. Imagine a simple scenario where two tasks, T1 and T2 work on an artifact consisting of ten name-value pairs. A business stakeholder could determine that the completion of T1 will require update of, say attributes 2-5 of the artifact and T2 requires update of attributes 6-10. Therefore, adding business value in this case can be clearly defined by the business stakeholder who thereby determines the boundaries of a task.

Notice that the condition of an artifact update is necessary from a modeling perspective, but not sufficient to truly determine the task boundaries. In the example above, the fact that attributes 2-5 are updated in task T1 is a business decision made by the analyst based on his insight into the work, and may be reflected in a condition that guards completion of T1. The artifact-centered approach does not prescribe this in any way, nor does it support modeling the execution of the task. The main reason to enforce artifact updates in tasks is that artifact-centered modeling is designed for creating accountability of work. Any work conducted should be traceable and hence be accounted for in a chunk of information in one or more artifacts.

*Ports* are the entry and exit points of tasks. We distinguish between input and output ports. A port can be associated with only one artifact type. Ports can have queues attached to them where artifacts wait to be read in or sent out. A port lives in the context of a task and an input port can have a trigger condition that instantiates the task. Usually a task is instantiated by an external (e.g. human) agent or a message.

A *repository* describes a waiting shelf or a buffer for an artifact. Tasks can push an artifact into a repository and pull it out of the repository. A *connector* connects an output port to an input port (task-task) or connects an output port to a repository (task-repository). Task-task connectors carry artifacts or simple messages. We support the use of messages mostly to allow for triggering of tasks by external agents, but do not encourage modeling message flows, as messages are not persistent entities and hence violate the design paradigm for accountability. Task-repository connectors carry artifacts when a task pushes an artifact into a repository. A task can either pull an artifact from a repository or read the content of an artifact in the repository.

# 3 Building Operational Models Using Patterns

## 3.1 Operational Patterns

During our practice with operational modeling, we designed nine operational patterns, which describe most common behaviors of business artifacts. These patterns are not exhaustive, but our customer engagements in the past five years convinced us that they are expressive enough to serve as the basic modeling constructs.

**Pattern 1 (Pipeline Pattern):** In a pipeline pattern, tasks are executed in sequence. An artifact is transported directly from an output port of a task to an input port of another. In this pattern, information processed by all sequential tasks is encapsulated by the same artifact. No new artifacts should be created within or from this pattern. An example of this pattern is shown in Fig. 2. In this example, task T3 "Analyze Results" is triggered right after the receipt of an artifact following the completion of task T2 "Perform Experiment". Each task updates the experiment record artifact (or EXP for short) and thus is considered to be a milestone in this artifact's lifecycle.

**Pattern 2 (Repository Pattern):** In a repository pattern, tasks are in sequence but execution is decoupled. After being processed by a task, an artifact is sent to a repository. The repository can respond to requests for this artifact. An example of this pattern is shown in Fig. 3. The main difference of this pattern from a pipeline pattern is that task T1 "Design Experiment" does not directly trigger the subsequent task T2 "Perform Experiments". Rather, task T2 is triggered asynchronously upon accessing artifacts from the repository for pending experiments.

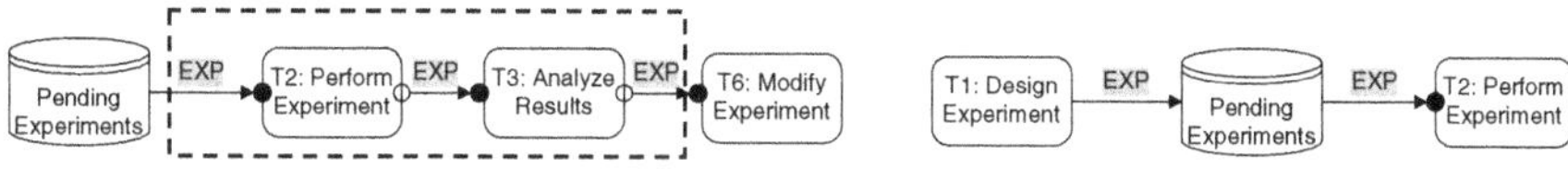


**Fig. 2.** Pipeline Pattern

**Fig. 3.** Repository Pattern

**Pattern 3 (Branch Pattern):** A branch pattern describes more than one option to process an artifact. Fig. 4 shows an example of this pattern. In this example, the results of an experiment are analyzed by task T3 "Analyze Results". Depending on the analysis result, one of the following three tasks can be executed: (1) T4 "Clone Experiment" (i.e., the experiment is repeated to test the reproducibility of the results), (2) task T5 "Update Protocol", and (3) task T6 "Modify Experiment". These options are exclusive and only one can be chosen, as required by the fact that each artifact is a unique entity and thus cannot be split to more than one location at any time.

**Pattern 4 (Convergence Pattern):** In a convergence pattern, a task or a repository can accept an artifact which may arrive from different sources. In general, a convergence pattern always happens together with branch patterns. When branch patterns create multiple possible ways to process an artifact, this artifact can follow different paths and then arrive at a common task through a convergence pattern. An example of this pattern is shown in Fig. 5. In this example, the artifact follows either the sequence: task T3 "Analyze results" → task T6 "Modify Experiment", or another sequence: task T3 "Analyze Results" → task T5 "Update Protocol" → task T6 "Modify Experiment", two exclusive paths to reach task T6.

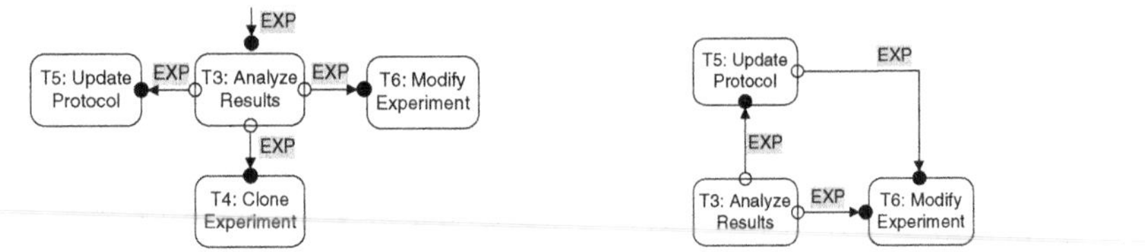


**Fig. 4.** Branch Pattern

**Fig. 5.** Convergence Pattern

**Pattern 5 (Project Pattern):** A project pattern is useful in collaborative scenarios where an artifact is worked on by many role players in an arbitrary order. An example is shown in Fig. 6. In this example, task T1 first creates an experiment and stores it in a repository. Then tasks T2 "Order Raw Material" and T3 "Request Supplies" are done in any order by pulling the artifact from and replacing it into the repository. T2 and T3 can be executed in an arbitrary order but while, e.g. T2 is working on the artifact, T3 has to wait for T2 to release the artifact back into the repository. Task T4 "Start Experiment" can be executed only after both T2 and T3 are completed, which would typically be realized by an appropriate guard condition on the input port of T4.

**Pattern 6 (Creation Pattern):** A creation pattern, as shown in Fig. 7, considers the correlation between different types of artifacts. Through a creation pattern, at least one new artifact is created. In Fig. 7, while an HTS (Candidate High Throughput Screening Protocol) artifact is processed through task T1 "Design Experiment", new EXP artifacts are created. In general, these two types of artifacts are correlated in some way. For example, an EXP artifact can have references to the HTS artifact.

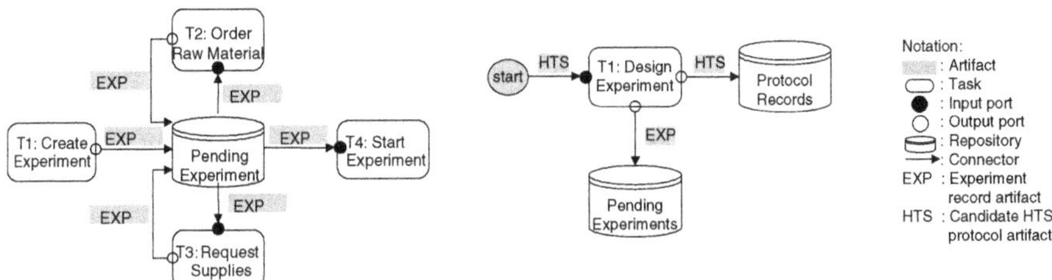


**Fig. 6.** Project Pattern

**Fig. 7.** Creation Pattern

**Pattern 7 (Synchronization Pattern):** A synchronization pattern considers the coordination between different types of artifacts. Through this pattern, a task acts on more than one artifact. The information content of one artifact is updated based on other artifacts. One example is shown in Fig. 8. In this example, the analysis of experiment results may indicate that the HTS protocol needs to be modified. Therefore, completing task T5 "Update protocol" requires two artifacts: an HTS artifact and an EXP artifact. After task T5, these two artifacts are synchronized and the updated HTS artifact is sent back to the repository. If multiple experiments are created, the HTS artifact may synchronize with each of them. Another example is shown in Fig. 9. When changes are made to a service order, a new artifact RFC (Request for Change) is created through task T2. Task T3 "Approve RFC" needs both the service order artifact and the RFC artifact as inputs. After T3, the service order is updated accordingly. In this example, synchronization happens only once.

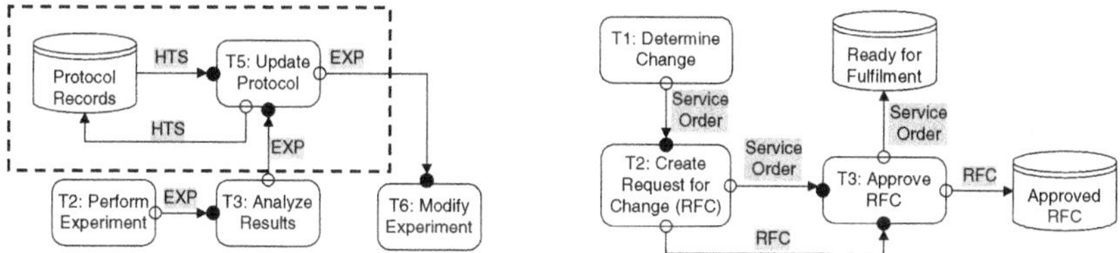


**Fig. 8.** Synchronization Pattern (Example 1) **Fig. 9.** Synchronization Pattern (Example 2)

**Pattern 8 (Rework Pattern):** A rework pattern is, in general, a loop. In this pattern, an artifact circulates in a set of tasks until an exit condition is satisfied. An example of such a pattern is shown Fig. 10. In this figure, after an experiment is performed and analyzed, if the results cannot be confirmed, the experiment needs to be repeated. Therefore, the experiment is cloned, sent to the repository "Pending Experiments", and then performed again.

**Pattern 9 (Disposal Pattern):** In some situations, an artifact may become unnecessary, for instance because of exceptions, and it drops from its lifecycle. An example is shown in Fig. 11. In this scenario, multiple experiments are created and each experiment is performed independently. When desirable results are achieved, all remaining pending experiments are disposed and sent to a repository, say "Disposed Artifacts", because there is no need for them.

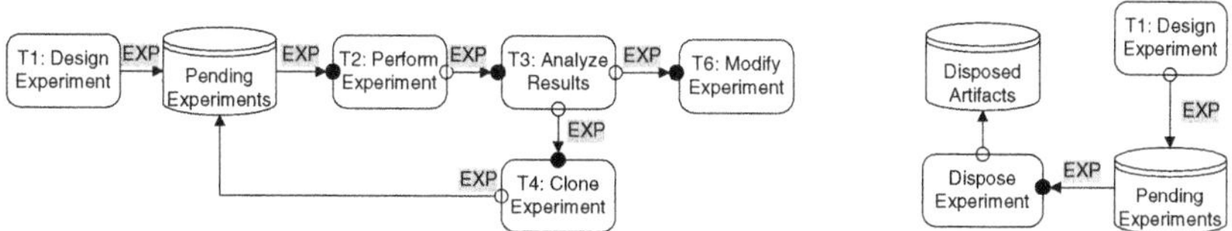


**Fig. 10.** Rework Pattern **Fig. 11.** Disposal Pattern

### 3.2 Putting Patterns Together – An Example

Having given nine operational patterns, next we continue the case study of Bayer pharmaceutical research and show how to use these patterns to build an operational model for industrializing drug discovery processes [6].

A drug discovery process starts with identifying and isolating the biological target– the biological structure associated with a specific disease. A very large number of chemical compounds that have the potential of inhibiting or neutralizing the malignant biological behavior of this target are selected during a procedure called high throughput screening (HTS). Experiments are conducted to further test chemical characteristics (ease of synthesis, solubility, reactivity, etc.) and biological characteristics (selectivity, toxicity, etc.) of these compounds. A HTS protocol gives precise and detailed instructions of performing these experiments. The protocol is evaluated and perhaps updated through a series of experiments. The target of this process is to generate an optimal HTS protocol which has maximum signal strength in the HTS apparatus in order to obtain unambiguous results. Two business artifacts, *candidate HTS protocol* (HTS) and *experiment records* (EXP), are identified.

Next, we describe business operation scenarios which create, process and archive the artifacts. Each scenario can be mapped to one or more operational patterns. Some matching patterns have been shown as examples in the previous section. We give the names of matching patterns at the end of each scenario.

(1) *Design experiment*: A lab head creates a candidate HTS protocol along with experiments. The protocol is stored in a repository and experiments are sent to the pending experiment repository (**Creation Pattern** (see Fig. 7)).

(2) *Perform experiment*: A lab technician performs an experiment from the pending experiment repository. Consecutively, the results are analyzed (**Repository Pattern** (see Fig. 3), **Pipeline Pattern** (see Fig. 2)).

(3) *Analyze results*: The lab technician and the lab head analyze the experiment results to determine one of the following options as the next step: (1) the experiment needs to be cloned and rerun; (2) the experiment needs to be modified and rerun; and (3) the protocol needs to be updated (**Branch Pattern** (see Fig. 4)).

(4) *Update protocol*: If a protocol needs to be updated, the protocol is retrieved from the protocol record repository, synchronized with experiment results, and sent back to this repository. After the update, either the experiment is determined to be complete and stored in a repository or to be modified and rerun (i.e. option (2) of Scenario (3)). Therefore, after the result analysis, the experiment can be modified directly or modified after the protocol update (**Synchronization Pattern** (see Fig. 8), **Branch Pattern, Convergence Pattern** (see Fig. 5)).

(5) *Rerun experiment*: a rerun experiment is first stored in repository "Pending experiments" and then processed as a new one (**Rework Pattern** (see Fig. 10)).

(6) *Prepare candidate protocol*: With experiment results, the lab head evaluates the protocol and archives completed experiments. Later, the lab head prepares to finalize the candidate protocol, requests pre-run, and stores it in a repository called "Candidate protocols". (**Synchronization Pattern**, **Pipeline Pattern**).

(7) *HTS lab*: The HTS lab retrieves the candidate HTS protocol for review. It may return the protocol and suggest further validation. Otherwise, the protocol is finalized and stored in an HTS Protocol repository (**Branch Pattern**).

(8) *Initiate additional experiments*: If further validation is needed, the lab head updates the candidate HTS protocol and creates additional experiments to the pending experiment repository (**Creation Pattern**, **Rework Pattern**).

It is very straightforward to formulate these scenarios after identifying their matching patterns. We can get a complete operational model shown in Fig. 12. Although an operational model is targeted at users at business levels, it also lends itself to formal analysis, verification, and simulation to ensure successful process execution. Next, we describe how to verify an operational model through Petri nets. We start with a brief introduction to Petri nets.

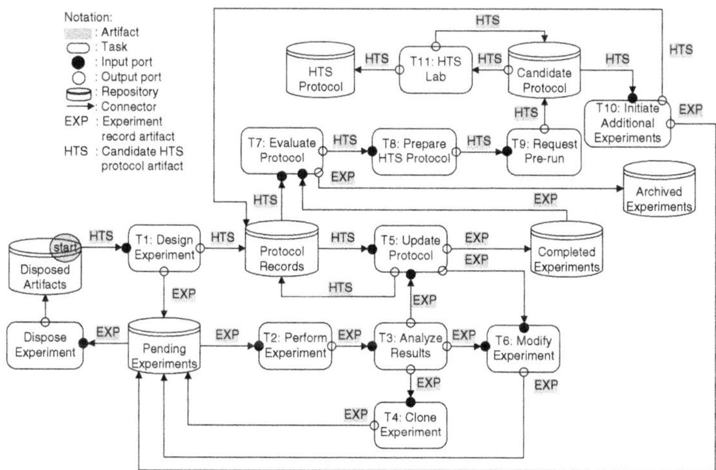


**Fig. 12.** Operational Model of Drug Discovery Process

# 4 Verifying Operational Models Using Petri Nets

## 4.1 Petri Net Preliminaries

Petri nets are a powerful tool for modeling the state transitions of systems in a variety of domains. A Petri net is a directed graph consisting of two kinds of nodes called *places* and *transitions*. In general, places are drawn as circles and transitions as boxes. Directed arcs connect transitions and places either from a transition to a place or from a place to a transition. *Arcs* are labeled with positive integers as their weight (the default weight is 1). Places may contain *tokens*. In Fig. 13(b), place P1 has a token, shown as a small disc. The firing rules of Petri nets are as follows [13]: (1)A transition $t$ is *enabled* if each input place of $t$ contains at least $w(p,t)$ tokens, where $w(p,t)$ is the weight of the arc from $p$ to $t$; and (2) The *firing* of an enabled transition $t$ removes $w(p,t)$ tokens from each input place $p$ of $t$, and adds $w(t,p)$ tokens to each output place $p$ of $t$, where $w(t,p)$ is the weight on the arc from $t$ to $p$.

In classical Petri nets, tokens are indistinguishable. A colored Petri net (CPN) is extended from the classical kind by tagging tokens with data values (i.e. *colors*) [10]. Moreover, in a colored Petri net, each place is associated with a type of data values (i.e., *color set*). For example, in Fig. 13(b), EXP is a color set and each color in this set stands for an experiment record artifact. In addition, each arc is attached with an arc expression specifying the tokens removed or added to a place. In Fig. 13(b), variable "exp" means that a token from color set EXP is required to fire transition T1 and after firing, the same token is put into place P3. The details of CPN can be found in [10,16].

## 4.2 Representing Operational Models as Petri Nets

We can transform operational models into colored Petri nets easily following several rules. First, each artifact type can be represented as a color set. For example, EXP in Fig. 13(b) is a color set for EXP artifacts. Accordingly, each artifact is represented as a token with a unique color in a color set. Second, a repository is transformed into a place tagged with a color set since it stores a particular type of artifacts. Third, each task is transformed into a transition and each of its output ports is represented as a place. Finally, each connector is converted to an arc and its associated artifact becomes a variable as an arc expression. Fig. 13 shows a transformation example. However, there are three exceptions to these general rules as follows.

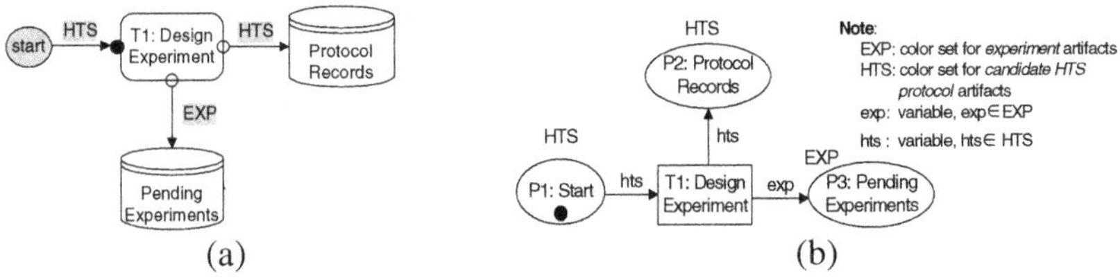


**Fig. 13.** An Operational Model and its Petri Net Representation

(1) *An output port is connected to a repository*. For example, task T1 is connected to the protocol record repository in Fig. 13(a). No Petri net representation for this port is needed. In Fig. 13(b), an arc directly connects transition T1 to place P2, which represents the repository.
(2) *Branch pattern*. The output ports of the branch task should be transformed into only one place, as shown in Fig. 14(b). After transition T5, a token is put into place P9 and it can fire either transition T6 or T12. If an output port is connected to a repository, a dummy transition, for example T12 in Fig. 14(b), is added in between two places.

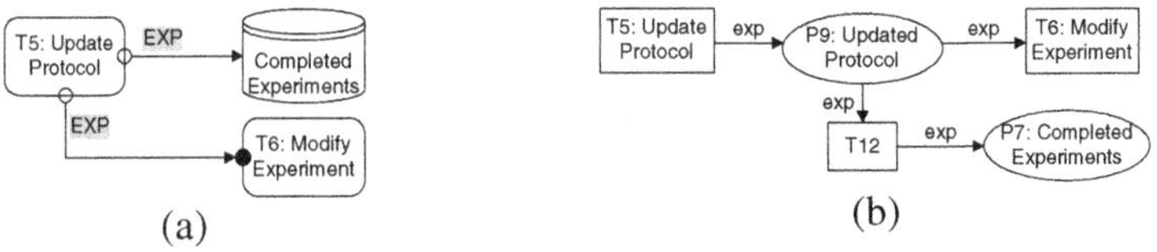


**Fig. 14.** Petri Net Representation of Branch Pattern

(3) *Convergence pattern*. The convergence task, for example T6 in Fig. 15(a), is duplicated so that each of its input ports belongs to one copy. Its Petri net representation is shown in Fig. 15(b).

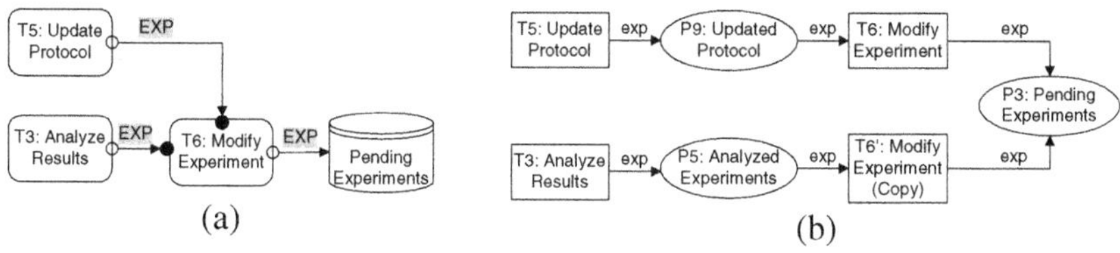


**Fig. 15.** Petri Net Representation of Convergence Pattern

Following these rules, the operational model of Fig. 12 can be transformed to a Petri net shown in Fig. 16. Next, we describe how to analyze and verify operational models using Petri nets.

### 4.3 Operational Model Verification

Operational models emphasize the uniqueness of business artifacts. Therefore, the objective of verification is to ensure the following important properties of artifacts: (1) *Persistence*: once created, an artifact cannot disappear; (2) *No split*: a business artifact can be at only one place at a time; and (3) *Reachability*: an artifact can reach any of its states (i.e. tasks or repositories in the operational model of this artifact).

These properties can be verified using a Petri net reachability graph [13]. A reachability graph shows the development of markings of a Petri net from an initial marking. A *marking* is denoted by a vector $M$, where $M(p)$ denotes the tokens in place $p$. For example, the initial marking of Fig. 13(b) is: $M_0$(P1)=1`"hts1" and $M_0$(P2) = $M_0$(P3) = $M_0$(P4) = 0 as only P1 has a token (denoted as "1`") with a color "hts1".

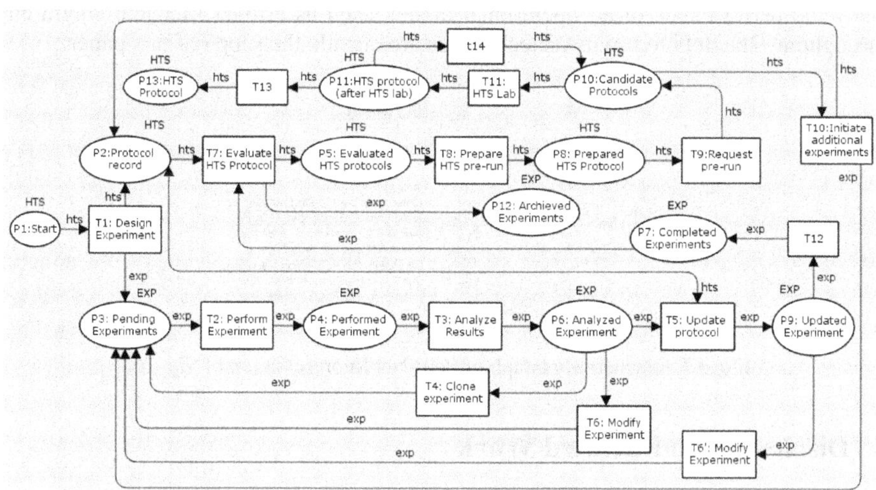


**Fig. 16.** Petri Net Representation of Operational Model in Fig. 12

After transition T1 fires, a new marking, say $M_1$, is generated: $M_1$(P2) = 1`"hts1", $M_1$(P3) =1`"exp1" (i.e., "exp1" is the new experiment created), $M_1$(P1) = $M_1$(P4) = 0.

Since an operational model describes an artifact's lifecycle and all artifacts of a type have exactly the same lifecycle, to verify the above artifact properties, imagine that we put one token of each color set (i.e., one artifact of each type) into its Petri net representation to play the token game. If the operational model is correct, for every marking *M* in the reachability graph of this Petri net, there is no more than one token in any place *p*, i.e., this Petri net is *safe* [13]. Moreover, since an artifact can be at only one state at a time, the maximum number of markings is $\prod n_i$ , where $n_i$ is the number of states of artifact type *i*. Therefore, the verification is very efficient. The artifact properties can be verified using reachability graphs as shown in Table 2.

Therefore, using the algorithm in [13], we can get a reachability graph, shown in Fig. 17, for the Petri net of Fig. 16. Note, the transition from node "P10, P12" to "P2, P3" shows a new experiment is initiated by task T10. The existing experiment has reached its final state and is removed. Minor modifications have been made to this algorithm to accommodate such a situation. Obviously, we can verify that these four artifact properties are guaranteed in this operational model.

**Table 2.** Verifying Artifact Properties Using Reachability Graphs

| Artifact Properties | Properties of Reachability Graph |
|---|---|
| *Persistence* | Any token in a marking must exist in all of its subsequent markings |
| *No split* | Places of a marking cannot have tokens with the same color |
| *Reachability* | Each transition is fireable, and for each place *p*, there exists at least a marking *M* such that \|*M*(*p*)\|=1. |

In addition, Petri nets allow simulation and other formal analyses. Simulation can be done using CPN Tools [16]. Moreover, we can perform theoretical analysis to study the performance of an operational model, such as artifact lifecycle length and throughput. The detailed analysis techniques are outside the scope of this paper.

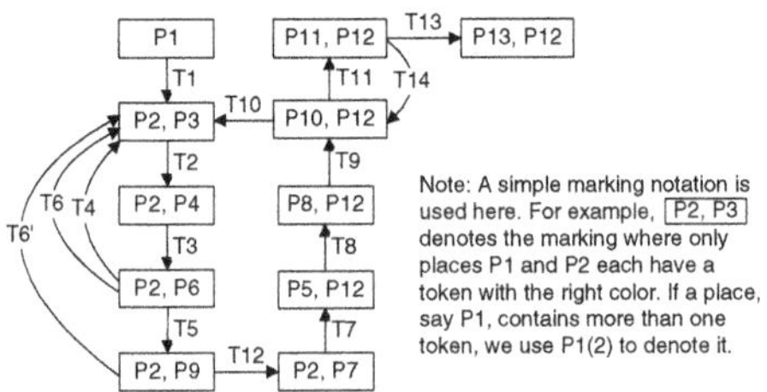


**Fig. 17.** Reachability Graph of Petri Net Representation of Fig. 16

## 5 Discussion and Related Work

Business operational modeling incorporates the contexture of a business as a first-class modeling primitive as manifested in business artifacts. The behavior of a business, described in the context of artifacts, models how business roles process artifacts to produce measurable business results. We found that the operational modeling approach can reduce the complexity of business problems significantly for two reasons. First, the artifact dimension tends to be manageable because there are typically just a few artifacts in any given business process. For example, we analyzed how IBM manages financial contracts, from the signing of a deal through creating and managing the contract. In spite of the complexity of IBM's business, we identified only 7 distinct artifacts. Second, the complexity of business processes is often exacerbated by a lack of guidelines for the granularity of activities. Standard process modeling approaches provide no criteria that prevent business analysts from including execution details. This lack of guidelines also leads to inconsistent granularity of activities, with some very detailed activities and some large chunks of the process represented by a single activity. Business artifacts, on the other hand, provide a context for the scope of a business task, which should be a distinct functional entity that updates one or more artifacts and produces a measurable business result.

We have shown the fundamental difference between operational modeling and traditional activity-centric process modeling, mainly workflow approaches with a focus on control flows [1,5,8], throughout this paper. Besides control-flow based workflows, recently, data-flow driven workflows have attracted increasing attention [12, 17]. The data-flow driven approach concerns the dependencies between data used by activities and derives control flows based on such dependencies. However, often the dependency information is insufficient for the generation of process models [12]. Moreover, it could be difficult to determine the dependencies of a large number of data objects. Operational modeling provides a framework to group data logically into a few unique entities and the modeling complexity is then greatly reduced.

Accordingly, operational patterns are also different from workflow patterns [2], which describe styles of control flows in workflow tools. Operational patterns should be understood as the styles of artifact behavior. We introduced several operational patterns, such as the creation and project patterns, which are unique in the context of operational modeling. Also, it is easy to understand that some workflow patterns such as "parallel split" cannot happen in operational models since an artifact is undividable.

Another related thread of work is the product-driven case handling approach [3], which addresses many concerns of workflows similar to ours especially with respect to the treatment of process context or data. A business artifact and a case are similar in many respects. Case-handling, however, details the structure of the case using data objects that can be managed and updated independently in various activities in the context of the case. Our approach treats business artifacts as unique entities that are updated within each task. To maintain proper granularity of business operations, we do not detail the data objects comprising artifacts and the activities that update these data objects. With significantly reduced complexity, operational models clearly omit execution details which are in the scope of case handling. Another interesting state-flow approach [4] uses process states as a means of driving a process towards its predefined objectives. Although it also emphasizes control over a process with respect to its business intent, this approach lacks an effective formalism for process goals and states. Rather, our approach clearly specify business intents as manifested in business artifacts and process objectives are achieved through business operations, which each make artifacts reach milestones in their lifecycles.

## 6 Conclusion and Future Work

In this paper, we presented the business artifact-centered operational modeling. Artifacts capture the contexture of a business, and operational models describe how a business goal is achieved by acting upon the business artifacts. We also showed how this approach fundamentally differs from traditional activity-centric process modeling.

This approach has been tested by a number of successful customer engagements. We summarized our best practices as nine operational patterns. These patterns can serve as basic constructs for developing operational models. Further, we transform operational models into colored Petri nets and verify the correctness of operational models through Petri net reachability analysis. Using operational models, a company is able to develop business process management IT solutions that are well aligned with its business intent. The MDBT Toolkit [14], which automatically generates IT solutions from operational models has been developed and tested in practice. In addition, a verification and analysis tool based on Petri nets is also under development.

As a future exercise, we plan to explore how operational models enable organizations to develop solutions based on Services-Oriented Architecture (SOA) [7]. Today's enterprises recognize the importance of SOA but struggle with the methodologies to implement SOA solutions. Operational models can provide insights into defining business relevant services at the appropriate level of granularity.

## References


1. van der Aalst, W.M.P.: The application of Petri nets to workflow management. The journal of Circuits, Systems and Computes 7(1), 21–66 (1997)
2. van der Aalst, W.M.P., ter Hofstede, A.H.M., Kiepuszewski, B., Barro, A.P.: Workflow patterns. Distributed and Parallel Databases 14(3), 5–51 (2003)
3. Aalst, W.M.P., Weske, M., Grünbauer, D.: Case handling: a new paradigm for business process support. Data and Knowledge Engineering 53, 129–162 (2005)
4. Andersson, T., Andersson-Ceder, A., Bider, I.: State Flow as a Way of Analyzing Business Processes-Case Studies. Logistics Information Management 15(1), 34–45 (2002)
5. Basu, A., Kumar, A.: Workflow Management Issues in e-Business. Information Systems Research 13(1), 1–14 (2002)
6. Bhattacharya, K., Guttman, R., Lyman, K., Heath III, F.F., Kumaran, S., Nandi, P., Wu, F., Athma, P., Freiberg, C., Johannsen, L., Staudt, A.: A model-driven approach to industrializing discovery processes in pharmaceutical research. IBM Systems Journal, 44(1): 145–162
7. Ferguson, D.F., Stockton, M.L.: Service-oriented architecture: programming model and product architecture. IBM Systems Journal archive 44(4), 753–780 (2005)
8. Georgakopoulos, D., Hornick, M., Sheth, A.: An Overview of Workflow Management From Process Modeling to Workflow Automation Infrastructure. Distributed and Parallel Database 3, 119–153 (1995)
9. Hammer, M.: Deep Change: How Operational Innovation can transform your Company, Havard Business Review, pp. 84-93 (2004)

10. Jensen, K.: Coloured Petri Nets: Basic Concepts, Analysis Methods and Practical Use, vol. 1. Springer, Heidelberg (1996)
11. Kumaran, S.: Model Driven Enterprise, Proceedings of Global Integration Summit 2004, Banff, Canada (2004)
12. Müller, D., Reichert, M.U., Herbst, J.: Flexibility of Data-driven Process Structures. In: Eder, J., Dustdar, S. (eds.) Business Process Management Workshops. LNCS, vol. 4103, pp. 179–190. Springer, Heidelberg (2006)
13. Murata, T.: Petri Nets: Properties, Analysis and Application. In: Proceedings of the Institute of Electrical and Electronics Engineers 77(4), 541–580 (1989)
14. Nandi, P., Kumaran, S.: Adaptive Business Objects - A new Component Model for Business Integration. In: Proceedings of the Seventh International Conference on Enterprise Information Systems (ICEIS 2005), Miami, USA (2005)
15. Nigam, A., Caswell, N.S.: Business artifacts: An approach to operational specification. IBM Systems Journal 42(3), 428–445 (2003)
16. Ratzer, V.A., Wells, L., Lassen, M.H., Laursen, M., Qvortrup, F.J., Stissing, S.M., Westergaard, M., Christensen, S., Jensen, K.: CPN Tools for Editing, Simulating, and Analysing Coloured Petri Nets. In: ICATPN 2003. LNCS, vol. 2679, pp. 450–462. Springer, Heidelberg (2003)
17. Sun, S., Zhao, J.L., Nunamaker, J.: On the Theoretical Foundation for Data Flow Analysis in Workflow Management, Americas Conference on InformationSystems 2005, Omaha, Nebraska, USA (2005)

# Policies and Aspects for the Supervision of BPEL Processes

Luciano Baresi, Sam Guinea, and Pierluigi Plebani

Dipartimento di Elettronica e Informazione – Politecnico di Milano
Piazza Leonardo da Vinci 32, 20133 Milano, Italy
{baresi,guinea,plebani}@elet.polimi.it

**Abstract.** The execution of business processes with BPEL relies on external Web services, which are not necessarily managed by the process owner. This implies the need to constantly verify the correctness of the interactions between the involved parties. This paper proposes a design process model for the definition of supervised processes, in which supervision rules are automatically generated starting from the policies that characterize the external services. These policies exploit WSCoL as a language for describing constraints on the messages exchanged with the business process. In addition, we also present a new version of Dynamo: a prototype of an aspect oriented execution environment that conjugates a BPEL engine and a supervision framework.

## 1 Introduction

The Service Oriented Computing paradigm is driving the development of a new generation of applications. Here, Web services expose their application logic and interoperate relying on XML-based protocols, such as SOAP, and descriptions, such as WSDL. A central node usually coordinates the interactions according to a predefined process specification in which Web services are used to perform the activities. BPEL represents the de-facto standard for specifying such processes. Among other things, BPEL processes specify when a Web service should be invoked and the data that must be exchanged; they do not, however, specify how the interaction occurs: i.e., if security needs to be considered, transactions enforced, or if messaging should be conducted reliabily.

Since the execution of a BPEL process relies on external Web services, not necessarily managed by the process owner, we need to constantly verify the corectness of the interactions among the invoked services. Moreover, if something goes wrong during the process execution, suitable recovery strategies must be performed. To this end, we propose policies as the means to specify how the interaction with external Web services must occur.

BPEL provides specific compensation handlers, but the supplied features —in their current version— are limited. In particular, all compensation activities are performed using a snapshot of the process state, which precludes the modification of "live" variables. Moreover, the decision to perform a compensation is a

J. Krogstie, A.L. Opdahl, and G. Sindre (Eds.): CAiSE 2007, LNCS 4495, pp. 340–354, 2007.

business-guided decision which, in contrast, must be hard-coded at design time and must be suitably implemented within the BPEL process.

At this stage, a number of monitoring approaches [1,2] have been proposed, to realize if a failure occurs during the invocation of the external Web services. Other approaches focus on the recovery problem with the goal of supporting self-healing processes [3]. However, no global solution (i) includes both monitoring and recovery, and (ii) considers design-time specification and run-time management.

The goal of this paper therefore is twofold. On the one hand, we propose a design process model for the definition of supervised BPEL processes. Supervision rules are automatically generated starting from the policies attached to the external Web services and defined by WS-Policy. Both policies and supervision rules exploit *WSCoL* (Web Service Constraint Language): a domain independent language to state monitoring assertions. We also introduce WSReL (Web Service Recovery Language) as the language to define the reaction strategy. On the other hand, we propose a new version of Dynamo (Dynamic Monitoring) [4]: an AOP-based framework for executing supervised BPEL processes, to monitor the execution and enact recovery strategies in case anomalous interactions take place. In this scenario, possible failures arise when: the service is not reachable, the service is down, or the service returns incorrect data. Analogously, possible recovery strategies could be to require a retry, a rollback, or a notification to the process manager.

To better clarify our approach, we introduce a running example. The example takes place in the field of automotive services. The business process, in fact, is intended to be executed on an automobile's onboard device. It provides users with the possibility to integrate a service for searching for parking lots with their navigation system.

When the on-board device is launched, it automatically retrieves —from the navigation system— the coordinates of the destination the user is driving towards, and the coordinates of the current position. We assume that the coordinates are given using the UTM (Universal Transverse Mercator) coordinate system[1]. The system also asks the user to define a maximum radius within which to look for parking lots. Once the system has obtained all the required data, it uses them to call an ActiveBPEL implementation of the process. We assume that the process interacts with a service similar to Microsoft's *Landmark* service to obtain the parking lots the user can choose from. Informally we can state that the Map Point service exposes the following policy: *"The service promises to provide a list of parking lots that are less than $x$ meters from a position indicated using UTM coordinates"*. How this policy is defined service-side in *WSCoL* will be shown in Section 3, while how it is used client-side to define monitoring and recovery will be demonstrated in Section 4.

The paper is structured as follows. Sections 2 discusses the actors and activities required to design a supervised BPEL process. Sections 3 and 4 detail the

[1] `http://en.wikipedia.org/wiki/Universal_Transverse_Mercator_coordinate_system`

approach at the service and the process side, respectively. Section 5 illustrates an aspect-oriented prototype environment.Finally, Section 6 compares our approach with existing ones, and Section 7conludes the paper.

## 2 Design Process Model

When defining a BPEL process, *designers* typically follow a standard design model. First, they search for partner services capable of guaranteeing the functionality and QoS needed to implement their systems. This is typically done by looking in UDDI-based service registries, where designers can find service descriptions that also contain policies that regulate how the interactions with that service have to take place. Common examples are directives regarding security (i.e., authentication, encryption, etc.).

Once the designer has found all the required services, the second step is to define the business logic itself, using the constructs offered by BPEL. In our process design model, the responsibility then passes over to to the *deployer*. This actor's main goal is to provide the business process with a valid descriptor for deployment. This role is usually played by a more tech-savvy person, someone who knows how the BPEL engine will have to be configured to comply with the policies attached to the services chosen by the designer. For example, ActiveBPEL uses its deployment descriptor to specify how the engine should behave when trying to contact certain endpoints, and how the message it intends to send them must be built (e.g., a message might need to be encrypted). The last step consists in deploying the process onto the execution engine.

In this paper we present a new kind of policy assertion, which can be used to define the functional and non-functional behavioral contract the client and the provider will have to comply with. We also present a client-side management framework that can be used in conjunction with a standard BPEL engine to monitor these policies and to react when they are not satisfied.

This leads to some necessary modifications in the standard process design model we just presented (see Figure 1). The first step still consists in searching for appropriate services in a UDDI registry (Step 1). The only difference is that we assume that the service specifications are augmented with our behavioral contracts specified using *WSCoL* (Web Service Constraint Language) [4]. In practice, the language is here used to define constraints on the messages the client and the service will exchange. Similarly to what happens in classical *design by contract* [5], the service provider promises a certain behavior (specified using post-conditions), if the client complies with certain requirements (specified using pre-conditions). If the client does not comply with the pre-conditions then the service will raise an exception. In contrast, if the service does not respect the post-conditions then the client should identify the violation and react properly.

Once a designer has evaluated the behavioral specifications, and chosen the partner services to use, the next step is to design the BPEL process (Step 2). (Step 3) consists in the deployer configuring the execution environment so that it can interact correctly with the service. This is achieved semi-automatically by

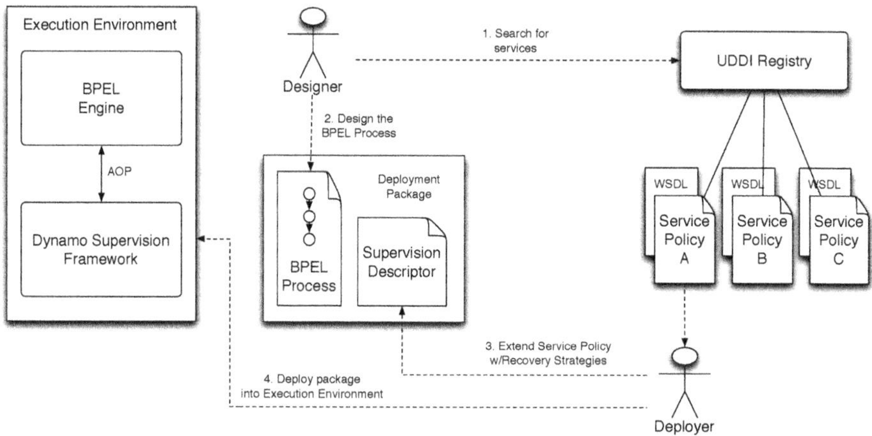


**Fig. 1.** Process Design Model

feeding the execution environment with an appropriate *descriptor*, containing a declarative specification of (1) the policies the system will monitor client-side, and (2) the recovery strategies the system will try to undertake in case of invalid interactions. This can happen for two reasons: it can be the process' fault (i.e., the pre-condition is not verified), or it can be the service's fault (i.e., the post-condition is not verified). The descriptor is created by extending the service-side policy definition. We currently only allow for two kinds of extensions. In the first, the designers can modify the conditions obtained from the service-side policy by strengthening the pre-condition and/or weakening the post-condition. Such modifications are considered acceptable, since they ensure that the partner service will receive a message which is compatible with the requirements it has expressed. In the second kind of extensions, the designer can define appropriate recovery strategies (Step 4). These strategies are performed client-side in case a pre-condition is violated (in this case the actions are performed prior to the interaction with the partner service), or in case a post-condition is violated. At this stage, we are only able to consider a set of recovery strategies that are related to a well-defined set of possible failures.

This approach results in a process that clearly separates the business logic (defined in BPEL) from the supervision activities (derived from the specified policy). This provides for greater flexibility, since both the recovery strategies and the modules enacting them can be customized without affecting the business process.

## 3 Service-Side Policies and WSCoL

According to the BPEL terminology and to the design process model introduced in the previous section, partners are Web services capable of performing one or more activities included in the process. These Web services are usually described by WSDL documents that define the operations, messages, and data-types involved during the invocation are defined. In this work, we aim at extending such a description by considering pre- and post-conditions on the incoming and outgoing messages. WSDL, indeed, cannot define, given a input parameter, which are the admissible values. In the same way, given an output parameter, the service client is not aware of the possible returning values.

For this reason, we assume that —along with the WSDL document— a WS-Policy document is provided. WS-Policy is a machine-readable language for representing the capabilities and requirements of a Web service. According to this specification both the service provider and the service user are able to argue about the behavioral aspects of a Web service. Briefly, a WS-Policy document is a composition of assertions, each of them representing an individual preference, requirement, capability or other property of the Web service. Assertions are organized according to two main operators: `ExactlyOne` and `All`. Given a set of assertions, the `ExactlyOne` operator states that only one assertion must hold at the same time, whereas, with the `All` operator, each assertion must hold.

WS-Policy is a part of the Web Services Policy Framework [6]. This framework also includes WS-PolicyAttachment [7] that specifies how a policy document can be attached to WSDL documents, UDDI entries, and generic XML files. In addition, the framework includes the guidelines for defining domain specific assertions. As listed in [8], some domain-specific assertions are now available to describe capabilities and requirements as security, reliable messaging, transactionality, and more. At this stage, no efforts have been done to describe pre- and post-conditions. For this reason, we aim at proposing such an assertions set, and to use *WSCoL* in a way that complies with the guidelines defined in [6]. These guidelines state that policy assertions representing opt-in, shared, and visible behaviors are useful pieces of metadata. In our case, pre- and post-conditions predicate on the incoming and outgoing messages of the Web services that are partners of our process (the client-side of the interaction). Pre-conditions obligate the service client to send correct data if it aims at obtaining useful results. In the same way, post-conditions make the service client aware of the range of possible values the Web service can send as a valid invocation response. Therefore, pre- and post-conditions defined in *WSCoL* affect the interaction among the parties. This means that *WSCoL* complies with the WS-Policy guidelines.

*WSCoL* has been previously introduced in [9] to express monitoring policies. In this work we aim at including an improved version of *WSCoL* to be used server-side. As discussed in the next section, the *WSCoL* expressions defined server side will inspire the supervision we provide with our client-side framework. *WSCoL* was originally intended for the monitoring of BPEL processes. It defines what should be monitored and how to collect the data required for such monitoring. *WSCoL* uses three different ways of collecting data: *internal variables* are part of the state of the running process, *external variables* are obtained externally by means of specific constructs for getting data from any remote component, which exposes a WSDL interface, and *historical variables* are obtained from previous process executions.

```
<wsp:Policy xml:base="http://www.microsoft.com/policies"
            wsu:Id="MapPointPolicy"
            xmlns:wsp="..."
            xmlns:wsu="...">
   <wsp:All xmlns:wscol="...">
      <wscol:MonitoredItems xmlns:wscol="...">
         <wscol:MonitoredItem type="precondition"
                path="//definitions/message[@name='parkingLotRequest']">
            <wscol:Expression>
               let $zone = "//definitions/message[@name='parkingLotRequest']
                   /part[@name='UTMZone']";
               let $northing = "//definitions/message[@name='parkingLotRequest']
                   /part[@name='UTMNorthing']";
               let $easting = "//definitions/message[@name='parkingLotRequest']
                   /part[@name='UTMEasting']";
               $zone >= 1 && $zone <= 60 &&
               $northing.ends-with("N") &&
               $easting.ends-with("E");
            </wscol:Expression>
         </wscol:MonitoredItem>
         <wscol:MonitoredItem type="postcondition"
                path="//definitions/message[@name='parkingLotResponse']">
            <wscol:Expression>
               let $parkings = "//definitions/message[@name='parkingLotResponse']
                   /part[@name='parking']";
               let $radius = "//definitions/message[@name='parkingLotRequest']
                   /part[@name='radius']";
               (forall $parking in $parkings;
                   ($parking/UTMEasting-$easting)^2 +
                   ($parking/UTMNorthing-$northing)^2 <= $radius^2);
            </wscol:Expression>
         </wscol:MonitoredItem>
      </wscol:MonitoredItems>
   </wsp:All>
</wsp:Policy>
```

**Fig. 2.** Landmark Ws-Policy example

When using *WSCoL* to define service-side policies, however, a few considerations must be made. First of all, instead of predicating on BPEL internal variables, we predicate on the messages being received and sent by the service. Secondly, both external and historical data are only considered at the client-side, to effectively monitor and enforce the constraints defined at the server-side.

To check whether collected data comply with defined constraints, *WSCoL* offers the typical boolean operators, such as `&&` (and), || (or), `!` (not), => (implies), and <=> (if and only if), relational operators, such as <, >, ==, <=, and >=, and mathematical operators such as +, −, ∗, /, and %. The language also supports predicate on sets of values through the use of universal and existential quantifiers, and other constructs, such as `max`, `min`, `avg`, `sum`, and `product`.

Considering our running example, Figure 2 shows the WS-Policy document attached to the Landmark Web service[2]. As pre-condition, the policy requires a valid UTM coordinate. A UTM coordinate is composed of three main information: *zone*, *easting*, and *northing*. Zone is a number from 1 to 60. Northing is a string consisting of six digits and an ending 'N' character. Finally, easting is similar to northing except it ends with an 'E'. The post-condition guarantees

[2] In this paper, for the sake of clarity, we use a simplified WSDL.

that the parking lots found are no more than $x$ meters away from the specified UTM location.

## 4 Client-Side Descriptors and WSReL

As previously stated, the client-side descriptor that instructs the Dynamo Supervision Framework is built as an extension of the service-side policies defined by the service provider. Two possible extensions are possible. In the first, the designer can take the policy defined in *WSCoL* and strengthen the pre-condition or weaken the post-condition. This guarantees run-time conformance to the original policy. In the second, the designer can add client-side recovery strategies to be performed if either the client or the service is not complying with the joint-behavioral contract. Due to lack of space we will not consider the first kind of extension but concentrate on how recovery is defined in *WSReL*.

### 4.1 Recovery Strategies

The recovery strategies in WSReL are based around the definition of a finite (but extensible) set of *Atomic Actions*. These actions are considered the building blocks we want to mix and match to define complex strategies. Our way of intending recovery is that these atomic actions work on a single process instance. They do not have access to the process definition. Therefore, the performed recovery is only valid for the life-span of a single process instance[3]. Moreover, the recovery strategies are performed synchronously (i.e., while the process is momentarily blocked).

The current set of *Atomic Actions* comprises: *ignore*, to simply ignore the anomaly, *notify*, to communicate to a user that something wrong happened, *halt*, to stop the process execution, *retry*, to impose that the system retry to execute the invocation a user-defined number of times, *rebind*, to indicate that the currently used Web service must be substituted with another service. At this stage we assume that the designer of the recovery strategy must indicate the endpoint of an equivalent service. For example, [10,11] discuss approaches for QoS-based Web service selection and Web service substitution with different interfaces are discussed. Additional Atomic Actions are: *changeSupervisionRules*, to modify how supervision is achieved and therefore to relax or tighten some constraints, *changeParams*, to modify the parameters associated with the considered supervision rule, *changeProcessParams*, to modify the parameters associated with the executing process, *call*, to call an external Web service, and *processCallback*, to directly invoke one of the event handlers supplied by the BPEL process.

Complex *Recovery Strategies* are not direct aggregations of atomic actions. Instead they are defined as multi-step processes, in which each step (i.e., each

[3] This allows us to have different client-side monitoring and recovery specifications for different stake-holders. However, it could be interesting to investigate recovery strategies, defined by the process provider itself, that have access to the process definition.

*Recovery Step*) attempts to fix the problem before giving up and passing on to the next. If a step is unsuccessful, it is rollbacked so that the next step can be attempted. If a step is successful, the system skips the others. A single step is defined as a conjunction of atomic actions that have to be executed. The way the system knows if a step was successful depends on the actions it contains. Some of the actions, in fact, require monitoring to be re-performed, while others are always successful (e.g., the *ignore* action). Another thing to keep in mind is that these actions have the power to modify the set of monitoring data being used, and the monitoring and recovery specifications themselves. Therefore, every time a step terminates unsuccessfully, all values are reverted to the original situation, as if no recovery had been attempted. This way, we deal with possible severe situations where the failure is not cause by the process to be monitored. On the other hand, the error might come by other software, such as operating systems, application servers, or BPEL engines.

WSReL allows more than one recovery strategy to be defined for a given erroneous situation. In fact, each strategy is always accompanied by a condition expressed in WSCoL and specified by the designer. Strategies also have an implicit prioritization, given by the order in which they are defined. The first strategy that can be applied is executed and the others are ignored. This is a simple way to avoid problems with multiple strategies enabled at the same time and to relax the constraint that *conditions* must be mutually exclusive.

### 4.2 Example Descriptor

Figure 3 illustrates an example of a client-side monitoring and recovery descriptor. It illustrates what is defined client-side for the server-side post-condition. First of all, the supervision rule contains the same WSCoL post-condition included in the server-side policy. This defines what the client-side framework will look out for.In case the post-condition fails, the strategies included in the descriptor are considered. Specifically, each strategy has a `<strategyCondition>` expressed in WSCoL. If the related expression holds then the specified `<step>`s are successively performed, until one of them results in an effective recovery. The `number` attribute indicates the order in which the steps are performed. In the example, there are two recovery strategies. The first, which is performed when the request is considered urgent by the user, consists of three recovery steps. The framework first tries to re-invoke the service, and then tries to dynamically bind to an equivalent service (i.e., to a service with the same WSDL interface). If neither is successful, its last resort is to notify the process provider via e-mail and halt the process execution. With the `<defaultstrategy>` we define a recovery strategy to be performed when all previous conditions do not hold. In this case, this strategy consists of only one step, which is to immediately notify the problem to the process provider and halt the execution. For the sake of simplicity, the strategy conditiosn are reported informally. Corresponding WSCoL expressions predicate on user context variables, which are external variables, stating the urgency of the request.

```
<wssup:SupervisionRule>
   <wssup:postcondition>
     <wscol:Expression id="postcond_1">
        let $parkings = "//definitions/message[@name='parkingLotResponse']
            /part[@name='parking']";
        let $radius = "//definitions/message[@name='parkingLotRequest']
            /part[@name='radius']";
        (forall $parking in $parkings;
            ($parking/UTMEasting-$easting)^2 +
            ($parking/UTMNorthing-$northing)^2 <= $radius^2);
      </wscol:Expression>
   </wssup:postcondition>
   <wssup:strategy>
      <wssup:strategycondition id="strategycond_1">
         <wscol:Expression>``Urgent request''</wscol:Expression>
      </wssup:strategycondition>
      <wssup:step number="1">
        <wssup:retry times="1"/>
      </wssup:step>
      <wssup:step number="2">
        <wssup:rebind url="http://..."/>
      </wssup:step>
      <wssup:step number="3">
        <wssup:notify>
            <wssup:message>...</wssup:message>
            <wssup:address>...</wssup:address>
         </wssup:notify>
         <wssup:halt/>
      </wssup:step>
   </wssup:strategy>
   <wssup:defaultstrategy>
      <wssup:step number="1">
        <wssup:notify>
            <wssup:message>...</wssup:message>
            <wssup:address>...</wssup:address>
         </wssup:notify>
         <wssup:halt/>
      </wssup:step>
   </wssup:defaultstrategy>
</wssup:MonitoringRule>
```

**Fig. 3.** WSReL example

## 5 Prototype

The prototype implementation we present in this section is based on AOP techniques. Its main goals are to provide BPEL process providers with the tools they need to deploy and manage processes that are aware of the kind of supervision rules presented in Section 4.

In this solution, business logic and supervision policy enforcement are defined and treated separately, since we advocate that *separation of concerns* facilitates both the design itself and later management. In this architecture, we augment —using AOP technology [12] (i.e., AspectJ [13])— a standard BPEL engine (i.e., ActiveBPEL) with notions on how to verify monitoring expressions and how to perform recovery. Business processes go unmodified and are deployed as usual, while supervision policies are deployed to a persistent component where they await activation.

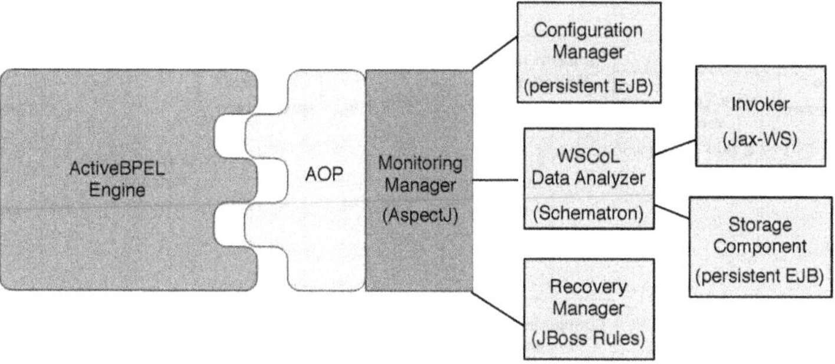


**Fig. 4.** The Architecture of the Dynamo Prototype

Figure 4 illustrates the overall design of the prototype. It is made up of seven main components.

1) The *ActiveBPEL Engine* is the BPEL engine we have chosen for our prototype, due to the fact that it is currently the most mature open-source engine available. Its implementation revolves around the run-time visit (using the *Visitor* pattern [14]) and management of an internal tree-based representation of the process. A thorough study of the platform led us to define our pointcuts[4] as (1) after the engine visits a *Receive* node, (2) before and after it visits an *Invoke* node, and (3) before and after it visits a *Pick* node. These were chosen since they represent the points in which the process interacts with the outside world.
2) The *Monitoring Manager* represents the main *advice*, that is to say the component that is weaved into the execution environment. The result is that —after the weaving— this component has direct access to the internal representation of the process in execution, and to its state (i.e., the set of instantiated BPEL variables). This allows it to collect data from the process itself, and provide them for analysis. This component is also responsible for managing all the steps in the monitoring process. We will defer a more in depth analysis of its behavior to Section 5.1.
3) The *Configuration Manager* is a persistent component in which we store all the supervision descriptions that have been devised, and that are waiting to be activated. The *Monitoring Manager* can query its contents by specifying the process it is executing, the unique id of the user of the business process, and the *Receive, Invoke*, or *Pick* activity being executed. As the reader can see, these allow the system to distinguish between different supervision policies for different users, and to guarantee personalized supervision. 4) The *WSCoL Data Analyzer* is the component responsible for actually verifying the monitoring expressions. The component takes the data collected from within the process, and the monitoring rules extracted from the *Configuration Manager*, and provides a monitoring result. If it needs extra data to perform its analysis (e.g., external or

[4] This term indicates —in the standard AOP terminology— the points in which we are interested in inserting our cross-cutting concern. In our case, it indicates the points in which we want to activate supervision.

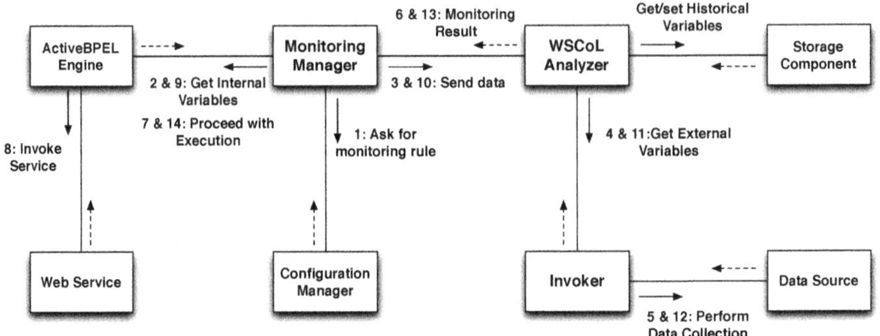


**Fig. 5.** Monitoring pre- and post-conditions

historical WSCoL variables), it can interact directly with the *Invoker* component to obtain data from external data sources, or with the *Storage Component* to obtain data pertaining to previous supervision activities.
5) The *Recovery Manager* is responsible for the execution of recovery strategies when monitoring has signaled an anomaly. It is based on the ECA rules paradigm [15] (i.e., *event-condition-action*), and was built using JBoss Rules [16] (formerly known as Drools). The *event* is implicit and consists in the anomaly itself being signaled. The *condition* consists of a two-level nesting of *if-then-else* clauses that allow the system to distinguish between different actions depending on the extent to which a monitoring expression is unverified. Both the clauses are expressed in WSCoL. The former reply the pre- or post-condition included in the service-side policy. The latter allows us to distinguish aong different reaction strategies. Finally, the *action* is a recovery strategy, as defined in Section 4.
6 and 7) The *Invoker* and the *Storage Component* are utility modules. The former allows to dynamically bind and invoke any Web service. The latter is used to store data collected during previous activations of the supervision framework.

### 5.1 Monitoring Manager

Figure 5 illustrates how all the aforementioned components come together to provide supervision. When the process execution is intercepted and the *Monitoring Manager* is activated, the first thing it does is to obtain the *processID*, the *userID*, and the *invokeID* needed to query the *Configuration Manager.* All these data are automatically provided by the execution engine, except for the *userID* which is provided by the user in the SOAP message that instantiates the process[5]. In our current implementation, the userID must be provided by the user when a new instance of the process is requested. However, in the future, the ID could be provided automatically by the system through authentication.

[5] This is the only modification that needs to be performed on the process definition to enable supervision.

The *Configuration Manager* is then queried for supervision rules that have been defined by that user, for that process, and in particular for that activity (Step 1). If none are found the execution is immediately returned to the engine.

However, if a rule exists, and it defines a pre-condition, the *Monitoring Manager* reads the monitoring expression to see what WSCoL internal variables have been defined, and need to be collected (Step 2). This is simplified by the fact that, thanks to the AOP weaving, the *Monitoring Manager* lives in the same execution space where the ActivbeBPEL variables are stored. Once all the data collection has been achieved, the data is formatted into XML and sent to the *WSCoL Analyzer* together with the monitoring expression itself (Step 3). The *WSCoL Analyzer* proceeds to finish data collection (external and historical variables) and perform verification. Once it has finished, the monitoring result is returned to the *Monitoring Manager* (Steps 4, 5, 6). At this point, if no error has been discovered the framework returns control to the execution engine which performs the service invocation (Steps 7, 8). When the supervision framework is re-activated, after completing the execution of the *Invoke* activity, it checks whether there is a post-condition. Its data collection and analysis are performed in the same way as for the pre-condition (Steps 9, 10, 11, 12, 13).

### 5.2 Recovery Manager

The current version of the recovery manager has been implemented using the JBossRules ECA rule engine. We use data collected during monitoring, and the monitoring results themselves, to produce JBoss rules that will "fire" according to the WSCoL strategy conditions in WSReL.

In order to guarantee the correct activation order for the recovery strategies, defined explicitly by the WSCoL conditions associated with the single strategies and implicitly by prioritization, we make use of: (1) the concept of *salience*, an integer value that gives a rule a certain priority (higher priority rules are executed before lower priority ones, while rules with the same priority are executed in a LIFO manner), (2) the concept of *activation-group*, a value which groups rules into sets in which only one rule can be activated, and (3) the concept of *agenda-group*, which allows the engine to discriminate between rule sets, and to execute only those actions that belong to the agenda-group that is said to be in "focus" (this can be set programmatically).

In our example we have two recovery strategies (see Figure 6). They are defined with the same agenda-group, meaning that the rule engine will try to activate them at the same time. However, they have two different salience values, meaning that `strategy_1` will be considered first. An activation-group is made explicit since the two strategies are mutually exclusive. If the first fails to "fix" the problem, then the second is activated.

Finally, the `recovery manager` performs the single recovery strategies by invoking a Java application (`recov_strategy_1()` and `recov_strategy_2()`) which contains the strategy steps and their atomic actions. In addition, the `Monitoring Manager` is also responsible for re-evaluating monitoring to see if

```
rule "Strategy_1"
   salience 2;
   agenda-group="postcond_1"
   activation-group="ag1"
   when
       strategycond_1
   then
       recov_strategy_1();
end
```

```
rule "Strategy_2"
   salience 1;
   agenda-group="postcond_1"
   activation-group="ag1"
   when
       strategycond_2
   then
       recov_strategy_2();
end
```

**Fig. 6.** JBoss rules example

recovery was successful. Monitoring being re-evaluated also translates into a complete cleansing of the JBoss Rules working space.

## 6 Related Work

Much work has been accomplished in the field of the specification and monitoring of service level agreements for Web services. Keller and Ludwig [17] advocate the need for a framework that can provide tools for the specification, measurement, and monitoring of QoS parameters. Ludwig et al. also present a revisitation of their work in [1], in which they adopt WS-Agreement [18] as their agreement language. They propose Cremona (Creation and Monitoring of Agreements) as an architecture that can facilitate the design and management of agreements through the use of templates. The architecture is mainly composed of two parts: an Agreement Protocol Role Management component, intended to help create and access agreements at run-time, and an agreement Service Role Management component, required to trigger agreement-driven provisioning of a service and to monitor their compliance. With respect to Cremona, which concentrates on QoS, our approach can be used to define more general properties. This guarantees a more widespread solution which can be adapted to many different needs.

Spanoudakis and Mahbub [2] have also developed a framework for monitoring requirements of BPEL-based service compositions. Their approach uses event-calculus for specifying the requirements that must be monitored. Requirements can be behavioral properties of the coordination process or assumptions about the atomic or joint behavior of the deployed services. The system observes system events during execution, and stores them in a database. Run-time checking is then interpreted as integrity constraint checking in a temporal deductive databases. Like our approach, they also provide reactive monitoring since erroneous situations can be found only after they have occurred. It is a less intrusive approach that proceeds in parallel to the execution of the business process. This leads to a lesser impact on performance but also to a lesser responsiveness in discovering run-time erroneous situations.

In our work, pre- and post-conditions are expressed using WSCoL since it provides compatibility with the rest of the proposed solution. Nevertheless, the policies can include conditions expressed according to different languages such as OCL (Object Constraint Language) or specific logics (e.g. temporal or descriptive). Even if some work, such as OWL-S [19] and WSDL-S [20], include pre- and post-conditions directly into the functional specification, we prefer to exploit WS-Policy. This way, we separate the technical details of the invocation from the constraints on exchanged data.

## 7 Conclusions and Future Work

In this paper we have presented an approach to supervise BPEL processes by exploiting policies and aspects. Policies are involved at design-time, when the process owner selects the external Web services to be invoked during the process execution. WS-Policy has been adopted as the language for expressing the behavioral aspects of the external Web services in term of pre- and post-conditions. In particular, WSCoL inspires a new set of assertions compliant with the WS-Policy framework. Policies also drive the process deployer during the configuration of the BPEL process. Using a WSReL descriptor the process is instructed to check the pre- and post-conditions during the service invocation and to properly react in case of violation. The descriptor is semi-automatically generated by starting from the policies attached to the external services. Finally, we present an AOP-based prototype, which is responsible for the execution of the business logic, and for its monitoring and recovery.

## References


1. Ludwig, H., Dan, A., Kearney, R.: Cremona: an architecture and library for creation and monitoring of ws-agreements. In: Proceedings of the 2nd International Conference on Service Oriented Computing, pp. 65–74. ACM, New York (2004)
2. Mahbub, K., Spanoudakis, G.: A framework for requirents monitoring of service based systems. In: Proceedings of the 2nd International Conference on Service Oriented Computing, pp. 84–93. ACM, New York (2004)
3. Modafferi, S., Mussi, E., Pernici, B.: SH-BPEL: a self-healing plug-in for Ws-BPEL engines. In: 1st workshop on Middleware for Service Oriented Computing (MW4SOC '06), Melbourne, Australia pp. 48–53 (2006)
4. Baresi, L., Guinea, S.: Towards dynamic monitoring of ws-bpel processes. In: Benatallah, B., Casati, F., Traverso, P. (eds.) ICSOC 2005. LNCS, vol. 3826, pp. 269–282. Springer, Heidelberg (2005)
5. Meyer, B.: Applying design by contract. IEEE Computer 25(10), 40–51 (1992)
6. Vedamuthu, A., Orchard, D., Hondo, M., Boubez, T., Yendluri, P.: Web Services Policy 1.5 - Primer (2006) http://www.w3.org/TR/2006/WD-ws-policy-primer-20061018
7. Sharp, C. (ed.): Web Services Policy 1.2 - Attachment (WS-PolicyAttachment) (2006) http://www.w3.org/Submission/WS-PolicyAttachment/
8. VV.AA.: Web Service Policy Framework (2006) http://www-128.ibm.com/developerworks/library/specification/ws-polfram/
9. Baresi, L., Guinea, S., Plebani, P.: WS-Policy for Service Monitoring. In: Bussler, C., Shan, M.-C. (eds.) TES 2005. LNCS, vol. 3811, pp. 72–83. Springer, Heidelberg (2005)

10. Antonellis, V.D., Melchiori, M., Santis, L.D., Mecella, M., Mussi, E., Pernici, B., Plebani, P.: A layered architecture for flexible web service invocation. Softw. Pract. Exper. 36(2), 191–223 (2006)
11. Fugini, M., Plebani, P., Ramoni, F.: A user driven policy selection model. In: Dan, A., Lamersdorf, W. (eds.) ICSOC 2006. LNCS, vol. 4294, pp. 427–433. Springer, Heidelberg (2006)
12. Kiczales, G., Lamping, J., Mendhekar, A., Maeda, C., Lopes, C.V., Loingtier, J., Irwin, J.: Aspect-oriented programming. In: ECOOP. pp. 220–242 (1997)
13. Kiczales, G., Hilsdale, E., Hugunin, J., Kersten, M., Palm, J., Griswold, W.G.: An overview of AspectJ. In: Knudsen, J.L. (ed.) ECOOP 2001. LNCS, vol. 2072, pp. 327–353. Springer, Heidelberg (2001)
14. Gamma, E., Helm, R., Johnson, R., Vlissides, J.: Design patterns: elements of reusable object-oriented software. Addison-Wesley Longman Publishing Co., Boston (1995)
15. McCarthy, D., Dayal, U.: The architecture of an active database management system. In: Proceedings of the 1989 ACM SIGMOD international conference on Management of data pp. 215–224 (1989)
16. Proctor, M., Neale, M., Lin, P., Frandsen, M.: Drools documentation. Technical report, JBoss.org (2006)
17. Keller, A., Ludwig, H.: The WSLA Framework: Specifying and Monitoring Service Level Agreements for Web Services. Journal of Network and Systems Management 11(1), 57–81 (2003)
18. Andrieux, A., Czajkowski, K., Dan, A., Keahey, K., Ludwig, H., Pruyne, J., Rofrano, J., Tuecke, S., Xu, M.: Web Services Agreement Specification (WS-Agreement). Global Grid Forum GRAAP-WG, Draft (August 2004)
19. Martin, D. (ed.): OWL-S: Semantic Markup for Web Services. W3C Submission (2004) http://www.w3.org/Submission/2004/SUBM-OWL-S-20041122/
20. Akkiraju, R., Farrell, J., Miller, J., Nagarajan, M., Schmidt, M.T., Shet, A., Verma, K.: Semantic Annotations for WSDL (2005) http://www.w3.org/Submission/WSDL-S/

# Goal Annotation of Process Models for Semantic Enrichment of Process Knowledge

Yun Lin and Arne Sølvberg

Dept. of Computer and Information System, Norwegian Univ. of Sci. &Tech.
Sem Sælands vei 7-9, NO-7491, Trondheim, Norway
{yunl,asolvber}@idi.ntnu.no

**Abstract.** A semantic annotation framework has been proposed to tackle the semantic heterogeneity problem of distributed process models in our earlier work. The goal annotation as part of the framework is further developed, in which goal ontology is annotated to process models to indicate the objectives or the capability of models. In the paper, we introduce a way to represent goal ontology, build relationships between goals and process models, and develop a goal annotation approach to process models. As an illustration, a case study is deployed with the proposed annotation approach. The results of the goal annotation enrich the semantics of process knowledge from stakeholders perspective in a cooperative goal-oriented manner. The ontology and the annotation results also facilitate the ontology-based queries for the semantic discovery and the reuse of heterogenous process models.

## 1 Introduction

As process knowledge, the distributed process models should be accessible and reusable when requesting them for achieving the cooperative business goals. However, those process models were originally created for achieving enterprises local goals. The local goals might be variously presented or even not presented explicitly which causes semantic heterogeneity problem in the goal representation. We need the consensual representations to specify the semantics of goals for the distributed process models, and to enable the machine to interpret and match them to the goal-oriented queries.

The process models involved in this research are the distributed enterprise models on the conceptual level, in which the semantic heterogeneity problem usually occurs. An ontology-based semantic annotation framework is developed to manage the semantic heterogeneity of process models in our previous work [1]. We have provided the approaches and strategies to deal with the semantic heterogeneity of meta-models and model contents. In this paper, the goal annotation based on goal ontology will be refined to complement the semantic annotation framework for facilitating the semantic management of process knowledge.

Goal annotation of process models is annotating process models and model fragments with goal ontology to specify the objectives of processes. Goal ontology is a set of concepts and relationships of semantic definitions about goals. Since the purpose of using goal ontology is to align the semantic representation of goals in a machine-understandable way, the goals should be represented formally in the goal ontology.

J. Krogstie, A.L. Opdahl, and G. Sindre (Eds.): CAiSE 2007, LNCS 4495, pp. 355–369, 2007.

Based on our objectives and the investigation of several goal modeling methods applied in requirements engineering and process modeling, we propose a representation of goal ontology serving goal-oriented semantic management of process knowledge. Therefore the principles of designing our goal ontology are i) the goal ontology should be process-achievable on the conceptual level, i.e. the goal can be targeted through the process modeling; ii) the corresponding relationships between goal ontology and process model should be easily built to facilitate the goal annotation; iii) the semantics of the goal ontology should be understandable and manipulable for both human and machine.

Goals have been used as an important mechanism for connecting requirements to design and supporting reuse [2]. Goal-driven search of design components [3] and discovery of services [4] uses such kind of mechanism, in which selecting the components or services capable of fulfilling the desired goals (requirements). The goal annotation is a pre-procedure to organize and define the process knowledge with goal ontology, i.e. building relationships between process models and pre-defined goal concepts. The definitions of those relationships are major tasks of the goal annotation, which indicate what relationships are supported in the annotation and how the annotation can be implemented. Consequently, the goal annotated models can be queried for the reuse in a goal-driven method with the goal ontology. The objective of the semantic annotation is to facilitate the knowledge management of heterogenous models via the semantic interoperability.

The goal annotation is accomplished together with other semantic annotations (profile annotation, meta-model annotation and model annotation). Thus, this research work on goal annotation of process models is presented as follows in this paper: we first present our semantic annotation framework to provide the research background of this work; our representation of goal ontology is introduced in section 3; then the corresponding relationships between process models and goal ontology are defined in section 4; the procedure of goal annotation accompanied with meta-model and model annotation is described in section 5; and a case study is deployed to illustrate the procedures of goal ontology building and the goal annotation; finally, we conclude the paper and outline the future work.

## 2 The Semantic Annotation Framework

Four main annotation sets constitute the framework: namely, profile annotation, meta-model annotation, model annotation and goal annotation. In the profile annotation, a set of metadata specify the significant characteristics of process models. In the meta-model annotation and the model annotation, we use ontologies to relate constructs across different modeling languages, as well as to align domain specific terminology used in models. Furthermore, the goal annotation is to specify the capacities of process models using a goal ontology. In this way we are able to solve semantic heterogeneity in model management.

A GPO (General Process Ontology) (see Figure 1) is proposed for annotating the process modeling languages in the meta model annotation. Applying the GPO in the meta model annotation, a process model is then described in a **PSAM** (process semantic annotation model).

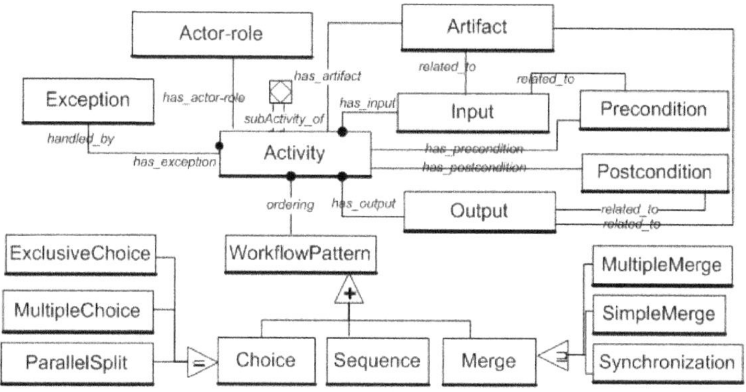


**Fig. 1.** The GPO

The model annotation and the goal annotation will be employed on the **PSAM**. The domain ontology and the goal ontology for a certain domain is built or selected by domain experts for the model annotation and the goal annotation respectively. Therefore, a **PSAM** contains concepts of GPO, domain specific ontology and goal ontology and is defined as follows.

$$PSAM = (AV, AR, AF, WP, I, O, \Theta^{pre}, \Theta^{pos} E, PD, PG)$$

Where $AV$ is a set of activities composing a process, $AR$ is a set of actor-roles interacting with a process, $AF$ is a set of artifacts participating in a process, $WP$ is a set of workflow patterns, and each workflow pattern denotes an ordering of activities. $I$ is a set of input parameters, $O$ is a set of output parameters, $\Theta^{pre}$ is pre-conditions when a process starts, $\Theta^{pos}$ is post-conditions when a process ends, $E$ is a set of possible exceptions occurring during a process. $PD$ is a subset of domain ontology ($D$) concepts, i.e. $PD \subseteq D$, including static ontology concepts and task ontology concepts. $PG$ is a subset of goal ontology ($G$), i.e. $PG \subseteq G$. Since ***Activity*** is the main concern in the goal annotation procedure, we provide the annotated activity structure as follows.

$$\begin{aligned} AV_i = (&id, model\_fragment, name, alternative\_name, has\_Actor-role, \\ &has\_Artifact, has\_Input, has\_Output, is\_in\_WorkflowPattern\_of, \\ &has\_Precondition, has\_Postcondition, has\_Exception, subActivity\_of, \\ &same\_as, different\_from, kind\_of, superConcept\_of, phase\_of, \\ &compositionConcept\_of, achieves\_Goal) \end{aligned}$$

We use $id$ and $model_fragment$ to locate the annotated model and the original model respectively. The $same\_as$, $different\_from$, $kind\_of$, $superConcept\_of$, $phase\_of$, $compositionConcept\_of$ are to annotate the activities with domain ontology, i.e. using semantic relationships mapping an activity with concepts defined in domain ontology. As the goal annotation, the goal ontology is referenced through $achieves\_Goal$. More details refer to [1].

## 3 Goal Ontology for Semantic Annotation

A process model represents how to do things not why to do things. Although a process must achieve certain goals, the relationships between goals and processes are not explicitly represented in many process models. With few process modelling language, goals can be modelled and linked to elements of process models, e.g. EEML process and goal models [5]. However, the representations of the goals and the relationships between goals and processes are variously presented in different models. In this work, goal ontology is applied to specify the capability of processes in a consensual way. The focus in this section is modeling goal ontology. We consider three principles to design goal ontology:

- The goal ontology should be process-achievable on the conceptual level, i.e. the goal can be targeted through the process modeling. Thus, the research will not include the goals related to technical factors such as the usage of computing resources or financial aims like reaching a certain amount of gross profit.
- The corresponding relationships between goal ontology and process model should be easily built to facilitate the goal annotation. The process modeling language constructs (e.g. actor, task, resource, ect.) are found overlapped in most existing goal modeling languages such as KAOS [6], i* [7], GBRAM [8] and etc. Accordingly, those goal modeling approaches could be referred to model goal ontology.
- The semantics of the goal ontology should be understandable and manipulable for both human and machine. We use Semantic Web technology — OWL to represent goal ontology. When they are well modeled in OWL Classes and OWL Properties, the model can be interpreted through OWL interpreters.

### 3.1 Semantic Representations of the Goal Ontology

Following the principles, we make a meta-model of the goal ontology considering the semantic expressivity of OWL as Figure 2. In this meta-model, goal ontology are defined based on the goal category and the goal target.

In general, goals can be classified into hard goals and soft goals [9]. Hard goals relate to functional requirements and they are obviously supported by process. Soft goals relate to non-functional requirements (colloquially "-ilities") which are about global qualities of a system. Soft goals are satisfied when there is sufficient positive and little negative evidence for this claim [9]. As the goal category, ***Hard Goal*** and ***Soft Goal*** are two upper level classes for the goals in all domains. Since hard goals are functional, for different domains they are described domain-specifically. Soft goals can be described generally in a set of "-ilities" (which are regarded as soft goal category), and then specified according to domains.

Checking the literatures of goal definitions, a goal is a condition or state of affairs in the world that stakeholders would like to achieve [10]. Pnina Soffer et al. defines a goal is a set of stable states in [11]. In [12], state of a thing is described as the vector of values for all property functions of a thing. In the context of process modeling, states must be represented as values for the properties of the process and the properties of the objects involved in the process. That is to say, a goal can be expressed as states of

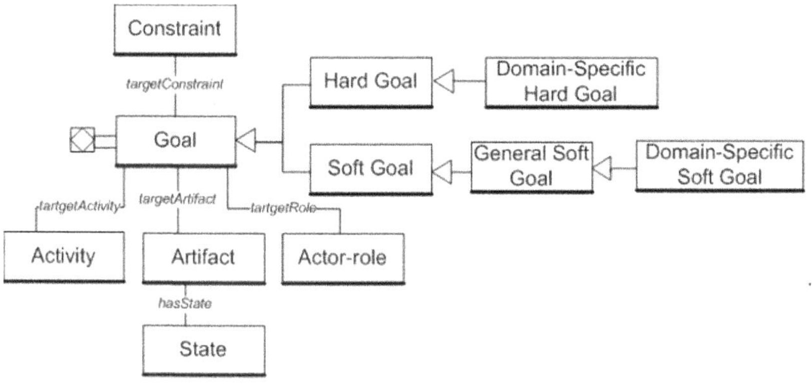


**Fig. 2.** The meta-model of the goal ontology

activities or states of artifacts. The goal target could be *Activity* or *Artifact*. Usually the 'accomplished' is regarded as the goal state of an activity, whilst the state of an artifact has to be specified for different goals. Goal is an organizational concept and goals are held by stakeholders. 'Actor' is defined to represent the goal owner in GRL and 'Agent' is applied in KAOS when analyzing the potential goal realizer. *Actor-role* is therefore the goal target in the goal ontology as well. In KAOS, goals are non-operational objectives and constraints are operational objectives. Although constraints are not goals, goals can be operated by constraints [6]. In this sense, *Constraint* is also the goal target. The relationships between *Goal* and those targets are simply defined because the purpose of the goal ontology is not to analyze the goal like those existing goal modeling methods. The targets show the different perspectives of viewing a goal. These targets are represented same as the concepts in the PSAM, which disclose the potential links between goal ontology and process models.

Decomposition relation – an important characteristic found in most goal analysis should be specified in the goal ontology. However, OWL does not provide any built-in primitives for part-whole relations (as it does for the subclass relation), but contains sufficient expressive power to capture most, but not all, of the common cases [13]. We therefore apply the simple part-whole relationship to represent the decomposition of goal concepts. The 'part' goals contribute the impacts to the 'whole' goal. The logic connections (OR, AND, XOR) between the parts are not considered in the goal ontology due to two reasons. One is the representation limits of OWL. The other one is that the concrete goal analysis mechanism is not necessary for a goal ontology. The goal ontology should be general to applications and how to decompose a goal depends on the specific applications. The goal ontology is more like a taxonomy of goal concepts serving for the semantic-aligned goal representation. The subsumption relationship (`owl:subClass`) represents the goal category hierarchy. The simple part-whole relationship provides the goal components hierarchy. The terminology presenting goal concepts in the goal ontology should be normalized. Further semantics of a goal are specified through the relationships (`owl:objectProperty` and `owl:dataProperty`) with the goal perspectives.

## 4 Relations Between Process Models and Goal Ontology

We study the relations between process models and goal ontology based on the PSAM model. As we have defined in the GPO in [14], an activity may be an atomic activity or a composite activity which is a synonym of process from this sense. We say that a process model comprises a set of activities (***AV***) and an activity can be decomposed into sub-activities. The activities are related to each other through flows according to our GPO definition [1]. If an activity in a process model is not an atomic activity [1] and it is also composed by a set of related activities, it is regarded as a process model fragment in this context. A goal can be linked to a whole process model or to a process model fragment. We assume that the process models are already organized into a decomposable activity hierarchy according to the task ontology in the model annotation phase. Each level activities in a process model can be considered as goal annotation targets.

**Definition 1.** *In the semantic annotation framework, a process model ($PM$) can be partitioned into several process model fragments ($PMF$). Each $PMF$ comprises a set of hierachically organized and decomposable $AV$.*

**Definition 2.** *Any goal concept ($g$) in the goal ontology ($G$) is possibly related to an activity ($av$) in a $PM$ or $PMF$:*

$$\forall(g, av) goalRelated(g, av) \qquad (a)$$

- *if the property* ***targetActivity*** *($av'$) of the $g$ is same or synonymic with $av$:*

$$\exists(av') targetActivity(g, av') \bigwedge av' = av \qquad (b)$$

- *if the property* ***targetArtifact*** *($af'$) of the $g$ is related to the output of $av$ and the* ***State*** *($s'$) of $af'$ is the value of the* ***Output*** *($o$) of the* ***Artifact*** *($af$):*

$$\exists(af, s', o) targetArtifact(g, af') \bigwedge hasState(af', s') \bigwedge s' = o \supset hasOutput(av, o) \bigwedge af' = af \supset relatedTo(o, af) \qquad (c)$$

- *if the property* ***targetRole*** *($ar'$) of the $g$ is related to an* ***Actor-role***$(ar)$ *involved in $av$:*

$$\exists(ar') targeRole(g, ar') \bigwedge ar' = ar \supset hasActor - role(av, ar) \qquad (d)$$

- *if the* ***targetConstraint*** *($c'$) expressed in the $g$ is involved in (involvedIn)* ***preCondition*** *($pre$),* ***postCondition*** *($post$) or* ***Exception*** *($e$) of $av$:*

$$\exists(c', pre, post, e) targetConstraint(g, c') \bigwedge (involvedIn(c', pre) \supset hasPrecondition(av, pre) \bigvee involvedIn(c', post) \supset hasPostcondition(av, post) \bigvee involvedIn(c', e) \supset hasException(av, e)) \qquad (e)$$

*Therefore,*

$$(a) \equiv (b) \bigvee (c) \bigvee (d) \bigvee (e)$$

[1] Note: An atomic activity can not be decomposed, but it is not an event either.

Checking above cases through matching algorithms can automatically provide a list of possible goal annotations. The decisions of the desired goal annotations are left to annotators.

**Definition 3.** *In the goal ontology (G), the hard goal set is notated as $G^h$ and the soft goal set is notated as $G^s$.*

The relations are further specified by the annotator based on the context of the process models and the goal ontology. We define two relations as follows:

**Definition 4.** *Hard goals can be achieved by an activity or activities. I.e. the relation between the activity (av) and the hard goal ($g^h$) is* ***achieves***$(av, g^h)$.

**Definition 5.** *Soft goals can be positively or negatively satisfied by an activity or activities. I.e. the relation between the activity (av) and the soft goal ($g^s$) is* ***positivelySatisfies***$(av, g^s)$ *or* ***negativelySatisfies***$(av, g^s)$.

Since the activities in the process models are decomposable, the relation between goals and a composite activity could be inferred based on the relations between goals and component activities.

**Definition 6.** *If an activity ($av$) is a component of another activity ($av'$) in a process model/model fragment, $av$ is the subactivity of $av'$, i.e.* **subActivityOf**$(av, av')$. *$av'$ is a* ***Composite Activity*** *in that model.*

Usually the effects of hard goals achieved by a subActivity can contribute to its composite activity. That is,

**Definition 7.** *If $av$ is the subactivity of $av'$ and $av$ achieves $g^h$, $av'$ achieves $g^h$:*

$$(\forall(av, g^h)\exists av')subActivityOf(av, av') \bigwedge achieves(av, g^h) \longrightarrow achieves(av', g^h)$$

However, the effects of soft goals can not be simply passed in the same way as hard goals. To a composite activity, the contribution of a soft goal from a subactivity might be enhanced or reduced by other subactivities which positively or negatively satisfy the same soft goal. The contribution could be calculated if the effects of soft goals are quantified. This issue is only simply considered in our current work by ***simple contribution calculation rules***. All effects of soft goals are regarded same. The contribution of soft goals to a composite activity is determined by comparing the numbers of subactivities which positively satisfy and negatively satisfy the same soft goals. That is,

**Definition 8.** *Let a $g^s$ is positively satisfied by $N$ subactivities, and is negatively satisfied by $M$ subactivities in a composite activity $av'$,*

- *if $N > M$,* ***positivelySatisfies*** $(av', g^s)$
- *if $N = M$, then the effect of $g^s$ is counteracted for $av'$.*
- *if $N < M$,* ***negativelySatisfies*** $(av', g^s)$

The relations between goals and activities defined in this section will be applied to build annotation links between goal ontology and process models in the goal annotation. That is to say, the meta data schema of the goal annotation for process models is:

```
ActivityID
   <achieves|positively satisfies|negatively satisfies>
GoalOntologyID
```

## 5 Goal Annotation Procedure

In the goal oriented requirements engineering, the goal analysis and modeling is a top to bottom procedure — decomposing high level goals down to lower level goals and operational activities. The goal annotation is a bottom to top procedure — first annotating low level subactivities and then annotating higher level activities and the whole process model with goal ontology. The goal annotation is employed based on the domain annotated PSAM models.

After the profile annotation, the PSAM is initially structured after the meta-model annotation, and then it is gradually filled with domain ontology in the model annotation. The goal annotation is employed as the final step of the whole annotation procedure, i.e. filling the PSAM with goal ontology. The semantic annotation procedure is illustrated in Figure 3.

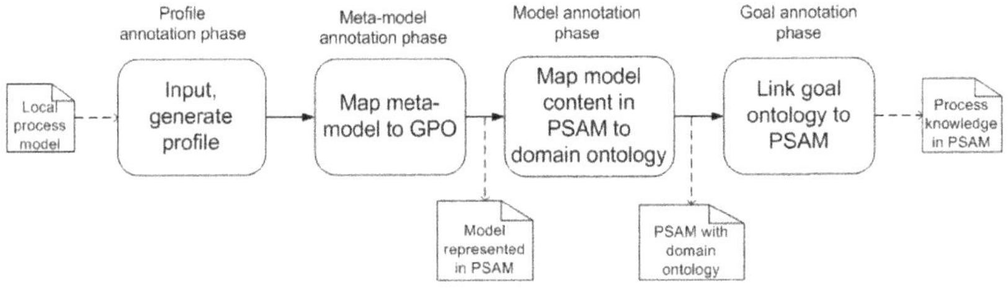


**Fig. 3.** The semantic annotation process based on PSAM

As the goal annotation, we update the PSAM with the goal annotation relations $achieves\_Goal/positively\_satisfies\_Goal/negatively\_satisfies\_Goal$ . Applying the `psam` as markup annotation language, an example of annotating a process model fragment with the goal ontology by representing an activity achieves a hard goal as follows:

```
<psam:Activity rdf:ID="ID">
  <psam:model_fragment rdf:resource="&MODEL_NAMESPACE#MODEL_ID">
  ...
  <psam:achieves rdf:resource="&GOAL_ONTOLOGY#GOAL_ONTOLOGY_CONCEPT"/>
```

We have discussed that the goal annotation of the process models and the model fragments is to link goal ontology to the activities identified in the models. Focusing on the activity, we describe the goal annotation procedure accompanied with the meta model annotation and the model annotation in a UML activity diagram in Figure 4.

Through the meta model annotation, activities are identified by the markup ***Activity*** in a **PSAM**. In the model annotation phase, if the domain task ontology is available

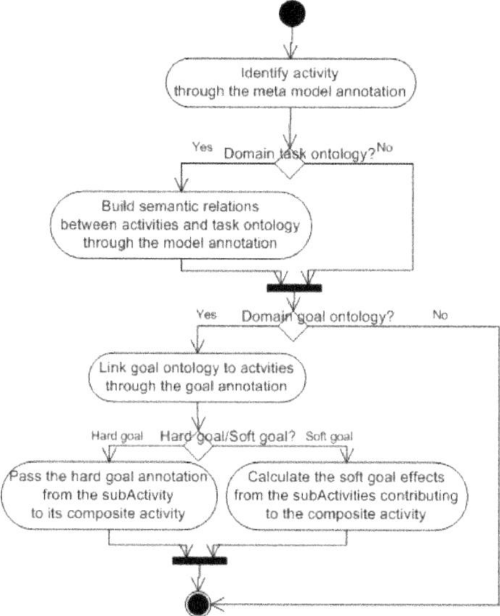


**Fig. 4.** The goal annotation procedure

as the activity references, the identified activities is annotated with the task ontology references via the semantic relations such as $same\_as$, $different\_from$, $kind\_of$, $superConcept\_of$, $phase\_of$, and $compositionConcept\_of$. If the domain goal ontology is available, the possible links between the activities and the goal ontology can be checked based on the relations described in section 4. We employ the annotation from the component activities to the composite activities. The contributions of the goals annotated to the low level component activities can be passed to or calculated for their upper level composite activities.

## 6 Case Study

In this section, we provide a case study to illustrate the goal ontology representation and the goal annotation procedure. For the sake of the brevity, we assume the models have already processed by the meta model annotation and the model annotation. Consequently, the goal annotation is employed on the **PSAM** model. We set our example in an industrial enterprise with supply-chain management systems. The SCOR (Supply Chain Operations Reference-model) [15] is applied as the domain ontology in this example. The goal ontology is also derived from the SCOR.

### 6.1 The TelCo Case Study

TelCo company is specialized in telecommunications, in the production and distribution of batteries, as well as in retail sales of everyday technology products. TelCo does not have its own warehouse but uses the services of logistics company Orbit Ltd. But

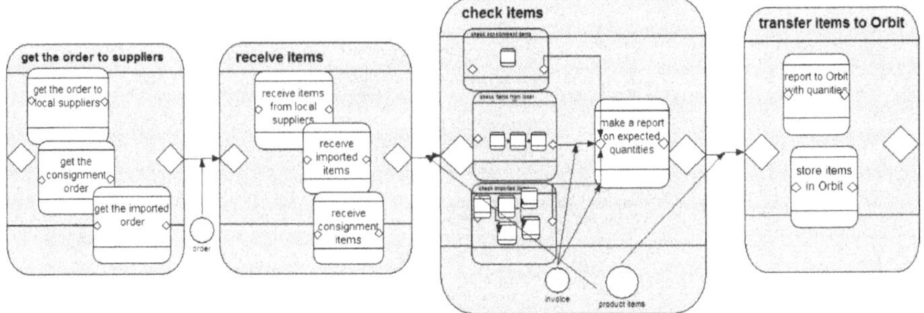


**Fig. 5.** TelCo item receiving process

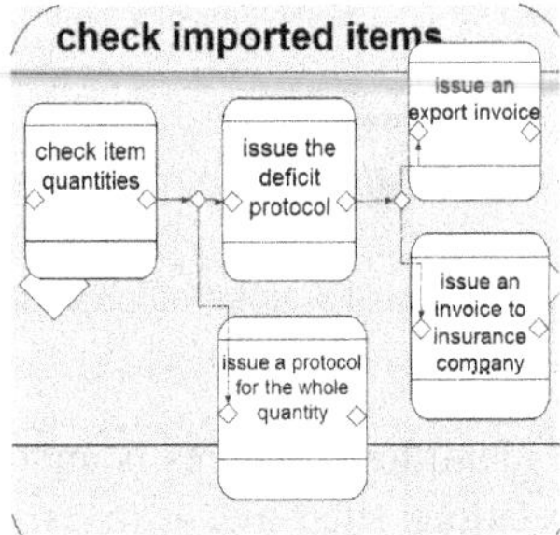


**Fig. 6.** Decomposition of the *check items*

TelCo has its logistics department who is responsible for items receiving and delivering. Thus the functions of logistics department are to order and control Orbit. The following main business processes of the logistics are items receiving, returns, orders from shops and franchisees, orders from dealers and inventories. In this case, we take the items receiving process as the annotation example. The process model of the items receiving is initially made in EEML. A simplified model mainly from the task viewpoint is displayed in Figure 5. The logistics department receives a report for expected quantities of items according to the order to the supplier, and prepares to receive the items. After receiving the items, the logistics department checks items. Sub-tasks are included in some tasks because there are three types of items receiving — regular orders to local suppliers, consignment and import. The process details are described in the decomposition of the task *check import items* in Figure 6. After checking the items, the received items will be transferred to Orbit. Meanwhile the department sends a report to Orbit in order to inform them what to expect.

The meta model annotation is to map the EEML meta model and the model structures to the GPO. Hereby, all the *tasks* in the EEML model are annotated as ***Activity*** in the **PSAM**. For the model annotation and the goal annotation, the SCOR is referenced as the domain ontology including the domain tasks and the domain goals.

## 6.2 The SCOR as Reference Ontology

The SCOR is a process reference model that has been developed and endorsed by the Supply-Chain Council as the cross-industry standard diagnostic tool for supply-chain management. The SCOR-model describes the business activities associated with all phases of satisfying a customer's demand.

For this case, we apply "S1 Source Stocked Product" model as task ontology for the item receiving process. The SCOR-model is depicted in Figure 7. Each process element is a task ontology concept which is referenced by the ***Activity*** in the model annotation (e.g. the ***Activity*** *check items* is annotated with the task ontology `Verify Product`). The inputs and outputs are the domain object concepts as ontology to annotate the ***Artifacts*** in the **PSAM** (e.g. the ***Artifact*** *product item* is annotated with the domain object ontology `Sourced Product On Order`). The details of the model annotation are not specified in this paper due to the space limitation.

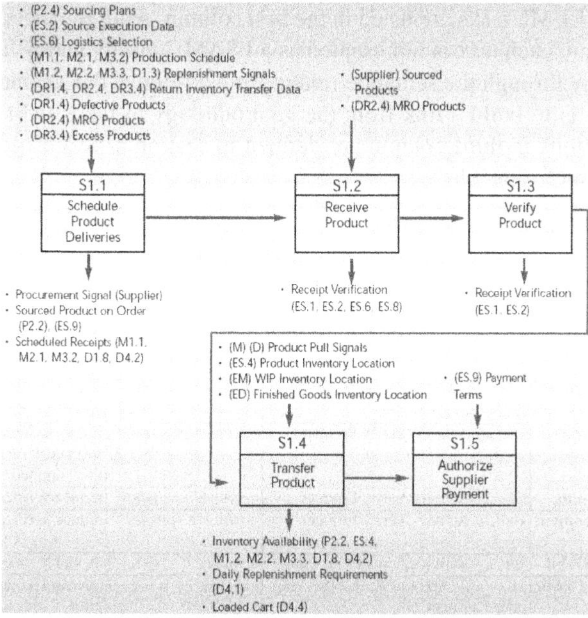


**Fig. 7.** S1 Source Stocked Product

The goal ontology in this domain is also from the SCOR. Usually the hard goals are derived from the level 3 process elements [15] and their inputs and outputs. The performance attributes defined in SCOR are *General Soft Goal Category* (generally in a set of "-ilities") such as **Reliability**,**Responsiveness**, **Flexibility**, **Cost**, and **Assets**. The domain specific soft goals are derived from the metrics of the performance attributes [11]. By analyzing the goal targets, we can identify the following goals derived from S1 (Table 1).

**Table 1.** Goal ontology derived from SCOR

| Goal Type | Goal Target | SCOR Goal Ontology |
|---|---|---|
| Hard Goal | *targetActivity* | sourced products are transferred; sourced products are verified. |
| | *targetArtifact and state* | sourced products on order; scheduled receipt; available inventory. |
| | *targetRole* | payment is authorized to supplier |
| Soft Goal | *targetConstraint* | improve supplier delivery to date performance (**Responsiveness**); invoices processed without error (**Reliability**). |
| | *targetArtifact* | improve delivery quantity performance (**Assets**); decrease % defective supplied (**Reliability**). |
| | *targetActivity* | reduce verification costs (**Costs**); reduce receiving & storage costs (**Costs**). |

## 6.3 Goal Annotation for Semantic Enrichment of Process Knowledge

Having a set of goal ontologies in the *Sourced Stocked Product* domain, we annotate the TelCo logistics department's item receiving process model. We consequently have the following annotation results listed in Table 2.

The orginial EEML tasks are listed in the first column. After the meta-model and the model annotation, each task is represented as a PSAM:Activity and is linked to SCOR domain ontology through the semantic relationships (refer to the second column). The goal annotation is to build a link from the goal ontology to the PSAM activities. The annotation relations in third column are defined in section 4. SCOR goal ontology in the fourth column from table 1.

**Table 2.** Goal annotation results

| EEML Tasks | PSAM Activities (*Model annotation with SCOR domain ontology*) | Goal Annotation Relations | SCOR Goal Ontology |
|---|---|---|---|
| **get the order to supplier** | *phase_of* Activity:Schedule Product Deliveries | *achieves* | sourced product on order |
| **get the order to supplier** | *phase_of* Activity:Schedule Product Deliveries | *positively_satisfies* | reduce receiving & storage costs |
| **check imported items** | *kind_of* Activity:Verify Product | *positively_satisfies* | improve delivery quantity performance; decrease % defective supplied |
| **check imported items** | *kind_of* Activity:Verify Product | *negatively_satisfies* | reduce verification costs |
| **check consignment items** | *kind_of* Activity:Verify Product | *positively_satisfies* | reduce verification costs |
| **transfer items to Orbit** | *kind_of* Activity:Transfer Product | *achieves* | available inventory |
| **issue the invoice; issue an export invoice** | *phase_of* Activity:Authorize Supplier Payment | *achieves* | payment is authorized to supplier |
| **issue the invoice; issue an export invoice; issue an invoice to insurance company** | *phase_of* Activity:Authorize Supplier Payment | *negatively_satisfies* | invoices processed without error |

The EEML task *get the order to supplier* can be regarded as a phase of the activity ontology `Schedule Product Deliveries` from SCOR. When annotating goals, we find two goals in the domain goal ontology related to this activity. It can achieve a hard goal `sourced product on order` and also positively satisfy a soft goal `reduce receiving & storage costs`. Three kinds of items receiving are checked through the activity *check items*. The procedure of *check imported items* is a little more complicated compared with the other two kinds of items receiving becuase it includes issuing deficit protocols and an invoice to insurance company. It therefore negatively satisfies the soft goal `reduce verification costs`. However, this check procedure can improve the delivery quantity performance and the deficit check can ensure the low % defective supplied. The consignment can simplify the check and item receiving procedure, so it positively satisfies the soft goal of `reduce verification costs`. To transfer items to Orbit is a procedure of `Transfer Product` and the result of the procedure is an `available inventory`. Issuing invoices are steps of `Authorize Supplier Payment`, in which *issue the invoice* to local suppliers and *issue an export invoice* are to authorize payment to suppliers. The soft goal of `invoices processed without error` will be risked by different invoice issue procedures.

After annotating the low level activity elements, the goal contributions can be calculated to the upper level activities. Taking the example of the composite activity *check items*, we have annotated its component activities with hard goals and soft goals. *Check imported items* negatively satisfies `reduce verification costs` and *check consignment items* positively satisfies `reduce verification costs`, so the effects are counteracted for the composite activity *check items* if we apply the ***simple contribution calculation rules***. Without negative counteraction, the soft goals `improve delivery quantity performance` and `decrease % defective supplied` contribute themselves to *check items*. The hard goals annotated to *issue the invoice* and *issue an export invoice* are also simply passed to *check items*. The soft goal `invoices processed without error` are negatively satisfied by *check items* and the effects are enhanced due to three component activities negatively satisfy this soft goal.

### 6.4 Process Knowledge Discovery and Reuse Based on the Semantic Annotation

Before the semantic annotation, the local process models are represented in EEML and the model contents are described with the concepts and terms locally defined, such as '*items*', '*supplier*', '*check items*'. The EEML might not be acquainted to other users who want to reuse the model. The users who want to search the model do not know the exact concepts and terms used in the model, if applying the keyword search. Usually, users want to find the model which can realize their goals. However, the goals usually do not be specified in the local process models or specified in the modeling languages strange to users.

The proposed semantic annotation approach makes the process knowledge of TelCo logistics department explicit and open to the third party who is interested in the model exchange, system integration and business cooperation in the SCOR domain. An external party would like to get the knowledge from the existing models, which include this EEML model and other heterogenous models in different modeling languages. She can make search based on the SCOR goal and domain ontology. If she provides a business goal which is represented in the SCOR goal ontology, the query will be matched to the goal annotated process models. For example, the user would like to check the existing models/model fragments which impact the soft goal reliability in the delivery process. In the goal ontology, the soft goal concept '`Reliability`' has some sub-classes such as 'Improve Delivery Performance', 'Improve Fill Rates' and 'Improve Perfect Order Fulfillment'. The user might decide to only take the goal 'Improve Delivery Performance' as the query. The mechine makes the match and the semantic inference between the query and the goal annotation in the process models. In the query results, we find not only the process model fragments annotated with the goal concept 'Improve Delivery Performance', also those annotated with the goals 'Improve supplier delivery to date performance' and 'Improve delivery quantity performance', because the goals 'Improve supplier delivery to date performance' and 'Improve delivery quantity performance' are part goals of 'Improve Delivery Performance' in the SCOR goal ontology. The models/model fragments annotated with the corresponding goal ontology will be returned to the user as a result.

The returned results include the original models and the annotation information. Suppose the users do not know EEML, but they know the GPO. Compared with a certain modeling language, GPO provides much simpler structures but enough core process semantics. The meta-model annotation provides the mapping relations between the EEML and the GPO. Also from the PSAM of a model, users can read the mapping between the local concepts in the original model and the domain ontology concepts. We assume that users well know the domain ontology as the domain standard. Consequently, the model annotation helps the users to understand the local concepts, which is required in the later model reuse and model transformation.

## 7 Conclusions and Future Work

As part of the semantic annotation framework, a goal annotation approach is elaborated in this paper. The semantic annotation framework is designed for managing the semantic heterogeneity of distributed process models on the conceptual level. The approach is based on the domain ontology, which provides common semantic representations of domain concepts. The semantic annotation framework aligns the semantic heterogeneity of models from different perspectives. The purposes of the goal annotation are two: 1) to enrich the semantics of the objectives of processes; 2) to provide a way to find the process knowledge based on business goals. Goal ontology is necessary for the goal annotation as the common semantic representations of process objectives. Based on the existing goal modeling methods, we proposed a way to describe the semantics of goals for the process model annotation purpose. We also discuss the relations and rules between process models and goal ontologies. The formulization of the relations can be implemented to facilitate the automatic or semi-automatic goal annotation procedure in practice. The proposed approaches are examined through a case study. Main work includes the establishment of a domain goal ontology from SCOR and the execution of the goal annotation based on the relations and rules. The possible applicability of the annotation result is also briefly presented.

The proposed goal ontology semantics and goal annotation relations can be further elaborated. More semantics are not explicitly represented and enriched, for instance, the *contribution* relationships between goals ('support', 'conflict', etc.), and the *impact degree* between activity and goal ('partially impact', 'totally impact'). The tools for the automatic annotation and the visualization of annotated models are also the challenges in this research.

## Acknowledgement

This work is partially supported by the Norwegian Research Foundation in the framework of Information and Communication Technology (IKT-2010) program. Also thanks the INTEROP project for the case study.

## References

1. Lin, Y., Strasunskas, D., Hakkarainen, S., Krogstie, J., Sølvberg, A.: Semantic annotation framework to manage semantic heterogeneity of process models. In: Dubois, E., Pohl, K. (eds.) CAiSE 2006. LNCS, vol. 4001, pp. 433–446. Springer, Heidelberg (2006)
2. Yu, E., Mylopoulos, J.: Why goal-oriented requirements engineering. In: Proc. of the 4th of International Workshop on Requirements Egnineering: Foundations of Software Quality (1998) http://www.cs.toronto.edu/pub/eric/REFSQ98.html
3. Hummel, K.A., Jochum, W., Leitich, S., Schandl, B.: Supporting meetings with a goal-driven service-oriented multimedia environment. In: MSC '05: Proceedings of the first ACM international workshop on Multimedia service composition, pp. 55–65. ACM Press, New York (2005)
4. Lin, M., Guo, H., Yin, J.: Goal description language for semantic web service automatic composition. In: Proc. of IEEE/IPSJ International Symposium on Applications and the Internet (SAINT 2005), 31 January - 4 February 2005, Trento, Italy. pp. 190–196 (2005)
5. Krogstie, J., Jørgensen, D.: Interactive models for supporting networked organisations. In: Persson, A., Stirna, J. (eds.) CAiSE 2004. LNCS, vol. 3084, pp. 550–562. Springer, Heidelberg (2004)
6. Dardenne, A., van Lamsweerde, A., Fickas, S.: Goal-directed requirements acquisition. Science of Computer Programming 20, 3–50 (1993)
7. Yu, E.: i*:an agent-oriented modelling framework (2006) http://www.cs.toronto.edu/km/istar/
8. Anton, A.I.: Goal based requirements analysis. In: Proc. Second Int. Conference on Requirements Engineering, ICRE 96. pp. 136–144 (1996)
9. Mylopoulos, J., Chung, L., Yu, E.: From object-oriented to goal-oriented requirements analysis. Commun. ACM 42, 31–37 (1999)
10. Grl, G.R.L.: ontology ( 2007) http://www.cs.toronto.edu/km/GRL/
11. Soffer, P., Wand, Y.: On the notion of soft-goals in business process modeling. Business Process Management Journal 11, 663–679 (2005)
12. Wand, Y., Weber, R.: On the deep structure of information systems. Information System Journal 5, 203–223 (1995)
13. W3C: Simple part-whole relations in owl ontologies (2005) http://www.w3.org/2001/sw/BestPractices/OEP/SimplePartWhole/index.html
14. Lin, Y., Strasunskas, D.: Ontology-based semantic annotation of process models. In: Proc. of 10th CAiSE/IFIP8.1/EUNO International Workshop on Evaluation of Modeling Methods in System Analysis and Design (EMMSAD05) Porto, Portugal (June 2005)
15. SCOR: Scor model (2006) http://www.supply-chain.org/page.ww?section=SCOR+Model&name= SCOR+Model

# Stakeholder Identification as an Issue in the Improvement of Software Requirements Quality

Carla Pacheco and Edmundo Tovar

Languages and Informatics Systems and Software Engineering Department
Faculty of Computer Science, Polytechnic University of Madrid. Spain
cpacheco@zipi.fi.ump.es, etovar@fi.upm.es

**Abstract.** Stakeholder identification together with its needs and expectations has been poorly realized in software projects. This is probably because the process is mistakenly viewed as a self-evident task in which direct users and the development team are the only stakeholders. It could also be due to the fact that the identification area can be substituted by opinions or knowledge from other more accessible sources of information. This paper provides a review of stakeholder identification literature and an overview of the state-of-the-art in methods for that purpose, which leads to a number of issues that are important in further research (e.g. developing a methodology). The paper findings are presented from two points of view: firstly, the impact of stakeholder identification on software requirements quality, and secondly, practices developed to carry out this task. Also, the present paper aims to describe the studies analyzed uniformly and show their contributions in this field.

**Keywords:** Stakeholder Identification Process, software process, effective practices, software requirements, elicitation.

## 1 Introduction

Requirements Engineering (RE) as a discipline was developed when the quality of requirements was recognized as a key factor in preventing many of the causes leading to software failure. Measures taken at an early stage of a project can have great repercussions, and they are also more beneficial than those taken at later stages. The problem of the "software crisis"[1] has, to a great degree, shifted to the area of requirements. Is there, then, some aspect of the requirements area that deserves particular attention? If so, this aspect should be taken into account at the initial stage of RE. Such is the case in requirements elicitation activities in which the problem to be solved is identified, and more importantly, the stakeholders must be identified. Relationships and ways of communicating between the development team and the customer are thereby established [1].

[1] The notion of a software crisis emerged at the end of the 1960s [24]. The term refers to difficulties in writing correct, intelligible and verifiable computer programs. The roots of the software crisis hinge around complexity, expectations, and change. Conflicting requirements have always been a hindrance to the software development process.

J. Krogstie, A.L. Opdahl, and G. Sindre (Eds.): CAiSE 2007, LNCS 4495, pp. 370–380, 2007.

However, identification of stakeholders as well as their needs and expectations are poorly done in software projects [2], probably because this process[2] is mistakenly viewed as a self-evident task in which direct users, clients, and the development team, are the only stakeholders. It could also be due to the fact that the identification area can be obviated or substituted by opinions or knowledge obtained from other more accessible sources of information. In the short term, this would create less conflict of interests resulting from different points of view.

The findings in the problem statement are presented from two points of view: firstly, the impact of stakeholder identification on software requirements quality, and secondly, practices developed to carry out this task. So far, there has not been any SIP framework or uniform description. In view of this, the present paper we offers a uniform description of SIP as a first step towards developing a methodology that would cover all necessary aspects of stakeholder identification. Hence, at this stage, the present paper does not represent a technical solution.

## 2 What Are the Consequences of Incorrect Stakeholder Identification on the Quality of Software Requirements?

To answer the question raised, it is necessary to consider what Software Requirement Specification Quality (SRSQ) involves. The IEEE Standard 830 [3] gives a summary of the properties that should ideally be part of software requirement specification. Any identification process that mistakenly recognizes someone as a stakeholder will probably include requirements which do not correspond to any real need (a feature of "Correctness" of the standard). On the other hand, when the identification task fails to detect participants who are needed for the software project, requirement specifications are no longer "Complete" due to the omission of relevant requirements for project success, and this could give rise to inconsistent specifications. Failing to obtain these properties can create risks that could affect the project. Completeness, correctness and consistency in the RSQ can be ensured by applying proper elicitation techniques such as scenarios, use cases, etc. All of these, however, require a previous Stakeholder Identification Process (SIP).

The review of relevant initiatives from the field of Software Engineering (SE) and Information Systems (IS), referred to in Section 3 "Stakeholder identification as seen in previous studies of Software Engineering", confirms how all of these distinguish different types of stakeholders involved in software development, each type having different priorities and interests. To the same extent, all initiatives take stakeholder identification for granted and confine their task to indicating "who" they can be. They do not, however, clarify "how" the SIP process must be carried out to ensure getting correct stakeholders and thereby obtain accurate and complete requirements.

The implications of SIP on the quality of requirements are evident. This has not gone unnoticed, not even by those who have created standards or done studies that have been used as guidelines for software improvement processes. The comparison of

[2] The CMMi model [2] defines a process as a set of activities that can be recognized as implementations of practices.

development studies shows that authors such as Wiegers [4] are more interested in identifying and applying effective practices[3] than in obtaining complete solutions for project development.

There have been calls for different initiatives based on collections of data and experiences with proposals for effective practices that successfully meet users' needs.

The CMMi [2] does not explicitly mention any specific stakeholder identification practice. According to the CMMi, stakeholder needs are fundamental in determining customer requirements. These needs must be analyzed together with stakeholders' expectations, constraints, interfaces, operational concepts, and product concepts and be harmonized, refined, and transformed into a set of customer requirements. This goal includes specific practices such as collecting basic needs and also eliciting unarticulated needs. This involves the previous specification of stakeholders.

Does the lack of mention of a specific stakeholder identification practice imply a lack of precision in the description of the standard, or does it imply that the importance of stakeholder identification with respect to the problem is reduced? To answer this question, other proposals of effective practices will now be analyzed in detail.

### 2.1 Effective Practices Recommended for Performing SIP and the Benefits of Proper Stakeholder Identification

- Hoffman's [5] studies identify some RE effective practices. These practices measure the effectiveness of a software project and are organized according to the different factors which have an impact on their objective. Within all of these practices, the ones related to SIP (that is, Identify and consult all likely sources of requirements) are framed within the knowledge factor under the argument that stakeholders should meet demands in terms of experience and expertise for effective team work. It's a question, then, of a) carefully selecting team members who are skilled in the application domain, IT, and RE processes, b) always assigning experienced, capable project managers to RE, and c) consulting domain experts and stakeholders at an early stage of the process to increase and validate the team's knowledge.

- Wiegers [4] identifies a group of effective practices related to the elicitation task. One of these practices (namely, Identify user classes and their characteristics) emphasizes the need for stakeholder identification. There may, in fact, be many groups of customers who use the product, and these can be classified in terms of frequency of use of the product, use characteristics, levels of privileges, or levels of skills. Since each type of project (for example, commercial applications, integrated systems, web developments, etc.) requires different experts, proper selection of stakeholders is recommended. This selection involves previous assessment of stakeholders in terms of risk and cost, and also taking into account standard types of

[3] Effective practices are activities that people with recognized expertise in a particular area have identified from experience as making significant contributions to project success. Approximating effective practices enables us to create a kit with a variety of practices that can be applied to different problems [23].

communication between users and developers. Communication, for example, in which developers can talk directly to potential users is more effective because it avoids loss of information generated by using intermediaries.

- Another approach to the specification of effective practices is the development of techniques and strategies which improve the RE processes. The REAIMS Project (8649) [6] set out to develop a maturity model for the RE similar to the SEI's CMM, that is, in terms of the scope to which an organization has defined the RE process based on effective practices. In this project, a set of basic guidelines is recommended for organizations that are at an initial stage of maturity. The guideline entitled "Identify and consult with the stakeholders of the system" recommends making a very specific list of stake-holders at an early stage of the RE process. It proposes a method of following the guidelines that ensures that only appropriate stakeholders are identified within each category of proposed stakeholder classifications. It is further suggested that an explicit list of stakeholders be drawn up and reasons given why the requirements will probably be important.

In summary, there is general agreement about the need to find effective practices relating to stakeholder identification in industry. So far, these practices have only defined different types of stakeholders on the basis of small groups of typical users and representatives of other people involved in the development of the project.

The benefits are evident: a proper selection of stakeholders improves the coverage of requirements, avoids overlapping of requirements in the user community, and allows for a more rational organization of requirements (Figure 1). In this way, people get involved more easily, and are less reluctant to implement the system and give information relating to requirements. However, effective practices, or standards such as CMMi, have the following limitations: they do not explain how to define the entire set of stakeholders. Furthermore, this process is not always self evident, and so the organization must be analyzed in order to encounter all possible stakeholders. Hence, the application of a stakeholder identification method sometimes becomes indispensable. This is developed in the following section.

## 3 Stakeholder Identification in Previous Software Engineering Studies

Software engineers need to identify, characterize, and handle all viewpoints of the different types of stakeholders [7]. Stakeholders, meaning all those involved in a project and have some interest in the software to be developed, may vary from one project to another. It is, therefore, always necessary to carry out an adaptation assessment of stakeholders' contributions and their vested interests in a project. In spite of the importance of identifying all the relevant parts (stakeholders) involved in a software project, the SIP area has received less attention than others in the SE.

Some of the main initiatives of SE recognize the existence of different types of stakeholders and who therefore need to be identified in each project.

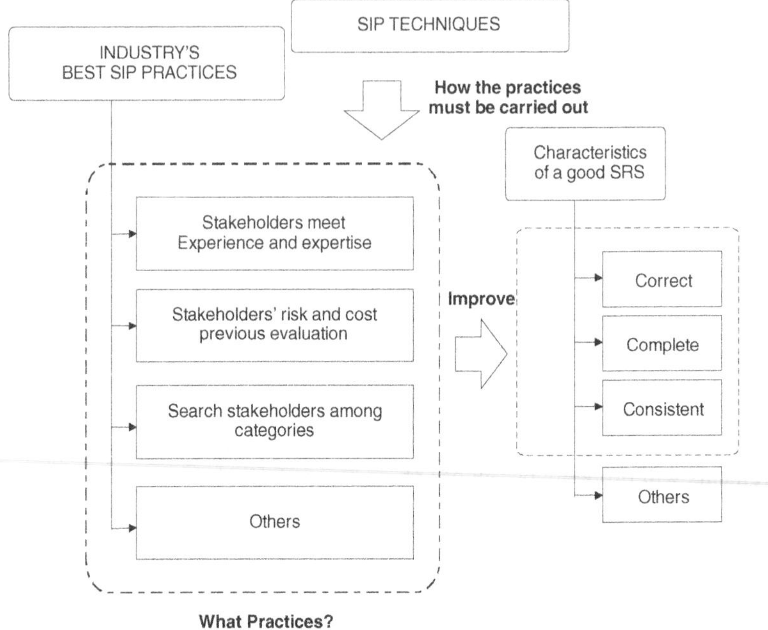


**Fig. 1.** Impact on the SIP effective practices on SRSQ

- The "Software Engineering Body Of Knowledge" (SWEBOK), emphasizes the description of the tasks related to each one of the proposed fields of knowledge, principally in what are called Requirements. For the SWEBOK, the process of requirements elicitation is a human activity in which stakeholders and relationships established between the development team and the client must be identified. [1]. RE is an interdisciplinary process in which all actors must become involved. Ignoring this can lead to the development of inappropriate systems. SWEBOK assigns the role of negotiator to the requirements engineer while other stakeholders are not taken into account.
- In an SEI Technical Report [8], stakeholders are identified at the requirements elicitation stage. They come from at least five communities involved in software development: clients/sponsors, users, developers, quality personnel, security personnel and the requirements analyst. The SEI's Capability Maturity Model Integration (CMMi- SW) [2] specifically proposes that stakeholders be selected from among customers, final users, developers, producers, test staff, suppliers, marketing staff, maintenance staff, and anyone who may affect or be affected by the software process and the final product. The CMMi defines two process areas related to RE: Requirements Management and Requirements Development. In each one of these areas, the different stakeholders involved are classified, and their different roles are defined on the basis of activities performed.
- In the standards developed by ISO and IEC, the ISO/IEC 12207 (software life cycle processes) provides a specific guideline to define the roles and responsibilities of some stakeholders in the life cycle of a software project, or product, or service. Some of the stakeholders mentioned are customers, quality personnel, software developers, etc. [9]

• Ian Sommerville [10] situates the identification of stakeholders within the stages of obtaining and analyzing the software requirements. Among the identified stakeholders are the final users who will interact with the system, and also anyone inside the organization who will be affected by such a system. Stakeholders also include engineers who develop or support other related systems, for example, business managers, the IT specialist, workers' representatives, etc.

• Roger S Pressman argues that stakeholders must be identified in the beginning of the RE process because many different participants are involved at this stage. Pressman identifies the following stakeholders as being the most common: business managers, brand managers, marketing staff, external and internal customers, consultants, product engineers, software engineers, sup-port and maintenance engineers, etc. [11].

• In the Rational Unified Process (RUP) within the SE process, stakeholder identification is carried out at the management requirements processing stage. The most obvious stakeholders in a software project are: the final user, the software developer, the purchaser, the project director, and anyone strongly interested in the project or those who need the project to solve their needs [12].

The aforementioned studies confirm the variety of existing stakeholders involved in software development, each having different priorities and interests. All of these studies take SIP for granted and don't go beyond indicating "who" the stakeholders may be. They do not mention "how" the process must be carried out to properly identify stakeholders as a prerequisite to obtaining exact and complete requirements.

## 4 What Is the Scope of Approximations in SIP Studies?

So far, we have explained how each software project may have different types of stakeholders, and how selecting them appropriately has a strong impact on software requirements quality, and consequently, on the success of the software project itself. The studies reviewed could give the impression that many attempts have been made to define and give detailed explanations of how the SIP is done. This, however, is not the case. Currently, stakeholder identification methods are few and since the process is described by each author the SIP lacks a common framework of study and a uniform description. The studies described so far in this paper seem to only partially solve the issue of stakeholder identification.

Based on RE literature relating to stakeholder identification processes, we propose to group them into three categories. Section 4.1 gives the first category of studies which limit themselves to only proposing a list of possible stakeholders. Section 4.2 presents the second category of studies which not only indicate who the stakeholders can be, but also studies their interactions. The third category, in section 4.3, deals with studies that include an assessment of stakeholders.

### 4.1 Studies That Exclusively Characterize Stakeholders

These studies provide a list of potential stakeholders from which it is possible to determine which ones are really relevant and how each one may be contacted. These, however, are incomplete because they only provide a helpful guide to establish a final list of stakeholders. What must not be overlooked is that stakeholders will normally have to contribute their effort, time and/or money, and they must therefore know what benefits can be gained in return. Potential stakeholders must therefore be characterized by gathering relevant information about them. This information may also be useful for evaluating a set of identified stakeholders, and for obtaining new and more appropriate configurations. Some examples are:

- Lauesen [13] only summarizes this information from responses to just three questions: what goals do they identify for the system? Why would they like to contribute? Or, what risks and costs do they envisage?
- Robertson's VOLERE template [14] is used during the whole requirements process for other purposes such as assessing quality and specifying business requirements. The template is a tool that helps to discover the difference between a stakeholder's wish and a real need, and helps to establish the range of a system. This fundamental aspect of the process ensures that all stakeholders know what is and what is not within their domain. However, it can also be used to get a general idea about the stakeholders who are participating in the development of a system. This method only differentiates between the client and the buyer, and thereby ignores other stakeholders. Other stakeholders may be found in the following categories: users, sponsors, test staff, business analysts, technology experts, system designers, marketing experts, legal experts, IT specialists, usability experts, representatives of external associations, etc. Each of the stakeholders taken into account is identified in terms of who he/she is and what role he/she will play. They are selected from a list of stakeholders initially proposed and on the basis of knowledge which may be necessary for the project.

In general, these studies cannot be regarded as identification of stakeholders because they only provide information that makes it easy to identify them. They do not ensure that all the necessary stakeholders are detected.

### 4.2 Studies Focusing on Interaction Between Stakeholders

Once we have an idea of who the main stakeholders are, the basic interactions between these actors should be identified. This enables stakeholders to clarify which part of the problem falls within each one's scope. The following range of studies deal with this aspect.

- Smith [15] proposes a context diagram to enable stakeholders to see what is happening in the system. This starts with a brainstorming activity in which all stakeholders must be taken into account. At the center of the diagram, an oval represents the project itself and a horizontal line divides the stakeholders. The upper section contains stakeholders who belong to the organization such as clients, functional departments, team members, etc. The lower part contains external staff such as advertising agencies (press, radio and television), competitors, citizens, government organizations or representatives.
- Coakes and Elliman [16] develop a method to identify stakeholders and their different viewpoints in a computer information system using a legacy system. The authors use a web that has a system of classification and stakeholders are grouped by using a holistic view of the situation. This facilitates an understanding and identification of agreements between stakeholder interests. The web not only

identifies stakeholders, it is also used to analyze relationships between activities which must be performed by the stakeholders and the members of the new system, with a view to prioritizing the proposed requirements. The system must be defined in terms of different boundaries: automation, technical, and total or human boundaries. Each boundary represents a wider view of the system and its impact. These limits are very generic and provide a general guideline as to who may be found within each limit. Stakeholder identification is a process that explores the web plane looking for interested parties. If stakeholders are identified, they are added to the web and thereby groups of stakeholders, needs and interests can be recognized.

- Sharp, Galal and Finkelstein [17] propose an approach to discover all stakeholders in the development of a specific software project. This identification is carried out by establishing a set of "baseline" stakeholders. From these, the "supplier" stakeholder (who provides information or supporting tasks to the baseline stakeholders) can be recognized, and also the "client" stakeholder (who examines products). Other stakeholders called "satellites" inter-act in various ways with the baseline stakeholders. "Interaction" may involve communicating, reading a set of rules or guidelines, searching for information, etc. Roles are assigned on the basis of an analysis of the interactions that can exist between different stakeholders and between the stakeholder and the system. Only the baseline stakeholder, however, is identified. The roles that they can perform are: users, developers, legislators, and decision makers.

- Preiss and Wegmann [18] adopt some fundamental principles of Systems Science to provide a generic, scientifically recognized basis that can aid stakeholder identification and classification. This method provides a framework that theoretically guarantees the identification of a complete set of relevant, abstract concepts and also all the stakeholders. The authors propose a generic stakeholder classification that is based on three principles: two systems, two viewpoints and two domains of enquiry. The software development life-cycle is divided into two stages: creation and operation, or in more concrete terms, the system creation stage (which includes conception, design and implementation), and the system operation stage (which includes system application in a real environment). Using this generic scheme, external and internal stakeholders can be identified.

## 4.3 Studies That Include Assessment of Stakeholders

- Mitchell, Agle and Wood [19] developed a theory of stakeholder identification based on the premise that identification is realized through an analysis of 'what' and 'who' affects the organization and also taking their salience into account. This identification is performed by analyzing stakeholders' interests in the project and considering three dynamic qualities, a) the power that the stakeholder has to propose requirements within the organization, b) legitimacy of the actions which a person performs within a certain social sys-tem constructed with the definition of norms, and c) urgency – meaning the degree of attention that a stakeholder claims from the project manager. Afterwards, possible stakeholders are grouped into three classes based on the degree of requirements priority: (a) latent or low salience, (b) expectant or moderate salience, and (c) definitive or high salience. Any stakeholder can become a definitive stakeholder by acquiring the missing attributes. Finally, the project manager determines who will be the 'feasible' stakeholder to include in the project, by analyzing a variety of existing stakeholder classes.

- The aim of the Method Engineering with Stakeholder Input and Collaboration, MEWSIC [20] (Young, McDonald, Edwards and Thompson), was to provide software developers with a practical tool to identify stakeholders. The method groups all the people involved in a project depending on the priorities of their interests. In stakeholder analysis, the MEWSIC approach suggests the identification of people related to the project and an assessment of their relevance to the project being developed to determine if he/she should or should not be taken into account. Appropriate stakeholders are those who have not only relevant knowledge and skills but also have suitable attitudes towards the process; some stakeholders, for example, are not suitable for negotiating. For this reason, this method proposes the use of personality tests to complement stakeholder analysis and make it possible to achieve an adequate fit between the stakeholder, the system context, and the characteristics of the development project. Furthermore, the method introduces group dynamics to observe stakeholder behavior within groups and how this affects a member's performance.

- McManus' approach analyzes the guide proposed by the World Bank (WB) model in 1996 for stakeholder identification [21]. The goal of this stakeholder analysis is to identify stakeholder categories using a WB questionnaire, to develop a strategic view of the situation and the relationship between different stakeholders and identified objectives, and to explain stakeholder interests and roles. The questionnaire results should provide information about stakeholders and their interests, the relationship between them, their motivations, and their ability to influence outcomes. The WB model proposes four groups of stakeholders. Primary stakeholders include those who, because of power, authority, responsibilities or claims over the resources, are primordial for any project initiative. Secondary stakeholders are those who have an indirect interest in the outcome. External stakeholders come from outside the project and expect something from the project team. And finally, the Extended stakeholders may often be of help to primary and secondary stakeholders to reach a unified vision and develop feasible plans of action. The roles of the four groups are established by means of the following activities: collecting and analyzing information, defining priorities and establishing goals, assessing available resources, project planning, designing strategies to implement these programs and dividing responsibilities among participants who manage a project, monitor progress, and evaluate results and impacts. However, since the identified stakeholders may not have enough capacity to participate in the project, each one's strengths and weaknesses is identified in order to form collaboration groups.

## 5 Conclusions

During software requirements elicitation we decide what exactly is to be produced. At this stage, the appropriate identification of stakeholders is vitally important as a means of understanding the environment in which the software project will be developed and operated, and also to identify which stakeholders will participate in the project. This is a key aspect in the process of obtaining the expected quality requirements specifications, in the sense that they must be appropriate, complete, and free of contradictions. This means that all stakeholders need to have appropriate knowledge and none stakeholder can be omitted. Good interaction is vital during the requirements-gathering process, and also between all stakeholders and the system to avoid conflicts and problems of communication arising from different points of view.

The state-of-the-art of SIP referred to in the present paper shows different interpretations of the scope of this process. All of the software initiatives referred to assume a way of contributing to the improvement of the software process by implementing a set of good industry practices for RE that have been identified, acknowledged, and disseminated, however they have not explained how to carry out the SIP. Some initiatives provide numerous examples of who can be stakeholders by establishing generic categories into which they may be grouped.

Other studies analyzed are more ambitious. However, the studies mentioned in this paper are not standardized and consequently the SIP is not standardized either. Not all of them, however, cover the same aspects and are not applicable to the same situations. This makes it difficult to select a correct stakeholder identification method be-cause some methods only characterize stakeholders but without assigning a stakeholder's role in a specific project (Lauesen [13] and Robertson's VOLERE template [14]); others like Smith [15], Coakes and Elliman [16], Sharp, Galal and Finkelstein [17] and Preiss and Wegmann [18] analyze stakeholder interaction but they do not cover human aspects of stakeholder identification (e.g. personality tests, human behavior). Only a few methods include stakeholder assessment (Mitchell, Agle and Wood [19], MEWSIC [20] and McManus [21]). Furthermore, not all the studies analyzed take into account aspects such as when and how we know that the stakeholders identified are sufficient for the project, and how all the information collected will be documented.

The SIP must take into account the impact of personality types and the roles they may play. Stakeholders are assessed in terms of their characteristics, the knowledge needed, their influence on a project, and the relationships between stakeholders. SIP must also develop schemes to characterize and evaluate appropriate relationships between all stakeholders. For example, labels such as "one person is in charge of", "this person is an assistant to", "he/she is crucial for", "he/she provides the information for" could be used.

Up to now, the SIP continues receiving very little attention from the different existing initiatives in software development (for example, CMMi, SWEBOK, the IEEE Standard 830), despite the fact that success in software products depends to a great deal on proper stakeholder identification in requirements specification. Therefore, on the basis of this literature survey, as further research we propose:

- The validation of all empirical studies analyzed in a specific project; this will enable us to determine its effectiveness.
- The development of a guide which would recommend the use of a specific method of stakeholder identification based on the particular characteristics of the project to be developed.
- Develop a new methodology to adequately perform the SIP by covering the shortcomings found in the methods analyzed in this paper.

## References

1. International of Electrical and Electronics and Electronics Engineers. Guide to the Software Engineering Body of Knowledge. New York (2004) www.swebok.org
2. Software Engineering Institute, Capability Maturity Model® Integration (CMMI®). Version 1.1, CMMI-SE/SW/IPPD/SS V1.1. Technical Report CMU/SEI-2002-TR-011 or ESC-TR-2002-011. Software Engineering Institute. Pittsburg, PA (March 2002)
3. IEEE. Guide to Software Requirements Specification, IEEE Std. 830-1998. IEEE Press. Piscataway, NJ (1998)
4. Wiegers, K.E.: Software Requirements. Microsoft Press. ISBN 0-7356-0631-5 (1999)
5. Hofmann, H.F., Lehner, F.: Requirements Engineering as a Success Factor in Software Projects. IEEE Software, July/August 2001 (2001)
6. Sommerville, P.S.: Requirements Engineering. A Good Practice Guide. J. Wiley & Sons Ltd. ISBN 0 471 97444 7 (1997)
7. Kotonya, G., Sommerville, I.: Requirements Engineering: process and techniques. John Wiley and Sons, New York (2000)
8. Software Institute Requirements Engineering Project. Issues in Requirements Elicitation. Technical Report CMU/SEI-92-TR-012 or ESC-TR-92-012. Software Engineering Institute, Pittsburg, PA (September 1992)
9. Lawson, H.W.: Defining Stakeholder Relationships. IEEE Computer Society, Washington (1999)
10. Sommerville, I.: Ingeniería de Software. In: 6a Edición, Addison-Wesley, London (2002)
11. Pressman Roger, S.: Software Engineering: A Practicioner's Approach, 6th edn. Mc Graw Hill, New York (2005)
12. Kruchten, P. (ed.): The Rational Unified Process an Introduction, 3rd edn. Addison Wesley, London (2003)
13. Lauesen, S.: Software Requirements: Styles and Techniques. In: Pearson Education, ISBN 0 201 74570 4, Addison-Wesley, London (2002)
14. Robertson, S., James, R.: Mastering the requirements process. Addison-Wesley, London (1999)
15. Smith, L.W.: Project clarity through stakeholders analysis. The Journal of Defense Software Engineering. Issue (December 2000) http://stsc.hill.af.mil/crosstalk/2000/12/smith.html
16. Elayne, C., Tony, E.: The role of the stakeholder in managing change. Communications of AIS, vol. 2, Article 4 (July 1999)
17. Sharp, H., Galal, G.H., Finkelstein, A.: Stakeholder Identification in the Requirements Engineering Process. In: Database & Expert System Applications, IEEE Press, New York (1999)
18. Preiss, O., Wegmann A.: Stakeholder Discovery and Classification Based on Systems Science Principles. Quality Software, 2001. IEEE Proceedings. Second Asia-Pacific Conference On, December 10-11, 2001. pp. 194–198 (2001)
19. Mitchell, R.K., Agle, B.R., Wood, D.J.: Toward a Theory of Stake-holder Identification and Salience: Defining the Principle of Whom and What Really Counts. Academy of Management Review. vol. 22(4) (1997)
20. Young, M., McDonald, S., Edwards, H.M., Thompson, J.B.: Quality & People in the Development of Situationally Specific Methods. Quality Software. 2001, Proceedings IEEE. Second Asia-Pacific Conference On December 10-11, 2001, pp. 199–203. (2001)
21. McManus, J.A: Stakeholder Perspective within Software Engineering Projects. In: Engineering Management Conference on Proceedings, IEEE International, October 18-21, 2004, vol. 2. pp. 880–884 ( 2004)

# The Impact of Task Structure and Negotiation Sequence on Distributed Requirements Negotiation Activity, Conflict, and Satisfaction

Bartel Van de Walle[1], Catherine Campbell[2], and Fadi P. Deek[2]

[1] Department of Information Systems and Management, Tilburg University, The Netherlands
[2] College of Computing Sciences, New Jersey Institute of Technology, USA
bartel@uvt.nl, {campbell, fadi.deek}@njit.edu

**Abstract.** This paper reports the findings of an experimental study of web-based negotiations among a group of distributed stakeholders involved in the design of a complex information system. Using a web-based communication system, the stakeholders had to reach agreement on a common set of software requirements taking into account their individual preferences as well as overall constraints of available time and budget. To support such complex negotiations, the objective of our study was to analyze the impact of providing structured task and explicit negotiation sequence support to the negotiating group with respect to their activity, conflict and satisfaction. Our results show that groups following a structured task are more active than groups lacking such structure. However, the absence of negotiation sequence and structured task support leads to greater satisfaction.

**Keywords:** distributed negotiation; software requirements; negotiation process support; group work; distributed group support systems.

## 1 Introduction

According to Weigand et al. [41], in a negotiation "*there are two or more participants in a situation of some kind of interdependence, each having some individual goals which may be partially incompatible, and in some form of the negotiation process, alternatives are investigated, of which one is mutually agreed upon as the acceptable outcome of the process.*" In today's global economy, distances between businesses' physical locations, possibly located in different time zones, often leave business partners no alternative but to conduct distributed negotiations over the internet, typically supported by dedicated software [19]. The use of software to support communication among distributed groups has been studied since a long time. Hiltz and Turoff were among the first to conduct empirical research on the use of computer-mediated communication systems for distributed groups [20], and the use of Decision Support Systems for negotiators was proposed around the same time [35]. These pioneering developments later evolved into dedicated Negotiation Support Systems (NSS) to aid negotiators, for instance by generating alternatives based on knowledge of the negotiators' preferences and utility functions [22]. NSS often draw

J. Krogstie, A.L. Opdahl, and G. Sindre (Eds.): CAiSE 2007, LNCS 4495, pp. 381–394, 2007.

from models and approaches in multi-criteria decision making, preference analysis or game theory [32, 33, 38].

Computer-mediated Communication (CMC) for software design teams has been investigated by Ocker and co-workers since the late 1990s [29, 30]. Ocker found that the creative task of deciding upon the initial specifications for the design of a software system can benefit from asynchronous CMC. Groups using CMC were found to be considerably more creative in their designs; the quality of the designs was also judged to be higher, though not significant [29]. In recent research on computer-mediated distributed software requirements engineering, Damian and co-workers studied the role of a facilitator in distributed group discussions. Faciliated meetings can help resolve disagreements among the group members, yet Damian found that "*a reduced richness of social behaviors in computer-mediated group settings made the groupfacilitation problematic*" [9]. Clearly, due to the nature of mediated communication in asynchronous interaction, it is relatively more difficult to coordinate distributed groups, and as an alternative to human faciliation, specific coordination structures must be arranged to overcome these difficulties. Kim, Hiltz and Turoff have examined the effect of system restrictiveness of coordination structures in an asynchronous environment. Their study found that less restrictive coordination structures are more appropriate to support asynchronously interacting distributed groups [26].

We continue in this paper along the lines of the CMC research on the impact of structured interaction, and introduce two types of interaction structures: a structured negotiation task, and a structured negotiation sequence. The effect of these structures on group activity, the group's satisfaction with the negotiation process, and conflict within the negotiating group is the focus of our investigation. Activity, satisfaction and conflict are key variables in Group Support Systems (GSS) research (see [14] for a detailed survey on experimental research in GSS, with an exhaustive list of experimental variables). We in particular focus on negotiations that take place during the phase of system requirements definition. Negotiation is typically needed in this phase to address and resolve conflicting system requirements which have been articulated by the different stakeholders at, or prior to, the start of the system design process [23].

In the following section, Section 2, we provide a concise theoretical overview of negotiation research and negotiation models on which we base our experimental study. We also introduce electronic negotiation systems (ENS) which provide information, communication and decision support to the negotiating members of a distributed group. In Section 3, we present our research model and formulate our research hypotheses. The experiment we have designed is presented in detail in Section 4, and the results of our experiment are given in Section 5. These results are discussed in Section 6, followed by our conclusions and a discussion of the limitations of this study in Section 7 where we also suggest avenues for future research.

## 2 Negotiation Research and Negotiation Support Systems

### 2.1 Negotiation Research

Drawing from a wide area of disciplines, in particular cognitive psychology and behavioral decision research, negotiation researchers have come to view participants in a negotiation as interdependent decision makers whose behavior is a result of choices based on often conflicting judgments about the negotiation situation [1]. In negotiated decisions, conflict and interdependence are indeed two key aspects. Conflicts may arise during negotiations because of the participants having different goals, yet conflict can also result from participants having common goals but different ideas on how to achieve them [31]. Decision makers in a negotiation are also interdependent, as one participant reaching his or her objective is influenced by others agreeing to it, and vice versa. In case the negotiation involves multiple participants, conflict among the participants can add considerable complexity and confusion as each participant has only limited information or knowledge about the other participants' preferences [34, 39]. According to the process model of negotiations first proposed by Gulliver [17] and later expanded by Kersten [23], the phases and activities of multiple participants during negotiations can be described as follows:

- *Search for arena and selection of the communication mode*: reaching agreement on where the decision process will take place. This can be a face-to-face or virtual meeting space for synchronous or asynchronous exchange of information.
- *Agenda setting*: agreeing to the issues and, if possible, partial problem representation and categorization.
- *Exploring the field*: further problem specification and analysis where parties try to establish limits, formulate best alternatives, assess their opponents, and decide on initial negotiation strategies.
- *Narrowing the differences and search for integration*: through exchange of information, the participants learn of the limitations of others, their aspirations and objectives, and knowledge about efficient solutions and their outcomes.
- *Search for agreement and improvements*: identify critical issues and areas of disagreement, try to develop joint proposals to come up with a limited number of acceptable compromises.

## 2.2 Distributed Group Support Systems and Electronic Negotiation Systems

The use of computer technology may help overcome constraints that are experienced in either face-to-face (same time – same place) or distributed negotiation situations. Group support systems (GSS) are designed to support decision makers in complex negotiation tasks, and have been proven to benefit the exchange of information among participants [11, 28]. It was found that groups interacting using a GSS exchanged about 50% more information than those interacting only verbally [11] or exchanged more unique, unshared information [34]. GSS structures or tools can be combined with basic communication support to enable decision making by groups distributed in space and time [7, 12] – a group support technology referred to as Distributed Group Support Systems (DGSS) [36]. Negotiation Support Systems (NSS) are GSS that specifically focus on negotiations and provide decision support to each negotiator [22, 27]. More advanced NSS feature group process structuring techniques, mediator support, and documentation of the negotiations [10, 15]. Electronic Negotiation Systems (ENS) in general employ the internet and computing technologies to support decision makers during negotiations, and have one or more of the following characteristics [24, 25]:

- Supports decision and concession making;
- Suggests offers and agreements;
- Assesses and criticizes offers and counteroffers;
- Structures and organizes the process;
- Provides information and expertise;
- Facilitates and organizes communication;
- Aids agreement preparation, and
- Provides access to negotiation knowledge, experts, mediators, or facilitators.

ENS range from straightforward web-based communication systems to sophisticated intelligent systems actively contributing to the negotiation process by assessing what is being negotiated while interacting with participants and providing offer assessment and critiques of counteroffers.

## 3 Distributed Software Requirements Negotiation: Research Model and Hypotheses

### 3.1 Software Requirements Negotiation Engineering

Negotiating conflicting requirements is an important part of the requirements phase of the software engineering life cycle [2, 8, 16]. Many studies have found that clear, consistent, and traceable requirements result in more robust, maintainable information systems [5, 6]. In order to provide this quality, consistency, and traceability, conflicts between users, analysts, and managers must be resolved during this important phase of software design and development [21]. Conflict in and of itself is not a negative activity at this stage of software development. Actually, conflict and its resolution at an early stage in the development process can have the result of saving both time and money further along in the development life cycle [18]. One of the most commonly known and studied approaches to resolving conflict and reaching agreement in requirements design is the spiral model proposed by Boehm and colleagues [3]. Boehm's spiral approach focuses on stakeholder collaboration and negotiation to achieve so-called 'win-win' software requirements where all stakeholders are winners in the negotiations. Boehm's model facilitates the uncovering and resolution of problems before they reach the implementation stage and therefore save time and money while at the same time enhancing software quality and reliability [4].

### 3.2 Research Model and Hypotheses

To support the negotiation process of groups whose members are distributed in space and time, an ENS can provide various levels of support to the negotiating group members. In this research, we focus on two basic types of support that the ENS can make available to negotiating parties: structured task and negotiation sequence support. We consider an ENS to provide structured task support when the ENS provides the negotiating group with clearly outlined specific steps on how to address their task, so that their discussions can proceed in a well-structured manner. In the absence of such a specific support within the ENS, we will consider the negotiation task as unstructured. Similarly, we consider the ENS to provide negotiation sequence support when the ENS provides the negotiating group with guidance on how to approach their negotiation, for instance by following the negotiation steps proposed by Gulliver and Kersten as described in Section 2.1 above. The objective of our research is to investigate how the presence or absence of structured negotiation task and negotiation sequence support in the ENS affects the activity during a distributed negotiation, as well as the satisfaction and conflict among participants.

Within GSS research, group effectiveness and productivity have often been measured in terms of number of ideas, alternatives or comments generated. Fjermestad and Hiltz identify not less than 167 productivity measures in their comprehensive review of about 200 experimental GSS studies published until 1998 [14]. Negotiation strategies, such as sequencing, can contribute to more collaboration and the convergence of different perspectives [3]. We therefore hypothesize:

**H1a.** Groups provided with negotiation sequence support will be more active compared to groups with no negotiation sequence support.

**H1b.** Groups provided with structured task support will be more active compared to groups with no structured task support.

Since structured task, as well as negotiation sequence support, provides guidance to the negotiating groups, we can hypothesize that their use in the ENS will reduce conflict in negotiating distributed groups:

**H2a.** Groups provided with negotiation sequence support will exhibit less conflict in their negotiations compared to groups without this support.

**H2b.** Groups provided with structured task support will exhibit less conflict in their negotiations compared to groups without this support.

Research on group decision making also investigates how satisfied groups are with the process to arrive at a decision. It has been shown that technology-supported groups can produce more democratic (individuals participating with equal value) or more fairly distributed decision-making processes [12, 40], which in turn leads to more satisfaction with the outcome. The more democratic the process, the more satisfying the group process and outcomes will be [13]. We therefore hypothesize:

**H3a.** Groups provided with negotiation sequence support will show more satisfaction with their negotiation process compared to groups with no negotiation sequence support.

**H3b.** Groups provided with structured task support will show more satisfaction with their negotiation process compared to groups with no structured task support.

# 4 Research Methodology

## 4.1 Experimental Design – Independent Variables

The independent variables in our experimental study are 'Structured Task' and 'Negotiation Sequence', leading to the 2 X 2 experimental design as shown in Table 1, which also lists the number of groups that were assigned to each experimental condition. Groups assigned in the 'structured task' condition are provided by the ENS with clearly outlined specific steps on how to approach their task. Groups assigned to the 'negotiation sequence' condition are aided through their negotiation by the ENS with guidelines according to the Gulliver and Kersten negotiation process [17, 23].

**Table 1.** Experimental design with number of groups and subjects per condition

<table>
<tr><td colspan="2" rowspan="2">CONDITION</td><td colspan="2">Task Structure Support</td><td></td></tr>
<tr><td>With</td><td>Without</td><td>Total</td></tr>
<tr><td rowspan="2">Negotiation Sequence Support</td><td>With</td><td>8 Groups<br>N = 43</td><td>8 Groups<br>N = 46</td><td>16 Groups<br>N = 89</td></tr>
<tr><td>Without</td><td>8 Groups<br>N = 39</td><td>8 Groups<br>N = 46</td><td>16 Groups<br>N = 85</td></tr>
<tr><td colspan="2">Total</td><td>16 Groups<br>N = 82</td><td>16 Groups<br>N = 92</td><td>32 Groups<br>N = 174</td></tr>
</table>

### 4.2 Dependent Variables and Measures

Three dependent variables are studied: Activity, Conflict, and Process Satisfaction. Conflict and Process Satisfaction are measured by a post-task questionnaire; the items used are listed in Table 2. All items are measured on a 7 point Likert-type scale with anchors from "Strongly disagree" to "Strongly agree". The questionnaire included items worded with proper negation and a shuffle of the items to reduce monotony of questions measuring the same construct. Activity is directly measured by the number of messages posted by the group.

**Table 2.** Conflict and Process Satisfaction question items used in the study

| Dependent variables | | Post-task questionnaire items |
|---|---|---|
| Conflict | 1 | To what extent did the group experience conflict? |
| | 2 | Did the group handle conflict effectively? |
| Process Satisfaction | 3 | How efficient would you describe your group's problem solving process? |
| | 4 | How coordinated would you describe your group's negotiation process? |
| | 5 | How fair would you describe your group's negotiation process? |
| | 6 | How understandable would you describe your group's negotiation process? |
| | 7 | How satisfying would you describe your group's negotiation process? |

### 4.3 Experimental Task

The experimental task offers a hypothetical scenario in which a distributed group of stakeholders is asked to negotiate the design requirements for a new information system. The information system described in the task is an Emergency Response Information System (ERIS) that is to be developed for use in a (fictitious) US County, and the stakeholders are organizations in the County which would use or rely on the ERIS in case of an emergency [37]. Five distinct roles were presented in the task: law enforcement, fire containment, public works, public health, and state home/civil defense unit liaison. Each participant in the experiment was provided with a specific role description according to the stakeholder organization he or she represented, and a set of organization-specific requirements the organization would like to see implemented in the ERIS. Participants only knew their personal requirements and were not informed of the requirements of the other stakeholders. Some requirements were common to all stakeholders while others were unique to the respective organizations, yet either unique or common requirements could only be 'disclosed' through the negotiation. The task description indicated that state and local government funds had been obtained to actually implement the ERIS, a task which was to be carried out by an independent software firm. However, since the funding was insufficient to implement all and everyone's requirements, the stakeholders were asked to negotiate to agree on a subset of requirements that will go forward for implementation in ERIS and would satisfy, in some way, the stakeholders most pressing needs. The stakeholders were told that the software firm's Project Manager would join their negotiations, whose role was described as to supply costs and resources to the group, to chair the meetings, and to provide facilitation.

### 4.4 Participants and Roles

The subjects for this study were graduate students enrolled in Systems Analysis and Design and Software Engineering courses at the New Jersey Institute of Technology. Demographics were collected with an online background questionnaire. The distribution by major was Computer Science 36%, Information Systems 15%, Engineering 44 %, Management 2%, and Other 3 %. The population included 51 females and 136 males. The education level of the students was 3.7 % undergraduate seniors, 39 % Information System Masters, 59% Computer Science Masters, 1% MBA, 2% PhD students, and 13% Other. The ages of the participants ranged from under 23 to over 40 with the majority in the 23-30 age range.

Subjects were randomly assigned to roles in groups of six, i.e., the five stakeholder roles and one Project Manager. The task was given to the students as part of their course work and they were graded individually on their participation. Initial participant training was completed in the classroom in a one-hour face-to-face training session. During this session the subjects were introduced to the domain topic and the general task to be completed, given a brief background of requirements engineering (this was covered in depth by their current or previous coursework), and introduced to the project roles and what was expected of them.

### 4.5 Electronic Negotiation Environment

The commercially available WebBoard software (a product of Akiva Corporation) is used as the web-based collaboration environment in which the distributed groups are conducting their negotiations. Although WebBoard is not specifically designed as an ENS, and does not have any standard negotiation support tools built-in, it does allow for an easy configuration of its group communication structures (threaded discussion forums) so that negotiations can take place as threaded discussions. WebBoard has a proven track record within education and corporate environments to create virtual classrooms, for internal project collaboration, for establishing standards and best practices and for customer support applications. The WebBoard interface was customized for the experimental conditions to accommodate for the presence or absence of structured task and negotiation sequence support.

### 4.6 Experimental Procedures

The experiment was conducted over two consecutive semesters during the academic year 2003-2004 at the New Jersey Institute of Technology (NJIT). During the fall 2003 semester, a pilot study was run to test the technology, group roles and interactions, task, and experimental procedures. Based on the results of the pilot study, some changes were made for the formal experiment which was run with graduate students in spring 2004. In the formal experiment in spring 2004, there were 32 Groups with eight groups per condition. The total subject population participating at the start of the experiment was 192; as a result of drop-out during the time of the experiment, the total number of participants was 174. After completion of a training task, the subjects were allowed to start the negotiations which were to be conducted during five consecutive days. At the end of the negotiations, the groups had four days to complete and comment on the final report before uploading it to WebBoard. Upon completion of the negotiations, the subjects were directed to a post-task online questionnaire. The subjects were encouraged to complete the questionnaire as soon as possible after they finished negotiating or by the sixth day of the experiment at the latest. Upon completion of the experiment and submission of all required documentation, each instructor was given a suggested grade for participation of each subject. Subjects were invited to an online debriefing conference which detailed the experimental design, procedures, and a brief theoretical background for the research.

## 5 Analysis of the Results

### 5.1 Negotiation Activity

Negotiation Activity variable was measured directly by analyzing the WebBoard conferences in which the group conducted their discussions, and count the number of messages posted by the groups.

**Table 3.** The ANOVA on messages posted

| Source | d.f. | MS | *F* | Significance |
|---|---|---|---|---|
| Negotiation Sequence Support | 1 | 195.031 | .248 | .623 |
| Task Structure Support | 1 | 4117.781 | 5.226 | **.030** |
| Negotiation Sequence Support * Task Structure Support | 1 | 1116.281 | 1.417 | .244 |
| Error | 28 | 787.942 | | |

Of the 32 groups that participated in the experiment, 29 did reach consensus on the list of requirements to be implemented. 21 groups reached consensus within 6 days. 11 groups took six days, 5 groups took seven days and one group each took eight and nine days respectively. 4 groups that did not have a negotiation sequence or task structure reached agreement within five days, with the most groups reporting consensus within that time. A two–way between-groups analysis of variance was conducted to explore the impact of negotiation sequence and task structure support on the group activity as measured by the number of messages generated by each group. From Table 3, it follows that there was a statistically significant effect for task structure ($p$ = .030). Therefore, **H1a** is supported. The main effect for negotiation sequence ($p$ = .623) did not reach statistical significance. Therefore Hypotheses **H1b** is not supported. There was no statistically significant interaction effect ($p$ = .244).

## 5.2 Conflict

The Conflict variable was measured by the two different post-task questionnaire items given in Table 2. For both items, a one–way between-groups analysis of variance was conducted, the results of which are shown in Table 4. For the first item, "*To what extent did the group experience conflict?*", no statistical significant effect was found ($F$ = 1.253, $p$ = 0.292). For the second item "*Did the group handle conflict effectively?*", a significant statistical effect was found ($F$ = 2.656, $p$ = 0.05) for the 'no negotiation sequence support' and 'no task structure support' conditions, meaning the groups in those conditions handled conflict the most effectively. The effect size, calculated using Eta Squared, was .04, and therefore the actual difference in means scores between the groups is quite small. Therefore, neither Hypothesis **H2a** (groups provided with negotiation sequence support will exhibit less conflict in their negotiations compared to groups without this support) nor Hypothesis **H2b** (groups provided with structured task support will exhibit less conflict in their negotiations compared to groups without this support) are supported.

**Table 4.** ANOVA on Conflict

**Conflict Measure 1**: To what extent did the group experience conflict?

| | Sum of Squares | df | Mean Square | F | Significance |
|---|---|---|---|---|---|
| Between Groups | 10.342 | 3 | 3.447 | 1.253 | .292 |
| Within Groups | 467.566 | 170 | 2.750 | | |

**Conflict Measure 2**: Did the group handle conflict effectively?

| | Sum of Squares | df | Mean Square | F | Significance |
|---|---|---|---|---|---|
| Between Groups | 14.335 | 3 | 4.778 | 2.656 | .050 |
| Within Groups | 305.854 | 170 | 1.799 | | |

## 5.3 Process Satisfaction

The Process Satisfaction variable was measured in the post-task questionnaire by questions 3-7 as given in Table 2. Chronbach's alpha of the process satisfaction scale is 0.89 which indicates that this measure is valid. The questions are summed and averaged to achieve a scale measure for this variable. Significance was tested using a two-way Analysis of Variance test. There was no main effect for negotiation sequence ($p = .46$). There was a statistically significant main effect for task structure ($p = .00$). The interaction effect ($p = .13$) did not reach statistical significance.

**Table 5.** ANOVA on process satisfaction

| Source | d.f. | MS | *F* | Significance |
|---|---|---|---|---|
| Negotiation Sequence Support | 1 | 26.355 | .560 | .455 |
| Task Structure Support | 1 | 855.697 | 18.176 | **.000** |
| Negotiation Sequence Support * Task Structure Support | 1 | 110.112 | 2.339 | .128 |
| Error | 170 | 47.078 | | |

The ANOVA results in Table 5 show that the Negotiation Sequence Support was not significant as a main effect ($F = 0.560$, $p = 0.455$). Therefore, Hypothesis **H3a** (groups provided with negotiation sequence support will show more satisfaction with their negotiation process compared to groups with no negotiation sequence support) is not supported. Structured Task Support on the other hand is significant as a main effect ($F = 18.17$, $p = 0.000$): those groups that did *not* follow a task structure had significantly *higher* process satisfaction than those that did. This implies that Hypothesis **H3b** (groups provided with structured task support will show more satisfaction with their negotiation process compared to groups with no structured task support) also is not supported. There is no significant interaction effect between Negotiation Sequence Support and Structured Task Support ($F = 2.339$, $p = 0.128$).

## 6 Discussion

The *negotiation activity* variable produced a significant effect for task structure. The groups using a task structure were significantly more active than those that were not provided with this structure. Negotiation sequence support however does not produce a significant effect, and the corresponding hypothesis is not supported. Also, there was no interaction effect for groups following a structured task and negotiation sequence. Using the task structure provided in the ENS may have helped the subjects focus on the task at hand and make the groups aware of all requirements available, leading to an increased communication activity.

The results of the items measuring the *conflict* experienced within the group show that the least conflict was in the groups that used no task structure and no negotiation sequence; this was a statistically significant finding. Therefore no conflict hypotheses were supported and our findings actually indicate the reverse of the experimental hypotheses for this variable. This may be explained by the fact that the lack of any support with respect to the task at hand and the negotiation process did allow the group members to communicate wherever and whenever they pleased within the electronic conference space. Since there was no guidance provided – no rules of the game – group members could not rely on 'objective' arguments to argue against this type of 'free' communication behavior. As such, the absence of structure may have lessened the pressure group members experience when participating in the negotiation. Conversely, it can be argued that the very presence of the structure and negotiation sequence support descriptions may have provided too much information for the students, leading to information overload.

The *process satisfaction* variable measured the negotiation and problem solving process as efficient, coordinated, fair, understandable, or satisfying. Satisfaction with the negotiation process of choosing a set of optimal requirements was found to be best when the groups did not follow either a negotiation sequence or structured task. Groups *not* following a structured task showed a significant result and felt the most satisfaction with their process, regardless of their negotiation support mode. Again, the absence of structure allowed the students to post freely, creating their own group process and communication structures in the WebBoard negotiation space. Subjects may prefer this type of communication mode and therefore may have felt less restricted in their group interactions. This perception may have been strengthened by most subjects' familiarity with the WebBoard software, which they have used for other purposes without task or negotiation support structures in place.

## 7 Conclusions and Limitations

Requirements negotiation has been well studied by several researchers, but to our knowledge this is the first study to address the negotiation of software requirements in an asynchronous and distributed communication mode. With global software engineering becoming the norm rather than the exception, interest in this area is increasing. Research interest has focused on the downstream phases of the software engineering process where workloads can more easily be partitioned and compartmentalized for distribution to different organizations or groups. The requirements engineering process has traditionally been conducted face-to-face. With the growth of global business and e-commerce, stakeholders are increasingly distributed, ensuring the contribution of all stakeholders is a crucial part of the requirements engineering process in order to obtain the most robust requirements for the proposed system. This research can contribute to encouraging and enabling distributed stakeholder groups to actively participate in the requirements process. The contribution of this research consists of a detailed analysis of two communication coordination mechanisms, the structured task and negotiation sequence. The results from our experiments were however less indicative than the authors had hoped for, which could be due to several limitations of this study.

Although there was an extensive pilot study prior to the formal experiment, several limitations still can be identified. Foremost perhaps, the experimental subjects were graduate students - a common experimental limitation that has affected several areas of the study. Most of the subjects were unfamiliar with the complex task domain of emergency response and therefore prior to participating, needed rather extensive training to participate in a meaningful manner. Subjects had to read and comprehend the roles and task; if these were not clear, it was difficult to negotiate from a position of strength and other stakeholders would be able to sway them more easily. In addition, many of the subjects had English as their second language, which also may have impacted their task and role comprehension. Another limitation of this study is the restrictive communication imposed on the participants, who could only communicate through the ENS (WebBoard). Nowadays, in most working environments, multiple modes of communication are concurrently used, such as chat or internet telephony. The single communication mode imposed by this experimental study may have been experienced as too rigid and restrictive by those subjects who had a more flexible experience in the work place.

These limitations also give us important directions for future research. The authors intend to conduct a follow-up experiment with professional emergency responders as subjects. Obviously, several restrictions regarding subject availability and ENS training and use have to be overcome. Preparations for this follow-up experiment are currently under way.

**Acknowledgments.** The first author's research is supported by the European Commission under the Sixth Framework Programme through a Marie Curie Intra-European Fellowship. We thank four anonymous referees for their helpful comments.

## References

1. Bazerman, M., Carroll, J.S.: Negotiator cognition. Research in Organizational Behavior 9, 247–288 (1987)
2. Blackburn, J., Scudder, C., Van Wassenhove, L.: Concurrent software development. Communications of the ACM 43(11), 200–214 (2000)

3. Boehm, B., Bose, P., Horowitz, E., Lee, M.J.: Software Requirements Negotiation and Renegotiation Aids: A Theory-W Based Spiral Approach. Communications of the ACM, pp. 243–253 ( 1995)
4. Boehm, B., Abi-Antoun, M., Port, D., Lynch, A.: Requirements Engineering, Expectations Management, and the Two Cultures. Center for Software Engineering, Technical Report, USC (1998)
5. Bray, I.K.: An Introduction to Requirements Engineering. Addison-Wesley, Essex, UK (2002)
6. Brooks, F.: No silver bullet: essence and accidents of software engineering. In: Kugler, H.J. (ed.) Information Processing, Elsevier Science Publishers, North-Holland (1986)
7. Bui, T., Jarke, M.: Communications requirements for group decision support systems. Journal of Management Information Systems 2(4), 8–20 (1986)
8. Conboy, K., Lang, M., Barry, C.: An investigation of the use of requirements prioritization in web-based information systems development. In: O'Toole et al. (eds.) Proceedings of 5th Irish Academy of Management Conference (2002)
9. Damian, D.E., Eberlein, A., Shaw, M.L.G., Gaines, B.R.: An Exploratory Study of Facilitation in Distributed Requirements Engineering. Requirements Engineering Journal 8(1) (2003)
10. Delaney, M.M., Foroughi, A., Perkins, W.C.: An empirical study of the efficacy of a computerized negotiation support system. Decision Support Systems 20, 185–197 (1997)
11. Dennis, A.R., Valacich, J.S., Connolly, T., Wynne, B.: Process structuring in electronic brainstorming. Information Systems Research 7(2), 268–277 (1996)
12. DeSanctis, G., Gallupe, R.B.: A foundation for the study of group decision support systems. Management Science 33(5), 589–609 (1987)
13. Dufner, D.: Effects of group support (listing and voting tools) and sequential procedures on group decision making using asynchronous computer conferences. Ph.D. Dissertation. Rutgers, The State University of New Jersey (1995)
14. Fjermestad, J., Hiltz, S.R.: An assessment of group support systems experimental research: methodology and results. Journal of Management Information Systems 15(3), 7–149 (1999)
15. Foroughi, A.: Minimizing negotiation process losses with computerized negotiation support systems. Journal of Applied Business Research 14(4), 15–26 (1996)
16. Grünbacher, P., Hofer, C.: Complementing XP with requirements negotiation. In: Proceedings of the 3rd International Conference on eXtreme Programming and Agile Processes in Software Engineering, pp. 105–108 (2002)
17. Gulliver, P.H.: Disputes and Negotiations: a Cross-Cultural Perspective. Academic Press, Orlando, Florida (1979)
18. Hall, T., Beecham, S., Rainer, A.: Requirements problems in twelve software companies: an empirical analysis. In: IEEE Proceedings of the Conference on Empirical Assessment in Software Engineering, pp.7–42 (2002)
19. Herbsleb, J.D, Moitra, D.: Global software development, IEEE Software, pp. 16–20 (March/June 2001)
20. Hiltz, S.R., Turoff, M.: The Network Nation: Human Communication via Computer. MIT Press, Cambridge (1993)
21. In, H., Roy, S.: Issues of visualized conflict resolution. In: Proceedings of the International Symposium on Requirements Engineering, Toronto, Canada, 10–15 (2001)
22. Jelassi, M.T., Foroughi, A.: Negotiation Support Systems: An Overview of Design Issues and Existing Software. Decision Support Systems 5, 167–181 (1989)

23. Kersten, G.E.: Support for group decisions and negotiations, an overview. INTERNEG Working Paper, pp.332–246 (1997)
24. Kersten, G.E.: Modeling Distributive and Integrative Negotiations. Review and Revised Characterization. Group Decision and Negotiation 10(6), 493–514 (2001)
25. Kersten, G.E.: E-negotiation systems: Interaction of people and technologies to resolve conflicts. INTERNEG Research paper presented at UNESCAP Third Annual Forum on Online Dispute Resolution, Melbourne, Australia (2004)
26. Kim, Y., Hiltz, S.R., Turoff, M.: Coordination Structures and System Restrictiveness in Distributed Group Support Systems. Group Decision and Negotiation 11(5), 379–404 (2002)
27. Lim, L.H., Benbasat, I.: A theoretical perspective of negotiation support systems. Journal of Management Information Systems 9(3), 27–44 (1992)
28. Nunamaker, J.F., Dennis, A.R., Valacich, J.S., Vogel, D.R., George, J.F.: Electronic meeting systems to support group work. Communications of the ACM 34(7), 40–61 (1991)
29. Ocker, R., Fjermestad, J., Hiltz, S.R., Johnson, K.: Effects of Four Modes of Group Communication on the Outcomes of Software Requirements Determination. Journal of Management Information Systems 15(1), 99–118 (1998)
30. Ocker, R., Hiltz, S.R., Turoff, M., Fjermestad, J.: The effects of distributed group support and process structuring on requirements development teams: results on creativity and quality. Journal of Management Information Systems 12(3), 127–153 (1996)
31. Pruitt, D.G., Carnevale, P.J.: Negotiation in Social Conflict, Open University Press (1993)
32. Raiffa, H., Richardson, J.: Negotiation Analysis. In: The Science and Art of Collaborative Decision Making, Harvard University Press, Cambridge (2003)
33. Rosenschein, J.S., Zlotkin, G.: Rules of Encounter. The MIT Press, Cambridge (1994)
34. Rutkowski, A.-F., Van de Walle, B., Van Den Eede, G.: The effect of group support systems on the emergence of unique information in a risk management process: a field study. In: Proceedings of the 39th Hawaii International Conference on System Sciences HICSS39 (2006)
35. Sprague, R.H., Carlson, E.D.: Building effective decision support systems. Prentice-Hall, Englewood Cliffs (1982)
36. Turoff, M., Hiltz, S.R., Baghat, A.N.F., Rana, A.R.: Distributed group support systems. MIS Quarterly, pp. 399–417 (1993)
37. Turoff, M., Chumer, M., Van de Walle, B., Yao, X.: The design of emergency response information systems. Journal of Information Technology Theory and Application 5(4), 1–36 (2004)
38. Van de Walle, B., Faratin, P.: Fuzzy preferences for multi-criteria negotiation. Position Paper for the American Association of Artificial Intelligence Fall, Symposium, Boston MA. Technical Report FS-01-03 (2001), pp. 116–119 (2001)
39. Van de Walle, B.: A relational analysis of decision makers' preferences. International Journal of Intelligent Systems 18, 775–791 (2003)
40. Watson, R.T.: A Study of Group Decision Support System Use in Three and Four-person Groups for a Preference Allocation Decision. Ph.D. Dissertation. University of Minnesota (1987)
41. Weigand, H., Schoop, M., de Moor, A., Dignum, F.: B2B Negotiation Support: The Need for a Communication Perspective. Group Decision and Negotiation 12, 3–29 (2003)

# Introducing Graphic Designers in a Web Development Process

Pedro Valderas, Vicente Pelechano, and Oscar Pastor

Department of Information System and Computation.
Technical University of Valencia, Spain
Cami de Vera s/n 46022
{pvalderas, pele, opastor}@dsic.upv.es

**Abstract.** Web development teams include not only software engineers but also graphic designers. In this work, we extend the OOWS method in order to introduce graphic designers into its development process. To do this, we extend the automatic code generation strategy of the OOWS method to obtain code that provides users with the information and functionality captured in the requirements model but without considering any kind of aesthetic aspect. We also propose a strategy to define domain-independent presentation templates. These templates can be applied to any web application developed by the OOWS method. These extensions allow us to define a web development process where graphic designers work together with analysts with a high degree of independence from each other but always in a coordinated way.

## 1 Introduction

When the Web Engineering was introduced at the beginning of the current decade [18], one of the problems that were presented to be solved was the multidisciplinary nature of Web development. This problem is related to the fact that Web applications handle information in its myriad forms (text, graphics, video, and audio) and it is very often published for worldwide access, publishing paradigm, and legal, social and ethical issues have to be taken on board. In this context, a correct understanding of additional disciplines such as usability, graphic design or information architecture is a key factor in the development of Web applications. As Powell [2] comments, web applications "involve a mixture between print publishing and software development, between marketing and computing, between internal communications and external relations, and between art and technology".

In this work, we present a first step in the definition of a whole multidisciplinary web development process. Taking into account the Powell's comment, we have started by considering the mixture between "art and technology" which led us to our central research question: "How can a coordinated work between graphic designers and analysts be achieved during the development of a Web application?"

As an answer to this research question we propose an approach that is based on the Web engineering method OOWS [8] [9]. OOWS follows the principles defined by the Model-Driven Development (MDD) [10] and it allows us to automatically obtain

J. Krogstie, A.L. Opdahl, and G. Sindre (Eds.): CAiSE 2007, LNCS 4495, pp. 395–408, 2007.

fully operative web application prototypes from a requirements specification. Then, considering that code can be automatically obtained, we extend the OOWS code generation strategy to obtain code which provides users with the information and functionality captured in the requirements model but without considering any kind of aesthetic property. We also propose a strategy to define domain-independent presentation templates which can be applied to any web application developed by the OOWS method.

Thus, the web development process that we propose introduces techniques and tools that (1) allow analysts to create a requirements model that precisely captures the user's needs and (2) allow graphic designers to define the look and feel of the web application (by means of domain-independent presentation templates) according to user preferences. These activities are performed with a high degree of independence between web professionals (analysts and graphic designers), but always in a coordinated way and directly interacting with the user. The OOWS code generation strategy is a key factor in achieving this. It allows us to automatically obtain code from the requirements model defined by analysts. This code is generated by following a strategy that allows us to associate it to the domain-independent presentation templates defined by graphic designers.

The main contributions of this work are:

- We propose a web engineering method that properly takes into account graphic designers in its development process.
- We clearly determine which activities each web professional (analysts and graphic designers) must perform.
- We provide web professionals with the techniques and tools needed to perform their tasks.

In order to develop this research work we have followed the research method described in [24]. This method is divided in five steps: (1) awareness of the problem, (2) suggestion, (3) development, (4) evaluation and (5) conclusion. The first step has been described in this section. Section 2 and 3 describe the second step. In these sections, we present first the related work and next we suggest a solution for the proposed research problem. This solution is a web development process based on the OOWS method. Sections 4, 5 and 6 support the third step in the research process: Section 4 presents the techniques and tools that support analysts in the performance of their activities. Section 5 introduces the OOWS method and the extended code generation strategy. Section 6 presents the techniques and tools that support graphic designers in the achievement of their activities. Section 7 evaluates the proposed solution by introducing the lessons learned after applying our approach in the development of several cases of study. Finally, conclusions and further work are presented in Section 8.

## 2 Related Work

Many web engineering methods, some partly or fully based on UML [3] (see e.g. OOHDM [4], UWE [7], WSDM [5], WebML [6], OOH [12]) and others based on formal foundations (see e.g. Schewe et al. [13]), have been proposed in order to give support to the development of Web applications. They provide solutions to both capture web application requirements and define web applications at conceptual level. Most of them also propose techniques to define the look and feel of web applications (OOHDM uses abstract data views, UWE uses presentation classes and framesets, WSDM uses an implementation model, WebML uses style sheets and OOH uses an abstract presentation diagram). Other approach to be considered in the field of Web engineering is the GX WebEngineering Method [22]. In this work, an assembly-based situational method engineering approach is used to develop a new Web engineering method by assembling elements of other three development methods: an old version of GX, UWE and the Unified Process. In all these cases, the published research works do not explicitly explain how these methods must be used by multidisciplinary web development teams. That is, its development processes are presented without considering the possibility that different web professionals work together in order to develop the Web application. In this context, the aim of our proposal is to complement current approaches by determining which tasks each professional (analysts and graphic designers) should achieve; which techniques and tools each professional should use; and how they should collaborate with each other.

In the context of Human-Computer Interaction, several approaches such as Campos et al. [15] or Granollers [16] take into account the multidisciplinary nature of web development. However, these approaches mainly focus on provide techniques that allow both graphic designers and experts in usability to correctly collaborate with end-users in order to design a usable web interface. Computer analysts who are in charge of the system functionality are not considered enough by these approaches.

To conclude, we want to mark that our work is a first step in the definition of a whole multidisciplinary web development process which follows the essence of methodologies such as Constantine et al. [14] but providing a more pragmatic solution.

## 3 The Web Development Process

In this section, we introduce a web application development process that takes into account development teams made up of analysts and graphic designers. To present this process, we use the notation and terms defined in the Software Process Engineering Metamodel (SPEM) proposed by the OMG [11]. First, we present the Disciplines that define our process as well as the Activities, WorkProducts, and ProcessRoles that are included in each Discipline. Next, we present the Sequencing of Activities that defines the development process.

### 3.1 Process Disciplines

According to the SPEM, an Activity is a piece of work performed by one ProcessRole in order to obtain a WorkProduct. A Discipline partitions the Activities within a process according to a common “theme”. The process that we propose is defined from three disciplines: Requirements Gathering, Graphic Design, and Code Generation.

#### 3.1.1 Requirements Gathering

This discipline includes those activities that are related to the handling of the user requirements. There are two of these activities: (1) elicit the user requirements and (2) specify requirements. The activities must be performed by analysts. The WorkProduct that analysts obtain after performing the activities is a requirements model.

Thus, the process proposes that analysts should interact with users and analyse their needs (activity 1). Then, analysts should accurately specify the users' needs into a requirements model (activity 2). In this context, analysts work on the problem space. They do not consider aspects related to implementation or graphic design.

### 3.1.2 Graphic Design

This discipline includes those activities that are related to the design of the look and feel of the web application. There are three of these activities: (1) interview the user, (2) design the look and feel and (3) associate the look and feel to code. The activities must be performed by graphic designers. The WorkProduct that they obtain is a presentation template.

Graphic designers work on the solution space. They design presentation templates (activity 2) that define aesthetic issues such as screen layout, colours, and font usage (the look and feel of the web application). To do this, they must interview the users (activity 1) in order to know their aesthetic preferences. Finally, graphic designers may be in charge of associating the presentation template to the code that implements the structural and behavioural aspects of the web application (activity 3). This activity is optional as explained in section 3.2.

### 3.1.3 Code Generation

This discipline includes those activities that are related to the automatic implementation of code. In order to define these activities we have based on both the development process of the OOWS method and its strategy of automatic code generation. OOWS is a web engineering method that allows us to automatically obtain web applications from a requirements model by following the principles defined by MDD (See [8] and [9] for detailed information). Thus, there are two activities for this discipline: (1) define a conceptual schema and (2) automatic code implementation. The activities must be performed by the OOWS method. The WorkProduct that the OOWS method obtains after performing these activities is either a web prototype or the final web application. We have defined the OOWS method as the ProcessRole due to the automation of its development process.

First, the OOWS method defines the web conceptual schema of the web application (from the requirements model defined by analysts, which is explained in the next subsection). To do this a model-to-model transformation is automatically applied [9]. Next, it generates code from the conceptual schema. To do this, a set of transformation patterns are applied [8] by means of the use of a generator tool [21]. This code implements either a web application prototype (if aesthetic aspects are not considered in the code generation activity) or the final web application (if code has been generated taking into account the aesthetic aspects). Aesthetic aspects are taken into account if graphic designers have defined a presentation template.

## 3.2 Sequencing of Activities

In order to define the sequencing of activities of a web development process, SPEM proposes the use of activity diagrams. Figure 1 shows the activity sequencing of the web development process proposed in this work. According to this figure, it is defined as follows: On the one hand, analysts create a requirements model after analyzing the user needs, and graphic designers define the look and feel of the web application (by means of a presentation template) after interviewing the users and determining their aesthetic preferences. As Figure 1 shows, there are no dependencies between the activities of the analysts and the activities of the graphic designers.

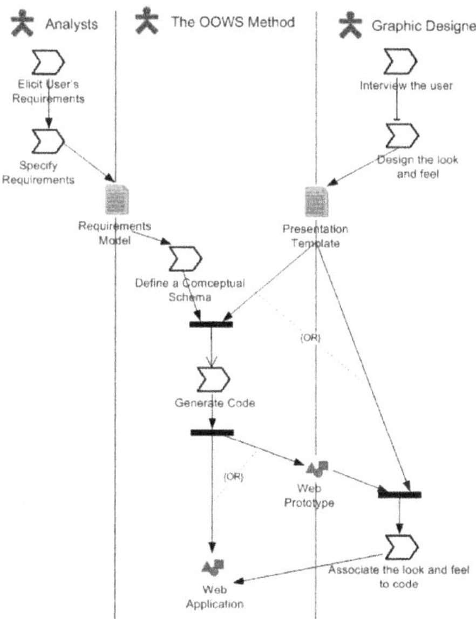


**Fig. 1.** Activity Sequencing

On the other hand, the OOWS method automatically creates the conceptual schema of the web application from the requirements model defined by the analysts. Then, OOWS can generate code in two ways: (1) If a presentation template is already defined, OOWS generates code and automatically associates the template to this code. Then, the final web application is obtained. (2) If a presentation template is not already defined, the code generated by the OOWS method implements a web application prototype. This prototype provides the user with the information and functionality captured in the requirements model but without any aesthetic properties (that is, in plain text). The aesthetic aspects of the web application prototype are defined in a later activity, where graphic designers associate a presentation template to the generated code.

## 4 Supporting the Activities of the Analyst

In this section, we present the techniques and tools that allow analysts to perform the activities included in the requirements discipline: elicit and specify requirements.

### 4.1 Requirements Engineering Techniques

In order to allow analysts to elicit web application requirements, we have defined an approach to interview the user by means of a wizard. We have defined several ontologies that characterize web applications of different types (e.g. e-commerce applications, web portals, directories, etc.). The wizard queries users to know the type of web application that they want to develop. Next, by following a question-guided process, the wizard queries users and prunes the proper ontology in order to obtain the user requirements. See [1] for further information.

In order to allow analysts to specify web application requirements we propose a technique that is based on the concept of task. This technique allows specifying not only requirements related to the structural and behavioural aspects (typical requirements of non-Web systems) but also requirements related to the navigational aspect (encouraged by Web applications). To do this, analysts must first define a task taxonomy for each kind of user that can interact with the web application. A task taxonomy specifies, in a hierarchical way, the tasks that a kind of user should perform when interacting with the web application.

Once the task taxonomy is defined, each leaf task is described by analyzing the interaction that users require from the web application. To do this, a strategy based on activity diagrams is proposed. Each activity diagram is defined from system actions (nodes depicted by dashed lines) or interaction points (nodes depicted by solid lines) that represent the moments during a task where the system and the user exchange information.

Finally, we must specify a set of information templates where the information that the system must store is described. We also use these templates to describe the information exchanged in each interaction point in detail. See [19] for more detailed information about the task-based requirements model.

### 4.2 Tool Support

In this section, we present a tool that supports analysts in the requirements specification of web applications. This tool is the Task Modeller. Figure 2 shows several snapshots of it.

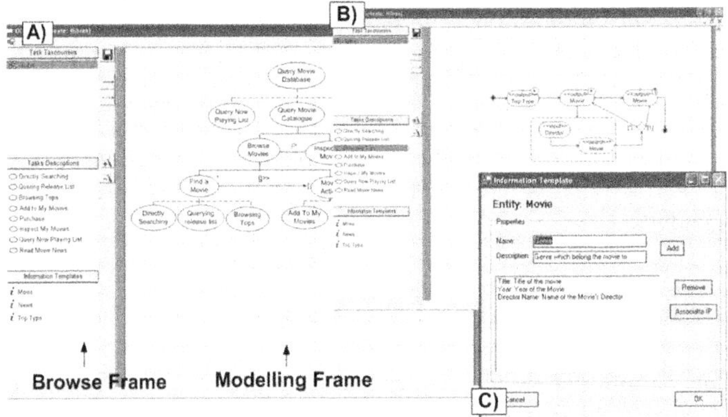


**Fig. 2.** The Task Modeller

Figures 2A and 2B show the main window of the task modeller. This window is divided into two frames: (1) The browse frame, which allows analysts to browse users and their associated task taxonomies (see upper side of Frame 1), tasks and their descriptions based on activity diagrams (see centre side) and information templates (see lower side). (2) The modelling frame, which allows analysts to define either a task taxonomy if a user is selected in the browse frame (see Figure 2A) or an activity diagram if a task is selected in the browse frame (see Figure 2B).

Furthermore, if analysts select an information template from the browse frame, they access a window such as the one shown in Figure 2C. This window allows analysts to define the information that the system must store as well as the information exchanged between the user and the system in each interaction point.

Once the analysts have built the task-based requirements model, the Task Modeller stores it in a XML document. According to the activity sequencing in Figure 2, the OOWS method takes this document as input in order to transform it into a web conceptual schema. Then, code is generated from the conceptual schema. The next section explains it.

## 5 Supporting the Activities of the OOWS Method

In this section, we present an overview of the OOWS method. Section 5.1 introduces the OOWS development process. Section 5.2 explains the code generation strategy.

### 5.1 The OOWS Development Process

As commented above, the OOWS method allows us to automatically obtain a web application from a requirements model. First, a model-to-model transformation is performed to derive the web application conceptual schema from the requirements model. In order to define this transformation we have first identify the set of mapping that allow us to obtain the elements of the OOWS conceptual model from the task-based requirements model (presented in section 4.1.1). Next, these mappings have been defined by following a strategy based on graph transformations. In order to automatically apply these transformations we use the AGG tool [23]. More information about the model-to-model transformation can be found in [9].

The OOWS conceptual schema is defined from several models that describe the different concerns of a web application: The system static structure and the system behaviour are described in three models (class diagram and dynamic-and functional models) that are borrowed from an object-oriented software production method called OO-Method [20]. The navigational aspects of a Web application are described in a navigational model [8].

Then, a strategy of automatic code generation is applied to the web conceptual schema in order to obtain a web application. This strategy implements web applications by following a three-tier architecture. The information and functionality (Application and Persistence tier) of the web application is generated by the OlivaNova tool [21] from the OO-Method models (structural and behavioural model). The navigational structure (Presentation tier) of the web application is generated by the OOWS case tool following directives specified in design templates [8].

Olivanova provides us with transformation engines for different platforms such as Visual Basic, .Net or Java. The OOWS case tool provides us with the generator engine that is next presented. If we have requirements concerning to the generated code we need to modify the transformation engine. This is not a difficult task if we consider that code is generated by applying transformation patterns. We just need to modify the code associated to each pattern in order to satisfy the requirements of the user. The generation strategy that we use is the same.

### 5.2 Implementing the Presentation Tier

OOWS generates web pages that provide users with specific information and functionality (according to the user's needs captured in the requirements model) but without any kind of aesthetic aspects. The information and functionality are obtained by requesting them to the Application and Persistent tiers (generated by the OlivaNova Tool). The aesthetic aspects are incorporated by means of presentation templates. To facilitate this, OOWS implements a web page as an aggregation of a set of logical content areas (see Figure 3A). Each logical content area provides a specific piece of data. Several areas are proposed by considering some aspect of usability defined in [17]:

- Information area (see Figures 3A, zone number as 1): presents the data and the functionality that users can access.
- Navigation area (see zone 2): provides links to the web pages that are available to users.
- Location area (see zone 3): shows the situation of the users. It indicates the page that is currently being shown and the navigational path that has been followed to reach the page.
- Application area (see zone 4): provides user facilities such as a link to the home page or access to the login form that are common to most web applications.
- Access-Structure area (see zone 5): provides users with mechanisms such as search engines or information indexes, which facilitates the access to the information.
- Corporative area (see zone 6): provides information about the organization such as the name, the e-mail, the logo, etc.

Figure 3A shows a page that provides information about a movie. Figure 3B shows the code that implement the information area of page in Figure 3A. This code is based on the <div> label. As we can see, this code does not include any kind of aesthetic aspect[1]. Each content area is defined by means of a div block. Each div block is also divided into sub-blocks in order to provide web designers with greater control for the definition of the web application aesthetics aspects (which is explained in the next section). The information_area block in Figure 3B is divided into two sub-blocks: data, which provides the properties of a selected movie; and operations, which provides users with the operations that they can activate. Each property is defined by two kinds of div blocks: img_att, which define a graphical property; and txt_att, which define a textual property. This distinction (according to the data type) provides web designers

[1] We have associated a default presentation template (similar to the IMDb web site) to the web page in Figure 3A in order to better visualize it.

with the possibility of defining different aesthetic properties for different data types. Finally, each property is implemented by means of two blocks: one that defines the property alias (e.g. "Year") and the other that defines the property value (e.g. "1960"). We explain next how graphic designers incorporate the aesthetic aspects.

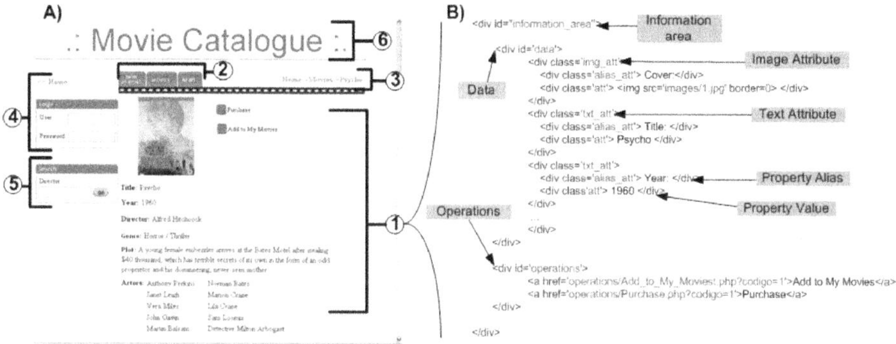


**Fig. 3.** Web page areas

# 6 Supporting the Activities of the Graphic Designer

In this section, we support graphic designers in the performance of the activities included in the graphic design discipline: interview the user; design the look and feel of the web application; and associate this template to the web application prototype.

In order to design the look and feel of a web application, graphic designers need to define a presentation template that is associated to the div-based code (presented in Section 5.2). Since this code is defined by means of the area-based strategy, graphic designers can define general-presentation templates. The general-presentation templates are templates whose presentation styles are defined without taking into account any specific web application domain. Styles are not defined by means of domain-specific terms such as CD, client, or invoice. Styles are defined by means of area-based terms such as information area, data, or operations. Then, these presentation templates can be applied to any web application developed following the OOWS method (without considering the Web application domain). This allows graphic designers to define presentation templates without having to take into account the requirements captured by analysts, providing thereby the desired level of independence between web professionals (see Section 3.2). For instance, graphic designers define the aesthetics properties of the information_area block and its sub-blocks. They do not need to consider whether the information provided in this area is related to books, clients or movies. Furthermore, because of these templates do not depend on a specific domain: (1) they do not need to be redesigned if requirements change (the code that supports new requirements will be also based on the <div> label) and (2) they can be reused in the development of several web applications (web applications are all implemented from the same areas).

Sections 6.1 and 6.2 present the technique proposed to define general-presentation templates and a tool that allows graphic designers to easily define these templates as well as to reuse predefined ones. Section 6.3 introduces a strategy that is based on the tool presented in Section 6.2 to perform the first activity (interviewing the user.

## 6.1 Defining General-Presentation Templates

In this section, we introduce a technique that allows graphic designers to define general-presentation templates. This technique is based on Cascading Style Sheets (CSS). General presentation templates are defined by creating a CSS style for each div block that implements a content area.

Figure 4B shows the CSS styles that define the aesthetic properties of the web page information area in Figure 4A. For instance, the *information_area* style defines the size, position and margin of the div block that implements this area as well as the font properties. The rest of CSS styles define properties for the sub-blocks defined inside the block that implements the information area. Thus, they inherit the aesthetic properties of the information_area style and only add or replace some properties. These aesthetic aspects are directly incorporated to the HMTL code by the Web browser. When the user accesses to a Web page the Web browser receives both the area-based HTML code and the area-based CSS styles. Then the browser composes both codes to create a Web page.

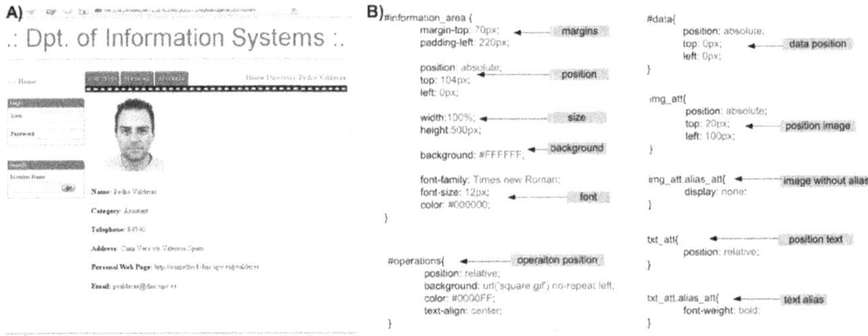


**Fig. 4.** Example of a general presentation template

Figure 4A shows a web page from a web site of a university department that provides information about a teacher. As we can see, this page presents the same look and feel design that the page in Figure 3A (which provide information about a different domain, movies): Content areas are located in the same places, and they share aesthetic properties such as colours, sizes, or font properties. Both pages share the same look and feel design because they are implemented by means of an area-based code and because they are associated to the same general-presentation template (this one that is partially presented in Figure 4B).

## 6.2 Tool Support

In this section, we present a tool that helps graphic designers in the definition and the reuse of general presentation templates. This tool also allows graphic designers to associate a presentation template to a specific web application prototype generated by the OOWS method. This tool is the Look and Feel Designer. It has two modes of work: (1) a basic mode that allows graphic designers to reuse general templates and (2) an advanced mode that allows graphic designers to define new templates.

Figure 5A shows a snapshot of the Look and Feel Designer in the basic mode. The tool is divided into three frames. Frame 1 shows the page tool. This tool provides users with the list of web pages (depicted by rectangles with the file name) that make up the loaded web application. To load a web application, the home page is selected from the file menu. The tool reads this file, and then follows the defined links to load the rest of web pages. Frame 2 shows the template tool. This tool provides users with the list of general-presentation templates that are stored in a repository. We have defined several general-presentation templates in order to provide graphic designers with default look and feel designs. However, as explained below, graphic designers can define their own general presentation templates, which are stored in the repository in order to allow their reuse in further developments. Frame 3 is the rendering zone. In this zone, graphic designers can see the page selected in the page tool with the aesthetic aspects defined in template selected in the template tool.

Graphic designers can reuse an already designed look and feel design as follows: (1) They load the new web application into the Look and Feel Designer. (2) They select the available pre-designed presentation templates from the template tool. To see the look and feel of each web page, they can select it from the page tool. (3) Once the graphic designers have decided on a presentation template, the tool automatically associates it to each web page. Figure 5B shows a web page, which provides a list of movies, with different look and feel designs.

To define new general presentation templates, graphic designers must use the advanced mode of the Look and Feel Designer. When this mode is selected, the rendering zone shows the main areas in which the web page (selected in the page tool) is divided. Graphic designers can click upon each area, and then a window for the definition of the aesthetic properties of the selected area is shown. This window provides an intuitive interface where graphic designers can define the CSS style associated to the selected area. For each area analysts can define properties such as visibility, position, size, margins, colors, etc. In this sense, graphic designers can define a new template by: (1) taking an existent general template and then modifying it to create a new one, or (2) selecting an empty template in order to create a new one from scratch.

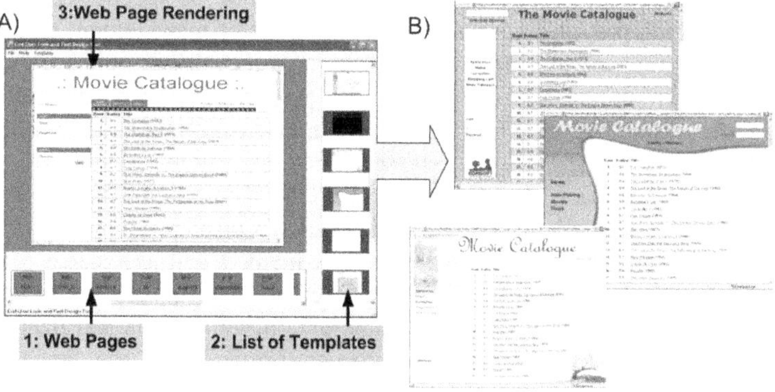


**Fig. 5.** Look and Feel Designer

### 6.3 Interviewing the User

In the previous section, we introduced the Look And Feel Designer, which allows graphic designers to define general-presentation templates and associate them to a web application. This tool can also be used to facilitate the interview with the user. Graphic designers can use this tool in two ways, depending on the moment in which users are interviewed:

- After generating the web application prototype. If a web application prototype has been generated, graphic designers can load it into the Look And Feel Designer. Then, graphic designers can apply the set of pre-defined templates to the prototype. This can guide users by allowing them to see the desired web application with different look and feel designs. In this sense, users can decide on a look and feel by directly seeing it on the final web application.
- Before generating the web application prototype. If code has not yet been generated, graphic designers can use the Look and Feel Designer in order to show the predefined presentation templates which are applied to a default web application. Thus, users can decide on a look and feel by seeing it in a real web application. Although this application is not the web application under development, the aesthetic properties will be applied in the same way (due to the area-based styles that define the presentation templates).

## 7 Lessons Learned

As a proof of concept, our approach has been successfully put into practice in the development of small and medium-size web applications, including the DSIC Department Web Site (http://www.dsic.upv.es), a rent a car company (http://www.rentacar-denia.com/), and a drinking water company (http://www.aguasdelbullent.com). We played the role of analysts and we asked to graphic designers of a software development company (CARE Technologies [21]) to collaborate with us. They were educated about our area-based implementation strategy and they were asked to define domain-independent presentation templates.

In the first two projects, Web applications are divided into two parts: (1) a public part which provides users with the information related to the specific domain and (2) a private part which provides administrators with a set of management tools. In these cases, we are a group of four analysts and we have to work in collaboration with a group of three graphic designers. In the last project, the Web application to be developed is a corporative site. In this case, we are two analysts and we have to work with one graphic designer.

The use of our approach provide us with several benefits: On the one hand, from the analyst perspective, clients got excited when we provide them with a software product (the prototype generated by the OOWS method and associated to a default presentation template) that partially supports their needs only one or two days after the first interview. This fact improves our interaction with clients and makes them to be more implicated in the development process. Furthermore, the fact that clients interact with a software product facilitated them to validate that requirements had been correctly captured. In the three web applications clients detected requirements that had not been correctly understood; changed their mind about some requirements when they saw the implementation; or added new requirements that arose interacting with the prototype. On the other hand, graphic designers could design several look and feel designs without interacting with us (analysts). Because of graphic designers are professionals which are familiarized with the CSS technology they have no problems to work with our presentation templates. Furthermore, they were very grateful with the fact that they just needed to focus on design activities. They did not need to make the effort of understanding the Web application domain. Graphic designers also found very useful the possibility of reusing presentation templates. For instance, in the first development project, the user selected a predefined presentation template and then graphic designers had not to create a new one. In the other projects, graphic designers just needed to personalize a predefined template to fit the user preferences. This made them to save a lot of time and effort.

Finally, we know that some drawbacks need to be improved: for instance, non-functional requirements such as usability or legal and social aspects are not properly considered. As commented at the introduction, our approach constitutes a first step in the definition of a whole multidisciplinary development process. We need to extend it in order to correctly introduce other web professionals such as usability experts, lawyers or domain specialists.

## 8 Conclusions and Further Work

We have presented a web development process which takes into account the multidisciplinary nature of web development teams. It has been defined by extending the OOWS method.

First, we have clearly identified and delimitated the tasks that each web professional must perform during the development of a web application. Then, we support web professionals in the development of these tasks by: (1) extending the automatic code generation strategy of the OOWS method in order to obtain code that gives support to the client's needs but without considering any kind of aesthetic property and (2) proposing a strategy to define general-presentation templates that can be applied to any web application developed by the OOWS method. These extensions allow web developers (analysts and graphic designers) to work with a high degree of independence from each other but in a coordinated way.

As further work, we want to extend this approach in order to take into account aspects related to personalized contents or designs. We also want to extend the generation code strategy in order to support new technologies such as Flash or Ajax.

## References

1. Valderas, P., Pelechano, V., Pastor, O.: Towards an End-User Development Approach for Web Engineering Methods. In: Dubois, E., Pohl, K. (eds.) CAiSE 2006. LNCS, vol. 4001, pp. 528–543. Springer, Heidelberg (2006)
2. Powell, T.A.: Web Site Engineering: Beyond Web Page Design. Prentice Hall, Englewood Cliffs (1998)

3. Object Management Group. Unified Modeling Language (UML) Specification Version 2.0 (2003) www.omg.org
4. Schwabe, D., Rossi, G., Barbosa, S.: Systematic Hypermedia Design with OOHDM. In: ACM Conf. on Hypertext, USA (1996)
5. De Troyer, O., Leune, C.: WSDM: A User-centered Design Method for Web sites. In: Proc. of the 7th WWW, pp. 85-94 (1997)
6. Ceri, S., Fraternali, P., Bongio, A., Brambilla, M., Comai, S., Matera, M.: Designing Data-intensive Web Applications. Morgan Kaufmann, Washington (2002)
7. Koch, N.: Software Engineering for Adaptive Hypermedia Applications. PhD thesis, Ludwig-Maximilians-University, Munich, Germany (2000)
8. Fons, J., Pelechano, V., Albert, M., Pastor, O.: Development of Web Applications from Web Enhanced Conceptual Schemas. In: Song, I.-Y., Liddle, S.W., Ling, T.-W., Scheuermann, P. (eds.) ER 2003. LNCS, vol. 2813, Springer, Heidelberg (2003)
9. Valderas, P., Fons, J., Pelechano, V.: Transforming Web Requirements into Navigational Models: An MDA Based Approach. In: Delcambre, L.M.L., Kop, C., Mayr, H.C., Mylopoulos, J., Pastor, Ó. (eds.) ER 2005. LNCS, vol. 3716, Springer, Heidelberg (2005)
10. Mellor, S.J., Clark, A.N., Futagami, T.: Model-driven development-Guest editor's intro. IEEE Software 20(5), 14–18 (2003)
11. Software Process Engineering Metamodel, version 1.1. Object Management Group. http://www.omg.org/technology/ documents/formal/spem.htm
12. Gomez, J., Cachero, C., Pastor, O.: Conceptual Modelling of Device Independent Web Applications. IEEE Multimedia Special Issue on Web Engineering, pp. 26–39, 04 (2001)
13. Schewe, K.-D., Thalheim, B.: Conceptual modelling of web information systems. Data and Knowledge Engineering 54(2), 147–188 (2005)
14. Constantine, L.L., Biddle, R., Noble, J.: Usage-Centered Design and Software Engineering: Models for Integration. In: Proc, ICSE (2003)
15. Campos, P.F., Nunes, N.J.: CanonSketch: A User-Centered Tool for Canonical Abstract Prototyping. In: Bastide, R., Palanque, P., Roth, J. (eds.) Engineering Human Computer Interaction and Interactive Systems. LNCS, vol. 3425, Springer, Heidelberg (2005)
16. Granollers, T.: User Centred Design Process Model. Integration of Usability Engineering and Software Engineering. INTERACT ' 03
17. Olsina, L., Rossi, G.: Measuring Web application quality with WebQEM. IEEE Multimedia Magazine 9(4), 20–29 (2002)
18. Deshpande, Y., Murugesan, s., Ginige, A., Hansen, S., Schwabe, D., Gaedke, M., White, B.: Web Engineering. Journal of Web. Engineering 1(1), 3–17 (2002)
19. Valderas, P., Fons, J., Pelechano, V.: Developing E-Commerce Application From Task-Based Descriptions. In: Bauknecht, K., Pröll, B., Werthner, H. (eds.) EC-Web 2005. LNCS, vol. 3590, pp. 65–75. Springer, Heidelberg (2005)
20. Pastor, O., Gomez, J., Insfran, E., Pelechano, V.: The OO-Method Approach for Information Systems Modelling: From Object-Oriented Conceptual Modeling to Automated Programming. Information Systems 26 (2001)
21. Olivanova Model Execution System. Care Technologies www.care-t.com
22. Van de Weerdl, I., Brinkkemper, S., Souer, J., Versendaal, J.: Situational Implementation Method for Web-based Content Management System-applications: Method Engineering and Validation in Practice. SPIP (2006)
23. Attribute Graph Grammar System. http://tfs.cs.tu-berlin.de/agg/
24. Vaishnavi, V., Kuechler, W.: Design Research in Information Systems (2004)

# Communication Abstractions for Distributed Business Processes

Lachlan Aldred[1], Wil M.P. van der Aalst[1,2], Marlon Dumas[1], and Arthur H.M. ter Hofstede[1]

[1] BPM Group, Queensland University of Technology, Australia
{l.aldred,m.dumas,a.terhofstede}@qut.edu.au
[2] Department of Mathematics and Computer Science, Eindhoven University of Technology, The Netherlands
w.m.p.v.d.aalst@tue.nl

**Abstract.** Languages for business process definition generally suffer from myopic approaches to capturing communication between distributed processes. Effective communication between processes requires: support for conversations involving interrelated interactions spread over time; ability to select and group messages based on their content, regardless of format and transport technology; and resolving contention between processes or tasks for common sets of messages. This paper presents a set of communication abstractions that provide a "glue" between the process layer and the middleware. The paper also reports on an implementation of these abstractions and an experimental evaluation.

**Keywords:** Business process integration, correlation patterns.

## 1 Introduction

At present, the definition of business processes that interact with one another in a distributed environment is hampered by a number of factors. Firstly, these processes are required to run on top of mainstream communication middleware which often does not support key features needed by applications in general [4,10,12,11], and process-oriented applications in particular. For instance message selectors in JMS[1] cannot filter messages based on their body. Secondly, there exist conceptual problems with state-of-the-art business process definition languages in regards to process-to-process communication abstractions. To receive message batches for example, Business Process Management Systems (BPMSs) force designers to incorporate dedicated code into the same scope as business process logic. Finally, apart from correlation mechanisms for routing messages to process instances, BPMSs accept messages sent to them without filtering, thus forcing unnecessary amounts of message selection code into the process definition. In summary, the distinction between process abstractions and communication abstractions is blurred in existing approaches.

[1] Java Message Service: java.sun.com/products/JMS accessed November 2006.

J. Krogstie, A.L. Opdahl, and G. Sindre (Eds.): CAiSE 2007, LNCS 4495, pp. 409–423, 2007.

This paper motivates and defines a set of communication abstractions at a layer in-between the business process and the middleware, that simplify the definition of interactions between distributed processes. The proposed abstractions have been implemented on top of a communication API, namely JCoupling [5], that abstracts away from the underlying middleware and protocols. Using this implementation, we have conducted preliminary experiments to compare this approach with respect to the approach embodied in the business process execution languages for Web services (WS-BPEL) [7].

The next section presents motivating scenarios and requirements for process-to-process communication. Section 3 introduces a communication model for distributed business processes addressing these requirements. Section 4 discusses the implementation and experimental evaluation of a prototype that implements these abstractions. Section 5 discusses related work and Section 6 concludes.

## 2 Motivating Requirements

Implementing interactions between business processes brings new issues on top of the traditional requirements found in distributed systems implementation (e.g. coupling over time and space [4,12], guaranteed delivery, encryption). In this section we distill some of these specialised requirements. These requirements are drawn from two studies on patterns in the area of integration of enterprise applications: the Enterprise Integration Patterns by Hohpe and Woolf [14], and the Correlation Patterns by Barros et. al. [9].

The correlation patterns described by Barros et. al. review conversation and message consumption patterns for distributed business processes. Most of those patterns led to requirements motivating our proposals. From Hohpe and Woolf we categorised the patterns into (1) Patterns forming requirements for a BPM messaging solution, (2) Patterns that are supported by most middleware, (3) Patterns composable from any data-aware, message-aware system, (4) Patterns concerning deployment and administration. The communication abstractions proposed in this paper focus on the first category. The technical report version of this paper [6] contains further details on this categorisation.

In the remainder of this section, we discuss the different requirements. For each requirement we cite the related patterns from [9,14].

**Conversations.** Support for conversations is necessary when business processes need to exchange more than one related message, and in particular when the processes are stateful, or execute over long periods. Furthermore typical process deployments, require many instances to share common channels and conversations enable messages finding their way to the correct instance. The technical report version of this paper [6] lists four forms of correlation. However this paper focuses on the most technically challenging – Property-based correlation – which assumes that many process instances share the same channel. Messages get routed to the appropriate instance by applying process-defined functions to incoming messages and then by matching the results to process instance values.

*Related to:* Key Based Correlation, Property Based Correlation, Reference Based Correlation, Conversation Overlap, Hierarchical Conversation, Initiator, Follower [9], Correlation Identifier [14].
*Scenario 1: Purchase Order* A purchase order is received. The process sends separate queries to different suppliers for each line item. Each response is correlated over the purchase order identifier, and over the line item number. This example demonstrates the need for nested conversations.

**Property-Based Message Selection.** Property-based message selection helps a process pick the best message off a channel. This significantly reduces the complexity logic within the process designed to iterate through an internal array of messages. There should be simpler abstractions for this.
*Related to:* Message Filter, Selective Consumer [14].
*Scenario 2: Line Items* A parts buyer wants to proceed with the best quote.

**Atomic Multiple Source Consumption.** When messages need to be joined, from more than one source, atomic multiple source consumption greatly reduces complexity in the process model. This is because the messages from each source may need to match over a certain field or property, and packaging them together removes the need to do this in the process. Furthermore it is much simpler to design exception handling if there is only one point of failure in the process. To demonstrate this, consider the possible outcomes of consuming them separately: (1) both message arrived, or (2) one failed to arrive, or (3) both failed to arrive, or (4) they both arrived however there was a property mismatch, or (5) there were two processes and each consumed only one message.
*Related to:* Atomic Consumption [9].
*Scenario 3: Purchase processing* Task "*ship-goods*" will not begin until there is a confirmation of credit from the accounts department, and all line items have been notified as being "in stock".

**Aggregated Consumption.** Consuming messages in one "pass" of a process loop greatly reduces complexity. There is no need to iterate through a set of receive actions, or to encode loop stop-conditions, which can often be a little arbitrary if the intent is to consume all available messages. Furthermore it may be necessary to choose messages only if their properties, taken collectively, satisfy certain criteria.
*Related to:* Aggregator [14].
*Scenario 4: Shipping Company* When at least 100 items have arrived destined for the same district, and a truck is available, the truck gets dispatched.

**Aggregated Consumption involving Time.** Business processes are in many cases very time sensitive, for example the hours of business (9am to 5pm, Monday to Friday, EST). Thus there exists the need to include the notion of time into the process layer, for example by allowing message selection based on temporal constraints. Temporal constraints can typically be either absolute or relative i.e. "15 – 20 November, 2006", or "within the last 7 days". Both styles are necessary with the latter being more challenging due to the group of eligible messages being in a continual state of flux with the passage of time.

*Related to:* Time-Based Correlation, Moving Time Window Correlation [9]
*Scenario 5: Time* If, over the last five working days, more than five percent of the incoming messages that arrive at the department contain complaints then an emergency quality control process gets launched.

**Contention.** Contention, or competition, for the same resources, is a natural phenomenon. For instance, auctions and goods tendering rely on contention between competitors. Messages, unlike auction items, have little or no intrinsic value and therefore contention over messages may not seem compelling. However, contention over messages is an enabling technique for load balancing of message consumption; wherein many instances of the same process share the workload by only consuming messages/events when they are ready.
*Related to:* Competing Consumers [14].
*Scenario 6: Competing Processes* Copies of a process are distributed onto different hosts. The first process instance to claim the message provides the service, thus distributing the processing workload.

**Interaction Cancelation.** The ability to cancel incomplete interactions.
*Related to:* Transactional Client [14], Cancelation [2].
*Scenario 7: Cancelation* A supplier having posted a receive purchase order request, discovers that the warehouse needs to be replenished first, and consequently cancels its earlier receive-request.

**Summary.** These requirements, being drawn from the respective patterns studies, ought to be supported by most state of the art solutions, however we found that this was not the case. For instance, despite WS-BPEL being possibly the most widely accepted standard in this domain, it only provides support for the first requirement. We also found some other motivating requirements in this domain. These include event handling, channel passing, and garbage collection. These are discussed in the technical report version of this paper [6]. It is one thing to support a wide range of problems in communicating business processes, however the language constructs exposed to the creators of these processes need to be suitable to their purpose, and they should be as intuitive, or conceptually aligned to their purpose as is possible.

## 3 Communication Model for Business Processes

In this section we introduce a model supporting interactions between processes addressing the requirements outlined above. In the proposed model, *instances* of a process model are hosted in a *process container*. The container uses a set of channels that are referred to by the process model. The channels are hosted in some form of message-oriented middleware. Channels can be used by a process container to send messages (outbound channel), receive messages (inbound channel), or perhaps both directions (bi-directional channel).

Outbound channels are composed of a unique name, possibly a type and/or an endpoint descriptor (the latter may be determined only at runtime). The proposed model abstracts away from the specific language used to describe message types. A particular embodiment of the concept of channel is one where message types are captured in XML Schema (or WSDL). Individual messages sent through the channel are then validated against its XML schema (or WSDL message type definition). Similarly, we abstract away from the mechanism used to describe destination endpoints. If using HTTP as a transport protocol for example, an endpoint can be described as a URL, while if using SOAP/HTTP, it can be described as a WSDL operation binding. Outbound channels support a range of message sending primitives which are described in detail in [5]. In the rest of the paper however, we focus on inbound channels.

Inbound channels have the same components as outbound ones but they additionally have a set of *properties*. A property is a function that takes as input a message and produces a literal value. This is similar to the concept of *property alias* defined in BPEL. However, as shown later, the scope of applicability of properties in our model is wider than that of BPEL. In BPEL, property aliases (and their composition in the form of *correlation sets*) are only used to correlate pairs of outbound and inbound messages. In contrast, in our model, properties can be used to perform other forms of message selection and aggregation.

A relation (i.e. a database table) is created for each inbound channel used in any process model. Each relation contains two predefined attributes: one of type `message identifier` and the other of type `timestamp`. Additionally, the relation contains one attribute (column) per property associated to the channel.

Properties are used to define *filters*. A filter is a function that is evaluated against the set of messages available for consumption over one or multiple channels. When the evaluation of a filter returns a non-empty set of messages, we say that the filter *matches* these messages. Filters can fulfill two purposes: (i) *garbage-collecting filters* are registered by a business process to discard unwanted or unnecessary messages over inbound channels; (ii) *message consumption filters* are used to consume one or multiple messages from one or multiple channels. Orthogonally, filters may be *one-off* or *persistent*. A one-off filter is immediately withdrawn after it has matched a message or set of messages, while persistent filters are preserved until explicitly withdrawn.

A filter is represented as a query over the relations(s) associated with the channel(s) it refers to. These queries are always constrained to produce a relation wherein each tuple contains `message identifier` attributes.

When a message arrives onto a channel, a tuple is inserted into the relation associated with that channel. This tuple always contains `message identifier` and `timestamp` attribute values, as well as attribute values obtained by applying each of the channel's properties to the incoming message. The inserted tuple represents the incoming message for the purpose of evaluating message filters. After being abstracted as a tuple, the incoming message is either:

- Immediately routed to a message receipt action if one has registered a filter that matches the messsage (possibly in combination with other messages).
- Discarded if the message matches any of the garbage-collecting filters registered for that channel.
- Queued until it matches a garbage-collecting or message consumption filter.

Conceptually, a filter is re-evaluated every time that a new message arrives to any of the channels it refers to (or continuously in the case of filters whose query depends on the current time). In practice however, the evaluation can be made incrementally and only when required.

Primitives for registering and withdrawing filters are provided as part of the communication framework. Registration and withdrawal of filters can be initiated either by the process container or by individual process instances. When a message consumption filter is registered, the filter is run once against the set of messages available in the channels referenced by the filter. If the filter matches one or several messages, these are removed atomically from their channel(s) and given back to the process container or process instance that registered the filter. If the filter is one-off, it is withdrawn. Should no match between a filter and the existing set of messages be found upon registration of the filter, the filter is maintained and re-evaluated whenever required as explained above, until the filter is either explicitly withdrawn or it matches a message (or set of messages). Once a match is found, the message(s) are routed to the corresponding process container or instance and the filter is removed.

Garbage-collecting filters work similarly: when registered, the filter is evaluated against the contents of the channels targeted by the filter. If a match is found, the matched message(s) are discarded. If the filter is one-off, it is withdrawn otherwise, it is preserved and it is re-evaluated when required.

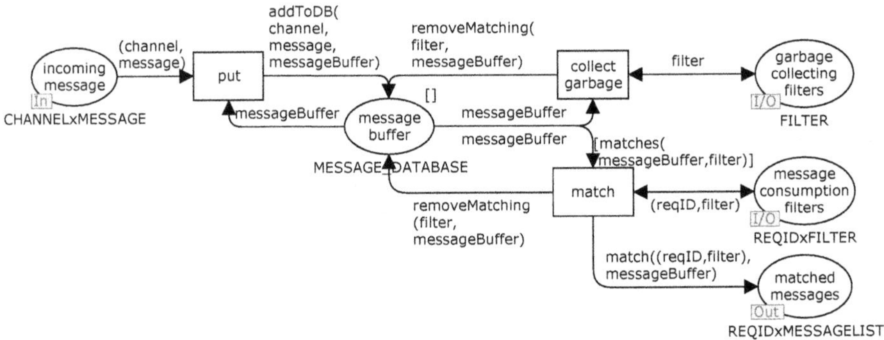


**Fig. 1.** Petri net capturing the treatment of inbound messages

The proposal is formally captured as a Coloured Petri net in Fig. 1. The net shows how inbound messages are stored and matched against filters. A token in place "*incoming message*" represents a message received by the communication layer. A transition called "*put*" moves tokens from this place to a place called "*message buffer*". This latter place holds a single token containing a list of all unmatched messages over all channels. Two of the places "*garbage collecting filters*" and "*message consumption filters*" are meant to contain one token per active filter. Transition "*collect garbage*" fires when there is a garbage-collecting filter that matches at least one of the messages in the message buffer. This transition puts back a modified message buffer in which all messages matching the garbage-collecting filter have been removed. Similarly, transition "*match*" fires if there is a consumption filter matching at least one message in the buffer. This transition also puts back a modified message buffer in which the matched messages are removed. In addition, it produces a tuple containing the filter and the set of matched messages into output place "*matched messages*". These tokens can then be routed to the process container or process instance that registered the filter in question. The latter is identified by a *request identifier* ("*reqID*"). For simplicity, the net only captures the case of "persistent filters", but it is easy to extend the net to deal with one-off filters: the only difference being that such filters should not be put back by transitions "*collect garbage*" and "*match*".

The proposed communication model abstracts away from the way channels and filters relate to business process activities or events. This way, the model can be integrated into a wide range of process definition languages. In BPEL, for example, inbound communication actions appear in two forms: as a standalone *receive* activity type and as the second leg of activities of type *invoke*, where the first leg corresponds to an outbound communication action. Thus, BPEL can be extended with the proposed communication primitives by enabling *receive* and *invoke* activities to refer to channels and filters as defined above. Channels can then be linked to *partner links* and *operations* in BPEL.

Similarly, the proposed model can be used to extend the YAWL process definition language to support a richer set of communication patterns. For example, we can define a type of message receipt task in YAWL such that: (i) upon enablement, the task registers a one-off message consumption filter defined as part of a task decomposition; (ii) the task then waits until the filter returns a match; (iii) should the task be cancelled before a match is found, the filter is withdrawn. Also, we can allow message consumption filters to be attached to the *initial condition* of a YAWL process model, to capture scenarios such as: "a new process instance should be started whenever a given type of message (or combination of messages) has been received." An integration of the proposed communication primitives into YAWL is left as future work.

## 4 Implementation and Evaluation

The implementation builds on a middleware service, and API, called JCoupling (available from `www.sourceforge.net/projects/jcoupling`). JCoupling supports a superset of the communication styles supported by mainstream communication middleware. It also abstracts away from transport protocol details, thus allowing us to concentrate on the core aspects of our proposal. It supports uni- and bi-directional communication, time, space, and synchronisation decoupling, and provides support for fault propagation. JCoupling can operate over open-JMS[2] and is able to use XML/HTTP or XML/TCP for message transport.

Figure 2 presents an architectural diagram of the prototype. It shows properties and filters being used during a simple interaction between two process tasks. In Fig. 2: (1) A message is sent by task "*T1*" over channel "*Ch1*" (denoted by

[2] Open JMS `www.openjms.sourceforge.net` accessed November 2006.

arrows labelled "1" going from the engine to the controller). (2) Properties "*P1*" and "*P2*" are used to extract values from the message. Next the message is put onto JCoupling (3.a), and a new tuple (row) is added to the relation (table) for channel "*Ch1*" (3.b). (4) The receiver task "*T2*" posts a filter over channel "*Ch1*". (5) Using the filter, the controller performs a query, over the relation for channel "*Ch1*". That query produces a tuple and the matching message is extracted from JCoupling (6). The callback to "*T2*" contains the message (7).

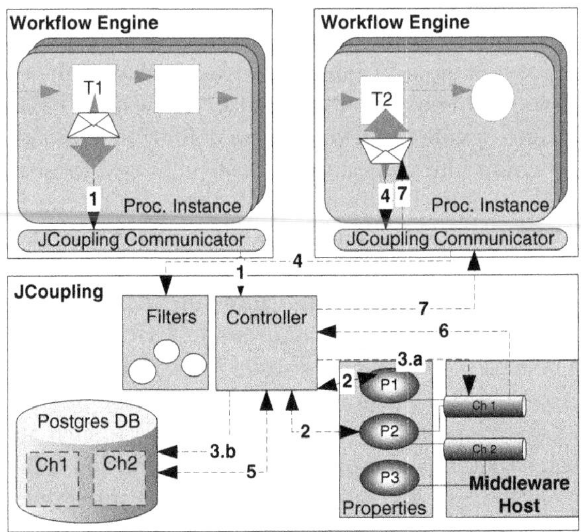


**Fig. 2.** Architecture of the proposal

This section gives further details on the implementation of properties, interactions and filters, and reports on an experimental evaluation of the prototype.

## 4.1 Implementing Properties

The prototype implementation contains an interface called `Property`. This interface and three implementing classes are presented in Fig. 3. A complete listing of this interface, and an explanation it's concrete classes is provided in the technical report [6]. Instances of `Property` extract various scalar values from messages.

We envision a configuration tool for creating property instances such that process designers do not need to write Java code. For example if the process designer wishes to create an XPath property, to extract "*PurchaseOrderID's*" from messages then it would only be necessary to supply a property name, a channel binding, an XPath expression, and a Type (e.g. `'PurchaseOrderId'`, `'PO_Chann'`, `'/order/@po-id/text()'`, `'Text'`).

As each new message arrives the prototype adds a tuple to the *property relation* for that message's channel using a PostgreSQL[3] database. As mentioned

[3] The PostgreSQL database system `www.postgresql.org` (accessed February 2007).

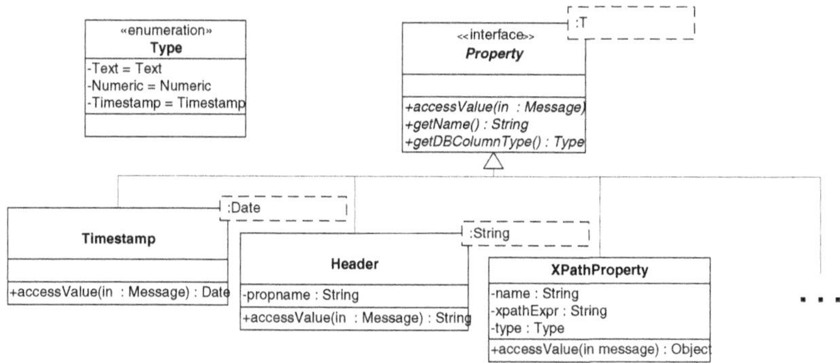


**Fig. 3.** UML of the property Interface

in Sect. 3, each channel's property relation has two default attributes (columns): `messageid` and `timestamp`. Table 1 presents a relation for channel `QuotesCh`.

**Table 1.** Relation corresponding to channel `QuotesCh`

| QuotesCh | | | |
|---|---|---|---|
| messageid | timestamp | quote | amount |
| C0000M0001 | 2006-11-15 16:35:50 | 4000 | 1100 |
| C1111M0002 | 2006-11-16 18:12:20 | 3000 | 1000 |
| C2222M0001 | 2006-11-16 20:42:53 | 2500 | 1001 |
| C3333M0003 | 2006-11-17 16:57:21 | 2000 | 999 |

### 4.2 Resolving Contention

A process may have to wait days before an appropriate message can be received. Hence, the prototype stores send/receive requests and performs callbacks when the interactions are complete. Indeed requests to send/receive, have their own lifecycle, including interaction cancelation (i.e. when process state change makes unnecessary an incomplete interaction). A race condition between two receive tasks is a perfect example of this. For example, a task "receive-bill-payment" and a task "receive-purchase-order-cancelation" operate such that the completion of one disables the other. The first message to arrive will make the other message unnecessary. This is a "pick" activity in BPEL, or a "deferred choice" in YAWL [1]. Accordingly the prototype (being a control layer between a process and the message layer), exposes a 'withdraw'/'cancel', primitive for send and receive interactions – also implementing the cancelation requirement (Sect. 2).

Another form of contention occurs where two tasks both want the same message/s. Contention may be a requirement (for example load balancing, Scenario 6), or it may be accidental, or perhaps even unavoidable due to the nature of the business process. Regardless contention, being a natural and sometimes necessary phenomenon, mandates a graceful approach to handling it[4].

[4] This is a distinguishing point from WS-BPEL, which throws runtime exceptions when contention between two receive-tasks occurs. Also, WS-BPEL's greedy routing of messages to process instances precludes the possibility to support contention.

To overcome the problems of contention within the context of our proposal we adopted the following algorithm in the prototype.

1. The receive request, containing its messages filter, is stored in the prototype.
2. Once the filter produces a non empty set ($\Omega$) of message-identifiers, it is locked to ensure that the receiver (task) cannot withdraw the request.
3. Then the message engine attempts to lock every message in $\Omega$.
4. If all messages were successfully locked, then for each message-id in $\Omega$, that property tuple is deleted, and each corresponding message gets removed from the buffer and sent to the requestor in a callback.
5. Finally the request filter is withdrawn from the engine.

Should the receiver be unable to lock the filter at step 2 the prototype withdraws the filter. At step 3, should any of the messages in $\Omega$ be already locked then the filter locked during step 2 is unlocked, and the filter is rescheduled. Hence the first locker of the message succeeds.

### 4.3 Implementing Filters

This section shows how the extensions to JCoupling outlined above, support the motivating requirements of Sect. 2. Property relations, abstracting from message content, enable the use of restricted SQL queries to produce relations containing message identifiers. The primary restriction we place over the queries is that their outer projections must only be over attributes of the domain `messageid`. i.e. for all attributes of a relation produced by applying any query to the property relations, the domain of that attribute must be `messageid`.

A single property, or combinations of properties can be used to select messages. We envision that the process modeller would have a library of configurable *filter templates*, each containing semi-complete queries. The filter templates would also allow a "design time" binding to process variables, enabling runtime data to be inserted into parameterised SQL. Alternatively, for those situations requiring sophisticated aggregate operations, or complex joining expressions, the process designer may instead write their own SQL.

**Property Based Selection.** Scenario 2 captured the need to select and proceed with the best quote, which is an example of selecting messages based on property values. The query in Listing 1 uses two properties defined over the channel "QuotesCh" (see Table 1). These are "price" and "quantity". When a receive request containing a query is invoked over the prototype, it will apply the query – returning messages in a callback to the requestor when results are found.

**Listing. 1.** This query combines 'price' and 'quantity' to find the best value offer

```
SELECT messageid
FROM QuotesCh
WHERE quantity >= 1000
AND price/quantity =
(  SELECT min (CostPerUnit)
   FROM (   SELECT price/amount AS CostPerUnit
            FROM QuotesCh ) AS Q1 )
```

**Conversations.** Scenario 1 outlined the need to correlate messages to an outer conversation for "purchase-order-ID" and an inner (nested) conversation for "line-item-id". Messages will only be correlated to a nested conversation if they satisfy correlation filters for the inner conversation, and all parent conversations. To achieve this we append an `AND-Clause` to the query. The following query (see Scenario 1) extracts runtime data from two process variables visible to the receive task: namely `PurchaseOrderID` and `LineItemID`.

**Listing. 2.** Achieving nested correlation through querying correlation properties

```
SELECT messageid
FROM PoResponseCh
WHERE PoID = $PurchaseOrderID$
AND ItemID = $LineItemID$
```

We envision that the process layer will generate queries for conversations from a "conversation" construct in the process model. This construct would declare which message properties are to be used for correlation, and whether each communication task involved initialises the conversation, or follows it. Furthermore if any task involved in the conversation wants to apply message filters we may have the task append more `AND` clauses to the generated query.

**Atomic Multiple Source, and Aggregate Consumption.** Achieving a combination of atomic multiple source consumption and aggregate consumption is possible by applying a query to two or more property relations. For example Scenario 4 required aggregate consumption of 100 packages destined for the same area, and a truck availability message from another channel. Listing 3 extracts runtime data from a process level variable called `deliveryDistrict`. The query either returns at least 100 tuples, or returns nothing.

**Listing. 3.** This query produces a relation of messageid's wherein the attribute (`Pack.messageid`) refers to messages from Channel `Packages`, and the attribute `Truck.messageid` refers to a message on Channel `TruckWaiting`. Related messages are linked, thus removing the need to relate messages in the process.

```
SELECT Pack.messageid, Truck.messageid
FROM Packages As Pack, (SELECT messageid FROM TruckWaiting LIMIT 1) AS Truck
WHERE Pack.deliveryDistrict = '$deliveryDistrict$'
AND 100 <= ( SELECT count(*)
             FROM Packages
             WHERE deliveryDistrict = '$deliveryDistrict$' )
```

**Aggregated Selection Involving Time.** Scenario 5 sought to find if at least 5% of messages that are less than 7 days old contain complaints; drawing on solutions to both time and aggregated selection. Additionally complaints were sought on *all channels*. A join between property relations will not work as every tuple from each relation represents one discrete event that needs to be considered separately. So unlike Listing 3, the result relation will have one attribute. A `view` over the `union` of source channels (property relations), solves this. In Listing 4 this view is named "Merged". It combines properties (`messageid, timestamp`, and `complaint`) from each of the source property relations.

**Listing. 4.** Using **union** and **view** to combine input sources for aggregate operations

```
CREATE VIEW Merged AS (
SELECT messageid, timestamp, complaint
FROM CustomerCh
UNION SELECT messageid, timestamp, complaint
FROM PartnerCh )
```

Using the above view, aggregate calculations over the messages, taken collectively, becomes feasible. Listing 5 demonstrates this. Lines 5 – 12 produce `false` unless 5%, or more, of the messages are complaints. Lines 3, 8, & 12 show the use of relative time expressions over the `timestamp` property. In cases like this we imagine that the process creator would write Listings 4 and 5 manually.

**Listing. 5.** This query, adapted from a continuous query in [8], produces a non-empty result of **messageid**'s when 5%, or more, of last week's messages contain complaints

```
1  SELECT messageid
2  FROM Merged
3  WHERE timestamp  > (CURRENT_TIMESTAMP - INTERVAL '7 days')
4  AND complaint = true
5  AND (    SELECT count(*)
6           FROM Merged
7           WHERE complaint = true
8           AND timestamp  > (CURRENT_TIMESTAMP - INTERVAL '7 days')
9       ) >= (
10          SELECT 0.05 * count(*)
11          FROM Merged
12          WHERE timestamp  > (CURRENT_TIMESTAMP - INTERVAL '7 days')
13      )
```

### 4.4 Performance Evaluation

To compare our correlation approach with that of WS-BPEL, we conducted experiments where up to ten thousand interactions were executed over our prototype and over a WS-BPEL simulator. Each experiment involved creating, in random order, at fixed intervals, a set of XML messages, all identical except for the value of one element which was mapped to a property. Receiving processes were spawned in the same way, each of which waited for one of the created messages. The code for the experiments is released with the JCoupling distribution.

The WS-BPEL correlation simulator was built using the same middleware as our prototype. This simulator receives messages off a designated channel. Each time a message arrives, an XPath expression is evaluated against it to extract a property value (this is called a `propertyAlias` in WS-BPEL). The extracted property value is then stored in a hash table together with the corresponding message identifier.[5] Concurrently, the simulator handles requests to consume incoming messages based on property values. When the simulator finds a match between a receive request and a message, the corresponding entry is deleted from the hash table, symbolising that the message has been correlated.

The test results are presented in Table 2. The table shows that our approach slightly underperforms that of WS-BPEL for small numbers of messages. This

[5] In the interest of fairness, each entry is written to disk after being added to the in-memory hash table, as our prototype stores property values in persistent tables.

**Table 2.** Results of performance tests

| Number of Interactions | | 50 | 100 | 250 | 500 | 1000 | 2500 | 5000 | 10000 |
|---|---|---|---|---|---|---|---|---|---|
| Time (ms) | Proposed Approach | 611 | 951 | 2230 | 5064 | 9003 | 25276 | 71409 | 234211 |
| | WS-BPEL Approach | 524 | 938 | 2103 | 4046 | 7825 | 20506 | 44204 | 133933 |
| Performance Difference | | 14% | 1% | 6% | 20% | 13% | 19% | 38% | 43% |

difference grows for larger numbers of messages. The accentuated difference can be explained by the fact that in the implementation of our approach, queries to match uncorrelated messages with pending receive requests are run against a persistent database, whereas in the WS-BPEL simulator, the corresponding lookup is done in-memory[6]. For larger numbers of messages, this leads to a performance penalty due to database cache management. A more consistent performance could be achieved by using an in-memory database system to implement our approach. Indeed, it is not necessary to make the correlation data structures persistent, only the messages themselves need to be persistent. In future, we plan to implement a more refined version of our approach and run a fairer and more detailed comparison against the WS-BPEL approach.

The performance penalty of our approach should be weighed against the additional functionality that it brings in. Indeed, our approach supports aggregate messaging, multi-source consumption and message contention. Moreover, as previously mentioned, there are opportunities to optimise the brute-force approach used in our implementation through incremental query evaluation.

## 5 Related Work

Communication in the context of distributed business processes has traditionally been researched from the perspective of *protocol* or *contract* definition. For example, the CrossFlow system [13] enables process designers to define contracts governing the communication between multiple workflows, possibly distributed across organisational boundaries. These contracts can be statically checked for consistency. Similarly, [3] proposes a method for capturing inter-workflow communication protocols and detecting deadlocks that can arise when inter-connecting processes with incompatible communication protocols. This body of work is complementary to our proposal, as we do not deal with static analysis, but rather with the routing of messages to processes at runtime.

WS-BPEL exhibits strong support for conversations with the exception of *Instance Channels*, which are not supported because WSDL *ports* are not generated during process execution. WS-BPEL, does not support the selection of messages based on their properties despite the use of properties in correlation sets. Nor does it support atomic-batched consumption. Contention over messages is not supported as messages are greedily consumed off channels and allocated to process instances immediately. Time-based message selection is not supported either. A WS-BPEL process consumes everything sent to it, thus forcing the modeller to select and throw away unwanted messages, as part of the core process

[6] Although entries are written to disk, lookups over the hash table are done in-memory.

logic. WS-CDL [15] has some distinguishing features, with respect to WS-BPEL, such as abstractions for channel passing and a global viewpoint over all actors in a choreography. However, in terms of the motivating requirements outlined in this paper, WS-CDL has very similar strengths and weaknesses to WS-BPEL.

Widom et al. [8,16] propose an approach to optimising the evaluation of continuous queries over one or many data streams. They address some of the problems associated with scalability of such queries and propose incremental evaluation techniques based on the type of query. We plan to apply some of their findings to enhance and optimise the evaluation of filters used in our proposal.

## 6 Conclusion

This paper proposed an inter-workflow communication and control layer, to lie between traditional workflow and messaging layers – providing an isolated area for the description and execution of communication. This would provide relief from using "spaghetti" solutions to achieve non-trivial interactions. The proposal is based on a strong and relatively simple set of abstractions – namely channels, properties, property relations and filters. Channels abstract from middleware topics, and queues. Properties abstract from message content and format, while property relations provide the foundation for property filters. Filters abstract from the business level requirements for choosing and selecting messages. These enable all forms of correlation, message selection, aggregated message consumption, and time based message consumption, over a single or multiple channels evaluated collectively. The possibility of contention between process/task instances is overcome by locking messages, and filter requests. The proposal has been implemented on top of a communication API, namely JDecouple, that abstracts away from the underlying middleware and communication protocols.

Future work will aim at integrating the proposed communication abstractions into process definition languages. Specifically, we aim to extend YAWL with the ability to attach communication actions to various elements of the notation.

**Acknowledgements.** This work is funded by ARC Discovery Grant DP0451092. The third author is funded by a Queensland Smart State Fellowship.

## References

1. van der Aalst, W., ter Hofstede, A.: YAWL: Yet Another Workflow Language. Information Systems 30(4), 245–275 (2005)
2. van der Aalst, W., ter Hofstede, A., Kiepuszewski, B., Barros, A.: Workflow patterns. Distributed and Parallel Databases 14(3), 5–51 (2003)
3. van der Aalst, W., Weske, M.: The P2P approach to Interorganizational Workflows. In: Dittrich, K.R., Geppert, A., Norrie, M.C. (eds.) CAiSE 2001. LNCS, vol. 2068, pp. 140–156. Springer, Heidelberg (2001)

4. Aldred, L., van der Aalst, W., Dumas, M., ter Hofstede, A.: On the Notion of Coupling in Communication Middleware. In: In Proceedings of the 7th International Symposium on Distributed Objects and Applications (DOA). Agia Napa, Cyprus, November 2005, pp. 1015–1033. Springer, Heidelberg (2005)
5. Aldred, L., van der Aalst, W., Dumas, M., ter Hofstede, A.: Understanding the challenges in getting together: The semantics of decoupling in middleware. Technical Report BPM-06-19, Business Process Management Center, Brisbane, Qld, Australia, 2006 ((February 2007), `http://www.bpmcenter.org` accessed
6. Aldred, L., van der Aalst, W.M.P., Dumas, M., ter Hofstede, A.H.M.: Abstractions for communication between distributed business processes. Technical Report BPM-06-28, Business Process Management Center, Brisbane, Qld, Australia, 2006, accessed (February 2007) `www.bpmcenter.org`
7. Alves, A., Arkin, A., Askary, S., Bloch, B., Curbera, F., Goland, Y., Kartha, N., Liu, C., König, D., Marin, M., Mehta, V., Thatte, S.,van der Rijn, D., Yendluri, P., Yiu, A.: Web Services Business Process Execution Language. Initial Draft of standards proposal by OASIS, June 2006, accessed (July 2006) `http://www.oasis-open.org/apps/org/workgroup/wsbpel/`
8. Babu, S., Widom, J.: Continuous queries over data streams. SIGMOD Record 30(3), 109–120 (2001)
9. Barros, A., Decker, G., Dumas, M., Weber, F.: Correlation Patterns in Service-Oriented Architectures. In (FASE). Proceedings of the 10th International Conference on Fundamental Approach to software Engineering, Braga, Portugal, March 2007, pp. 245–259. Springer, Heidelberg (2007)
10. Beugnard, A., Fiege, L., Filman, R., Jul, E., Sadou, S.: Communication Abstractions for Distributed Systems. In: Buschmann, F., Buchmann, A.P., Cilia, M.A. (eds.) Object-Oriented Technology. ECOOP 2003 Workshop Reader. LNCS, vol. 3013, pp. 17–29. Springer, Heidelberg (2004)
11. Cypher, R., Leu, E.: The semantics of blocking and nonblocking send and receive primitives. In: Siegel, H (ed.) Proceedings of 8th International parallel processing symposium (IPPS), pp. 729–735 (April 1994)
12. Eugster, P., Felber, P., Guerraoui, R., Kermarrec, A.: The Many Faces of Publish/Subscribe. ACM Computing Surveys 35(2), 114–131 (2003)
13. Grefen, P., Aberer, K., Hoffner, Y., Ludwig, H.: CrossFlow: Cross-organizational Workflow Management in Dynamic Virtual Enterprises. International Journal of Computer Systems, Science, and Engineering 15(5), 277–290 (2001)
14. Hohpe, G., Woolf, B.: Enterprise Integration Patterns: Designing, Building, and Deploying Messaging Solutions. Addison-Wesley, Boston (2003)
15. Kavantzas, N., Burdett, D., Ritzinger, G., Fletcher, T., Lafon, Y., Barreto, C.: Web Services Choreography Description Language Version 1.0. Candidate Recommendation, (November 2005) `http://www.w3.org/TR/ws-cdl-10/`
16. Olston, C., Jiang, J., Widom, J.: Adaptive filters for continuous queries over distributed data streams. In: SIGMOD '03: Proceedings of the 2003 ACM SIGMOD international conference on Management of data, pp. 563–574. ACM Press, New York (2003)

# Questionnaire-driven Configuration of Reference Process Models

Marcello La Rosa, Johannes Lux, Stefan Seidel, Marlon Dumas, and Arthur H. M. ter Hofstede

BPM Group, Queensland University of Technology, Australia
{m.larosa,j.lux,s.seidel,m.dumas,a.terhofstede}@qut.edu.au

**Abstract.** Reference models are a widely accepted means to facilitate reusable information system and organizational design. At present, besides domain knowledge, the configuration of a reference model requires a thorough understanding of the notation it is captured in. This hinders the involvement of domain experts without specialized modeling background, in the configuration of reference models. In this paper, we propose a questionnaire-driven approach to reference model configuration which abstracts away from the modeling language. For illustration, we show how this approach can be applied to reference process models captured in the Configurable EPC notation. To demonstrate its applicability, the proposal has been implemented as a toolset that guides users through the configuration process by means of a form-based interface.

## 1 Introduction

A *reference process model* is a model of day-to-day operations in a given domain such as supply chain management, logistics, human resource management or film production. Reference process models are intended to be configured in a specific context (e.g. for a given organization or project) leading to individualized process models. A major benefit of configuring a reference process model for a given project, as opposed to building a new model from scratch, is the ability to reuse and build upon proven practices [5].

Reference process models in commercial use, such as the IT Infrastructure Library (ITIL) [18] or the Supply Chain Operations Reference model (SCOR) [17], lack a representation of configuration alternatives, configuration decisions, and relationships between these decisions and alternatives. This hinders the configuration of these models. Notations for representing configuration alternatives in process models have been put forward to address this shortcoming. An example is the Configurable Event-driven Process Chains (C-EPCs) notation [13] which extends EPCs, a generic process modeling notation, with the ability to capture variation points, constraints that restrict the allowed variations and guidelines for configuring variation points.

However, these approaches suffer from two major limitations. Firstly, in notations such as C-EPCs, which are designed to capture individual variation points, it is difficult to understand which of these points are affected by a high-level

J. Krogstie, A.L. Opdahl, and G. Sindre (Eds.): CAiSE 2007, LNCS 4495, pp. 424–438, 2007.

configuration decision. Secondly, these approaches require that the stakeholders involved in the configuration of a reference process model have a thorough understanding of both the application domain and the process modeling notation. While it is normal to assume that the stakeholders who produce the reference process model itself are familiar with the notation in question, it is less realistic to assume that those who provide input for configuring these models (e.g. a logistics expert) are sufficiently proficient with the notation.

To address these shortcomings, we propose a questionnaire-driven approach to reference model configuration that directly captures configuration choices and their dependencies. Via so-called *facts* that represent answers, *questions* are linked to variation points in reference models. Questions are expressed in natural language and can be answered by domain experts without extensive knowledge of the underlying reference model. The only major assumption made is that questions have a finite or discretized domain of possible answers. This assumption is reasonable, given that the number of configuration alternatives in a reference process model (e.g. in a C-EPC) is also finite. This assumption allows the models to be efficiently analyzed to prevent the user from entering inconsistent answers.

In this paper, we show how the proposed questionnaire-driven configuration approach can be applied to reference process models defined as C-EPCs. We also illustrate how this approach, linked with C-EPCs, can be applied to a reference process model with non-trivial interdependencies. We also show how the configuration process can be supported by a toolset that guides users through the configuration questions by means of a form-based interface.

The rest of the paper is organized as follows. In the next section we briefly describe C-EPCs. We then introduce a working example and demonstrate how the formalized approach to questionnaire-driven configuration can be applied. We further show how the automated configuration has been implemented including the mapping between C-EPC and interactive questionnaires. The paper concludes with related work and an outlook including our future research agenda.

## 2 Background: Configurable Event-driven Process Chains

Event Driven Process Chains (EPCs) [6] are a widely used modeling language whose main components are *events*, *functions*, *connectors* and *arcs* linking these elements. Events represent triggers or conditions, functions correspond to tasks, and connectors denote splits and joins of type *AND*, *OR* or *XOR*.

C-EPCs extend EPCs by providing a means to represent variability in EPC reference process models. This is achieved by identifying a set of variation points (*configurable nodes*) in the model, to which possible values (*alternatives*) can be assigned, as well as constraints to restrict the combination of allowed values. By configuring each variation point to exactly one value among the ones allowed, it is possible to derive an EPC model from the starting C-EPC.

Variation points are nodes of type *function* or *connector*, highlighted in bold in the model. *Configurable functions* can be set as included (*ON*), excluded

(*OFF*) or conditionally skipped (*OPT*). The first two alternatives allow one to decide a priori whether to keep or permanently discard the function; the last option permits the deferral of this choice to run-time, where the execution of the function can be skipped on an instance-by-instance basis. *Configurable connectors* can only be mapped to equally or less expressive connector types. Consequently, a configurable *AND*-connector can only be mapped to a regular *AND*-connector. A configurable *XOR* can be set to a regular *XOR* or to an outgoing/incoming sequence $SEQ_n$ of events and functions (where $n$ is the node starting the sequence). A configurable *OR* can be mapped to a regular *OR*, *XOR*, *AND* or to a single sequence. Moreover, *configuration requirements* formalize constraints over the values of variation points, whilst *configuration guidelines* express advices and industry best practices to aid the configuration process. They are both expressed in the form of logical predicates and depicted as notes attached to the variation points involved. Only requirements are mandatory and must hold for a configuration to be valid. Finally, a partial order over variation points can be defined as a suggested order for configuring the nodes of the model.

The following definitions formalize the above concepts and closely follow [13]:

**Definition 1 (Configurable EPC).** *A configurable EPC is a ten-tuple C-EPC* $= (E, F, C, l, A, F^C, C^C, O^C, R^C, G^C)$ *where:*

- $E, F, C, l$ *and* $A$ *refer to standard EPC sets of* events, functions, connectors, *a mapping to define a* label AND, XOR, *or* OR *for each connector, and* arcs,
- $F^C \subseteq F$ *is the set of* configurable functions,
- $C^C \subseteq C$ *is the set of* configurable connectors,
- $O^C \subseteq (F^C \cup C^C) \times (F^C \cup C^C)$ *is a* partial order *over the configurable nodes,*
- $R^C$ *is the set of* configuration requirements,
- $G^C$ *is the set of* configuration guidelines.

**Definition 2 (Partial Order for Connectors).** *The partial order* $\leq^C$ *is defined on* $CT \cup CTS$ *where* $CT = \{AND, OR, XOR\}$ *is the set of connector types and* $CTS = \{SEQ_n \mid n \in E \cup F \cup C\}$ *is the set of sequence operators.* $\leq^C = \{(n, n) \mid n \in CT\} \cup \{(XOR, OR), (AND, OR)\} \cup CTS \times \{XOR, OR\}$.

The partial order $\leq^C$ is used to determine, by restriction, the set of values each configurable connector can be mapped to. For example $XOR \leq^C OR$ implies the second configurable connector type can be mapped to the first connector type.

A *configuration* is a mapping that links a configurable node to an allowed value, according to the node type. It also ensures that a sequence can be chosen as value, only if it is an incoming branch for a configurable join-connector, or an outgoing branch for a configurable split-connector.

**Definition 3 (Configuration $l^C$).** *Let C-EPC* $= (E, F, C, l, A, F^C, C^C, O^C, R^C, G^C)$ *be a configurable EPC. The mapping* $l^C \in (F^C \to \{ON, OFF, OPT\}) \cup (C^C \to CT \cup CTS)$ *is a* configuration *of C-EPC iff for each* $c \in C^C$*:*

- $l^C(c) \leq^C l(c)$,
- *if* $c \in C_J$ *and* $l^C(c) = SEQ_n$ *for some* $n \in E \cup F \cup C$*, then* $(n, c) \in A$*, where* $C_J$ *is the set of* join connectors,

– *if* $c \in C_S$ *and* $l^C(c) = SEQ_n$ *for some* $n \in E \cup F \cup C$, *then* $(c, n) \in A$, *where* $C_S$ *is the set of* split connectors.

When the above requirements are met, we write $valid_{C-EPC}(l^C)$, or simply $valid(l^C)$ if from the context the C-EPC involved is clear.

## 3 Working Example

As part of a research effort focusing on business process management for the Post-Production phase in the screen business, we developed a C-EPC reference process model [15]. An extract of this model is presented in Fig. 1 and will be used as working example throughout the paper.

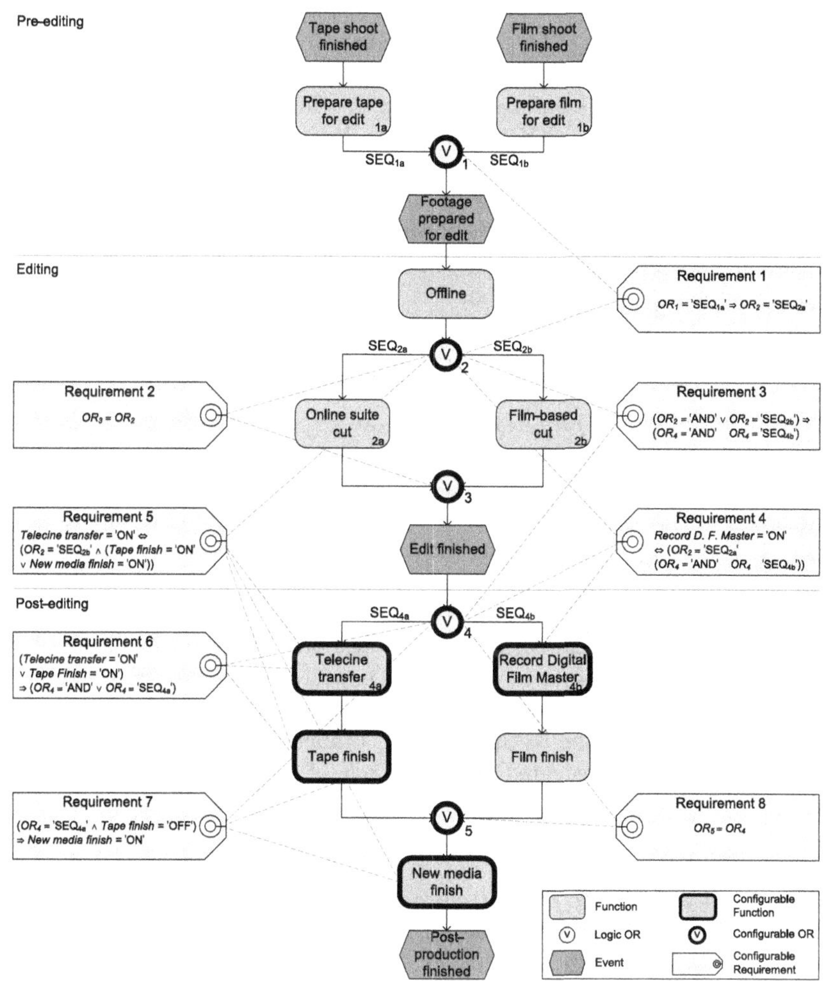


**Fig. 1.** The Post-Production reference process model example

Post-Production aims at the edit and technical completion of a screen business project and encompasses three main steps: pre-edit, edit and post-edit. In the first phase the footage arriving from the shooting is prepared for editing by synchronizing audio and video. The shooting format can be tape, film, or both media. Of the two, film results in a more costly operation as special treatments are required for making it visible and permanent. The choice of the medium is modeled in the C-EPC example via configurable connector $OR_1$, which can be set to the value $SEQ_{1a}$ for tape, $SEQ_{1b}$ for film, or $AND$ for both.

The first activity of the edit phase is the Offline (function *Offline* in the model), where the main creative editing part is carried out on a low resolution medium. This is followed by the cut stage, where the editing decisions previously taken are committed in a high quality format. The cut can be *Film-based* or carried out in an *Online Suite*, according to the format of the shooting media. This variability is achieved in the model by configuring $OR_2$ and $OR_3$ to just one of the two branches ($SEQ_{2a,} SEQ_{2b}$) or to their parallel execution ($AND$). However, the film-based variant can be selected only if at least parts of the project were shot on film. This is enforced by *Req. 1*, while *Req. 2* ensures $OR_3$ to be configured the same way as $OR_2$, since they share the same incoming/outgoing branches. As a result, we can impose requirements only on the split, as the join connector will be configured accordingly. Note that we could have defined $(OR_2, OR_3) \in O^C$ to specify that $OR_2$ should be configured before $OR_3$. However the partial order is just a suggestion, and does not compel $OR_3$ to behave as $OR_2$. Therefore *Req. 2* is needed.

In post-edit, the project is finished for delivery on tape, film, new media or any combination of these formats. The overall finishing process varies on the basis of the delivery media and may involve further tasks, according to the configuration choices made before. A *Film-based cut* is an expensive activity. Therefore, if performed, it must lead at least to a *Film finish.* This is guaranteed by *Req. 3* attached to connector $OR_4$. In fact, if the Film-based branch is enabled by $OR_2$, then function *Film finish* will be executed either if $OR_4$ is equal to $AND$ or $SEQ_{4b}$. However, if the cut has been done only in the online suite, then a further task, modeled by configurable function *Record Digital Film Master*, is needed to transfer the editing results to the so-called 'film master'. This is stated in *Req. 4*. Analogously, *Telecine transfer* is used only if the cut is film-based and if at least a finish on tape or new media is expected. This behavior is enforced by *Req. 5*, attached to configurable functions *Tape finish* and *New media finish.* These two functions belong to the outgoing branch $SEQ_{4a}$ of $OR_4$; thus *Req. 6* ensures that $SEQ_{4a}$ is activated if at least one of these two functions has been set to $ON$. *Req. 7* guarantees that at least one finish medium is selected, as *New media Finish* must be set to $ON$, if no film nor tape finish is desired (i.e. if $OR_4 = SEQ_{4a}$). Finally, *Req. 8* imposes $OR_5$ to take the same value as $OR_4$.

For simplicity's sake we did not consider those configuration alternatives involving run-time choices ($XOR$, $OR$, $OPT$) and implied the existence of further requirements to avoid such alternatives. Nonetheless, this is an example of how interdependencies over configurable nodes can be complex and intricate when the model refers to a real configuration scenario. In such cases, model-based configurations may turn out to be unacceptably arduous. Moreover, domain experts – supposed to be in charge of the configuration – are likely to be unaware of business process notations, as in the screen business case. In order to tackle these issues, in the next section we propose a new approach to configuration.

# 4 Approach

## 4.1 Questionnaire-Driven Configuration

We propose to represent choices independently of specific notations or languages, by means of a set of facts, representing the space of possible answers to a set of questions. Questions can be answered solely requiring domain expertise, via an interactive questionnaire that guides the configuration, by posing only the relevant questions in an order consistent with the interdependencies.

Making a choice corresponds to setting a *fact* within a *question*. Facts are simply statements such as "tape shooting" or "film finish". Initially they are *unset* while at run-time they can be asserted or negated by setting their value to *true*, resp. *false*. For example, setting "tape shooting" = *false*, would mean that we are not interested in shooting on tape. Each fact features a default value and can be marked as 'mandatory' if it needs to be set explicitly. Under certain restrictions, a non-mandatory fact can be left *unset* at configuration-time. In this case its default value is used instead.

Facts are grouped in questions according to their content, so that all the facts of the same group can be set at once by answering the associated question. Each question features at least one fact and the set of questions must cover all the facts. Although a fact can appear in more than one question, its value can be set only the first time, and must be preserved in all the subsequent questions that contain it. However, the value of a fact previously set can still be changed by rolling back the question.

A *facts setting* is any combination of facts values such that all the facts have been set, either explicitly by answering questions or by using their default.

Fig. 2 depicts a possible structure of questions/facts for representing variability in Post-Production. All questions and facts are assigned a unique id and a description. For example, facts $f_1$ to $f_3$ refer to typical budget ranges for a Post-Production project, so they all are grouped in question 1 asking for the estimated project budget. Also, these facts are mandatory as we want users to explicitly answer $q_1$. Indeed, the choice of budget is rather important as it affects the Post-Production phase overall. Default values have been assigned in order to reflect the typical choices made in a medium budget project, pitched for cinema and home video distribution. Hence, $f_2$ in $q_1$ has default value equal to *true* as well as $f_4$ and $f_6$ in $q_2$, which relates to the primary distribution channels, and so on for the other facts. Other questions would allow users to choose the shooting media ($q_3$) and the shooting format ($q_6$, $q_7$), the type of cut ($q_4$) and the expected deliverables ($q_5$).

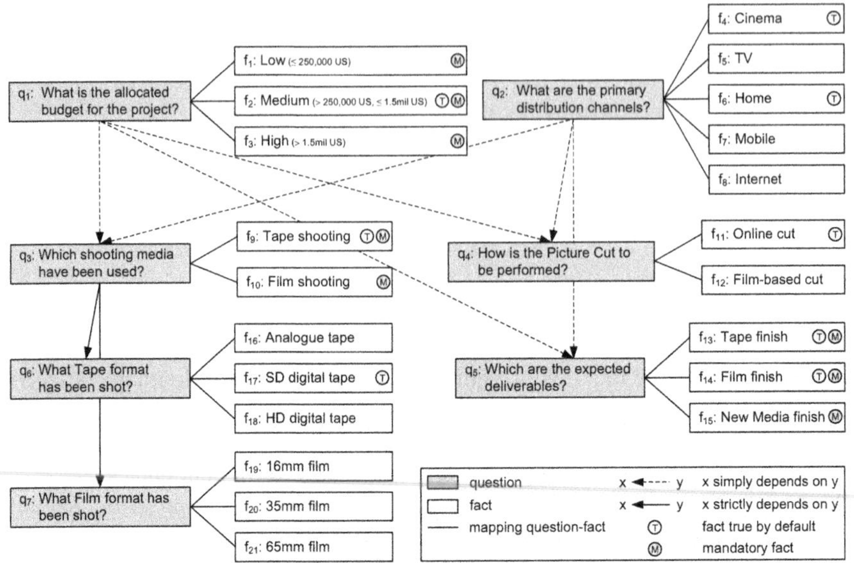


**Fig. 2.** A possible structure of questions/facts for the Post-Production example

Questions can be connected via two different types of dependencies. Dependencies determine a partial order for posing questions to users, and can be arbitrary as long as undesired cycles are avoided. A *simple dependency* (dashed arrow in Fig. 2) is used when a question "may" depend on another, whilst a *strict dependency* (plain arrow) is used to model a compulsory order over two questions. For example in Fig. 2, $q_3$ allows users to choose the shooting media between tape ($f_9$) and film ($f_{10}$). This question "simply" depends on $q_1$ and $q_2$, viz., it can be posed only after answering at least one of $q_1$ and $q_2$. On the other hand, $q_6$ – where the tape format is determined – strictly depends on $q_3$, as it is reasonable to make this choice only after deciding on the shooting media which includes tape as possible alternative ($f_9$). Although not shown in this example, dependencies can also be defined over facts, by following the same rules.

Dependencies provide a means for ordering questions but do not affect facts values. Answering a question in a given way may restrict the allowed answers to subsequent questions, and not all combinations of answers may lead to valid facts settings. We model interdependencies over facts values as a set of constraints in propositional logic, used to restrict the number of possibilities. A facts setting is thus a *configuration* if and only if it complies with the constraints.

The following constraints, drawn from an analysis of the parameters leading to variations in the Post-Production reference process model, are defined over the facts identified in Fig. 2:[1]

[1] $\dot{\vee}$ indicates the exclusive disjunction ($XOR$), a commutative and associative relation.

C1: $f_1 \dot{\vee} f_2 \dot{\vee} f_3$
C2: $f_1 \Rightarrow \neg(f_{10} \vee f_{14})$
C3: $f_2 \Rightarrow \neg f_{10}$
C4: $f_4 \vee f_5 \vee f_6 \vee f_7 \vee f_8$
C5: $f_4 \Rightarrow f_{14}$
C6: $f_5 \Rightarrow f_{13}$
C7: $f_6 \Rightarrow (f_{13} \vee f_{15})$
C8: $(f_7 \vee f_8) \Rightarrow f_{15}$
C9: $f_9 \vee f_{10}$
C10: $f_{11} \vee f_{12}$
C11: $\neg f_{10} \Rightarrow \neg f_{12}$
C12: $f_{13} \vee f_{14} \vee f_{15}$
C13: $(f_{16} \dot{\vee} f_{17} \dot{\vee} f_{18}) \Leftrightarrow f_9$
C14: $\neg(f_{16} \vee f_{17} \vee f_{18}) \Leftrightarrow \neg f_9$
C15: $f_{12} \Rightarrow f_{14}$
C16: $(f_{19} \dot{\vee} f_{20} \dot{\vee} f_{21}) \Leftrightarrow f_{10}$
C17: $\neg(f_{19} \vee f_{20} \vee f_{21}) \Leftrightarrow \neg f_{10}$.

For example, C1 ensures that exactly one fact is asserted in $q_1$, as a project places itself only in a specific budget range. On the other hand, due to C4, more than one distribution channel can be selected in $q_2$, as it makes sense for a project to have multiple releases (e.g. *TV* and *Home*).

We said that due to the costs involved, a film-based cut would be worthwhile if it implied a subsequent finish on film. This is captured by C15, affecting the way $q_4$ and $q_5$ can be answered. In truth, as per C2, a low budget choice ($f_1 = true$) implies no shooting on film nor release on film is possible ($f_{10}, f_{14} = false$). As a result, for low budgets a film-based cut is not allowed either (C11).

Dependencies and constraints are not overlapping concepts. They rather complement each other, as shown by C11 over $f_{10}$ and $f_{12}$. These two facts occur in $q_3$ resp. $q_4$, but the questions do not depend on each other. Thus, by setting $f_{10}$ to *false*, $f_{12}$ is forced to *false* too, although we could have already negated the latter by answering $q_4$ before $q_3$. In this case, only a constraint is used to achieve the desired behavior. On the other hand, as per C13, (exactly) one tape format can be chosen in $q_6$, if and only if tape has been selected as shooting medium in $q3$ ($f_9 = true$). Otherwise, no tape format can be specified (C14). Anyhow, $q_6$ cannot be answered before $q_3$. In such a case, dependencies and constraints work together to ensure the shooting format being decided only after the shooting medium, and according to its type.

It is not in the scope of this paper to provide a method for identifying a set of dependencies and constraints such as the above ones. Rather, the focus is on defining a (meta-)model for capturing questions, facts, constraints and dependencies, and then linking facts to variation points in a C-EPCs.

## 4.2 Formal Definition of Configuration Models

Due to space limitations, this section presents only a reduced definition of *Configuration Model (CM)*, which is the formal underpinning to our approach. The complete definitions, technical details and proofs can be found in [7].

**Definition 4 (Reduced Configuration Model).** *A reduced configuration model is a six-tuple $rCM = (F, F_D, F_M, Q, map_{QF}, CS)$ where:*

- *$F$ is a finite, non-empty set of* facts *where a* fact *is a boolean variable,*
- *$F_D \subseteq F$ is the default setting, i.e. the set of* facts *whose default is asserted,*
- *$F_M \subseteq F$ is the set of* mandatory facts,
- *$Q$ is a finite (non-empty) set of* questions,
- *$map_{QF} \in Q \to \mathcal{P}(F) \setminus \{\varnothing\}$ is a function mapping questions onto sets of facts, such that $\bigcup_{q \in Q} map_{QF}(q) = F$,*[2]

[2] $\mathcal{P}$ indicates a power set.

– *$CS \subseteq \mathcal{P}(F)$ is the set of the* allowed settings *of the facts in $F$, such that $F_D \in CS$, i.e. the default setting is always allowed.*

Elements of $CS$ are those facts settings that satisfy all the constraints, where only the facts asserted are present in each element. Hence, if a fact is not contained in a clause of $CS$, it follows that the fact is negated in that setting. Also, as the default setting must be always allowed, set $CS$ is non-empty. The definition of set $CS$ has been used to construct a set of functions for detecting possible conflicts over facts constraints at design-time, and for dynamically restricting the space of available configurations at run-time. Those definitions can be found in [7]. An implementation of these concepts is discussed in Section 5.

A *configuration* $\sigma$ of $CM$ is thus a facts setting whose true values form exactly an element of $CS$, i.e. a facts setting that does not violate the constraints.

### 4.3 Mapping C-EPCs to Configuration Models

We propose a simple method to define a mapping between a C-EPC and a questionnaire-driven configuration model. The idea is to assign a boolean function over the facts of the configuration model, to each configuration a variation point can assume in the C-EPC. For example, the first configurable node in the C-EPC process of Fig. 1, $OR_1$, according to the type of shooting medium, can be set to $AND$ (for both tape and film), $SEQ_{1a}$ (for tape only) and $SEQ_{1b}$ (for film only). In the configuration model of Fig. 2 this would correspond to answer $q_3$ (*Which shooting media have been used?*) with both $f_9, f_{10} = true$ in the first case, with only $f_9 = true$ in the second case, and with only $f_{10} = true$ in the third (where $f_9$ is *Tape shooting* and $f_{10}$ is *Film shooting*). This is equivalent to checking whether $f_9 \wedge f_{10}$, or $f_9 \wedge \neg f_{10}$, or $\neg f_9 \wedge f_{10}$ holds against a given configuration over facts, obtained by answering the questions shown in Fig. 2. Thus, we assign each of these functions to the corresponding configuration of $OR_1$. The remaining configuration alternatives of $OR_1$, not allowed by the model (i.e. $OR, XOR$), are simply given a *false* function.

A requirement for a mapping between a C-EPC and configuration model to be unambiguous is that, for any facts configuration, exactly one boolean function should evaluate to *true* for each variation point in the model. This way we avoid a facts configuration that may lead to zero or more than one alternative for a variation point. In the above example this is enforced by C9, which excludes the fourth combination $\neg f_9 \wedge \neg f_{10}$ as a possible condition of facts values. Once each variation point has been configured with a proper configuration value, an action, attached to that value, has to be performed on the C-EPC net so as to reflect the values chosen for facts. A formalization of the mapping is given below.

**Definition 5 (CF-Mapping, Valid CF-Mapping, Actions, CA-Mapping).** *Let C-EPC $= (E, F, C, l, A, F^C, C^C, O^C, R^C, G^C)$ be a configurable EPC, $l^C \in (F^C \to \{ON, OFF, OPT\}) \cup (C^C \to CT \cup CTS)$ a configuration of C-EPC, and let $rCM = (F, F_D, F_M, Q, map_{QF}, CS)$ be a reduced configuration model. For each configurable node $cn \in C^C \cup F^C$:*

- $L^{cn} = \{l^C(cn) \mid valid(l^C)\}$ *is the set of all the configurations of C-EPC for a given cn,*
- $map^{cn}_{CF} \in L^{cn} \rightarrow \mathbb{B}_F$ *is a* CF-Mapping, *i.e. a function mapping a configuration of cn to a boolean function defined over F,*
- $\varphi \equiv \bigvee_{v \in CS}(\bigwedge_{f \in v} f \ \wedge \ \bigwedge_{f \in F \setminus v} \neg f)$ *is a boolean function evaluating to* true *if there exist constraints over F,*
- $valid(map^{cn}_{CF})$ *holds iff, under the assumption that constraints over F exist, for every configurable node exactly one configuration holds, i.e. iff* $\varphi \Rightarrow \dot{\bigvee}_{e \in L^{cn}}(map^{cn}_{CF}(e))$ *is a* tautology*,*
- *Act is a finite set of* actions*, corresponding to modifications in the C-EPC model in order to reflect choices made over facts,*
- $map^{cn}_{CA} \in L^{cn} \rightarrow Act$ *is a function assigning an action to each configuration* $l^C$ *of cn.*

Given a configuration $\sigma$ of $CM$, for each configurable node $cn$ of C-EPC, an action $a$ related to a configuration $l^C$ of $cn$ is performed if the boolean function associated to $l^C$ evaluates to *true* given the values assigned to facts in $\sigma$. As functions exclude each other, only one action per variation point is executed.

The following table shows for each configurable node in the C-EPC example of Fig. 1, the associated boolean function defined over the facts of Fig 2.

We can see that some facts have a 1-1 mapping with C-EPC variation points (e.g. $f_{13}, f_{15}$). In general though, a fact can have a wider impact on the process model (1-N). Consider for example $f_1$ (low budget). Although this fact does not appear in any function in the above table, if asserted, it would affect a number of configuration nodes due to C2 and C11. Namely all the variation points whose boolean functions feature $\neg f_{10}, \neg f_{12}, \neg f_{14}$. This would lead to the following configuration: $OR_1 = SEQ_{1a}, OR_2 = OR_3 = SEQ_{2a}, OR_4 = OR_5 = SEQ_{4a}$, where all the branches of the C-EPC model involving an activity related to film have been denied.

In general, the more impact a fact has on successive facts, the more variation points in the process model are likely to be affected. This depends on the way both constraints over facts and boolean functions have been defined.

Facts 16 to 21 – regarding the shooting formats for tape and film – are the only ones that do not affect any variation point, neither directly nor indirectly. Thus, they do not appear in any boolean function of Tab. 1. Indeed these facts influence those variation nodes occurring in the sub-processes of functions *Prepare Tape for edit* and *Prepare Film for edit*, that are not shown in Fig. 1.

## 5 Tool Support

With the purpose of validating our questionnaire-driven approach from a practical perspective, we implemented a set of tools during the course of this research. Each tool is a stand-alone application responsible for specific tasks in the configuration process. However, when combined, the tools provide end-to-end support for reference process model configuration, from the collection of the answers via

**Table 1.** Mapping configuration alternatives for the Post-Production example

| Configurable node $cn$ | Configuration $l^C(cn)$ | Boolean function $map_{CF}^{cn}(l^C(cn))$ |
|---|---|---|
| $OR_1$ | AND | $f_9 \wedge f_{10}$ |
| | $\text{SEQ}_{1a}$ | $f_9 \wedge \neg f_{10}$ |
| | $\text{SEQ}_{1b}$ | $\neg f_9 \wedge f_{10}$ |
| | OR | *false* |
| | XOR | *false* |
| $OR_2$ | AND | $f_{11} \wedge f_{12}$ |
| | $\text{SEQ}_{2a}$ | $f_{11} \wedge \neg f_{12}$ |
| | $\text{SEQ}_{2b}$ | $\neg f_{11} \wedge f_{12}$ |
| | OR | *false* |
| | XOR | *false* |
| $OR_3$ | same as $OR_2$ | |
| $OR_4$ | AND | $(f_{13} \wedge f_{14}) \vee (f_{12} \wedge \neg f_{13} \wedge f_{15})$ |
| | $\text{SEQ}_{4a}$ | $(f_{13} \wedge \neg f_{14}) \vee (\neg f_{13} \wedge \neg f_{14} \wedge f_{15})$ |
| | $\text{SEQ}_{4b}$ | $(\neg f_{13} \wedge f_{14} \wedge \neg f_{15}) \vee (\neg f_{12} \wedge \neg f_{13} \wedge f_{14})$ |
| | OR | *false* |
| | XOR | *false* |
| $OR_5$ | same as $OR_4$ | |
| *Telecine transfer* | ON | $(\neg f_{11} \wedge f_{13}) \vee (\neg f_{11} \wedge f_{15})$ |
| | OFF | $\neg((\neg f_{11} \wedge f_{13}) \vee (\neg f_{11} \wedge f_{15}))$ |
| | OPT | *false* |
| *Record Digital Film Master* | ON | $\neg f_{12} \wedge f_{14}$ |
| | OFF | $\neg(\neg f_{12} \wedge f_{14})$ |
| | OPT | *false* |
| *Tape finish* | ON | $f_{13}$ |
| | OFF | $\neg f_{13}$ |
| | OPT | *false* |
| *New media finish* | ON | $f_{15}$ |
| | OFF | $\neg f_{15}$ |
| | OPT | *false* |

questionnaires, to the release of a configured process model. So far we only support the C-EPC language, but the architecture is such that new modules for new configurable process notations can be easily plugged in. Due to space limitations we just present an overview of the architecture, without entering into the details of the implementation.[3] Fig. 3 depicts the architecture.

The *Quaestio* tool takes an XML serialization of a configuration model as input and guides the configuration interactively by posing only the relevant questions in an order consistent with the interdependencies. Questions can be answered by users or automatically by the system (using defaults), and they can be rolled back. To deal with the complexity of dynamically checking propositional logic constraints, Quaestio embodies a propositional constraint solver[4]

[3] Downloadable from `http://sky.fit.qut.edu.au/~dumas/ConfigurationTool.zip`

[4] Downloadable from `http://www-verimag.imag.fr/~raymond/tools/bddc-manual`

based on Shared Binary Decision Diagrams (SBDDs) [10]. It has been proven that algorithms based on SBDDs can efficiently deal with systems made up of around one million of possibilities [10]. Accordingly, Quaestio can scale with configuration scenarios made up of thousands of facts and around one million of possible configurations.

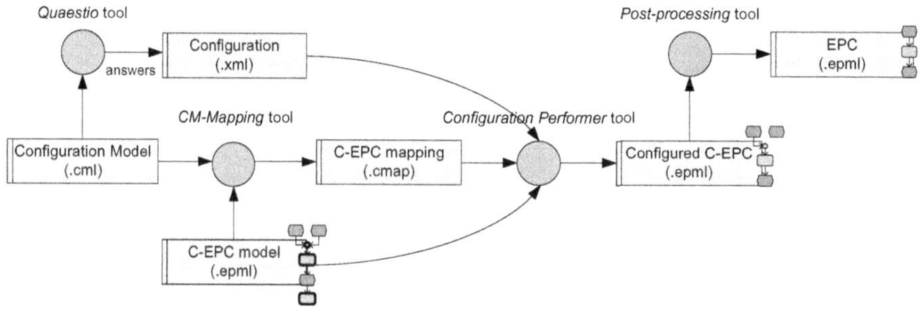


**Fig. 3.** The software architecture of the tools implemented

The *CM-Mapping* tool allows designers to define boolean functions over the facts of a configuration model, and to link them to the variation points of a C-EPC net, whose EPML[5] serialization is taken as input. This tool uses the SBDD calculator to check whether the constraints over the facts are consistent, and whether the functions assigned to the configuration values of each variation point are in mutual exclusion, so as to generate only valid mappings.

The *Configuration Performer* tool takes as input a configuration over facts generated by *Quaestio*, the EPML of the C-EPC model, and the mapping linking the C-EPC to the corresponding configuration model. It gives as output an EPML representation of a configured C-EPC, where each configurable node has been marked with a configuration value, according to the EPML syntax for C-EPC presented in [8]. This artifact is then post-processed by a tool implementing the derivation algorithm presented in [9]. The output of the latter is a syntactically correct EPC model.

Fig. 4 depicts the EPC model resulting from the application of configuration $\sigma = \{f_1, \neg f_2, \neg f_3, \neg f_4, \neg f_5, \neg f_6, f_7, f_8, f_9, \neg f_{10}, f_{11}, \neg f_{12}, \neg f_{13}, \neg f_{14}, f_{15}, ...\}$ to the Post-Production C-EPC model.[6] This configuration corresponds to a low budget project shooting on tape, performing an online cut and releasing on the new media *Mobile* and *Internet*. According to the mapping of Tab. 1, the C-EPC variation points assume the following configuration values: $OR_1 = SEQ_{1a}$, $OR_2 = OR_3 = SEQ_{2a}$, $OR_4 = OR_5 = SEQ_{4a}$, *Telecine transfer* = *OFF*, *Record Digital Film Master* = *OFF*, *Tape finish* = *OFF*, *New media finish* = *ON*.

[5] `http://wi.wu-wien.ac.at/~mendling/EPML`

[6] The model is shown in *EPC Tools*, `http://wwwcs.upb.de/cs/kindler/Forschung/EPCTools`

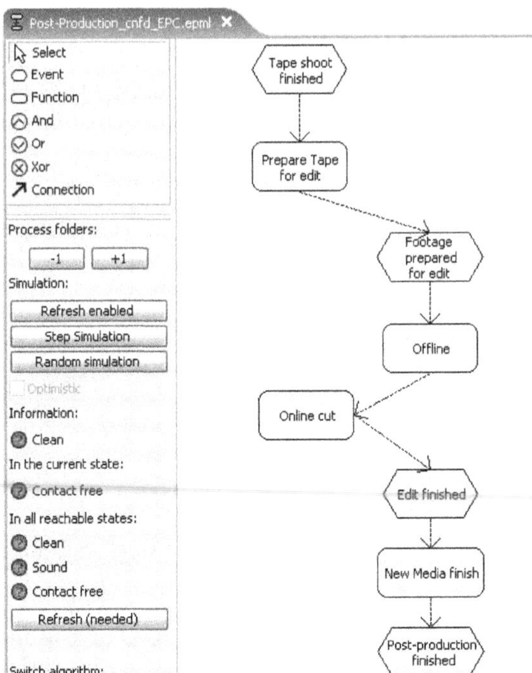


**Fig. 4.** The EPC for low-budget Post-Productions, obtained from the C-EPC of Fig. 1

## 6 Related Work

Conceptual support for adapting reference models is not prevalent in the field of IS. Existing approaches exhibit a heterogenous set of methods for reusing reference models but do not provide formalized support to abstract from the model during the configuration process. Process alternatives are depicted as process specializations in [16]. For their activation, these specializations are linked to conditions expressed in questions. However, dependencies and constraints between questions cannot be expressed. Following a domain engineering approach, [12] defines so-called stereotypes to specify the multiple appearance of model elements. The model instantiation is not supported by any abstraction from the actual model. There are also no means to depict dependencies and constraints.

Reference model adaptation mechanisms have been introduced in [3]. Among others, these include logical terms and attributes that are linked to model elements to indicate which sections are not relevant for a specific application scenario and have to be removed. The approach supports the configuration apart from the actual model. However, it does not address the definition of management questions and their dependencies. Consequently, existing approaches to reference model adaptation support configuration but still require users to be model experts in order to perform it. However, as we shown in this paper, there are scenarios where this is unfeasible, due to the complexity of interdependenciesof configuration decisions and the degree of variability.

Our work is also related to questionnaire systems. A range of commercial products supports the creation of online questionnaires. They rely on the notion of question flows and support dependencies among questions but lack support to capture constraints among them [11]. Form definition languages like XForms [4] support both dependencies and constraints but are unfit for our purpose as they rely on first-order logic constraints which can not be efficiently analyzed.

The field of Software Configuration Management deals with models and languages to capture how a collection of available options impact the way a software system is built from a set of components. For a comparison between our proposal and existing work in this area, refer to [7].

## 7 Conclusion

In this paper we showed how a questionnaire-driven model configuration can be applied to the configuration of reference process models. We do not claim that questionnaire-driven configuration will substitute configurable reference models. Rather, we suggest the combined use of both. While languages for configurable reference modeling such as C-EPCs support the specification of variation points and alternatives, interactive questionnaires provide a front-end allowing users to configure these models by reasoning directly in terms of the concepts of the domain in question (e.g. film production). To demonstrate this, we have implemented a toolset that generates interactive questionnaires for model configuration, and a mapping between C-EPCs and facts gathered by these questionnaires.

Future research has to show that our approach can be applied to other modeling languages as well. As an example, we are developing a configurable process definition language based on the YAWL environment [1]. This is a suitable area to apply the results of our research, due to YAWL's comprehensive support for the workflow control-flow patterns [2]. Also, there is an opportunity in future to apply the questionnaire-driven approach beyond the realm of reference process models, e.g. to configure data models.

The need for abstracting the configuration from actual models can be argued as follows. First, the model user is not required to have extensive knowledge of both the domain and the reference model. Second, when variation points are added to a model, configuration complexity increases dramatically and, thus, configuring the model without any means of abstraction gets close to unmanageable. Regarding this research, some limitations have to be pointed out. First, the actual impact of questionnaire-based configuration on the modeling process has not been empirically investigated. To this end we are conducting focus groups and surveys with a selected group of screen business experts, so as to evaluate the perceived usefulness and ease of use of the approach. Second, in this paper we have not elaborated on the relationship between the construction of configurable reference models and the construction of interactive questionnaires. The question still remains whether the questionnaire or the reference model should be constructed first. Alternatively, this could be an iterative process in which the reference model construction influences the questionnaire and vice versa.

**Acknowledgments.** The authors wish to thank Wil van der Aalst, Florian Gottschalk and Michael Rosemann for their valuable feedback.

## References

1. van der Aalst, W.M.P., Aldred, L., Dumas, M., ter Hofstede, A.H.M.: Design and Implementation of the YAWL System. In: Persson, A., Stirna, J. (eds.) CAiSE 2004. LNCS, vol. 3084, pp. 142–159. Springer, Heidelberg (2004)
2. van der Aalst, W.M.P., ter Hofstede, A.H.M., Kiepuszewski, B., Barros, A.P.: Workflow Patterns. Distributed and Parallel Databases 14(1), 5–51 (2003)
3. Becker, J., Delfmann, P., Knackstedt, R.: Adaptive Reference Modeling: Integrating Configurative and Generic Adaptation Techniques for Information Models. In: Reference Modeling Conference 2006, Passau (2006)
4. Boyer, J., Landwehr, D., Merrick, R., Raman, T., Dubinko, M., Klotz, L.: XForms 1.0 se, W3C Recommendation (2006) `http://www.w3.org/MarkUp/Forms`
5. Fettke, P., Loos, P.: Classification of reference models: a methodology and its application. Information Systems and e-Business Management 1(1), 35–53 (2003)
6. Keller, G., Nüttgens, M., Scheer, A.W.: Semantische Processmodellierung auf der Grundlage Ereignisgesteuerter Processketten (EPK). Veröffentlichungen des Instituts für Wirtschaftsinformatik, University of Saarland, Saarbrücken (1992)
7. La Rosa, M., van der Aalst, W.M.P., Dumas, M., ter Hofstede, A.H.M., Gottschalk, F.: Generating Interactive Questionnaires From Configuration Models (2006) Available at QUT ePrints `http://eprints.qut.edu.au/archive/00006325`
8. Mendling, J., Recker, J., Rosemann, M., van der Aalst, W.M.P.: Towards the Interchange of Configurable EPCs. In: EMISA, pp. 8–21 (2005)
9. Mendling, J., Recker, J., Rosemann, M., van der Aalst, W. M. P.: Generating correct EPCs from configured C-EPCs In: Proc. of the 2006 ACM (SAC), April 23-27, 2006, France, pp. 1505–1510 (2006)
10. Minato, S., Ishiura, N., Yajima, S.: Shared Binary Decision Diagram with Attributed Edges for Efficient Boolean function Manipulation. In: DAC, pp. 52–57 (1990)
11. Morton, K., Carey-Smith, C., Carey-Smith, K.: The QUEST Questionnaire System. In: Proc. of the 2nd ANNES, pp. 214–217. IEEE Computer Society, Washington (1995)
12. Reinhartz-Berger, I., Soffer, P., Sturm, A.: A Domain Engineering Approach to Specifying and Applying Reference Models. In: Desel, J., Frank, U. (eds) Workshop EMISA, LNI vol. 75, pp. 50–63. German Informatics Society (2005)
13. Rosemann, M., van der Aalst, W.M.P.: A Configurable Reference Modelling Language. Information Systems 32(1), 1–23 (2007)
14. Scheer, A.-W., Nüttgens, M.: ARIS Architecture and Reference Models for Business Process Management. In: van der Aalst, W.M.P., Desel, J., Oberweis, A. (eds.) Business Process Management. LNCS, vol. 1806, pp. 376–389. Springer, Heidelberg (2000)
15. Seidel, S., Rosemann, M., ter Hofstede, A.H.M., Bradford, L.: Developing a Business Process Reference Model for the Screen Business - A Design Science Research Case Study. In: 17th ACIS, Adelaide (2006) `www.screenbusiness.org`
16. Soffer, P., Golany, B., Dori, D.: ERP modeling: a comprehensive approach. Information Systems 28(6), 673–690 (2003)
17. Stephens, S.: The Supply Chain Council and the SCOR Reference Model. Supply Chain Management - An International Journal 1(1), 9–13 (2001)
18. Taylor, C., Probst, C.: Business Process Reference Model Languages: Experiences from BPI Projects. In: Proc. of INFORMATIK, pp. 259–263 (2003)

# Formalization and Verification of EPCs with OR-Joins Based on State and Context

Jan Mendling[1] and Wil van der Aalst[2]

[1] Vienna University of Economics and Business Administration
Augasse 2-6, 1090 Vienna, Austria
jan.mendling@wu-wien.ac.at
[2] Eindhoven University of Technology
P.O. Box 513, 5600 MB Eindhoven, The Netherlands
w.m.p.v.d.aalst@tue.nl

**Abstract.** The semantics of the OR-join in business process modeling languages like EPCs or YAWL have been discussed for a while. Still, the existing solutions suffer from at least one of two major problems. First, several formalizations depend upon restrictions of the EPC to a subset. Second, several approaches contradict the modeling intuition since the structuredness of the process does not guarantee soundness. In this paper, we present a novel semantical definition of EPCs that addresses these aspects yielding a formalization that is applicable for all EPCs and for which structuredness is a sufficient condition for soundness. Furthermore, we introduce a set of reduction rules for the verification of an EPC-specific soundness criterion and present a respective implementation.

## 1 Introduction

The Event-driven Process Chain (EPC) is a business process modeling language for the represention of temporal and logical dependencies of activities in a business process (see [1]). EPCs offer *function type* elements to capture the activities of a process and *event type* elements describing pre- and post-conditions of functions. Furthermore, there are three kinds of *connector types* (i.e. AND, OR, and XOR) for the definition of complex routing rules. Connectors have either multiple incoming and one outgoing arc (join connectors) or one incoming and multiple outgoing arcs (split connectors). As a syntax rule, functions and events have to alternate, either directly or indirectly when they are linked via one or more connectors. *Control flow arcs* are used to link elements.

The informal (or intended) semantics of an EPC can be described as follows. The AND-split activates all subsequent branches in a concurrent fashion. The XOR-split represents a choice between exclusive alternative branches. The OR-split triggers one, two or up to all of multiple branches based on conditions. In both cases of the XOR- and OR-split, the activation conditions are given in events subsequent to the connector. Accordingly, splits from events to functions are forbidden with XOR and OR since the activation conditions do not become clear in the model. The AND-join waits for all incoming branches to complete,

J. Krogstie, A.L. Opdahl, and G. Sindre (Eds.): CAiSE 2007, LNCS 4495, pp. 439–453, 2007.

then it propagates control to the subsequent EPC element. The XOR-join merges alternative branches. The *OR-join* synchronizes all active incoming branches, i.e., it needs to know whether the incoming branches may receive tokens in the future. This feature is called *non-locality* since the state of all (transitive) predecessor nodes has to be considered.

Since the informal description cannot be directly translated into proper semantics (see [2]), EPCs arguably belong to those process modeling languages for which state based correctness criteria such as soundness are not directly applicable. Instead, several authors have proposed to consider structuredness of the process graph as an alternative criterion for correctness (see e.g. [3,4,5]). Essentially, in a structured process model each split connector matches a join connector of the same type and loops have one XOR-join as entry and one XOR-split as exit point. These building blocks can be nested and extended with sequences of functions and events. The structuredness of a process model can be tested by repeatedly applying reduction rules that collapse several nodes of the respective building blocks. If the reduction yields a single node for the whole process, the model is structured. While structuredness represents a sufficient condition for soundness of Petri nets (see [6,7]), the application of reduction rules to EPCs such as proposed in [5] rather represents a heuristic. Figure 1 gives an example of a structured EPC that can be reduced to a single node by first collapsing the two OR-blocks, the AND-block, and then the loop. The EPC of Figure 2 extends this model with two additional start events $e4$ and $e5$. Due to the introduction of the OR-joins $c9$ and $c10$, there are only two structured blocks left between $c3$ and $c4$ and between $c5$ and $c6$. Still, if we assume that the start event $e1$ is always triggered, there is no problem to execute this unstructured EPC. If the start event $e4$ is triggered, it will synchronize with the first loop entry at $c1$.

Against this background, we present a novel EPC semantics definition that has the following qualities. First, it is applicable for all EPCs that are syntactically

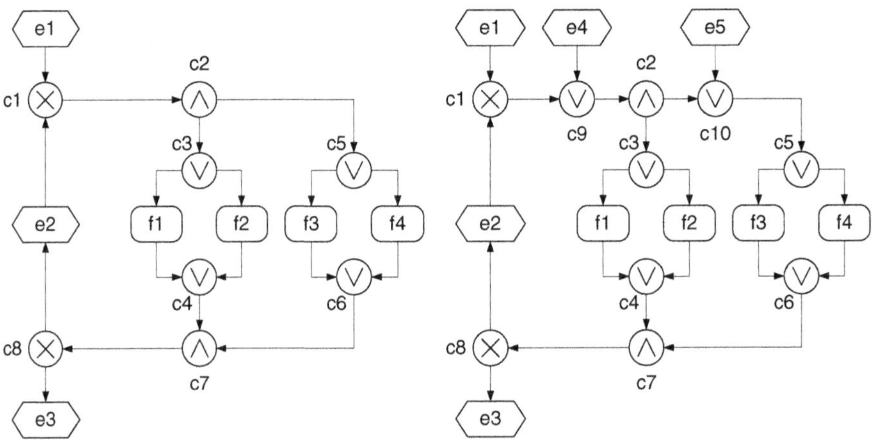


**Fig. 1.** A structured EPC with two OR-blocks $c2 - c5$ and $c3 - c4$ on a loop

**Fig. 2.** An unstructured EPC with one OR-block $c3 - c4$ and an OR loop entry

correct while several existing proposals restrict themselves only to a subset. Second, for this new semantics structuredness is a sufficient condition for soundness. This aspect is of central importance both for the intuition of the semantics and for the efficient verification based on reduction rules. The remainder of this paper is structured as follows. In Section 2 we use the EPCs of Figures 1 and 2 as a running example to discuss related work on process modeling languages with OR-joins, i.e. EPCs and YAWL, in particular. This discussion reveals that none of the existing formalizations captures both the complete set of syntactically correct EPCs and at the same time supports the intuition that structured models are sound. In Section 3 we give an EPC syntax definition and present our novel EPC semantics definition based on state and context. In Section 4 we elaborate on the relationship of structuredness and soundness showing that a structured EPC is indeed sound according to the new semantics. This is an important result that makes the semantics also a candidate for the formalization of YAWL nets without cancellation areas. Section 5 concludes the paper and gives an outlook on future research.

## 2 Related Research

The transformation to Petri nets plays an important role in early EPC formalizations. A problem of these approaches is their restriction to a subset of EPCs. The first concept is presented by *Chen and Scheer* [8] who define a mapping of structured EPCs with OR-blocks to colored Petri nets. A similar proposal is repeated by *Rittgen* [9]. Yet, while these first Petri net semantics provide a formalization for structured EPCs such as in Figure 1, it does not provide semantics for OR-joins in unstructured EPCs.

The transformation approach by *Langner, Schneider, and Wehler* [10] maps EPCs to Boolean nets, a variant of colored Petri nets whose token colors are 0 (negative token) and 1 (positive token). A connector propagates both negative and positive tokens according to its logical type. This mechanism is able to capture the non-local synchronization semantics of the OR-join similar to dead-path elimination (see [11]). A drawback is that the EPC syntax has to be restricted: arbitrary structures are not allowed. If there is a loop it must have an XOR-join as entry point and an XOR-split as exit point which are both mapped to one place in the resulting Boolean net. As a consequence, this approach does not provide semantics for the unstructured EPC in Figure 2.

*Van der Aalst* [12] presents an approach to derive Petri nets from EPCs. While this mapping provides clear semantics for XOR- and AND-connectors as well as for the OR-split, it does not cover the OR-join. Dehnert presents an extension of this approach by mapping the OR-join to a Petri net block [13]. Since the resulting Petri net block may not necessarily synchronize multiple tokens at runtime (i.e., a non-deterministic choice), its state space is larger than the actual state space with synchronization. Based on the so-called relaxed soundness criterion it is possible to check whether a join should synchronize (cf. [13]).

*Nüttgens and Rump* [14] define a transition relation for EPCs that addresses also the non-local semantics of the OR-join, yet with a problem: the transition relation for the OR-join refers to itself under negation. *Van der Aalst, Desel, and Kindler* show, that a fixed point for this transition relation does not always exist [15]. They present an example to prove the opposite: an EPC with two OR-joins on a circle waiting for each other. This vicious circle is the starting point for the work of *Kindler* towards a sound mathematical framework for the definition of non-local semantics for EPCs [2]. The technical problem is that for the OR-join transition relation $R$ depends upon $R$ itself in negation. Instead of defining one transition relation, he considers a pair of transition relations $(P, Q)$ on the state space $\Sigma$ of an EPC and a monotonously decreasing function $R$. Then, a function $\varphi((P, Q)) = (R(Q), R(P))$ has a least and a greatest fixed point. $P$ is called pessimistic transition relation and $Q$ optimistic transition relation. An EPC is called *clean*, if $P = Q$. For most EPCs, this is the case. Some EPCs such as the vicious circle EPC are *unclean* since the pessimistic and the optimistic semantics do not coincide. The EPC of Figure 2 belongs to the class of unclean EPCs.

*Van der Aalst and Ter Hofstede* define a workflow language called YAWL [16] which also offers an OR-join with non-local semantics. The authors propose a definition of the transition relation $R(P)$ with a reference to a second transition relation $P$ that ignores all OR-joins. A similar semantics that is calculated on history-logs of the process is proposed by *Van Hee et al.* in [17]. *Mendling, Moser, and Neumann* relate EPCs to YAWL by the help of a transformation [18]. Even though this definition provides semantics for the full set of models, it yields a deadlock if the OR-joins $c4$ and $c6$ are activated. In cases of chained OR-joins, there might be a lack of synchronization (see [19]). Motivated by these problems *Wynn et al.*, present a novel approach based on a mapping to Reset nets. Whether an OR-join can fire (i.e. $R(P)$) is decided depending on (a) a corresponding Reset net (i.e. $P$) that treats all OR-joins as XOR-joins and (b) a predicate called *superM* that hinders firing if an OR-join is on a directed path from another enabled OR-join. In particular, the Reset net is evaluated using backward search techniques that grant coverability to be decidable (see [21,22]). A respective verification approach for YAWL nets is presented in [23]. The approach based on Reset nets provides interesting semantics but in some cases also leads to deadlocks, e.g. if the OR-joins $c4$ and $c6$ are activated.

Table 1 summarizes existing work on the formalization of the OR-join. Several early approaches define syntactical restrictions such as OR-splits to match corresponding OR-joins or models to be acyclic (see [8,10]). Newer approaches impose little or even no restrictions (see [2,16,23]), but exhibit unexpected behavior for OR-block refinements on loops with further OR-joins on it.

In the following section, we propose a novel semantics definition that provides soundness for structured EPCs without restricting the set of models based on the concepts reported in [19]. For verification we follow a reduction rules approach similar to the one proposed in *Sadiq & Orlowska* [3]. Unfortunately, the verification algorithm presented in [3] turned out to be incorrect since the set of reduction rules provided was shown to be incomplete [24,7]. In [24] there was an attempt to repair this by adding additional reduction rules. In [7] it was shown that the considered class of process models coincides with the well-known class

**Table 1.** Overview of OR-join semantics and their limitations

| OR-join semantics | Restricted to | Correctness of structured models |
|---|---|---|
| Chen et al. [8] | structured EPCs | correct |
| Langner et al. [10] | structured EPCs | correct |
| Kindler [2] | clean EPCs | correct (no proof available) |
| van der Aalst et al. [16] | no restriction | potential deadlock, lack of synchronization |
| Wynn et al. [23] | no restriction | potential deadlock |

of free-choice nets for which a compact and complete set of reduction rules exist [6]. Moreover, using the well-known Rank Theorem for free-choice nets it is possible to find any errors in polynomial time for the class of workflow considered in [3] extended with loops. A set of reduction rules for EPCs was first mentioned in *van Dongen, van der Aalst, and Verbeek* [5]. Still, their reduction rules are not related to a semantics definition of EPCs, but rather given as heuristics. In this paper, we extend this work by relating reduction rules to EPC soundness and provide specific rules to deal with multiple start and end events.

# 3 EPC Syntax and Semantics

## 3.1 EPC Syntax

There is not only one, but there are several approaches towards the formalization of EPC syntax because the original paper introduces them only in an informal way (see [1]). The subsequent syntax definition of EPCs is an abbreviation of a more elaborate definition given in [19] that consolidates prior work.

**Definition 1 (EPC Syntax).** A flat $EPC = (E, F, C, l, A)$ consists of four pairwise disjoint and finite sets $E, F, C$, a mapping $l : C \to \{and, or, xor\}$, and a binary relation $A \subseteq (E \cup F \cup C) \times (E \cup F \cup C)$ such that

- An element of $E$ is called *event*. $E \neq \emptyset$.
- An element of $F$ is called *function*. $F \neq \emptyset$.
- An element of $C$ is called *connector*.
- The mapping $l$ specifies the type of a connector $c \in C$ as *and*, *or*, or *xor*.
- $A$ defines the control flow as a coherent, directed graph. An element of $A$ is called an *arc*. An element of the union $N = E \cup F \cup C$ is called a *node*.

In order to allow for a more concise characterization of EPCs, notations are introduced for incoming and outgoing arcs, paths, and several subsets.

**Definition 2 (Incoming and Outgoing Arcs, Path).** Let $N$ be a set of *nodes* and $A \subseteq N \times N$ a binary relation over $N$ defining the arcs. For each *node* $n \in N$, we define the set of incoming arcs $n_{in} = \{(x, n) | x \in N \wedge (x, n) \in A\}$, and the set of outgoing arcs $n_{out} = \{(n, y) | y \in N \wedge (n, y) \in A\}$. A *path* $a \hookrightarrow b$ refers to a sequence of nodes $n_1, \ldots, n_k \in N$ with $a = n_1$ and $b = n_k$ such that for all $i \in 1, \ldots, k$ holds: $(n_1, n_2), \ldots, (n_i, n_{i+1}), \ldots, (n_{k-1}, n_k) \in A$. This includes the empty path of length zero, i.e., for any node $a : a \hookrightarrow a$.

**Definition 3 (Subsets).** For an *EPC*, we define the following subsets of its nodes and arcs:

- $E_s = \{e \in E \mid |e_{in}| = 0\}$ being the set of start-events,
  $E_{int} = \{e \in E \mid |e_{in}| = 1 \wedge |e_{out}| = 1\}$ being the set of intermediate-events,
  $E_e = \{e \in E \mid |e_{out}| = 0\}$ being the set of end-events.
- $A_s \subseteq \{(x, y) \in A \mid x \in E_s\}$ as the set of start-arcs,
  $A_{int} \subseteq \{(x, y) \in A \mid x \notin E_s \wedge y \notin E_e\}$ as the set of intermediate-arcs, and
  $A_e \subseteq \{(x, y) \in A \mid y \in E_e\}$ as the set of end-arcs.

In contrast to other approaches, we assume only a very limited set of constraints for a EPC to be correct. For an extensive set of constraints see e.g. [19].

**Definition 4 (Syntactically Correct EPC).** An $EPC = (E, F, C, l, A)$ is called syntactically correct, if it fulfills the requirements:

1. $EPC$ is a directed and coherent graph such that $\forall n \in N : \exists e_1 \in E_s, e_2 \in E_e$ such that $e_1 \hookrightarrow n \hookrightarrow e_e$
2. $|E| \geq 2$. There are at least two events in an EPC.
3. Events have at most one incoming and one outgoing arc.
   $\forall e \in E : |e_{in}| \leq 1 \wedge |e_{out}| \leq 1$.
4. Functions have exactly one incoming and one outgoing arcs.
   $\forall f \in F : |f_{in}| = 1 \wedge |f_{out}| = 1$.
5. Connectors have one incoming and multiple outgoing arcs or multiple incoming and one outgoing arc. $\forall c \in C : (|c_{in}| = 1 \wedge |c_{out}| > 1) \vee (|c_{in}| > 1 \wedge |c_{out}| = 1)$. If a connector does not have multiple incoming or multiple outgoing arcs, it is treated as if it was an event.

### 3.2 EPC Semantics Based on State and Context

In this subsection, we introduce a novel formalization of the EPC semantics. The principal idea of these semantics lends some concepts from *Langner, Schneider, and Wehler* [10] and adapts the idea of Boolean nets with true and false tokens in an appropriate manner. The reachability graph that we will formalize afterwards depends on the state and the context of an EPC. The *state* of an EPC is basically an assignment of positive and negative tokens to the arcs. Positive tokens signal which functions have to be carried out in the process, negative tokens indicate which functions are to be ignored. In order to signal OR-joins that it is not possible to have a positive token on an incoming branch, we define the *context* of an EPC. The context assigns a status of *wait* or *dead* to each arc of an EPC. A wait context indicates that it is still possible that a positive token might arrive; a dead context status means that no positive token can arrive anymore. For example, XOR-splits produce a dead context on those output branches that are not taken and a wait context on the output branch that receives a positive token. A dead context at an input arc is then used by an OR-join to determine whether it has to synchronize with further positive tokens or not.

**Definition 5 (State and Context).** For an $EPC = (E, F, C, l, A)$ the mapping $\sigma : A \rightarrow \{-1, 0, +1\}$ is called a state of an $EPC$. The positive token captures the state as it is observed from outside the process. It is represented by a black circle. The negative token depicted by a white circle with a minus on it has a similar semantics as the negative token in the Boolean nets formalization. Arcs with no state tokens on them have no circle depicted. Furthermore, the mapping $\kappa : A \rightarrow \{wait, dead\}$ is called a context of an $EPC$. A wait context is represented by a w and a dead context by a d next to the arc.

In contrast to Petri nets we distinguish the terms marking and state: the term marking refers to state $\sigma$ and context $\kappa$ collectively.

**Definition 6 (Marking of an EPC).** For a syntactically correct $EPC$ the mapping $m : A \rightarrow \{-1, 0, +1\} \times \{wait, dead\}$ is called a marking. The set of all markings $M$ of an EPC is called marking space with $M = A \times \{-1, 0, +1\} \times \{wait, dead\}$. The projection of a given marking $m$ to a subset of arcs $S \subseteq A$ is referred to as $m_S$. If we refer to the $\kappa$- or the $\sigma$-part of $m$, we write $\kappa_m$ and $\sigma_m$, respectively, i.e. $m(a) = (\sigma_m(a), \kappa_m(a))$.

The propagation of context status and state tokens is arranged in a four phase cycle: (1) dead context, (2) wait context, (3) negative token, and (4) positive token propagation. Whether a node is enabled and how it fires is illustrated in Figure 3. A formalization of the transitions for each phase is presented in [25].

1. In the first phase, all *dead context* information is propagated in the EPC until no new dead context can be derived.
2. Then, all *wait context* information is propagated until no new wait context can be derived. It is necessary to have two phases (i.e., first the dead context propagation and then the wait context propagation) in order to avoid infinite cycles of context changes (see [25]).
3. After that, all *negative tokens* are propagated until no negative token can be propagated anymore. This phase cannot run into an endless loop (see [25]).
4. Finally, one of the enabled nodes is selected and propagates *positive tokens* leading to a new iteration of the four phase cycle.

In order to set the start and the end point of the four phases, we define the initial and the final marking of an EPC similar to the definition in *Rump* [26].

**Definition 7 (Initial Marking of an EPC).** For an $EPC$ $I \subseteq M$ is defined as the set of all possible initial markings, i.e. $m \in I$ if and only if [1]:

- $\exists a_s \in A_s : \sigma_m(a_s) = +1$,
- $\forall a_s \in A_s$: $\sigma_m(a_s) \in \{-1, +1\}$,
- $\forall a_s \in A_s$: $\kappa_m(a_s) = wait$ if $\sigma_m(a_s) = +1$ and $\kappa_m(a_s) = dead$ if $\sigma_m(a_s) = -1$, and
- $\forall a \in A_{int} \cup A_e : \kappa_m(a) = wait$ and $\sigma_m(a) = 0$.

[1] Note that the marking is given in terms of arcs.

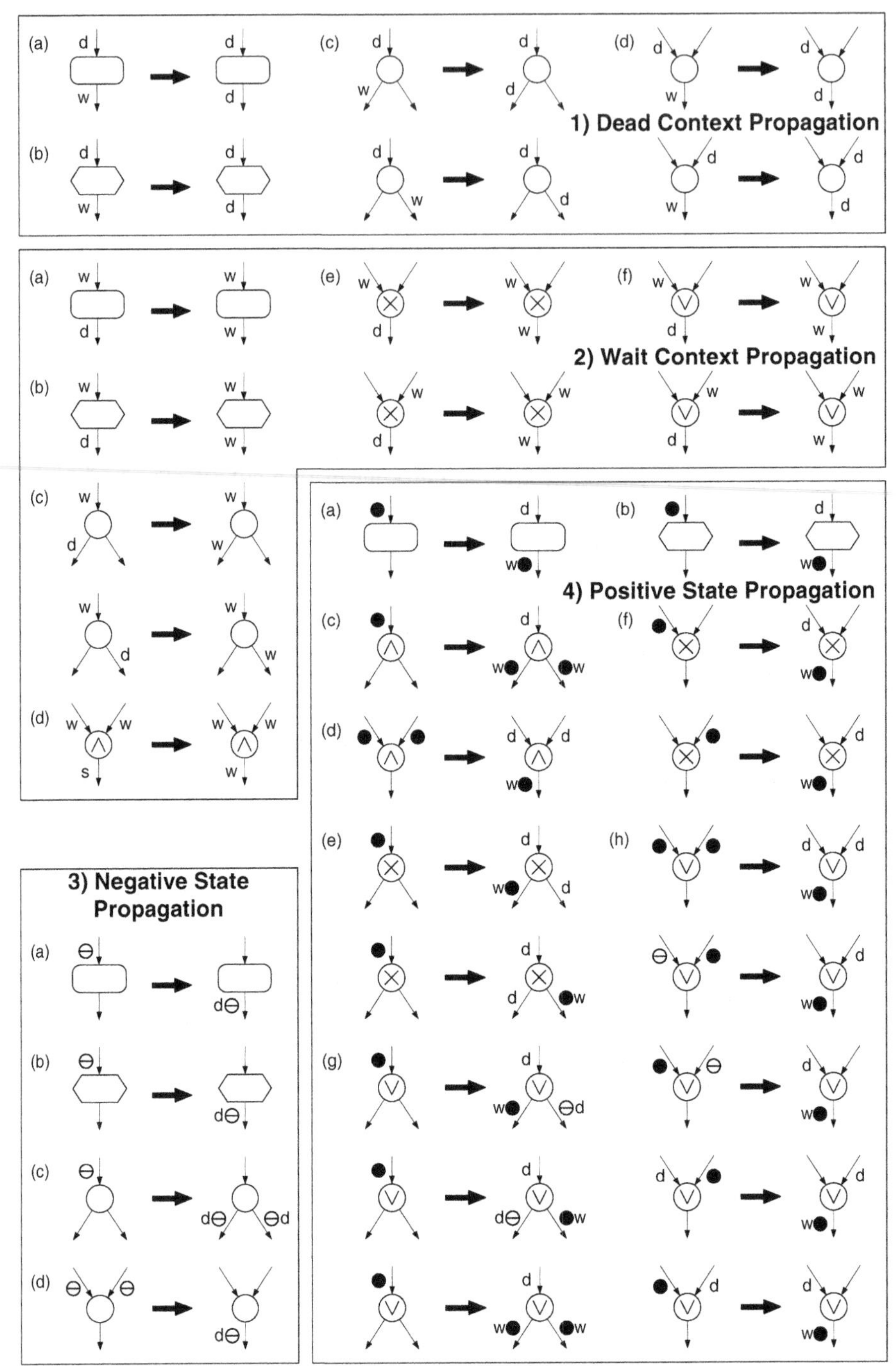


**Fig. 3.** Propagation of context and state in four phases

**Definition 8 (Final Marking of an EPC).** For an *EPC* $O \subseteq M$ is defined as the set of all possible final markings, i.e. $m \in O$ if and only if:

- $\exists a_e \in A_e$: $\sigma_m(a_e) = +1$ and
- $\forall a \in A_{int} \cup A_s : \sigma_m(a) \leq 0$.

Initial and final markings are the start and end points for calculating the reachability graph of an EPC. In this context a marking $m'$ is called reachable[2] from another marking $m$ if and only if after applying the phases of dead and wait context and negative token propagation on $m$, there exists a node $n$ whose firing in the positive token propagation phase produces $m'$. Then, we write $m \stackrel{n}{\rightarrow} m'$, or only $m \rightarrow m'$ if there exists some node $n$ such that $m \stackrel{n}{\rightarrow} m'$. Furthermore, we write $m_1 \stackrel{\tau}{\rightarrow} m_q$ if there is a firing sequence $\tau = n_1 n_2 ... n_{q-1}$ that produces from marking $m_1$ the new marking $m_q$ with $m_1 \stackrel{n_1}{\rightarrow} m_2, m_2 \stackrel{n_2}{\rightarrow} ... \stackrel{n_{q-1}}{\rightarrow} m_q$. If there exists a sequence $\tau$ exists such that $m_1 \stackrel{\tau}{\rightarrow} m_q$, we write $m_1 \stackrel{*}{\rightarrow} m_q$. Accordingly, we define the reachability graph *RG* as follows.

**Definition 9 (Reachability Graph of an EPC).** $RG \subseteq M_{RG} \rightarrow N \times M_{RG}$ is called the reachability graph of an EPC if and only if:

(i) $\forall i \in I : i \in M_{RG}$.
(ii) $\forall m, m' \in RG : (m, n, m') \Leftrightarrow m \stackrel{n}{\rightarrow} m'$.

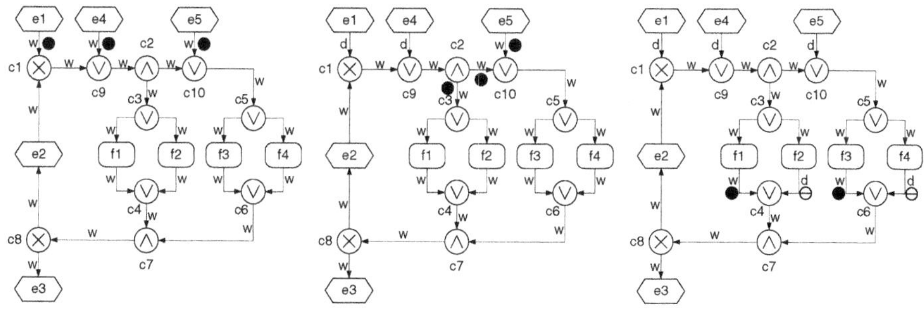


**Fig. 4.** Applying the Transition Relations

Based on the previous definitions we can discuss the behavior of the unstructured example EPC of Figure 2. This EPC and three markings are depicted in Figure 4. The *first marking* shows the example EPC in an initial marking with all start arcs carrying a positive token represented by a black circle. In this marking only the XOR-join *c*1 is allowed to fire – the other OR-joins have a *wait* context on one of their incoming arcs; therefore, they are not allowed to fire. In the *second marking* a token is propagated from *c*1, via synchronizing with the second token at *c*9, to the AND-split *c*2. The context of the start arcs has changed to *dead*, but

[2] A formalization of *reachability* is given in [25].

the arcs between the connectors $c1 - c9$ and $c9 - c2$ are still in *wait* since a token may arrive here via $e2$ on the loop. In order to arrive at the *third marking*, first the connector $c10$ has to fire. After that both OR-splits $c3$ and $c5$ are activated and fire a positive token to the left branch and a negative token to the right branch. After passing functions $f1$ to $f4$ we achieve the current marking with the OR-joins $c4$ and $c6$ being activated since both input arcs carry a token (a positive and a negative). *After this marking*, the two positive tokens generated by $c4$ and $c6$ synchronize at the AND-join $c7$. Then the loop can be run again, or the end arc can be reached. The loop can be executed without a problem since both OR-joins $c9$ and $c10$ have a *dead* context on the arcs coming from the start events. Therefore, the OR-join can fire using the last transition rule of (h) in positive state propagation.

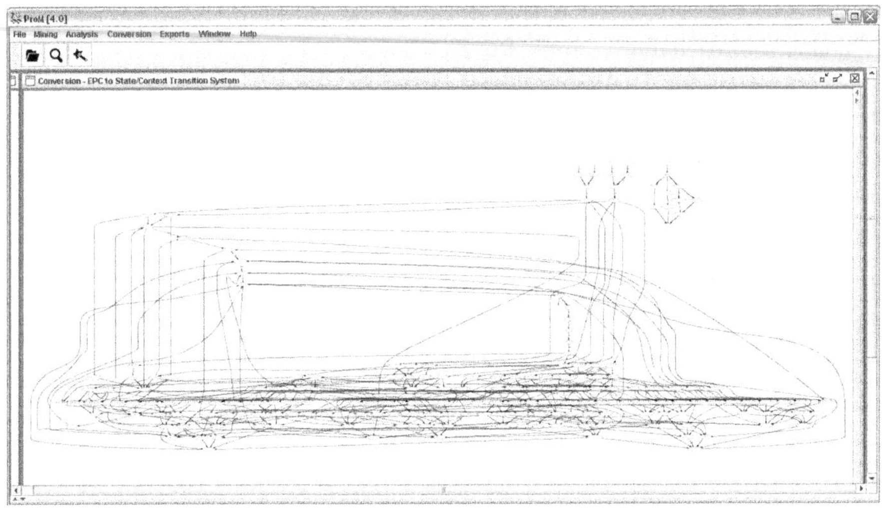


**Fig. 5.** Reachability Graph for the unstructured example EPC

We have implemented the reachability graph calculation as a conversion plug-in for the ProM framework [27]. Figure 5 displays the reachability graph of the unstructured example EPC that we used to illustrate the behavioral semantics. It can be seen that this graph is already quite complex for a small EPC. The complexity of this example basically stems from three facts. First, there are seven different initial markings. Second, parts of the loop can be executed in concurrency. Third, there are two OR-splits that both can activate either one or the other or both output arcs. Similar to the state explosion in Petri nets, the calculation of the reachability (or coverability) graph can turn out to be very inefficient for verification. Therefore, we discuss an EPC-specific variant of soundness and it verification using reduction rules in the following section.

## 4 EPC Verification Based on Reduction Rules

Soundness is an important correctness criterion for business process models introduced in [28]. The original soundness property is defined for a Workflow net, a Petri net with one source and one sink, and requires that (i) for every state reachable from the source, there exists a firing sequence to the sink (option to complete); (ii) the state with a token in the sink is the only state reachable from the initial state with at least one token in it (proper completion); and (iii) there are no dead transitions [28]. For EPCs, this definition cannot be used directly since EPCs may have multiple start and end events. Based on the definitions of the initial and final marking of an EPC, we define soundness of an EPC analogously to soundness of Workflow nets.

**Definition 10 (Soundness of an EPC).** An *EPC* is sound if there is a set of initial markings $I$ such that:

(i) For each start-arc $a_s$ there exists an initial marking $i \in I$ where the arc (and hence the corresponding start event) holds a positive token. Formally: $\forall a_s \in A_s : \exists i \in I : \sigma_i(a_s) = +1$

(ii) For every marking $m$ reachable from an initial state $i \in I$, there exists a firing sequence leading from marking $m$ to a final marking $o \in O$. Formally: $\forall i \in I : \forall m \in M \ (i \xrightarrow{*} m) \Rightarrow \exists o \in O \ (m \xrightarrow{*} o)$

(iii) The final markings $o \in O$ are the only markings reachable from a marking $i \in I$ such that there is no node that can fire. Formally: $\forall m \in M : \nexists m'(m \rightarrow m') \Rightarrow m \in O$

Given this definition, the EPCs of Figures 1 and Figure 2 are sound, and any initial marking of the second must include the state $\sigma_i(e_1, c_1) = +1$ for all $i \in I$.

Related to this soundness definition, we identify a set of reduction rules that is soundness preserving. A *reduction rule* $T$ is a binary relation that transforms a source $EPC_1$ to a simpler target $EPC_2$ that has less nodes and/or arcs (cf. e.g. [6]). A reduction rule is bound to a *condition* that defines for which arcs and nodes it is applicable. The reduction rules for sound EPCs include (1) Sequence Elimination, (2) Block Reduction, (3) Simple Loop Reduction, (4) Join Reduction, (5) Split Reduction, (6) Start-Join Reduction, and (7) Split-End Reduction (see Figure 6). Some of these rules (i.e.,1-5) were defined in previous work by [5]. In the following we sketch why these rules are soundness preserving for the given EPC semantics definition.

**(1) Sequence Elimination:** An element $n$ with one input and one output arc can be eliminated. This rule is applicable for functions and intermediate events, but also connectors with such cardinality can be produced by the other rules. As mentioned before in Def. 4, these connectors are treated as if they were events. The idea for proving that the rule preserves soundness can be sketched as follows. Based on the soundness of the unreduced $EPC_1$ we have to show that the reduced $EPC_2$ is also sound. In order to meet (i) we consider $I_2 = I_1$ of $EPC_1$. For (ii) we consider the node $x$ that enables $n$, i.e. $m_1 \xrightarrow{x} m_2$, and the firing of $n$, i.e.

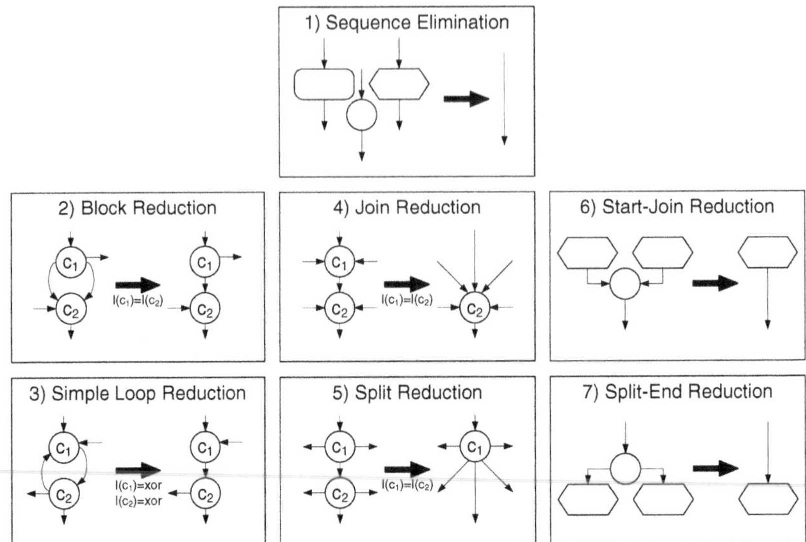


**Fig. 6.** Soundness preserving reduction rules for EPCs

$m_2 \xrightarrow{n} m_3$ of $EPC_1$. Obviously, in $EPC_2$ every marking that corresponds to $m_3$ is reachable from $m_1$ by firing $x$. Therefore, still for all markings that can be reached from some initial marking, some final marking is reachable. Since no new transitions are introduced, the final markings are still the only markings that meet (iii). Therefore, $EPC_2$ is also sound.

**(2) Block Reduction:** Multiple arcs from split- to join-connectors of the same type can be fused to a single arc. This might result in connectors with one input and one output arc. The above argument also holds for this reduction, but it must be adapted to cover all states that might be produced by firing $c_1$.

**(3) Simple Loop Reduction:** The arc from an XOR-split to an XOR-join can be deleted if there is also an arc from the join to the split. This rule might produce connectors with one input and one output arc. The above argument also holds for this rule.

**(4) Join Reduction:** Multiple join connectors having the same label are merged to one join. The above argument on soundness can be adapted here.

**(5) Split Reduction:** Multiple split connectors are reduced to one split. The above argument can be adapted for this rule.

**(6) Start-Join Reduction:** Multiple start events that are merged to one start event. We replace the two joined start events of $EPC_1$ in each initial marking by the merged start event such that (i) is met for $EPC_2$. Since any marking that is reachable by firing the join in $EPC_1$ is also reachable directly from the start event in $EPC_2$, but no additional marking is reached, (ii) and (iii) hold respectively for $EPC_2$. Therefore, $EPC_2$ is sound.

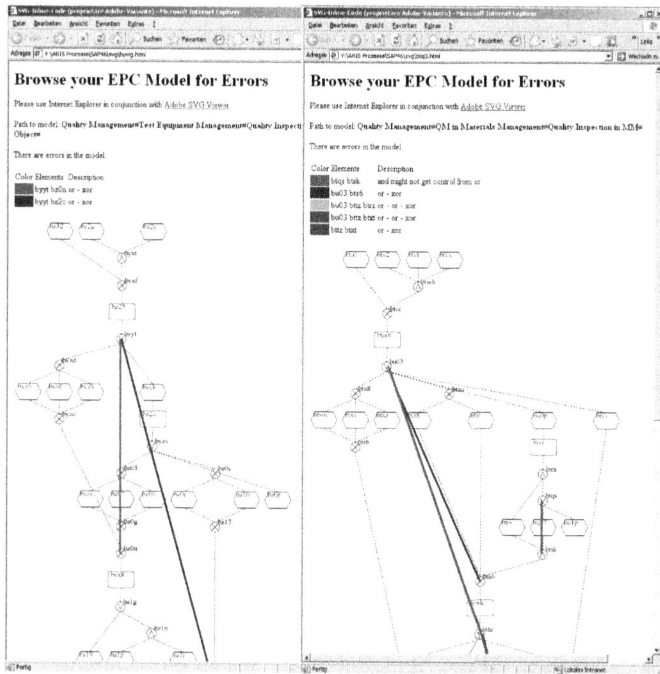


**Fig. 7.** Examples from the SAP reference model

**(7) Split-End Reduction:** Splits to multiple end events can be reduced to one end event. The above argument can be adapted for this rule.

Based on these reduction rules, it can be shown that the structured EPC of Figure 1 is indeed sound. Beyond that, we have implemented the reduction rules for EPCs that are available as ARIS XML files. Figure 7 shows two EPCs from the SAP reference model [29]. Both these models were analyzed with an existing verification approach based on the relaxed soundness criterion [13]. Even though they are relaxed sound, they still have structural problems. Using reduction rules we found that the EPCs are not sound according to the definition reported in this paper. In both models there are OR-splits that are joined with an XOR. The website http://wi.wu-wien.ac.at/epc offers an interface to the implementation of the reduction rules. Uploading an ARIS XML file generates an error report such as shown in Figure 7.

## 5 Contribution and Limitations

In this paper we presented a novel semantics definition for EPCs covering also the behavior of the OR-join. In contrast to existing semantical proposals for business process modeling languages with OR-joins, our definition provides semantics that are (1) applicable for any EPC without imposing a restriction on the syntax, and (2) intuitive since structuredness of the process model yields sound behavior. This is an important finding because there is up to now no solution reported that covers both aspects (1) and (2) in a formalization of the OR-join. Furthermore, the reduction rules that we presented and their implementation as a web interface are a useful tool for the verification of EPCs. In particular, the start-join and the split-end reduction rule directly address the definition of a soundness notion for EPCs. Moreover, they provide a novel solution for the problem of multiple start and end events in an EPC which is not appropriately covered by existing approaches so far. Still, our approach has a limitation with respect to the completeness of the reduction rules. While for free-choice Petri nets there is a complete set of reduction rules, this completeness is not achieved by the seven rules for EPCs. In future work, we aim to enhance our set by adding further rules in order to provide for an efficient verification of EPC soundness.

## References


1. Keller, G., Nüttgens, M., Scheer, A.W.: Semantische Prozessmodellierung auf der Grundlage Ereignisgesteuerter Prozessketten (EPK). Heft 89, Institut für Wirtschaftsinformatik, Saarbrücken, Germany (1992)
2. Kindler, E.: On the semantics of EPCs: Resolving the vicious circle. Data Knowl. Eng. 56, 23–40 (2006)
3. Sadiq, W., Orlowska, M.E.: Applying graph reduction techniques for identifying structural conflicts in process models. In: Jarke, M., Oberweis, A. (eds.) CAiSE 1999. LNCS, vol. 1626, pp. 195–209. Springer, Heidelberg (1999)
4. Kiepuszewski, B., ter Hofstede, A.H.M., Bussler, C.: On structured workflow modelling. In: Wangler, B., Bergman, L. (eds.) CAiSE 2000. LNCS, vol. 1789, pp. 431–445. Springer, Heidelberg (2000)
5. van Dongen, B., van der Aalst, W., Verbeek, H.M.W.: Verification of EPCs: Using reduction rules and petri nets. In: Pastor, Ó., Falcão e Cunha, J.F. (eds.) CAiSE 2005. LNCS, vol. 3520, pp. 372–386. Springer, Heidelberg (2005)
6. Esparza, J.: Reduction and synthesis of live and bounded free choice petri nets. Information and Computation 114, 50–87 (1994)
7. van der Aalst, W., Hirnschall, A., Verbeek, H.: An Alternative Way to Analyze Workflow Graphs. In: Banks-Pidduck, A., Mylopoulos, J., Woo, C., Ozsu, M. (eds.) CAiSE 2002. LNCS, vol. 2348, pp. 535–552. Springer, Heidelberg (2002)
8. Chen, R., Scheer, A.W.: Modellierung von Prozessketten mittels Petri-Netz-Theorie. Heft 107, Institut für Wirtschaftsinformatik, Saarbrücken (1994)
9. Rittgen, P.: Paving the Road to Business Process Automation. In: Proc. of ECIS 2000. pp. 313–319 (2000)
10. Langner, P., Schneider, C., Wehler, J.: Petri Net Based Certification of Event driven Process Chains. In: Desel, J., Silva, M. (eds.) ICATPN 1998. LNCS, vol. 1420, Springer, Heidelberg (1998)
11. Leymann, F., Altenhuber, W.: Managing business processes as an information resource. IBM Systems Journal 33, 326–348 (1994)
12. van der Aalst, W.: Formalization and Verification of Event-driven Process Chains. Information and Software Technology 41, 639–650 (1999)
13. Dehnert, J., Rittgen, P.: Relaxed Soundness of Business Processes. In: Dittrich, K.R., Geppert, A., Norrie, M.C. (eds.) CAiSE 2001. LNCS, vol. 2068, pp. 151–170. Springer, Heidelberg (2001)

14. Nüttgens, M., Rump, F.J.: Syntax und Semantik Ereignisgesteuerter Prozessketten (EPK). In: Desel, J., Weske, M. (eds.): Promise'02. Vol. 21 of LNI. pp. 64–77 (2002)
15. van der Aalst, W., Desel, J., Kindler, E.: On the semantics of EPCs: A vicious circle. In: Nüttgens, M., Rump, F. J. (eds.): Proc. of EPK'02. pp. 71–79 (2002)
16. van der Aalst, W., ter Hofstede, A.: YAWL: Yet Another Workflow Language. Information Systems 30, 245–275 (2005)
17. Hee, K., Oanea, O., Serebrenik, A., Sidorova, N., Voorhoeve, M.: Workflow model compositions perserving relaxed soundness. In: Dustdar, S., Fiadeiro, J.L., Sheth, A. (eds.) BPM 2006. LNCS, vol. 4102, pp. 225–240. Springer, Heidelberg (2006)
18. Mendling, J., Moser, M., Neumann, G.: Transformation of yEPC Business Process Models to YAWL. In: Proc. of ACM SAC, 2, 1262–1267 (2006)
19. Mendling, J., van der Aalst, W.: Towards EPC Semantics based on State and Context. In: Nüttgens, M., Rump, F. J., Mendling, J. (eds.): Proc. of EPK'06 pp. 25–48 (2006)
20. Wynn, M., Edmond, D., van der Aalst, W., ter Hofstede, A.: Achieving a General, Formal and Decidable Approach to the OR-join in Workflow using Reset nets. In: Ciardo, G., Darondeau, P. (eds.) ICATPN 2005. LNCS, vol. 3536, pp. 423–443. Springer, Heidelberg (2005)
21. Leuschel, M., Lehmann, H.: Coverability of reset petri nets and other well-structured transition systems by partial deduction. In: Palamidessi, C., Moniz Pereira, L., Lloyd, J.W., Dahl, V., Furbach, U., Kerber, M., Lau, K.-K., Sagiv, Y., Stuckey, P.J. (eds.) CL 2000. LNCS (LNAI), vol. 1861, pp. 101–115. Springer, Heidelberg (2000)
22. Finkel, A., Schnoebelen, P.: Well-structured Transition Systems everywhere! Theoretical Computer Science 256, 63–92 (2001)
23. Wynn, M., van der Aalst, W., ter Hofstede, A., Edmond, D.: Verifying Workflows with Cancellation Regions and OR-joins: An Approach Based on Reset Nets and Reachability Analysis. In: Dustdar, S., Fiadeiro, J.L., Sheth, A. (eds.) BPM 2006. LNCS, vol. 4102, pp. 389–394. Springer, Heidelberg (2006)
24. Lin, H., Zhao, Z., Li, H., Chen, Z.: A novel graph reduction algorithm to identify structural conflicts. In: Proc. of HICSS. 289 (2002)
25. Mendling, J.: Detection and Prediction of Errors in EPC Business Process Models. Ph.D. Thesis, Vienna University of Economics and Business Administration (2007)
26. Rump, F.J.: Geschäftsprozessmanagement auf der Basis ereignisgesteuerter Prozessketten - Formalisierung, Analyse und Ausführung von EPKs. Teubner (1999)
27. van Dongen, B., Medeiros, A., Verbeek, H., Weijters, A., van der Aalst, W.: The ProM framework: A New Era in Process Mining Tool Support. In: Ciardo, G., Darondeau, P. (eds.) ICATPN 2005. LNCS, vol. 3536, pp. 444–454. Springer, Heidelberg (2005)
28. van der Aalst, W.: Verification of Workflow Nets. In: Azéma, P., Balbo, G. (eds.) ICATPN 1997. LNCS, vol. 1248, pp. 407–426. Springer, Heidelberg (1997)
29. Keller, G., Teufel, T.: SAP(R) R/3 Process Oriented Implementation: Iterative Process Prototyping. Addison-Wesley, London (1998)

# Towards More Extensible MetaCASE Tools

Vincent Englebert and Patrick Heymans

University of Namur
Computer Science Department
PRECISE Research Center in Information Systems Engineering
Rue Grandgagnage 21, B-5000 Namur, Belgium

**Abstract.** In this paper, we suggest a solution to several limitations of current metaCASE technology: (i) the limited number of modelling levels, (ii) the rigid separation between those levels, (iii) the limited bootstrapping possibilities, (iv) the hardcoding of various types of information (e.g. GUI related information), and (v) the inability to record links between semantically related (e.g. referrentially redundant) constructs.

Our proposal is centered around a 2-layer metamodelling language called MetaL. MetaL is characterised by ubiquitous reflexivity (metacircularity) and extended reification capabilities. The language is presented and applied to illustrative examples. Its pros and cons are discussed and an on-going prototypical metaCASE implementation is reported.

## 1 Introduction

Conceptual modelling languages (CML) have long been recognized as powerful means to reduce complexity during the development of information systems (IS). CMLs are usually supported by CASE (computer-assisted software engineering) tools primarily used to edit, browse and record models. CASE tools can also have a broad range of additional features: code generation, model analysis, documentation generation, traceability, version control, etc. Nowadays, CASE tools have become essential assets in IS development projects, especially with the MD* (MDA, MDD, MDE,...) initiatives. MD* advocates the pervasive use of models and CASE tools in order to automate the generation of runtime artefacts (programs, databases, interfaces, etc.).

The ubiquity of models goes together with an increased need for flexibility. The users of CMLs (e.g. UML) often want to turn them into, or complement them with, domain specific languages (DSL) designed for a specific usage and/or technological domain. Unfortunately, the CML adaptation support offered by the large majority of CASE tools is very limited. For example, the most one can expect from a UML tool is usually support for *profiles*, i.e. using stereotypes, tagged values and constraints to alter the UML notation. Profiling is widely recognized as a very poor extension mechanism which only enables to superficially change some aspects of the syntax of CMLs [10] . Metamodel and semantics changes are out of its scope.

J. Krogstie, A.L. Opdahl, and G. Sindre (Eds.): CAiSE 2007, LNCS 4495, pp. 454–468, 2007.

Specific support is therefore needed to define DSLs and quickly obtain supporting tools. A technical solution to those problems has been explored during the last twenty years with metaCASE technology [1,6,11,26,14,19,16]. MetaCASE tools allow method engineers (a) to define new metamodels, (b) to define concrete notations (either textual or graphical) and (c) to edit, browse and record the models. They also sometimes feature import/export facilities or accomodate plugins (e.g., for specific model analyses).

In more recent works, the term "domain specific modelling (DSM) environment" has emerged. Newer proposals [3,17] are based on the same ideas as metaCASE but show an emphasis on model transformations and code generation, in accordance with the MD* vision. One of the ideas is that more effective code generation can be devised from DSLs than from general purpose languages. DSM environments should thus enable the quick and iterative development of DSLs and transformations.

To define DSLs, both metaCASE and DSM environments rely on metamodelling languages (MMLs) such as EBNF, MOF [23], Kermeta [21], OWL [20], GOPPRR [18], or KM3 [17]. Unfortunately, each of them faces at least one of the following limitations:

- the inability or difficulty to record links between semantically related (e.g. referentially redundant [24]) constructs in different (meta)models,
- the limited number of editable modelling levels (or M-levels), usually restricted to $M_2$ (metamodel) and $M_1$ (model)[1],
- the rigid separation between M-levels, i.e. the impossibility for relationships to traverse them, which is known as *strict metamodelling* [2,12],
- the hardcoding of various types of information, including the MML itself, parts of concrete syntax, the tool's GUI, version control (if any), etc.

For now, we illustrate the first and third limitations with a small example taken from [24]. In Section 3, we make the case for the other limitations.

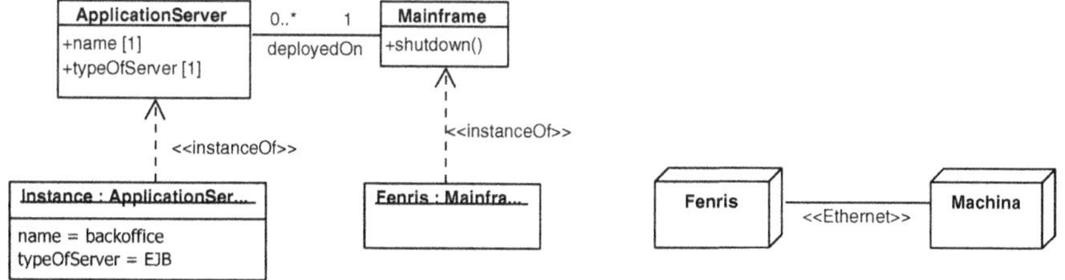


**Fig. 1.** Implicit referrential redundancy between UML diagrams (adapted from [24])

The example is about a mainframe called "Fenris", modelled both as a node in a UML deployment diagram, and as an object in a UML object model (see Fig. 1). We assume that the modeller is using an editing environment generated

[1] We refer to the standard OMG 4-layer architecture.

by a metaCASE tool to support the two languages. If she now wants to change the name of the mainframe from "Fenris" to "George" while editing the object diagram, what will happen? Typically, the name of the object will change, but the modification will not be propagated to the deployment diagram. This is because the two pieces of information are not formally linked within the tool. The reason for this usually lies in the limitations of the MML. Without multiple specialisation or multiple instantiation, it is difficult to cope with the sharing of properties between instances of constructs in different metamodels [24]. But another complication appears: the two name properties (of deployment nodes and objects, respectively) belong to distinct abstraction levels. This is impossible in most tools (with notable exceptions, see Section 5).

This example illustrates the need for more flexible MMLs. Similar problems are acknowledged by other authors [2,24,16,12] who also suggested solutions. In this paper, we introduce a new MML called MetaL. It integrates and extends ideas from these authors and from existing languages (RDF-Schema [4], Telos [22] and GXL [13]) to provide enhanced flexibility: besides multiple instantiation and multiple specialisation, it supports meta-circularity (reflexity of instantiation), non-strict metamodelling, and adaptable granularity of constructs. As we will see, these features yield reification capabilities beyond current state of the art, which allow for elegant and powerful bootstrapping mechanisms.

The structure of the paper goes as follows. Section 2 describes $\text{MetaL}_1$ and $\text{MetaL}_2$, the two fundamental layers of MetaL. Section 3 illustrates the benefits of MetaL based on examples. Section 4 outlines the features and architecture of a metaCASE tool based on MetaL currently under development. A comparison with related approaches appears in Section 5. The limitations and future works are discussed in Section 6, before Section 7 concludes the paper.

## 2 The MetaL Language

MetaL consists of two distinct *layers* (see Fig. 2) — not to be confused with *M-levels*, although these are supported too. The first layer, $\text{MetaL}_1$, is a formally defined minimal kernel language that serves to give the second layer, $\text{MetaL}_2$, a formal foundation. This way, both languages are given unambiguous definitions, and can be interpreted by tools. Of the two, $\text{MetaL}_2$ is the only language to be manipulated directly by end-users (metamodellers or method engineers). Therefore it proposes abstractions familiar to them: 'metamodel', 'metaobject', 'metaproperty', etc. The end-users can structure the (meta)models into 'compartments', e.g. to reflect M-levels. However, an important characteristic of $\text{MetaL}_2$ is that it does neither enforce a strict separation between levels, nor is it limited to a fixed number of levels (as illustrated in Fig. 2).

The formal definition of $\text{MetaL}_1$ and $\text{MetaL}_2$ can be found in [9]. In Section 2.1, we briefly introduce the main concepts of $\text{MetaL}_1$ and provide graphical representation conventions, also used for $\text{MetaL}_2$. $\text{MetaL}_2$ is described in Section 2.2. The reader should understand that $\text{MetaL}_2$ is not necessarily the terminal, nor the only, user language that will be defined on top of $\text{MetaL}_1$. New

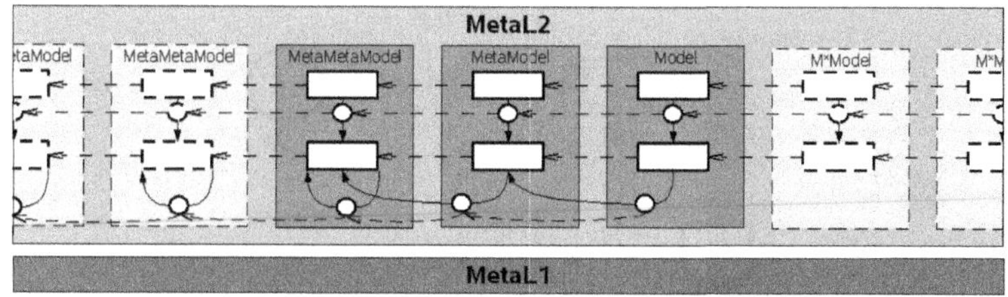


**Fig. 2.** Overview of $\text{MetaL}_1$ and $\text{MetaL}_2$

modelling purposes, or experience gained in using $\text{MetaL}_2$, are likely to lead us to define new or improved MMLs on top of $\text{MetaL}_1$.

### 2.1 Layer 1: The $\text{MetaL}_1$ Language

We consider a metaCASE tool's repository to contain *dataitems* (*items*, for short). This set of items is called $D$. Every item is either an *object* (from $O$), a *property* (from $P$), or both: $D = O \cup P$, where $O$ and $P$ are not disjoint. Graphically, we will represent objects by rectangles and properties by ovals (see Fig. 2, 3, 4, 5 and 6). When an item is both an object and a property, we represent the item by superimposing the two shapes (like for `MetaRole`, `transition` or `wp♯1` in Fig. 4).

$O_t$, a subset of $O$, denotes the set of *object types*. Similarly, $P_t$, a subset of $P$, denotes the set of *property types*. $D_t = O_t \cup P_t$ is the set of types. When an item is a type, this is represented by thickening the border of its shape. Hence, bold rectangles (e.g. $\omega$ in Fig. 3) denote object types and bold ovals (e.g. $\pi$ in Fig. 3) denote property types. An item that is both an object and a property needs not necessarily be a type in these two *roles* (e.g. `wp♯1` in Fig. 4).

Every item must have at least one type. Properties have exactly one. We thus allow *multiple instantiation* for object types. This is recorded by the instanceOf relations: $\mathsf{instanceOf}_O \subseteq O \times O_t$ and $\mathsf{instanceOf}_P : P \rightarrow P_t$. instanceOf relations are depicted with dashed arrows drawn from the instance to the type. Specialisation happens through the isa relation (graphically, a thick arrow) defined between object types: $\mathsf{isa} \subseteq O_t \times O_t$. Unlike common practice in similar languages, e.g. [22], we do not require isa to be acyclic, not to forbid multiple specialisation. However, we have the usual constraint that instantiation is closed by specialisation: $\forall x \in O \cdot \forall a, b \in O_t \cdot (x, a) \in \mathsf{instanceOf}_O \ \wedge \ (a, b) \in \mathsf{isa}^* \Rightarrow (x, b) \in \mathsf{instanceOf}_O$, where $\mathsf{isa}^*$ denotes the transitive closure of isa.

A property possesses a *domain* and a *range* which are both objects: $dom, ran : P \rightarrow O$. Graphically, the domain and the range of a property can be identified by a plain line arrow originating from the domain, crossing the property, and ending at the range (see e.g. $\pi$ from $\omega$ to $\omega$ in Fig. 3, or `wp♯2` from `transition` to `statechart` in Fig. 4). In the case of a property type, the domain and the range are object types: $\forall p \in P_t \cdot dom(p) \in O_t \ \wedge \ ran(p) \in O_t$. For a given property, a type of its domain (resp. range) must necessarily be the domain (resp. range)

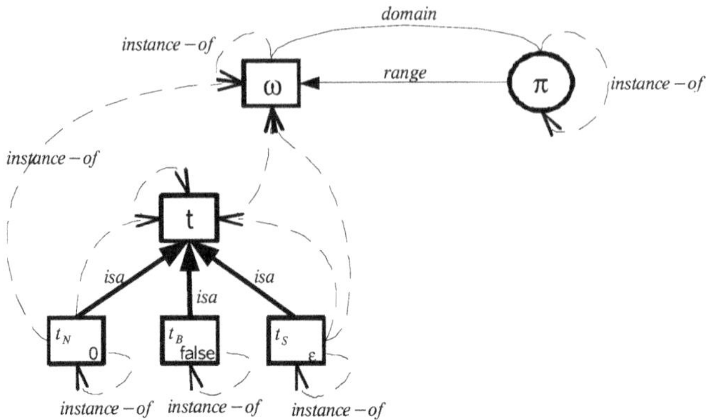


**Fig. 3.** The seminal items of $\text{MetaL}_1$

of its type: $\forall p \in P \cdot \forall pt \in P_t \cdot (p, p_t) \in \mathsf{instanceOf}_P \Rightarrow (dom(p), dom(p_t)) \in \mathsf{instanceOf}_O \wedge (ran(p), ran(p_t)) \in \mathsf{instanceOf}_O$.

The repository is initialized with some seminal items such as the object type $\omega$, and the property type $\pi$. Object type $t$ is an instance of $\omega$. $t$ is the super type of all *elementary types* such as $t_{\mathbb{B}}$ (booleans), $t_{\mathbb{N}}$ (natural numbers), and $t_{\mathbb{S}}$ (strings of characters)[2]. All instances of $t_{\mathbb{B}}$ are in $O_{\mathbb{B}}$ ($O_{\mathbb{B}} = \{b \mid (b, t_{\mathbb{B}}) \in \mathsf{instanceOf}_O\}$). $O_{\mathbb{B}}$ is the subset of $O$ representing the boolean *value objects*. Similarly, all instances of $t_{\mathbb{N}}$ are in $O_{\mathbb{N}}$, and so on. For each value object, the function *val* returns a *value*: $val : (O_{\mathbb{B}} \rightarrow \mathbb{B}) \cup (O_{\mathbb{N}} \rightarrow \mathbb{N}) \cup (O_{\mathbb{S}} \rightarrow \mathbb{S}) \cup \ldots$

A major difference between MetaL and other similar MMLs is its full-fledged support for reification. This is intended, among other things, to counter two usual limitations: the fixed number of levels and the strict separation between them. This is also intended to open the way for powerful bootstrapping mechanisms discussed further in this paper. In $\text{MetaL}_1$, reification is materialized mainly by (i) the non disjointness of $P$ and $O$ (see above), (ii) the absence of too rigid axioms (for example, there is no non-acyclicity axiom for $\mathsf{isa}$), (iii) multiple instantiation of objects, and, most prominently, (iv) the *reflexivity* of the $\mathsf{instanceOf}$ relation for the seminal items. In clear: every seminal item is an instance of itself. For the elementary types, this entails that they should all be both object types and values objects (i.e. $t_{\mathbb{B}} \in O_{\mathbb{B}}$, $t_{\mathbb{N}} \in O_{\mathbb{N}}$, $t_{\mathbb{S}} \in O_{\mathbb{S}}$, etc.). To this end, we also need to give them default values: $val(t_{\mathbb{B}}) = \mathsf{false}$, $val(t_{\mathbb{N}}) = 0$, and $val(t_{\mathbb{S}}) = \epsilon$[3] (Fig. 3).

### 2.2 Layer 2: The $\text{MetaL}_2$ Language

In the second layer, we define a metametamodel on top of $\text{MetaL}_1$ to provide appropriate abstractions to metamodellers and method engineers. The most

[2] We only present the types that are used in the remainder for this paper.

[3] $\epsilon$ denotes the empty string.

important constructs are presented here and illustrated in the left compartment of Fig. 4. A more complete description can be found in [9].

The core construct is the `MetaObject` object type. It is characterized by a name (a property type between `MetaObject` and $t_{\mathbb{S}}$). A metaobject `has` a `MetaProperty`, an object type that is also a subtype of `MetaObject`. Metaproperties are characterized by a type (the generic type $t$) and a cardinality (`cardP`). The second subtype of `MetaObject` is `MetaRole`, an object- and property type. As a property type, `MetaRole` takes `MetaObject` both as its domain and its range. As an object type, it possesses a cardinality property (type), `cardR`, indicating whether the metarole denotes a one-to-many, a many-to-many,... relationship[4].

The `MetaModel` object type is yet another subtype of `MetaObject` and denotes an aggregate composed of other metaobjects. The `wp` (whole-part) object- and property type defines this aggregation. As an object type, `wp` is the domain for three other property types (not shown in Fig. 4):

- `rename` (with range $t_{\mathbb{S}}$) allows us to name a metaobject differently for each aggregate it pertains to. For instance, the `class` metaobject could be renamed `entity-type` within the ER metamodel, or `table` within the relational metamodel;
- `canNotBeShared` (with range $t_{\mathbb{B}}$) specifies whether the *instances* of the metaobject at the model level can be shared. For example, can a class belong both to a UML class diagram and an ER model?
- `notDependentOn` (with range $t_{\mathbb{B}}$): can the aggregate still exist if one of its components is removed[5].

### 2.3 MetaL$_2$ Example

The metametamodel is to be used to create metamodels. The compartement in the middle of Fig. 4 represents a portion of a metamodel of statecharts in MetaL$_2$. `statechart` is a metamodel defined as the aggregation of the metaobjects `state` and `transition` (the latter being a metarole). Both `state` and `transition` possess names. Not shown is that, as an aggregate, a statechart is not dependent on these components (`wp.notDependentOn=true`), and that these components can be shared with other models (`wp.canNotBeShared=false`).

In turn, the `statechart` metamodel can be used as a 'pattern' to create concrete models such as the `coffeemachine` statechart (see right compartment in Fig. 4). This concrete model is created as an instance of the statechart metamodel. It is an aggregate consisting of concrete objects such as states `idle` and `busy`, and the transition going from the former to the latter.

The graphical conventions used in Fig. 4 and throughout this paper to represent the items in a MetaL repository are by no means prescriptive of the concrete syntax of the metamodelling environment. This is still an open question. Currently, only a Java API is available to create items (see Section 4).

---

[4] A special value in $t_{\mathbb{N}}$ is used to encode cardinality values: 1 for one-to-many, 2 for many-to-many, etc.

[5] If set to true, it makes `wp` semantically equivalent to composition (strong aggregation) in UML class diagrams.

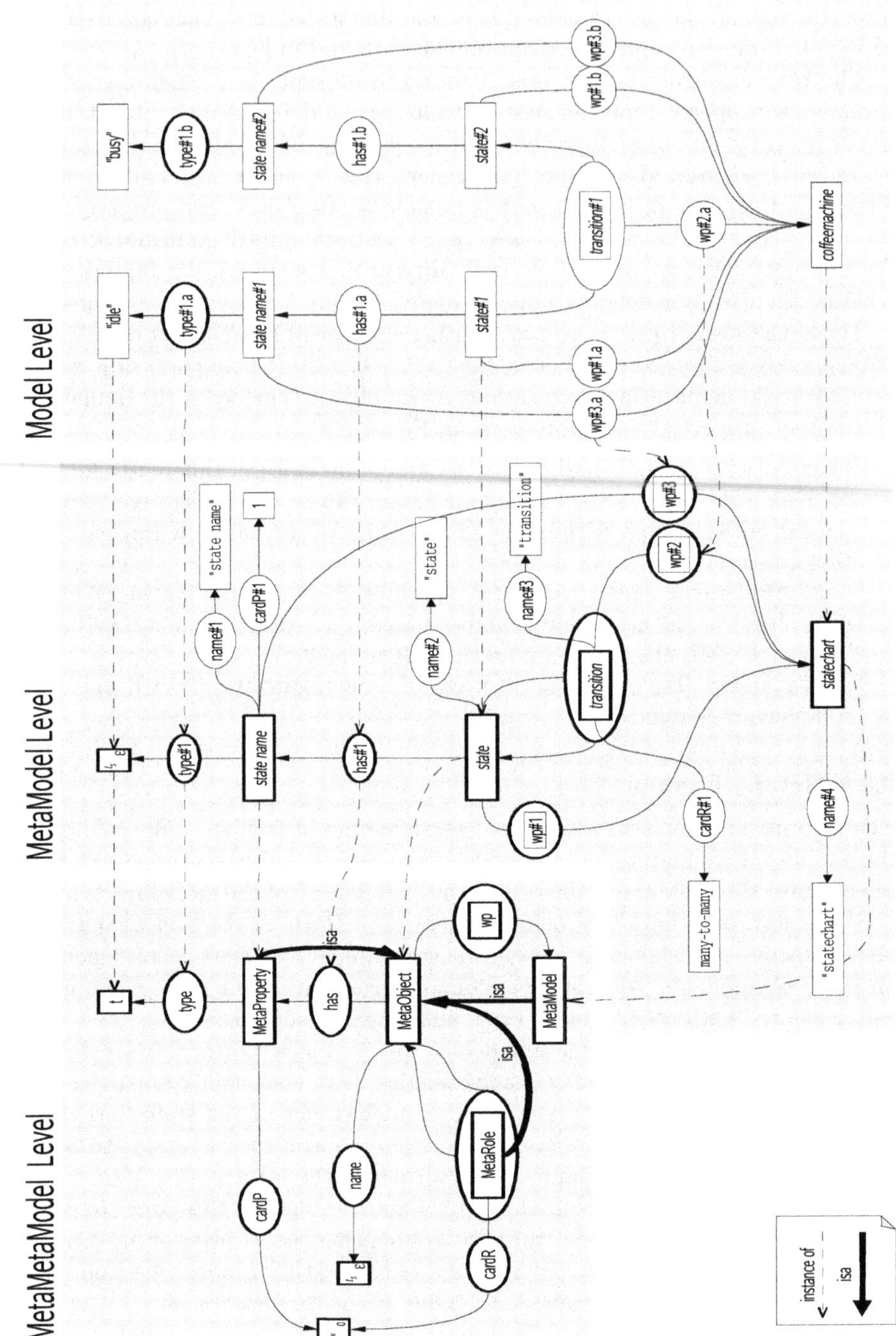


**Fig. 4.** MetaL applied to statecharts, with 3 M-levels: the metametamodel ($M_3$), the statechart metamodel ($M_2$), and a concrete statechart model ($M_1$). (Not all items appear).

## 3 Benefits

This section presents the main benefits of a metaCASE based on MetaL. We first expose the general benefits through an illustrative example (Section 3.1). We then focus on the reification capabilities (Section 3.2).

### 3.1 General Benefits

The example from [24] used in the introduction motivated the need for maintaining explicit links between (meta)models. Thus, if the same concept (e.g., "mainframe") is represented in several (meta)models, these referential redundancies are explicitely recorded and (meta)models can be kept easily in-sync at all times. How one can achieve this in MetaL is illustrated in Fig. 5. Three metamodels are defined: one for deployment diagrams (`Deployment`), one for class diagrams (`Static`), and one for a new kind of diagram (or DSL) supposedly defined by the user (`Infrastructure`).

As in Fig. 4, metamodels (i.e. instances of the `MetaModel` object type) are shown as (coloured) panes. The same convention is used for models and the metametamodel. The fact that an item is graphically represented within a pane stands for an instance of the `wp` property type between the (meta(meta))model and the item. Moreover, some panes overlap. Items found at the interesection of panes are those shared (though `wp`) between (meta)models. For convenience, we have numbered some items in Fig. 5. These numbers are used in the following text to help identify items more quickly.

For instance, the metaobjects `Deployment.node`*(1)* and `Static.class`*(2)* have been defined to share the same metaproperty (`name`*(3)*). `Deployment` is instantiated into a model*(4)* that contains a node*(5)* whose name is "Fenris"*(6)*. `Static` is instantiated into a model*(7)* that denotes a class*(8)* named "Mainframe" which owns a method named "shutdown". A new metamodel*(9)* is also created to model infrastructures. It borrows the `mainframe` class*(8)* from the static diagram, and promotes it as a metaobject which can have a specific name. An infrastructure model*(10)* is created, that borrows `Mainframe Fenris`*(5)* from the deployment model. It defines its infrastructure name*(11)* as the same value as its name*(12)* in the deployment model, but another name could be used.

The above repository exploits several features of MetaL to counter the usual limitations of meta-repositories:

- First, since their aggregations are not necessarily disjoint, *(meta)models can overlap*: `Deployment MetaModel` and `Static MetaModel` share the `name` metaproperty; `Deployment Model #1` and `Infrastructure Model #1` share the `Mainframe Fenris` concept but with distinct types; and finally `Static Model #1` and `Infrastructure MetaModel` share the `Mainframe` item.
- Second, *sharing between (meta)models can be arbitrarily fine-grained*: the elementary units of sharing are metaobjects. Since every item is a metaobject, the language allows great flexibility: one of more (meta)objects, (meta)roles,

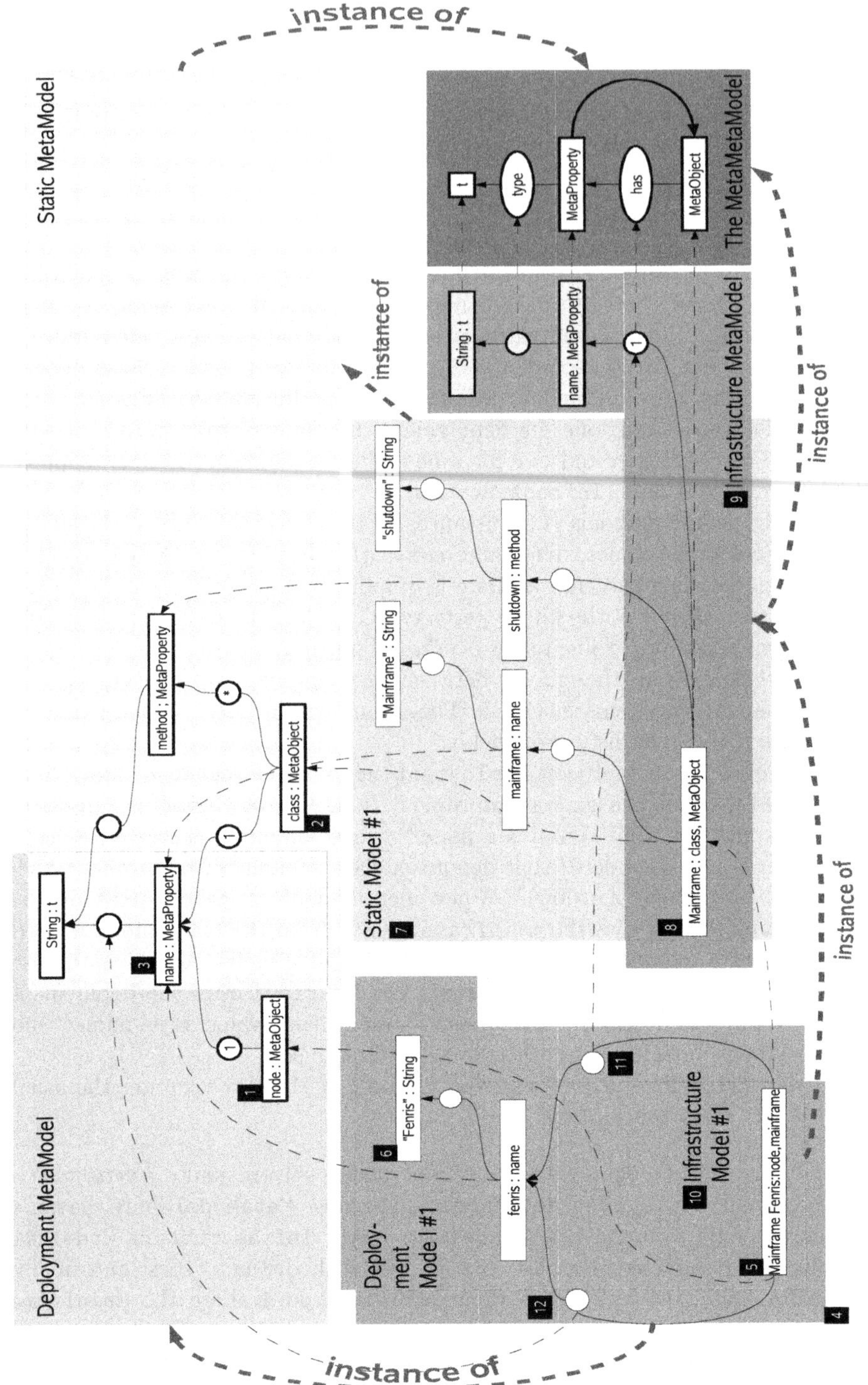


**Fig. 5.** Crosscutting models and metamodels in MetaL. (Not all items appear).

(meta)properties, or even (meta)models can be shared. Moreover, the sharing of a(n) (meta)object does not entail the sharing of its (meta)properties or (meta)roles.

- Third, *an object can be an instance of several object types*, i.e. we have *multiple instantiation*. In the example, `Mainframe Fenris`(5) is both a `node` and a `mainframe`. Multiple instantiation is not supported by many tools, but it is a powerful means of implementing overlapping modelling views. It can be used for example to approximate the *facet language* imagined in [25] (see Section 5).
- Fourth, *the language is freed from the instance/type delimitation* and *relationships can cross M-levels*. In the example, this allows `node` and `mainframe` to share the same property (`fenris`) although they belong to distinct M-levels. Most approaches (with the notable exceptions of [2,16,12]) do not allow cross-level (non instance-of) relationships, a.k.a. *strict metamodelling*.
- Fifth, since metamodels are metaobjects, *the same item can be either considered as an atomic concept (a metaobject) or a complex aggregate (a metamodel)*. In the latter case, distinct definitions (sets of member items) can even be possible. For instance, the mainframe can be modelled as a node; this node could in turn be a model which decomposition would show its inner parts (e.g. motherboard, chip, devices, ...); the same node could be decomposed according to another type that would show its software components (process, libraries, ...). Because of space limitations, this aspect could not be illustrated in Fig. 5.
- Finally, another benefit that we cannot show here is *metamodel refinement*. Since they are metaobjects, metamodels can also specialise each other. For example, a metamodel `m1` could be defined as a specialisation of `m2`. So, `m1` can inherit from the constructs previously defined in `m2`, but possibly rename them, add new constructs, and propagate refinements to lower levels, e.g. to models. This type of metamodel refinement is much more powerful than profile-based customization.

### 3.2 Reification, Meta-circularity and Bootstrapping

The definition of $\text{MetaL}_2$ on top of $\text{MetaL}_1$, and the fact that every $\text{MetaL}_1$ item is an instance of itself (see Fig. 3), allow for a reflexive definition of the metametamodel. This opens the way for elegant and powerful bootstrapping, an example of which is found in Fig. 6.

Being a reflexive object type, `MetaObject` (see Fig.4) possesses at least one instance at the metamodel level: itself. The same happens with the other elements of the metametamodel (`MetaProperty`, `MetaModel`, `MetaRole`, etc.).

With this reflexivity, we can reify some previously 'hardcoded' constructs. As an example, consider the property type `name` found in the metametamodel of Fig.4. Being a metaobject, `MetaObject` can have a metaproperty, e.g. $P$ pointing to $t_{\mathbb{S}}$. If we further instantiate `MetaObject` into itself, at the model level `MetaObject` can now be linked to $P_1$, an instance of $P$ leading to the string "MetaObject". The old `name` property type has thus become obsolete.

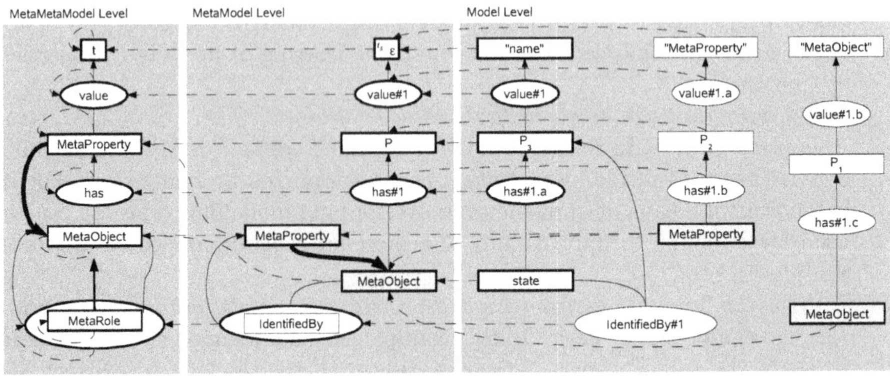


**Fig. 6.** An excerpt of the metametamodel with explicit reflexive instanceOf relations allowing reification of the `name` property type into a first-class metaproperty $P$

The conceptual elegance of reflexivity is relatively unimportant wrt to its ability to bootstrap advanced features into the MML. Fig. 6 illustrates this possibility with the addition of the metarole `IdentifiedBy` between `MetaObject` and `MetaProperty`, making it possible now for metaobjects such as `state` to be *identified by* a subset of their metaproperties ($P_3$ in this case). A similar process can be followed to enrich the (meta)languages with integrity constraints, version control, traceability, etc. among many others. This could not be illustrated here because of space limitations.

## 4 The metaCASE Architecture

A metaCASE prototype based on MetaL is currently being implemented as a Java application. At this stage, we have completed the implementation of a transactional repository whose architecture is presented in Fig. 7. It has been made independent from a specific persistence technology and is able to manage several projects at the same time in different formats: JDO, RDF, XML,... The programme is currently 20 KLOC and offers an API as well as some basic input forms to create and edit dataitems, both in $\text{MetaL}_1$ and $\text{MetaL}_2$.

The general philosophy of the metaCASE tool we are building on top of this repository is to be completely model-driven. That is, every aspect of the tool is a model compliant with some metamodel. Thus, all the dimensions of its observable characteristics are editable, refinable and extensible. We will do so for the GUI (menus, look-and-feel, etc.), the specification of the concrete user notations, version management, collaborative aspects, etc.

For a previous metaCASE tool, we developed GRASYLA [8], a powerful declarative language for describing the concrete syntax of metamodel items (metaobjects, metaproperties, etc.) through equations and symbolic expressions on these items. Taking advantage of the bootstrapping capabilities of the repository, GRASYLA is being redefined as a metamodel. This way, not only arbitrarily many graphical representations could be associated to the same metamodel

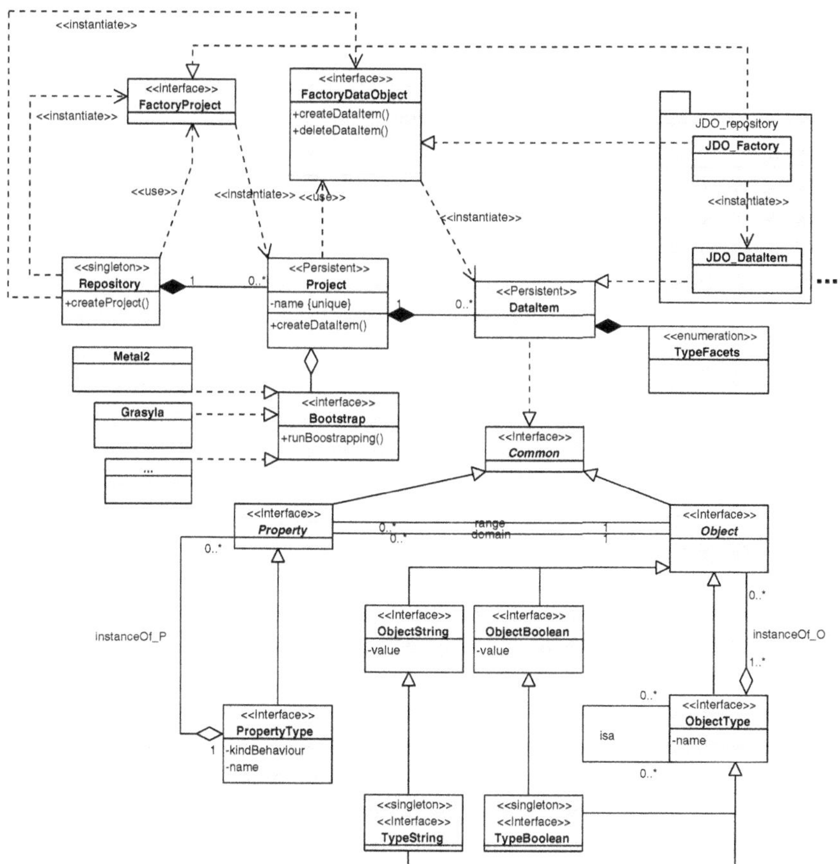


**Fig. 7.** Class diagram of the repository. A project holds a set of dataitems and has a specific persistence manager. A dataitem implements a set of interfaces as described in Section 2.

items, but it will also be possible to generate graphical editors for all languages, at any level, including the MML and GRASYLA themselves.

## 5 Related Work

The current de facto industry standards for modelling and MD* are UML and its supporting CASE tools. Unfortunately UML's extensibility mechanisms are based on the MOF [23] and therefore very limited. These limitations are being tackled by metaCASE tools and metamodelling frameworks. A complete survey is impossible in to fit in here, so we concentrate on the approaches that are the closest in spirit, and we address only the characteristics discussed in this paper.

In the tool arena, the closest are MetaEdit+ [26], Metis [16], ConceptBase [15] and GME [19]. MetaEdit+ allows the sharing of properties between objects, and the sharing of objects between models, but does not allow them to span several of its 3 M-levels ($M_1$ to $M_3$, where $M_3$ is fixed). Metis circumvents this limitation with a fully reflexive metamodelling facility. However, the absence of technical details on the MML [16] make it hard to evaluate Metis' limits. ConceptBase supports Telos [22], a very flexible MML. It has unlimited M-levels and non strict metamodelling, but has more rigid axioms (see Section 2.1) and graphical limitations wrt GRASYLA. A notable strength of ConceptBase is a declarative language for constraints, rules and queries. GME [19] is also a fulfledged metaCASE tool but, like MOF, imposes disjoint metatypes (e.g. a classifier can not be a package) which prevents adequate treatment of referrential redundancy, and is limited to 3 M-levels.

We also find related approaches in metamodelling frameworks. MMF [5] proposes a reflexive MML with constructs similar to ours. In [2], the concept of 'clabject' (half-class, half-object) is introduced to circumvent strict metamodelling. This approach is extended with the concept of powertype in [12]. We support these views (all our elementary and metamodel items are 'clabjects' or reflexive) but extend them with multiple instantiation, generalized reflexivity and unlimited M-levels.

## 6 Limitations and Future Work

Despite their enhanced flexibility, our language and tool currently suffer several limitations.

First, although we build on the experience of previous prototypes [7,8], we are still at a relatively early (re)development stage. The features we currently support are quite basic: a format-independent transactional repository based on MetaL, with an API and basic input forms (see Section 4). Nevertheless, we are confident that bootstrapping will significantly accelerate the development of the remaining features, and especially the re-implementation of GRASYLA.

A concrete user syntax for MetaL still needs to be defined. The one used in this paper is not prescriptive but could be a default. Another could be based on class diagrams. Editors will be bootstrapped with GRASYLA.

By definition, the extension possibilities of a metaCASE tool are endless, so we cannot hope to enumerate them here. However, additional features that we envisage to bootstrap are version control, traceability and process integration. The flexibility of the tool makes it possible to develop or tune those features at later times without the need to reconsider architectural decisions, which is a great advantage. An uncertainty is the performance that the full tool will have since the cost of flexibility is the multiplication of indirection layers. The current results obtained by manipulating metamodels such as those of various UML diagrams are very satisfactory, but performance will need to be carefully evaluated for advanced usages.

A missing feature, and a major advantage of ConceptBase over our approach, is a declarative high-level language for constraints, rules and queries. At the moment, the high-level API can be used for this, but entails using Java. Again, bootstrapping might be an elegant way to introduce such a language. This option will be explored in future work.

Finally, our objectives and design will have to be validated in the light of empirical results. When the maturity of the tool will allow, experimentation with users will be carried out.

## 7 Conclusion

The need for customizable and extensible CASE environments has become ubiquitous. Despite the progress made by metaCASE tools, some limitations may continue to discourage their use. In this paper, we recalled some problems and proposed an integrated solution: MetaL. MetaL is a formal, fine-grained metamodelling language with relaxed axioms and enhanced reflexivity. It allows to easily master referrential redundancy, to span abstraction levels, and to escape the three layers limit. Moreover, its meta-circularity allows metametamodel extension and virtually infinite bootstrapping of additional features.

## References

1. Alderson, A.: Meta-CASE technology. In: Endres, A., Weber, H. (eds.) Software Development Environments and CASE Technology. LNCS, vol. 509, pp. 81–91. Springer, Heidelberg (1991)
2. Atkinson, C., Kühne, T.: Meta-level independent modeling. In: International Workshop Model Engineering (in Conjunction with ECOOP'2000). Cannes, France (June 2000)
3. Atkinson, C., Kühne, T.: The role of meta-modeling in mda. In: Bezivin, J., France, R. editors, Workshop in Software Model Engineering (2002)
4. Brickley, D., Guha, R.V.: RDF vocabulary description language 1.0: RDF Schema. Technical report, W3C (February 2004)
5. Clark, T., Evans, A., Kent, S.: Engineering modelling languages: A precise metamodelling approach. In: FASE. LNCS, vol. 2306, pp. 59–173. Springer, Heidelberg (2002)
6. Ebert, J., Süttenbach, R., Uhe, I.: Meta-CASE in practice: a case for KOGGE. In: Olivé, A., Pastor, J.A. (eds.) Advanced Information Systems Engineering. In: $9^{th}$ International Conference CAiSE'97, Barcelona, Catalonia, Spain. LNCS, vol. 1250, pp. 203–216. Springer, Heidelberg (1997)
7. Englebert, V., Hainaut, J.-L.: DB-MAIN: A next generation meta-CASE. Information Systems (Special issue on meta-modelling and methodology engineering) 24(2), 99–112 (1999)
8. Englebert, V., Hainaut, J.-L.: GRASYLA: Modelling CASE tool GUIs in Meta-CASEs. In: Vanderdonckt, J., Puerta, A. (eds.) Proceedings of the 3rd International Conference on Computer-Aided Design of User Interfaces (CADUI'99), Louvain-la-Neuve, Kluwer, Dordrecht (1999)
9. Englebert, V., Heymans, P.: MetaL: a formal specification. Technical Report PRECISE-06-01, University of Namur - PRECISE Research Centre, Rue grandgagnage 21, 5000 Namur, Belgium (2006)

10. France, R.B., Ghosh, S., Dinh-Trong, T., Solberg, A.: Model-driven development using uml 2.0: Promises and pitfalls. em Computer 39(2), 59–66 (2006)
11. Froehlich, G., Tremblay, J.-P., Sorenson, P.: Providing support for process model enaction in the Metaview metasystem. In: $7^{th}$ International Workshop Computer-Aided Software Engineering (CASE'95), Toronto, Ontario, Canada, IEEE Computer Society Press, Washington (1995)
12. Gonzalez-Perez, C., Henderson-Sellers, B.: A powertype-based metamodelling framework. Software and System Modeling 5(1), 72–90 (2006)
13. Holt., Schürr., Sim., Winter.: GXL: A graph-based standard exchange format for reengineering. Science of Computer Programming 60(2), 149–170 (2006)
14. Honeywell. DOME Guide, Version 5.2.1 (1999)
15. Jeusfeld, M.A., Quix, C.: Meta modeling with conceptbase. In: Proceedings 1st Workshop on Meta-Modelling and Corresponding Tools (WoMM'05), Essen, Germany (7-8 March 2005)
16. Jørgensen, H.D., Karlsen, D., Lillehagen, F.: Collaborative modeling and metamodeling with the enterprise knowledge architecture. Enterprise Modeling and Information Systems Architectures, An International Journal, 1(1) (2005)
17. Jouault, F., Bézivin, J.: KM3: A DSL for metamodel specification. In: Gorrieri, R., Wehrheim, H. (eds.) FMOODS, Bologna, Italy. LNCS, vol. 4037, pp. 171–185. Springer, Heidelberg (2006)
18. Kelly, S., Lyytinen, K., Rossi, M.: MetaEdit+: A Fully Configurable Multi-User and Multi-Tool CASE and CAME Environment. In: Constantopoulos, P., Mylopoulos, J., Vassiliou, Y. (eds.) Proceedings of the $8^{th}$ International Conference CAiSE'96 on Advanced Information Systems Engineering, Heraklion, Crete, Greece. LNCS, vol. 1080, pp. 1–21. Springer, Heidelberg (1996)
19. Ledeczi, A., Maroti, M., Bakay, A., Karsai, G.: The generic modeling environment. In: WISP'2001, Budapest, Hungary, IEEE Computer Society Press, Washington (2001)
20. Mcguinness, D.L., van Harmelen, F.: OWL web ontology language overview (February 2004)
21. Muller, P.-A., Fleurey, F., Jézéquel, J.-M.: Weaving executability into object-oriented meta-languages. In: Kent, S., Briand, L. (eds.) Proceedings of MODELS/UML'2005, Montego Bay, Jamaica. LNCS, vol. 3713, pp. 264–278. Springer, Heidelberg (2005)
22. Mylopoulos, J., Borgida, A., Jarke, M., Koubarakis, M.: Telos: Representing knowledge about information systems. Information Systems 8(4), 325–362 (1990)
23. OMG. Meta Object Facility (MOF) 2.0 Core Specification, ptc/04-10-15 edition (2004)
24. Opdahl, A.L., Henderson-Sellers, B.: A unified modelling language without referential redundancy. Data Knowl. Eng. 55(3), 277–300 (2005)
25. Opdahl, A.L., Sindre, G.: Facet modelling: An approach to flexible and integrated conceptual modelling. Information Systems 22(5), 291–323 (1997)
26. Rossi, M., Kelly, S.: Construction of a CASE tool: The case for MetaEdit+. In: First International Symposium on Constructing Software Engineering Tools (CoSET'99), Los Angeles, USA (May 1999)

# Concepts for Incremental Method Evolution: Empirical Exploration and Validation in Requirements Management

Inge van de Weerd, Sjaak Brinkkemper, and Johan Versendaal

Department of Information and Computing Sciences
Utrecht University, Utrecht, The Netherlands
{i.vandeweerd, s.brinkkemper, j.versendaal}@cs.uu.nl

**Abstract.** Product software companies are confronted with performance failures in their processes for which standard theories on situational method engineering need to be revisited. By developing a knowledge infrastructure, we support these companies with their method evolution by increasing the maturity of their processes incrementally. We first identify and formalize general method increments that are found in an exploratory case study. Then, we formalize common process needs, by developing a root-cause map for software product management and by identifying the root causes and process alternatives that are related to them. We validate the formalized method increments, and process needs by applying them to an extensive case study conducted at Infor Global Solutions. The results show that the formalized method increment types cover all increments that were found in the exploratory case study, and that the root-cause map is a useful technique to model the root causes encountered in product software companies.

**Keywords:** method engineering, meta-modeling, software process improvement, incremental method evolution, root cause analysis.

## 1 Introduction: Incremental Method Evolution

Many organizations are struggling with the evolution of their information systems development methods [6]. To control this, several software process improvement methods have been proposed (e.g. [8] [14]), which can be implemented in different ways and which are evolutionary in nature. In our research, we focus on such an evolutionary approach instead of a mere revolutionary approach for several reasons: a) it is a fundamental way to reduce risk on complex improvement projects [10]; and b) we observe in practice that this is the natural way for method evolution [26] [27].

This evolutionary approach has been subject of research in various scientific studies: methods have been developed to measure and to increase a company's maturity [8] [14]; studies have been carried out to find the best approach to instigate a process improvement [17] [22]; and research has been done on the key success factors that influence software process improvement [15]. However, in 2002, it was estimated that still 70% of software process improvement projects failed [21].

J. Krogstie, A.L. Opdahl, and G. Sindre (Eds.): CAiSE 2007, LNCS 4495, pp. 469–484, 2007.

In this work, we choose to take the existing research on software process improvement a step further. Our aim is to develop a knowledge infrastructure that supports product software (PS) companies that build off-the-shelf software products for a market [28] in the *incremental evolution* of their methods, by dealing with their process needs and guiding them to higher maturity levels. We keep the increments local (i.e. one process at a time is changed) and small (in comparison to existing incremental approaches with larger increments like CMM [14] and SPICE [8]).

In the next section, we describe our research approach, introduce process-deliverable diagrams for modeling methods, and describe the context of this research. In section 3, we define and formalize the process needs. In section 4, we validate the formalized method increments by carrying out a case study at Infor Global Solutions. Finally, in section 5, we describe our conclusions and future research.

## 2 Research Approach

Our aim is to support PS companies in their method evolution, by improving parts, or fragments, of their existing methods in an automated way. Method engineering [3] has been used successfully to engineer (parts of) methods for specific situations [1] [16]; to serve as an instrument in software process improvement [27]; and to use as an approach to manage evolutionary method development by integrating formal meta-models with an informal method rationale [19].

For scoping reasons we limit our research to the software product management domain of PS companies, covering requirements management, release planning, product roadmapping, and portfolio management. In industry, software product management is a clearly defined function, but in science research is fragmented [24].

### 2.1 Research Question and Methodology Outline

We define the following research question:

*"How can product software companies improve their software product management methods in an evolutionary way, using method fragment increments?"*

We address this question by applying method engineering theory. Incremental method engineering has been subject to research by e.g. [10] and [23]. However, a definition of method increment seems not to be available. Therefore, we define a *method increment* as: a method adaptation, in order to improve the overall performance of a method. Note that adaptation can mean insertion, editing or removal of method fragments.

Actual method increments in industry are explored in an explorative case study at a HRM software vendor (from now on: HRM case study), in order to derive a list of method increment *types* that occur during the evolution. By formalizing and generalizing the increments, we model incremental evolution of a product software company's processes. The formalized increments are then validated in an ERP case study. Using Root Cause Analysis (RCA, [18]) techniques we determine an initial set of root causes of process needs that PS companies may encounter in the software product management domain. RCA has been applied to process improvement and incident prevention in software and non-software industries; see for example [11].

With respect to the HRM case study we determine an initial set of root causes that may lead to process improvement alternatives. This set and our RCA application are also validated in the ERP case study.

## 2.2 Meta-modeling with Process-Deliverable Diagrams

For the analysis of method increments, we use process-deliverable diagrams (PDDs), a meta-modeling technique that is based on UML activity diagrams and UML class diagrams [25]. The resulting PDDs model the processes on the left-hand side and deliverables on the right-hand side (see Figure 1). Examples of PDDs can be found in Figure 5 and 6.

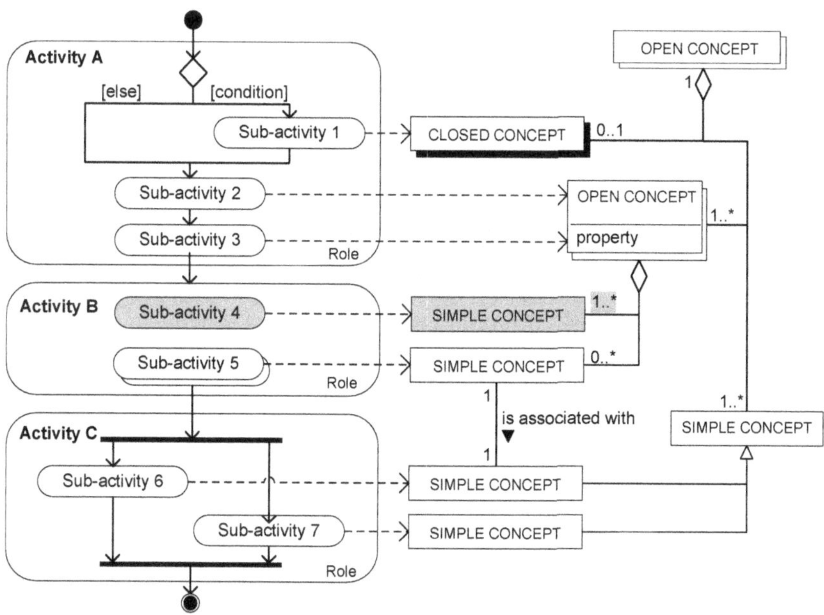


**Fig. 1.** Process-data diagram

We follow standard UML [13] conventions, but some minor adjustments have been made for modeling development processes. Firstly, deliverables can be **simple** or **compound**. Simple deliverables do not contain any sub deliverables and are visualized with a rectangle. Compound deliverables contain one or more sub deliverables. Compound deliverables can be **open**, visualized with an open shadow, to indicate that it contains sub deliverables. The sub deliverables can be shown in the same diagram, by using aggregation, or in another diagram (for example for space saving). **Closed** compound deliverables, visualized with a closed shadow, indicate that that sub deliverables exist, but are not relevant in this context. Similarly, open en closed activities are used in the diagram. The dotted arrows indicate which deliverables result from the activities. More details on this modeling technique can be found in [25] and [27].

The PDD, visualized in Figure 1, is called a **snapshot**, a model of the process as it was at a certain moment in time [27]. The evolution of a method over time exists of a number of these snapshots. By comparing snapshots, method increments can be analyzed. In Figure 1, we marked sub activity 4 and its corresponding concept. We use this notation to show the method increment of this snapshot compared to a snapshot earlier in time.

## 2.3 A Knowledge Infrastructure for Incremental Method Evolution

The context in which we want to support PS companies with the incremental evolution of their processes is described in [27], where we propose the Product Software Knowledge Infrastructure (PSKI, [27]). Several knowledge repositories for software development methods have been proposed and developed (e.g. the OPEN Process Framework [9]). However, the PSKI is not only a knowledge repository, but it also analyzes the process need of a company in order to deliver meaningful advice. In Figure 2, the PSKI is illustrated as well as the PS company that interacts with it. The PSKI contains a method base, in which method fragments, situational factors, maturity capabilities and assembly rules are stored.

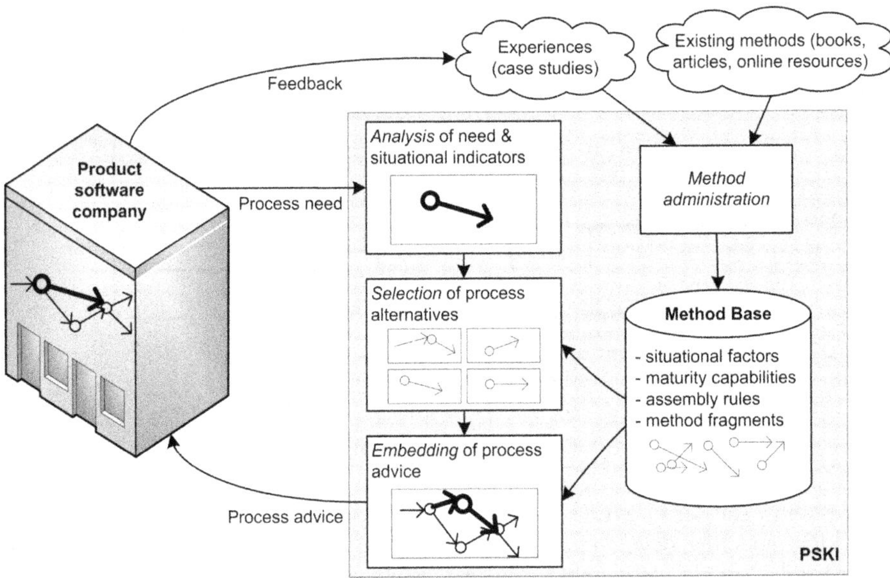


**Fig. 2.** Product Software Knowledge Infrastructure

*Analysis of need and situational indicators*

The first step is the analysis of the process need and situational indicators. The process need is analyzed using Root Cause Analysis, (RCA). Through RCA the root causes of a process need are determined using the following sequence, see also [11] and [18]: 1) which *process difficulties* actually occur; 2) what are the so-called *causal factors* of the difficulties; and 3) what are the actual *root causes* per causal factor, using a root cause map designed for PS companies. We define a root cause as (one of) the underlying reasons of a process need, solving one or more causal factors, and relating to one or more actors, activities and deliverable concepts (referring to figure 1). Situational indicators contain information about the process and the company. Examples are company size, development platform and sector.

*Selection of process alternatives*
Once the root causes are known for a process need, directions for software process improvement can be sought taking into account situational factors. For this, the method base is used. Links between maturity capabilities and root causes are available in the method base in order to identify possible process improvement alternatives. Examples of maturity capabilities are listed in [26]. We define a process alternative as a method fragment of a particular maturity capacity that settles one or multiple root causes of the process need.

*Embedding of process advice*
The last step is embedding the process advice in the company's existing processes. A process advice, which contains a process description, templates and examples, is sent back to the company. The person responsible for process improvement at the company will then start the organizational deployment of the process advice. This roll-out process also includes the insertion of the increment in the existing processes.

## 3 Definition and Formalization

This section defines and formalizes method increments and the development problems that lead to these increments. The rationale for this formalization is twofold: First, we use it to analyze the method increments that we found in the HRM case study (see Section 4.4). Secondly, the formalization is used as a first step to develop a formal structure for the method base of the PSKI in which method fragments can be edited. Firstly, we define method evolution, snapshot and method increments. Then, based on the meta-meta model of PDDs we present a list of all possible increment types with some method fragment insertion rules. Thirdly, we analyze problems that lead to the method increments and develop a root cause map for software product management (RCM for SPM).

### 3.1 Definitions of Incremental Method Evolution

As the PDD technique is based on UML, we can utilize the available formalizations in the literature. There appears to be two kinds of formalizations: those based on the formal language **Z**, e.g. [5] and [20] and those using first order predicate logic, e.g. [2], of which we chose the latter due to its concise presentation.

We start the formalization with the assumption that there is some kind of universe of consistent methods, called $\mathsf{M}$. We assume furthermore, that these methods in $\mathsf{M}$ can be executed by project members, i.e. the method descriptions are available, complete, and consistent. The evolution of the method in a particular company can then be seen as a series of methods $m_1, m_2, \ldots, m_n \in \mathsf{M}$. For reasoning about time we introduce the time dimension $\mathsf{T}$. The set of method fragments is called $\mathsf{F}$.

**Definition 3.1.** The mapping $\mathsf{method}: \mathsf{T} \rightarrow \mathsf{M}$, where $m = \mathsf{method}(t)$ means that the method $m \in \mathsf{M}$ is the valid method at time $t$.

The methods change in the course of time, and this allows us to define the notion of snapshot of a method.

**Definition 3.2.** A *method adaptation time* is a point of time where the method has been adapted. Let $\mathcal{T}$ be the set of method adaptation times, i.e. $\mathcal{T} = \{t_1, t_2, t_3, \ldots, t_n\}$ such that $\forall i\ \forall t$: $\mathsf{method}(t_i) = \mathsf{method}(t) \neq \mathsf{method}(t_{i+1})$.

**Definition 3.3.** A *method snapshot* is a method $m \in \mathsf{M}$ that was valid at a particular time, i.e. $\exists t_i \in \mathcal{T}$; $m = \mathsf{method}(t_i)$.

**Definition 3.4.** A *method evolution* is a set $\mathsf{S} \subseteq \mathsf{M}$ consisting of the method snapshots, i.e. $\mathsf{S} = \{\mathsf{method}(t_i) | \ t_i \in \check{\mathsf{T}}\}$. So $\mathsf{S}$ is the set of methods that have been valid in the course of time.

We are now able to define method increments. As in common method engineering practices a method is seen as being composed of method fragments or method chunks [3] [16]. Such a method is consistently created using well-formedness rules of process composition and deliverable configuration. These rules are not elaborated here, as they can be found in [4].

**Definition 3.5.** The predicate $\mathsf{contains}$: $\mathsf{F} \times \mathsf{S}$ : $\mathsf{contains}(f,s) \equiv$ fragment $f$ is contained in snapshot $s$.

Then we can define an method increment as a method fragment that is part of $\mathsf{method}(t_i)$ but not in $\mathsf{method}(t_{i-1})$.

**Definition 3.6.** A *method increment* is a method fragment $f \in \mathsf{F}$ such that $\exists i$ $\mathsf{contains}(f, \mathsf{method}(t_i)) \wedge \neg \mathsf{contains}(f, \mathsf{method}(t_{i-1}))$

This means that the method increments are a collection of method fragments that have been introduced in the method during the method adaptations between $t_i$ and $t_{i-1}$. In the following section we will then formalize the various types of increments

## 3.2 Formalization of Method Increments

In Figure 3 the meta-meta model of PDD is given, denoted in (again!) a UML Class diagram.

The meta-meta model is a simplified view of the full UML definition of Class diagrams and Activity diagrams [13] with special emphasis on the adaptations discussed in Section 2.2 and the definitions in 3.1. Figure 3 shows that a method consists of method fragments, that we distinguish as process fragments for the process part of a method and deliverable fragments similarly. Note that the creation of deliverables is modelled in the association *edits* between Activities and Concepts.

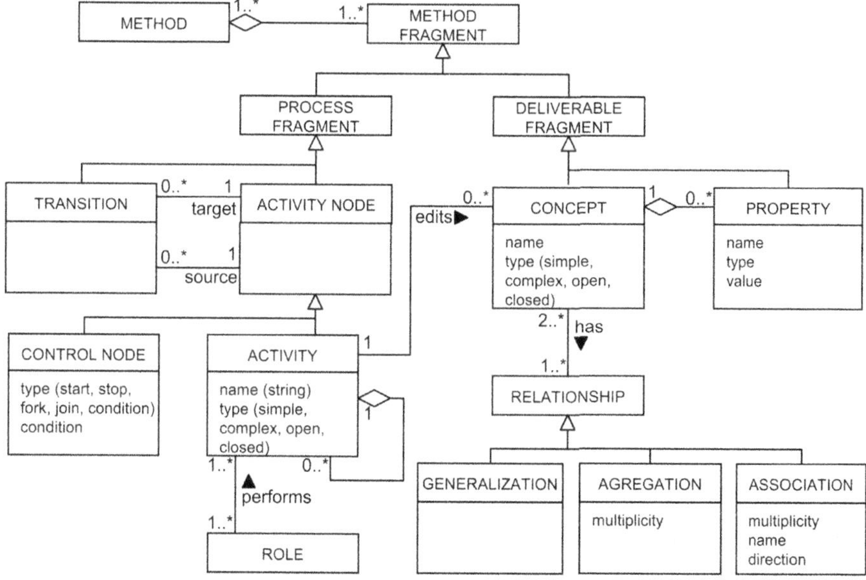


**Fig. 3.** Meta-meta model of PDD

The structure of the meta-meta-model and the earlier case studies [27] to method evolution revealed that 18 elementary increment types can be distinguished:

- *insertion* of a concept, property, relationship, activity node, transition, role
- *modification* of a concept, property, relationship, activity node, transition, role
- *deletion* of a concept, property, relationship, activity node, transition, role

The complete method increments from one snapshot to another can then be seen as a composition of elementary increment types.

The UML formalization of [2] postulates the existence of unary predicates for each class in a class diagram, e.g. concept($c$) means that $c$ is a concept in the model. However, in our research we require evolution of methods over the various snapshots, so we enhance these unary predicates to binary predicates with the method as an additional parameter. So concept($c$,$m$) means that $c$ is a concept in the method $m$. Method increments can now be defined as polymorphic mappings on the set of method fragments and methods.

**Definition 3.7.** The mapping insert: F x M → M: insert($f$,$m_1$) = $m_2$ means that the method fragment $f$ has been inserted in the method $m_1$ resulting into method $m_2$.

**Definition 3.8.** The mapping modify: F x F x M → M: modify($f_1$,$f_2$,$m_1$) = $m_2$ means that the method fragment $f_1$ in the method $m_1$ has been modified to the fragment $f_2$ in method $m_2$.

**Definition 3.9.** The mapping delete: F x M → M: delete($f$,$m_1$) = $m_2$ means that the method fragment $f$ has been deleted from the method $m_1$ resulting into method $m_2$.

The rules for the elementary increments can then be formulated. For the sake of brevity we list the rules for the insertion of concepts and properties. Both rules are illustrated with an example that is taken from the increment example in Section 4.3.

**Rule 3.1.** Insertion of concepts:

$$\text{insert}(c,m_i) = m_{i+1} \Rightarrow \neg\text{concept}(c,m_i) \wedge \text{concept}(c,m_{i+1})$$

Rule 3.1 states when a concept has been inserted into method mi to get method mi+1. So, for instance:

insert(RELEASE TABLE,BaanIncr2) = BaanIncr3 ⇒ ¬concept(RELEASE TABLE,BaanIncr2) ∧ concept(RELEASE TABLE,BaanIncr3)

This means that when the concept RELEASE TABLE is inserted into BaanIncr2 resulting into BaanIncr3, then RELEASE TABLE is not a concept present in BaanIncr2 and is present as concept in BaanIncr3.

**Rule 3.2.** Insertion of properties:

$$\text{insert}(p,m_i) = m_{i+1} \wedge \text{property}(p,m_{i+1}) \Rightarrow [\forall c: \text{concept}(c,m_i) \wedge \neg\text{contains}(p,c)] \wedge [\exists 1c: \text{concept}(c,m_{i+1}) \wedge \text{contains}(p,c)]$$

Rule 3.2 tells that when property $p$ is inserted into snapshot $m_i$ resulting into snapshot $m_{i+1}$, then $p$ is not a property of any concept in $m_i$ and there is just one concept in $m_{i+1}$ of which $p$ is the property. So, for instance:

insert(topic,BaanIncr2) = BaanIncr3 ∧ property(topic,BaanIncr3) ⇒ [∀c: concept(REQUIREMENT,BaanIncr2) ∧ ¬contains(topic,REQUIREMENT)] ∧ [∃1c: concept(REQUIREMENT,BaanIncr3) ∧ contains(topic,REQUIREMENT)]

This means that when the property topic is inserted into snapshot BaanIncr2, resulting into BaanIncr3, then topic is not a property of any concept in BaanIncr2 and there is just one concept, namely REQUIREMENT, in BaanIncr3 of which is topic the property.

Analogously, rules for the other 16 elementary method increments can be formulated, while taking the method assembly rules in [4] into account. Based on our earlier work on method assembly this formalization is extremely straightforward and will support the construction of the PSKI currently under development.

### 3.3 Root Cause Analysis for Product Software

Based on the general Root Cause Map (RCM) [18], the reference framework for software product management (SPM) [24], and the HRM case study [27], we are able to construct an initial RCM for SPM, as is depicted in Figure 4.

During the interviews conducted in the HRM case study, two major process difficulties for requirements management were recognized:

A. Customers do not see that their required features and software improvement wishes are implemented in new releases.
B. The company finds its requirements gathering process for new features not productive.

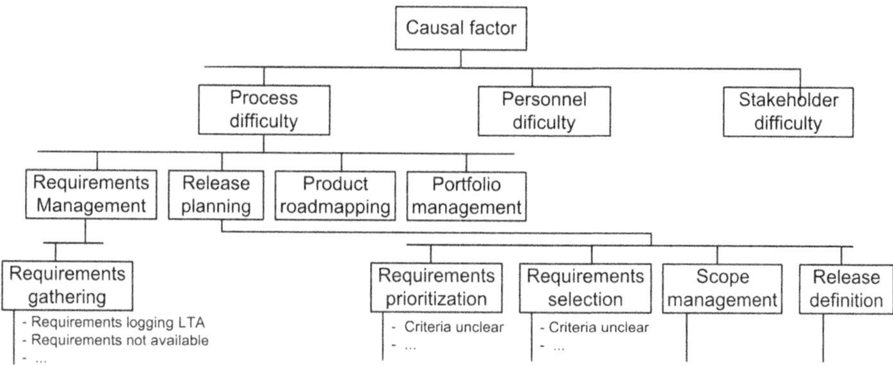


**Fig. 4.** The explorative case root-cause map

When we apply RCA to these process difficulties, we identify a number of causal factors: To communicate a suggestion for improvement, a customer can contact the sales representative; in some cases the sales representative replies that suggestions should be posted to the helpdesk; in other cases the sales representative forwards the suggestion to the helpdesk; and some suggestions are not logged at all.

As for the second process difficulty, when a new release is defined, the helpdesk, the development manager and the software engineers are consulted. Rather arbitrary, but fitting a defined planning schedule, the development of a new release is triggered. Consequently, we identify three causal factors:

*C1. Customers have difficulty in making their wishes known*
*C2. Customer requirements are not registered effectively*
*C3. Scoping of releases is rather arbitrary*

The following root causes can be identified (indicated are the corresponding causal factors):

*R1. Requirement logging is less than adequate (LTA) (root cause for C1 & C2)*
*R2. Requirements are not available (root cause for C3)*
*R3. Criteria for requirements prioritization are unclear (root cause for C1 & C3)*
*R4. Criteria for requirements selection are unclear (root cause for C3)*

In [27] a threefold solution for the two major process difficulties(A & B) is described:

*S1. Introduction of a separate activity for receiving and logging new requirements;*
*S2. Introduction of a wish list (requirements database) with wishes (requirements) containing a priority attribute;*
*S3. Introduction of a separate activity for prioritizing wishes.*

When we map this process on the PSKI, this threefold solution would be described in a process advice, containing process descriptions, templates and examples. Note that RCA was not the basis for the solution finding at the HRM case study. However, if we do take into account the RCA and the resulting root causes we find that solution S1 addresses R1 and R2, solution S2 addresses R2 and partly R3, solution S3 addresses R3. Note that R4 has not been properly addressed in the solution. We conclude that in the HRM case study, RCA was a useful approach for finding process improvements alternatives. This will be further validated in Section 4.

# 4 ERP Case Study

We carried out a case study at Infor Global Solutions (specifically the former Baan company business unit), a vendor of ERP (Enterprise Resource Planning) software (see for example [12]). The goal of the ERP case study is to validate the increment types defined in Section 3.3 and the root-cause map in Section 3.4. In 1978, Baan was established as a book-keeping consulting company. Over the years, the company changed from a consultant company to a software developer for businesses. Baan was quoted on the Nasdaq stock exchange as an independent company from 1995 to 2000.

## 4.1 Case Study Design

Different sources are used to collect information. Firstly, several interviews have been conducted with six former employees of Baan. Two explorative 3-hour interviews were conducted with the Process Engineer of Baan. Based on this interview, the method evolution between 1997 and 2002 was modeled. This information was cross-checked by conducting 2-hour follow-up interviews with five other employees of Baan, consisting of two former (Senior) Product Managers, a Director ERP Development, a Manager ERP Product Ownership and a Software Engineering Process Group Manager for Baan Development. In these interviews, also the snapshots of 1994, 1996, 2003, 2004 and 2006 were identified and modeled.

Secondly, a document study was carried out. Documentation provided by the Process Engineer was used to complement and validate the results from the interviews. This documentation consisted of process descriptions, templates and examples of methods and work products used at Baan in the period 1997 until 2006. From the period before 1997 no documentation was available. We focused on the following case study questions, related to software product management:

- Which snapshots can you identify in the method evolution?
- Which methods were used per stage? Which activities can be distinguished?
- Which deliverables resulted from these methods?
- Which process difficulties arose in this stage? Why was an increment needed?

With the information gathered in the case study, we modeled 14 snapshots in PDDs, each representing a method that was used in a particular moment in time [26].

## 4.2 Method Snapshots

We analyzed 14 snapshots of the evolution of the software development process at Baan, with emphasis on product management activities. The time period that is covered in the ERP case study ranges from 1994 to 2006.

Note that, although some method increments entail the removal of a method fragment, we still describe them as increments, as described in Section 2.1. In the

**Table 1.** Overview of method increments at Baan

| # | Increment | Date |
|---|---|---|
| 0 | Introduction requirements document | 1994 |
| 1 | Introduction design document | 1996 |
| 2 | Introduction version definition | 1998, May |
| 3 | Introduction conceptual solution | 1998, November |
| 4 | Introduction requirements database, division market and business requirements, and introduction of product families | 1999, May |
| 5 | Introduction tracing sheet | 1999, July |
| 6 | Introduction product definition | 2000, March |
| 7 | Introduction customer commitment process | 2000, April |
| 8 | Introduction enhancement request process | 2000, May |
| 9 | Introduction roadmap process | 2000, September |
| 10 | Introduction process metrics | 2002, August |
| 11 | Removal of product families & customer commitment | 2003, May |
| 12 | Introduction customer voting process | 2004, November |
| 13 | Introduction master planning | 2006, October |

following section, one of these increments, namely the increment between snapshot 2 and 3, is further elaborated on. The other increments are described in [26].

### 4.3 Increment Example: Introduction of the Conceptual Solution

In Figure 5, increment # 2 of the ERP case study is visualized. Looking at the process-side of the diagram, we can distinguish one main activity, i.e. 'Requirements', and three sub-activities.

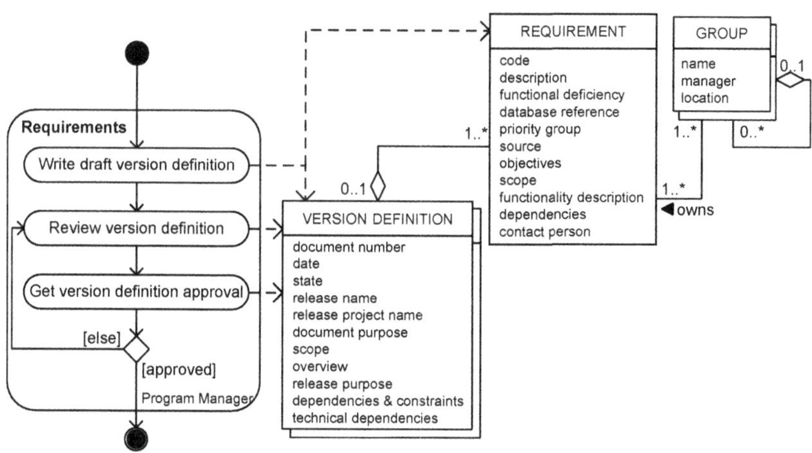


**Fig. 5.** Snapshot of increment #2

The first sub-activity, 'Write draft version definition', results in the concepts VERSION DEFINITION and REQUIREMENT. The latter is connected to VERSION DEFINITION by means of aggregation. Both have a number of attributes, and finally, a REQUIREMENT is owned by a GROUP, that has the responsibility for this REQUIREMENT. The next sub-activity is to review the VERSION DEFINITION. If the approval is obtained, the next activity can be started; otherwise the VERSION DEFINITION has to be reviewed again.

In Figure 6, increment #3 is visualized. In this snapshot, one extra activity is included. Note, however, that this activity is open, i.e. this activity contains further sub activities that are elaborated elsewhere. Due to space limitations, the elaboration on this activity is not included in this paper.

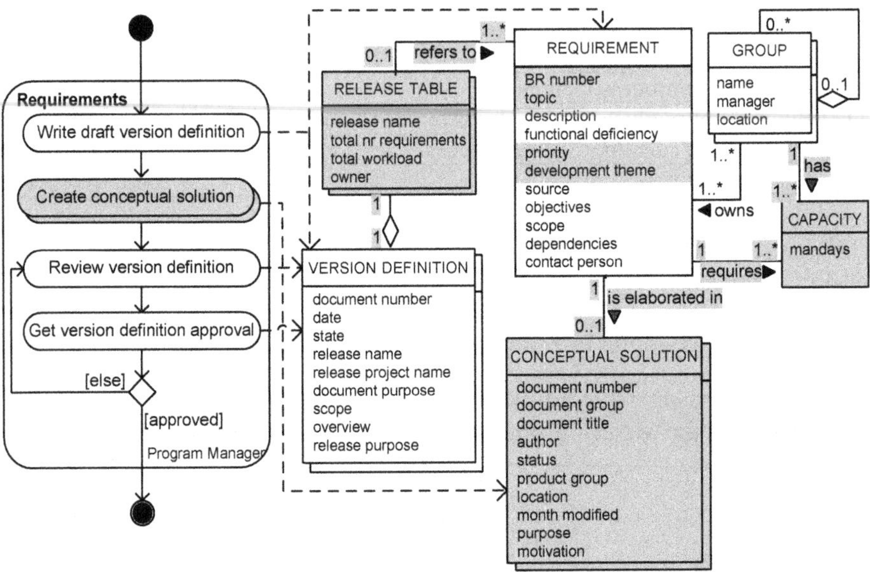


**Fig. 6.** Snapshot of increment #3

## 4.4 Root Cause Analysis of Method Increments

In increment 3 (Figure 6) we distinguish the following increment types, based on the formalization in Section 3.2:

I1. Insertion of an activity node, i.e. 'Create conceptual solution'
I2. Insertion of a concept, i.e. RELEASE TABLE, CONCEPTUAL SOLUTION and CAPACITY
I3. Insertion of a property, i.e. the properties added to REQUIREMENT
I4. Insertion of a relationship, i.e. the relationships connecting the introduced concepts to the existing concepts

Now we focus on RCA. The increments are included to solve one or more problems. Based on the interviews, several process needs were identified in the snapshot of increment #2. The most important ones were:

A. Development managers find it hard if not impossible to determine a VERSION DEFINITION that is feasible with available resources, and consequently makes sub-optimal scoping decisions. In detail: signals from the market, as well as from internal stakeholders, indicate that a new release should be developed. The development managers ask the program managers and architects to establish the version definition for the different software modules. The program managers and architects collect, with some difficulty, the features and requirement from different sources. They select a set of features to be developed according to their own opinions. The program managers and architects discuss the draft version definition with the development managers, and make changes to the selection of features.

B. Software engineers find it hard to read the VERSION DEFINITION in order to built what is requested, and consequently do not build the precise features that were intended to be build. In detail: in the version definition each new software product REQUIREMENT is elaborated by the program manager and/or product architect. They describe dependencies with other REQUIREMENTS in the text associated with a requirement. The software engineers read the (often badly written) requirements, interpret requirement texts, possibly asking their program managers and architects for explanations. Subsequently, the requirements are built in the software product.

We identify the following causal factors:

*C1. Requirement collection is difficult*
*C2. Text elaborations of requirements have different authors*
*C3. Requirements dependency descriptions are unstructured*
*C4. Interpretation of requirement is ambiguous*

If we apply the earlier constructed root cause map for product software to this particular increment we choose to extend it accordingly in order to address all identified causal factors (see Figure 7). Note that, although we had to extend the root cause map with the (bold) root causes, it fits the constructed structure as derived in Section 3 very well.

The root causes of the four identified causal factors are fourfold:

*R1. Requirements are scattered throughout the company in different documents (root cause for C1 & C2)*

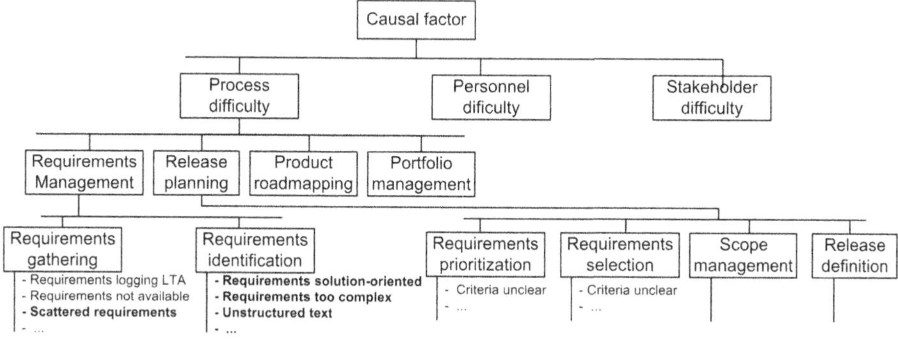


**Fig. 7.** Baan increment extended root cause map for product software

*R2. Some requirements are written in a solution-oriented way (root cause for C2 & C4)*
*R3. Requirements are too complex (root cause for C4)*
*R4. Requirements are written in unstructured text (root cause for C3 & C4)*

R2 and R3 led to the introduction of the CONCEPTUAL SOLUTION in increment #3. This document was used to write a solution on conceptual level for the particular REQUIREMENT. In this way, solution-oriented texts are also kept out the requirements themselves. R4 partly led to the decomposition of the VERSION DEFINITION and REQUIREMENTS. A RELEASE TABLE is used, in which information on the separate REQUIREMENTS is summarized. No (full) solution is implemented for R1 and R4. These are taken into account in the subsequent increment, which is described in [26].

We note that the RCM for SPM has been extended on the lowest level, but that higher levels were untouched, indicating that for two different companies, the RCM for SPM is a useful tool. We conclude that RCA can be used in software development improvements (as in [8]) and more specifically software product management. The root causes showed in the RCM for SPM provide means for the PSKI to determine process improvement alternatives.

### 4.5 Validity Threats

In exploratory research, three types of validity are important [29]. Firstly, construct validity concerns the validity of the research method. We satisfy this type of validity by using multiple sources of data (interviewees and documents) and by maintaining a chain of evidence. Furthermore, we had key informants review the draft case study report. Secondly, the external validity concerns the domain to which the results can be generalized. We carried out the case study in the software product management domain in PS companies. The same protocol is followed as in earlier case studies in PS companies. Finally, to guarantee the reliability of the case study, all information should be recorded. This is done by maintaining a case study database which contains all relevant information used in the case study. This case study database consists of interview notes, documentation and process-data diagrams of all modelled methods.

## 5 Conclusion

By presenting a formal approach to incremental process improvement, we provided PS companies with an instrument to improve their software product management methods in an evolutionary way. Firstly, we formalized the method increments that occur during method evolution. Doing this provided insight in the evolution process, which can be used when assembling a method advice. Secondly, we presented an approach for the structural analysis of process needs, by using root cause analysis. By applying this analysis in a case study, we found that this approach and the corresponding root cause map can be of great value in the support of incremental method evolution.

Currently, we are working on the realization of the PSKI. We aim to further integrate the root cause analysis approach in the PSKI in order to map root causes to maturity capabilities and method fragments. The formalization of method increments is used to implement assembly rules. In the future, we plan to fill the method base with situational factors, method fragments and assembly rules. Finally, we plan to test the PSKI at PS companies of different sizes and in different sectors, in order to test the mapping between situational factors, maturity capabilities and method fragments.

## References


1. Aydin, M.N., Harmsen, F.: Making a Method Work for a Project Situation in the Context of CMM. In: Proceedings of the 14th International Conference on Product Focused Software Process Improvement, Rovaniemi, Finland, pp. 158–171 (2002)
2. Berardi, D., Cali, A., Calvanese, D., De Giacomo, G.: Reasoning on UML class diagrams. Artificial Intelligence 168, 70–118 (2005)
3. Brinkkemper, S.: Method Engineering: Engineering of Information Systems Development Methods and Tools. In: Information and Software Techn, vol. 38, pp. 275–280. Elsevier, Amsterdam (1996)
4. Brinkkemper, S., Saeki, M., Harmsen, F.: Meta-modelling Based Assembly Techniques for Situational Method Engineering. Information Systems 24(3), 209–228 (1999)
5. Clark, T., Evans, A., Kent, S.: The Metamodelling Language Calculus: Foundation Semantics for UML. In: LNCS, vol. 2029, pp. 17 –31Springer, Heidelberg (2001)
6. Conradi, R., Fernström, C., Fuggetta, A.: A Conceptual Framework for Evolving Software Processes. In: ACM SIGSOFT Software Eng. Notes 18(4), 26–35 (1993)
7. Cronholm, S., Ågerfalk, P.J.: On the Concept of Method in Information Systems Development. In: Proceedings of the 22nd Information Systems Research Seminar in Scandinavia 1, 229–236 (1999)
8. El, E.K., Melo, W., Drouin, J.-N. (eds.): SPICE: The Theory and Practice of Software Process Improvement and Capability Determination. IEEE Computer Soc. Press, Los Alamitos (1997)
9. Henderson-Sellers, B.: Process Metamodelling and Process Construction: Examples Using the OPEN Process Framework (OPF). Annals of Software Eng. 14, 341–362 (2002)
10. Krzanik, L., Simila, J.: Is my Software Process Improvement Suitable for Incremental Deployment? 8th International Workshop on Software Technology and Engineering Practice (STEP'97) p. 76 (1997)
11. Leszak, M., Perry, D.E., Stoll, D.: A Case Study in Root Cause Defect Analysis, ICSE p. 428 (2000)
12. Natt och Dag, J., Gervasi, V., Brinkkemper, S., Regnell, B.: Speeding up Requirements Management in a Product Software Company: Linking Customer Wishes to Product Requirements through Linguistic Engineering. In: Proceedings of the 12th IEEE International Requirements Engineering Conference pp. 283–294 (2004)
13. Object Management Group: UML 2.0 Superstructure Specification. Technical Report ptc/04-10-02 (2004)
14. Paulk, M.C., Curtis, B., Chrissis, M.B., Weber, C.V.: Capability Maturity Model for Software (Version 1.1) (SEI/CMU-93-TR-24, ADA263403). Pittsburgh, Pa.: Software Engineering Institute, Carnegie Mellon University (1993)
15. Rainer, A., Hall, T.: Key Success Factors for Implementing Software Process Improvement: a Maturity-Based Analysis. Journal of Systems and Software 62(2), 71–84 (2002)

16. Ralyté, J., Rolland, C.: An Assembly Process Model for Method Engineering. In: Advanced Information Systems Engineering. In: CAiSE 2001. LNCS, vol. 2068, pp. 267–283. Springer, Heidelberg (2001)
17. Richardson, I., Ryan, K.: Software Process Improvements in a Very Small Company. Software Quality Professional 3(2), 23–35 (2001)
18. Root Cause Analysis Handbook: A Guide to Effective Incident Investigation, ABS Group Consulting, Inc, Houston, TX (1999)
19. Rossi, M., Ramesh, B., Lyytinen, K., Tolvanen, J.-P.: Managing evolutionary method engineering by method rationale. Journal of the Association for Information Systems 5(9), 356–391 (2004)
20. Saeki, M.: Toward Formal Semantics of Meta Models. In: International Workshop on Model Engineering, Nice, France (2000)
21. SEI: Process maturity profile of the software community. Software Engineering Institute, Carnegie Mellon University (2002)
22. Stelzer, D., Mellis, W.: Success Factors of Organizational Change in Software Process Improvement. In: Software Process: Improvement and Practice, vol. 4(4), pp. 227–250. John Wiley & Sons, New York (1998)
23. Tolvanen, J.-P.: Incremental method engineering with modeling tools: theoretical principles and empirical evidence. Jyväskylä Studies in Computer Science, Economics and Statistics 47, University of Jyväskylä, PhD Dissertation thesis (1998)
24. Weerd, I., van de Brinkkemper, S., Nieuwenhuis, R., Versendaal, J., Bijlsma, L.: Towards a Reference Framework for Software Product Management. In: Proc. of the 14th International Requirements Engineering Conference, Minneapolis, Minnesota, USA pp. 312-315 (2006)
25. Weerd, I., van de Brinkkemper, S., Souer, J., Versendaal, J.: A Situational Implementation Method for Web-based Content Management System-applications. In: Software Process: Improvement and Practice. Vol. 11(5), pp. 521–538. John Wiley & Sons, New York (2006)
26. Weerd, I., van de Brinkkemper, S., Versendaal, J.: Incremental Method Evolution in Requirements Management: A Case Study at Baan 1994-2006. Institute of Computing and Information Sciences, Utrecht University. Technical report UU-CS-2006-057 (2006)
27. Weerd, I., van de Versendaal, J., Brinkkemper, S.: A Product Software Knowledge Infrastructure for Situational Capability Maturation: Vision and Case Studies in Product Management. In: Proceedings of the 12th Working Conference on Requirements Engineering: Foundation for Software Quality (REFSQ'06), Luxembourg (2006)
28. Xu, L., Brinkkemper, S.: Concepts for Product Software. To appear in: European Journal of Information Systems (2007)
29. Yin, R.K.: Case study research: Design and methods (3rd edn.). Beverly Hills, CA: Sage Publishing (2003)

# ReeF: Defining a Customizable Reengineering Framework

Gemma Grau and Xavier Franch

Universitat Politècnica de Catalunya (UPC)
c/ Jordi Girona 1-3, Barcelona E-08034, Spain
{ggrau, franch}@lsi.upc.edu

**Abstract.** During their life span, organizations must adapt continuously to an always evolving context and so have to do their Information Systems and the processes around them. The scope of these changes ranges from small-scale maintenance modifications or the redefinition of some business processes to the full deployment of a new system. In all cases, the resulting Information System will seldom be built from the scratch; as even when deploying it for the first time, we may consider that it starts from the description of the current human processes. For that reason, we may consider Information System development and its evolution as a reengineering process. In this paper, we present a framework that defines the generic activity of reengineering using Method Engineering techniques. The framework is built upon existing reengineering methods from different disciplines and provides six generic phases that can be instantiated with the purpose of defining new reengineering methods.

**Keywords:** Method Engineering, Reengineering Framework, Business Process Reengineering, Software Process Reengineering, *i** Modelling and Analysis.

## 1 Introduction

Information Systems (IS) are in continuous change for various reasons. Changes that affect the system over time include requirements, technology and business processes [30]. All these changes are diverse in nature and may require different treatments according to their impact over the IS. On the one hand, the current software may have to be rebuilt, in order to create a product with added functionality, better performance and reliability, and improved maintainability [26]. On the other hand, if the changes on the business are too profound, a new IS may have to be deployed by adapting an already existing legacy system or by building a new one. Therefore, in all these situations, there are processes, artefacts and knowledge that can be taken as a starting point.

According to [31], traditional reengineering activities include: identifying, delineating, and modelling the existent process; analysing it for deficiencies; proposing new solutions; and implementing the new design in terms of new technical systems and new organizational structures. It is possible to observe that most of the methods proposed for the specification, development or acquisition of IS already support some of these activities. For instance, some IS methods explicitly mention the term reengineering in their proposal, as in [1], [2], [3], [4], [20], [21], [26], [30], [32].

J. Krogstie, A.L. Opdahl, and G. Sindre (Eds.): CAiSE 2007, LNCS 4495, pp. 485–500, 2007.

On the other hand, some other methods not defined in the reengineering context, tackle with some of those activities, among them we mention [7], [8], [10], [11], [16], [17], [19], [24], [28]. Therefore, we may consider that changes on ISs are all part of a reengineering activity, which supports our claim of IS development being treatable similarly to IS reengineering.

An observation that can be made is that each above-mentioned reengineering-related approach focuses on a particular discipline: business processes [2], [20]; software architectures [3], [26]; or software platforms [4], [32]. Despite of this diversity, there are a lot of similarities when the methods are deeply analysed. Actually, some of the differences lie more in the detail (e.g., using this or that technique) than in the rationale or the rough reengineering process. However, in some proposals, some of the reengineering activities and artefacts are not mentioned and the lack of a generic framework makes difficult to apply them through a complete reengineering process.

In order to address this problem, we propose ReeF, a customizable **Ree**ngineering **F**ramework which is based on the principles of Method Engineering [5], [23], [25], [28] with the aim of assisting on the construction of new processes based on the existing ones. ReeF has been built in two steps: first, abstracting the phases and method artefacts from the existing method PR*i*M, a Process Reengineering *i** Method [13], by using the Approach for Method Reengineering [28]; and second, generalizing the obtained phases and method artefacts by analysing other existing reengineering techniques from different domains [1], [2], [3], [4], [7], [20], [21], [26], [30], [32]. Once obtained and validated, we show an example of framework customisation by defining SAR*i*M, a method for software architectures reengineering based on *i**.

The benefits resulting from this process are twofold. In the one hand, the definition of ReeF may help to understand, reconcile, and analyse existing reengineering methods, and also to formulate new specific ones. With this aim, ReeF clearly establishes the reengineering phases and the method artefacts involved in each phase (techniques needed, modelling languages used, tool support provided, and roles). On the other hand, the abstraction and generalization mechanisms used for abstracting and generalizing ReeF from other methods (such as PR*i*M), may be applied to generate other customizable frameworks based on a different development point of view (as we have done with process reengineering).

The rest of the paper is organized as follows. In section 2 we outline the research method followed to define the framework. The PR*i*M method, upon which ReeF is based, is presented in section 3. The proposed framework is detailed in section 4 and customized in section 5 for obtaining SAR*i*M. Finally section 6 presents the conclusions and future work.

## 2 Research Method

The main purpose of this research is to define a generic framework in which existing reengineering techniques can be reconciled, adapted and analysed. As a result, new reengineering methods in different disciplines and domains can be created by derivation and combination of reusable fragments. As a result, this work is related with Method Engineering, which is the discipline that constructs new methods from parts of existing methods [5]. There are several proposals that address Method Engineering [5], [23], [25], [28], among which we remark:

- The OPEN Process Framework (OPF) [25] is a generic framework that provides a repository with a wide range of Method Components, which are different parts of existing methods described at different levels of detail that can be used for defining other methods in different domains. A Method Component can be specialized into Endeavour, Language, Producer, Stage, Work Product or Work Unit, which, in turn, can all be specialized forming a complete hierarchy of elements. The OPF repository of Method Components is very complete, thus enabling the selection of those components more suitable for the specific purposes of the method.
- The Approach for Method Reengineering [28] proposes a bottom-up process for transforming already existing methods into several pieces of *method chunks* which are stored in a *method base*. From the stored *method chunks*, assembly-based construction of methods is done by applying the following three steps [23]: method requirements specification, *method chunks* selection and *method chunks* assembly.

We have considered using the OPF approach for generating ReeF; more precisely we have studied the customizations for a Business Reengineering Project and for a Framework Project. However, in both cases, the level of detail provided in OPF is too broad for our purposes. For instance, the OPF reengineering phase description includes aspects such as management, quality, and testing; but does not include all the basic activities that we have identified in reengineering methods. Because of that, we decided to use another approach for defining our reengineering framework, but we still using OPF for assessing the analysis of existing reengineering methods, as a kind of classification schema. On the other hand, *method chunks* are specific of the method reengineered and, so, its granularity level is too detailed for being part of the generic framework. However, we can observer that it is possible to abstract and generalize the concepts of the specific *method chunks* into a set of generic *method chunks*. There are several approaches on how to document, store and reuse the different method parts [4], [5], [23], [25], [27], [28] that could be used to define and customize ReeF. However, as we use *method chunks* during the definition of the method, we keep on using them for illustrating its customization, as it is done in [23], [27], [28]. Consequently, we assume that *method chunks* are stored in a *method base*.

Taking those aspects into account, we have adopted a research method that, given an existing reengineering method, abstracts and generalizes its *method chunks*. As a result we have ReeF, a generic reengineering framework, which can be further customized by using other *method chunks* previously stored in the *method base*.

In order to abstract the initial set of method chunks using method Reengineering, we analyse PR*i*M, a Process Reengineering *i** Method [13]. We consider this method adequate as starting point because, as detailed in Section 3: 1) it is constructed after a rigorous state of the art of business process reengineering techniques; 2) it makes use of widespread techniques and artefacts in its definition instead of proposing ad-hoc ones; 3) some of the underlying ideas are applicable to contexts other than business process reengineering; 4) as authors, we have experience in applying the method and, so, access to all the components that we want to abstract onto the customizable framework which is an information sometimes difficult to obtain whilst analysing other methods.

The definition of ReeF is done in two steps: abstraction and generalization. Abstraction is the process of extracting common features from specific examples, whereas generalization is the process of formulating general concepts by abstracting common properties of instances. During the abstraction process, the phases of PR*i*M are analysed in order to synthesize its *method chunks*, following the principles given in [28]. PR*i*M is a method specific for the process reengineering domain. Thus, for obtaining a generic framework, we need to apply a generalization process over other reengineering methods from different domains. As a result, a new set of *method chunks* is obtained, with the particularity that the method artefacts (namely, the techniques, modelling languages, tool support and roles involved) are specified by stating its generic definitions instead of their particular ones. Also, special emphasis is given on the generic intention (the goal) that each *method chunk* pursues. The generic framework is then defined by analysing and reconciling all the obtained elements.

The customization of ReeF is done by applying the following steps: refinement, operationalization and combination. During refinement, the generic definitions stated in the *method chunks* of ReeF, are refined into specific ones for the domain of application. During the operationalization step, the refined statements of ReeF are used for selecting from the *method base* those *method chunks* that better accomplish a certain purpose. In order to facilitate this step, the *method chunks* can be classified according to a set of criteria [23], [27]. Finally, during combination, the selected method chunks are combined in order to obtain the new method. As we have mentioned, these steps can also be done by using other methods [5], [25].

Fig.1 presents an overview of the research method. We observe that the validation of ReeF is twofold. On the one hand, the proposed research method used for the definition of ReeF ensures that the different reengineering methods analysed can be successfully defined as instances of the framework. On the other hand, we define a new method for the domain of software architectures with the objective of validating its customization. The new method, called SAR*i*M, is then defined by customizing

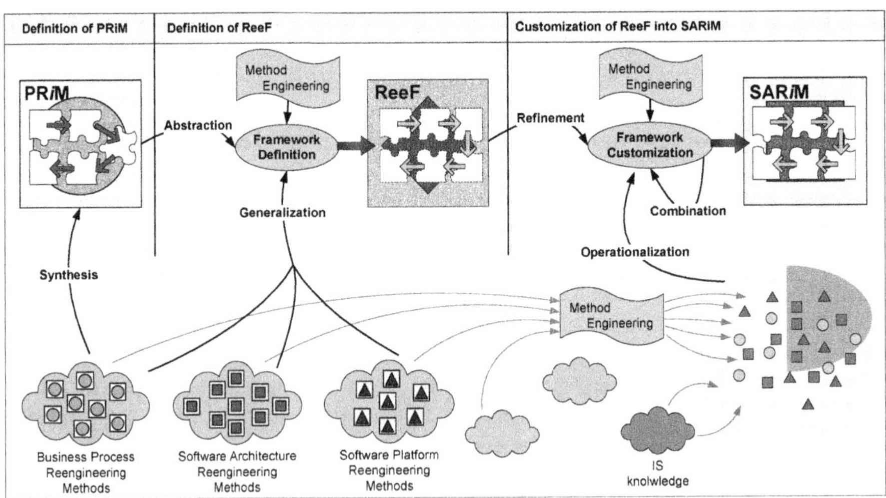


**Fig. 1.** Overview of the Research Method used for defining ReeF

ReeF, and combines *method chunks* from existing reengineering methods with specific techniques from the software architectures domain.

## 3 PR*i*M: A Process Reengineering *i** Method

In our previous research we defined PR*i*M [13], a Process Reengineering *i** Method that addresses the specification of Information Systems from the process reengineering perspective. The *i** framework [31] is a well consolidated goal-oriented approach that allows to model Information Systems in a graphical way, in terms of actors and dependencies among them. The use of the *i** framework in this context provides an appropriate milieu where the current process rationale is modelled by means of intentional concepts and the evaluation of the alternatives is done by analyzing the rationale behind the modelled intentional concepts.

Analysis and evaluation of *i** models is commonly done in a qualitative manner by using the analysis capabilities provided by the Strategic Rational Model. Instead, a goal of PR*i*M is to address the evaluation of alternatives from a quantitative point of view by applying structural metrics over the *i** models as proposed in [10], [11]. According to [9], one of the problems of the *i** framework is the repeatability when constructing the models. As repeatability is a fundamental property when applying structural metrics and it is not ensured by other *i** modelling techniques [14], the main motivation behind PR*i*M definition has been to ensure this property. Because of that, during the definition of PR*i*M we analysed several well-known business process reengineering and requirements engineering methods [12] in order to incorporate in the new method the adequate techniques, roles and artefacts. We highlight these included elements in the description of the method provided below, and summarize them in Table 1. We also remark that PR*i*M is defined upon the business process reengineering phases presented in [31] but adding a first preliminary phase for obtaining the information of the current processes.

The first phase of PR*i*M involves capturing and recording the information about the current process in order to inform further phases. The approach adopted is based on the RESCUE method [19] and, as a result, requirements engineers produce Human Activity Models (hereafter, HAM). During the second phase, the *i** model is build. In order to ensure repeatability when constructing the models, PR*i*M provides concrete guidelines that transform the information in Detailed Interaction Scripts (hereafter, DIS) to *i** elements. Thus, one of the activities the first phase is to adapt the information on the HAM to DIS. As both approaches share a common structure, simple transformation rules are provided in order to do it, and consistency checks are defined latter on the method for checking that they have been correctly applied.

In the second phase, the *i** model is built in two differentiated steps in order to distinguish the functionality performed by the stakeholders from their strategic intentionality. This approach is based on the semantically distinction of descriptive goals and prescriptive goals given in [2]. Therefore descriptive goals are modelled on the *operational i* model* by using the information in the DIS, and prescriptive goals are modelled on the *intentional i* model*. As a result of this process a *complete i* model* of the current process is obtained.

**Table 1.** Phases of PR*i*M, detailing the techniques, activities, inputs and outputs involved

| Phase | Activity | Input | Techniques | Roles | Output |
|---|---|---|---|---|---|
| **Phase 1: Analysis of the current process** | | | | | |
| | Analysis of the current process | Current process | Observation | Process analyst | Human Activity Diagrams (HAM) |
| **Phase 2: Construction of the *i** model of the current process** | | | | | |
| | Transformation | HAM | Transformation rules | *i** modeller | DIS |
| | Actor Identification and modelling | DIS | Analysis of HAM | *i** modeller | *i** model actors |
| | Building the Operational *i** model | DIS | Transformation Rules | *i** modeller | Operational *i** model |
| | Building the Intentional *i** model | Operational *i** model | Provided Guidelines | Process analyst | Intentional *i** model |
| | Checking the Complete *i** model | Intentional *i** model | Consistency checks | *i** modeller | Complete *i** model |
| **Phase 3: Generation of alternatives for the new process** | | | | | |
| | Reengineering the current process | Complete *i** model | Requirements Elicitation Patterns | Requirements engineer | Enriched *i** model |
| | Adding new actors to the process | Enriched *i** model | Analysis of the market | Process designer | Actors for an *i** alternative (one) |
| | Reallocating responsibilities | Enriched *i** model, Actors | Provided Guidelines | Process designer | Alternative *i** model (one) |
| | Checking the consistency | Alternative *i** models (all) | Consistency Checks | *i** modeller | Consistent *i** alternatives |
| **Phase 4: Evaluation of alternatives for the new process** | | | | | |
| | Choosing suitable properties | Extended *i** model | Observation of needs from model | Process analyst | Properties |
| | Defining property metrics | Properties | Definition guidelines | Process analyst | Property metrics |
| | Evaluating alternative models | Consistent *i** alt. Metrics | Evaluation principles | *i** modeller | Evaluation Results |
| | Evaluation Trade-off analysis | Evaluation results | Trade-off analysis | Process analyst | Suitable *i** model solution |
| **Phase 5: Specification of the new Information System** | | | | | |
| | Specification of the new IS. | Suitable *i** model solution | Transformation guidelines | *i** and Use Case modellers | Use Case model of new IS. |

The first activity of the third phase is to obtain the goals of the new process, which is done by using the *complete i* model* of the current process and applying KAOS [8] for analysing it. As KAOS and *i** are both goal-oriented, the acquired goals are added to the *complete i* model*, yielding to the *enriched i* model*. With the aim of satisfying these goals, several process alternatives are systematically generated by adding new *i** actors (which are mainly human, software or hardware), removing some of the existing ones and reallocating the responsibilities between them. This process is guided by the aim to satisfy the different new goals on the *enriched i* model*, which is done by applying the techniques proposed in [20]. As a result, several *alternative i* models* are produced.

In the fourth phase, the different *alternative i* models* are evaluated by applying structural metrics over them [10], [11]. Trade-off analysis is needed in order to select the most suitable solution. Finally, in the fifth phase, PR*i*M proposes the generation of the new Information System specification from the *i** model of the chosen alternative which follows the work proposed by [29].

The PR*i*M method is based on an exhaustive state-of-the-art on business process reengineering methods [12] complemented with well established requirements engineering techniques such as KAOS [8]. The use of these techniques provides an additional strength to all the phases defined on the method, and they have facilitated the development of J-PR*i*M [15], a tool that supports the application of the method. These are arguments that support using PR*i*M as starting point for formulating the framework.

Also we would like to remark the benefits of the use of *i** in PR*i*M. On the one hand, *i** supports all the phases of the method, allowing an *assembly of methods by association* [23], because no connection between the product models has to be done when combining the different *method chunks*. Actually, this also facilitates the substitution of most of the techniques applied on the phases for other *i** techniques with the same aims, without great modifications and without altering the result (e.g., the generation of alternatives can be done by using the organizational patterns proposed in [22]). On the other hand, as *i** is goal-oriented and agent-oriented, it allows reasoning at the goal and agent levels, which aligns with the strategic nature of reengineering processes. Consequently, in the *assembly of methods by integration* [23], goal-oriented and agent-oriented *method chunks* are easily adapted to represent the concepts in a unique *i** model (e.g., in phase 3, KAOS goals are represented in the *i** model).

## 4 Defining ReeF, a Customizable Reengineering Framework

In this section we explain the construction of ReeF in two differentiated processes, abstraction and generalization, starting from the PR*i*M method.

### 4.1 The Abstraction Process

In the Abstraction process we extract common reengineering features from the specific method PR*i*M. Thus, we use the Approach for Method Reengineering proposed in [28] over PR*i*M for achieving the proposed four main intentions: Define a section, Define a guideline, Identify a *method chunk*, and Define a *method chunk*. Due to the lack of space, we present directly the application of the method in our context; a complete description of the foundations can be found in [27], [28].

The PR*i*M method has a well defined process model and, so, in order to identify its sections we use the *functional strategy* in order to establish the *method map sections* from its phases. The intentions (or goals) of each phase of PR*i*M are identified and documented using the Method Reengineering suggested notation, as follows: Analyse the current process using Human Activity Modelling; Conceptualize the current process into an *i** model, Elicit requirements for the new process and explore different process alternatives based on them; Assess the generated process alternatives using evaluation techniques; and Create the specification of the new Information System.

When reviewing the guidelines associated to these intentions, we realize that the section "Elicit requirements for the new process and explore different process alternatives based on them" contains two different products that could be treated independently. Thus, we apply the *progression discovery strategy* and, as a result, the section is divided into two different ones: "Elicit requirements for the new process using a goal-oriented approach" and "Explore new process alternatives using process generation heuristics". Once the sections are defined, the guidelines indicating how to proceed to achieve the objective of each identified section are also defined by applying Method Reengineering. For instance, the *method chunk* "Explore new process alternatives using process generation heuristics" has the strategic guideline that it is shown at the bottom of Fig. 2.

The *method chunks* are identified by using a *section-based discovery strategy*. We consider that each of the identified sections represents a *method chunk* because they can be reused separately outside its original method. Actually, as PR*i*M does so, we do not consider to apply any other strategy to identify more *method chunks*. Therefore we may define them already. At the top of Fig. 2 we present the descriptor for the *method chunk* "Explore new process alternatives using process generation heuristics".

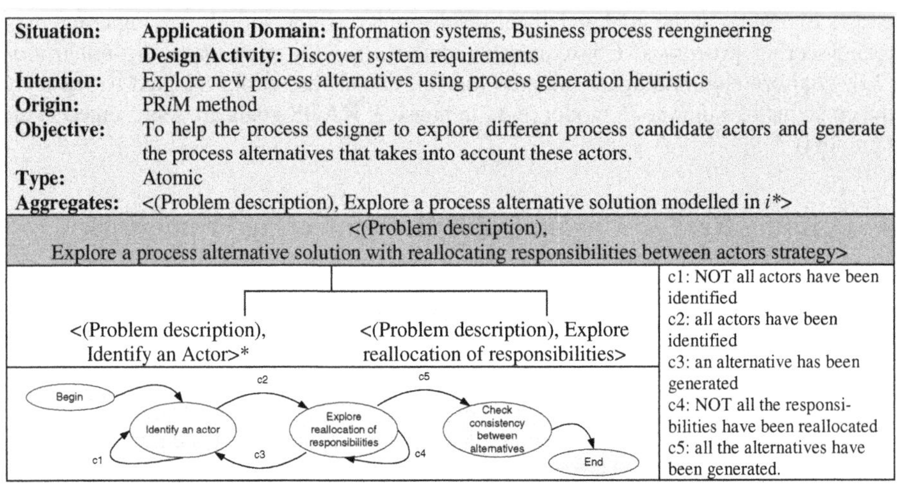


**Fig. 2.** *Method chunk* "Explore new process alternatives using generation heuristics"

Once all the PR*i*M *method chunks* are identified, we abstract their intentions and the method artefacts used and, as a result, we obtain a set of *abstract method chunks*. Table 2 shows the results of these abstractions, where we can observe that the intentions of the PR*i*M *method chunks* are written in an abstract manner in order to help further customization of the method. This is done by substituting the PR*i*M specific artefacts (techniques, modelling languages, tool support and roles) for its equivalent generic artefacts, which are written between the symbols <>. The flow of the artefacts involved in the *abstracted method chunks* shows that they are treated in a specific order, hence establishing that they are sequential. In the fourth *abstracted method chunk* of Table 2, we show how the abstracted method artefacts are documented by stating a description and some of the examples of the analysis techniques, modelling languages, tool support and roles involved. The rest of the method artefacts abstraction is straightforward. A more formal documentation of the framework could be stated by using [6]. A complete catalogue of method artefacts can be found at the OPF repository [25].

**Table 2.** Generic notation for the intentions of the *abstracted method chunks* (abridged as *amc*)

| | |
|---|---|
| **amc 1:** | **Analyse** [source] <domain artefact> **using** <analysis techniques> **obtaining** <analysed artefact> |
| **amc 2:** | **Conceptualize** <analysed artefact> **into** <model artefact> |
| **amc 3:** | **Elicit** <requirements artefact> **for the** [final] <domain artefact> **using** <elicitation techniques> |
| **amc 4:** | **Explore** [candidate] <domain artefact> **using** <generation techniques> **obtaining** [generated] <domain artefact> |
| | **Techniques:** Techniques and heuristics used to explore candidate solution artefacts (e.g., application of organizational patterns, application of architectural patterns, heuristics and guidelines for the generation of alternatives). |
| | **Modelling language:** Formalisms used to conceptualize the candidate solution artefacts (e.g., business process reengineering models, conceptual models, scenarios, architecture description languages, goal hierarchies, actor-dependency models such as *i**) |
| | **Tool Support:** Tools that aims at supporting the exploration of candidate solutions using an specific formalism (e.g., scenario generation tools, generation of alternative architectures tools) |
| | **Roles Involved:** Analyst, which is domain expert, responsible of exploring the solution artefacts (e.g., process analyst, software architectures analyst, systems analyst). |
| **amc 5:** | **Assess** [generated] <domain artefact> **using** <evaluation techniques> |
| **amc 6:** | **Create** [final]<specification artefact> **for the** [new] <domain artefact> **using** <model transformation techniques> |

## 4.2 The Generalization Process

During the generalization process we formulate general concepts by analysing the common properties of other reengineering methods. Once the initial set of *method chunks* are identified, we apply again the Approach for Method Reengineering [28] to analyse more reengineering methods in order to obtain a generalization of the process. The undertaken review includes the methods used in the definition of PR*i*M (now studied from a Method Reengineering perspective) [2], [20]; business process reengineering methods [1], [24]; architecture reengineering methods [3], [21]; and platform reengineering methods [4], [32]. As a result, we obtain the *method chunks* of these processes. In Table 3 we present an excerpt of it by showing the intentions obtained from analysing the Scenario-based Software Architecture Reengineering method [3]. We observe that each intention corresponds to an *abstracted method chunk* with only one exception: after the elicitation of the functional requirements, the method assesses the current software architecture.

**Table 3.** Intentions proposed by the Scenario-based Software Architecture Reengineering [3]

<table>
<tr><th>Method</th><th>Scenario-based Software Architecture Reengineering [3]</th></tr>
<tr><td>amc 1:</td><td rowspan="2">These method chunks are not defined, as the method establishes as its input: the source <software architecture> conceptualized into <scenarios></td></tr>
<tr><td>amc 2:</td></tr>
<tr><td>amc 3:</td><td>Elicit <functional requirements> for the final <software architecture></td></tr>
<tr><td>amc 5:</td><td>Assess <current software architecture> using <scenario-based evaluation></td></tr>
<tr><td>amc 4:</td><td>Explore candidate <software architecture> using <QA-optimizing architecture transformations></td></tr>
<tr><td>amc 5:</td><td>Assess generated <software architecture> using <scenario-based evaluation></td></tr>
<tr><td>amc 6:</td><td>This method chunk is not defined. The output of the method is: <improved architecture design></td></tr>
</table>

When analysing the *method chunks* obtained from applying Method Reengineering over all the previously mentioned reengineering methods, we observe the following:

- The analysed *methods chunks* present intentions that can be considered equivalent to the *abstracted method chunks*. For instance, all the methods share the intention of "Explore new solution artefacts", although they propose different guidelines to satisfy it.
- Not all the analysed methods present a sequential instantiation of the *abstracted method chunks*, as most of them omit some intentions. We remark that usually the omitted phases are the ones at the beginning or at the end of the process. For instance, in [3], [7], [21], the first two intentions are not mentioned as they assume that the information of the current situation is already studied and modelled for their purposes, but they all generate and evaluate candidate software architectures.
- Some of the studied methods propose a preliminary evaluation of the modelled process before the elicitation of new requirements. For instance, [3], [24], [32].
- Some of the methods allow iteration between the phases, allowing eliciting new requirements, exploring new solutions and evaluating them several times before choosing the final solution [3], [4], [7], [24].
- All the analysed methods have the *abstracted method chunks* for exploring and assessing the solution artefacts. However, in some of the methods the assessment is implicit in the exploration of the solutions as if it were a cycle between both phases. For instance, in [4] and [21] the designer generates the solutions according to its own criteria, which means an implicit evaluation of the current solution.
- All the studied methods have their intentions executed in the sequential order established by the *abstracted method chunks*. An extreme example of this is the work proposed in [24] where different reengineering processes can be generated from applying a set of map strategies, and the generated methods are compliant with ReeF.
- The method artefacts obtained in the studied *method chunks* are equivalent to those abstracted in ReeF and, although the proposed techniques come from different domains, their intentions and roles are an instance of the ones abstracted.
- All the methods use a modelling language for communicating between its phases. The common modelling languages are visual models (e.g., Use Case Maps [7], enterprise business process models [24]) and structured text (e.g., scenarios [3]).

Taking those considerations into account, we generalize the *abstracted method chunks* obtained in ReeF and we establish the following restrictions:

- There is a sequential order within the different *abstracted method chunks*, but it is possible to omit the ones at the beginning or at the end, as some methods do.
- It is possible to assess the source artefact after it is modelled, in order to inform the elicitation of requirements.
- It is possible to iterate between the phases: the evaluation of alternatives can inform a new elicitation of requirements; new alternatives are generated and evaluated; and so on and so forth, until a final solution is found.

As a result, the ReeF framework is composed by six phases, which are shown in Fig. 3. The blue arrows show the sequence of execution of the phases according to the *abstracted method chunks* allowing the diversions and iterations previously mentioned. The framework defines, for each of these phases, the work products needed (inputs) and produced (outputs) during the phases, the techniques (including the activities for obtaining the work products, the transformations between models and the tool support used) and the roles that are involved. As the framework is generic, customization has to be applied in order to instantiated it. We remark that during customization it is possible to define the new method by using different techniques for each of its iterations if needed.

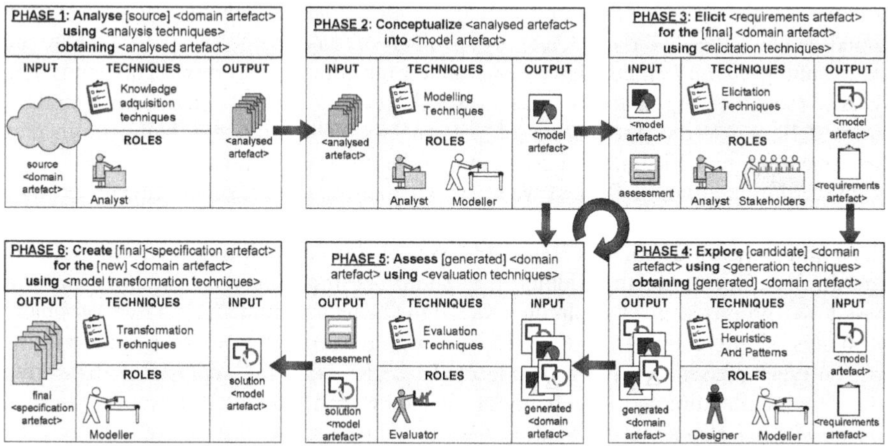


**Fig. 3.** Phases, inputs, outputs, techniques and roles abstracted in ReeF

## 5 Customizing ReeF into SAR*i*M

As an example of application of the framework we propose the definition of SAR*i*M, a Software Architecture Reengineering *i** Method. The aim of SAR*i*M is to adapt the experience in using PR*i*M to the domain of software architectures. The use of *i** as a modelling language has several advantages. On the one hand, *i** allows to represent functional and non-functional requirements as well as business goals at the same level, thus bridging the gap that is usually found between requirements and architectures [16]. On the other hand, *i** has already been successfully used for the representation of software architectures [18]. As a result, the customization strategy followed in the SAR*i*M case has prioritized operationalization over refinement and combination (see Fig. 1).

**Refinement.** The generic intentions (or goals) defined in the *abstract method chunks* of ReeF are refined for the particular domain of software architecture in order to establish the main objectives to be satisfied in the new method, see Table 4. We observe that only the desired artefacts are refined and that the precise technique may still be undefined. As we do in PR*i*M, we do not consider the evaluation of the current software architecture before the elicitation of the new requirements.

**Operationalization**. Once the intentions are defined, we search into the *method base* those *method chunks* that better accomplish the intention. We propose to classify the *method chunks* in the database according to three dimensions: intention they support, domain they are designed for, and modelling language used. The reason for taking considering the modelling language is that, if the *method chunks* do not share the same modelling language, a transformation technique has to be applied between them, so it is recommended to take this aspect into account in order to facilitate further steps. However, other classification criteria for the *method base* can be used [3], [23].

Based on the refined intentions, the search for the appropriate *method chunks* in the *method base* is facilitated, as the set of candidate elements is delimited. We remark that the *method base* is not complete and not all the *method chunks* required may be found there. If this is the case, a study of other suitable methods has to be done and the resulting *method chunks* have to be added into the *method base*. This study may include reengineering methods but also well-know requirements engineering methods or guidelines for the application of patterns that, although not being defined as reengineering methods, may support some of the proposed phases.

In the third column of Table 4 we show whether the *method chunks* available in the method base are supported or not by PR*i*M. In the example presented in Table 3 we show the intentions of the *method chunks* for the Scenario-based Software Architecture Reengineering Method [3]. There, the fourth *method chunk* is scenario-based and proposed a set of architecture transformation guidelines based on quality attributes. As this intention satisfies the one we have refined in SAR*i*M, we use it.

The other phases that are not supported by PR*i*M are the analysis of the current software architecture and the elicitation of requirements for the future one. As there are no *method chunks* in the method base to support those phases, we analyze other methods for doing it. More precisely, we have searched in the field of requirements engineering and we have selected the Architecture Reconstruction Method [17] for the recovery and analysis of the current architecture, and the CBSP method [16] to be adapted to the *i** notation for the elicitation of the new requirements.

**Combination.** Once the *method chunks* are selected, method engineering techniques for assembling can be applied [5], [23], [25], [27], [28] in order to obtain the final method. The combination of the *method chunks* is out of the scope of this work, as it

**Table 4.** Refinement step, customazing ReeF in the domain of Software Architectures

| Generic Intention in ReeF | Refinement into SAR*i*M | *Method chunks* Operationalization |
|---|---|---|
| **Analyse** [source] <domain artefact> **using** <analysis techniques> **obtaining** <analysed artefact> | **Analyze source** software architecture **using** *<architecture analysis technique>* | Not supported by PR*i*M: operationalized by the Architecture Reconstruction Method. |
| **Conceptualize** <analysed artefact> **into** <model artefact> | **Conceptualize** the software architecture **into** an *i** model | Supported by PR*i*M: needs previous transformation of the results into DIS |
| **Elicit** <requirements artefact> **for the** [final] <domain artefact> **using** <elicitation techniques> | **Elicit** quality requirements **for the final** software architecture **using** *<elicitation technique>* | Not supported by PR*i*M: operationalized by the CBSP method. |
| **Explore** [candidate] <domain artefact> **using** <generation techniques> | **Explore candidate** software architectures **using** *<generation techniques>* | Not supported by PR*i*M: use of the Scenario-based Software Architecture Reengineering Method. |
| **Assess** [generated] <domain artefact> **using** <evaluation techniques> | **Assess generated** software architecture **using** *i** structural evaluation techniques. | Supported by PR*i*M: needs the generated architectures to be represented as *i** models. |
| **Create** [final]<specification artefact> **for the** [new] <domain artefact> **using** <model transformation techniques> | **Create final** specification **for the new** software architecture **using** *i** to use cases transformation techniques. | Supported by PR*i*M: can be applied directly from the previous phase. |

has already been addressed in [23], [27]. We just remark that, following the criteria in [23] all the *method chunks* are combined following the established order and using the *assembly by association*, where transformation techniques are applied in order to transform *i** models to scenarios. In the *method chunks* for requirements elicitation and architectures generation, we apply an *assembly by integration*, as the tight link between *i** and requirements engineering techniques, facilitates it.

## 6 Conclusions and Future Work

In this paper we have argued that the evolution of Information Systems very often leads to a reengineering activity. There are a lot of methods proposed in the literature at different levels (business processes, software processes, software architectures, etc.). This methods support reengineering both consciously, by applying the term reengineering in its proposal; and unconsciously, by mentioning the phases that characterize reengineering. However, as far as we know, there is not a common framework to reason about reengineering and this has been the motivation of our proposal. ReeF has been defined following the principles of Method Engineering because this technique is specially well-suited when defining new methods based on existing ones. As a result, the advantage of applying the framework is twofold:

- It provides a common umbrella under which the different existing reengineering proposals may be analysed, compared for possible adoption, customized to particular contexts and even composed to deal with reengineering at different levels. In particular, an existing method could be enlarged to deal with some activity not covered in its definition, or some technique may be changed with some other identified as similar.
- It allows formulating new reengineering approaches starting from that framework, not only facilitating that task, but also providing an ontology of reference and the possibility of reusing methods, techniques, models and tools from a common experience base.

ReeF is not intended to deliver an exhaustive catalogue with all the possible phases and techniques, but instead it serves as a generic, customizable framework, which provides, among other things, different levels of abstraction and the possibility of choosing between different characteristics. More precisely, we argue that the framework satisfies the following guiding principles proposed by the OPF [25]:

- **Flexibility.** In order to allow maximum flexibility when customizing, the phases of ReeF provide: atomicity, in the way that the activities it proposes are related to only one concept of the reengineering activities; optionality, certain phases can be avoided if the customization requires so; and iteration, in those methods that require several iterations of some of the phases.
- **Standardization.** Reef uses the most common terminology in the business process reengineering field. For techniques, roles and activities it uses the already standardized terminology and concepts coming from the OPF.
- **Completeness.** ReeF is complete in the sense that it includes all the elements that may be needed in a reengineering process. Although it not provides a complete

repository of elements for instantiate the framework, it provides techniques for constructing this repository.

- **Openness.** ReeF remains open in the sense that there is not a closed list of elements and also because it is not necessary to instantiate all those elements, allowing the method engineer to customize them accordingly to its goals.
- **Reengineering Best Practices.** ReeF is based on the abstraction and generalization of well-know reengineering methods and related requirements engineering techniques.
- **Usability.** ReeF facilitates usability by providing guidelines for its customization, as it is shown in the customization of ReeF into SAR*i*M.
- **Reuse.** The framework supports reuse of methods providing the context where to customize the method and a set of elements as examples.

Further work will involve the application of ReeF on the combination of reengineering methods that work in different domains (e.g., business process reengineering and architecture reengineering). This includes the definition of more *method chunks* and method artefacts into the *method base* and how to document and classify them in order to facilitate their customization. We are mainly interested in the use of *i** and the *method chunks* proposed in PR*i*M as a basis for this process and we want to adapt J-PR*i*M [15] in order to provide tool support for the whole process.

## Acknowledgements

This work has been partially supported by the CICYT programme, project TIN2004-07461-C02-01. Gemma Grau work is supported by an UPC research scholarship.

## References

1. van der Aalst, W.M.P., van Hee, K.M.: Framework for Business Process Redesign. In: Proceedings of the Fourth Workshop on Enabling Technologies: Infrastructure for Collaborative Enterprises, pp. 36–45 (1995)
2. Antón, A.I., McCracken, W.M., Potts, C.: Goal Decomposition and Scenario Analysis in Business Process Reengineering. In: Proceedings of CAiSE 1994. LNCS, vol. 811, pp. 94–104. Springer, Heidelberg (1994)
3. Bengtsson, P., Bosch, J.: Scenario-based Software Architecture Reengineering. In: Proceedings of the 5th International Conference on Software Reuse, pp. 308–317 (1998)
4. Bouillon, L., Vanderdonckt, J., Chow, K.C.: Flexible Re-engineering of Web Sites. In: Proceedings of the 9th International Conference on Intelligent user interface (2004)
5. Brinkkemper, S., Saeki, M., Harmsen, F.: Assembly Techniques for Method Engineering. In: Proceedings of CAiSE 1998. LNCS, vol. 1413, pp. 381–400. Springer, Heidelberg (1998)
6. Brinkkemper, S., Saeki, M., Harmsen, F.: A Method Engineering Language for the Description of Systems Development Methods. In: Proceedings of CAiSE 2001. LNCS, vol. 2068, pp. 473–476. Springer, Heidelberg (2001)
7. de Bruin, H., van Vliet, H.: Scenario-based Generation and Evaluation of Sofware Architectures. In: Proceedings of the Third International Conference on Generative and Component-Based Software Engineering, LNCS, vol. 2186, pp. 128–139. Springer, Heidelberg (2001)

8. Dardenne, A., van Lamsweerde, A., Fickas, S.: Goal-directed Requirements Acquisition. Science of Computer Programming 20(1-2), 3–50 (1993)
9. Estrada, H., Martínez, A., Rebollar, O., Pastor, J.: An Empirical Evaluation of the i* in a Model-Based Software Generation Environment. In: Proceedings of CAiSE 2006. LNCS, vol. 4001, pp. 513–527. Springer, Heidelberg (2006)
10. Franch, X.: On the Quantitative Analysis of Agent-Oriented Models. In: Proceedings of CAiSE 2006. LNCS, vol. 4001, pp. 495–509. Springer, Heidelberg (2006)
11. Franch, X., Grau, G., Quer, C.: A Framework for the Definition of Metrics for Actor-Dependency Models. In: Proceedings of RE 2004, pp. 348–349
12. Grau, G.: State of the Art for the Systematic Construction and Analysis of i* Models for assessing COTS-Based Systems Development. Research Report LSI-06-38-R. Available at: http://www.lsi.upc.edu/ techreps/files/R06-38.zip
13. Grau, G., Franch, X., Maiden, N.A.M.: A Goal Based Round-Trip Method for System Development. In: Proceedings of REFSQ 2005, pp. 71–86 (2005)
14. Grau, G., Cares, C., Franch, X., Navarrete, F.J.: A Comparative Analysis of i* Agent-Oriented Modelling Techniques. In: Proceedings of SEKE 2006, pp. 657–663 (2006)
15. Grau, G., Franch, X., Ávila, S.: J-PRiM: A Java Tool for a Process Reengineering i* Methodology. In: Proceedings of RE 2006, pp. 352–353 (2006)
16. Grünbacher, P., Egyed, A., Medvidovic, N.: Reconciling software requirements and architectures with intermediate models. Software and Systems Modeling 3(3), 235–253 (2004)
17. Guo, G.Y., Atlee, J.M., Kazman, R.: A Software Architecture Reconstruction Method. In: Proceedings of WICSA 1999, pp. 15–34 (1999)
18. The i* wiki at: http://istar.rwth-aachen.de/ Last Accessed : November 2006
19. Jones, S., Maiden, N.A.M.: RESCUE: An Integrated Method for Specifying Requirements for Complex Socio-Technical Systems. Book chapter in Requirements Engineering for Sociotechnical Systems, Idea Group Inc. (2004)
20. Katzenstein, G., Lerch, F.J.: Beneath the Surface of Organizational Processes: A Social Representation Framework for Business Process Redesign. ACM Transactions on Information Systems 18(4), 383–422 (October 2000)
21. Kim, M., Lee, J., Kang, K.C., Hong, Y., Bang, S.: Re-engineering Software Architecture of Home Service Robots: A Case Study. In: Proceedings of ICSE 2005, pp. 505–513 (2005)
22. Kolp, M., Giorgini, P., Mylopoulos, J.: Organizational Patterns for Early Requirements Analysis. In: Proceedings of CAiSE 2003. LNCS, vol. 2681, pp. 617–632. Springer, Heidelberg (2003)
23. Mirbel, I., Ralyté, J.: Situational method engineering: combining assembly-based and roadmap-driven approaches. Requirements Engineering, 11(1) (2005)
24. Nurcan, S., Rolland, C.: A multi-method for defining the organizational change. Information and Software Technology 45(2), 61–82 (February 2003)
25. The OPEN Process Framework (OPF) at: www.opfro.org. Last accessed: November 2006.
26. Pressman, R.S.: Software Engineering: a Practitioner's Approach. In: International Edition, 6th edn. McGraw-Hill, New York (2005)
27. Ralyté, J.: Ingénierie des méthodes par assemblage de composants. Thèse de doctorat en informatique de l'Université Paris 1 (Janvier 2001)
28. Ralyté, J., Rolland, C.: An Approach for Method Reengineering. In: ER 2001. LNCS, vol. 2224, pp. 471–484. Springer, Heidelberg (2001)
29. Santander, V.F.A., Castro, J.F.B.: Deriving Use Cases from Organizational Modeling. In: Proceedings of RE 2002, pp. 32–39 (2002)

30. Smith, J.D., Hybertson, D.: Implementing Large-Scale COTS Reengineering within the United States Department of Defense. In:Proceedings of ICCBSS 2002. LNCS, vol. 2255, pp. 243–256. Springer, Heidelberg (2002)
31. Yu, E.: Modelling Strategic Relationships for Process Reengineering, PhD. thesis, University of Toronto (1995)
32. Zhang, W., Jarzabeg, S., Loughran, N., Rashid, A.: Reengineering a PC-based System into the Mobile Device Product Line. In: Proceedings of the 6th International Workshop on Principles of Software Evolution, pp. 149–160 (2003)

# Publishing and Discovering Information and Services for Tagged Products

Christof Roduner and Marc Langheinrich

Institute for Pervasive Computing, ETH Zurich, 8092 Zurich, Switzerland
roduner@inf.ethz.ch, langhein@inf.ethz.ch

**Abstract.** Radio frequency identification (RFID), and more recently the development of Near Field Communication (NFC) technology, have popularized the idea of linking real-world products with online information and services. Apart from early prototypes, however, the benefits of such automated identification technologies have so far been mostly available to industry, rather than consumers. With the next generation of mobile phones capable of reading both traditional bar codes through their integrated cameras, as well as RFID tags using the NFC standard, end-users themselves could take full advantage of such ubiquitous identification labels, given novel information architectures that go beyond simple web pages or industrial enterprise resource planning (ERP) systems. This paper presents an open lookup infrastructure that allows commercial, public, and private entities to easily provide information and services associated with tagged items, thus facilitating the rapid development and deployment of applications based on everyday products.

## 1 Introduction

The idea of linking information and services to physical objects has been investigated in many research projects, either using printed markers [1,2,3,4,5], embedding RFID tags [6,7,8], or even by attaching small infrared beacons [9]. In its simplest form, product identification technology has been widely used as early as in the mid-1970s, when bar code labels began to speed up the checkout process in supermarkets. Today, bar codes have become truly ubiquitous, forming the backbone of many automated processes, such as in airline ticketing and baggage handling, in libraries and video rental shops, in hospitals, and – most of all – in industrial supply chain management.

During the past few years, radio frequency identification (RFID) labels have gradually begun to replace traditional bar code labels, as they offer two distinct advantages: Firstly, RFID labels do not require a line of sight between a reader and a tag, thereby allowing large numbers of tags to be read quickly. Secondly, traditional one-dimensional product bar codes (so-called EAN or UPC bar codes)[1] can only be used to identify products at class-level, while RFID

[1] EAN stands for *European Article Number*, UPC for *Universal Product Code*. They represent the official product identification codes for European and North American products, respectively.

J. Krogstie, A.L. Opdahl, and G. Sindre (Eds.): CAiSE 2007, LNCS 4495, pp. 501–515, 2007.

tags allow the identification of individual items (as RFID offers more digits in a smaller area), thus allowing for more detailed product tracking capabilities.

However, so far the benefits of such identification technologies – be it bar codes or RFID tags – have mainly been limited to industry, i.e., manufacturers, distributors, wholesalers, and retailers, who were able to automate many of their logistical processes. This situation may change in the near future, as modern mobile phones are increasingly able to directly read identification tags found on consumer products: Mobile phones equipped with a *Near Field Communication* (NFC, see `www.nfc-forum.org`) module are already able to read specific types of RFID tags, while recent camera-equipped models can easily decode traditional EAN or UPC bar code symbols found on virtually all consumer goods [10].

While the above-cited projects certainly offer a wide variety of applications for tagged products, they nevertheless assume monolithic, centrally administrated services – such as calling up an online dictionary when putting a bound dictionary on the desk [8], opening a product's web page [9], or launching an application-specific user interface [4]. However, the general availability of information tags and corresponding reader devices opens up the possibility for novel and innovative applications that cannot be planned for. Having to install and run separate applications and infrastructures for each of the potentially available services for a product (e.g., a price comparison service, allergy warnings, a calorie calculator, or warranty information) would soon overburden users, application developers, and system administrators.

Ideally, an open service infrastructure would allow any party, e.g., manufacturers, consumer interest groups, governmental agencies, or even enthusiastic end-users, to dynamically add services to a specific product or product group, which could then be presented to and selected by consumers right when they scan a product. This paper presents our *open lookup infrastructure* for tagged items, which allows

1. manufacturers, third-parties, and individuals to publish product-specific resources (i.e., information and services), and
2. consumers to dynamically find and use these resources.

We begin by describing a set of envisioned scenarios and an analysis of the corresponding requirements for our lookup infrastructure in Section 2. The overall architecture of our system is presented in Section 3, with implementation details given in Section 4. Section 5 concludes with a discussion of three prototype applications that we built on top of our infrastructure, demonstrating the value and feasibility of our approach.

## 2 Application Scenarios and Requirements

Augmenting physical products for end-user lookup is an attractive option, especially for manufacturers. A frozen food company could for example differentiate its frozen spinach by providing an instant "recipe of the week"-suggestion, which customers would be able to access simply by pointing their mobile phone at the product. At the same time, a consumer interest organization could provide background information about the product's health benefits, e.g., praising it for the organic fertilizer that was used for it, or maybe warning consumers of genetically modified ingredients. Another example scenario for end-user lookup could be a faulty appliance, such as a printer or a coffee maker, which would provide a diagnostic code over an integrated NFC interface. By touching the appliance with a mobile phone, users could pick up this diagnostic information and receive instructions on how to resolve the problem. Alternatively, a list of nearby repair centers could be displayed. For missing consumables (e.g., printing paper, coffee capsules, or filter units), third party stores could offer users one-click reordering options. Last but not least, tagged products could also provide machine-readable instructions for other appliances, which would, e.g., allow a microwave oven to prepare a frozen meal, or a washing machine to warn users when the wrong temperature for a certain garment is selected.

Based on these scenarios, we can derive a number of high-level requirements for an infrastructure that should facilitate the discovery of resources (i.e., information and services) associated with a physical product.

**Publication, search, and retrieval of resources.** In general, many resources may be linked to a single tagged product at the same time. Our infrastructure must therefore provide mechanisms to store and find these resources. At the same time, an application might want to limit or focus the resources returned when looking up information and services associated with a given tagged product. Such search restrictions could be based on certain topics (e.g., "health aspects") or certain types of resources to look up (e.g., "expiration date"). Furthermore, the concept of context [11] (e.g., the location or status of the appliance the user interacts with) should be supported as a search criterion.

**Openness.** As the example scenarios illustrate, a range of stakeholders could have a potential interest in associating resources with a product. We therefore want to allow both product manufacturers and third-parties (such as advocacy organizations, competitors, or individuals) to be able to publish resources for a particular product. Similarly, our system should also allow the sharing of publicly available resources. In line with Lessig's "Creative Commons" approach [12], we expect the general availability of resources and a supporting infrastructure to prompt the development of numerous innovative applications, as is the case with current Web 2.0 mashups.

**Extensibility.** In our application scenarios above, resources can be of very different types. Our infrastructure should thus not try to define a limited set of foreseen resource types and their uses. Rather, it should provide extension points that allow third-parties, be it individuals or industries, to come up with their own resource types that can be shared using our system. Additionally, it must be possible to easily integrate existing information systems, such as enterprise resource planning (ERP) systems, to make their data accessible as resources.

**Lightweight and secure architecture.** Most of the discussed applications run on either mobile devices or embedded systems. Due to the resource constraints

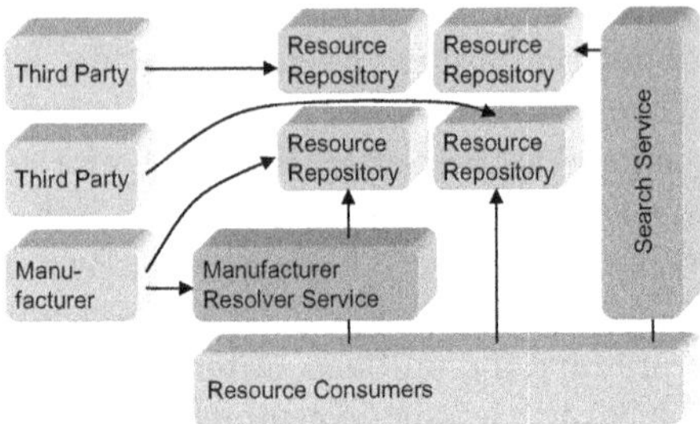


**Fig. 1.** *Open lookup infrastructure.* The center of our architecture are *resource repositories* containing *resource descriptions.* In order to find one or more repositories, given a particular product tag, mobile devices or stationary appliances either access a known repository (e.g., of their favorite consumer interest group), use the *manufacturer resolver service,* or query generic *search services.*

typically found on these platforms, we need to employ lightweight protocols in our lookup infrastructure. The open nature of our system mandates the use of security mechanisms. For example, in certain applications users must be able to determine the authenticity of a party providing a resource.

Resource discovery, i.e., finding resources by specifying a set of desired attributes in a distributed environment, has been an active field of research. While many of the protocols in this domain, such as Jini [13], UPnP [14], SLP [15], and Konark [16], focus on ad-hoc networks and do not scale beyond single communities, our application scenarios span larger networks (i.e., the Internet). While INS/Twine [17] seems to be well suited for large networks, it gives resource providers no control over where their resources are stored, which would be a problem for tagged product resolving, as it gives resource providers no control over service quality and cost sharing. UDDI [18] is a discovery service used in the domain of web services, so it should in principle scale well and offer finer control options. However, UDDI is not well suited for the lightweight lookup of simple and small bits of information. Very much related to our work is EPCglobal's EPCIS [19] – a standard for sharing product related information. However, its applicability is limited to the logistics domain. In summary, none of these existing solutions meet our requirements as discussed above.

## 3 Architecture

In this section, we discuss the architecture and four core concepts of our lookup service: resources and their descriptions, resource registries, a manufacturer resolver service, and search services (see Figure 1).

### 3.1 Resources and Resource Descriptions

Resources are at the core of our system. They offer information on, or services for, a physical product. Typical examples for resources range from a simple website to complex web services. Resources can be provided by the original product manufacturer or any other party. Resource consumers can be product owners, business partners, or appliances. For every resource, a resource provider must create a *resource description* that specifies all the metadata that is needed to consume the information or service.

Figure 2 shows an example of a resource description. A resource description includes the following main elements:

- The *resource ID* element is a pseudo-random value that serves as a globally unique identifier (GUID) for the resource.
- The *tag ID* element denotes the identifiers of those tags on physical products that a resource is associated with. The tag ID can specify a product at an item- or class-level. Different numbering schemes, such as EPC[2] and EAN/UPC, are supported. Note that a resource can carry several tag IDs and thus apply to several products. This can be helpful, e.g., when the same (or similar) product is sold under different identifiers.
- The *profile* element can be used to express that the resource adheres to the syntax and semantics that are defined in a certain profile. Typically, a profile will be defined by an industry (e.g., in a standardization group). The food industry could, for example, specify in a profile how the expiration date of a product is to be represented in a resource. Profiles are essential in cases where a resource is not interpreted by humans, but processed by an appliance.[3]
- The *url* element points to the actual resource (e.g., a website). Alternatively, the resource can be stored directly in the *data* element if it is relatively small (e.g., a product's expiration date), which avoids an additional round-trip. The syntax and semantics of the data available via either the *url* or *data* element are defined by the resource's profile as indicated in the *profile* element.
- If specified, the *context* field defines in which situation the resource is relevant. In order to enable interoperability, we predefined the following context elements that can be used to restrict a resource's applicability: time (date, time, weekday), location (coordinates, city, country), and status (expressed as a simple string) of the appliance the user interacts with. Note that this list is easily extensible by resource providers. Exact values, value ranges, and regular expressions are supported for each context element.
- The *title* and *description* elements describe the resource in natural language.
- Finally, the resource provider can digitally sign the resource description using the optional *signature* element.

Figure 2 shows an example of a resource description, in this case describing the expiration date of a particular bottle of milk. Notice that the example is

[2] EPC stands for *Electronic Product Code* and is the designated, global successor to both the EAN and UPC numbering scheme. It is administered by EPCglobal.

[3] Note that this element does not actually contain a syntactical or semantical description, but merely serves as an identifier for a format agreed upon by participants, similar to the *Content-Type* field in HTTP.

```
resource id:  f5f7305bf097af39c68b790d817d7889f788f222
     tag id:  urn:ean.ucc:7610200337481
    profile:  http://foodindustry.org/profiles/expiration-date/
        url:  (empty)
       data:  2007-05-31
    context:  (empty)
      title:  Expiration date
description:  Expiration date for OrganicMilk, 1 liter
  signature:  (empty)
```

**Fig. 2.** *Example resource description.* Descriptions can be expressed in various formats, e.g., XML or even binary, depending on the particular communication and storage needs of a product (example given in an abstract format).

given in a generic format, which in practice can be instantiated in a number of formats, such as XML or even binary form, depending on the particular use case. Also, resource descriptions for food products might equally well be entire data sets (e.g., expiration date, allergy information, country of origin) instead of just a single data item (e.g., expiration date) as in the above example – this can be standardized as needed by the various standard bodies.

### 3.2 Resource Repository

The resource repository is responsible for storing resource descriptions and making them available to resource consumers. Resource repositories can be deployed by any party interested in offering resources, such as a manufacturer or an advocacy group. In this way, a single resource repository typically contains the descriptions of resources that are thematically related. Operators can flexibly configure access restrictions to their resource repositories. For example, a manufacturer will in most cases run a read-only repository, while a community-operated product reviews repository might allow anyone to add or even update resource descriptions (very much like today's Wikis). The same applies to the querying side, where a consumer reviews publisher might limit access to its repository to paying customers only.

The three basic operations offered by a repository are `RegisterResource`, which is used to publish a resource description, `RemoveResource` to delete a published description, and `LookupResource`, which returns the descriptions of those resources matching the query conditions provided by the caller. A query can consist of up to four elements:

- The *tag ID* element must be provided to denote the product for which resources are looked up. A lookup can be performed at both class- and item-level.
- A *profile* element can be indicated to only fetch resources adhering to it.
- A *search term* element can be specified to restrict the resulting resources based on their textual description.

```
     tag id:  urn:epc:id:sgtin:0652642.800031.400
    profile:  http://appliances.org/troubleshooting-hints/
search term:  (empty)
    context:  status=E683[hint]
```

**Fig. 3.** *Example lookup request sent to a resource repository.* Based on a particular printer status (as sent through the printer's NFC interface), a user could query directly for information on a particular printer in the context of a "status=E683" code.

- Using the *context* element, the caller can specify an arbitrary number of context values. Each value must be marked as either a *hint* (favoring resources with a matching context element) or a *requirement* (excluding resources with no matching context element).

An typical lookup request is shown in Figure 3. It shows a request as it could, for example, be sent to a printer manufacturer's resource repository in order to obtain troubleshooting instructions when the printer is in a malfunctioning state. The printer's status code is read by the mobile phone's NFC module and used as context information to narrow down the lookup.

A resource repository can also be configured to allow user feedback on resources. The incorporation of feedback allows community-based applications where the quality of content is controlled by users submitting confidence values for resources. At the moment, we only provide the `SendBinaryFeedback` operation, which can be called by users to express their approval or disapproval of a resource. The order in which resource descriptions are returned by the repository depends on these ratings. Finally, the resource repository can be configured to synthesize resource descriptions of a specific profile using custom-built *wrappers*. Wrappers can be used to integrate existing information systems, such as an ERP, into the lookup infrastructure.

Note that resource repositories are in principle no different from traditional Web servers. Therefore, the same well-established mechanisms for achieving security, reliability, and scalability can be used. For example, a repository could be replicated and made accessible through a load-balancer that routes traffic according to the individual repositories' availability and load.

### 3.3 Manufacturer Resolver Service and Search Service

In order to make use of resource descriptions, users must be able to locate the resource repositories containing them. This is the task of the *manufacturer resolver services* and *search service*. They connect a product EPC or EAN/UPC to a resource repository where this product's resource descriptions can be found.

The use cases in which the various deployed resource repositories are accessed by potential resource consumers can be divided into four groups. In the first group, only the product manufacturer's repository is of interest. An example for such a case is a washing machine that checks the handling instructions of every piece of clothing put into it. In the second group, there is a single repository that is used for every lookup. An example for this case is an application that allows a user to check prices offered by other dealers for a physical product at hand. In the third group, a lookup is performed in several repositories at the same time. An example for such a case is a browser application that lets users specify a number of repositories operated by interest groups (e.g., environmental, political, etc.) they care about. The browser would then, for example, display all reviews regarding a product that can be found in the repositories relevant to the user's interests. We envision repository directories similar to the Dmoz Open Directory Project (`www.dmoz.org`) from which users can pick the repositories they find interesting. In the fourth group, a user wants to search all repositories for resources associated with a given product. This case comes into play when no relevant resources can be found in the repositories the user has registered in his or her browser. In this case, the consumer would simply query his or her favorite search service for relevant repositories. It is clear from these considerations that the architecture needs to include both a *manufacturer resolver service* that links a tag ID to the manufacturer's resource repository and a *search service* to find resources across the boundaries of single repositories.

Why is there only a resolver service for manufacturers? Why not for distributors, vendors, or consumer interest groups? After all, the example scenarios in Section 2 above illustrated that a wide variety of parties might want to offer their descriptions to consumers, each for an equally valid reason. The question of who gets to supply information to a product, i.e., who gets to "define" its properties, is actually highly political. Our system uses a pragmatic approach, inspired both by technology and legal realities. Manufacturers already play a special role in the life of a product. They are responsible for its safety, they supply manuals, organize warranties and repairs, and often also handle its recycling. In many scenarios, manufacturers thus will be legally the main, if not the only, authoritative source for information. From a technical point of view, manufacturers are also much easier to localize, given their (industrial) ID. This is because the current EPC standard (and, to some extent, also the EAN/UPC standard) contain special mechanisms to quickly identify a product's manufacturer from an EPC or EAN/UPC code. Our manufacturer resolver service makes use of this mechanism (see Section 4 below for details), thus ensuring that users can always locate the repository of a product's manufacturer.

All other information and service providers are harder to identify and find. While one could conceive a central registry where all repositories would be registered, this would violate our *openness* and *extensibility* principles set forth in Section 2. Instead, we decided to complement our manufacturer resolver with an orthogonal, decentralized, search-based approach, building on existing web search technology. Just as today's web spiders and robots, specific resource search services would crawl repositories, create an index, and answer search queries. A query passed to a search service's `Search` operation consists of the same four elements (tag ID, profile, search term, context – see above) as a `LookupResource` request sent to a single resource repository.[4] Of course, users can also directly access repositories, e.g., of their favorite product review magazine, by manually entering its address, by receiving the address via Bluetooth or SMS, or by finding it in a directory of resource repositories.

### 3.4 Deployment and Use

How would these architectural parts be used to deploy and/or make use of individual product descriptions? This depends on the individual stakeholder.

A *manufacturer* would begin with setting up a public, read-only resource repository, e.g., using an add-in to a standard web server. It would then create resources for each of its products – either informational resources such as web pages or user manuals, or service resources, such as a recipe service or a diagnostics program – and prepare corresponding resource descriptions for each of these resources. These would be entered into its resource repository, which in turn would be registered with the manufacturer resolver service.[5]

A *third-party* wishing to provide information for a certain product (e.g., an advocacy organization or even a governmental agency) would start out similarly. After setting up a repository, creating a number of resources and publishing their descriptions in the repository, however, a third party would need to advertise this repository to potential users (as it cannot make use of the manufacturer resolver service). Instead, it would register its resource repository with a search service or repository directory (similar to Yahoo or the Dmoz Open Directory Project), and/or communicate its repository URL to end-users through traditional advertising, e.g., TV, SMS, and print media.

Without any special configuration, *end-users* can always contact the manufacturer's resource repository, which can be found via the manufacturer resolver service, in order to retrieve a list of "official" resources offered for a product. Similarly, they can use a search service to find resources available from third parties that have registered their repository with the search service. Alternatively, they can manually configure resource repositories that they find especially interesting, using the above mentioned out-of-band advertising mechanisms.

As with the World Wide Web, the cost of running our infrastructure is borne by those publishing resources. Parties interested in participating must either set up their own resource repository (i.e., a dedicated machine with a 24/7 Internet connection), or find someone to do so on their behalf (e.g., a hosting company). Just like the Web, our repository infrastructure can be built gradually and without central coordination.

Since anyone can publish arbitrary resources, data quality will become an issue. Until sophisticated search engines that can provide ranked results are

[4] While this mechanism could in principle be also applied to the manufacturer's repository, thus eliminating the need for a special manufacturer resolver, we decided to make use of existing resolution mechanisms in order to guarantee users that at least the manufacturer information can be located.

[5] See Section 4 for details on how the manufacturer's resolver service is registered.

available, we expect that word-of-mouth recommendation and independent editorial review (e.g., popular press) will lead to the emergence of a set of resource repositories that are known to provide quality content. Just as it has become standard with websites today, manufacturers and third parties will eventually run and advertise their repositories in both print and electronic media, treating them as yet another means for differentiating their products and services.

Given these characteristics, our approach differs from automatic service discovery as implemented in, e.g., UDDI [18] or E-Speak [20]. In the scenarios addressed by these technologies, selecting the right services is a matter of semantic description and automated matching. In the use cases presented above, however, selecting resources is simpler, as the search scope is limited to entries linked to a certain physical product at hand. We therefore believe that the adoption process for our infrastructure would be considerably faster.

## 4 Implementation

Based on the concepts described above, we implemented a prototype of our resource lookup infrastructure. For each of its three building blocks, the implementation is reviewed in this section.

**Resource Repository.** Resource repositories are implemented using Java Servlets and a relational database for resource, feedback, and user management. Fulltext search capabilities are implemented using the Apache Lucene search engine. The implementation provides bindings to SOAP, XML-RPC, and REST [21]. Optionally, TLS can be used for increased security.

**Manufacturer Resolver Service.** Resolving the manufacturer's resource repository is implemented using the Object Naming Service (ONS) [22]. ONS is a global infrastructure that is used as part of EPCglobal's EPC Network to find the EPCIS[6] of a product's manufacturer. It resolves a product's identifier (its EPC number) to a URL pointing to the corresponding EPCIS by leveraging the existing Domain Name System (DNS) infrastructure. The basic principle of ONS is to append ".sgtin.id.onsepc.com" to the EPC's string representation. Using standard DNS infrastructure, the resulting domain name (e.g., `000024.0614141.sgtin.id.onsepc.com`) is then queried for "NAPTR" records (a type of record as defined by the DNS specifications), which contain the URL to the manufacturer. We use a custom value (`EPC+ResRep`) in the `service` field of the NAPTR record in order to distinguish our URLs pointing to the manufacturer's resource repository from other data in the ONS (typically URLs pointing to an EPCIS).

**Search Service.** We believe that indexing of resource repositories is a task that could be best done by already existing web search services such as Google. In our prototype system, we developed a simple search service based on the

[6] EPCIS stands for *EPC Information Services* and is an integral part of the EPC Network. The EPCIS holds logistical information on a product in the EPC-enabled industrial supply chain.

Apache Lucene search engine. Our search service crawls all registered resource repositories, creates an index, and can be queried using the `Search` operation.

In addition to this, we extended our search service implementation beyond crawling resource repositories. The Internet is full of standard web pages containing information that pertain to physical products. Such information range from product reviews to user guides and blog entries. If we consider such standard web pages as potential resources linked to physical products, we can easily build a search service for these particular resources. Similarly to the Technorati blog search service, we use an empty anchor-tag (i.e., an `<a/>` HTML-element) to mark a web page as being a resource belonging to a certain physical product. A weblog author could for example link a posting to a physical product with EAN number 7610200337481 by including the element `<a href="http://tagged.example.org/tagid/ean/7610200337481"/>` into the HTML source code.[7] As most search engines support a *link* operator to find all web pages linking to a given URL, it is possible to leverage these systems to easily find pages marked with such an `<a/>` element. Our original intention was to implement the search service around one of the large Internet search engines. However, as this turned out not to work reliably, we again used Apache Lucene as the underlying search technology. When the search service receives a `Search` request, it internally queries the Lucene search engine, converts the search results into resource descriptions with the *profile* element set to "webpage" and the *url* element set to the respective web page's address, and returns these resource descriptions to the caller.

Depending on the client's request, matching resource descriptions found in resource repositories and synthesized from web pages are returned either separately or aggregated. Our search service implementation provides bindings to both XML-RPC and REST.

## 5 Prototype Applications

To illustrate the value that our lookup infrastructure offers to the development of applications around tagged products, and to validate our architectural design choices, we built three demonstrator applications. All prototypes were implemented as Java MIDlets on a Nokia 3220 mobile phone. The MIDlets use the REST binding to connect to both the resource repositories and the search service. XML parsing of service responses is implemented using kXML, a lightweight parser for J2ME with minimal memory footprint. Our demonstrators rely on the Nokia 3220's integrated NFC reader, even though conventional EAN/UPC barcode symbols could be equally used as tagging technology.

### 5.1 Calorie Tracker

The first demonstrator allows users to track their daily calorie intake (see Figure 4). Calorie information on products is fetched from a user-extendable

[7] Notice how this link does *not* enclose any text, which is how traditional hyperlinks work. Instead, this singular anchor indicates that this entire page applies to the referenced resource.

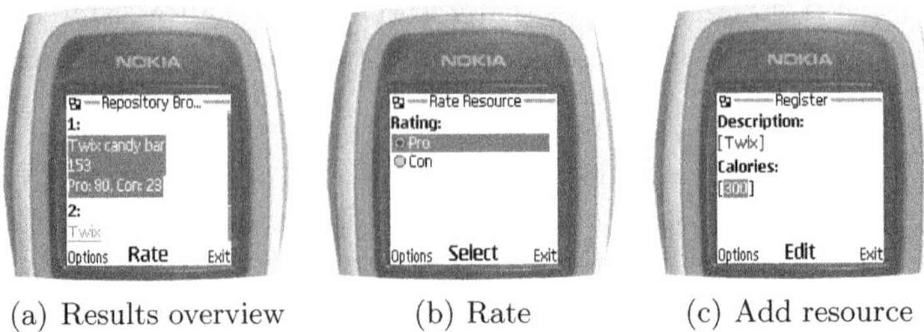


(a) Results overview (b) Rate (c) Add resource

**Fig. 4.** *"Calorie Tracker" application.* An example for a community-built and -maintained product repository.

resource repository. The application demonstrates the possibility of a community-built and -maintained resource repository, by creating new resources and adding feedback to them directly on a mobile phone. To ensure basic quality control, we borrow a concept from community websites and let users approve or disapprove resources created by other users. For every resource, the number of positive and negative votes is recorded and taken into account when resources are ranked in response to a query. If there are no entries for a product, or if a user does not agree with any of the returned values, a new resource can be created. When a user touches a product with the NFC phone, a `LookupResource` request with the acquired tag ID is performed on the "calories repository". The result contains a list of resource descriptions, consisting of the textual description, the calorie number, and feedback, as partly shown in Figure 4(a). While browsing through the results, the user has the possibility to rate a result. Figure 4(b) shows the form for entering a rating for a resource. If none of the suggestions are correct, the user can add a new resource as shown in Figure 4(c).

### 5.2 Shopping Assistant

A second example application provides users with background information on products. Upon scanning a tagged product, the "shopping assistant" contacts 3 resource repositories: First, the manufacturer to obtain allergy information according to the "allergy" profile that we assume has been defined by the food industry. Second, a repository implementing a wrapper to the product's price information at Amazon.com. Third, a repository offering information on environmental issues of a given product. Based on the resources obtained from these

**Table 1.** *Resource repository queries.* Three examples for a shopping assistant (see Section 5.2), trying to find information pertaining to an identified product.

| Repository | `LookupResource` elements |
|---|---|
| manufacturer | `tagid=urn:ean.ucc:9783540240037, profile=allergy` |
| price information | `tagid=urn:ean.ucc:9783540240037, profile=price` |
| env. information | `tagid=urn:ean.ucc:9783540240037, profile=review` |

```
<resDescriptions repository="http://repos.allergy.org/">
  <item resId="b5fe3a5bf077af32c68b790d817d7339f724f209">
    <profile>allergy</profile> <title>Allergy Information</title>
    <data><almonds/></data>
  </item>
</resDescriptions>

<resDescriptions repository="http://repos.envprot.org/">
  <item resId="73cd125bf097af69c64b790d817d7899f788ffa7">
    <profile>review</profile> <title>Environmental Information</title>
    <data>Acme Crop. has repeatedly distributed its toxic waste...</data>
  </item>
</resDescriptions>
```

**Fig. 5.** *Responses from resource repositories.* These (abbreviated) replies illustrate potential replies to the queries shown in Table 1.

repositories, the assistant informs the user if the product conflicts with his or her allergy profile, if it is available from Amazon and for what price, and if there are any environmental issues with it. Table 1 shows queries for an example product sent to the 3 repositories, while Figure 5 illustrates two received responses. All results are aggregated and displayed as shown in Figures 6(a) and 6(b). The Amazon book price resources are automatically created by a custom wrapper that leverages the Amazon Web Services to fetch the current price of books.

### 5.3 Appliance Support

Our last application uses context in the form of a status code obtained from a malfunctioning appliance, such as a printer, to find information that can help solve the problem. We use the search service to locate web pages, blog entries, or other sources of information that are marked as relevant to the product at hand in the status encountered. Figure 6(c) shows an overview of the results found for a printer in a certain status. By selecting "Goto", the user can launch the device's web browser and open the web page (Figure 6(d)). The two special links that mark the pages that were found as relevant for a product with EAN tag ID "6420256000052" and context "status=3762" were <a href="http://tagged.example.org/tagid/ean/6420256000052"/⟩ and <a href="http://tagged.example.org/context/status/3762"/⟩, respectively.

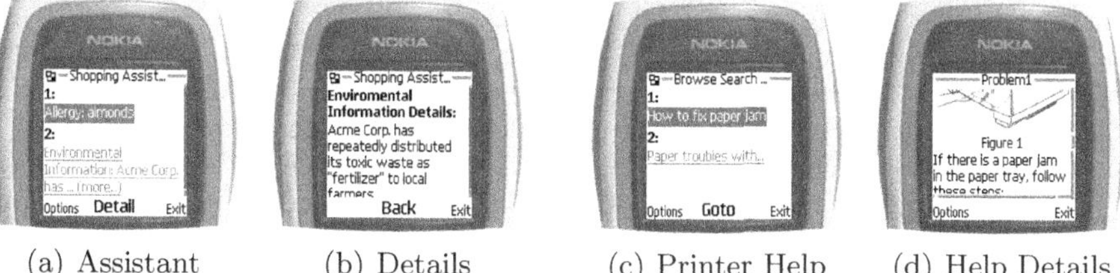


(a) Assistant (b) Details (c) Printer Help (d) Help Details

**Fig. 6.** "Shopping Assistant" and "Appliance Support" applications

## 6 Conclusion

The idea of linking information and services with physical objects is a powerful concept, especially when we are able to augment millions of everyday products with such resources. Realizing the vision of every product being augmentable raises the question of how interested parties can flexibly associate information and services with a product. We address this issue by presenting the concept and architecture of an *open lookup infrastructure* for resource descriptions that fulfills the requirements derived from a range of example application scenarios. We validated the infrastructure by implementing its key components prototypically. We also implemented three demonstrator applications to illustrate how it facilitates the development of novel applications involving digitally augmented, tagged products. In a corresponding user study [23], our demonstrators received very positive reviews from our test subjects. In summary, our open lookup infrastructure offers four key benefits to the various stakeholders involved. Firstly, it allows users to find out what information and services are available for a physical product. Secondly, it gives resource providers access to potential consumers. Thirdly, it enables manufacturers to increase the value of their products by adding information and services to them. And finally, it provides application developers with concepts and services that facilitate the implementation of novel applications.

## References

1. Ishii, H., Ullmer, B.: Tangible Bits: Towards Seamless Interfaces between People, Bits and Atoms. In: CHI '97: Proc. of the SIGCHI conference on Human factors in computing systems, pp. 234–241. ACM Press, New York (1997)
2. Ljungstrand, P., Redström, J., Holmquist, L.E.: WebStickers: Using Physical Tokens to Access, Manage and Share Bookmarks to the Web. In: DARE '00: Proc. of DARE 2000 on Designing augmented reality environments, pp. 23–31. ACM Press, New York (2000)
3. Rekimoto, J., Nagao, K.: The World through the Computer: Computer Augmented Interaction with Real World Environments. In: UIST '95: Proc. of the 8th annual ACM symposium on User interface and software technology, pp. 29–36. ACM Press, New York (1995)
4. Rohs, M., Bohn, J.: Entry Points into a Smart Campus Environment – Overview of the ETHOC System. In: IWSAWC '03: Proc. of the 23rd International Conference on Distributed Computing Systems, pp. 260–266. IEEE Computer Society Press, Los Alamitos (2003)
5. Smith, M.A., Davenport, D., Hwa, H., Turner, T.: Object AURAs: A Mobile Retail and Product Annotation System. In: EC '04: Proc. of the 5th ACM conference on Electronic commerce, ACM Press, pp. 240–241. ACM Press, New York (2004)
6. Lampe, M., Metzger, C., Fleisch, E., Zweifel, O.: Digitally Augmented Collectibles. In: Adjunct Proc. of 8th Annual ACM Symposium on User Interface Software and Technology (UIST), Seattle (2005)
7. Römer, K., Schoch, T., Mattern, F., Dübendorfer, T.: Smart Identification Frameworks for Ubiquitous Computing Applications. Wireless Networks 10(6), 689–700 (2004)

8. Want, R., Fishkin, K.P., Gujar, A., Harrison, B.L.: Bridging Physical and Virtual Worlds with Electronic Tags. In: CHI '99: Proc. of the SIGCHI conference on Human Factors in Computing Systems, Pittsburgh, PA, USA, pp.370–377 ( (1999)
9. Kindberg, T., Barton, J., Morgan, J., Becker, G., Caswell, D., Debaty, P., Gopal, G., Frid, M., Krishnan, V., Morris, H., Schettino, J., Serra, B., Spasojevic, M.: People, Places, Things: Web Presence for the Real World. Mob. Netw. Appl. 7(5), 365–376 (2002)
10. Adelmann, R., Langheinrich, M., Floerkemeier, C.: Toolkit for Bar Code Recognition and Resolving on Camera Phones – Jump Starting the Internet of Things. In: Informatik 2006 workshop on Mobile and Embedded Interactive Systems (MEIS'06) (2006)
11. Dey, A.K.: Understanding and Using Context. Personal Ubiquitous Comput. 5(1), 4–7 (2001)
12. Lessig, L.: The Future of Ideas: The Fate of the Commons in a Connected World. Random House Inc., New York, USA (2001)
13. Sun Microsystems: Jini Architectural Overview (1999) `www.sun.com/software/jini/whitepapers/architecture.pdf`
14. UPnP Forum: UPnP Device Architecture (2000) `www.upnp.org`
15. Guttman, E.: Service Location Protocol: Automatic Discovery of IP Network Services. IEEE Internet Computing 3(4), 71–80 (1999)
16. Helal, S., Desai, N., Verma, V., Lee, C.: Konark - A Service Discovery and Delivery Protocol for Ad-Hoc Networks. In: IEEE Wireless Communications and Networking Conference (WCNC 2003) vol.3, pp. 2107–2113 (2003)
17. Balazinska, M., Balakrishnan, H., Karger, D.: INS/Twine: A Scalable Peer-to-Peer Architecture for Intentional Resource Discovery. In: Proc. of the First International Conference on Pervasive Computing. LNCS, vol. 2414, pp. 195–210. Springer, Heidelberg (2002)
18. UDDI: UDDI Technical White Paper (2000) `www.uddi.org/pubs/Iru_UDDI_Technical_White_Paper.pdf`
19. EPCglobal: EPCglobal Architecture Framework Version 1.0 (2005)
20. Kim, W., Graupner, S., Sahai, A., Lenkov, D., Chudasama, C., Whedbee, S., Luo, Y., Desai, B., Mullings, H., Wong, P.: Web E-Speak: Facilitating Web-Based E-Services. IEEE MultiMedia 9(1), 43–55 (2002)
21. Fielding, R.T., Taylor, R.N.: Principled Design of the Modern Web Architecture. ACM Trans. Inter. Tech. 2(2), 115–150 (2002)
22. EPCglobal: Object Naming Service (ONS) Specification Version 1.0 (2005)
23. Roduner, C., Langheinrich, M., Floerkemeier, C., Schwarzentrub, B.: Operating Appliances with Mobile Phones – Strengths and Limits of a Universal Interaction Device. In: Proc. of Pervasive 2007. LNCS, Springer, Heidelberg (2007)

# Automating Standard Operating Procedures in Intensive Care

Martin Sedlmayr[1], Thomas Rose[1,2], Torben Greiser[1], Rainer Röhrig[3], Markus Meister[3], and Achim Michel-Backofen[3]

[1] Fraunhofer FIT, Schloss Birlinghoven, 53754 Sankt Augustin, Germany
martin.sedlmayr,thomas.rose@fit.fraunhofer.de
[2] Informatik V, RWTH Aachen University, 52056 Aachen, Germany
[3] Department of Anaesthesology, Intensive Care Medicine, Pain Therapy, Justus-Liebig-University Giessen, 35392 Giessen, Germany
rainer.roehrig@chiru.med.uni-giessen.de

**Abstract.** Supporting medical processes is among the most difficult endeavors. In contrast to uniform and unvaried workflows, the complexity and dynamics of patient treatment processes prevents the application of standard methodologies and tools, such as workflow systems. Despite long-term research in flexible and adaptive workflows as well as computerized clinical guidelines there are hardly any applications used in clinical routine. However, Standard Operation Procedures are a key element for any hospital to continuously improve their processes with regard to quality of patient care as well as resources required. Based on a three-level representation of know-how about patient care and treatment, we present a methodology for a stepwise formalisation and automation of clinical guidelines embedded into a patient data management system.

## 1 Introduction

Patient care and treatment certainly constitutes a knowledge intensive endeavor. Knowledge about indications and treatment options is overwhelming and each member of an intensive care unit is confronted with a continuous and dynamic flow of information about the patient [1]. Increasing numbers of implementations of quality management procedures in addition to cost pressure foster the application of best practices according to evaluated standards [2]. Hence, a methodology for capturing the know-how about processes and the use of this know-how for process guidance is an essential ingredient for any hospital management.

Several knowledge and process management approaches have been proposed. A *clinical pathway* describes a complete diagnosis and treatment plan during hospitalization of a patient related to a diagnosis or a symptom [2]. Such structured care plans are aimed to reduce variability, to reduce cost and to increase quality of care by describing essential steps in the care of patients with a specific clinical problem while each step may be formalized in terms of a *Standard Operating Procedure (SOP)*.

J. Krogstie, A.L. Opdahl, and G. Sindre (Eds.): CAiSE 2007, LNCS 4495, pp. 516–530, 2007.

*Clinical practice guidelines* are "systematically developed statements to assist practitioner decisions about appropriate health actions for specific clinical circumstances" (Field/Lohr, Institute of Medicine, 1990). In Germany, "Leitlinien" are medical and scientific recommendations for medical treatment in specific situations. They are issued by national and medical associations. Notably, they do not deal with any economic issues and have no legal implications such as liability (http://www.bda.de).

This body of information about patient care and treatment is increasing due to its promotion by organizations around the world: Internet portals such as http://www.guidelines.gov and http://www.leinlinien.de offer access to guideline repositories in text format such as HTML, Microsoft Word or PDF.

Once providing a uniform modeling methodology, repositories of formalized clinical and medical processes will establish reference processes and hence enable the reuse of evidence-based knowledge. Thus, health care providers can share their knowledge about treatment eventually leading to the establishment of standards across hospitals.

However, formal representations of this know-how are scarce and often limited to well structured processes such as the treatment of chronic diseases [3,4]. However, there is little support in complex and dynamic cases.

To start with, one first has to reach a consensus among clinical experts about best practices. From a technology point of view, there are four major obstacles for a formalisation and automation of guidelines [5]:

- Diversity of different control patterns for structuring activities, i.e. some follow well defined paths whereas others need to be re-arranged on the fly according to the patient's status;
- Representation of ad-hoc decisions that are based on personal experience rather than on text-book knowledge;
- Monitoring of the patient's state of health (in terms of a patient data management system);
- Integration into an operational environment.

Project OLGA (Online Guideline Assist) has been conducted as an interdisciplinary endeavor of physicians and computer scientists to develop a methodology for the capture of processes in the clinical domain and the utilization of these processes for online assistance in terms of guidelines. The pivotal objective of OLGA is the translation of mainly text-based guidelines into executable workflow (fragments) with a full-fledged integration into patient data management systems (PDMS). In the course of this translation GLIF (Guideline Interchange Format) [6] is employed as intermediary between medical and technical specifications. Moreover, awareness concepts are included in the PDMS to make physicians aware of recommendations for treatment or any change of a patient's condition.

This paper is organized as follows. Section 2 reviews the state-of-the-art of computerized support for medical processes. Then, section 3 presents our approach for online guideline assistance for hospitals with a focus on intensive care units, which can be characterized by their high level of process complexity and vitality. A particular focus will be put on workflow support that builds upon an existing PDMS. Section 4 describes the implementation of our system at the level of executable guidelines, while section 5 concludes with an outlook on future services that are enabled by our core implementation.

## 2 State of the Art: Process Support

Research on the support of medical processes can be traced back as early as the 70s when expert systems offered support in the execution of clinical guidelines such as ONCOCIN [7]. At that time, those systems were standalone expert shells requiring manual input of patient data due to a lack of electronic patient records. In addition, as stand-alone tools they were not integrated into the clinical workflow, but external applications that required massive user attendance.

Today, recent patient data management systems (PDMS) offer electronic access to the complete dataset of a patient, enabling the development of online support systems.

On the one hand, workflow management approaches are moving to the medical domain. Flexible solutions are investigated and some have been applied to the medical applications. On the other hand, experts in the medical domain studied computerized clinical guidelines, yet rather independently from classical workflow management.

### 2.1 Workflow Management

For many years, automating office and production processes has fostered the development of various kinds of workflow systems. A model of processes, activities and resources is specified which is interpreted by a runtime engine to distribute tasks and data among users and applications. Numerous standards exist (see e.g. www.ebpml.org or www.bpm-guide.de), which differ by support for different standards, products, vendors or domains [8].

Workflow systems come to their full potential when standardized, fixed process flows exist. However, in many cases and domains this condition is not met, because processes cannot be fully specified a-priori (complexity) or too many context specific dependencies alter the flow (dynamics) [9]. Thus, means to increase flexibility in workflows are a topic of intensive research including:

- Workflow models are changed during execution: the runtime engine reacts on pre-defined events and conditions and alters the process according to a list of rules. Well known representatives of this approach are Event Condition Action rules (ECA) used e.g. by AgentWork [10];
- Generic place-holder processes will be replaced during runtime by fragments from a repository of alternative implementations (e.g. Worklets in YAWL [11]);
- Instead of using a fixed, predefined process model a set of process fragments are combined during runtime according to rules and constraints [12,13] or its configuration is interactively supported [14];

- The case handling approach replaces the explicit model of activity flow by implicit routing, i.e. activities are enabled if its preconditions are met. Conditions are based on the data of the process [15,16].

Flexibility is mainly achieved by two techniques [17], viz flexibility by selection or flexibility by adaptation. Selection is supported for example by the Worklet approach [11] and allows to build a hierarchical repertoire of alternative process implementations which are selected during runtime based on the specific context of the executing instance. This allows to modify and extend alternatives independently from the overall process specification even during runtime. However, this kind of selection is useful mainly when the requirements for flexibility in the workflow process can be identified in advance. In other cases, flexibility by adaptation is needed. Adaptation means the modification of either the workflow schema or the current instance. This adaptation can be based for example on rules [10], constraints [13] or user decisions [14]. However, current enterprise workflow standards and engines do not have broad support for this kind of flexibility.

Besides the many approaches for making classical workflows more flexible by techniques such as ECA rules, innovative approaches, such as case handling and ad-hoc workflows, exist. However, despite naming the medical domain as a model environment, the mapping of clinical guidelines to classical or adaptive workflows still needs further research.

### 2.2 Computerized Clinical Guidelines

A number of methods have been developed during the last decade to support the computerization of guidelines [3,18]. These vary from XML-based document formats over rule-based decision support systems to workflow-like execution systems (see http://www.openclinical.org for a comprehensive list).

The intention in which a guideline modeling system has been developed significantly influences the scope of the model: support for

- different types of guidelines,
- different modes of use,
- adaptation of guidelines for local use,
- integration with institutional systems,
- revision tracking, and
- managing complexity.

Hence, different representation formalisms and computational techniques, such as rule-based, logic-based, networke-based and workflow-like, are used and each system puts a specific emphasis only on certain aspects such as specification of intentions or formal languages [6].

Despite differences in the technical realization, each approach models care delivery as a process, a chain of activities over time, decisions and condition states.

In the end, the process and decision models shall be automatically executable by execution engines. However, existing execution engines are mainly proprietary implementations by the authors of the models as a proof-of-concept and only a few have been commercialized (e.g. PROForma [19]). None of the guideline systems except GUIDE [20] has been built on top of enterprise standards or tools.

The authors of GLIF and PROforma investigated an approach of Guideline Execution by Semantic Decomposition of Representation (GESDOR) [21]: while developing a common standard modeling language for guideline specification seems to be impractical, the various models share common tasks and elements on an implementation level in an execution engine.

Primitive tasks such as data collection, decision making, branching, and synchronization as well as auxiliary tasks, such as criterion evaluation and event management, are common to all models. Each of these tasks can be described by a set of input and output elements, subtasks and constraints. As such, the GLIF and PROforma models have been mapped to these generic tasks. The GESDOR engine – which is based on GLEE (see below) – was able to successfully execute guidelines of either models. However, no attempt has been made to map these execution primitives to standard workflow concepts.

Clinical guidelines and SOPs do not represent process models as defined by workflow management. They are a set of rules that gather knowledge about patient care and medical treatment. Only some of them include process fragments. Most projects focus on the modeling of expert knowledge. In doing so, the models are used for planning tasks or certain analysis [22]. So far, hardly any considers an automated execution of a guideline in terms of a workflow system.

Recent approaches take workflow systems into account but abandon the top-down specification of guidelines. Instead, activities will be triggered by events in the workflow system (bottom up) [23]. However, typical workflow properties such as reliability, security and scalability, are still out of consideration although they are of importance for any operational system.

## 3 Our Approach

Our approach is distinguished from related projects in two regards:

- online access to patient data in the course of process execution; and
- interdisciplinary translation of text-based guidelines into workflows that can be executed on top of PDMS.

It is founded in the HL7-centric patient data management system (PDMS) ICUdata (http://www.imeso.de) that is operational in Giessen since 1999. GLIF is employed as intermediary language for capturing know-how about medical processes.

OLGA obeys a three-step approach towards the capture of know-how for the automation of clinical guidelines. Figure 1 depicts the different degrees of formalisation and presents the process of stepwise translation from text-based representations towards an executable workflow representation.

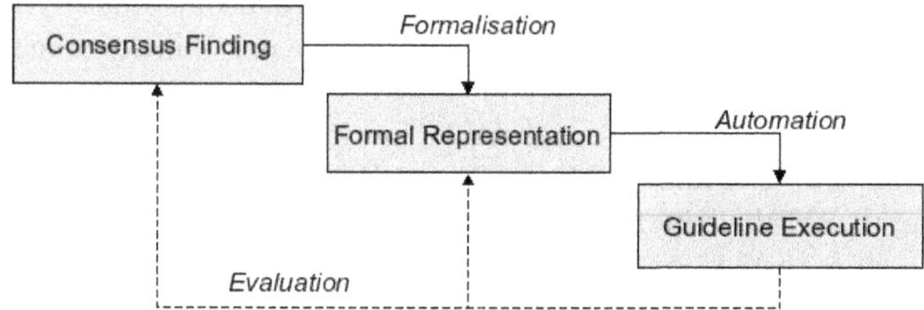


**Fig. 1.** Stepwise formalization of guidelines in OLGA

1. First, a consensus has to be reached on the medical process flow. The definition is to be based on relevant literature and knowledge sources and has to be adapted to the specific settings in the ward by experienced domain experts. This consensus process yields simple flowcharts and textual descriptions. This first step is in line with guideline definition procedures in the medical domain.
2. Second, these charts and texts have to be formalized, i.e. have to be translated into a process modeling language. We use GLIF as an intermediary language between medical and technical specifications since GLIF is an acknowledged format for guideline modeling.
3. Third, the formal GLIF model is translated into executable workflow languages with the help of mapping assistants. These workflows then can be executed by workflow engines like YAWL or JBPM. Despite differences of the modeling languages in expressiveness, control constructs and the like, the mapping process shares the same issues: not all domain relevant concepts can unambiguously be mapped to workflow elements.

### 3.1 Formalization

In Germany, the Society of German Anesthetists (Bund Deutscher Anästhesisten, BDA) and the Society of Anesthesiology and Intensive Care (DGAI) operate a SOP exchange platform in the Internet. We have analyzed 23 SOPs from this repository according to formalization in GLIF and automation [24].

Even the most detailed treatment procedures require additional annotations if they are to be automatically executed. Often the description of temporal aspects and priorities of steps in the care process are too vague, taking interpretation by the human expert user for granted. For example, the diagnosis of a heart stroke depends not on a single event but on a complex combination of findings which have a certain temporal relationship.

Looking into the control patterns of the guideline models, one can find sequences, loops and the like as in any other process models [5]. The expressiveness of Event-driven Process Chains (EPC) has been compared with process samples collected in hospitals and concludes with the open issues of expressing intuitive, non-deterministic decisions, modeling the inner dynamics of the patient and its effect on the process flow, the complexity of time relations of activities and the many kinds of iterations. These issues go beyond the possibilities of many process languages [25].

The Guideline Interchange Format GLIF has been developed as a joint effort of research groups at Columbia, Stanford, and Harvard (Intermed Collaboratory) and was first published in 1998 [6]. Its main purpose has been the sharing of guidelines across institutions. GLIF models are not limited to knowledge about the treatment process, but also include applicability criteria, strength of recommendation (for scientific assessment), links to further knowledge sources and authoring maintenance information.

Specifications of guidelines take place on three layers of abstraction in GLIF. At the *conceptual level*, a guideline is drawn as flowchart to catch the generic chain of activities without having to concentrate on details. These details, e.g., data definition, decision criteria and control flow, are specified at the *computable level*. The mapping of tasks and data to any institutional system is represented at the *implementable level* (which has not yet been defined in GLIF).

The flowchart graph contains nodes for actions, decisions, routing (branch and synchronization), and patient states. The data model uses the HL7 Reference Information Model (RIM, see http://www.hl7.org). The expression language GELLO [26] is based on the Object Constraint Language (OCL) and has become an accepted standard by HL7 and ANSI.

The ontology editor Protege (http://protege.stanford.edu) is used as standard editor for GLIF guidelines. A number of add-ons such as the GraphWiz for the flowchart or the InstanceTree for editing are available. GLIF guidelines for vaccination and chronic cough treatment have been executed and evaluated with about 2000 patient cases using by the GLIF Guidelines Execution Engine (GLEE) [27].

### 3.2 Automation and Support

We have been looking into implementation options based on workflow techniques and tools after deciding to support GLIF in OLGA. Two options exist for automating the formalized guidelines: direct execution by a kind of GLIF interpreter or indirect execution by mapping GLIF models into another process language.

For *direct execution*, a runtime engine such as the Guideline Execution Engine (GLEE) for GLIF [27] manages execution states of the model elements (Guideline_Steps) such as active, prepared, started and finished and allows for user-defined schedules between the states. Such a proprietary implementation allows developers to concentrate on the automatable subset of the modeling standard, but is bound to it. GESDOR [21] introduced an interesting approach by aiming at a kind of virtual machine for guideline execution engines providing common functions of GLIF and PROforma.

*Indirect execution* – by translating a GLIF guideline into another workflow format – allows for utilization of established business standards. It may utilize already existing infrastructure and inter-operate with other applications. Though not all concepts of GLIF can be directly mapped to standard workflow elements (e.g. events, see below), careful selection and modeling of guidelines can prevent the usage of not-mappable structures. This approach has been chosen for the implementation of OLGA.

# 4 Implementation

OLGA provides tools and services for supporting care providers in the selection and execution of guidelines. These are integrated into the patient data management system (PDMS) at the anaesthesiological intensive care unit of the university hospital in Giessen. Next to each of the 14 beds on the ward is a terminal showing the electronic patient chart.

The PDMS uses HL7 messages to communicate among clients and the server. The OLGA execution engine filters relevant messages and triggers activities in the guideline repository. Activities and recommendations are communicated back using HL7 and become part of the electronic health record. Visualisation towards the user is inside the PDMS as part of the patient's chart.

## 4.1 System Design

The patient data management system PDMS used in the ICU in Giessen supports modular information processing. Data are exchanged through HL7 messages routed by master services between modules [28]. Any change to patient data is propagated to the database as well as to all clients that operate on this patient's data (event-based communication). The communication network and event mechanism is open not only to PDMS modules but also to external applications.

The OLGA execution engine implements such an external module. It uses the notification services to listen to HL7 messages and filters relevant data to trigger guideline execution. Automatically generated advise, recommendations and request for further input (observations) are sent as HL7 messages and become part of the patient record. There is no direct interaction between the OLGA tools and the end-user: medical recommendations as result of a guideline execution are part of the patient record shown in the user interface of the PDMS (see Figure 3). Therefore, OLGA integrates with the normal tools and workflow of physicians and nurses and does not appear as another tool on top to handle.

Two options exist for the architecture of OLGA: client/bed-side installation or as central server. Bed-side terminals have the most complete and current record of patient data. If the network communication fails, new data about the patient arrives from manual interaction with the chart and from connected devices (ventilator, patient monitor). This means, guidelines can be executed locally even if there is no connection with the network. Synchronization with the database and other clients is handled by the local communication master. Therefore, client-side seems advantageous with respect to reliability, safety and scalability and hence has been chosen for the implementation of OLGA.

### 4.2 Formalization Using GLIF

We have modeled several guidelines in GLIF using Protege. The guidelines comprise weaning from long-term ventilation, SIRS/sepsis therapy, acute coronary syndrome and scoring. The examples ought to cover all kinds of control structures, interaction patterns and data patterns. Our experience using GLIF needs to be distinguished between GLIF itself and the modeling tool Protege.

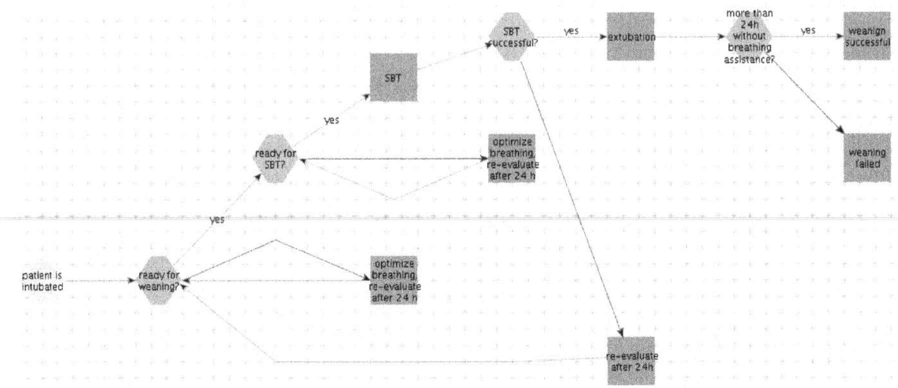


**Fig. 2.** Simplified weaning protocol modelled in GLIF

Using a flowchart at the conceptual layer is perceived very well among users. It is an intuitive graphical representation of the flow of activities. The simple flow can be successively orchestrated by events, conditions and decision rules. However, it is not easy to decide which should go on either level. In our experience, users tend to put as much as possible on the graphical level: for example decision criteria are modeled as a chain of `Decision_Steps` instead of a single decision containing a set of rules. It is argued that although the flowchart becomes more complex, the reasoning about the algorithm as well as comprehension is felt to be better.

Partly, the confusion is caused by the Protege modeling tool. While Protege is a well known and often used tool for modeling ontologies, it offers little support specific to GLIF modeling. Its generic view on ontologies as classes, slots and instances makes editing more complicated. Usually more than ten windows are open when modeling a GLIF decision criterion. In addition, some functions seem to misbehave, for example the arrows in the graphical widget and the corresponding attributes in the model (`next_step`, branches) are not synchronised leaving the model prone to constraint errors.

As a consequence, we decided to build a graphical editor for OLGA guidelines more focused on the medical end-user. While being inspired by GLIF, Protege and other workflow editors will allow us for more flexibility towards visualization, simulation and advanced modeling features not yet specified [25].

### 4.3 Workflow Mapping

We have implemented a mapping assistance tool to be used by the knowledge engineer to translate GLIF models into jPDL to be executed by jBPM. The JBoss Business Management Framework jBPM by Jboss is a flexible and extensible open-source environment for executing workflows specified in the jBPM Process Definition Language jPDL (http://www.jbpm.org). The graph oriented programming model can be easily extended by own node types and therefore looks promising for implementing GLIF execution. Further, jBPM can easily be integrated with the leading Jboss application server as well as the rule engine Drools (for rule-based decision support).

Based on the Protege API, OLGA reads the GLIF model and translates step-by-step to jPDL. In principle, each guideline step maps to a node in the workflow model and relations are mapped to transitions. Advanced concepts such as interpretation of decision rules are delegated to specific classes (implementations of the interface `Action_Handler` in jBPM). Also, task specifications in GLIF (`Action_Specification`) are mapped to `Action_Handlers` in jBPM. It can be observed, that it is the first of the three conceptual layers in GLIF (conceptual layer) that is mapped straight forward to the workflow whereas the details on layer two and three are implemented by extensions to the workflow engine. A more in-depth discussion on the mapping can be found in [29].

The following excerpt shows the beginning of the weaning protocol mapped to jPDL workflow language in XML. The workflow is used to implement the transitions between action states whereas the implementation of the action is delegated to external action handlers (e.g. `GetInputAction` to read the result of the spontaneous breathing test (parameter `sbt_status`) of the current patient).

```
...
<node name="order SBT">
  <action class="jpdl.handler.WriteOrderAction">
    ...
  </action>
  <transition to="wait for SBT" />
</node>
<node name="wait for SBT">
  <action class="jpdl.handler.GetInputAction">
    <parameter>sbt_status</parameter>
  </action>
  <transition to="SBT ok?"/>
</node>
<decision name="SBT ok?">
  <transition name="no" to="wait 4h"/>
  <transition name="yes" to="order sedation reduction">
    <condition>#{sbt_status == "ok"}</condition>
  </transition>
</decision>
...
```

The converter was tested successfully first for process-oriented test cases specified in GLIF and then using simpflified clinical guidelines. `Guideline_Steps`, `Action_Specifications` and decision conditions are mapped whereas iterations and events are to be done.

### 4.4 Automation

The current state of the implementations allows for the automation of the selected guidelines. For example, a simplified version of the SOP for weaning of a patient from automated ventilation has been formalized and converted into an executable workflow.

Whenever a ventilated patient arrives on ward, the guideline/workflow is triggered. It enters an order into the chart for additional findings such as "Is patient status ok?" and "Is patient awake?", if not already present in the record, to trigger the next steps ("reduce sedation", "extubate").

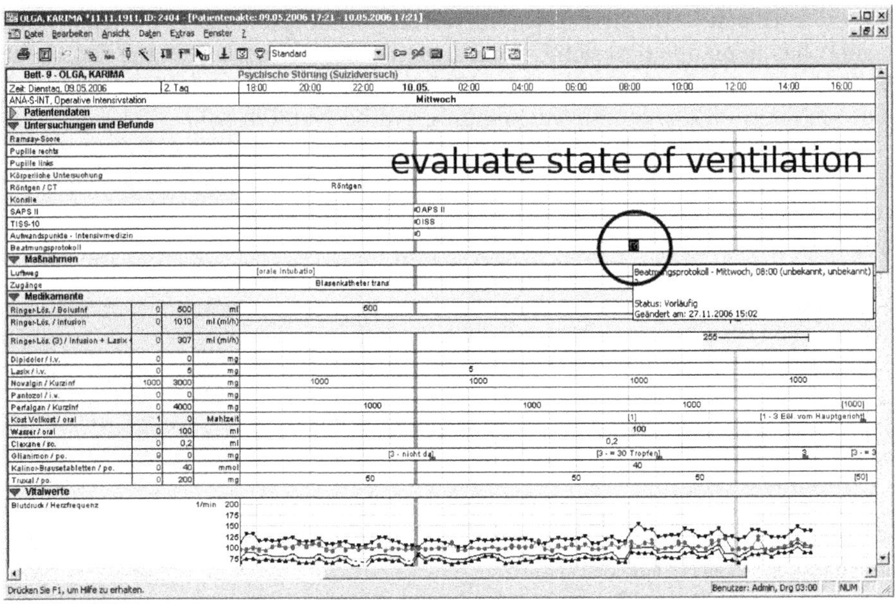


**Fig. 3.** Patient chart of the PDMS showing an order entry triggered by the weaning protocol

In Figure 3 one can see the entry to reassess the state of ventilation of the patient as entered by the automated workflow. The order was triggered by the patient being intubated and ventilated afterwards ("orale Intubation").

Although rather simple in its current state, the process shows all patterns of [5]. Whereas activities in each of the ventilation and sedation threads are sequential, both threads happen in parallel. Exceptions occur, if a patient extubates himself. Advanced iteration specification is necessary, because patients should not be extubated or woken up (sedation) during night time.

Our tools will implement a framework that will allow us to try various approaches of process mapping to overcome the described problems based on real patient data in a PDMS. The iterative, bottom-up approach starting with simplified guidelines enables early and continuous feedback by having operational guidelines applied in daily routine.

## 5 Conclusion

Computerized support of medical processes has surfaced as challenging research question. This challenge is founded in the fact, that process management concepts appear beneficiary for process improvements for hospitals at first glimpse. However, once leaving the domain of administrative processes (where hospitals currently witness a widespread deployment of EPC-based modelling) and the domain of well-definable chronic diseases, one is confronted with an increased complexity and flexibility of medical processes. Although the process management community has developed an array of concepts for more flexible workflows, such approaches still fell short for accurately representing medical processes. On the other hand, the medical community has developed knowledge representation languages for the capture of medical processes without accounting for the execution of processes. Such languages merely allow for the representation of medical know-how, but they do not support the execution of processes. Our approach strives to bridge this gap by an interdisciplinary development methodology. This methodology allows a stepwise formalization of medical processes from flowchart-oriented representations in GLIF, via mapping assistants towards executable process fragments for workflow management systems.

Our approach contributes to two research issues:

1. *Formalization of clinical guidelines* – Our formalization of medical processes is based on the knowledge representation language GLIF, which is dedicated to the capture of processes by medical experts. Various interaction tools support medical experts in the capture and maintenance of their know-how. Due to the ambiguous semantics of representations concepts, mapping assistants have been implemented for the translation of GLIF specifications into workflow specifications.
2. *Automation of medical workflows* – Once translated into a workflow specification, the PDMS hosts the workflow processes. Rather than interacting with a workflow management system, the physician uses the PDMS as interface to the workflow system.

Currently OLGA supports the specification and execution of medical processes on top of an operational PDMS. Until now, the system has been developed in a testbed environment. Once a reasonable number of medical pathways have been specified and operationalized, field trials will be conducted in a hospital environment to evaluate system performance and acceptance. OLGA will be tested in daily routine on the ICU ward in Giessen. The permission of the ethics commission and experience of the personnel in using research prototypes for electronic patient records make this rare opportunity possible.

In addition, awareness and notification concepts have been studied to explore options on how to make physicians aware of incoming messages, alerts and certain changes in a patients record [30]. The primary design goal is to stick with the interface of the PDMS for workflow interaction.

Future work will address the topics of:

- *Evidence-based medicine* – Once medical pathways have been used in daily routine, the electronic patient records managed by the PDMS serve as valuable source for data- and process mining of patient care and treatment. Thus, impacts of specific medical pathways can be traced and evaluated leading to evidence about medical processes.
- *Control patterns* – Control patterns have until now only been studied for business processes [8]. Medical processes have given birth to new types of control patterns, yet more research is needed to identify typical control patterns and design appropriate means for interaction and support. Initial experiences with the GLIF environment have unveiled such control patterns, but corresponding workflow counterparts have to be designed.

**Acknowledgements.** This work is partly supported by the Deutsche Forschungsgemeinschaft within the scope of project Online Guidance Assist (RO3053/1-1; RO2030/2-1). OLGA is a joint project of Fraunhofer FIT, Germany, and the Klinik für Anästhesiologie, Intensivmedizin, Schmerztherapie und Palliativmedizin, Universitätsklinikum Giessen und Marburg, Standort Giessen, Germany.

## References

1. Imhoff, M., Webb, A., Goldschmidt, A.: of Intensive Care Medicine. ESCIM, E.S.: Health informatics. Intensive Care Med. 27(1), 179–186 (January 2001)
2. Campbell, H., Hotchkiss, R., Bradshaw, N., Porteous, M.: Integrated care pathways. BMJ 316(7125), 133–137 (January 1998)
3. Peleg, M., Tu, S., Bury, J., Ciccarese, P., Fox, J., Greenes, R.A., Hall, R., Johnson, P.D., Jones, N., Kumar, A., Miksch, S., Quaglini, S., Seyfang, A., Shortliffe, E.H., Stefanelli, M.: Comparing computer-interpretable guideline models: a case-study approach. J Am. Med. Inform. Assoc. 10(1), 52–68 (2003)
4. Shahar, Y., Young, O., Shalom, E., Galperin, M., Mayaffit, A., Moskovitch, R., Hessing, A.: A framework for a distributed, hybrid, multiple-ontology clinical-guideline library, and automated guideline-support tools. J Biomed. Inform. 37(5), 325–344 (2004)
5. Sarshar, K., Dominitzki, P., Loos, P.: Einsatz von Ereignisgesteuerten Prozessketten zur Modellierung von Prozessen in der Krankenhausdomme - eine empirische Methodenevaluation. In: Nttgens, M., Rump, F., (eds.) Proceedings of EPK 2005, Hamburg, Germany, CEUR Workshop Proceedings, vol. 167, pp. 97–116 (2005)
6. Peleg, M., Boxwala, A.A., Tu, S., Zeng, Q., Ogunyemi, O., Wang, D., Patel, V.L., Greenes, R.A., Shortliffe, E.H.: The InterMed approach to sharable computer-interpretable guidelines: a review. J Am. Med. Inform. Assoc. 11(1), 1–10 (2004)

7. Shortliffe, E.H., Scott, A.C., Bischoff, M.B., Campbell, A.B., van Melle, W., Jacobs, C.D.: ONCOCIN: An Expert System for Oncology Protocol Management. In: Proceedings of the 7th International Joint Conference on Artificial Intelligence, pp. 876–881(1981)
8. van der Aalst, W.M.P., ter Hofstede, A.H.M., Kiepuszewski, B., Barros, A.P.: Workflow Patterns. Distributed and Parallel Databases 14, 5–51 (2003)
9. Heinl, P., Horn, S., Jablonski, S., Neeb, J., Stein, K., Teschke, M.: A comprehensive approach to flexibility in workflow management systems. SIGSOFT Softw. Eng. Notes 24(2), 79–88 295675 79-88 (1999)
10. Müller, R., Greiner, U., Rahm, E.: AGENT WORK: a workflow system supporting rule-based workflow adaptation. Data & Knowledge Engineering 51(2), 223–256 (2004)
11. Adams, M., ter Hofstede, A., Edmond, D., van der Aalst, W.: Facilitating Flexibility and Dynamic Exception Handling in Workflows through Worklets. In: The 17th Conference on Advanced Information Systems Engineering-CAiSE Short Paper Proceedings. CEUR Workshop Proceedings vol.161 (2005)
12. Deng, S.G., Yu, Z., Wu, Z.H., Huang, L.C.: Enhancement of workflow flexibility by composing activities at run-time. In: Proceedings of the 2004 ACM symposium on Applied computing, pp. 667–673 (2004)
13. Mangan, P.J., Sadiq, S.: A Constraint Specification Approach to Building Flexible Workflows. Journal of Research and Practice in Information Technology 35(1), 21–39 (2002)
14. Rupprecht, C., Peter, G., Rose, T.: Ein modellgestuetzter Ansatz zur kontextspezifischen Individualisierung von Prozessmodellen. Wirtschaftsinformatik 41(3), 226–236 (1999)
15. Schuschel, H., Weske, M.: Fallbehandlung: Ein Neuer Ansatz zur Unterstützung Prozessorientierter Informationssysteme. In: Desel, J., Weske, M., eds.: Prozessorientierte Methoden und Werkzeuge für die Entwicklung von Informationssystemen - Promise. vol.21, pp.52–63 Potsdam, GI (2002)
16. van der Aalst, W.M.P., Weske, M., Grunbauer, D.: Case Handling: A New Paradigm for Business Process Support. Data. and Knowledge Engineering 53(2), 129–162 (2005)
17. Halliday, J., Shrivastava, S., Wheater, S.: Flexible workflow management in the OPENflow system. In: Enterprise Distributed Object Computing Conference, 2001. EDOC '01. Proceedings. Fifth IEEE International (September 4-7, 2001) pp. 82–92 (2001)
18. Wang, D., Peleg, M., Tu, S.W., Boxwala, A.A., Greenes, R.A., Patel, V.L., Shortliffe, E.H.: Representation primitives, process models and patient data in computer-interpretable clinical practice guidelines: a literature review of guideline representation models. Int. J. Med. Inform. 68(1-3), 59–70 (2002)
19. Sutton, D.R., Fox, J.: The syntax and semantics of the PROforma guideline modeling language. J. Am. Med. Inform. Assoc. 10(5), 433–443 (2003)
20. Quaglini, S., Stefanelli, M., Lanzola, G., Caporusso, V., Panzarasa, S.: Flexible guideline-based patient careflow systems. Artif. Intell. Med. 22(1), 65–80 (2001)
21. Wang, D., Peleg, M., Bu, D., Cantor, M., Landesberg, G., Lunenfeld, E., Tu, S.W., Kaiser, G.E., Hripcsak, G., Patel, V.L., Shortliffe, E.H.: Gesdor - a generic execution model for sharing of computer-interpretable clinical practice guidelines. AMIA Annu Symp Proc. pp. 694–698 (2003)
22. Peleg, M., Tu, S., Manindroo, A., Altman, R.B.: Modeling and analyzing biomedical processes using workflow/Petri Net models and tools. Medinfo. Journal. Article. 11(Pt 1), 74–78 (2004)

23. Tu, S., Musen, M., Shankar, R., Campbell, J., Hrabak, K., McClay, J., Huff, S., McClure, R., Parker, C., Rocha, R.: Modeling Guidelines for Integration into Clinical Workflow. Medinfo. 11, 174–178 (2004)
24. Meister, M., Michel-Backofen, A., Jost, A., Röhrig, R., Junger, A., Hempelmann, G.: Evaluation der Abbildbarkeit und Automatisierbarkeit von Standard Operating Procedures mit dem Guideline Interchange Format (GLIF). In: 53. Jahrestagung der DGAI, Leipzig, Abstraktband DAC 2006 (2006)
25. Sedlmayr, M., Rose, T., Röhrig, R., Meister, M., Michel-Backofen, A.: Ansatz zur Automatisierung klinischer Guidelines in GLIF mit Workflow Techniken. In: 51. Jahrestagung der Deutschen Gesellschaft für Medizinische Informatik, Biometrie und Epidemiologie (GMDS). Klinische Forschung vernetzen. 10.-14. September 2006. Leipzig: Universität Leipzig (2006)
26. Sordo, M., Boxwala, A.A., Ogunyemi, O., Greenes, R.A.: Description and status update on GELLO: a proposed standardized object-oriented expression language for clinical decision support. Medinfo. 11(Pt 1), 164–168 (2004)
27. Wang, D., Peleg, M., Tu, S.W., Boxwala, A.A., Ogunyemi, O., Zeng, Q., Greenes, R.A., Patel, V.L., Shortliffe, E.H.: Design and implementation of the GLIF3 guideline execution engine. J. Biomed. Inform. 37(5), 305–318 (2004)
28. Michel-Backofen, A., Demming, R., Röhrig, R., Benson, M., Marquardt, K., Hempelmann, G.: Realizing a Realtime Shared Patient Chart using a Universal Message Forwarding Architecture. Stud. Health. Technol. Inform. 116, 509–514 (2005)
29. Sedlmayr, M., Rose, T., Röhrig, R., Meister, M., Michel-Backofen, A.: Formalisierung und Automatisierung von SOPs in der Intensivmedizin. In: 8. Internationale Tagung Wirtschaftsinformatik; Karlsruhe, Germany (2007)
30. Röhrig, R., Meister, M., Michel-Backofen, A., Sedlmayr, M., Uphus, D., Katzer, C., Rose, T.: Online Guideline Assist in Intensive Care Medicine - Is the login-authentication a sufficient trigger for reminders? Stud. Health. Technol. Inform. 124, 561–568 (2006)

# Composing Data-Providing Web Services in P2P-Based Collaboration Environments

Mahmoud Barhamgi[1], Pierre-Antoine Champin[1], Djamal Benslimane[1], and Aris M. Ouksel[2]

[1] LIRIS Laboratory, Claude Bernard Lyon1 University
69622 Villeurbanne, France
{mahmoud.barhamgi,pierre-antoine.champin,djamal.benslimane}@liris.cnrs.fr
[2] The University of Illinois at Chicago
Depts. Of Information and Decision Sciences and Computer Science
aris@uic.edu

**Abstract.** P2P Data Management Systems (PDMSs) have traditionally focused on the integration of data sources to support information processing on the Web. Recent trends suggest that the same problem may be viewed through the lens of data-providing Web services instead. In this paper, we propose a method to supplement current PDMSs with capability to handle data sources exposed as services. In our solution, Data-Providing services are modeled as RDF parameterized views. An algorithm is devised to compute those data-providing services' compositions that are capable of answering a given query. The conducted experiments, though preliminary, show encouraging results, the algorithm scales up to 100 views in 6 seconds. Our data-driven composition approach applies special data treatment techniques between the composed services prior to query. This may include, merging, differencing and intersecting the returned results of two or more of Data-Providing services.

**keywords:** Data-Driven Services composition, Data Integration, Views.

## 1 Introduction

The advent of Web services technology has increasingly enabled to expose information resources as services reachable to partners engaged in e-collaboration applications (e.g., e-Gov, e-Entreprise, e-Health ...etc). This is motivated by the need to circumvent proprietory implementations when accessing and retrieving data items from autonomous and heterogeneous collaborating systems, regardless of the proprietary systems employed. Thus, peers have access to each others' set of data-providing services. Data-Providing Services, henceforth referred to as DP services, are different from Effect-Providing services (e.g. fulfilling a car reservation, charging a credit card,...etc) in that their invocation only returns data without changing the state of the system. In the eHealth domain, for example, Data-Providing services could involve one or more of the processes: 1). A healthcare peer may require data from several legacy applications, sensors and devices equipped with proprietary interfaces. Web services provide a means to bridge

J. Krogstie, A.L. Opdahl, and G. Sindre (Eds.): CAiSE 2007, LNCS 4495, pp. 531–545, 2007.

the proprietary data sources during integration. 2). Health-care peers may outsource the management of their own data to other peers. For example, a hospital may rely on an independent laboratory to manage some information about its patients. The outsourcing peer, the hospital in this example, may then be able to access its data via a set of data Web services provided by the peer recipient of outsourcing, which in the case is the laboratory. 3). Peers often support may only a limited set of queries (modeled as services) over their schema to constrain the way data is accessed (by the partners) because of privacy constraints [9,17]. 4). Access rights (e.g. a doctor has an access right higher than a nurse) are fairly simple to implement with Web services. Readers are referred to [5] for further information on the use of Web services in the eHealth domain.

Obviously, a query issued at a peer, say peer $P_1$ (see figure 1), which requests information about a patient admitted at $P_1$, may use local DP services at $P_1$ as well as DP services offered by its partners. For example, a query posted at a hospital will invoke services at other peers, such as laboratories, to whom all or partial management patient data was outsourced.

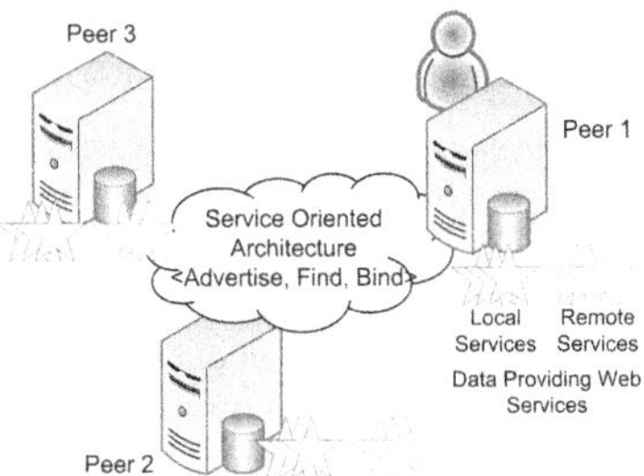


**Fig. 1.** Data sharing via Data-Providing Web services in a loosely coupled environment

Data management and integration in loosely coupled P2P-based information systems have been investigated intensively in recent years. Several systems and architectures have been proposed, including for example, Edutella [11], Piazza[8], PEPSEINT[3], SEWASIE[1], Hyperion [18], PeerDB [14], SDQNET [19], and SCOPES [15]. However, in these systems, without exception, data at peers is presented either in a syntactic form (XML) or in semantic forms, a combination of OWL instances and inferencing capabilities. Queries are applied directly to data answers. None of the systems has examined query processing in environments, where Web services are increasingly being adopted to share data. Data accessibility in this case postpones direct access to data until query execution, as data is hidden by the services. Query resolution necessitates a priori rewriting of the query in terms of available services in a way that guarantees successful execution of the query. This may require coordination of the query to insure desirable outcome.

In this paper we propose an approach for supporting service-accessed data sources (or DP services) in e-collaboration environments. Our approach relies on modeling DP Web services as *RDF Parameterized Views* over the domain ontology. In comparison with the standards of *semantic web services* like OWL-S[1] and WSDL-S[2], our description permits to represent the semantic relationship between In/Out of described services. Individual peers in this environment describe their services and exchange these descriptions with one another. Upon receiving a query, a peer makes use of its stored RDF PVs (for its local services, or remote services) in the query resolution process. A query is resolved by a composition of DP services that would satisfy fully the query.

Observe that this is different from the traditional Web services composition [12] in two ways. First, the traditional composition is "task-driven" since composed services collaborate to achieve a more complex task (or functionality ), e.g. a full-package journey reservation service out of hotel, plane and car reservation Web services. In our framework, the composition is "data-driven" (it does lead to an augmented functionality), where composed services collaborate to achieve as much complete answer as possible. Second, while in traditional composition a composite service is defined and implemented to run in sustainable way, in our framework a composition serves only to answer a received query.

Thus, the key contribution in this paper is a novel approach, which enables peer-to-peer systems to handle the case where data portions are behind services (they are not readily available). To our knowledge this was not addressed yet in these systems. In addition, in our system, peers which still expose their data in forms such as (OWL, RDFS instances) can still inter-operate with the others since our queries are issued in SPARQL[3], a standard query language suited for querying data in the semantic Web.

The main contributions of this paper are: 1). A new approach for answering queries using Data-Providing services, this includes modeling services as RDF parameterized views, and a service-based query rewriting algorithm that is capable to compute the possible services compositions answering a given query. 2. A P2P system (under development) that supports the needs encountered in e-Systems that make an extensive use of Data-Providing Web services.

The rest of the paper is organized as follows. In section 2 we model DP services as RDF views over the domain ontology. Section 3, is devoted to the query rewriting in terms of services (or DP services composition). In section 4 we present the implementation status of our approach. In section 5 we review related works. Finally, in section 6 we conclude the paper and present our future works.

## 2 Modeling Data-Providing Web Services as Views

This section is devoted to model both our queries and DP Web services. Based on this modeling we devise an algorithm to rewrite queries in terms of services.

### 2.1 Motivation Example

This example is extracted from the healthcare domain. Assume a physician conducting a research study about the harmful effects of some medications. In this

[1] http://www.w3.org/Submission/OWL-S/
[2] http://www.w3.org/Submission/WSDL-S/
[3] http://www.w3.org/TR/rdf-sparql-query/

task, the physician needs to examine the test results for patients who have been administered the studied medication. The phisician can express his query as *Q1: what are the tests performed by patients who have been administered a medication termed as "Some Stuff"?*. Also suppose that the peer holding the query has references to the services shown in table 1 (both local and remote services). In

**Table 1.** Local and remote services

| Service Location | Service | Description |
|---|---|---|
| remote | $WS_1$ | retrieves Test A (specialization of Test (see figure2)) |
| remote | $WS_2$ | retrieves Test B (specialization of Test (see figure2)) |
| local | $WS_3$ | returns the medications list taken by a given patient |
| local | $WS_4$ | returns patients (their names) who have been administered a given medication |

order to answer $Q_1$, several services should be composed, they are in particular $WS_1$, $WS_2$ and $WS_4$ (local and remote services). Notice that it does not suffice to rely on the services' inputs and outputs to decide whether they can answer the query or not, rather the semantic relation between the service's input and output must be taken in consideration. In section (2.3), we capture this relation by modeling a service as a *RDF parameterized view.* Queries are rewritten in terms of services by exploiting these views.

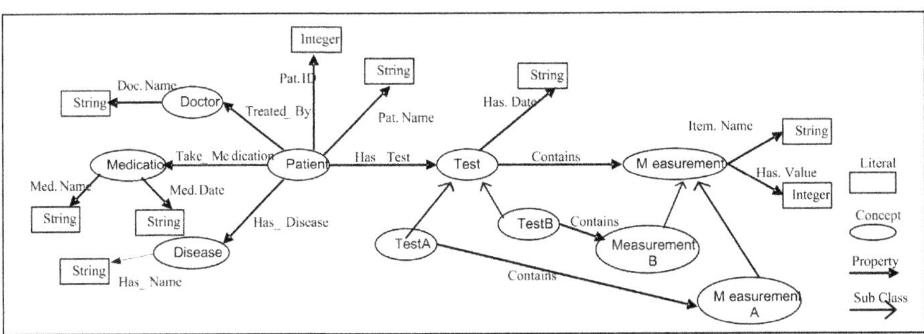


**Fig. 2.** An example of OWL ontology modeling the peer's local data items and the items provided by its partners

## 2.2 Queries

OWL has became the de facto standard for modeling Web resources. OWL primitives include classes, properties and Datatypes. Properties break up in two types; Object properties relating classes and Datatype properties relating classes to datatypes.

**Definition 1.** *In our context an OWL ontology O is a 6-tuple (C, L, DP, OP, SC, SP) where:*

1. *C is the set of classes.*
2. *L is the set of datatypes.*
3. *DP is the set of datatype properties. Each datatype property has a domain in C and a range in L.*
4. *OP is the set of object properties. Each object property has its domain and range in C.*
5. *SC is a relation over $C \times C$, representing the sub-class relationship between classes.*
6. *SP is a relation over $OP \times OP \cup DP \times DP$, representing the sub-property relationship between homogeneous properties.*

Materialized instances of an OWL ontology form a graph where nodes are labeled by a class (instance nodes) or a datatype (literal nodes), and edges are labelled by a property consistent with the linked nodes' labels. Users in our framework are allowed to issue queries on that graph.

Given the previous definition, a query on the instance graph has the following definition.

**Definition 2.** *A query on the instance graph of an ontology O is a 3-tuple (Backbone, Ct, Out) such that:*

1. *Backbone is a sequence of the form:*
   $?c_1(C_1).\Psi.p_{1.2}.?c_2(C_2).\Psi.p_{2.3}\ \ldots\ \Psi.p_{n-1.n}.?c_n(C_n)$, *where*
   - $?c_i$ *is a variable of type* $C_i$, *and* $C_i$ *is a class of the ontology.*
   - $p_{i.j}$ *are object properties holding between* $?c_i$ $(1 \leq i \leq n)$.
   - $\Psi$ *is a linking operator and it is used when one variable* $?c_i$ *is linked to more than one other variable such that each of these variables pose a condition on the selection of* $?c_i$. *The semantics of this operator is that instances of* $?c_i$ *must satisfy all of the conditions specified by the* $\Psi$*'s outgoing paths.*
2. *Ct is the constraints set imposed on datatype properties of* $?c_{i:1\rightarrow n}$.
3. *Out is the output set, it comprises output variables (and their projected datatype properties).*

We implement this form of queries with SPARQL query language. This form is suitable when matching queries against the services as we shall see next. In the spirit of this definition our query became:

$Q_1$:

Backbone=
?T1(type:Test) . $[Has-Test]^{-1}$ . ?P1(type:Patient) . [Take-Medication] . ?M1(type:Medication)
Ct= {$M1(Name="Some Stuff")},
Out= {?T1(Result)},

## 2.3 DP Services as Views

Web services are usually modeled with the de facto standard for service description OWL-S. In particular, OWL-S's *Service Profile* permits to model the service's effect (operation in terms of WSDL), inputs and outputs. On the other hand, Data Providing Services have no explicit effects, instead, the semantic relation holding between their inputs and outputs must be captured. Therefore OWL-S may not be the best choice for describing them since it does not allow to capture this relation. We model Data-Providing Services in our approach as *RDF Parameterized Views (PVs)* over OWL ontology as they necessitate a particular set of inputs (the parameters values) for their invocation and return a particular set of outputs. Initially a parameterized view is a technique that has been used to describe content and access methods in the widely used Global-as-View (GaV) integration architectures [7], and also recently to describe privacy constraints in [17].

*Each PV is a predicate* $WS_i(c_i)$*:- 4-**tuple** <**Backbone, Ct, In, Out**>* where, $WS_i(c_i)$ is called the *view head* and it comprises the name of corresponding service and its returned results. The rest is called the *view body* and it has the following contents:

1. ***Backbone** it comprises both the variables set **C** (of classes types) linking the input and the output of the service, and the object properties set **OP** relating the different variables in C.*
2. ***Ct** is the constraints set imposed on the datatype properties of C without being required inputs of the service.*
3. ***In** is the necessary literals for the service invocation.*
4. ***Out** is the output literals.*

According to this definition, the parameterized views for our example have the form presented in figure 3. Concretely we establish these views with RDF triples as showed in figure 9 (in the appendix)

WS1(TestA (result)):-
{ {?TestA . [Has_Test]$^{-1}$ . ?Patient },
Ct: {øø}
In: {$Patient(Name)}
Out: {?TestA(Result)} }

WS2(Measurement(Name), TestB(Result)):-
{ { ?TestB . Ψ { . [Has_Test]$^{-1}$ ?Patient
. [Contains]$^{-1}$ . ?Measurment },
Ct: {øø}
In: {$Patient(Name)}
Out: {?TestB(Result), ?Measurment(Name, Value)} }

WS3(Medication(Name)):-
{ {?Medication . [Take_Medication]$^{-1}$ . ?Patient}
Ct: {øø}
In: {$Patient(Name)}
Out: {?Medication(Name)} }

WS4(Patient(Name)):-
{ {?Patient . [Take_Medication] . ?Medication},
Ct: {øø}
In: {$Medication(Name)},
Out: {?Patient(Name)} }

**Fig. 3.** The defined Parameterized Views for the DP services in the running example

Notice that the parameterized view not only indicates the output and the input of the service, but also how they are semantically related with respect to the underlying ontology. Schematic representations of our services' PVs are shown in figure 4, where circles, triangles and squares represent respectively variables of types defined by the ontology classes, mandated inputs (literals) for the service invocation process, and the literals returned by the services.

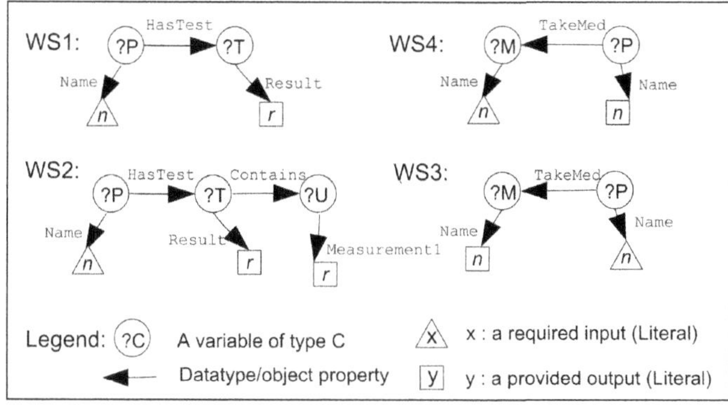


**Fig. 4.** The schematic representations of services's parameterized views PVs

## 3 Composing DP Web Services

Note that peers do not expose data portions as OWL instances, rather data is behind services. Thus queries cannot be applied squarely to data, instead the targeted peer must analyze both the received query and the defined *PVs* to settle on services whose composition (data-driven composition) can return relevant results. However, before proceeding with the composition process, the defined views must be preprocessed.

### 3.1 Preprocessing the Defined RDF Parameterized Views

Before the rewriting process, the parameterized views should be preprocessed. This preprocessing includes the following steps.

**Step 1. Extending the obtained PVs to reflect OWL "explicit" subclassing statements.** For those peers which have not the capability to apply some reasoning while matching the query with available PVs, obviously a query making reference to the concept "Test" cannot be answered with a *PV* if this makes reference to another concept to define the same data item (e.g. the concept "TestA", a specialization of Test) although this *PV* (or service) returns relevant information. To remedy this, there are two possible solutions. The first is to include in the algorithm the capability to verify whether a concept is a super/sub class of another (based on the ontology definition) while matching both query and services backbones. This is expensive in terms of the time necessary for the rewritings computation with large ontologies. The other solution is to extend previously defined *PVs* with the constraints *subClassOf, subPropertyOf* that are explicitly declared in the ontology. For example in (figure 10, case A) a new triple was added to the *PV* of $WS_1$ indicating that an instance of "TestA" is also an instance of "Test".

**Step 2. Skolemizing triples.** Variables denoting classes in PVs need to be *skolemized* [2], that is to replace each variable by a *skolem* function helpful to merge instances stemming from different services, e.g. the variable ?Patient (of type Patient) is replaced by the function SF1(Name), that is to say if two instances have the same name then they are considered as being denoting the same entity and thus can be merged. An example of a skolemized $PV$ is shown in (figure 10, case B). The properties of a skolem function for a particular class are chosen by the domain expert.

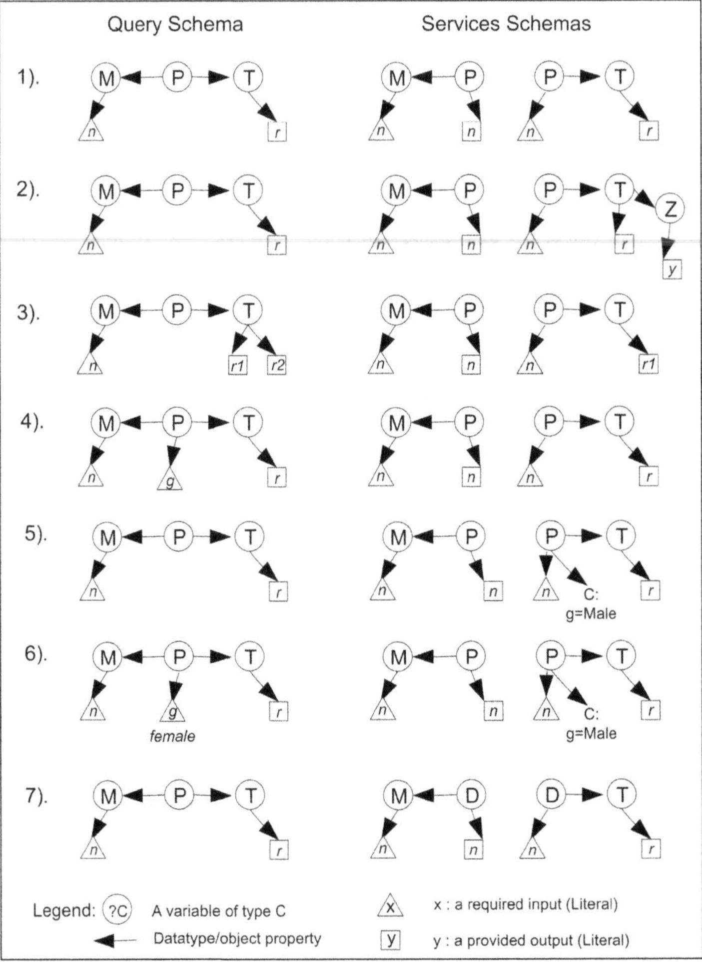


**Fig. 5.** Matching query's schema and Web Services' parameterized views (For simplicity, properties labels were left out)

## 3.2 Composing DP Services Based on Their Views

For formal discussion assume a query $Q$ *(backbone*$_Q$*,* $Ct_Q$*,* $Out_Q$*)* and a set of services, each has a $PV_i(backbone_{S_i}, Ct_i, In_i, Out_i)$. In order to satisfy Q, backbones union of selected services has to cover the query's backbone and the final output of the composition satisfies $Out_Q$. As can be seen in the figure 5, several cases may occur while the query rewriting process; they are as follows:

1. Case (1). The union of the services' backbones covers $backbone_Q$, all of the requested data items (datatypes) are provided (depicted as "r" in the figure), and the Q's constraints list $Ct_Q$ is satisfied with the union of $Ct_i$. In this case services fully satisfy Q and thus their combination returns a valid rewriting.
2. Case (2). The union is larger than the query backbone with provision to all of the asked outputs. In this case it should be verified whether the additional concepts (e.g. the variable Z in figure 5) pose additional constraints (thus returning more specific results) or if they have a corresponding input parameter necessary for the service invocation (in the last case the service cannot be invoked as a necessary input will not be available).
3. Case (3). The union provides a partial result as some literals do not appear in the output. Herein if one of the missing outputs is mandated then the combination of these services will be considered as an invalid rewriting.
4. Case (4). The union of the services' backbones covers $backbone_Q$ but a constraint in the $Ct_Q$ was dropped (e.g. patient gender must be male). Here if dropped constraints were mandated then these services will be rejected. Otherwise these constraints can be enforced on data flow between services.
5. Case (5). The union of the services backbones covers $backbone_Q$, but with enforcing an additional constraint that was not specified in the query's $Ct_Q$. Herein obtained results will be specific, however they are still relevant ones.
6. Case (6). The union of the services backbones covers $backbone_Q$, but there is a conflicting constraint between Q and one of the services (e.g. the gender property has conflicting values male vs. female). The rewriting here is invalid.
7. Case (7). The union of the services backbones does not cover the query backbone. In this case these services must be rejected even if they return similar outputs to the demanded ones (e.g. in (figure 5, case (7)) the first service returns the doctors names who have prescribed a medication, and the second returns the test results which were verified by a particular doctor).

All of these observations were dealt with in our Web services-based query rewriting algorithm presented next.

**Query resolution algorithm.**

Inputs:
-A query Q $< Q_{backbone}, Ct_Q, OS_Q >$.
-The service List L, where each service $S_i \in L$ has a $PV_i(Ser_{i\ backbone}, Ct_i, In_i, Out_i)$.

Populate the list RSL (Relevant Services List) where $S_i \in$ RSL iff ($\exists\ ?c_i \in Out_i$, $\wedge\ \exists\ ?c_j \in OS_Q$) such that both $?c_i$ and $?c_j \in O : c_i$
**for** each $S_i \in L$ **do**
    **for** each variable $?c_i \in OS_Q$ where $c_i$ is its corresponding type class in O **do**
        **if** $c_i$ appears in the $Out_i$ **then**
            Add $S_i$ to RSL
        **else if** $c_i$ is a subclass to one or more of used classes in $OS_Q$ **then**
            //$S_i$ returns generic result.
            Reject $S_i$ unless otherwise specified by the user.
        **else if** $c_i$ is a superclass to one or more of used classes in $OS_Q$ **then**
            Add $S_i$ to RSL
**if** $RSL$ is not empty **then**

```
      for each S_i in RSL do
      //If Ser_backbone i contains a variable v_i of type not used in
         Q_backbone such that v_i poses a constraint in the Ct_i ∪ In_i then reject
         S_i;
            Construct the types schema (sub graph of O) QTS used in Q;
            Construct the types schema (sub graph of O) STS used in S_i;
            if ∃c_i such that c_i ∈ STS and c_i ∉QTS then
               if∃ ?v ∈ c_i such that ?v poses a constraint in Ct_i ∪ In_i then
                  Reject the service
       //Verify if the S_i's backbone is covered by Q's backbone
          Take the backbones of S_i and the Q
          Let ?c_1 be the common output variable between S_i and Q
          Let ?c_in be a variable enclosing some literals necessary for
          the service's invocation
          Let ?c_i be a variable varying from ?c_1 till ?c_in in the service backbone
            for (?c_i=?c_1 till ?c_i=?c_in) in S_i's backbone do
             Let O : c_i be the corresponding class type of ?c_i
             Let O : c_j be the corresponding class type of ?c_j, where
             ?c_j is ?c_i's analogous variable in Q's backbone
               if (¬(c_i ≡ c_j) or ¬(c_i subclass c_j) )then
                  Reject S_i
               Let Ct_S and Ct_Q are the constraints sets pertaining
            to variables involved in compared backbones
               if exist a conflicting constraint between Ct_S and Ct_Q then
                  S_i is rejected
               else if Ct_S > Ct_Q then
                  //S_i returns more restricted results, the user has
               the choice to whether or not accept specific results
               else if Ct_S < Ct_Q then
            S_i returns more general results, the user
            has the choice to whether or not accept general results.
               if S_i is not rejected yet then
                  Insert the service predicate in the query
                  Eliminate the service backbone from the query backbone
                  Eliminate the service output from the query output
                  Insert the service's inputs in the Q's outputs and mark
                  them as mandated ones
                  if Q:OS is not satisfied with the Q:Ct then
                     Repeat the algorithm on the new Query
   else
      Q cannot be resolved
Output: The rewritings list.
```

The possible rewritings of our query (in the running example) are shown in figure 6. Our algorithm starts with looking for services which provide at least one of the asked outputs. It finds that two services do provide relevant results ($WS_1$ and $WS_2$) since these services return the results of "TestA" and "TestB" respectively (subclasses of "Test" (subClassing constraints are added in the views definitions)). Each of these services corresponds to an independent rewriting. The backbones of $WS_1$ and $WS_2$ match part of Q's backbone, and their constraints lists satisfy the involved constraints in the query's Ct. Next, in each rewriting the algorithm eliminates the service's backbone from the query's backbone and its provided output from the Q's output set. Then, it inserts the needed inputs for the service invocation as mandated outputs in the new query Q's OS (Output List). The obtained result after this iteration is shown in figure 6. Then the same algorithm is applied again on the new yielded query in each rewriting. This time, it turned out that $WS_4$ satisfies the new query as it

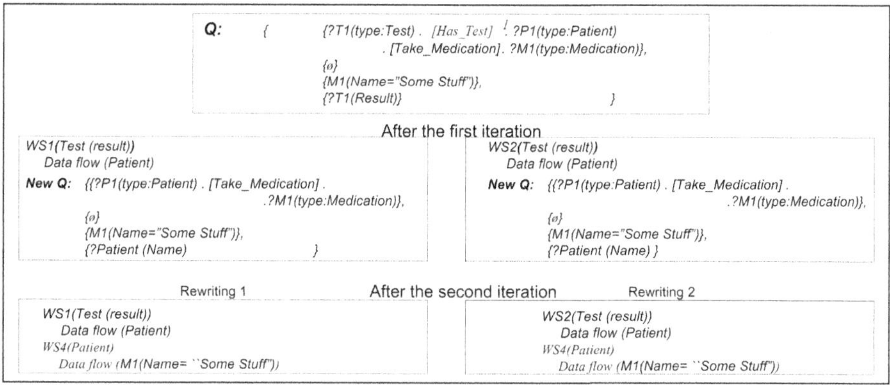


**Fig. 6.** The query rewriting process for the running example. There were two yielded rewritings for Q.

provides the required outputs, its backbone matches exactly the query backbone, and its inputs are satisfied with query constraints list.

### 3.3 The Composition Execution

The previous algorithm yields a certain number of query rewritings (set of DP services compositions). These compositions encode the invocation order of the combined services. Before executing these compositions, we apply an extra algorithm (this was left out for space limitation) in order to superimpose these compositions (if possible). The intent of this algorithm is to avoid the duplicate invocation of the same service across several compositions. For instance, instead of invoking $WS_4$ twice, it suffices to invoke it one time and then to use the obtained results to invoke both of $WS_1$ and $WS_2$ (see figure 7, case a ). In our system the results of each service invocation are automatically transformed in the form of OWL instances (skolem functions are exploited during this process) before being used to invoke subsequent services or sent to the requester . This helps in merging and aggregating results stemming from different services and enable us to detect, if needed, data inconsistences (between services) based on some semantic reasoners (e.g. two instances of patient with the same value of a functional property (e.g. national ID) but with different names, probably misspelled, are detected).

We need to apply extra treatment and processing over data flow among combined services. In general, three semantic operators can be applied. They are as follows.

- **Semantic Union** ($WS_i \cup WS_j$): This operator is used to semantically combine the outputs (OWL instances) of the services $WS_i$, $WS_j$. The outcome includes the disjoint instances provided by $WS_i$ and $WS_j$ and the semantically equivalent instances provided by both only once. For example suppose that there is a second service ( $WS^*$) equivalent to $WS_4$ (depicted with doted

lines in figure 7, case b). In this case, obtained results of both $WS^*$ and $WS_4$ must be combined with the union operator before invoking the subsequent services.
- **Semantic Intersection** ($WS_i \cap WS_j$): This operator can be used to return the semantically equivalent instances provided by both $WS_i$ and $WS_i$.
- **Semantic Difference** ($WS_i \oslash WS_j$): It can be used to return the instances provided by $WS_i$ excluding the equivalent instances provided by $WS_j$.

Note that treatments on data flow are applied in the execution time of the composition. Returning back to our example, the execution of the compositions is done as follows. First, $WS_4$ is invoked with medication name. The returned patients' information is put automatically in the form of OWL instances. Then, for each obtained instance of patient we invoke both $WS_1$ and $WS_2$ and the results are materialized as OWL instances then sent to the requester.

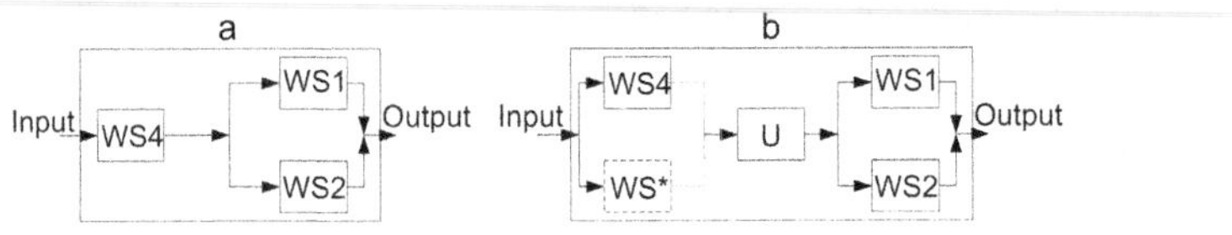


**Fig. 7.** Data flow in Data driven Web services composition

## 4 Implementation

We are focusing on integrating our framework for supporting DP services within a P2P-based data management system that is being developed for the purpose of integrating proprietary data resources of some private health centers. In this system, each peer has its own proper ontology, and it maps it to the neighboring ontologies (in a pair-wise manner) via *OWL mapping constructs*. When a peer receives a query from neighboring peers (or from the peer's user) it tries to resolve it by composing its DP services. We have implemented our algorithm using Jena Framework [6]. Currently we are conducting some experimental tests on our algorithm to measure the impact of PVs number and the ontology volume on the rewritings computation time. Figure 8 shows some preliminary results of our experimentation. The algorithm can scale up to 100 views under 6 seconds. This test was conducted in the following context: 1). The used ontology contains 32 classes and 223 properties. 2). Test performed on a PC with a single 3.06 GHz and 512MB RAM.

## 5 Related Works

The work presented in this article is closely related to research in several areas.

Previous research works in P2P data integration and interoperation have focused on traditional data sources (with a direct data accessibility) and overlooked

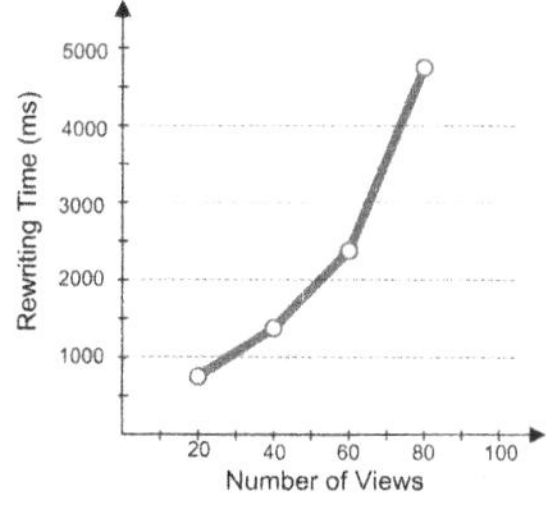


**Fig. 8.** Preliminary results of evaluating the performance of the algorithm as a function of the PV number

an enormous number of data sources exposed as Web services (or Data-Providing services) in the P2P environment. We cite for example P2P systems like Hyperion [18], Piazza [8], Edutella [11] PEPSINT [3], SEWASIE [1]. These systems attempt to integrate relational databases in the environment by establishing either *mapping tables* [18] or GaV/LaV-based mappings [10] among the different data sources. Some other recent systems like [19,4] suppose that data sources are fully transformed into forms like (OWL or RDFS) before applying, in a subsequent step, queries to data in these forms. In general, queries in these systems are forwarded from peer to another and applied to data on each site then results are sent to the requester. In our system, results can be obtained in two ways. Either queries are matched with DP services on each peer and then matched services are sent back to the requester, or matched services are composed on each site then executed before sending the final result to the requester. In addition, our queries are formulated against OWL instances before being resolved in terms of services, consequently any peer that still exposes its resources in the traditional way can participate and respond to the others' requests in the collaboration environment.

Another interesting research area is the combination of Publish/Subscribe mechanisms and P2P systems [21,13]. In these systems individual peers establish their *acquaintances groups* based on a *subscription/publication model.* Peers describe their data as *publications* and express their needs as *subscriptions*, then while subscriptions are forwarded from peer to another they get matched against publications. Matched publications are sent back to the original peer along with information about their hosting peers which will become part of the original peer's acquaintances group. We intend to employ this paradigm in advertising and matching DP services. The RDF views of DP services will be encapsulated within publications then disseminated on the network. Also, several algorithms for publications/subscriptions matching were proposed. Notably the one proposed in [16] for checking whether or not a subscription is covered by a set of similar subscriptions can be used to check whether or not a composition of DP services satisfies the constraints set of the treated query. In the near future, we intend to incorporate this algorithm in our framework.

Previous works in the area of web services composition have focused on constructing a composite service fulfilling a sophisticated task (or functionality) out

```
WS1(?TestA<Result>):-
  Backbone = { ?Patient . rdf:type .        Op1: Patient
               ?Patient . Op1:has_Test . ?Test A
               ?Test A . rdf:type .         Op1: Test A }
  Ct       = { øø }
  In       = { ?Patient . Op1: Pat_Name . $Name }
  Out      = { ?Test A . Op1:has_result . ?A_Result }

WS2(?TestB<Result>, ?Measurment<Value>):-
  Backbone = { ?Patient . rdf:type . Op1: Patient
               ?Patient . Op1:has_Test . ?Test B
               ?Test B . rdf:type . Op1: Test B
               ?Test B . Op1:contains . ?Measurement }
  Ct       = { øø }
  In       = { ?Patient .Op1: Pat_Name. $Name }
  Out      = { ?Test B . Op1:has_result . ?B_Result
               ?Measurement . Op1:hasValue . ?Value }

WS3(?Medication<Name>):-
  Backbone = { ?Patient . rdf:type . Op1: Patient
               ?Patient . Op1:Take_Medication . ?Medication
               ?Medication . rdf:type . Op1: Medication }
  Ct       = { øø }
  In       = { ?Patient . Op1: Pat_Name . $Name }
  Out      = { ?Medication . Op1:Med_Name . ?Med_Name }

WS4 (?Patient<Name>):-
  Backbone = { ?Patient . rdf:type . Op1: Patient
               ?Patient . Op1:Take_Medication . ?Medication
               ?Medication . rdf:type . Op1: Medication }
  Ct       = { øø }
  In       = { ?Medication . Op1:Med_Name . $Name }
  Out      = { ?Patient . Op1: Has_Name . ?Patient_Name }
```

**Fig. 9.** The Parameterized views defined for the running example's services

```
WS1(?TestA<Result>):-
  Backbone = { ?Patient . rdf:type . Op1: Patient
               ?Patient . Op1:has_Test . SF2(?Id)
               ?Test A.   rdf:type . Op1: Test A
               ?Test A.   rdf:type . Op1: Test }
  Ct       = { øø }
  In       = { ?Patient . Op1: Pat_Name . $Name }
  Out      = { ?Test A . Op1:has_result . ?A_Result }

  A. Extending PVs with subClassing
  constraints

WS1(?TestA<Result>):-
  Backbone = { SF1(?n) .  rdf:type . Op1: Patient
               SF1(?n) .  Op1:has_Test . SF2(?Id)
               SF2(?Id).  rdf:type . Op1: Test A
               SF2(?Id).  rdf:type . Op1: Test }
  Ct       = { øø }
  In       = { SF1(?n) .  Op1: Pat_Name . $Name }
  Out      = { SF2(?Id) . Op1:has_result . ?A_Result }

  B. Skolemizing Pvs
```

**Fig. 10.** An extended PV for WS1. A new triple was added to reflect the relation between Test and Test A.

of a set of primitive services fulfilling simpler tasks each. For example authors in [20], describe a SHOP2 based system to automatically compose Web services. In addition to the input and output constraints, their system can also handle web services with preconditions and effects. The key difference between Web services composition here and the composition in our work is that while composition in these approaches is task-driven, it is data-driven in our work. That is to say, the ultimate objective of the composition is to provide as much complete answer as possible to the user queries. Also in our composition we need to apply extra treatment on data flow between services (data aggregation, redundancy elimination...etc).

## 6 Conclusions and Future Works

In this paper we have modeled Data-Providing services as RDF parameterized views over the domain ontology. When individual peers get interrogated, they resolve their received queries in terms of the available Data-Providing services. In the near future, we intend to incorporate some efficient algorithms [16] for constraints satisfaction checking while composing DP services. Also we need to deal with the efficiency issues of the composition. Another research direction, is to include data-mediating services in our algorithm (to convert data values if a conversion is needed e.g. the conversion of a measurement unit).

## References


1. Bergamaschi, S., Fillottrani, P.R., Gelati, G.: The sewasie multi-agent system. In: AP2PC, pp. 120–131 (2004)
2. Chen, H., Wu, Z., Wang, H., Mao, Y.: Rdf/rdfs-based relational database integration. In: ICDE, p. 94 (2006)
3. Cruz, I.F., Xiao, H., Hsu, F.: Peer-to-peer semantic integration of xml and rdf data sources. In: AP2PC, pp. 108–119 (2004)
4. Dimitrov, D.A., Heflin, J., Qasem, A., Wang, N.: Information integration via an end-to-end distributed semantic web system. In: International Semantic Web Conference, pp. 764–777 (2006)
5. Dogac, A., Laleci, G., Kirbas, S., Kabak, Y., Sinir, S., Yildiz, A., Gurcan, Y.: Artemis: Deploying semantically enriched web services in the healthcare domain. Information Systems Journal (Elsevier) (2006)
6. Jena Framwork. `http://jena.sourceforge.net/`
7. Halevy, A.Y.: Answering queries using views: A survey. VLDB J. 10(4), 270–294 (2001)
8. Halevy, A.Y., Ives, Z.G., Madhavan, J., Mork, P., Suciu, D., Tatarinov, I.: The piazza peer data management system. IEEE Trans. Knowl. Data Eng. 16(7), 787–798 (2004)
9. LeFevre, K., Agrawal, R., Ercegovac, V., Ramakrishnan, R., Xu, Y., DeWitt, D.J.: Limiting disclosure in hippocratic databases. In: VLDB, pp. 108–119 (2004)
10. Lenzerini, M.: Data integration: A theoretical perspective. In: PODS, pp. 233–246 (2002)
11. Löser, A., Siberski, W., Wolpers, M., Nejdl, W.: Information integration in schema-based peer-to-peer networks. In: CAiSE, pp. 258–272 (2003)
12. Maamar, Z., Benslimane, D., Ghedira, C., Mrissa, M.: Views in composite web services. IEEE Internet Computing 9(4), 79–84 (2005)
13. Muthusamy, V., Jacobsen, H.-A.: Small scale peer-to-peer publish/subscribe. In: P2PKM (2005)
14. Ooi, B.C., Shu, Y., Tan, K.-L.: Relational data sharing in peer-based data management systems. SIGMOD Record 32(3), 59–64 (2003)
15. Ouksel, A.M.: In-context peer-to-peer information filtering on the web. SIGMOD Record 32(3), 65–70 (2003)
16. Ouksel, A.M., Jurca, O., Podnar, I., Aberer, K.: Efficient probabilistic subsumption checking for content-based publish/subscribe systems. In: Middleware, pp. 121–140 (2006)
17. Rizvi, S., Mendelzon, A.O., Sudarshan, S., Roy, P.: Extending query rewriting techniques for fine-grained access control. In: SIGMOD Conference, pp. 551–562 (2004)
18. Rodríguez-Gianolli, P., Garzetti, M., Jiang, L., Kementsietsidis, A., Kiringa, I., Masud, M., Miller, R.J., Mylopoulos, J.: Data sharing in the hyperion peer database system. In: VLDB, pp. 1291–1294 (2005)
19. Spyropoulou, E., Dalamagas, T.: Sdqnet: Semantic distributed querying in loosely coupled data sources. In: ADBIS, pp. 55–70 (2006)
20. Wu, D., Parsia, B., Sirin, E., Hendler, J.A., Nau, D.S.: Automating daml-s web services composition using shop2. In: International Semantic Web Conference, pp. 195–210 (2003)
21. Yang, J., Papazoglou, M.P., Krämer, B.J.: A publish/subscribe scheme for peer-to-peer database networks. In: CoopIS/DOA/ODBASE, pp. 244–262 (2003)

# Participative Enterprise Modeling: Experiences and Recommendations

Janis Stirna[1], Anne Persson[2], and Kurt Sandkuhl[1]

[1] Jönköping University, PO Box 1026, SE-551 11, Jönköping, Sweden
{janis.stirna,kurt.sandkuhl}@ing.hj.se
[2] University of Skövde, P.O. Box 408, SE-541 28 Skövde, Sweden
anne.persson@his.se

**Abstract.** The objective of this paper is to report a set of experiences of applying participative enterprise modeling in different organizational contexts. While the authors have successfully applied the approach in many organizations, the paper primarily concentrates on three cases. On the basis of these experiences the paper presents a set of generic principles for applying participative enterprise modeling.

**Keywords:** Enterprise modeling, participative modeling.

## 1 Introduction

Enterprise Modeling (EM) is an activity where an integrated and commonly shared model describing different aspects of an enterprise is created. An Enterprise Model comprises a number of related "sub-models", each focusing on a particular aspect of the problem domain, such e.g. processes, business rules, concepts/information/data, vision/goals, and actors. EM is often used for developing organization's strategies, business process restructuring, business process orientation, communication of work procedures, eliciting information system requirements, etc. More about the applicability of EM is available in [1]. In all these activities the development team addresses frequent challenges of how to discover the domain knowledge, how to consolidate different stakeholder views, and how to represent this knowledge in a coherent and comprehensive model. Additionally, there is a need to ensure that the decisions made and reflected in the models are taken-up and implemented in reality. Addressing these challenges by the traditional consulting approach of fact gathering, analysis and then delivering an expert opinion is not efficient when dealing with "ill-structured" or "wicked" problems [2] typically occurring in organizations. As a result participative EM, i.e. modeling in facilitated group sessions, has been established as a valuable and practicable instrument for solving organizational design problems (c.f., e.g. [3, 4, 5, 6, 7]).

Therefore, the objective of this paper is to *report a set of experiences of applying participative enterprise modeling in different organizational contexts.* While the approach has been successfully applied in many organizations this paper primarily concentrates on three cases, namely a healthcare organization, a firm developing

J. Krogstie, A.L. Opdahl, and G. Sindre (Eds.): CAiSE 2007, LNCS 4495, pp. 546–560, 2007.

components for the automotive industry, and a municipality. On the basis of these experiences we also present a set of generic participative EM principles.

The research approach is conceptual and argumentative based on a number of case studies that were carried out in public and private organizations [8, 9, 10] and an interview study with experienced practitioners [11].

The remainder of the paper is organized as follows. In section 2 we provide a background to EM methods and ways of working. Section 3 presents three case studies, namely, Kongberg Automotive, the Riga City Council and Skaraborgs Sjukhus. Section 4 presents a set of best practices of participative EM. The recommendations are illustrated using the three cases. Conclusions and future work are, finally, discussed in section 5.

## 2 Background to Enterprise Modeling

In Scandinavia, Business or Enterprise Modeling were initially developed in the eighties by Plandata, Sweden [12], and later refined by the Swedish Institute for System Development (SISU). A significant innovation in this strand of EM was the notion of business goals as part of an Enterprise Model, enriching traditional model component types such as entities, and business processes. The SISU framework was further developed in the ESPRIT projects F3 – "From Fuzzy to Formal" and ELEKTRA – "Electrical Enterprise Knowledge for Transforming Applications". The current framework is denoted EKD – "Enterprise Knowledge Development" [4, 6]. Apart from the "Scandinavian" strand of EM, a variety of other methods have been suggested (c.f., e.g. [13], [14], [15], [16], [17], [18]).

[1] show that EM can be used for two main types of objectives – (1) developing the business, e.g. developing business vision, strategies, redesigning the way the business operates, developing the supporting information systems, or (2) ensuring the quality of the business, e.g. sharing the knowledge about the business, its vision, the way it operates, or ensuring the acceptance of business decisions through committing the stake-holders to the decisions made.

### 2.1 EKD

EKD – Enterprise Knowledge Development method [6] is a representative of the Scandinavian strand of EM methods. It defines the modeling process as a set of guidelines for participative way of working and the modeling product in terms of six sub-models each focusing on a specific aspect of an organization (see table 1).

The ability to trace decisions, components and other aspects throughout the enterprise is dependent on the use and understanding of the relationships between the different sub-models addressing the issues in table 1. When developing a full enterprise model, these relationships between components of the different sub-models play an essential role. E.g. statements in GM allow different concepts to be defined more clearly in the CM. A link is then specified between the corresponding GM component and concepts in CM. In the same way, goals in the GM motivate the existence of processes in the BPM. Links between models make the model traceable. They show, for instance, why certain rules, processes and information system requirements have been introduced.

**Table 1.** Overview of the sub-models of the EKD method

| | **Goals Model (GM)** | **Business Rules Model (BRM)** | **Concepts Model (CM)** | **Business Process Model (BPM)** | **Actors and Resources Model (ARM)** | **Technical Component & Requirements Model(TCRM)** |
|---|---|---|---|---|---|---|
| Focus | Vision and strategy | Policies and rules | Business ontology | Business operations | Organizational structure | Information system needs |
| Issues | What does the organization want to achieve or to avoid and why? | What are the business rules, how do they support organization's goals? | What are the things and "phenomena" addressed in other sub-models? | What are the business processes? How do they handle information and material? | Who are responsible for goals and process? How are the actors interrelated? | What are the business requirements to the IS? How are they related to other models? |
| Components | Goal, problem, external constraint, opportunity | Business rule | Concept, attribute | Process, external proc., information set, material set | Actor, role, organizational unit, individual | IS goal, IS problem, IS requirement, IS component |

## 2.2 C3S3P

C3S3P is based on work in several EU projects from the area of networked and extended enterprises. An extended enterprise is a dynamic networked organization, which is created ad-hoc to reach a certain objective using the resources of the participating cooperating enterprises. In order to support solutions development for extended enterprises, the EXTERNAL project [19] developed a methodology for extended enterprise modeling [20], which initially was named SGAMSIDOER based on the abbreviations of the different modeling steps proposed: Scoping of the extended enterprise, Gather existing partner information, Analyze extended enterprise potential, Model extended enterprise, Simulate and analyze model scenarios, Implement Model, Deploy extended enterprise, Operate, as well as Evaluate, Re-engineer extended enterprise. This methodology was further developed towards a complete customer delivery process denoted C3S3P, which is used in the ATHENA[1] and MAPPER[2] projects. C3S3P, like SGAMSIDOER, aims at executable solutions based on visual EM. The seven C3S3P phases are:

- Concept Testing: pre-studies are performed to investigate whether EM is a suitable and accepted way of developing executable solutions for the networked enterprise
- Scaffolding aims at creating shared knowledge and understanding among the participants of the project about the scope and challenges of the project.
- Scenario Modeling: creation of executable models supporting the networked enterprise in the defined scope including all relevant dimensions required, like process, product, organization or IT-systems
- Solutions Modeling: refining the scenario model by integration personnel, product structures, document templates and IT systems required for using the enterprise model in an actual project
- Platform Configuration: configure the solution models for use in the networked or extended enterprise by connecting the enterprise model to the platform used

[1] http://www.athena-ip.org/
[2] http://mapper.troux.com/

- Platform Delivery: encompasses the roll-out of model-configured solutions
- Performance Improvement by capturing indicators for process and product quality and using adequate management instruments.

# 3 Cases of Applying Participative EM in Practice

This section presents three EM application cases at Kongberg Automotive, the Riga City Council and Skaraborgs Sjukhus.

## 3.1 Participative EM at Kongsberg Automotive (KA)

Kongsberg Automotive is a first tier supplier to the worldwide automotive industry developing and manufacturing gearshift controls and seat comfort systems. EM at KA was performed during the EU FP6 project MAPPER, which aims at supporting collaboration between different actors in the automotive supply chain by using reconfigurable enterprise models and providing an infrastructure with collaboration services and for executing these models. The enterprise models developed follow the POPS* approach [21]; they include dimensions like processes, organization structures, product structures, IT-systems and any other dimension relevant for the modeling purpose in one single visual enterprise model.

The development of enterprise models roughly followed the C3S3P approach (section 2.2) and was developed in different phases. In this paper we focus on the scaffolding phase, aiming to create a joint understanding in the modeling team about current situation, challenges and way of working of the modeling team. The group consisted of 3 to 7 employees from KA representing different roles (material specialist, electrical engineer, product manager, etc.), 3 to 4 participants from research organizations and 1 or 2 experts from the developer of the modeling tool used. All modeling sessions added to 6 days of work for the team. During the first modeling workshop the team agreed to use the following roles:

- *Process owner* being responsible for establishing the modeling activity within the enterprise, selecting the right personnel resources, arranging meetings, etc.
- *Facilitator* providing expertise in using the selected modeling process and tool as well as providing supports the modeling process and model development by coaching the modelers. This role facilitates model construction and development.
- *Modeling expert* having in-depth knowledge in the modeling method and tools.
- *Tool operator* responsible for documenting the enterprise models in the computerized tool during the modeling process
- *Domain expert* providing knowledge about the domain under consideration, which is basis for modeling.

The work was done during joint modeling workshops with about 10 participants. We used the visual EM tool METIS[3]. All aspects of modeling were jointly performed in a model projected on a large screen. Each modeling workshop started with defining objectives of the session. The scaffolding focused on the case under discussion – the

[3] Information about the METIS product is available at http://www.trouxmetis.com/

process of innovation at KA. The main task was to establish a process for creating new and innovative products in the advanced engineering department.

After the workshops, the produced models were consolidated by the facilitator. This task aimed at improving the visual structure of the model, completing textual descriptions of the model elements and identifying open questions, inconsistencies and needs for refinement. The workshops that followed always started with a walk-through of the current model version, in order to update all partners about the current status, introduce changes that have been made as well as raise issues for discussion.

The use of a visual modeling language supported the participatory approach providing a means for instant discussion of the modeling results, checking accuracy, and correcting potential shortcomings. It also helped updating all partners on the current status after a period of offline work. Visual modeling was equally efficient and useful in the initial sessions devoted to brainstorming and in final modeling sessions used for refining models. While the participative approach might have appeared to consume a lot of resources it gave the desired result of stakeholder involvement and correctness of the model. With a facilitator and tool operator managing the modeling tool skillfully the delays for updating the model were rather small. We have to recognize that with increasing level of detail, the visual models became quite complex and large, which makes it difficult to provide paper-based versions for stakeholders preferring to work with paper printouts. In our case there was no such stakeholder, but in other projects this would have been a likely situation.

### 3.2 Paticipative EM at the Riga City Council (RCC)

Riga City Council (RCC) is a municipality responsible for administration of public affairs of the city of Riga. In the past ten years the RCC has developed a large amount of information systems supporting its various functions. However, these systems did not address the growing need for managing RCC's organizational knowledge and competence. To answer this challenge RCC participated in FP5 IST project "Hypermedia and Pattern Based Knowledge Management for Smart Organizations" with the objective of developing and adopting a system for collecting and disseminating knowledge regarding strategic issues of capital importance for decision making at different levels in the city's administration [10].

The initial phase of the project was devoted to setting the Knowledge Management (KM) strategy, KM processes, and requirements for KM systems. We used participative EM and the EKD method. The way of working consisted of:

- Interviews about the current state and future vision for KM with ca 50 high ranking politicians and managers working in various committees and departments of the RCC. The additional intangible effect of these interviews was increased awareness about and popularity of the project.
- Two modeling sessions with the top level management of the RCC in order to decide the KM vision and outline the KM adoption process
- Selecting three KM pilot applications – at the Riga Drug Abuse Prevention Centre, at the School Board of Riga City, and at the Traffic Department.

- A series of modeling sessions in each pilot area targeting specific issues of these departments. In total the sessions at this stage added to ca 10 days of work for several modeling teams of 5-10 participants. Each session started with a review of the work previously done.
- One modeling session about integrating the pilot cases and developing overall KM processes for the RCC.

The roles of participants were similar to the ones described in section 3.1. During the modeling sessions we used a large plastic wall and post-it notes to document the model. This approach proved to be useful because it does not require the modeling participants to "channel" their input to the model through an operator of a computerized tool, which often slows down the creative process. After the modeling session the facilitator documented the resulting models in the Visio tool. At the final stages of the project the modelers refined models by using the tool directly.

The participants accepted the participative way of working because it was apparent to them that it helps them to discuss issues openly and to agree on decisions. The role of facilitator was appreciated. In the beginning of the project the method providers acted as modeling facilitators, but as the project progressed the two people from the RCC developed the competency of facilitation and took over this role.

### 3.3 Participative EM at Skaraborgs Sjukhus (SKaS)

Skaraborgs Sjukhus (SKaS) is a cluster of hospitals in Western Sweden working together with primary care centers and municipal home care to provide high-quality healthcare to the citizens in the region within which they act. Some medical specialties have a higher degree of collaboration between the hospital, primary care centers and municipalities than others. An example of this is the treatment and prevention of leg ulcers. To decrease the healing time for various types of leg ulcers e.g. with diabetic patients large efforts are made by all actors involved to e.g. develop new and more efficient treatment methods and care routines. To address the challenge of efficient knowledge sharing among various actors in the healthcare process (e.g. nurses in primary care and municipal home care) all three healthcare organizations participated in a project to build a knowledge repository for learning and sharing of best practices with regard to treatment and prevention methods for leg ulcers [8].

In the project, participative EM has mainly been used as a means to develop a knowledge map that describes the content and structure of the knowledge repository. The knowledge map is in the form of an EKD concepts model. The roles of participants were similar to the ones described in section 3.1 and 3.2. The domain experts in this case were doctors and specialist nurses at the hospital. An initial version of the model was developed early on in the project using the "plastic wall" approach described in section 3.2. iGrafx Flowcharter was used to document the model. Throughout the project the model was refined a number of times as the understanding of the problem domain improved, among all actors involved. Changes to the model were made directly in the computer tool. In this case the concepts model functioned as a detailed "blueprint" for creating the knowledge repository. Therefore, particular attention was given to developing a model that was precise and correct.

# 4 Recommendations of Using Participative EM

This section presents our recommendations for conducting participative EM in practice. While in the course of discussion we will refer to the three application cases outlined in section 3, the knowledge is also grounded into application cases in organizations such as e.g. British Aerospace (UK), Capital Bank (UK), Public Power Corporation (Greece), Sema Group (France), Telia (Sweden), Vattenfall (Sweden), Volvo (Sweden), Verbundplan (Austria), RRC College (Latvia). It is also built on an interview study mainly targeting experienced practitioners [11].

## 4.1 Assess the Organizational Context

In any project, understanding the organization's *power and decision-making structure* is essential. It is within the boundaries of these structures that the stakeholders create their Enterprise Model. Having access to and having the trust of the relevant decision-makers is especially critical for participative EM project managers when it comes to obtaining enough effort from domain experts. The planning of participative EM sessions will openly reveal that the stakeholders involved will need to allocate their time and effort to modeling work. If participative EM is relatively new to the organization, the amount of man-hours will seem unnecessarily large, which may cause some reluctance with decision-makers. If other activities in the organization at any time are given a higher priority than the EM project, the resources allocated for modeling sessions will most likely be reduced. This will have a strong negative impact on the modeling result.

In the SKaS case the project had high priority with management and this priority was not changed during the project. Management also seemed to trust the judgments of project management. The group of domain experts and method experts was stable for the duration of the project. Certain days were allocated for the project in the domain experts' weekly schedules, which made it easy to plan for modeling sessions.

*Organizational culture* has significant impact on the results and effects of participative EM [1, 11]. In fact, it seems that failure to properly understand the culture of an organization is perhaps one of the most critical risks in participative EM. Participative EM requires that the participants consider themselves authorized to state their opinions and to suggest solutions. This approach is therefore only suitable in consensus-oriented organizations. In authoritative cultures, it will be extremely difficult to achieve consensus-driven participation in the modeling groups.

Official documents/systems (policy documents, strategy documents, internal instructions, web-site etc.) often reveal some of the organizational culture. Ask direct questions about how the organization looks upon the concepts of responsibility, co-operation and participation. This will give an idea of the management philosophy in the organization. Also, ask questions about how people in the organization will be informed about the project. If very strong restrictions are put on the involvement of a circle of people outside the modeling group, this may either indicate an authoritative culture or a hidden agenda. Ask questions about how any additional stakeholders may be contacted and involved. A strong enforcement of the official decision-making structure indicates that the modeling team will not be free to contact people without talking to their superiors. This may also indicate an authoritative culture.

Attitudes towards participation are often revealed in the way people talk about the problem at hand, other people in the organization, etc. Observe how people act when talking with each other and with the method provider. Look for attitudes towards different types of actors (superiors, subordinates, opposite sex, etc.). E.g. in a group of people, it can be observed by looking at the faces of people whether or not they agree with what is said or whether they approve of another person or not. Exaggerated agreement with a superior may indicate a need to always express opinions that are in line with those of a superior. This may also indicate an authoritative culture.

In a consensus-oriented culture subordinates can question superiors, the dialogue between levels of the organization is open and direct and reward systems encourage initiatives from all levels of the organization. In an authoritative culture management is by directives only, the dialogue is indirect, and where there are no reward systems for initiatives from different levels of the organization. Note that in an organization, different types of cultures can reside in departments, divisions, subsidiaries etc. This mixed organizational culture may be an effect of mergers between organizations. Note also that organizational culture may be amplified by the official decision-making structure. If the organizational culture seems to be authoritative, do *not* use the participative approach to EM. Try other approaches such as e.g. interviewing.

If the organizational culture is undecided, try to negotiate that modeling will be done in two steps. The first step will function as an initial test of whether or not a participative approach is suitable. If active participation is not achieved in the modeling team, use traditional interviewing for the remainder of the project. If the culture is mixed, use the participative approach for work in the consensus-oriented part of the organization and some non-participative approach in the more authoritative parts of the organization. However, do not mix the two groups of people.

The organizational culture at KA is characterized by distributed working groups from different cultures as design and manufacturing facilities around the world require an awareness of how to integrate different ways of working. This formed an excellent basis for a consensus oriented way of working where inclusion of different opinions and equality regarding expressing contradictory viewpoints was accepted as a natural way of working. In the RCC the culture was mixed. Some organizational units had an authoritative culture and some had a consensus oriented culture. For the trial applications we chose three with consensus oriented culture.

In the SKaS case the culture was mixed. In general, the sense of hierarchy between different professions such as doctors and nurses in a healthcare organization is very strong. However, in the modeling group, where both doctors and nurses were represented, the culture was consensus oriented. This could be explained by the strong common dedication to solve the problem at hand.

Hidden agendas will decrease the possibility of achieving the project goals, since different stakeholders will try to steer the project towards their own goals. The project definition states the official goals of the project. It serves as important input to detecting hidden agendas. If the organization has hidden agendas, it may be reluctant to give the necessary authority to stakeholders, which could be "suspected" of jeopardizing that agenda. There can be hidden agendas as a part of a project, and the whole project itself could be a hidden agenda. The latter is the most fatal one.

Interviews with stakeholders before starting the project may reveal hidden agendas, but in that case they need to be carried out by an experienced person with good social skills. Questions about how the project was initiated, how the project is anchored in the organization and how the result will be used afterwards are useful as probes. Hidden agenda may also and will often surface during the project, which calls for open discussions with the customer.

In RCC the project itself did not have hidden agendas, but on the other hand it dealt with an issue – knowledge sharing – which related to some hidden agendas mostly concerning reluctance of information sharing. We identified these issues during interviews and as a result prepared modeling objectives addressing them.

In the SKaS case there were no hidden agendas since all stakeholders had complete agreement with regard to both the problem definition and the problem solution.

### 4.2 Assess the Problem at Hand

There are two views among practitioners when it comes to defining the problem at hand. Some stress the importance of obtaining a clear problem definition and seem to believe that it is possible to acquire such a clear definition. Others, often the most experienced practitioners, claim that clearly defined problems in most cases are illusions and that they rather are detected as the project progresses. The objective of the project is negotiated with the customer or process/project owner. There are two main approaches to this end – (1) interview the key decision maker(s) on the customer side about the objective, or (2) conduct short participative EM session to identify the objective, preferably involving other stakeholders than the key decision maker(s). Approach (2) can be used when it is difficult for the customer to pinpoint the problem, which normally means that the customer is uncertain about what exactly she/he wants to achieve. If the uncertainty still remains after the short EM session, this may indicate that the problem at hand is a "wicked problem".

Assessing the complexity of a problem definition is an essential part of the project negotiation. Problem complexity influences the project planning in terms of activities and resources. For resources, the complexity of the problem influences the requirements of the actor/s responsible for carrying out the EM project. Three types of problems can be observed:

- Fairly "simple" problems have a clear definition and a perceivable solution, and which do not require the co-ordination of a large number of different preconditions, activities, actors and resources.
- "Complex" problems have a fairly clear definition and a perceivable solution, but which require the co-ordination of a large number of different preconditions, activities, actors, and resources.
- "Wicked problems" are ill-structured problems, which have no clear problem definition and where there is no way of measuring that the problem is solved.

For "simple" and "complex" problems, proceed to the planning phase. If the problem is considered to be "complex", ensure that a highly skilled person will lead the project. If the problem is considered to be "wicked", negotiate that the project will be carried out in three steps:

1. A pre-study phase where modeling is the approach to obtaining agreement to the main scope of the project.

2. A negotiation phase, where the actual project is negotiated and planned. Since a "wicked" problem comprises many unknown factors, the customer must be made aware of this. Preferably, the project planning should contain a number of evaluation steps, where the results of the project are evaluated and new decisions are made regarding the continuation of the project.
3. A completion phase, where the defined problem is solved as best can be done.

In the KA case, interviews with the project owner and the manager of the department under consideration gave a good impression of the problem at hand: the process of innovation had to be refined and probably restructured. As the complexity of this task even after the interviews was not fully clear, the decision was made to conduct several EM phases addressing different scopes and levels of detail. The selected C3S3P approach supported this intention quite nicely. However, it has to be noted that time plan and budget of the EU project MAPPER provided adequate frame conditions in terms of resources. In a commercial project, the frame conditions would have been subject of negotiation after every phase.

In the RCC case the project was initially seen as complex and the problem as unclear. After pre-interviewing (see section 3.2) the overall vision of KM at RCC and project's focus was modeled in two sessions with top management representatives. Based on the outcome of these sessions we planned the pilot cases.

In the SKaS case the problem was perceived to be fairly "simple" at the outset of the project. This assumption was not changed during the project. Hence, there was no need for re-planning.

### 4.3 Assign Roles in the Modeling Process

We recommend assigning the typical roles used in project management such as project owner, steering group and quality manager and in addition the following roles specifically related to participative EM projects.

The *modeling facilitator* is responsible for choosing the modeling language used in the sessions, conducting modeling sessions, assisting the modeling participants to discuss, capture and structure ideas, as well as helping to develop the model during the session. The facilitator is *only* there to moderate the problem solving process among the domain experts, not to solve the problem. The ownership of a problem and its solution should always remain with the stakeholders. We recommend using two modeling facilitators if possible. Two facilitators should always be used if the modeling group is larger than 8 people and/or if the duration is planned for more than 6 hours. In smaller projects the modeling facilitator may also act as tool operator.

The *tool operator* is responsible for drawing the model into a computerized tool. This can be done either after the modeling session or during the session. In the latter case the tool operator has to work in tandem with the facilitator to ensure that all the ideas and wishes of all participants are reflected in the model correctly.

The *modeling participants*, also called domain experts, are responsible for providing correct knowledge about the problem domain and making sure that it is reflected in the model. They are the problem solvers.

Allocating the relevant domain competency profiles to the project is a task that should not be taken lightly, since domain knowledge is the most critical resource in an EM project. One way of ensuring that the participating domain experts will contribute their knowledge in the modeling sessions is to interview them in advance. They also need to be prepared for what will happen during the sessions. This is particularly critical in organizations where the employees are not used to modeling in general and particularly to modeling in a group. Before the modeling session each participant has to: (1) understand the objective of the modeling session, (2) agree upon the importance of this objective, (3) feel personally capable to contribute to a positive result, and (4) be comfortable with the rest of the team (including the facilitator).

The best way of preparing the participants is to carry out individual interviews. In general, experienced facilitators do not face problems to get the customer to accept interviewing the modeling participants in advance as part of the project. In contrast, the less experienced claim that they seldom are given the opportunity to carry out interviews. They are in general aware of the importance of interviews and say that they need to improve their ability to negotiate the resources to carry them out.

At KA, interviews with the participants were done in a series of meetings with 2-3 persons. In these meetings, the area under consideration was discussed in order to prepare for the EM sessions and to investigate whether additional stakeholders should contribute specific domain knowledge. One result of these meetings was a textual description of the issues addressed. Furthermore, all participants of the EM sessions got basic training in the visual modeling language used and in the modeling tool.

In the RCC case the modeling participants were interviewed before the modeling session, which allowed us to plan the session, e.g. identify specific objectives, select participants, questions, and plan the course of the seminar.

In the SKaS case the participants in the modeling group were used to participative modeling from a previous project and stated positive experiences from that. The previous project had the same modeling facilitator. The participants in the group had all been involved in defining the project. Therefore, the project leader decided that interviewing the participants was not necessary. The positive outcome of the project supported this view.

The competency of the others than domain experts is equally critical. Three aspects of the project determine the needed EM competency in the team of method experts:

1. The degree of problem complexity. A wicked problem will need a more experienced and skilled modeling team than a simple problem.
2. The degree of creativity needed for the solution. Designing the future state or radically changing the current state needs more method competency than describing the current state.
3. The size of the project and the needed co-ordination effort. A large modeling project will need a leader and experienced and skilled method expert with a holistic view. Furthermore, a large project might also require someone being responsible only for documenting and managing the modeling results.

In the RCC case the problem was complex and the problem definition was abstract allowing many alternative solutions involving different stakeholder types. To answer these challenges three pilot projects addressed the problem from different perspectives. They each had a pilot owner, a modeling facilitator and a tool operator. The owner of the whole project at RCC coordinated the efforts of the pilot cases. In summary the complexity and the size of the project required experienced modeling facilitators with project management skills.

In the SKaS case the problem was fairly "simple" and the project definition gave very little need for creativity. The size of the project when it comes to modeling activities was rather small. Since the project leader and facilitators were experienced the project was carried out in a very controlled manner.

### 4.4 Acquire Resources for the Project in General and for Preparation Efforts in Particular

One important insight that characterizes experienced EM practitioners is that it is unprofessional to assume responsibility for a project without the necessary resources [11]. A professional attitude, although drastic, is to refuse projects without the proper resources. This seems in fact to be part of the professional ethics of expert practitioners.

Management support is essential for a project to be successful. It is a critical precondition for obtaining the necessary resources and for motivating stakeholders to commit to the modeling work. It will also facilitate the involvement of skilled method experts even if it may be costly. Management should also give the modeling team the authority to act and make decisions within the boundaries of the project. A critical issue is the persistence with which management keeps supporting the project even if more resources are needed later on, or if the project runs into other types of problems. This is particularly important if the project is larger than 1-2 modeling sessions.

The KA case is an excellent example for the positive effects of full management support. After having made the decision to conduct the project, which was based on a clear description of goals and the planned process, the responsible manager did not only arrange for the required personnel resources and facilities, but also promoted the project in the organization and opened all desirable information sources.

In the RCC case management provided enough resources to carry out the pilot applications. On the other hand the top management representatives sometimes canceled their participation at a modeling seminar or sent a replacement instead.

In the SKaS case support from management was strong. The modeling team was given the needed modeling resources and the authority to carry out the project as they deemed fit. This shows that management had great confidence in the team and that the priority of the project was high.

Our case studies and interview study indicate that the effort spent on preparation is in direct relation to the quality of the project results. Experienced practitioners claim that they do not accept projects where the resources for preparation are too small, while the inexperienced report severe problems that are related to lacking preparation. We suggest distributing effort in a modeling project according to: preparations (assessing the organization, project definition, interviews, etc) ~40%, modeling seminars ~30%, and documenting and reporting ~30% of the total effort. The figures are mainly based on interviews with practitioners with 10-30 years of participative EM experience. This distribution of resources is only given as an indication; depending on the project aim and duration they may actually vary within ca 10%. E.g. some very short projects might not require extensive documentation.

The KA and RCC cases followed this distribution of effort in general. In the SKaS case the distribution of effort was, however, different. The need for preparation was much smaller due to the fact that the problem was simple and well defined and that the modeling team knew each other from before. Also, the domain experts in the modeling group were involved in defining the project.

### 4.5 Conduct Modeling Sessions

Carrying out a modeling session needs concentration and dedication from all participants involved. A detailed discussion about how a session should be managed is beyond the scope of this paper. However, we consider the following issues to be of utmost importance for the quality of the outcome of a modeling session:

- Each modeling session should have set clear objectives of practical value to the organization.
- Use a modeling notation that everyone understands. Participation will be severely hampered if too much attention is put into understanding the notation used. If the participants are not used to modeling, use a relatively simple and intuitive notation.
- Do not "train" the modeling participants in method knowledge. It is the responsibility of the modeling facilitator that the chosen method/notation is correctly used. Too much attention to the method/notation used will distract the modeling participants from solving the problem at hand. Our experience is that hands-on practice is the best way of becoming acquainted with a method/notation.
- Keep everyone involved and focused on the problem at hand. Avoid side discussions that will distract attention from the problem at hand.
- Do not accept unknown participants in the modeling session. In the best case they will keep silent and leave early, because they do not have the background knowledge that allows them to participate efficiently. In the worst case they might try to sabotage the modeling effort, to fulfill their own agenda – something that should have been discovered in the pre-interviewing stage.
- The problem owner and/or the insiders of the problem area should not dominate the seminar – the point of having a broader modeling group is to extend the view.
- Establish a common vocabulary – developing a CM might help to achieve this.
- Develop models in parallel – e.g. decide on a business goal, then switch to modeling a business process that would fulfill the goal, and then model the necessary roles performing and being responsible for the process. How to shift the group's attention between the sub-models depends on the project objectives, the situation in the organization, and the findings in the pre-interviews. More guidance about this is available in [6] and [11].
- Make concrete decisions in the session – attach roles and responsibilities to goals, processes and change actions.
- The same model might need to be improved in several modeling sessions because the group's understanding of the modeling issue tends to change during the project.
- The result of the session, the model, should deliver a solution – a common situation is that the model is too "polite", addressing only general and well known issues without tackling some of the hard problems of the organization.
- Make sure that everyone knows what will happen after the seminar – and whether they should carry out some of the actions decided and documented in the model.

In the RCC case we followed these guidelines. On a few occasions of top level managers sent replacements to modeling seminars. These people were either unable to contribute and left early or started to investigate how is this project related to other projects which they knew.

In the SKaS case modeling successfully proceeded according to the above recommendations. Since the participants were familiar with modeling from before, there was need to train them. In the previous project they were involved, however, there was a conscious decision to train the participants by hands on experience, which proved successful. The EKD notation was perceived to be easy to understand by the participants, even if they had no previous experience from modeling. The modeling team was stable throughout the project, which facilitated the constant refinement of the EKD concepts model, which was at the center of the modeling activities.

## 5 Concluding Remarks

This paper has presented a number of generic recommendations for carrying out participative EM in diverse organizational contexts. The main elements of EM are the notation and the modeling process. In participative EM, most of the critical success factors pertain to the modeling process. The positive effects of participative EM are: (1) Enhanced quality of the Enterprise Model, (2) Consensus among stakeholders, and (3) Acceptance and commitment to the modeling result [11]. However, to achieve these effects substantial knowledge and understanding of the modeling process is needed, as well as experience and skills with regard to managing people in a modeling session. Hence, successfully carrying out participative EM is a task that is far from trivial, which requires skillful and experienced professionals both when it comes to the roles of facilitator and project manager. In summary, this emphasizes the need for developing effective training programmes for facilitators and EM project managers.

## References

1. Persson, A., Stirna, J.: An explorative study into the influence of business goals on the practical use of Enterprise Modelling methods and tools. In: Proceedings of the 10th International Conference on Information Systems Development (ISD 2001), Kluwer, London (2001)
2. Rittel, H.W.J., Webber, M.M.: Planning Problems are Wicked Problems. In: Cross (ed.) Developments in Design Methodology, John Wiley & Sons, Chichester (1984)
3. F3-Consortium. F3 Reference Manual, ESPRIT III Project 6612, SISU, Sweden (1994)
4. Loucopoulos, P., Kavakli, V., Prekas, N., Rolland, C., Grosz, G., Nurcan, S.: Using the EKD Approach: The Modelling Component, UMIST, Manchester, UK (1997)
5. Nilsson, A.G., Tolis, C., Nellborn, C. (eds.): Perspectives on Business Modelling: Understanding and Changing Organisations. Springer-Verlag, Heidelberg (1999)
6. Bubenko, J.A., j., P.A., Stirna, J.: User Guide of the Knowledge Management Approach Using Enterprise Knowledge Patterns, IST Programme project Hypermedia and Pattern Based Knowl-edge Management for Smart Organisations, no. IST-2000-28401, KTH, Sweden, (2001) http://www.dsv.su.se/~js/ekd_user_guide.html

7. Niehaves, B., Stirna, J.: Participative Enterprise Modelling for Balanced Scorecard Implementation. In: 14th European Conference on Information Systems (ECIS 2006), Gothenburg, Sweden (2006)
8. Stirna, J., Persson, A., Aggestam, L.: Building Knowledge Repositories with Enterprise Modelling and Patterns - from Theory to Practice. In: proceedings of the 14th European Conference on Information Systems (ECIS), Gothenburg, Sweden (June 2006)
9. Carstensen, A., Högberg, P., Holmberg, L., Johnsen, S., Karlsen, D., Lillehagen, F., Lundqvist, M., Ohren, O., Sandkuhl, K., Wallin, A.: Kongsberg Automotive Requirements Model, deliverable D6, MAPPER, IST proj. no 016527 (2006)
10. Mikelsons, J., Stirna, J., Kalnins, J.R., Kapenieks, A., Kazakovs, M., Vanaga, I., Sinka, A., Persson, A., Kaindl, H.: Trial Application in the Riga City Council, deliverable D6, IST Programme project Hypermedia and Pattern Based Knowledge Management for Smart Organisations, project no. IST-2000-28401. Riga, Latvia (2002)
11. Persson, A.: Enterprise Modelling in Practice: Situational Factors and their Influence on Adopting a Participative Approach, PhD thesis, Dept. of Computer and Systems Sciences, Stockholm University, No 01-020, (2001) ISSN 1101-8526
12. Willars, H.: Handbok i ABC-metoden. Plandata Strategi (1988)
13. Bajec, M., Krisper, M.: A methodology and tool support for managing business rules in organisations. Information Systems 30(6), 423–443 (2005)
14. Castro, J., Kolp, M., Mylopoulos, J.: A Requirements-Driven Software Development Meth-odology. In: CAiSE 2001. LNCS, vol. 2068, pp. 108–123. Springer, Heidelberg (2001)
15. Dobson, J., Blyth, J., Strens, R.: Organisational Requirements Definition for Information Technology. In: Proceedings of the International Conference on Requirements Engineering 1994, Denver/CO (1994)
16. Fox, M.S., Chionglo, J.F., Fadel, F.G.: A common-sense model of the enterprise. In: Proceedings of the 2nd Industrial Engineering Research Conference, Institute for Industrial Engineers, Norcross/GA (1993)
17. Yu, E.S.K., Mylopoulos, J.: From E-R to A-R - Modelling Strategic Actor Relationships for Business Process Reengineering. In: Proceedings of the 13th International Conference on the Entity-Relationship Approach, Manchester, England (1994)
18. Zorgios, Y., (ed.): Enterprise State of the Art Survey, Part 3, Enterprise Modelling Methods, DTI ISIP Project Number 8032, AIAI, The University of Edinburgh (1994)
19. Krogstie, J., Jørgensen, H.D.: Interactive Models for Supporting Networked Organizations. In: Proceedings of CAiSE'2004. LNCS, Springer, Heidelberg (2004)
20. Krogstie, J., Lillehagen, F., Karlsen, D., Ohren, O., Strømseng, K., Thue Lie, F.: Extended Enterprise Methodology. Deliverable 2 in the EXTERNAL project, available at (2000) http://research.dnv.com/external/deliverables.html
21. Lillehagen, F.: The Foundations of AKM Technology. In: Proceedings 10th International Conference on Concurrent Engineering (CE), Madeira, Portugal (2003)

# Negotiating Models

Peter Rittgen

School of Business and Informatics, University College of Borås, 501 90 Borås, Sweden
peter.rittgen@hb.se

**Abstract.** We investigate the process of collaborative modeling by analyzing conversations and loud thinking during modeling sessions and the resulting models themselves. We discovered the basic activities of the modeling teams on the social, pragmatic, semantic and syntactic levels and derived a schema for the pragmatic level. Our main conclusion is that team-based modeling can be characterized as a negotiation process. Drawing on these results we suggest a tool support for modeling.

## 1 Introduction

With the abundance of literature on modeling one would expect that the process of modeling itself, i.e. how models are actually created, is well understood. And indeed, methods for developing, e.g., business process models can be found with ease [1-9]. Most of these methods operate on a coarse-grain level by specifying in which order diagrams should be developed, for example. Some provide also guidelines on how to create a specific diagram, especially in object-oriented modeling [10]. But in practice the use of a method is often reduced to the use of its notation as the former is fraught with principal problems [11]. We therefore claim that there is a need to study the modeling process from a descriptive rather than from a prescriptive perspective to find out what actually happens when people model. The ultimate aim is to support these activities with appropriate tools.

Descriptive approaches to understanding the modeling process are scarce. Only a few deal with collaborative modeling (see section 2). The others assume a scenario where a single expert modeler creates a formal model of some part of a business [12-15]. These studies identify sets of general heuristics for successful modeling without going down to the level of the concrete steps that are performed in creating models. Their results are hardly applicable to business modeling in general for a number of reasons. Firstly, a business model is rarely developed by an expert alone but rather by a team involving representatives of the respective business(es) and externals. Secondly, the problem domain of general business modeling is often less well-structured and formal languages are of limited use. Thirdly and last, the goal of providing tool support for collaborative modeling requires the identification of detailed steps.

The objective of this paper is to discover the elementary activities and the structure of the modeling process, i.e. a meta-model of the modeling process. This is done by studying, in a descriptive way, the work performed by small groups of modelers that were assigned the same task: To develop business process models for a hospital based

J. Krogstie, A.L. Opdahl, and G. Sindre (Eds.): CAiSE 2007, LNCS 4495, pp. 561–573, 2007.

on a detailed description of the processes in natural language. The group members were homogeneous concerning their modeling experience and their roles, i.e. there was no a-priori assignment of a group leader or modeling expert. The scope of this research is therefore limited to text-based modeling.

## 2 Related Research

Collaborative modeling processes have been studied by [16-20]. In [16] modeling involves domain experts, modeling mediators and model builders. It is viewed as a form of information gathering dialogue where knowledge is elicited from the domain experts. This view can be challenged because modeling is a social and communicative process where much of the information is created by and through the process rather than gathered from domain experts. We have therefore studied a situation where the participants had no a priori roles but all started from a similar position of having very little domain knowledge and collectively tried to make sense of the case described in a text document. [16] goes on developing meta model-based strategies which are of a prescriptive nature. Contrary to that and as outlined above our aim is to investigate what is actually done during modeling, i.e., we take a descriptive approach.

[17] emphasizes the importance of natural language as the primary medium and identifies two principal activities and associated roles: the domain expert who concretizes an informal model and a system analyst who abstracts a formal model. A detailed process model of both activities is given that again aims at prescribing steps to be performed to achieve a "good" analysis model. [18] distinguishes between an elicitation and a formalization dialogue and develops a modeling procedure by generalizing existing procedures for specific modeling languages. The authors claim that these procedures are descriptions of the modeling process (i.e., "documented procedures") but the focus is again clearly on prescription (see, e.g., their use of the term "guidebook"). In this sense all other approaches can also be seen as descriptive but we use this term in a more direct sense, i.e. meaning the direct observation of modeling behavior with the purpose of getting a richer and more detailed description of the modeling process.

[19] acknowledges that modeling is not only a knowledge elicitation process but also a knowledge creation and dissemination process. It is viewed as a structured conversation where sub-conversations are associated with goals and strategies (the latter are elaborated in [16]). We fully agree that modeling is a conversation but we claim that it is a specific type of conversation, namely a negotiation. This idea is implicitly present in [19] where the dialogue structure contains negotiation elements such as propose and accept. We elaborate this point in the following sections and deliver a more detailed negotiation model. [19] also advocates the use of controlled language and validation. We consider the latter as problematic as it has often been observed that domain experts falsely agree with a model not being fully aware of all its implications.

[20] studies the influence of situational factors on modeling (in particular, enterprise modeling). The author's aim is to create an environment that facilitates and supports participative modeling.

## 3 Research Method

Keeping this background in mind, we set out to study a situation where groups of modelers worked on a textual description of a business case with the purpose of deriving business process models. To understand the modeling process, we assumed that two factors are predominant in model creation:

- The internal mental processes of each modeler, and
- The conversations between modelers and within the group.

To get access to the former we used a think-aloud process-tracing methodology [13, 21] where the observants speak out what they are currently thinking. The utterances were then transcribed yielding the think-aloud protocols. The same is done with the conversations. In addition to that we also considered the product of the modeling process, the models themselves, to fill the gaps in the protocols and to help with interpreting ambiguous phrases in them. Open issues that could not be dealt with in this way were marked on the coding scheme and clarified by ex-post interviews with the respective groups.

To develop a preliminary coarse-grain categorization we turned to theories in the pertinent literature, particularly in organizational semiotics. We used the upper four 'rungs' of the semiotic ladder [22]: syntactic, semantic, pragmatic, and social. They refer to the structure of sign systems (e.g., a language), the meaning of the signs, their use, and the norms of a community, respectively. An initial coding phase within this framework revealed that the syntactic and semantic levels, which together make up the language level, are divided into the natural language domain and the modeling language domain depending on the kind of language used to describe the business.

The activities on the pragmatic level were classified as 'Understanding' and 'Organizing the Modeling Process'. The former term was then further refined into 'Undestanding the language' and 'Understanding the text', the latter can be divided into 'Setting the agenda' and 'Negotiation'. The social level consists of rules for acceptance and rejection in the negotiation. A detailed discussion of these categories can be found in the respective sections. The results are summarized in fig. 1.

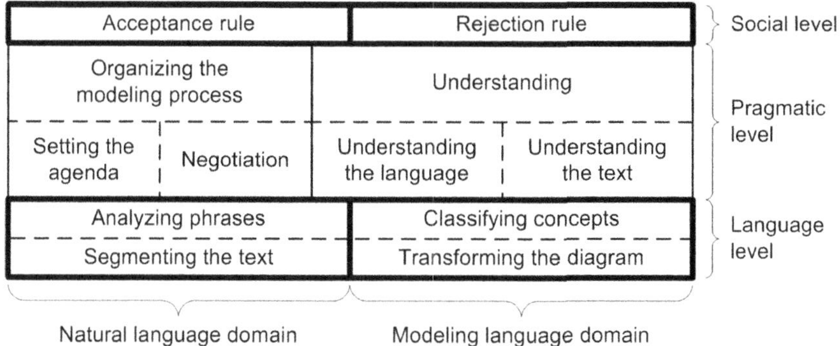


**Fig. 1.** Levels and domains

We conducted 3 experiments that involved a total of 26 groups of 2-3 students in informatics over a period of 3 years. The students were provided with a textual description of four business processes in a hospital. They were asked to model these processes with the help of two different modeling languages that they could choose freely from a set of four languages: ARIS-EPC [8], FMC-Petri nets [23], UML [24, 25], and DEMO [26]. Based on the results of these experiments we derived a layered meta-model of the modeling process that includes a model of the negotiation process.

# 4 Results

We carried out the main coding of the material within the framework stipulated by fig. 1. Examples of that procedure are shown in the respective section. The results are presented here in the order of the levels from top to bottom.

## 4.1 Social Level

The social norms within a modeling team are mainly made up of rules for determining whether a proposal is accepted or rejected. We observed that these rules do not have to be logical complements which allows for situations where a proposal can be neither rejected nor accepted but requires further convincing to decide one way or the other. A termination rule was applied occasionally to force a decision if a negotiation got stuck, i.e., when there were no more changes in the individuals' convictions over an extended period of time. We witnessed two types of rules:

- *Rules of majority*, where a certain number of group members had to support or oppose a proposal in order for the whole group to accept or reject it (e.g., more than half). A tie-break rule was sometimes specified (e.g., for the case of an equal number of supporters and opponents). The tie-break could involve seniority issues.
- *Rules of seniority*, where the weight of a group member's support or opposition was related to his or her status within the group. This status could be acquired (e.g., by experience) or associated with a position to which the member was appointed. A frequent example of this was the case of a more experienced modeler who was considered as the leader by the group and took decisions on their behalf. The other members filled the role of consultants in such a case.

These rules were sometimes set up explicitly before the group began their work, or in an early phase of this work. But in most cases they rather emerged as the result of each member's behavior. Individuals making regular contributions of high quality were likely to acquire seniority. In homogeneous teams majority rules were used more often.

## 4.2 Pragmatic Level

On the pragmatic level we discovered two distinct types of behavior, each of which can be classified in two sub-categories (the abbreviations of the categories are used as indices of the respective coded terms later on):

- *Understanding*, which concerns the text of the case description (index UT) or the (modeling) language (index UL), and
- Organizing the modeling process, which involves two types of activities: *setting the agenda* (index SA) and *negotiation* (index N).

Understanding was established by questions and answers. If the respondent could not provide clarification, an assumption was made. For details see table 1.

Agendas have been used by the participants in our study as an instrument for roughly structuring the modeling session. They were introduced in the beginning and then adapted during the session if necessary. On the whole most groups started by reading the case description completely and then organized their work around the flow of the text. For further details refer to table 1.

**Table 1.** Generic activities on the pragmatic level

| Activity | Coding | Example |
|---|---|---|
| Modeler *m* makes proposal *p*. | $\mathbf{propose}_{N}$ $(m, p)$ | "I suggest that ..." |
| Modeler *m* withdraws his/her proposal *p*. | $\mathbf{withdraw}_{N}$ $(m, p)$ | "My idea does not work; let us forget it." |
| Modeler *m* expresses consent to proposal *p*. | $\mathbf{support}_{N}$ $(m, p)$ | "I agree with that." |
| Modeler *m* expresses objection to proposal *p*. | $\mathbf{challenge}_{N}$ $(m, p)$ | "I think you are wrong." |
| Modeler *m* delivers argument *a* to support *p*. | $\mathbf{argue\_for}_{N}$ $(m, p, a)$ | "Yes, because that's an operation you do, so that must be a transaction." |
| Modeler *m* delivers argument *a* to challenge *p*. | $\mathbf{argue\_against}_{N}$ $(m, p, a)$ | "We cannot do this because an and-connector cannot have two inputs and two outputs at the same time." |
| Modeler *m* proposes *p'* instead of *p*. | $\mathbf{counter}_{N}$ $(m, p, p')$ | "We should have the records after the evaluation and not at the same time." |
| Modeler *m* needs clarification on issue *q*. | $\mathbf{ask}_{UT/UL}$ $(m, q)$ | "Does the patient just come to the hospital?" (UT) |
| Modeler *m* provides a possible answer *a* to question *q*. | $\mathbf{assume}_{UT/UL}$ $(m, q, a)$ | "I think the sticky labels are required for tests." |
| Modeler *m* gives a definite answer *a* to question *q*. | $\mathbf{clarify}_{UT/UL}$ $(m, q, a)$ | "An XOR connector cannot follow an event." (UL) |
| Add activity *a* to the agenda as item number *n*. | $\mathbf{add}_{SA}$ $(a, n)$ | "First we will read the case completely." |
| Perform next activity on agenda. | $\mathbf{perform}_{SA}$ | "Next is patient care" |

The majority of the activities on the pragmatic level were associated with negotiation, though (see also table 1). This is surprising as modeling is typically rather pictured as an intuitive act that is largely the product of a creative brain (e.g., a consultant) that possibly receives some input from other stakeholders in the modeling process (e.g., domain experts from the respective departments).

From these results we can draw interesting conclusions for the design of a system that supports modeling (see section "Architecture of a Modeling Support System". This concludes the pragmatic level. The next section proceeds with the semantic level.

### 4.3 Semantic Level

The semantic level is concerned with the concepts of the modeled domain, in our case business processes. It is therefore also called the conceptual level. These concepts are

**Table 2.** Generic activities and concepts on the semantic level

| Activity/concept | Coding | Example |
|---|---|---|
| Phrase $t_1$ is considered equivalent to phrase $t_2$. | **interpret**$_{AP}$ $(t_1, t_2)$ | "An anamnesis is the same as a case history of the patient." |
| Phrase $t$ is considered an instance of foundational concept $c$. | **classify**$_{CC}$ $(t, c)$ | "The 'nurse' is an actor." |
| Actor $x$ performs action $a$. | **perform**$_{CC}$ $(x, a)$ | "The nurse treats the patient." |
| Action $a$ is triggered by the conjunction of events $e_1, e_2, \ldots$ | **wait-for**$_{CC}$ $(a, e_1, e_2, \ldots)$ | "When the certificate of discharge and the transfer report are ready, the nurse copies the transfer documents." |
| Event $e$ is raised by all of the actions $a_1, \ldots$ being finished. | **define**$_{CC}$ $(e, a_1, \ldots)$ | "A document is ready when it has been filled in and signed." |
| Action $a$ is triggered by any of the events $e_1, e_2, \ldots$ | **merge**$_{CC}$ $(a, e_1, e_2, \ldots)$ | "If the patient is new or the treatment is ineffectual, the physician examines him." |
| Event $e$ is raised by any of the actions $a_1, a_2, \ldots$ being finished. | **merge**$_{CC}$ $(e, a_1, a_2, \ldots)$ | "If the nurse has recorded the patient's data or the clerk has registerered him, the patient is admitted." |
| Actions $a_1, \ldots$ are triggered by event $e$ simultaneously. | **trigger**$_{CC}$ $(e, a_1, \ldots)$ | "The arrival of the patient starts the admission process." |
| Action $a_2$ follows action $a_1$. | **after**$_{CC}$ $(a_1, a_2)$ | "First, the nurse hands the physician the admission papers, then he performs the admission examination." |
| The completion of action $a$ raises any of the events $e_1, e_2, \ldots$ | **branch**$_{CC}$ $(a, e_1, e_2, \ldots)$ | "The nurse determines whether the admission time is before 3 p.m. or not. |
| The completion of action $a$ raises all of the events $e_1, e_2, \ldots$ | **fork**$_{CC}$ $(a, e_1, e_2, \ldots)$ | "When the examination has been completed, the tentative diagnosis is available and lab tests are requested." |
| Action $a$ creates info object $o$. | **create**$_{CC}$ $(a, o)$ | "During registration a patient's data is recorded." |
| Action $a$ removes info object $o$. | **remove**$_{CC}$ $(a, o)$ | "Discharge implies deletion of the entry in the bed register." |
| Action $a$ uses info object $o$. | **use**$_{CC}$ $(a, o)$ | "To assign a bed the nurse consults the bed register." |
| Action $a$ changes info object $o$. | **update**$_{CC}$ $(a, o)$ | "The ward books are updated." |

often closely linked to the ones that are found in the language used for modeling the domain. The modeler expresses the perceived or constructed reality in terms of the concepts that the language provides, be it a natural language or a modeling language. This implies that the chosen language both enables and restricts the modeler in having and expressing certain thoughts. To prevent that these restrictions affect the number and types of concepts we identified in the modeling language domain (= *Classifying Concepts*, index *CC*) we treated them regardless of the language they were expressed in. As a consequence the results of the coding are not divided by language. Not all of the generic concepts are found in all languages, though. The foundational concepts are: actor, action/function, event and (information) object. Between them relational concepts are defined. They are listed in table 2 without the surrounding **classify** activity, for convenience. By phrase we mean a text fragment (e.g., from the case description). The index *AP* refers to *Analyzing Phrases*, the main activity of the natural language domain).

This completes the activities on the semantic level. The next section deals with the syntactic level.

## 4.4 Syntactic Level

On the syntactic level we distinguish again between the natural language and the modeling language domain. In the latter the diagrams of the respective language are built (*Transforming Diagrams*, *TD*). They consist of nodes, edges that connect them, and labels attached to both. In general the foundational concepts are represented by nodes and the relational concepts by edges but this is not necessarily true for all languages (e.g., Petri nets). In the natural language domain the text is segmented into useful units for the analysis on the semantic level (*Segmenting the Text*, *ST*). The generic activities on this level are listed in table 3.

**Table 3.** Generic activities on the syntactic level

| Activity | Coding | Example |
|---|---|---|
| Text fragment $t$ is used as the unit of analysis. The fragment can be: text, section, sentence, nominal phrase, verbal phrase or word. | **focus**$_{ST}$ $(t)$ | "What does this sentence mean?" |
| Introduce a new node $n$ of type $t$. | **introduce**$_{TD}$ $(n, t)$ | "We insert an ellipse for the nurse." |
| Attach label $l$ to node $n$. | **label**$_{TD}$ $(n, l)$ | "The function is called 'Transport patient'." |
| Remove node $n$. | **remove**$_{TD}$ $(n)$ | "This place is not needed." |
| Connect node $n_1$ to node $n_2$ with edge $e$ of type $t$. | **connect**$_{TD}$ $(n_1, n_2, e, t)$ | "We need a dashed line from 'Nurse' to 'Update ward books'." |
| Attach label $l$ to edge $e$ at place $p$. | **label**$_{TD}$ $(e, l, p)$ | "Write 'Before 3 p.m.' on the arrow." |
| Remove edge $e$. | **remove**$_{TD}$ $(e)$ | "The arrow should not point that way." |

This completes the list of levels. In the following section we show an example of how a part of a model is negotiated.

## 5 Example

Somewhere in the middle of the modeling session concerning the admission process of the hospital, group B encountered a difficult situation. They were just focusing the following sentence in the case description (the part in square brackets is intended for the reader and was not present in the case text):

"The lab results are evaluated and the results [of the evaluation] are put in the medical record."

Table 4 shows the associated discussion of the group and the respective coding. The members of this group are called A and B. We have left out the syntactic level for convenience.

**Table 4.** An example discussion

| Utterance | Coding |
|---|---|
| A: "I think we should introduce a function 'Evaluate lab results' first." | $p_1$ = **propose**$_{\mathrm{N}}$ (A, **classify**$_{\mathrm{CC}}$ ('Evaluate lab results', function)) |
| B: "O.K." | **support**$_{\mathrm{N}}$ (B, $p_1$), **accept** ($p_1$) |
| B: "And then another function 'Put results in medical record'." | $p_2$ = **propose**$_{\mathrm{N}}$ (B, **classify**$_{\mathrm{CC}}$ ('Put results in medical record', function)) |
| A: "Wait! Is that not rather an output of the first function?" | $p_3$ = **counter**$_{\mathrm{N}}$ (A, $p_2$, **classify**$_{\mathrm{CC}}$ ('Medical record', object) + **update**$_{\mathrm{CC}}$ (Evaluate lab results, Medical record)) |
| B: "You are right! That makes more sense." | **support**$_{\mathrm{N}}$ (B, $p_3$), **accept** ($p_3$), **reject** ($p_2$) |

This small example is supposed to give an impression of how the modeling process is structured. In the next section we discuss a potential tool support for this process.

## 6 Tool Support for the Modeling Process

Our analyses of the modeling sessions showed us that modeling is a complex process involving issues such as collective sense-making, negotiations and group decisions. It is therefore worthwhile to consider tool support for this process. This is particularly true in an interorganizational setting where participants are often geographically distributed. The tool we envision helps group members in understanding the modeling situation, creating and discussing modeling alternatives, and deciding on the best one, all in a shared internet-based environment. The following paragraphs elaborate on the components that such a tool should provide.

According to our results modeling is a relatively well-structured process. It consists of a limited number of well-defined activities on all levels of the semiotic ladder. We are aware that further research will reveal more activities but from the experience of the three experiments that yielded a decreasing number of new ones, we are confident that the total number of activities will converge with respect to a given domain, e.g., business processes. The activities identified so far can therefore be assumed to be relatively stable in that domain. To a certain extent this is even true across different business process modeling languages, although the terminology of concepts may vary and not every concept is realized in each of the languages. But the findings will not carry over to another domain due to the domain specificity of the language level. The other levels are likely to work, though.

An analysis of the workflows on the pragmatic level revealed a structure that goes beyond the mere identification of generic activities. We found out that the negotation process actually follows a certain pattern. This pattern is shown in fig. 2.

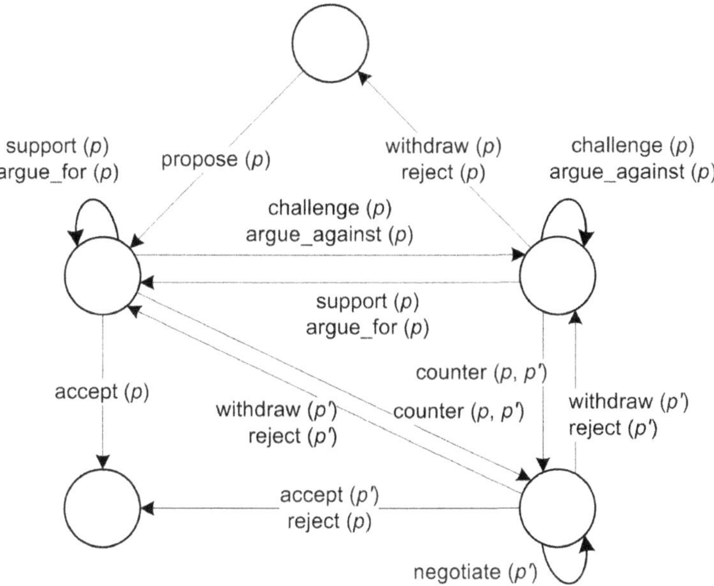


**Fig. 2.** Negotiation pattern

It consists of an initial and reject state at the top, a state where acceptance is favored (upper left-hand corner), a state where rejection is favored (upper right-hand corner), a recursive sub-state for negotiating a counter-proposal (lower right-hand corner) and an accept state (lower left-hand corner). Each of the states allows for a set of certain pragmatic activities that take the negotiation to a different state. We have left out the parameters concerning the modeler who performs the activity and the argument (if present). In general any modeler can perform any activity but there are a few rules to be observed. A modeler making a proposal is implicitly assumed to support it. He is the only one who may withdraw it. A counter-argument is brought up by a different modeler but a counter-proposal can also be made by the proponent of the original proposal, e.g., to accommodate counter-arguments. With the help of the pattern of fig. 2 we can control the negotiation component of a modeling support system. On the other levels we were not able to discover an equally strong pattern of activities. This will affect the kind of support a tool can provide at the language level.

The architecture of a modeling support system, i.e., a system that supports a group in developing models, is still under investigation. Some authors have suggested groupware systems that help teams in collective sense-making [18, 27-29] which is an important part of the modeling process. [29] reports on an approach, Compendium, that is the result of 15 years of experience. Compendium combines three different areas: meeting facilitation, graphical hypertext and conceptual frameworks. To make them work, facilitation is viewed as essential to remove the cognitive overhead for the group members, i.e., the necessity to develop hypertext literacy, which cannot be assumed in all participants. On the technology side, the critical elements are question-based templates, metadata and maps. They allow participants to move freely between different levels of abstraction and formalization as the need dictates. The question-based templates guide the process by supplying relevant questions, the answers to which will lead the group towards a better understanding of the problem and towards the development of appropriate solutions (e.g., models). The metadata is used to provide additional information that is also considered relevant but was not anticipated in the templates or lies at the intersection of templates. The maps have a hierarchical structure and the same concept can appear in different maps so that its use in different contexts can be understood. This feature is called transclusion.

Groupware systems for collective sense-making, as the one mentioned, address an important issue in collaborative modeling. They can therefore be used as the core of a modeling support system (MSS). So far these systems are typically tailored for specific modeling languages though (in the case of Compendium, World Modeling Framework and Issue-Based Information System). For an MSS they need to be more modular so that any modeling language can be "plugged in" (e.g., other enterprise or information systems modeling languages). In addition, there is also the need for a negotiation component that facilitates structured arguments and decisions regarding modeling choices. The model shown in Fig. 2 can function as an initial workflow template controlling such a negotiation component. Once instantiated the actual workflow can then be adjusted to the concrete modeling situation.

## 7 Conclusions

We studied group modeling sessions in detail, both regarding conversations between the group members and the mental processes within each individual. By doing this we derive a sub-categorization of the upper four levels of the semiotic ladder, generic activities of business process modeling at all of these levels and a negotiation pattern at the pragmatic level. On the basis of these results we suggest a tentative architecture of a system that supports group modeling. Our aim with this research is two-fold. On the one hand we want to develop a better understanding of the modeling process that has been largely neglected by researchers so far. Such an improved understanding can lead to better modeling methods and thereby ultimately to higher quality of models.

On the other hand we are also interested in providing computer support to those modelers that work in a group environment. Modeling is a highly demanding task that is further complicated by the dynamics of group work. Effective support is therefore essential, especially if some of the group members are inexperienced as is often the case in business modeling sessions, where typically a majority of the participants does not have any modeling background. But it is precisely this latter type of participant that contributes most to the actual design of the model with his or her knowledge of the relevant business domain. Both the speed and quality of the models can therefore benefit tremendously if we can manage to involve these people directly as modelers instead of relying on the bottleneck of the modeling expert for all communication within the group. The suggested tool support can accommodate this by giving the expert seniority (i.e., the right to make the final decision) and turning the domain experts into effective consultants that make proposals (thereby reversing the traditional roles in IT consulting).

The modeling support system can also be seen as a special kind of group decision support system (GDSS, [30]) if we consider that the accept and reject decisions in the negotiation process are the key to model design. There is significant empirical support for the claim that GDSS are beneficial [30-39], particularly for larger groups and/or complex tasks. Many of these benefits carry over to modeling support systems, e.g., reduced meeting time, higher quality of the decisions, broader involvement of all participants, higher effectiveness of decisions, etc.

Our research studied text-based modeling only. This is not a realistic scenario for practical modeling situations. We are confident though that our results are relevant for real-world modeling to some extent. The social level is fairly independent of the way in which a modeling alternative was derived (text-based or other) as the decision rule rather depends on the alternatives themselves. The same is true for the language level as we can safely assume that natural language and modeling languages will play an important role in any modeling endeavor. We therefore expect differences primarily on the pragmatic level, and here especially in the areas "setting the agenda" and "understanding". Whether modelers just interpret a text or communicate with domain experts will have considerable impact on the way the agenda is determined. Likewise the issue of understanding has to be extended to cover forms of communication other than analyzing text.

So far we have only looked at business process modeling. Other domains in the business and information systems areas remain to be explored. It should also be noted that our study has been performed in a contrived setting albeit with a realistic case. Further confirmation, and especially consolidation, is therefore required, preferably by means of a field study. In addition to this it seems reasonable to build a prototype of a modeling support system, and to test it in a realistic modeling scenario. We are confident that these measures will contribute to a better understanding of the process of modeling, both from a cognitive and a collaborative perspective, and they will eventually help us to better support modelers in their challenging task.

## References


1. AMICE: CIMOSA: Open System Architecture for CIM, 2nd revised and extended version. Springer, Berlin (1993)
2. Barker, R.: CASE*Method: Entity-Relationship Modelling. Addison-Wesley, Wokingham (1990)
3. Bernus, P., Nemes, L.: A Framework to Define a Generic Enterprise Reference Architecture and Methodology. Computer Integrated Manufacturing Systems 9, 179–191 (1996)
4. Goldkuhl, G., Röstlinger, A.: Joint elicitation of problems: An important aspect of change analysis. In: Avison, D.E., Kendall, J.E., DeGross, J.I. (eds.) Human, organizational and social dimensions of Information systems development. North-Holland, Amsterdam, pp. 107–125 (1993)

5. Menzel, C., Mayer, R.J.: The IDEF Family of Languages. In: Bernus, P., Mertins, K., Schmidt, G. (eds.) Handbook on Architectures for Information Systems, pp. 209–241. Springer, Berlin (1998)
6. Ould, M.: Business Process Management: A Rigorous Approach. Meghan-Kiffer Press, Tampa, FL (2005)
7. Roboam, M., Zanettin, M., Pun, L.: GRAI-IDEF0-Merise (GIM): integrated methodology to analyse and design manufacturing systems. Computer-Integrated Manufacturing Systems 2, 82–98 (1989)
8. Scheer, A.-W.: ARIS - Business Process Modeling. Springer, Berlin (1999)
9. Williams, T.J.: The Purdue enterprise reference architecture. Computers in Industry 24, 141–158 (1994)
10. Bennet, S., McRobb, S., Farmer, R.: Object-Oriented Systems Analysis and Design. McGraw-Hill, Maidenhead (1999)
11. Introna, L.D., Whitley, E.A.: Against method: exploring the limits of method. Logistics Information Management 10, 235–245 (1997)
12. Morris, W.T.: On the Art of Modeling. Management Science 13, B-707–B717 (1967)
13. Srinivasan, A., Te´eni, D.: Modeling as Constrained Problem Solving: An Empirical Study of the Data Modeling Process. Management Science 41, 419–434 (1995)
14. Willemain, T.R.: Insights on Modeling from a Dozen Experts. Operations Research 42, 213–222 (1994)
15. Willemain, T.R.: Model Formulation: What Experts Think about and When. Operations Research 43, 916–932 (1995)
16. Bommel, P., Hoppenbrouwers, S.J.B.A., Proper, H.A.E., Weide, T.P.v.d.: Exploring Modelling Strategies in a Meta-modelling Context. In: Meersman, R., Tari, Z., Herrero, P. (eds.) On the Move to Meaningful Internet Systems 2006: OTM 2006 Workshops - OTM Confederated International Workshops and Posters, AWESOMe, CAMS, COMINF, IS, KSinBIT, MIOS-CIAO, MONET, OnToContent, ORM, PerSys, OTM Academy Doctoral Consortium, RDDS, SWWS, and SebGIS, Proceedings, Part II. vol.4275, pp. 1128–1137. Springer, Berlin, Germany (2006)
17. Frederiks, P.J.M., Weide, T.P.v.d, Weide, T.P.v.d.: Information Modeling: the process and the required competencies of its participants. Data. & Knowledge Engineering 58, 4–20 (2006)
18. Hoppenbrouwers, S.J.B.A., Lindeman, L., Proper, H.A.: Capturing Modeling Processes - Towards the MoDial Modeling Laboratory. In: Meersman, R., Tari, Z., Herrero, P. (eds.) On the Move to Meaningful Internet Systems 2006: OTM 2006 Workshops - OTM Confederated International Workshops and Posters, AWESOMe, CAMS, COMINF, IS, KSinBIT, MIOS-CIAO, MONET, OnToContent, ORM, PerSys, OTM Academy Doctoral Consortium, RDDS, SWWS, and SebGIS, Proceedings, Part II. vol.4275, pp. 1242–1252. Springer, Berlin, Germany (2006)
19. Hoppenbrouwers, S.J.B.A., Proper, H.A., Weide, T.P.: v.d.: Formal Modelling as a Grounded Conversation. In: Goldkuhl, G., Lind, M., Haraldson, S. (eds.) Proceedings of the 10th International Working Conference on the Language Action Perspective on Communication Modelling (LAP'05), Kiruna, Sweden. Linköpings Universitet and Högskolan i Borås, Linköping and Borås, pp. 139-155 ( (2005)
20. Persson, A.: Enterprise Modelling in Practice: Situational Factors and their Influence on Adopting a Participative Approach. Department of Computer and Systems Sciences, Stockholm University (2001)
21. Ericsson, K., Simon, H.: Protocol Analysis: Verbal Reports as Data. MIT Press, Boston (1993)

22. Stamper, R.: The Semiotic Framework for Information Systems Research. In: Nissen, H., Klein, H., Hirschheim, R. (eds.) Information Systems Research: Contemporary Approaches and Emergent Traditions. North-Holland, Amsterdam, pp. 515-517 (1991)
23. Keller, F., Wendt, S.: FMC: An Approach Towards Architecture-Centric System Development. In: Keller, F., Wendt, S. (eds.) 10th IEEE Symposium and Workshop on Engineering of Computer Based Systems, pp. 173–182. IEEE Computer Society, Pasadena, CA (2003)
24. OMG: UML 2.0 Superstructure Specification. OMG, Needham, MA (2004)
25. OMG: Unified Modeling Language: Infrastructure. OMG, Needham, MA (2006)
26. Dietz, J.L.G.: Understanding and modeling business processes with DEMO. In: Akoka, J., Bouzeghoub, M., Comyn-Wattiau, I., Métais, E. (eds.) Proceedings of the 18th International Conference on Conceptual Modeling ER '99, pp. 188–202. Springer, Berlin (1999)
27. Boehm, B., Grunbacher, P., Briggs, R.O.: Developing Groupware for Requirements Negotiation: Lessons Learned. IEEE Software 18, 46–55 (2001)
28. Briggs, R.O., de Vreede, G.J., Nunamaker, J.: Collaboration Engineering with Thinklets to Pursue Sustained Success with Group Support Systems. Journal of MIS 19, 31–63 (2003)
29. Conklin, J., Selvin, A., Buckingham Shum, S., Sierhuis, M.: Facilitated Hypertext for Collective Sensemaking: 15 Years on from gIBIS. In: Weigand, H., Goldkuhl, G., de Moor, A. (eds.) Proceedings of the 8th International Working Conference on the Language-Action Perspective on Communication Modeling (LAP'03), Tilburg, The Netherlands (2003)
30. Aiken, M., Vanjani, M., Krosp, J.: Group decision support systems. Review of Business 16, 38–42 (1995)
31. Bamber, E.M., Watson, R.T., Hill, M.C.: The effects of group support system technology on audit group decision-making. Auditing: A Journal of Practice & Theory 15, 122–134 (1996)
32. Benbasat, I., Lim, L.H.: The effects of group, task, context, and technology variables on the usefulness of group support systems: A meta-analysis of experimental studies. Small Group Research 24, 430–462 (1993)
33. Bidgoli, H.: A new productivity tool for the 90's: Group support systems. Journal of Systems Management 47, 56–62 (1996)
34. Burke, K., Chidambaram, L., Lock, J.: Evolution of relational factors over time: A study of distributed and non-distributed meetings.In: Proceedings of the Twenty-Eighth Hawaii International Conference on System Sciences vol. 4, pp. 14–23 ( (1995)
35. Cass, K., Heintz, T.J., Kaiser, K.M.: Using a voice-synchronous GDSS in dispersed locations: A preliminary analysis of participant satisfaction. In: Proceedings of the Twenty-Fourth Hawaii International Conference on System Sciences vol. 3, 555-563 (1991)
36. Chudoba, K.M.: Appropriations and patterns in the use of group support systems. Database for Advances in Information Systems 30, 131–148 (1999)
37. Fjermestad, J., Hiltz, S.R.: An assessment of group support systems experiment research: Methodology and results. Journal of Management Information Systems 15, 7–149 (1998/1999)
38. Jackson, N.F., Aiken, M.W.V., Mahesh, B.H., Bassam, S.: Support group decisions via computer systems. Quality Progress 28, 75–78 (1995)
39. Townsend, A.M., Whitman, M.E., Hendrickson, A.R.: Computer support system adds power to group processes. HRMagazine 40, 87–91 (1995)

# Change Patterns and Change Support Features in Process-Aware Information Systems

Barbara Weber[1,⋆], Stefanie Rinderle[2], and Manfred Reichert[3]

[1] Quality Engineering Research Group, University of Innsbruck, Austria
Barbara.Weber@uibk.ac.at

[2] Inst. Databases and Information Systems, Ulm University, Germany
stefanie.rinderle@uni-ulm.de

[3] Information Systems Group, University of Twente, The Netherlands
m.u.reichert@cs.utwente.nl

**Abstract.** In order to provide effective support, the introduction of process-aware information systems (PAIS) must not freeze existing business processes. Instead PAIS should allow authorized users to flexibly deviate from the predefined processes if required and to evolve business processes in a controlled manner over time. Many software vendors promise flexible system solutions for realizing such adaptive PAIS, but are often unable to cope with fundamental issues related to process change (e.g., correctness and robustness). The existence of different process support paradigms and the lack of methods for comparing existing change approaches makes it difficult for PAIS engineers to choose the adequate technology. In this paper we suggest a set of changes patterns and change support features to foster systematic comparison of existing process management technology with respect to change support. Based on these change patterns and features, we provide an evaluation of selected systems.

## 1 Introduction

Contemporary information systems (IS) more and more have to be aligned in a process-oriented way. This new generation of IS is often referred to as Process-Aware IS (PAIS) [1]. In order to provide effective process support, PAIS should capture real-world processes adequately, i.e., there should be no mismatch between the computerized processes and those in reality. In order to achieve this, the introduction of PAIS must not lead to rigidity and freeze existing business processes. Instead PAIS should allow authorized users to flexibly deviate from the predefined processes as required (e.g., to deal with exceptions) and to evolve PAIS implementations over time (e.g., due to process optimizations or legal changes). Such process changes should be enabled at a high level of abstraction and without affecting the robustness of the PAIS [2].

The increasing demand for process change support poses new challenges for IS engineers and requires the use of change enabling technologies. Contemporary

⋆ This work was done during a postdoctoral fellowship at the University of Twente.

J. Krogstie, A.L. Opdahl, and G. Sindre (Eds.): CAiSE 2007, LNCS 4495, pp. 574–588, 2007.

PAIS, in combination with service-oriented computing, offer promising perspectives in this context. Many vendors promise flexible software solutions for realizing adaptive PAIS, but are often unable to cope with fundamental issues related to process change (e.g., correctness and robustness). This problem is further aggravated by the fact that several competing process support paradigms exist, all trying to tackle the need for more process flexibility (e.g., adaptive processes [3,4,5] or case handling [6]). Furthermore, there exists no method for systematically comparing the change frameworks provided by existing process-support technologies. This, in turn, makes it difficult for PAIS engineers to assess the maturity and change capabilities of those technologies. Consequently, this often leads to wrong decisions and misinvestments.

During the last years we have studied processes from different application domains and elaborated the flexibility and change support features of numerous tools and approaches. Based on these experiences, in this paper we suggest a set of *changes patterns* and *change support features* to foster the comparison of existing approaches with respect to process change support. Change patterns allow for high-level process adaptations at the process type as well as the process instance level. Change support features ensure that changes are performed in a correct and consistent way, traceability is provided, and changes are facilitated for users. Both change patterns and change support features are fundamental to make changes applicable in practice. Finally, another contribution of this paper is the evaluation of selected approaches/systems based on the presented change patterns and change support features.

Section 2 summarizes background information needed for the understanding of this paper. Section 3 describes 17 change patterns and Section 4 deals with 6 crucial change support features. Based on this, Section 5 evaluates different approaches from both academia and industry. Section 6 discusses related work and Section 7 concludes with a summary.

## 2 Backgrounds

A PAIS is a specific type of information system which allows for the separation of process logic and application code. At run-time the PAIS orchestrates the processes according to their defined logic. Workflow Management Systems (e.g., Staffware [1], ADEPT [3], WASA [5]) and Case-Handling Systems (e.g., Flower [1,6]) are typical technologies enabling PAIS.

For each business process to be supported a process type represented by a *process schema* $S$ has to be defined. In the following, a process schema is represented by a directed graph, which defines a set of *activities* – the process steps – and control connections between them (i.e., the precedence relations between these activities). Activities can either be atomic or contain a sub process (i.e., a reference to a process schema $S'$) allowing for the hierarchical decomposition of a process schema. In Fig. 1a, for example, process schema $S1$ consists of six activities: Activity `A` is followed by activity `B` in the flow of control, whereas `C` and `D` can be processed in parallel. Activities `A` to `E` are atomic, and activity `F` constitutes a sub process with own process schema $S2$. Based on a process schema $S$, at run-time new *process instances* $I_1, \ldots, I_n$ can be created and executed. Regarding process instance $I_1$ from Fig. 1a, for example, activity A is completed and activity B is activated (i.e., offered in user worklists). Generally, a large number of process instances might run on a particular process schema.

PAIS must be able to cope with change. In general, changes can be triggered and performed at two levels – the process type and the process instance level (cf. Fig. 1b) [2]. Schema changes at the type level become necessary to deal with the evolving nature of real-world processes (e.g., to adapt to legal changes). Ad-hoc changes of single instances are usually performed to deal with exceptions, resulting in an adapted *instance-specific* process schema.

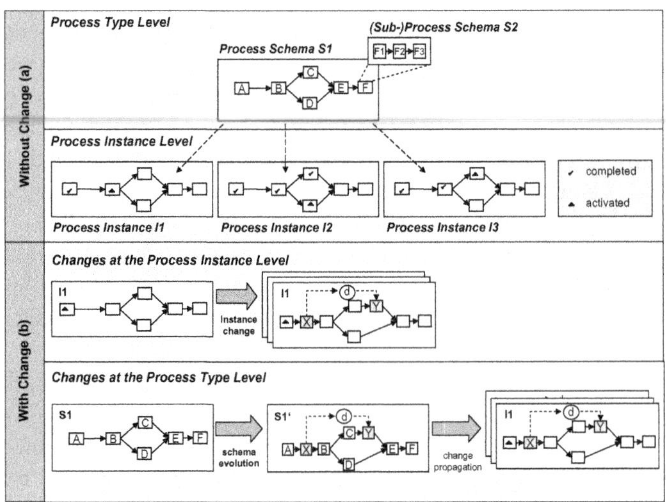


**Fig. 1.** Core Concepts

## 3 Change Patterns

In this section we describe 17 characteristic patterns we identified as relevant for *control flow changes* (cf. Fig. 2). Adaptations of other process aspects (e.g., data or resources) are outside the scope of this paper. Change patterns reduce the complexity of process change (like design patterns in software engineering reduce system complexity [7]) and raise the level for expressing changes by providing abstractions which are above the level of single node and edge operations. Consequently, due to their lack of abstraction, low level change primitives (add node, delete edge, etc.) are not considered to be change patterns and thus are not covered in this section.

As illustrated in Fig. 2, we divide our change patterns into *adaptation patterns* and *patterns for predefined changes*. Adaptation patterns allow modifying the schema of a process type (type level) or a process instance (instance level) using high-level change operations. Generally, adaptation patterns can be applied to the whole process schema or process instance schema respectively; they do not have to be pre-planned, i.e., the region to which the adaptation pattern is applied can be chosen dynamically. By contrast, for predefined changes, at build-time, the process engineer defines regions in the process schema where potential changes may be performed during run-time.

For each pattern we provide a name, a brief description, an illustrating example, a description of the problem it addresses, a couple of design choices, remarks regarding its implementation, and a reference to related patterns. *Design Choices* allow for parametrization of patterns keeping the number of distinct patterns manageable. Design choices which are not only relevant for particular patterns, but for a whole pattern category, are described only once at the category level. Typically, existing approaches only support a subset of the design choices in the context of a particular pattern. We denote the combination of design choices supported by a particular approach as a *pattern variant.*

| CHANGE PATTERNS | | | |
|---|---|---|---|
| ADAPTATION PATTERNS (AP) | | | |
| **Pattern Name** | **Scope** | **Pattern Name** | **Scope** |
| **AP1:** Insert Process Fragment(*) | I / T | **AP8:** Embed Process Fragment in Loop | I / T |
| **AP2:** Delete Process Fragment | I / T | **AP9:** Parallelize Process Fragment | I / T |
| **AP3:** Move Process Fragment | I / T | **AP10:** Embed Process Fragment in Conditional Branch | I / T |
| **AP4:** Replace Process Fragment | I / T | **AP11:** Add Control Dependency | I / T |
| **AP5**: Swap Process Fragment | I / T | **AP12:** Remove Control Dependency | I / T |
| **AP6**: Extract Sub Process | I / T | **AP13:** Update Condition | I / T |
| **AP7**: Inline Sub Process | I / T | | |
| PATTERNS FOR PREDEFINED CHANGES (PP) | | | |
| **Pattern Name** | **Scope** | **Pattern Name** | **Scope** |
| **PP1:** Late Selection of Process Fragments | I / T | **PP3:** Late Composition of Process Fragments | I / T |
| **PP2:** Late Modeling of Process Fragments | I / T | **PP4:** Multi-Instance Activity | I / T |

I... Instance Level, T ... Type Level
(*) A process fragment can either be an atomic activity, an encapsulated sub process or a process (sub) graph

**Fig. 2.** Change Patterns Overview

### 3.1 Adaptation Patterns

Adaptation patterns allow to structurally change process schemes. Examples include the insertion, deletion and re-ordering of activities (cf. Fig. 2). Fig. 3 describes general design choices valid for all adaptation patterns. First, each adaptation pattern can be applied at the process type or process instance level (cf. Fig. 1b). Second, adaptation patterns can operate on an atomic activity, an encapsulated sub process or a process (sub-)graph (cf. Fig. 3). We abstract from this distinction and use the generic concept *process fragment* instead. Third, the effects resulting from the use of an adaptation pattern at the instance level can be permanent or temporary. A *permanent instance change* remains valid until completion of the instance (unless it is undone by a user). By contrast, a *temporary instance change* is only valid for a certain period of time (e.g., one loop iteration) (cf. Fig. 3).

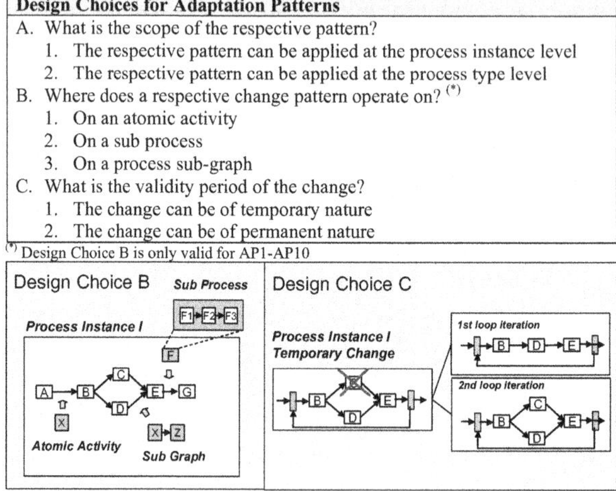

**Design Choices for Adaptation Patterns**

A. What is the scope of the respective pattern?
   1. The respective pattern can be applied at the process instance level
   2. The respective pattern can be applied at the process type level

B. Where does a respective change pattern operate on? (*)
   1. On an atomic activity
   2. On a sub process
   3. On a process sub-graph

C. What is the validity period of the change?
   1. The change can be of temporary nature
   2. The change can be of permanent nature

(*) Design Choice B is only valid for AP1-AP10



**Fig. 3.** Design Choices for Adaptation Patterns

We describe four selected adaptation patterns in more detail. These four patterns allow for the insertion, deletion, movement, and replacement of process fragments in a given process schema. The *Insert Process Fragment* pattern (cf. Fig. 4a) can be used to add process fragments to a process schema. In addition to the general options described in Fig. 3, one major design choice for this pattern (Design Choice D) describes the way the new process fragment is embedded in the respective schema. There are systems which only allow to serially insert a fragment between two directly succeeding activities. By contrast, other systems follow a more general approach allowing the user to insert new fragments between two arbitrary sets of activities [3]. Special cases of the latter variant include the insertion of a fragment in parallel to another one or the association of the newly added fragment with an execution condition (*conditional insert*). The *Delete Process Fragment* pattern, in turn, can be used to remove a process fragment (cf. Fig 4b). No additional design choices exist for this pattern. Fig. 4b depicts alternative ways in which this pattern can be implemented.

The *Move Process Fragment* pattern (cf. Fig. 5a) allows to shift a process fragment from its current position to a new one. Like for the *Insert Process Fragment* pattern, an additional design choice specifies the way the fragment can be embedded in the process schema afterwards. Though the *Move Process Fragment* pattern could be realized by the combined use of AP1 and AP2 (*Insert/Delete Process Fragment*), we introduce it as separate pattern as it provides a higher level of abstraction to users. The latter also applies when a process fragment has to be replaced by another one. This change is captured by the *Replace Process Fragment* pattern (cf. Fig. 5b).

We have only described the most relevant adaptation patterns. Additional patterns we identified are: swapping of activities (AP5), extraction of a sub process from a process schema (AP6), inclusion of a sub process into a process schema (AP7), embedding of an existing process fragment in a loop (AP8),

**a) Pattern AP1: Insert Process Fragment**

**Description**: A process fragment is added to a process schema.
**Example**: For a particular patient an allergy test has to be added due to a drug incompatibility.
**Problem**: In a real world process a task has to be accomplished which has not been modeled in the process schema so far.
**Design Choices (in addition to the ones in Fig. 3):**

D. How is the additional process fragment X embedded in the process schema?
  1. X is inserted between 2 directly succeeding activities (serial insert)
  2. X is inserted between 2 activity sets (insert between node sets)
     a) Without additional condition (parallel insert)
     b) With additional condition (conditional insert)

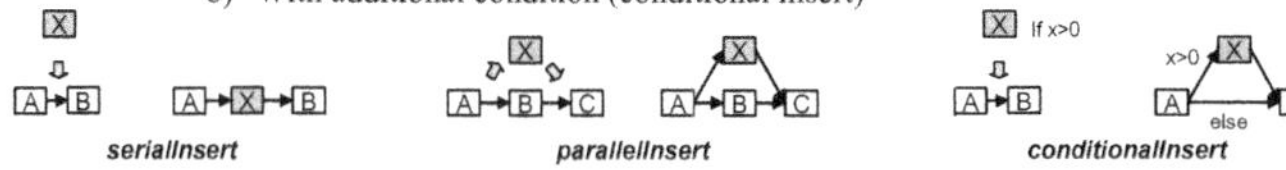


**Implementation**: The *insert* adaptation pattern can be realized by transforming the high level insertion operation into a sequence of low level change primitives (e.g., add node, add control dependency).

**b) Pattern AP2: Delete Process Fragment**

**Description**: A process fragment is deleted from a process schema.
**Example**: For a particular patient no computer tomography is performed due to the fact that he has a cardiac pacemaker (i.e., the computer tomography activity is deleted).
**Problem**: In a real world process a task has to be skipped or deleted.

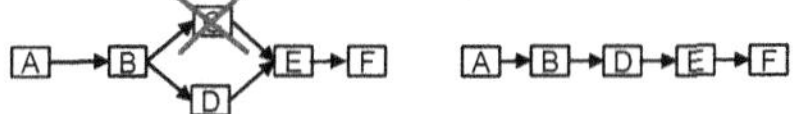

**Implementation**: Several options for implementing the *delete* pattern exist: (1) The fragment is physically deleted (i.e., corresponding activities and control edges are removed from the process schema), (2) the fragment is replaced by one or more null activities (i.e., activities without associated activity program) or (3) the fragment is embedded in a conditional branch with condition *false* (i.e., the fragment remains part of the schema, but is not executed).

**Fig. 4.** Insert (AP1) and Delete (AP2) Process Fragment patterns

**a) Pattern AP3: Move Process Fragment**

**Description**: A process fragment is moved from its current position in the process schema to another position.
**Example**: Usually employees are only allowed to book a flight, after getting approval from the manager. For a particular process instance the booking of a flight is exceptionally done in parallel to the approval activity (i.e., the book flight activity is moved from its current position to a position parallel to the approval activity).
**Problem**: Predefined ordering constraints cannot be completely satisfied for a set of activities.

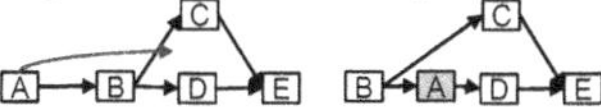

**Design Choices:**

D. How is the process fragment X embedded in the process schema?
  1. X is inserted between 2 directly succeeding activities (serial move)
  2. X is inserted between 2 activity sets (move between node sets)
     a) Without additional condition (parallel move)
     b) With additional condition (conditional move)

**Implementation**: This adaptation pattern can be implemented based on Pattern AP1 and AP2 (insert / delete process fragment).
**Related Patterns:** *Swap* adaptation pattern (AP5) (not detailed in the paper)

**b) Pattern AP4: Replace Process Fragment**

**Description**: A process fragment is replaced by another process fragment.
**Example**: Instead of the computer tomography activity, the X-ray activity shall be performed for a particular patient.
**Problem**: A process fragment is no longer adequate, but can be replaced by another one.

**Implementation**: This adaptation pattern can be implemented based on Pattern AP1 and AP2 (insert / delete process fragment).

**Fig. 5.** Move (AP3) and Replace (AP4) Process Fragment patterns

parallelization of process fragments (AP9), embedding of a process fragment in a conditional branch (AP10), addition of control dependencies (AP11), removal of control dependencies (AP12), and update of transition conditions (AP13). A description of these patterns can be found in [8].

### 3.2 Patterns for Predefined Changes

The applicability of adaptation patterns is not restricted to a particular process part a priori. By contrast, the following patterns predefine constraints concerning the parts that can be changed. At run-time changes are only permitted within these parts. In this category we have identified 4 patterns, *Late Selection of Proces Fragments* (PP1), *Late Modeling of Process Fragments* (PP2), *Late Composition of Process Fragments* (PP3) and *Multi-Instance Activity* (PP4) (cf. Fig. 6). The *Late Selection of Process Fragments* pattern (cf. Fig. 7) allows to select the implementation for a particular process step at run-time either based on predefined rules or user decisions. The *Late Modeling of Process Fragments* pattern (cf. Fig. 8a) offers more freedom and allows to model selected parts of the process schema at run-time. Furthermore the *Late Composition of Process Fragments* pattern (cf. Fig. 8b) enables the on-the fly composition of process fragments (e.g., by dynamically introducing control dependencies between a set of fragments).

In case of *Multi-Instance Activities* the number of instances created for a particular activity is determined at run-time. We do not consider multi-instance activity patterns in detail as they constitute some of the workflow patterns described in [9]. Multi-instance activities enable the creation of a particular process activity during run-time. The decision how many activity instances are created can be based either on knowledge available at build-time or on some run-time knowledge. We do not consider multi-instances of the former kind as change pattern since their use does not lead to change. For all other types of multi-instance activities the number of instances is determined based on run-time knowledge which can or cannot be available a-priori to the execution of the multi-instance activity. While in the former case the number of instances can be determined at some point during run-time, this is not possible for the latter case. We consider multi-instance activities as change patterns too, since their dynamic creation works like a dynamic schema expansion.

## 4 Change Support Features

So far, we have introduced a set of change patterns, which can be used to accomplish changes at the process type and/or process instance level. However, simply counting the number of supported patterns is not sufficient to analyze how well a system can deal with process change. In addition, change support features must be considered to make change patterns useful in practice (cf. Fig. 9). Relevant change support features include *process schema evolution* and *version control*,

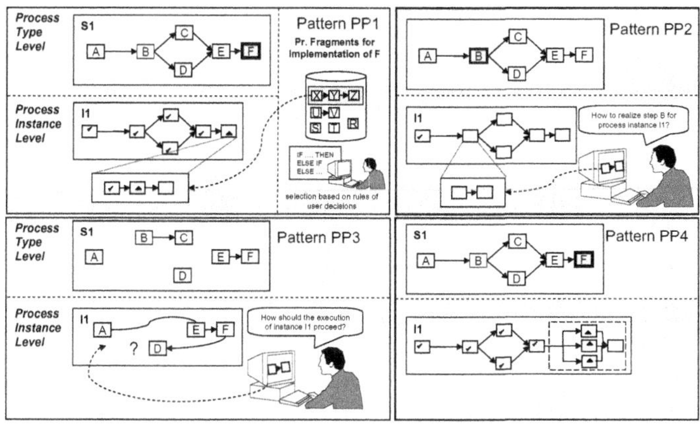


**Fig. 6.** Patterns for Predefined Changes (Overview)

**Pattern PP1: Late Selection of Process Fragments**

**Description**: For particular activities the corresponding implementation (activity program or sub process model) can be selected during run-time. At build time only a placeholder is provided, which is substituted by a concrete implementation during run-time (cf. Fig. 6).

**Example**: For the treatment of a particular patient one of several different sub-processes can be selected depending on the patient's disease.

**Problem**: There exist different implementations for an activity (including sub-processes), but for the selection of the respective implementation run-time information is required.

**Design Choices:**

A. How is the selection process done?
   1. Automatically based on predefined rules
   2. Manually by an authorized user
   3. Semi-automatically: options are reduced by applying some predefined rules; user can select among the remaining options

B. What object can be selected?
   1. Atomic activity
   2. Sub process

C. When does late selection take place?
   1. Before the placeholder activity is enabled
   2. When enabling the placeholder activity

**Implementation**: By selecting the respective sub process or activity program, a reference to it is dynamically set and the selected sub-process or activity program is invoked.

**Related Patterns**: Prerequisite for Pattern *Late Modeling of Process Fragment* (PP2)

**Fig. 7.** Late Selection of Process Fragments (PP1)

change correctness, change traceability, access control and change reuse[1]. As illustrated in Fig. 9 the described change support features are not equally important for both process type level and process instance level changes. Version control, for instance, is primarily relevant for changes at the type level, while change reuse is particularly useful at the instance level [10].

### 4.1 Schema Evolution, Version Control and Instance Migration

In order to support changes at the process type level, version control for process schemes should be supported (cf. Fig. 9). In case of long-running processes, in

[1] Again we restrict ourselves to the most relevant change support features. Additional change support features not covered in this paper are change concurrency control and change visualization.

**a) Pattern PP2: Late Modeling of Process Fragments**

**Description**: Parts of the process schema have not been defined at build-time, but are modeled during run-time for each process instance (cf. Fig. 6). For this purpose, placeholder activities are provided, which are modeled and executed during run-time. The modeling of the placeholder activity must be completed before the modeled process fragment can be executed.
**Example**: The exact treatment process of a particular patient is composed out of existing process fragments at run-time.
**Problem**: Not all parts of the process schema can be completely specified at build time.
**Design Choices:**

A. What are the basic building blocks for late modeling?
   1. All process fragments (including activities) from the repository can be chosen
   2. A constraint-based subset of the process fragments from the repository can be chosen
   3. New activities or process fragments can be defined
B. What is the degree of freedom regarding late modeling?
   1. Same modeling constructs and change patterns can be applied as for modeling at the process type level (*)
   2. More restrictions apply for late modeling than for modeling at the process type level
C. When does late modeling take place?
   1. When a new process instance is created
   2. When the placeholder activity is instantiated
   3. When a particular state in the process is reached (which must precede the instantiation of the placeholder activity)
D. Does the modeling start from scratch?
   1. Late modeling may start with an empty template
   2. Late modeling may start with a predefined template which can then be adapted

**Implementation**: After having modeled the placeholder activity with the editor, the fragment is stored in the repository and deployed. Finally, the process fragment is dynamically invoked as an encapsulated sub-process. The assignment of the respective process fragment to the placeholder activity is done through late binding.
**Related Patterns**: necessitates *Late Selection of Process Fragments* (PP1) of the dynamically modified fragment

(*) Which of the adaptation patterns are supported within the placeholder activity is determined by the expressiveness of the used modeling language.

**b) Pattern PP3: Late Composition of Process Fragments**

**Description**: At build time a set of process fragments is defined out of which a concrete process instance can be composed at run time. This can be achieved by dynamically selecting fragments and adding control dependencies on the fly (cf. Fig. 6).
**Example**: Several medical examinations can be applied for a particular patient. The exact examinations and the order in which they are performed are defined for each patient individually.
**Problem**: There exist several variants of how process fragments can be composed. In order to reduce the number of process variants to be specified by the process engineer during build time, process instances are dynamically composed out of fragments.

**Fig. 8.** Late Modeling (PP2) and Late Composition of Process Fragments (PP3)

| **Change Support Features** | | | |
|---|---|---|---|
| **Change Support Feature** | **Scope** | **Change Support Feature** | **Scope** |
| **F1: Schema Evolution, Version Control and Instance Migration** | T | 2. By change primitives | |
| | | **F3: Correct Behavior of Instances After Change** | I + T |
| No version control – Old schema is overwritten | | **F4: Traceability & Analysis** | I + T |
| 1. Running instances are canceled | | 1. Traceability of changes | |
| 2. Running instances remain in the system | | 2. Annotation of changes | |
| Version control | | 3. Change Mining | |
| 3. Co-existence of old/new instances, no instance migration | | **F5: Access Control for Changes** | I+T |
| 4. Uncontrolled migration of all process instances | | 1. Changes in general can be restricted to authorized users | |
| 5. Controlled migration of compliant process instances | | 2. Application of single change patterns can be restricted | |
| **F2: Support for Ad-hoc Changes** | I | 3. Authorizations can depend on the object to be changed | |
| 1. By change patterns | | **F6: Change Reuse** | I |

T ... Type Level, I ... Instance Level

**Fig. 9.** Change Support Features

addition, controlled migration of already running instances, from the old process schema version to the new one, might be required. In this subsection we describe different existing options in this context (cf. Fig. 10).

If a PAIS provides no version control feature, either the process designer can manually create a copy of the process schema (to be changed) or this schema is overwritten (cf. Fig. 10a). In the latter case running process instances can either be withdrawn from the run-time environment or, as illustrated in Fig. 10a, they remain associated with the modified schema. Depending on the execution state of the instances and depending on how changes are propagated to instances which have already progressed too far, this missing version control can lead to inconsistent states and, in a worst case scenario, to deadlocks or other errors [2]. As illustrated in Fig. 10a process schema $S1$ has been modified by inserting activities X and Y with a data dependency between them. For instance $I1$ the change is uncritical, as $I1$ has not yet entered the change region. However, $I2$ and $I3$ would be both in an inconsistent state afterwards as instance schema and execution history do not match (see [2]). Regarding $I2$, worst case, deadlocks or activity invocations with missing input data might occur.

By contrast, if a PAIS provides explicit version control two support features can be differentiated: running process instances remain associated with the old schema version, while new instances will be created on the new schema version. This approach leads to the co-existence of process instances of different schema versions (cf. Fig. 10b). Alternatively a migration of a selected collection of process instances to the new process schema version is supported (in a controlled way) (cf. Fig. 10c). The first option is shown in Fig. 10b where the already running instances $I1$, $I2$ and $I3$ remain associated with schema S1, while new instances ($I4$-$I5$) are created from schema $S1'$ (co-existence of process instances of different schema versions). By contrast, Fig. 10c illustrates the controlled migration of process instances. Only those instances are migrated which are *compliant*[2] with $S1'$ ($I1$). All other instances ($I2$ and $I3$) remain running according to $S1$. If instance migration is uncontrolled (as it is not restricted to *compliant* process instances) this will lead to inconsistencies or errors. Nevertheless, we treat the uncontrolled migration of process instances as a separate design choice since this functionality can be found in several existing systems (cf. Section 5).

### 4.2 Other Change Support Features

**Support for Ad-hoc Changes:** In order to deal with exceptions PAIS must support changes at the process instance level either through high level changes in the form of patterns (cf. Section 3) or through low level primitives. Although changes can be expressed in both ways, change patterns allow to define changes at a higher level of abstraction making change definition easier.

**Correctness of Change:** The application of change patterns must not lead to run-time errors (e.g., activity program crashes due to missing input data, deadlocks, or inconsistencies due to lost updates or vanishing of instances).

[2] A process instance $I$ is compliant with process schema $S$, if the current execution history of $I$ can be created based on $S$ (for details see [2]).

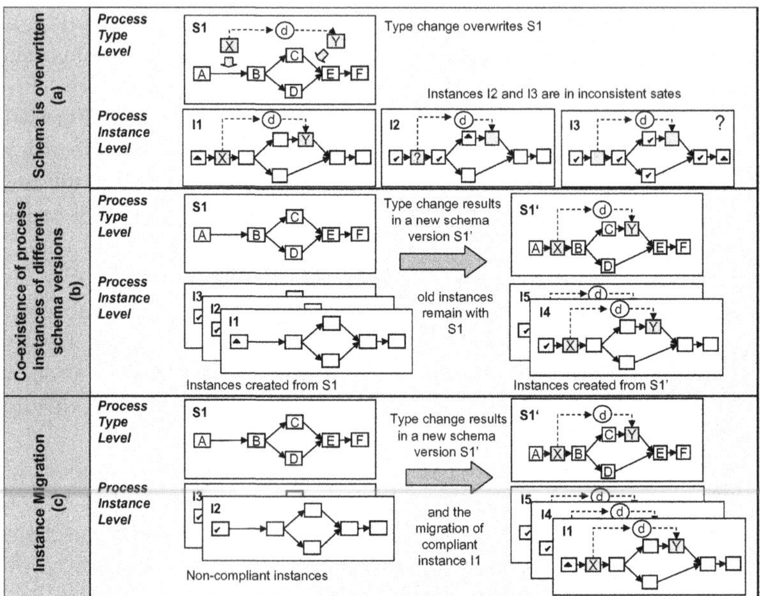


**Fig. 10.** Version Control

Different criteria (see [2]) have been introduced to ensure that instances can only be updated to a new schema if they are compliant with it.

**Traceability and Analysis:** To ensure traceability of changes, they have to be logged. For adaptation patterns the applied changes have to be stored in a change log as change patterns and/or change primitives. While both options allow for traceability, change mining [11] becomes easier when the change log contains high-level information about the changes as well. Regarding patterns for predefined changes, an execution log is usually sufficient to enable traceability. In addition, logs can be enriched with more semantical information, e.g., about the reasons and context of the changes [10]. Finally, change mining allows for the analysis of changes (e.g., to support continuous process improvement) [11].

**Access Control for Changes:** The support of change patterns leads to increased PAIS flexibility. This, in turn, imposes security issues as the PAIS becomes more vulnerable to misuse. Therefore, the application of changes at the process type and the process instance level must be restricted to authorized users. Access control features differ significantly in their degree of granularity. In the simplest case, changes are restricted to a particular group of people (e.g., to process engineers). More advanced access control components allow to define restrictions at the level of single change patterns (e.g., a certain user is only allowed to insert additional activities, but not to delete activities). In addition, authorizations can depend on the object to be changed, e.g., the process schema.

**Change Reuse:** In the context of ad-hoc changes "similar" deviations (i.e., combination of one or more adaptation patterns) can occur more than once. As it requires significant user experience to define changes from scratch change reuse should be supported. To reuse changes they must be annotated with contextual information (e.g., about the reasons for the deviation) and be memorized by the PAIS. This contextual information can be used for retrieving similar problem situations and therefore ensures that only changes relevant for the current situation are presented to the user [12,10]. Regarding patterns for predefined changes, reuse can be supported by making historical cases available to the user and by saving frequently re-occurring instances as templates.

## 5 Change Patterns and Change Support in Practice

In this section we evaluate approaches from both academia and industry regarding their support for change patterns as well as change features. For academic approaches the evaluation has been mainly based on literature. In cases where it was unclear whether a particular change pattern or change feature is supported or not, the respective research groups were additionally contacted. The evaluated academic approaches are ADEPT[3], WIDE [13], Pockets of Flexibility [14], Worklets/Exlets [4,15], CBRFlow [12,10], MOVE [16], HOON [17], and WASA [5]. In respect to commercial systems only such systems have been considered for which we have hands on experience as well as a running system installed. This allowed us to test the change patterns and change features. As commercial systems Staffware [1] and Flower [6] were considered. Evaluation results are summarized in Fig. 11. A detailed description of the evaluated approaches can be found in [8].

If a change pattern or change support feature is not supported at all, the respective table entry will be labeled with "-". Otherwise, it describes the exact pattern variants as supported by listing all available design choices. In case no design choices exist for a particular change pattern, which is supported, the respective table entry is simply labeled with "+". Partial support is labeled with "o". As an example take change pattern PP1 of the Worklet/Exlet approach [4,15]. The string "A[1,2], B[1,2], C[2]" indicates that design choices A, B and C are supported. Further, it shows for every design choice the exact options available (e.g., for design choice A, Options 1 and 2 are supported).

In particular an adaptation pattern will be only considered as being provided, if the respective system supports the pattern directly, i.e., based on one high-level change operation. Of course, adaptation patterns can be always expressed by means of a set of basic change primitives (like add node, delete node, add edge, etc.). However, this is not the idea behind adaptation patterns. Since process schema changes (at the type level) based on these modification primitives are supported by almost each process editor, this is not sufficient to qualify for pattern support. By contrast, the support of high-level change operations allows introducing changes at a higher level of abstraction and consequently hides a lot of the complexity from the user. Therefore changes can be performed in a more efficient and less error prone way. In addition, in order to qualify as an adaptation pattern the application of the respective change operations must not be restricted to predefined regions in the process.

Several of the adaptation patterns (e.g., AP3 or AP4) can be implemented by applying a combination of the more basic patterns AP1, AP2, AP10 and AP11. However, a given approach will only qualify for a particular adaptation pattern, if it supports this pattern directly (i.e., it offers one respective change operation).

Note that missing support for adaptation patterns does not necessarily mean that no run-time changes can be performed. As long as feature F2 is supported ad-hoc changes to running process instances are possible (for details see [8]). In general, if a respective approach provides support for predefined change patterns like for instance late modeling of process fragments (PP1) or late selection of process fragments (PP2) the need for structural changes of the process schema can be decreased making feature F3 less crucial.

The evaluation of selected approaches shows that there exists no single system which supports all change patterns and features (cf. Table 11). In particular, none of the approaches provides both adaptation patterns and predefined change patterns, which would allow addressing a much broader process spectrum. While predefined change patterns allow to reduce the need for structural changes during run-time by providing more flexible models, adaptation patterns allow for structural changes which cannot be pre-planned. In addition, they make changes more efficient, less complex and less error-prone through providing high-level change operations.

| Change Patterns and Change Support | | | | | | | | | |
|---|---|---|---|---|---|---|---|---|---|
| Pattern/ Feature | Academic | | | | | | | Commercial | |
| | ADEPT / CBRFlow | WIDE | Pockets of Flexibility | Worklets / Exlets | MOVE | HOON | WASA | Staffware | Flower |
| **Change Patterns** | | | | | | | | | |
| **Adaptation Patterns** | | | | | | | | | |
| AP1 | A[1, 2], B[1,2,3], C[1,2], D[1, 2] | A[2], B[1], C[2], D[1,2] | – | – | – | – | – | – | – |
| AP2 | A[1, 2], B[1,2,3], C[1,2] | A[2], B[1], C[2] | – | – | – | – | – | – | A[2], B[1], C[2] |
| AP3 | A[1, 2], B[1,2,3], C[1,2], D[1, 2] | – | – | – | – | – | – | – | – |
| AP4 | – | A[2], B[1], C[2] | – | A[1], B[2], C[1,2] | – | – | – | – | – |
| **Preplanned Change Patterns** | | | | | | | | | |
| PP1 | – | – | – | A[1,2], B[1,2], C[2] | – | A[1,2], B[1,2], C[2] | – | A[1,2], B[1,2], C[2] | – |
| PP2 | – | – | A[1,2], B[2], C[2], D[1,2] | – | A[1], B[1], C[3], D[1,2] | – | – | – | – |
| PP3 | – | – | – | – | – | – | – | – | – |
| PP4 | – | + | – | – | – | – | – | + | + |
| **Change Features** | | | | | | | | | |
| F1 | 3, 5 | 3, 5 | – | 3 | – | – | 3, 5 | 3, 4 | 1, 2, 3 |
| F2 | 1 | – | 2 | 2 | 2 | 2 | 2 | 2 | 1 |
| F3 | + | + | + | ○ | + | + | + | – | – |
| F4 | 1, 2, 3 | 1 | 1 | 1 | 1 | 1 | 1 | 1 | 1 |
| F5 | 1, 2, 3 | 1, 3 | 1, 2, 3 | 1, 2, 3 | 1, 3 | 1, 2, 3 | 1 | 1, 2, 3 | 1, 2, 3* |
| F6 | + | – | + | + | – | – | – | – | – |

(*) Flower supports Option 2 and 3 of feature F4 only for process instance changes, but not for process type changes

**Fig. 11.** Change Patterns and Change Support Features in Practice

## 6 Related Work

Patterns were first used to describe solutions to recurring problems by Ch. Alexander, who applied patterns to descibe best practices in architecture [18]. Patterns also have a long tradition in computer science. Gamma et al. applied the same concepts to software engineering and described 23 patterns in [7].

In the area of workflow management, patterns have been introduced for analyzing the expressiveness of process modeling languages (i.e., control flow patterns [9]). In addition, workflow data patterns [19] describe different ways for modeling the data aspect in PAIS and workflow resource patterns [20] describe how resources can be represented and utilized in workflows. The introduction of workflow patterns has significant impact on the design of PAIS and has contributed to the systematic evaluation of PAIS and process modeling standards. However, to evaluate the powerfulness of a PAIS regarding its ability to deal with changes, the existing patterns are important, but not sufficient. In addition, a set of patterns for the aspect of workflow change is needed. Further, the degree to which control flow patterns are supported provides an indication of how complex the change framework under evaluation is. In general, the more expressive the process modeling language is (i.e., the more control flow and data patterns are supported), the more difficult and complex changes become.

In [21] exception handling patterns which describe different ways for coping with exceptions are proposed. In contrast to change patterns, exception handling patterns like *Rollback* only change the state of a process instance (i.e., its behavior), but not its schema. The patterns described in this paper do not only change the observable behavior of a process instance, but additionally adapt the process structure. For a complete evaluation of flexibility, both change patterns and exception handling patterns must be evaluated.

## 7 Summary and Outlook

In this paper we proposed 17 change patterns (and described 8 of them in detail) and 6 change support features, which in combination allow to assess the power of a particular change framework. In addition, we evaluated selected approaches and systems regarding their ability to deal with process changes. We believe that the introduction of change patterns complements existing workflow patterns and allows for more meaningful evaluations of existing systems and approaches. In combination with workflow patterns the presented change framework will enable (PA)IS engineers to choose process management technologies

Future work will include change patterns for aspects other than control flow (e.g., data or resources) and patterns for more advanced adaptation policies (e.g., the accompanying adaptation of the data flow when introducing control flow changes) as well as the evaluation of additional systems and approaches.

**Acknowledgements.** We would like to thank S. Shadiq, M. Adams, M. Weske and Y. Han for their valuable feedback and the many fruitful discussions, which helped us to significantly improve this paper.

## References

1. Dumas, M., ter Hofstede, A., van der Aalst, W. (eds.): Process Aware Information Systems. Wiley Publishing, Chichester (2005)
2. Rinderle, S., Reichert, M., Dadam, P.: Correctness criteria for dynamic changes in workflow systems – a survey. Data and Knowledge Engineering 50, 9–34 (2004)
3. Reichert, M., Dadam, P.: ADEPT$_{flex}$ – supporting dynamic changes of workflows without losing control. JIIS 10, 93–129 (1998)
4. Adams, M., ter Hofstede, A.H.M., Edmond, D., v.d.Aalst, W.M.: A service-oriented implementation of dynamic flexibility in workflows. In: Coopis'06 (2006)
5. Weske, M.: Workflow management systems: Formal foundation, conceptual design, implementation aspects. University of Münster, Germany, Habil Thesis (2000)
6. van der Aalst, W., Weske, M., Grünbauer, D.: Case handling: A new paradigm for business process support. Data and Knowledge Engineering. 53, 129–162 (2005)
7. Gamma, E., Helm, R., Johnson, R., Vlissides, J.: Design Patterns: Elements of Reusable Object-Oriented Software. Addison-Wesley, New York (1995)
8. Weber, B., Rinderle, S., Reichert, M.: Identifying and evaluating change patterns and change support features in process-aware information systems. Technical Report Report No. TR-CTIT-07-22, CTIT, Univ. of Twente, The Netherlands (2007)
9. van der Aalst, W.M.P., ter Hofstede, A.H.M., Kiepuszewski, B., Barros, A.P.: Workflow patterns. Distributed and Parallel Databases 14, 5–51 (2003)
10. Rinderle, S., Weber, B., Reichert, M., Wild, W.: Integrating process learning and process evolution - a semantics based approach. In: BPM 2005, pp. 252–267 (2005)
11. Günther, C., Rinderle, S., Reichert, M., van der Aalst, W.: Change mining in adaptive process management systems. In: CoopIS'06, pp. 309–326 (2006)
12. Weber, B., Wild, W., Breu, R.: CBRFlow: Enabling adaptive workflow management through conversational cbr. In: ECCBR'04, Madrid, pp. 434–448 (2004)
13. Casati, F.: Models, Semantics, and Formal Methods for the design of Workflows and their Exceptions. PhD thesis, Milano (1998)
14. Sadiq, S., Sadiq, W., Orlowska, M.: A framework for constraint specification and validation in flexible workflows. Information Systems 30, 349–378 (2005)
15. Adams, M., ter Hofstede, A.H.M., Edmond, D., v. d. Aalst, W.M.: Dynamic and extensible exception handling for workflows: A service-oriented implementation. Technical Report BPM Center Report BPM-07-03, BPMcenter.org (2007)
16. Th. Herrmann, A.-W., Scheer, H.W. (eds.): Verbesserung von Geschftsprozessen mit flexiblen Workflow-Management-Systemen - Verffentlichungen des Forschungsprojektes MOVE. Bd. 1 - 4. Physica Verlag, Heidelberg (1998)
17. Han, Y.: Software Infrastructure for Configurable Workflow Systems. PhD thesis, Univ. of Berlin (1997)
18. Alexander, C., Ishikawa, S., Silverstein, M. (eds.): A Pattern Language. Oxford University Press, New York (1977)
19. Russell, N., ter Hofstede, A., Edmond, D., van der Aalst, W.: Workflow data patterns. Technical Report FIT-TR-2004-01, Queensland Univ. of Techn. (2004)
20. Russell, N., ter Hofstede, A., Edmond, D., van der Aalst, W.: Workflow resource patterns. Technical Report WP 127, Eindhoven Univ. of Technology (2004)
21. Russell, N., van der Aalst, W.M., ter Hofstede, A.H.: Exception handling patterns in process-aware information systems. In: CAiSE'06 (2006)

# Analyzing the Dynamic Cost Factors of Process-Aware Information Systems: A Model-Based Approach

Bela Mutschler[1,2], Manfred Reichert[2], and Stefanie Rinderle[3]

[1] DaimlerChrysler Group Research, P.O. Box 2360, 89013 Ulm, Germany
bela.mutschler@daimlerchrysler.com
[2] Information Systems Group, University of Twente, The Netherlands
m.u.reichert@utwente.nl
[3] Institute Databases and Information Systems, University of Ulm, Germany
stefanie.rinderle@uni-ulm.de

**Abstract.** Introducing *process-aware information systems* (PAIS) in enterprises is usually associated with high costs. It is therefore crucial to understand those factors that determine these costs. Though software cost estimation has received considerable attention during the last decades, it is difficult to apply existing approaches to PAIS. This difficulty particularly stems from the inability of these techniques to deal with the dynamic interactions of the many technological, organizational and project-driven cost factors which specifically arise in the context of PAIS. Picking up this problem, this paper presents an approach to investigate the complex cost structures of PAIS engineering projects based on evaluation models. We present a formalism to design such evaluation models, discuss one characteristic evaluation model and its derivation in detail (based on the outcome of an empirical study), and introduce the notion of *value-based evaluation patterns* to enable the reuse of evaluation models.

## 1 Introduction

*Process-aware information systems* (PAIS) separate process logic from application code [1], and orchestrate the processes at run-time according to their defined logic [2]. For implementing PAIS, numerous process support paradigms (e.g., workflow management, service flows, case handling), process modeling standards (e.g., WS-BPEL, BPML), and tools (e.g., ARIS Toolset, Tibco Staffware) have been introduced [3].

While the benefits of PAIS are typically justified by improved process performance [4,5] and cheaper process implementation [6], there exist no approaches for systematically analyzing related costs. Though software cost estimation has received considerable attention during the last decades and has become an essential task in information system engineering, it is difficult to apply existing approaches to PAIS. This difficulty stems from the inability of these approaches to cope with the numerous technological, organizational and project-driven cost factors which have to be considered in the context of a PAIS (and which do only partly exist in data- or function-centered information systems) [7]. As an example consider the costs for redesigning processes. Another challenge deals with the dependencies between these factors. Activities for *business process redesign* [8], for example, can be influenced by intangible impact factors like *process*

J. Krogstie, A.L. Opdahl, and G. Sindre (Eds.): CAiSE 2007, LNCS 4495, pp. 589–603, 2007.

*knowledge* or *end user fears*. These dependencies result in dynamic economic effects which influence the overall costs of a PAIS engineering project. Existing techniques [9] are typically not able to deal with such dynamic effects as they rely on static models based upon snapshots of the considered software system.

What is needed is a comprehensive approach that enables engineers to model and investigate the complex interplay between the cost and impact factors that arise in the context of PAIS. For this purpose, this paper[1] introduces a sophisticated and practically approved, model-based methodology to better understand and systematically investigate the complex cost structures of PAIS engineering projects. We present a formalism to design qualitative evaluation models and discuss one characteristic evaluation model and its derivation in detail (based on the outcome of an empirical study). In response to the problems observed during the exploratory use of our methodology in practice, we additionally introduce the notion of *value-based evaluation patterns*.

The remainder of this paper is organized as follows. Section 2 describes our qualitative cost analysis methodology. Section 3 introduces value-based evaluation patterns. Section 4 discusses related work, and Section 5 concludes with a summary.

## 2 The EcoPOST Cost Analysis Methodology

This section describes the main steps of our approach for modeling, analyzing and understanding those factors and their complex interplay that determine the dynamic costs of PAIS engineering projects. Section 2.1 describes the terminology used in our approach. Section 2.2 introduces the notation of our evaluation models. Section 2.3 gives an illustrating example. Section 2.4 motivates the use of simulation for analyzing the dynamic implications as described by our evaluation models. Section 2.5 deals with the specification of simulation models. Finally, Section 2.6 summarizes major lessons learned from a (pilot) case study in the automotive domain.

### 2.1 Terminology

Basically, we distinguish between different kinds of evaluation factors. *Static Cost Factors* (SCF) represent costs that can be precisely quantified in terms of money. The value of a SCF does not considerably change during a PAIS engineering project (except for its time value, which is not further considered in this paper). Thus, the value of a SCF can be considered as constant. As typical examples of SCF consider software license costs, hardware costs, or costs for external consultants.

*Dynamic Cost Factors* (DCF), in turn, represent costs that are determined by activities related to a PAIS engineering project. The (re)design of business processes prior to the introduction of PAIS, for example, constitutes such an activity. These activities cause measurable efforts, which, as they are influenced by other, often intangible factors, can vary. A DCF "Costs for Business Process Redesign", for instance, may be influenced by an intangible factor "Willingness of Staff Members to support Redesign

[1] Our research has been conducted in the EcoPOST project [7,10] which deals with the development of an evaluation framework for analyzing PAIS from a value-based perspective (see http://is.ewi.utwente.nl/research/).

Activities". Obviously, if staff members do not contribute to a redesign project by providing needed information (e.g., about process details), any redesign effort will be ineffective and will increase costs. If staff willingness is additionally varying during the redesign activity (e.g., due to a changing communication policy), the DCF "Costs for Business Process Redesign" will be subject to more complex effects. In our framework, intangible factors like "Willingness of Staff Members to support Redesign Activities" can be represented by so called *impact factors*.

*Impact Factors* (ImF) are intangible evaluation factors that influence DCF, i.e., the activities underlying a DCF. In particular, ImF lead to the evolution of DCF, which, in turn, makes the estimation and analysis of DCF a difficult task to accomplish. As examples consider factors such as "End User Fears", "Availability of Process Knowledge", or "Ability to redesign Business Processes". Opposed to SCF and DCF, the values of ImF are not quantified in monetary terms, but in a qualitative manner. "End User Fears", for example, can be quantified by means of a "Degree of End User Fears" (which can be "low" or "high"). Also, ImF can be either *static* or *dynamic*. The value of a static ImF ($ImF_S$) does not considerably evolve during a PAIS engineering project and can be considered as constant (like the value of a SCF). The value of a dynamic ImF ($ImF_D$), by contrast, may be changing. Like the evolution of DCF, the evolution of dynamic ImF is caused by (both static and dynamic) ImF.

## 2.2 Evaluation Models

To better understand the evolution of DCF in PAIS engineering projects as well as DCF interference through ImF, we utilize *evaluation models*. In particular, each DCF is represented and analyzed by exactly one evaluation model. These models are specified using the System Dynamics [11,12] notation[2] (cf. Fig. 1A) [7].

**Model Notation.** An evaluation model comprises a set of *model variables* which are denoted as *evaluation factors*. In our context SCF, DCF, and ImF correspond to evaluation factors. Different types of variables exist. *State variables* can be used to represent dynamic factors, i.e., to capture changing values of DCF (e.g., the "Costs for Business Process Redesign"; cf. Fig. 1B) and dynamic ImF (e.g., a certain degree of "Process Knowledge"). A state variable is graphically denoted as rectangle (cf. Fig. 1B), and its value at time $t$ is determined by the accumulated changes of this variable from starting point $t_0$ to present moment $t$ ($t > t_0$); similar to a bathtub which accumulates – at a defined moment $t$ – the amount of water which has been poured into it in the past. Each state variable needs to be connected to at least one *source* or *sink*. Both sources and sinks are graphically denoted as cloud-like symbols (cf. Fig. 1B).

[2] We decided to use System Dynamics based on a literature review of potential modeling formalisms. Out of the investigated formalisms, *System Dynamics* (SD) and *Bayesian Networks* (BN) promised to be of particular usefulness in our context. Both formalisms allow to explicitly model networks of evaluation factors. However, BN deal with uncertainty and focus on determining probabilities of events. SD, by contrast, neglects the issue of "(un)certainty" and deals with the analysis of dynamic effects which occur in networks of interacting factors. As we can typically determine whether a certain factor is relevant in a given scenario, we decided to use SD.

Values of state variables change through inflows and outflows. Graphically, both flow types are depicted by twin-arrows which either point to (in the case of an inflow) or out of (in the case of an outflow) the state variable (cf. Fig. 1B). Picking up again the bathtub image, an *inflow* is a pipe that adds water to the bathtub, i.e., inflows increase the value of a state variable. An *outflow*, by contrast, is a pipe that purges water from the bathtub, i.e., outflows decrease the value of a state variable. The DCF "Costs for Business Process Redesign" shown in Fig. 1C, for example, increases through its inflow "Cost Increase" and decreases through its outflow "Cost Decrease".

Returning to the bathtub image, we further need "water taps" to control the amount of water flowing into the bathtub, and "drains" to specify the amount of water flowing out. For this purpose, a *rate variable* is assigned to each flow (graphically depicted by a valve; cf. Fig. 1B).

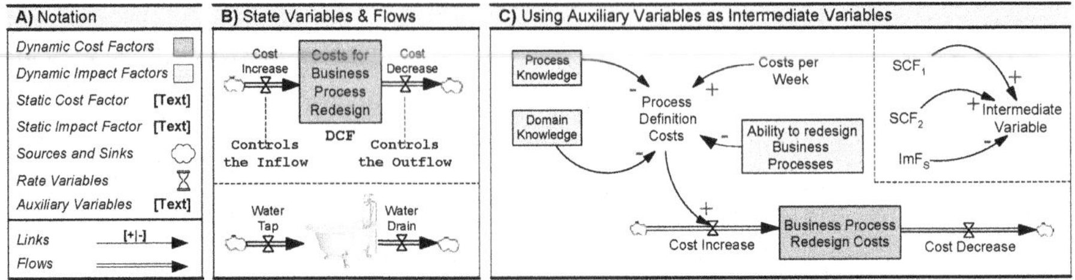


**Fig. 1.** Evaluation Model Notation and initial Examples

Besides state variables, evaluation models may comprise *constants* and *auxiliary variables* (which are both graphically represented by their name). Constants are used to represent static evaluation factors, i.e., SCF and static ImF in our context. Auxiliary variables, in turn, represent intermediate variables. As an example consider the auxiliary variable "Process Definition Costs" in Fig. 1C. Both constants and auxiliary variables are integrated into an evaluation model with *links* (not flows), i.e., labeled arrows. A *positive link* (labeled with "+") between x and y (with y as dependent variable) indicates that y will tend in the same direction if a change occurs in x. A *negative link* (labeled with "-") expresses that the dependent variable y will tend in the opposite direction if the value of x changes. Relationships as expressed by links either can be linear or non-linear (cf. Section 2.5 for details). Altogether, we can define:

***Definition (Evaluation Model).*** *A graph EM = (V, F, L) is called evaluation model, if the following holds:*

- $V := S \dot\cup X \dot\cup R \dot\cup C \dot\cup A$ *is a set of model variables with*
  - *S is a set of state variables,*
  - *X is a set of sources and sinks,*
  - *R is a set of rate variables,*
  - *C is a set of constants,*
  - *A is a set of auxiliary variables,*

- $F \subseteq ((S \times S) \cup (S \times X) \cup (X \times S))$ *is a set of edges called flows,*
- $L \subseteq ((S \times A \times Lab) \cup (S \times R \times Lab) \cup (A \times A \times Lab) \cup (A \times R \times Lab) \cup$ $(C \times A \times Lab) \cup (C \times R \times Lab))$ *is a set of edges called links with* $Lab := \{+, -\}$ *being the set of link labels, where*
  - $(q_i, q_j, +) \in L$ *with* $q_i \in (S \dot\cup A \dot\cup C)$ *and* $q_j \in (A \dot\cup R)$ *denotes a positive link,*
  - $(q_i, q_j, -) \in L$ *with* $q_i \in (S \dot\cup A \dot\cup C)$ *and* $q_j \in (A \dot\cup R)$ *denotes a negative link.*

**Model Correctness.** For defining evaluation models we introduce additional constraints (*model design rules*) to be taken into account: (1) DCF and dynamic ImF have to be represented by state variables, (2) SCF and static ImF must be represented as constants, (3) every state variable $v$ must be connected to at least one source or sink $q$, i.e., $\forall v \in S$ : $\exists(q,v) \in F \vee \exists(v,q) \in F$ with $q \in X$, (4) every model variable must be used in at least one binary relation, i.e., $\forall v, q \in (S \dot\cup X) : \exists(v,q) \in F \vee \exists(q,v) \in F$ and $\forall q \in (A \dot\cup C) \wedge$ $\forall v \in (A \dot\cup R) : \exists(q, v, [+|-]) \in L$, (5) every rate variable of the evaluation model is influenced by at least one link, i.e., $\forall v \in R \wedge q \in (S \dot\cup A \dot\cup C) : \exists(q, v, [+|-]) \in L$, and (6) there exist no cycles solely consisting of auxiliary variables, i.e., $\neg\exists < q_0, q_1, ..., q_r > \in A^{r+1}$ with $q_0 = q_r$ and $q_k \neq q_l$ for $k, l = 1, \ldots, r; k \neq l$.

Rules for the correct use of flows and links are illustrated in Fig. 2A and Fig. 2B. By contrast, Fig. 2C - Fig. 2H show examples of incorrect models. DCF and $ImF_D$, for example, may be only influenced by flows, and not by links as shown in Fig. 2C. Flows may be only connected to DCF and $ImF_D$, but not to auxiliary variables or constants as depicted in Fig. 2D. Links pointing to constants (e.g., SCF, $ImF_S$) as denoted in Fig. 2E and Fig. 2F are also not valid. Finally, flows and links connecting DCF to $ImF_D$ (and vice versa) are also not considered as correct (cf. Fig. 2G and Fig. 2H).

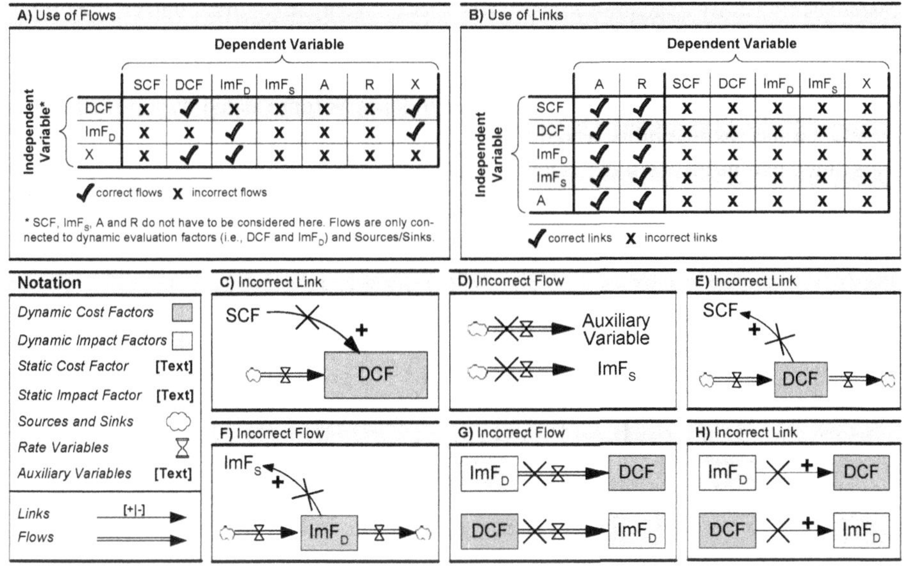


**Fig. 2.** Model Design Rules and Examples of Incorrect Modeling

Model correctness does not only presume compliance with existing model design rules. It also deals with the development of models that are suitable to represent real-world settings. Therefore, we accomplished user surveys and case studies (see below).

## 2.3 Illustrating Example

Fig. 3 shows a model which describes the influence of the dynamic ImF "End User Fears" on the DCF "Costs for Business Process Redesign". More specifically, this model reflects the assumption that the introduction of a PAIS may cause end user fears, e.g., due to a high degree of job redesign and due to changed social clues. Such end user fears can lead to emotional user resistance. This, in turn, can make it difficult to get support from the users while introducing a PAIS. Such models are of significant value for PAIS engineers, e.g., due to their suitability to serve as a conscious-raising tool about basic economic effects in PAIS engineering projects.

**Model Details.** Basic to this evaluation model is a cyclic structure connecting the four ImF "End User Fears", "Emotional Resistance", "Ability to acquire Process Knowledge", and "Ability to redesign Business Processes". Their arrangement (cf. Fig. 3) illustrates the following coherence: Increasing end user fears result in increased emotional resistance of end users. This dependency is represented by a positive link from the ImF "End User Fears" to the "Resistance Growth Rate" (which controls the inflow of the ImF "Emotional Resistance"). An increasing emotional resistance of end users, in turn, results in a decreasing ability to acquire process knowledge. Reason is that an increasing emotional resistance makes profound process analysis (e.g., based on interviews with process participants) a difficult task to accomplish. This dependency is represented by a negative link from the ImF "Emotional Resistance" to the rate variable "Decreasing Ability to acquire Process Knowledge" (which, in turn, controls the inflow of the ImF "Ability to acquire Process Knowledge").

A decreasing ability to acquire process knowledge results in a decreasing ability to redesign business processes. Again, this dependency is represented by a positive link. Finally, an increasing ability to redesign business processes can even enforce end user fears since end users often consider business process redesign activities as a potential

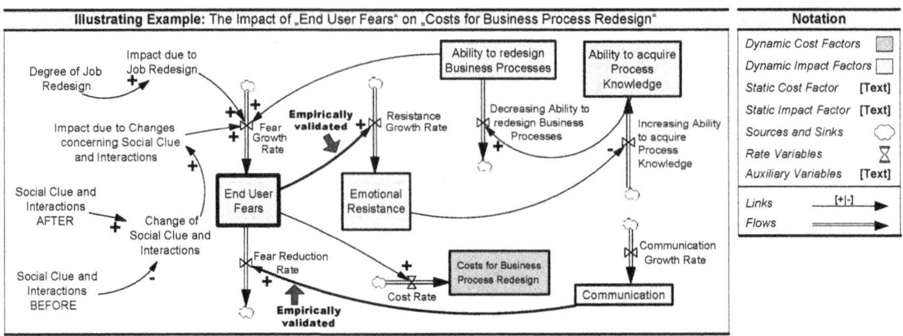


**Fig. 3.** Dealing with the Impact of End User Fears

threat for their own job. This dependency is represented by another positive link. Note that the "Fear Growth Rate" is not only biased by this link. It is also influenced by the "Degree of Job Redesign" and the "Change of Social Clue and Interactions" (which is calculated from the social clues and interactions before and after the business process redesign). Finally, "Communication" is considered as well. This ImF deals with the information of end users about the goals of introducing a PAIS.

**Empirical Validation.** To empirically confirm our assumptions as represented by this evaluation model (and the many other ones we have developed for PAIS engineers) we conducted an online survey[3] among 70 business process management experts. Regarding the above example, we have analyzed the ImF "End User Fears" and the ImF "Communication" in more detail. First, we asked for the *relevance* of the factor (Question 1). Second, we asked whether there are potential *dependencies* between this factor and other ones (Question 2). Only those survey participants who confirmed the existence of dependencies were directed to two additional questions which dealt with the further specification of the confirmed dependency. Question 3 addressed the *semantic specification* of the dependency, whereas Question 4 asked for the *strength* of the dependency. Note that we interpret our survey results from a qualitative viewpoint, i.e., our results do not allow for precise quantifications of the investigated effects.

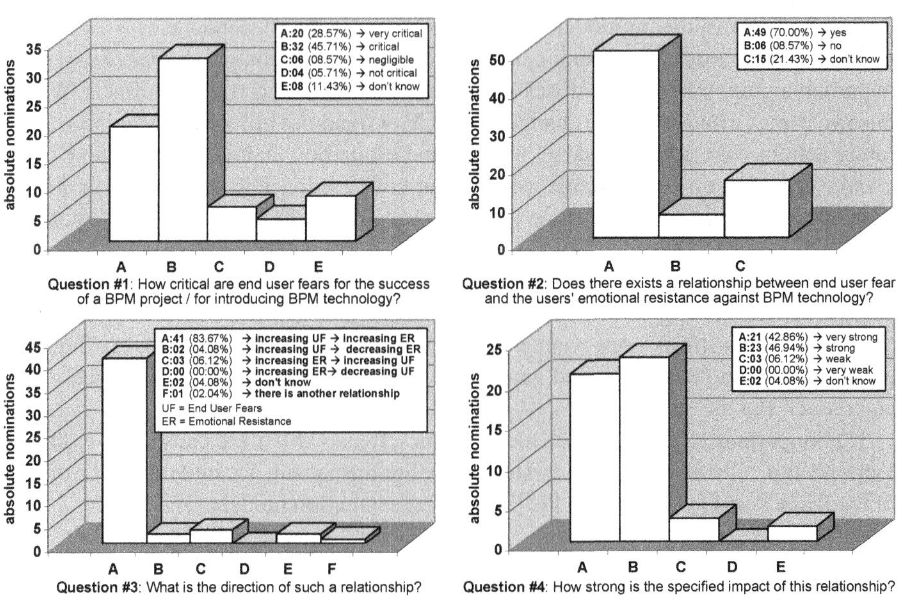


**Fig. 4.** Validating the Impact of End User Fears

Consider Fig. 4. The majority of 74.28% of the survey participants considers end user fears as "very critical" (28.57%) or "critical" (45.71%) for the overall success of

[3] We have summarized the complete results of this survey in [13].

*business process management* (BPM) projects (cf. Question 1 in Fig. 4). More specifically, 70% of the survey respondents confirm that there is a relationship between end user fears and the emotional resistance of end users against BPM technology (cf. Question 2 in Fig. 4). This particularly confirms the positive link connecting these two variables in Fig. 3. Out of these respondents, 83.67% share the opinion that increasing end user fears result in increasing emotional resistance (cf. Question 3 in Fig. 4). Finally, 89.8% of the respondents state (cf. Question 4 in Fig. 4) that the impact of end user fears on emotional resistance either is "very strong" (42.86%) or "strong" (46.94%).

The evaluation model shown in Fig. 3 also considers the ImF "Communication". The majority of 92.86% of the survey participants consider communication between a BPM project's stakeholders as an "essential" (47.14%), "very important" (35.71%) or "important" (10%) factor. Furthermore, 78.57% of the respondents confirm that there is a relationship between communication and end user fears (cf. Fig. 3). Out of these, 74.55% share the opinion that an increasing communication results in decreasing end user fears. Finally, 85,45% of the respondents state that the impact of communication on end user fears either is "very strong" (29.09%) or "strong" (56.36%).

### 2.4 Investigating the Evolution of DCF and Dynamic ImF Through Simulation

The change of DCF and dynamic ImF is caused by the interplay between the different elements of an evaluation model, i.e., by the complex interdependencies between dynamic and static evaluation factors, flows, and links. In this context, *feedback loops* are of particular importance. A feedback loop is a *closed cycle* of causes and effects. Within this cycle, past events (like the change of a DCF or dynamic ImF) are utilized to control future actions (like another change of the same evaluation factor). In other words, if a *change* occurs in a model variable, which is part of a feedback loop, this change will be propagated around the loop [12]. As an example consider the feedback loop depicted in Fig. 3 (cf. Section 2.3).

We distinguish between two types of *loop polarities*. First, *positive* (or *self-reinforcing*) loops generate growth of DCF and dynamic ImF (cf. Fig. 5A). Second, *negative* (or *self-correcting*) loops counteract and oppose growth (cf. Fig. 5B). If evaluation models contain both positive and negative feedback loops, more complex effects may emerge (cf. Fig. 5C and Fig. 5D).

It is important to mention that the dynamic effects which are caused by feedback loops are typically not easy to understand [14]. For this reason, we investigate the effects of feedback loops by simulating [15] respective evaluation models.

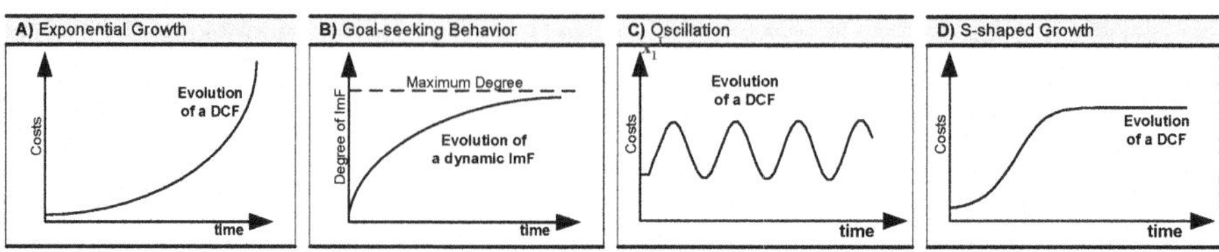


**Fig. 5.** Feedback in Evaluation Models: Overview of potential dynamic Effects

## 2.5 Specifying Simulation Models

In our EcoPOST framework, a *simulation model* consists of a number of *algebraic equations* — one for each model variable (i.e., dynamic and static evaluation factors as well as rate variables and auxiliary variables). The basic components of these algebraic equations are the model variables. In our approach, we use different types of algebraic equations for the different variables of an evaluation model (cf. Fig. 6A):

- ***Static Evaluation Factors***: Static evaluation factors (i.e., SCF and static ImF) are specified using a numerical value in a *constant equation* (e.g., "`Business Process Redesign Costs = 1000 $/Week`"). A specific variant of a constant equation is an *initially computed constant*. In fact, it will often become necessary to specify a constant in terms of another constant if the former depends on the latter and the former should change in any simulation run where the latter is given a new value [14]. As an example consider the following equation: `Process Redesign Costs = 1000 $/Week * Risk Factor`. Note that initially computed constants need to be evaluated only once at the beginning of a simulation.
- ***Dynamic Evaluation Factors***: Dynamic evaluation factors (i.e., DCF and dynamic ImF) are specified by *integral equations* in our approach [14]. Such equations specify the accumulation of a dynamic evaluation factor from a starting point $t_0$ to the present moment $t$ (cf. Fig. 6B). More specifically, DCF and dynamic ImF integrate their *net flow*. The net flow during any interval $[t_1, t_2]$ is the area bounded by the graph of the *net rate* between the start and the end of the interval (cf. Fig. 6C). Thus, the value of a dynamic evaluation factor at $t_2$ can be calculated as the sum of its value at $t_1$ and the area under the net rate curve between $t_1$ and $t_2$. In Fig. 6C, the value at $t_1$ is $S_1$. Adding the area under the net rate curve between $t_1$ and $t_2$ increases the value to $S_2$. The net flow is determined by one or several rate variables.

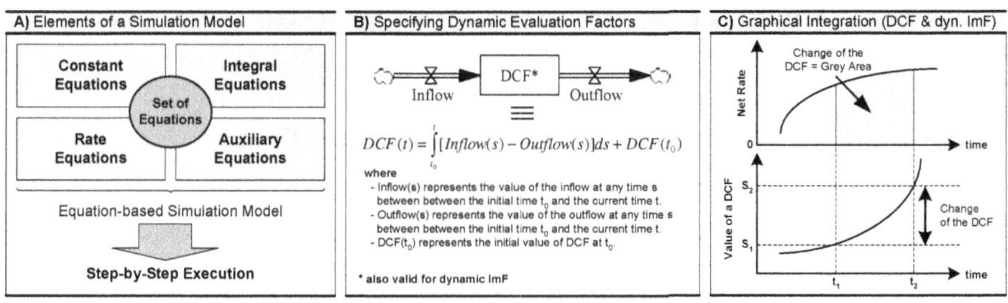


**Fig. 6.** Integration of Flows for Dynamic Evaluation Factors

- ***Rate Variables***: Are expressed by *rate equations*. Rate equations specify the change of DCF or dynamic ImF between two computed conditions. Flows connected to DCF are specified by rate equations describing the amount of costs flowing to, from, or between DCF. Rate equations for flows connected to dynamic ImF specify the impact flowing to, from, or between dynamic ImF. In any case, a rate equation uses information (i.e., values) from other model variables (SCF, DCF, dynamic ImF, and auxiliary variables) to calculate a specific change. In the context of a specific rate variable, the relevant information is represented by those model variables that are connected to the rate variable by links (cf. Section 2.2).

- ***Auxiliary Variables***: Are specified by *auxiliary equations*. Constituting elements of these equations may be SCF, DCF, ImF, rate variables, and auxiliary variables. Note that auxiliary equations are evaluated after the integral equations on which they depend, and before the rate equations of which they are part.

Often, an ImF has a nonlinear impact on DCF. Such nonlinearities have to be represented in our simulation models as well. For this purpose, we use a specific kind of auxiliary equation (implying that nonlinearities require the introduction of additional auxiliary variables in our evaluation models). Specifically, we use *table functions* transferring an input value (e.g., a certain degree of process knowledge) into a corresponding output value (e.g., expressing a specific effect on a DCF) through a *lookup function f* [16]. Linear interpolation is used for values lying between the specified table values. Fig. 7, for example, shows typical table functions. Dependent on the degree of an ImF a specific *impact rating* is derived. An impact rating less than 1 results in decreasing costs (cf. Fig. 7A). A rating equal to 1 neither does increase nor decrease costs. A rating larger than 1 results in increasing costs (cf. Fig. 7B and Fig. 7C). Quantifications based on such impact ratings are also known from software cost models like COCOMO [17].

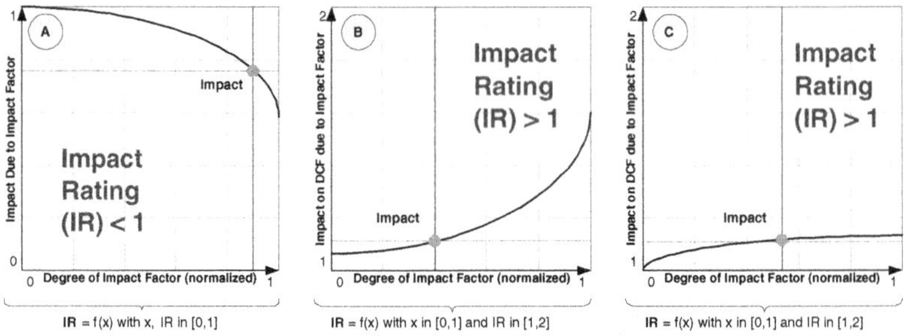


**Fig. 7.** Table Functions for quantifying Impact Factors

For the design of our evaluation models as well as their simulation existing System Dynamics modeling and simulation tools can be used (e.g., *Vensim* [18]). To support administrative tasks related to our framework, we have implemented the *EcoPOST Cost Benefit Analyzer*. Among other things, this tool comprises a knowledge base module for storing and managing VBEP as well as entire evaluation scenarios and a module for visualizing EcoPOST evaluations.

## 2.6 Using the Methodology in Practice - Lessons Learned

We have applied our methodology in an exploratory case study in the automotive domain. In this case study, we have analyzed cost overruns observed during the introduction of a large PDM system for the integrated support of engineering processes.

The initial business case for this project comprised seven major cost categories. In our case study, we analyzed three of them in more detail: *process management costs*, *IT system realization costs*, and *specification and test costs*. In particular, we analyzed whether the observed cost overruns in these cost categories could have been predicted using our cost analysis method. Based on real project data, interviews with project members, two user surveys, and practical experiences, we developed a set of evaluation models using our methodology and analyzed the effects described by these models using simulation. Taking our evaluation models, we were able to explain the observed cost overruns. Moreover, our models helped project members to better understand the complex cost structure of the analyzed project.

Our case study has also revealed several difficulties. In particular, it has turned out that the design of evaluation models constitutes a complex and time-consuming task. Evaluation models tend to become rather complex due to the large number of potential SCF, DCF, ImF and causal dependencies between them, and each evaluation model had to be designed from scratch. This resulted in the loss of valuable modeling experiences. In response to these issues the following section introduces the notion of *value-based evaluation patterns* (VBEP).

## 3 Value-Based Evaluation Patterns (VBEP)

The introduction of PAIS in production environments often exhibits similarities. Picking up these similarities by means of customizable generic evaluation patterns would be a useful step to simplify the use of our methodology and to increase model reuse. Therefore, we introduce a VBEP as a predefined, but customizable evaluation model.

**Characterization.** Basically, VBEP use the same elements as introduced in Section 2, i.e., they consist of an *evaluation model* and an associated *simulation model*. More precisely, each VBEP constitutes a template for specific sets of DCF and/or ImF we encounter when introducing PAIS.

Our approach distinguishes between primary and secondary VBEP. A *primary* VBEP describes a particular DCF, and a *secondary* VBEP describes an ImF. Characteristic VBEP are summarized in Table 1 (primary VBEP) and Table 2 (secondary VBEP).

**Table 1.** Primary VBEPs

| Name | Description |
|---|---|
| *Process (Re)Design* | This VBEP deals with the costs of business process redesign activities prior to and during the development of a PAIS. Such a redesign may become necessary for several reasons, e.g., to increase the degree of automated process activities or to eliminate process performance flaws. |
| *Organizational Change* | This VBEP deals with the costs of changing an organization due to the introduction of a PAIS. As examples consider the adaptation of organizational structures like team structures and single jobs. |
| *Process Evolution* | This VBEP deals with the costs caused by the adaptation of business process changes. In fact, many business processes are continuously changing due to evolving business requirements. Any process change necessitates the adaptation of the supporting PAIS. |
| *Enterprise Architecture* | This VBEP deals with the costs caused by preparing an enterprise architecture for the introduction of a PAIS (e.g., costs for implementing interfaces to other legacy systems). |
| *Work Profile Change* | This VBEP deals with the costs related to changes in work profiles of end users of a PAIS. In particular, costs are caused by simultaneously holding up the new and the old work profile for some time. |

**Table 2.** Secondary VBEPs

| Name | Description |
|---|---|
| *Process Knowledge* | This VBEP deals with the effects of process knowledge, e.g., about data and control flows. Acquiring process knowledge causes efforts, e.g., for conducting interviews with process participants. However, process knowledge can also have a positive impact on other activities such as business process redesign. |
| *Domain Knowledge* | This VBEP deals with the impact of domain knowledge, e.g., of the experience of project members, on the costs of introducing a PAIS. Acquiring domain knowledge causes efforts, e.g., for the time needed to understand a complex domain. However, domain knowledge can also have a positive impact. |
| *Process Ownership* | This VBEP deals with the effects of a clear or unclear ownership of the business processes to be supported by a PAIS. The definition of explicit process ownerships typically implies efforts. However, clear process ownerships will have a positive impact if they are well-defined. |
| *Process Transparency* | This VBEP deals with the effects of process transparency during the introduction of a PAIS. A high process transparency has a positive impact on other activities such as the redesign of business processes. |
| *End User Fears* | This VBEP deals with the impact of end user fears on the ability to redesign business processes. We have already discussed this VBEP (cf. Fig. 3) in Section 2.3. |

All these VBEP have been systematically derived based on the results of case studies we conducted in several information system engineering projects in the automotive domain (e.g., in projects dealing with the introduction of PDM and ERP systems). Furthermore, we rely on results of several online surveys. However, we do not claim for completeness here, and we are continuously working on the extension of our pattern collection.

Generally, VBEP enable the reuse of historical evaluation data. This reduces the need for designing evaluation models from scratch. Moreover, VBEP are useful as a means to increase the awareness for cost effects in PAIS engineering projects.

**Customization.** Customization becomes necessary as VBEP are applied in different evaluation context. Thereby, we distinguish between the customization of the *evaluation model* (Step I in Fig. 8) and the *simulation model* (Step II in Fig. 8) of a VBEP. The former always requires the subsequent adaptation of the underlying simulation model, while the latter is also possible without adapting the assigned evaluation model.

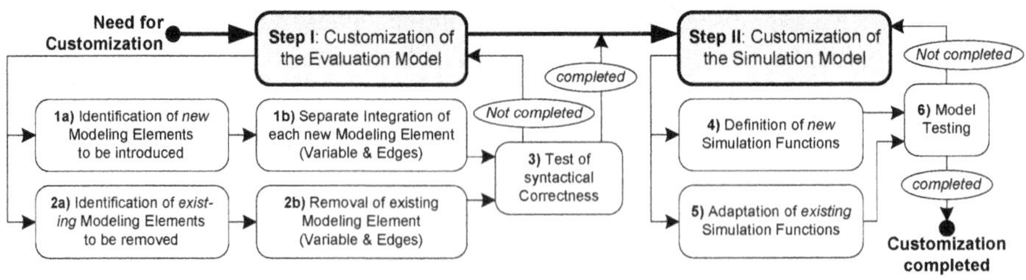


**Fig. 8.** Step-by-Step Customization of VBEP

Adapting an evaluation model can be achieved by adding or removing model variables, flows, or links (Step 1a/b and Step 2a/b in Fig. 8). The VBEP "End User Fears" (cf. Fig. 3), for example, could be customized by introducing an ImF "Management Commitment" to take into account the impact of this factor on end user fears. Therefore, the new ImF "Management Commitment" is connected to the ImF "End User Fears".

In our example this can be achieved with a negative link to denote that increasing management commitment results in decreased user fears. The correctness of a customized VBEP is ensured through the design rules discussed in Section 2.2.

Customizing a simulation model, by contrast, requires adaptations of the equations of the simulation model (Step 4 and Step 5 in Fig. 8). As examples of potential customizations consider changes of SCF values or adaptations of rate functions.

**Merging VBEP.** Customization becomes also necessary when VBEP are merged. Assume, for example, that an ImF "End User Fears" (cf. Fig. 9B) has to be considered in the context of a DCF "Costs for Business Process Redesign" (cf. Fig. 1C). This can be realized by merging a secondary VBEP (specifying the additional ImF) with a primary VBEP (specifying the DCF). Regarding our evaluation models, this merge can be (partially) automated. As input, a respective algorithm needs two evaluation models EM1 and EM2. The merge of EM1 and EM2 is then accomplished through a systematic comparison of all model variables from EM1 with all model variables from EM2. If a model variable from EM1 (e.g., a DCF) has the same name and type as a model variable from EM2, both variables (and their links) will be merged.

Applying this algorithm requires that the evaluation models to be merged exhibit some overlap, i.e., both models have to contain at least one identical model variable. In our example, the ImF "Ability to Redesign Business Processes" has been the mixing

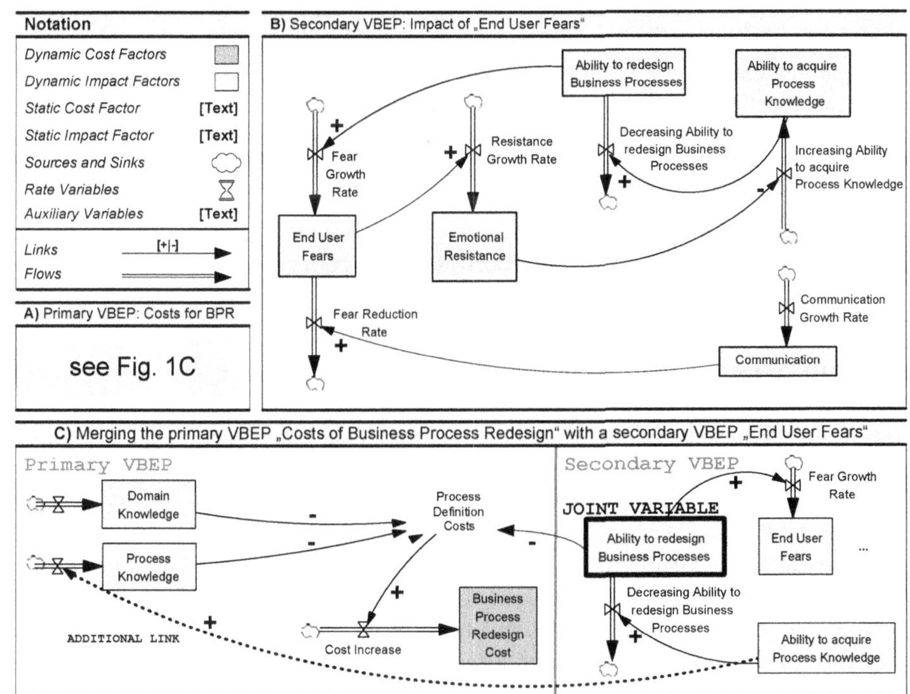


**Fig. 9.** Combining primary and secondary VBEP

point (cf. Fig. 9C). If there exist no identical variables, evaluation models can be merged manually. Besides, any merge typically requires an additional editing of the newly derived model (regardless whether the merge has been automatically conducted or not). In Fig. 9), for example, we introduce an additional link between the ImF "Ability to acquire Process Knowledge" and the ImF "Process Knowledge".

## 4 Related Work

Boehm et. al [19] propose a classification of cost estimation techniques into six major categories. In particular, they distinguish between *model-based approaches* (e.g., COCOMO, SLIM), *expertise-based approaches* (e.g., the Delphi method), *learning-oriented approaches* (using neural networks or case based reasoning), *regression-based approaches* (e.g., the ordinary least squares method), *composite approaches* (e.g., the Bayesian approach), and *dynamic-based approaches* (which explicitly acknowledge that cost factors change over the duration of the system development). Picking up this classification, our methodology can be considered as an example of a dynamic-based approach (the other five categories rely on static analysis models).

The use of patterns has been widely discussed since the advent of computer science research. At present, the software community is using numerous variations of patterns largely for software architecture (*conceptual patterns*), design (*design patterns*), programming (*XML schema patterns*, *J2EE patterns*, etc.), as well as for software development processes. Recently, the idea of using patterns has been also applied to more specific domains like workflow management [20] or inter-organizational control [21].

Regarding the reuse of (System Dynamics) models, one has to distinguish between two basic directions. On the one hand, authors like Senge [22], Eberlein and Hines [23], Liehr [24], and Myrtveit [25] introduce predefined generic structures (with slightly different semantics). All these approaches satisfy the capability of defining "components". On the other hand, Winch [26] proposes a more restrictive approach which is only based on the parameterization of generic structures (without providing standardized modeling components). Our approach picks up ideas from both directions, i.e. we address both the definition of generic components and customization.

## 5 Summary

This paper has presented a qualitative cost analysis methodology to investigate the complex dependencies and interactions of those factors that determine the costs of PAIS engineering projects. We have presented a formalism to design evaluation models and exemplarily discussed one evaluation model and its derivation based on the results of an empirical study. Finally, we have introduced the notion of *value-based evaluation patterns* (VBEP) as a means to enable the reuse of evaluation data in different context.

Note that the expressiveness of simulation always depends on the plausibility and resilience of the underlying simulation models. Therefore, we have additionally accomplished various empirical and experimental research activities (e.g., software experiments, online surveys, case studies) in order to put the quantifications gained from our simulation models on a more reliable basis (cf. [27] for examples).

## References


1. Dehnert, J., van der Aalst, W.: Bridging the Gap between Business Models and Workflow Specification. Int'l. Journal of Cooperative Information Systems (2004)
2. Reichert, M., Rinderle, S., Kreher, U., Dadam, P.: Adaptive Process Management with ADEPT2. Proc. 21th ICDE, pp.1113-1114 (2005)
3. Dumas, M., van der Aalst, W.M.P., ter Hofstede, A.H. (eds.): Process-aware Information Systems: Bridging People and Software through Process Technology. Wiley, Chichester (2005)
4. Reijers, H.A., van der Aalst, W.M.P.: The Effectiveness of Workflow Management Systems - Predictions and Lessons Learned. Int'l. J. of Inf. Manag. 25(5), 457–471 (2005)
5. Choenni, S., Bakkera, R., Baetsa, W.: On the Evaluation of Workflow Systems in Business Processes. Electronic Journal of IS Evaluation (EJISE) vol.6(2) (2003)
6. Kleiner, N.: Can Business Process Changes Be Cheaper Implemented with Workflow-Management-Systems?. In: Proc. IRMA 2004, pp. 529–532 (2004)
7. Mutschler, B., Reichert, M., Bumiller, J.: Designing an Economic-driven Evaluation Framework for Process-oriented Software Technologies. In: Proc. 28th ICSE, pp. 885–888 (2006)
8. Yu, E.: Modelling Strategic Relationships for Process Reengineering. PhD Thesis, University of Toronto (1995)
9. Mutschler, B., Zarvic, N., Reichert, M.: A Survey on Economic-driven Evaluations of Information Technology. Technical Report, TR-CTIT-07, University of Twente (2007)
10. Mutschler, B., Reichert, M., Bumiller, J.: An Approach for Evaluating Workflow Management Systems from a Value-Based Perspective. In: Proc. 10th IEEE EDOC, pp. 477–482 (2006)
11. Richardson, G.P., Pugh, A.L.: System Dynamics - Modeling with DYNAMO (1981)
12. Ogata, K.: SD. Prentice-Hall, Englewood Cliffs (2003)
13. Mutschler, B., Reichert, M.: A Survey on Evaluation Factors for Business Process Management Technology. Technical Report, TR-CTIT-06-63, University of Twente (2006)
14. Forrester, J.W.: Industrial Dynamics. Productivity Press, Cambridge, London (1961)
15. Vangheluwe, H., de Lara, J., Mosterman, P.J.: An Introduction to Multi-Paradigm and Simulation. In: Proc. AIS 2002, pp. 9–20 (2002)
16. Mutschler, B., Reichert, M.: Simulation Models for Analyzing the Dynamic Costs of Process-aware IS. Technical Report, TR-CTIT-07-14, University of Twente (2007)
17. Boehm, B., Abts, C., Brown, A.W., Chulani, S., Clark, B.K., Horowitz, E., Madachy, R., Reifer, D., Steece, B.: Software Cost Estimation with Cocomo 2. Prentice-Hall, Englewood Cliffs (2000)
18. Vensim: Ventana Systems (2006) `http://www.vensim.com/`
19. Boehm, B., Abts, C., Chulani, S.: Software Development Cost Estimation Approaches - A Survey. Technical Report, USC-CSE-2000-505 (2000)
20. van der Aalst, W.M.P., ter Hofstede, A.H.M., Kiepuszewski, B., Barros, A.P.: Advanced Workflow Patterns. In: Proc. 7th CoopIS, LNCS 1901, pp. 18–29 (2000)
21. Kartseva, V., Hulstijn, J., Tan, Y.H., Gordijn, J.: Towards Value-based Design Patterns for Inter-Organizational Control. In: Proc. 19th Bled Electronic Commerce Conference (2006)
22. Senge, P.M.: The 5th Discipline: The Art and Practice of the Learning Organization (1990)
23. Eberlein, R.J., Hines, J.H.: Molecules for Modelers. In: Proc. 14th System Dyn. Conf. (1996)
24. Liehr, M.: A Platform for SD Modeling – Methodologies for the Use of predefined Model Components. In: Proc. 20th System Dynamics Conf. (2002)
25. Myrtveit, M.: Object-oriented Extensions to SD. In: Proc. 18th System Dynamics Conf. (2000)
26. Winch, G.: User-parameterized generic Models: A Solution to the Conundrum of Modelling Access for SMEs? SD Review 18(3), 339–357 (2003)
27. Mutschler, B., Reichert, M., Bumiller, J.: Why Process-Orientation is Scarce: An Emp. St. of Process-oriented IS in the Automotive Industry. In: Proc. 10th IEEE EDOC, pp. 433–438 (2006)

# Author Index

# Lecture Notes in Computer Science

For information about Vols. 1–4424

please contact your bookseller or Springer

Vol. 4539: N.H. Bshouty, C. Gentile (Eds.), Learning Theory. XII, 634 pages. 2007. (Sublibrary LNAI).

Vol. 4538: F. Escolano, M. Vento (Eds.), Graph-Based Representations in Pattern Recognition. XII, 416 pages. 2007.

Vol. 4534: I. Tomkos, F. Neri, J. Solé Pareta, X. Masip Bruin, S. Sánchez Lopez (Eds.), Optical Network Design and Modeling. XI, 460 pages. 2007.

Vol. 4531: J. Indulska, K. Raymond (Eds.), Distributed Applications and Interoperable Systems. XI, 337 pages. 2007.

Vol. 4530: D.H. Akehurst, R. Vogel, R.F. Paige (Eds.), Model Driven Architecture- Foundations and Applications. X, 219 pages. 2007.

Vol. 4529: P. Melin, O. Castillo, L.T. Aguilar, J. Kacprzyk, W. Pedrycz (Eds.), Foundations of Fuzzy Logic and Soft Computing. XIX, 830 pages. 2007. (Sublibrary LNAI).

Vol. 4526: M. Malek, M. Reitenspieß, A. van Moorsel (Eds.), Service Availability. X, 155 pages. 2007.

Vol. 4525: C. Demetrescu (Ed.), Experimental and Efficient Algorithms. XIII, 448 pages. 2007.

Vol. 4524: M. Marchiori, J.Z. Pan, C.d.S. Marie (Eds.), Web Reasoning and Rule Systems. XI, 382 pages. 2007.

Vol. 4523: Y.-H. Lee, H.-N. Kim, J. Kim, Y. Park, L.T. Yang, S.W. Kim (Eds.), Embedded Software and Systems. XIX, 829 pages. 2007.

Vol. 4521: J. Katz, M. Yung (Eds.), Applied Cryptography and Network Security. XIII, 498 pages. 2007.

Vol. 4519: E. Franconi, M. Kifer, W. May (Eds.), The Semantic Web: Research and Applications. XVIII, 830 pages. 2007.

Vol. 4517: F. Boavida, E. Monteiro, S. Mascolo, Y. Koucheryavy (Eds.), Wired/Wireless Internet Communications. XIV, 382 pages. 2007.

Vol. 4515: M. Naor (Ed.), Advances in Cryptology - EUROCRYPT 2007. XIII, 591 pages. 2007.

Vol. 4514: S.N. Artemov, A. Nerode (Eds.), Logical Foundations of Computer Science. XI, 513 pages. 2007.

Vol. 4510: P. Van Hentenryck, L. Wolsey (Eds.), Integration of AI and OR Techniques in Constraint Programming for Combinatorial Optimization Problems. X, 391 pages. 2007.

Vol. 4509: Z. Kobti, D. Wu (Eds.), Advances in Artificial Intelligence. XII, 552 pages. 2007. (Sublibrary LNAI).

Vol. 4508: M.-Y. Kao, X.-Y. Li (Eds.), Algorithmic Applications in Management. VIII, 428 pages. 2007.

Vol. 4506: D. Zeng, I. Gotham, K. Komatsu, C. Lynch, M. Thurmond, D. Madigan, B. Lober, J. Kvach, H. Chen (Eds.), Intelligence and Security Informatics: Biosurveillance. XI, 234 pages. 2007.

Vol. 4505: G. Dong, X. Lin, W. Wang, Y. Yang, J.X. Yu (Eds.), Advances in Data and Web Management. XXII, 896 pages. 2007.

Vol. 4504: J. Huang, R. Kowalczyk, Z. Maamar, D. Martin, I. Müller, S. Stoutenburg, K.P. Sycara (Eds.), Service-Oriented Computing: Agents, Semantics, and Engineering. X, 175 pages. 2007.

Vol. 4501: J. Marques-Silva, K.A. Sakallah (Eds.), Theory and Applications of Satisfiability Testing – SAT 2007. XI, 384 pages. 2007.

Vol. 4500: N. Streitz, A. Kameas, I. Mavrommati (Eds.), The Disappearing Computer. XVIII, 304 pages. 2007.

Vol. 4496: N.T. Nguyen, A. Grzech, R.J. Howlett, L.C. Jain (Eds.), Agent and Multi-Agent Systems: Technologies and Applications. XXI, 1046 pages. 2007. (Sublibrary LNAI).

Vol. 4495: J. Krogstie, A. Opdahl, G. Sindre (Eds.), Advanced Information Systems Engineering. XVI, 606 pages. 2007.

Vol. 4493: D. Liu, S. Fei, Z. Hou, H. Zhang, C. Sun (Eds.), Advances in Neural Networks – ISNN 2007, Part III. XXVI, 1215 pages. 2007.

Vol. 4492: D. Liu, S. Fei, Z. Hou, H. Zhang, C. Sun (Eds.), Advances in Neural Networks – ISNN 2007, Part II. XXVII, 1321 pages. 2007.

Vol. 4491: D. Liu, S. Fei, Z.-G. Hou, H. Zhang, C. Sun (Eds.), Advances in Neural Networks – ISNN 2007, Part I. LIV, 1365 pages. 2007.

Vol. 4490: Y. Shi, G.D. van Albada, J. Dongarra, P.M.A. Sloot (Eds.), Computational Science – ICCS 2007, Part IV. XXXVII, 1211 pages. 2007.

Vol. 4489: Y. Shi, G.D. van Albada, J. Dongarra, P.M.A. Sloot (Eds.), Computational Science – ICCS 2007, Part III. XXXVII, 1257 pages. 2007.

Vol. 4488: Y. Shi, G.D. van Albada, J. Dongarra, P.M.A. Sloot (Eds.), Computational Science – ICCS 2007, Part II. XXXV, 1251 pages. 2007.

Vol. 4487: Y. Shi, G.D. van Albada, J. Dongarra, P.M.A. Sloot (Eds.), Computational Science – ICCS 2007, Part I. LXXXI, 1275 pages. 2007.

Vol. 4486: M. Bernardo, J. Hillston (Eds.), Formal Methods for Performance Evaluation. VII, 469 pages. 2007.

Vol. 4485: F. Sgallari, A. Murli, N. Paragios (Eds.), Scale Space and Variational Methods in Computer Vision. XV, 931 pages. 2007.

Vol. 4484: J.-Y. Cai, S.B. Cooper, H. Zhu (Eds.), Theory and Applications of Models of Computation. XIII, 772 pages. 2007.

Vol. 4483: C. Baral, G. Brewka, J. Schlipf (Eds.), Logic Programming and Nonmonotonic Reasoning. IX, 327 pages. 2007. (Sublibrary LNAI).

Vol. 4482: A. An, J. Stefanowski, S. Ramanna, C.J. Butz, W. Pedrycz, G. Wang (Eds.), Rough Sets, Fuzzy Sets, Data Mining and Granular Computing. XIV, 585 pages. 2007. (Sublibrary LNAI).

Vol. 4481: J. Yao, P. Lingras, W.-Z. Wu, M. Szczuka, N.J. Cercone, D. Ślęzak (Eds.), Rough Sets and Knowledge Technology. XIV, 576 pages. 2007. (Sublibrary LNAI).

Vol. 4480: A. LaMarca, M. Langheinrich, K.N. Truong (Eds.), Pervasive Computing. XIII, 369 pages. 2007.

Vol. 4479: I.F. Akyildiz, R. Sivakumar, E. Ekici, J.C.d. Oliveira, J. McNair (Eds.), NETWORKING 2007. Ad Hoc and Sensor Networks, Wireless Networks, Next Generation Internet. XXVII, 1252 pages. 2007.

Vol. 4478: J. Martí, J.M. Benedí, A.M. Mendonça, J. Serrat (Eds.), Pattern Recognition and Image Analysis, Part II. XXVII, 657 pages. 2007.

Vol. 4477: J. Martí, J.M. Benedí, A.M. Mendonça, J. Serrat (Eds.), Pattern Recognition and Image Analysis, Part I. XXVII, 625 pages. 2007.

Vol. 4476: V. Gorodetsky, C. Zhang, V.A. Skormin, L. Cao (Eds.), Autonomous Intelligent Systems: Multi-Agents and Data Mining. XIII, 323 pages. 2007. (Sublibrary LNAI).

Vol. 4475: P. Crescenzi, G. Prencipe, G. Pucci (Eds.), Fun with Algorithms. X, 273 pages. 2007.

Vol. 4474: G. Prencipe, S. Zaks (Eds.), Structural Information and Communication Complexity. XI, 342 pages. 2007.

Vol. 4472: M. Haindl, J. Kittler, F. Roli (Eds.), Multiple Classifier Systems. XI, 524 pages. 2007.

Vol. 4471: P. Cesar, K. Chorianopoulos, J.F. Jensen (Eds.), Interactive TV: a Shared Experience. XIII, 236 pages. 2007.

Vol. 4470: Q. Wang, D. Pfahl, D.M. Raffo (Eds.), Software Process Dynamics and Agility. XI, 346 pages. 2007.

Vol. 4468: M.M. Bonsangue, E.B. Johnsen (Eds.), Formal Methods for Open Object-Based Distributed Systems. X, 317 pages. 2007.

Vol. 4467: A.L. Murphy, J. Vitek (Eds.), Coordination Models and Languages. X, 235 pages. 2007.

Vol. 4465: T. Chahed, B. Tuffin (Eds.), Network Control and Optimization. XIII, 305 pages. 2007.

Vol. 4464: E. Dawson, D.S. Wong (Eds.), Information Security Practice and Experience. XIII, 361 pages. 2007.

Vol. 4463: I. Măndoiu, A. Zelikovsky (Eds.), Bioinformatics Research and Applications. XV, 653 pages. 2007. (Sublibrary LNBI).

Vol. 4462: D. Sauveron, K. Markantonakis, A. Bilas, J.-J. Quisquater (Eds.), Information Security Theory and Practices. XII, 255 pages. 2007.

Vol. 4459: C. Cérin, K.-C. Li (Eds.), Advances in Grid and Pervasive Computing. XVI, 759 pages. 2007.

Vol. 4453: T. Speed, H. Huang (Eds.), Research in Computational Molecular Biology. XVI, 550 pages. 2007. (Sublibrary LNBI).

Vol. 4452: M. Fasli, O. Shehory (Eds.), Agent-Mediated Electronic Commerce. VIII, 249 pages. 2007. (Sublibrary LNAI).

Vol. 4451: T.S. Huang, A. Nijholt, M. Pantic, A. Pentland (Eds.), Artifical Intelligence for Human Computing. XVI, 359 pages. 2007. (Sublibrary LNAI).

Vol. 4450: T. Okamoto, X. Wang (Eds.), Public Key Cryptography – PKC 2007. XIII, 491 pages. 2007.

Vol. 4448: M. Giacobini et al. (Ed.), Applications of Evolutionary Computing. XXIII, 755 pages. 2007.

Vol. 4447: E. Marchiori, J.H. Moore, J.C. Rajapakse (Eds.), Evolutionary Computation,Machine Learning and Data Mining in Bioinformatics. XI, 302 pages. 2007.

Vol. 4446: C. Cotta, J. van Hemert (Eds.), Evolutionary Computation in Combinatorial Optimization. XII, 241 pages. 2007.

Vol. 4445: M. Ebner, M. O'Neill, A. Ekárt, L. Vanneschi, A.I. Esparcia-Alcázar (Eds.), Genetic Programming. XI, 382 pages. 2007.

Vol. 4444: T. Reps, M. Sagiv, J. Bauer (Eds.), Program Analysis and Compilation, Theory and Practice. X, 361 pages. 2007.

Vol. 4443: R. Kotagiri, P.R. Krishna, M. Mohania, E. Nantajeewarawat (Eds.), Advances in Databases: Concepts, Systems and Applications. XXI, 1126 pages. 2007.

Vol. 4440: B. Liblit, Cooperative Bug Isolation. XV, 101 pages. 2007.

Vol. 4439: W. Abramowicz (Ed.), Business Information Systems. XV, 654 pages. 2007.

Vol. 4438: L. Maicher, A. Sigel, L.M. Garshol (Eds.), Leveraging the Semantics of Topic Maps. X, 257 pages. 2007. (Sublibrary LNAI).

Vol. 4433: E. Şahin, W.M. Spears, A.F.T. Winfield (Eds.), Swarm Robotics. XII, 221 pages. 2007.

Vol. 4432: B. Beliczynski, A. Dzielinski, M. Iwanowski, B. Ribeiro (Eds.), Adaptive and Natural Computing Algorithms, Part II. XXVI, 761 pages. 2007.

Vol. 4431: B. Beliczynski, A. Dzielinski, M. Iwanowski, B. Ribeiro (Eds.), Adaptive and Natural Computing Algorithms, Part I. XXV, 851 pages. 2007.

Vol. 4430: C.C. Yang, D. Zeng, M. Chau, K. Chang, Q. Yang, X. Cheng, J. Wang, F.-Y. Wang, H. Chen (Eds.), Intelligence and Security Informatics. XII, 330 pages. 2007.

Vol. 4429: R. Lu, J.H. Siekmann, C. Ullrich (Eds.), Cognitive Systems. X, 161 pages. 2007. (Sublibrary LNAI).

Vol. 4427: S. Uhlig, K. Papagiannaki, O. Bonaventure (Eds.), Passive and Active Network Measurement. XI, 274 pages. 2007.

Vol. 4426: Z.-H. Zhou, H. Li, Q. Yang (Eds.), Advances in Knowledge Discovery and Data Mining. XXV, 1161 pages. 2007. (Sublibrary LNAI).

Vol. 4425: G. Amati, C. Carpineto, G. Romano (Eds.), Advances in Information Retrieval. XIX, 759 pages. 2007.